P9-BYA-261

R
616
V

11150

Van Amerongen C.

The way things work
book of the body

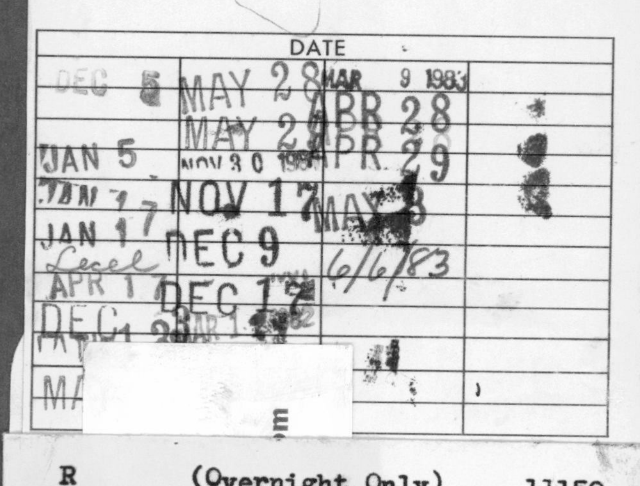

R
616
V

(Overnight Only)

11150

Van Amerongen

The way things work book
of the body

© THE BAKER & TAYLOR CO.

THE WAY THINGS WORK BOOK OF THE BODY

THE WAY THINGS WORK BOOK OF THE BODY

LIBRARY
Santa Venetia School
San Rafael, Calif.
WITHDRAWN

SIMON AND SCHUSTER • NEW YORK

11150

Original German language edition entitled *Wie Funktioniert das? Der Mensch und seine Krankheiten*
© 1973 Bibliographisches Institut AG, Mannheim. This translation and adaptation by C. van Amerongen, M. Sc., A.M.I.C.E. Copyright © 1979 by George Allen & Unwin, Ltd.
All rights reserved
including the right of reproduction
in whole or in part in any form
Published by Simon and Schuster
A Division of Gulf & Western Corporation
Simon & Schuster Building
Rockefeller Center
1230 Avenue of the Americas
New York, New York 10020

Designed by Helen Barrow
Manufactured in the United States of America

1 2 3 4 5 6 7 8 9 10

Library of Congress Cataloging in Publication Data

Van Amerongen, C
The way things work book of the body.

Translation and adaptation of Der Mensch und seine Krankheiten.
Includes index.
1. Medicine, Popular. I. Bibliographisches Institut A. G., Mannheim. Fachredaktion für Naturwissenschaft und Medizin. Der Mensch und seine Krankheiten. II. Title. III. Title: Book of the body.
RC81.V29 616 78-23944

ISBN 0-671-22454-9

A Foreword From the Publishers

The Way Things Work series has been one of the most remarkable phenomena of modern publishing, both because it presents a unique way to explain the most complicated concepts and artifacts of modern technology to the ordinary, nonscientific reader, and because it achieved a success so great that it can best be described as a nationwide explosion of enthusiasm, almost instantly making the original book, and later its succeeding volumes, into a record-breaking best seller and an institution. Well over 1,000,000 copies have been sold of the first book alone, in its various editions, and the worldwide sale of the complete series is astronomical.

Now the same technique of explanation and illustration has been applied to the human body, to health and to medicine, in this extraordinary and unique volume you hold in your hands. Until now, illustrated books about the body have tended to resemble *Gray's Anatomy,* splendid and detailed works for medical students, but of very little real use to the lay reader, for whom they were mainly a curiosity. Unlike these books, *The Way Things Work Book of the Body* applies to even the most complex phenomena and physiological processes the tried and true method of explaining things by diagrams. In these pages you will not only learn about the human body, and your health, but discover, in clear, simple and graphic form, *how it works*.

We believe that this book represents a major breakthrough in human knowledge and understanding. We cannot do better than quote the words of Isaac Asimov about the original volume of *The Way Things Work:*

Once upon a time, there were primitive priesthoods of magic, and members of those priesthoods cast spells, muttered runes, made intricate diagrams on the floor with powders of arcane composition. All this was intended to make the future clear, or bring the rain, or ward off evil, or lure an enemy to an agonizing end. Onlookers, when there were any, would watch with awe and no little fear, believing utterly in the efficacy of all this and in the existence of powerful and dangerous forces they could not themselves control.

Nowadays, there is a modern priesthood of science that calls on the power of expanding steam, of shifting electrons or drifting neutrons, of exploding gasoline or uranium, and does so without spells, runes, powders or even any visible change of expression. In response, onlookers are without awe, for, indeed, they seem to participate in the magic. By moving a lever, interrupting a light beam, or closing a contact, they can accelerate a car, open a door, or blow up a city.

Yet have we the right to sneer at the centuries of "ignorance" when witches were feared and gloat at our own age of "enlightenment" when scientists are respected? Do we understand more about the scientific principles and technological details of today than our ancestors did of the witches' spells of yester-century?

What it amounts to is that we live in a world of black boxes—all of us do, however well we may be educated. We drive automobiles that are complete mysteries to us, and, for that matter, the lock on our front door may be equally mysterious.

It is to this problem that The Way Things Work *addresses itself. The aim is to take any device you can think of, from the ball-point pen, through lightning conductors and gunpowder, to gearbox transmission and the nuclear bomb, and to explain how each of them works. . . .*

On the right-hand side of each spread is a diagram or a series of diagrams in red and black, drawn and labeled neatly and clearly. On the left-hand side of the spread is descriptive matter. . . .

*The book is exhaustive, and has an excellent index. There is little you can wonder about and not find. . . . For instance, I use an electric typewriter with a typing head instead of a keyboard. It had never seriously occurred to me to wonder how the keys control the typing head. With the book in my hand, it did occur to me to be curious and I turned to the two double-page spreads devoted to typewriters (one to mechanical and electrical keyboard typewriters and the second to the typing head). Well, I know now—in general, anyway—even though I still wouldn't dream of trying to repair the machine myself.**

—THE PUBLISHERS

*From the review of *The Way Things Work* by Isaac Asimov, *New York Times Book Review,* November 19, 1967; © 1967 by The New York Times Company. Reprinted by permission.

Contents

PART THREE: THE BLOOD

PART FOUR: THE RESPIRATORY ORGANS

PART FIVE: THE DIGESTIVE SYSTEM

PART EIGHT: THE ENDOCRINE GLANDS

PART NINE: MUSCLES AND BONES

PART TEN: THE TEETH AND THE PALATE

PART ELEVEN: THE SKIN

PART TWELVE: THE SENSORY ORGANS

PART THIRTEEN: INFECTIOUS DISEASES

PART FOURTEEN: TUMORS

Part One

GENERAL CONSIDERATIONS

CELLS, TISSUES, ORGANS

The various organs of the human body—heart, liver, brain, etc.—are composed of tiny units, only a few thousandths of a millimeter in size, called *cells*. Groups of cells, all of the same kind and together performing a particular function, are referred to as *tissues*. Thus, muscular tissue consists of assemblies of cells elongated side by side and capable of contracting in response to a stimulus. Cells that can transform external stimuli into electric signals and transmit them to the brain are associated with one another to form nerve tissue. Another kind of tissue is particularly well suited for supporting or holding together large assemblies of other tissues and is, for this reason, known as connective tissue; it consists of relatively few cells, but forms strong elastic fibers. Often various kinds of tissue, associated in larger units, together perform certain physical functions; such structural and functional units are called *organs*. Some organs, acting in combination with others, perform a vital main function essential to the very existence and survival of the organism as a whole. They thus constitute various organic systems such as the circulatory system, the respiratory system and the alimentary system (the latter comprising the mouth, tongue, teeth, esophagus, stomach, intestine and the associated digestive glands, e.g., pancreas).

The organism performs its many and varied functions more particularly with the aid of an advanced division of labor. This is possible only by specialization, which is manifested at all levels, from the major organic systems down to the various tissues and the individual cells that compose them. For instance, in a muscle, in a gland or in a nerve we find highly specialized cells which can contract or can secrete a particular substance or can transmit a stimulus, and which are equipped for their respective special functions by having an appropriate characteristic structure. Yet all these so widely differing types of cells have, in the course of biological evolution, developed from remote unicellular ancestors, and the approximately 60 million million cells in the human body are also all derived from a single cell, the fertilized egg cell. Hence it follows that, despite advanced specialization, even the cells in which an extremely high degree of modification has occurred must have preserved certain basic features in common with all the other cells in the body. These features, shared by all the cells, are distinctly visible under the microscope, especially at high magnifications such as the electron microscope can attain.

The *cell* comprises a nucleus and a layer of jellylike substance called protoplasm enveloping the nucleus (Fig. 1-1). The protoplasm is enclosed within the *cell membrane,* which forms the boundary between the cell and its surroundings and which, more particularly in multicellular organisms, separates it from other cells and from the extracellular space. The *protoplasm* (or *cytoplasm*) contains a number of small specialized structures, so-called *organelles,* which are of various kinds, e.g., the endoplasmic reticulum, ribosomes, the Golgi apparatus, mitochondria and lysosomes. The *nucleus* of the cell contains a strongly acid compound called *deoxyribonucleic acid* (DNA), formerly known as chromatin because of its capacity for staining with basic dyes. This compound is the material of inheritance which, when a cell divides, is transmitted to the two daughter cells.

The mechanism of ordinary cell division (involving division of the nucleus) is called *mitosis;* it commences with a change in the nuclear material: more particularly, special structures called *chromosomes* develop in it. These are the carriers of the *genes* (hereditary factors), which consist of DNA and which actually transmit the hereditary characteristics of the dividing cell to its daughter cells. In the chromosomes, which occur in pairs, the DNA is present in the form of double strands, the two halves of which constitute the pattern that is copied in the nuclear material of the daughter

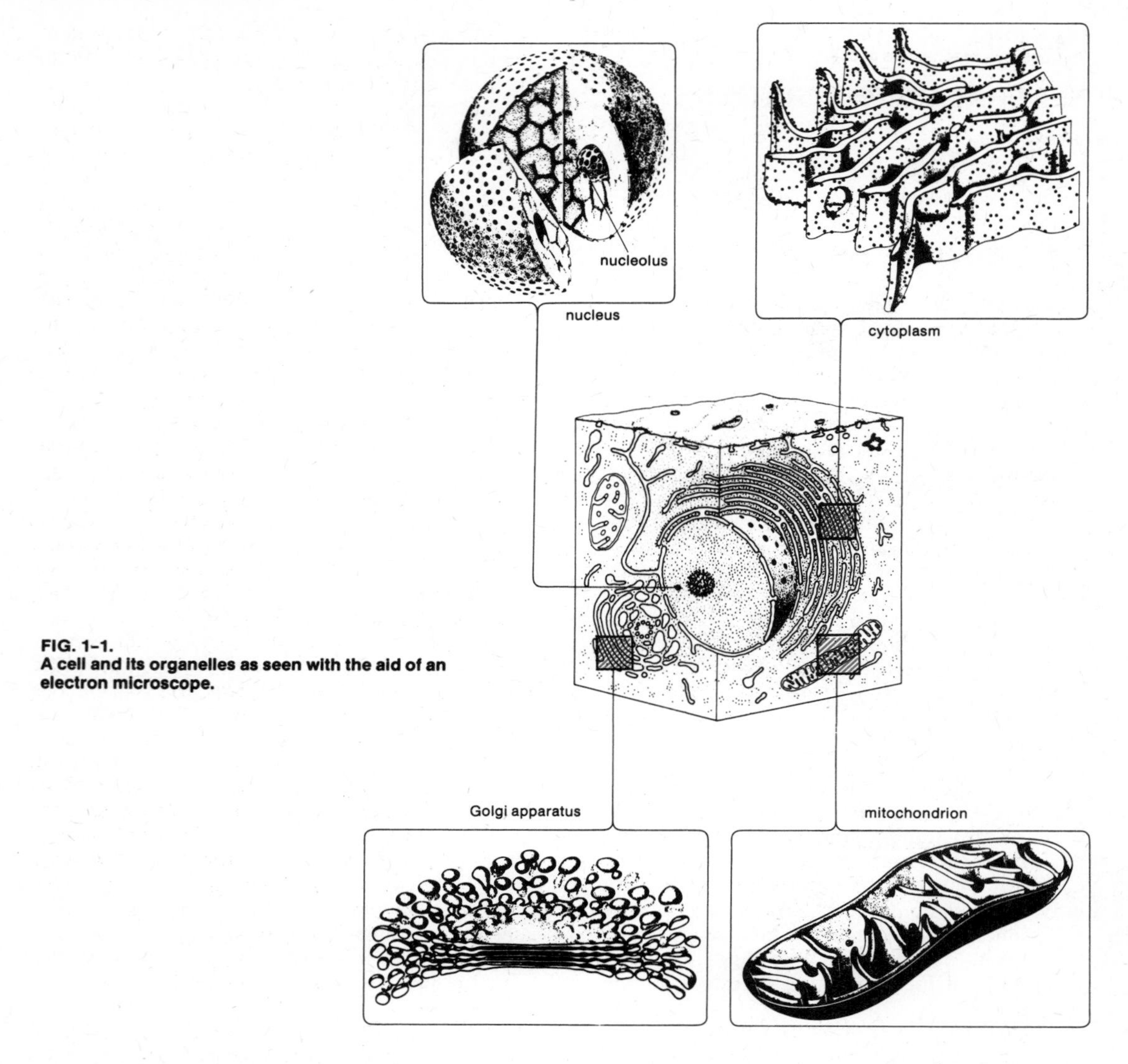

FIG. 1-1.
A cell and its organelles as seen with the aid of an electron microscope.

cells. In mitosis the two halves of each DNA double strand become separated, and each immediately produces a new matching strand. Thus there are two identical sets of chromosomes with their corresponding DNA strands present in the cell's nucleus shortly before it divides. Division of the nucleus occurs directly after division of the protoplasm. In this way each daughter cell receives one complete set of chromosomes with the hereditary factors they contain.

In the nucleus a small dense body, the *nucleolus,* is normally discernible, but disappears during mitosis (Fig. 1-1).

The nucleus of the cell is not just the carrier and transmitter of the hereditary factors from one generation of cells to the next; the DNA, i.e., the material of inheritance contained in the cells, also holds all the information and instructions for the functioning of the individual cell for the duration of its existence. It is ensured that in each cell, according to its specialized character and functional condition, of the complete set of instructions contained in the DNA, only a specific small number of instructions—appropriate only to that type of cell and to any particular instant of time—are in fact carried out. This is achieved in a highly

FIG. 1-2.
Instructional mechanism of cell.

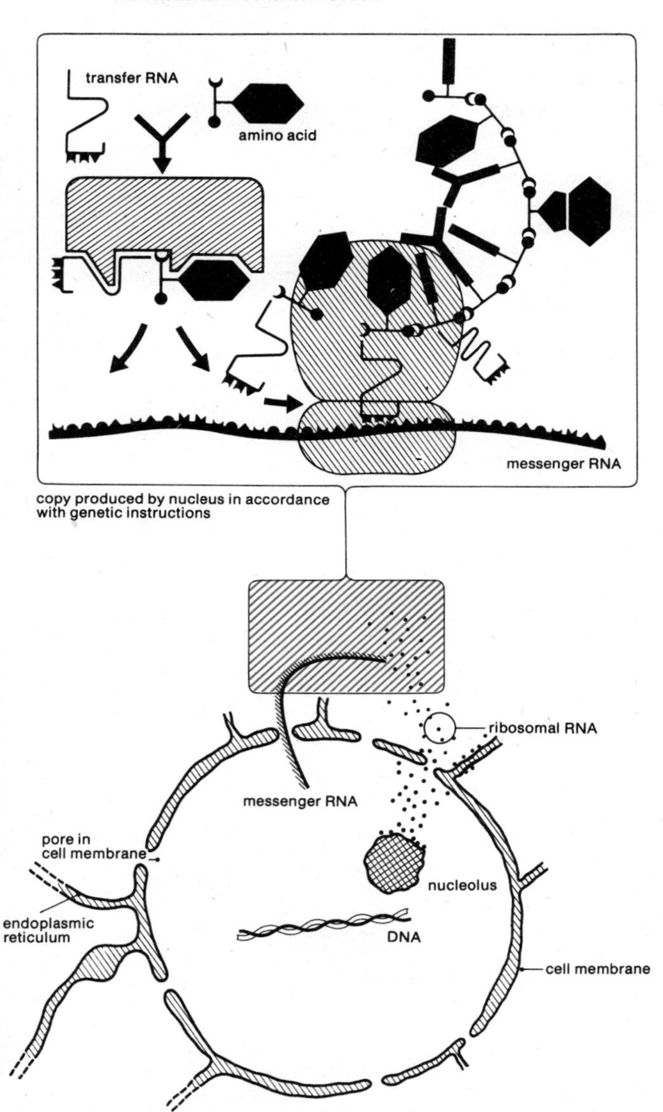

Courtesy Wittmann and Jockusch

ingenious and complex manner, the precise details of which are still under investigation. Of those instructions that are to be carried out in a particular cell and at a particular time (i.e., the instructions that are not "repressed"), a special copy is produced, but now not for transmission to daughter cells: it is intended for the cell's own use. This copy, composed of ribonucleic acids (RNA), makes its way out of the nucleus through minute pores in the nuclear membrane into the protoplasm surrounding the nucleus and there serves as the pattern for the execution of the instructions it conveys.

Each gene (i.e., each unit of the material of inheritance), in carrying out its hereditary instruction, brings about the synthesis of a particular enzyme protein. The enzymes in turn control and regulate the metabolism of the cell (cf. page 142). Therefore the nucleus is the cell's control device which copies parts of the set of hereditary instructions and passes them to the protoplasm for execution (Fig. 1-2).

Protein synthesis in the cell takes place in the *endoplasmic reticulum,* a complex system of pairs of membranes in the cytoplasm (the term *cytoplasm* refers to all the protoplasm excluding the nucleus), and is achieved with the aid of the *ribosomes.* The latter are granules consisting of protein and, for the greater part, of ribosomal RNA (r-RNA). Assisted by other nucleic acids—called transfer RNA and messenger RNA—they are able to link amino acids together, in a sequence specified by the cell's nucleus, to form long protein chains. The synthesized enzyme proteins perform essential functions in the metabolism of every cell. In addition, many cells specialize in producing large quantities of particular proteins (such as digestive proteins and protein hormones) for the benefit of the organism as a whole. These products, synthesized in the rough-surfaced endoplasmic reticulum (so called because of the presence of ribosomes as a granular covering), are packed in the form of tiny vesicles by the *Golgi apparatus* and are removed from the cell through the cell membrane. The Golgi apparatus, located close to the nucleus of the cell, consists of a system of tiny membranous tubes which probably communicate with the endoplasmic reticulum (Fig. 1-1).

All these chemical reactions of the cell require not only materials but also chemical energy to keep them going. This energy is, in general, made available in the form of high-energy phosphate compounds, usually adenosine triphosphate (ATP). The energy needed for building up the ATP itself is obtainable from the anaerobic breakdown of carbohydrates (the term *anaerobic* refers to a biochemical process which takes place in the absence of free oxygen). Much more ATP is formed if the carbohydrates, especially sugar, are not just partly broken down under anaerobic conditions, but are instead fully decomposed ("burned") aerobically, i.e., with the aid of oxygen, to produce carbon dioxide and water as the final products. This process of complete oxidation takes place in the *mitochondria* (chondrosomes), the highly specialized biochemical "power stations" and "combustion furnaces" of the living cell (Fig. 1-3). Each of these microscopic bodies consists of a dense matrix enclosed within a double membrane whose inner component is folded inward in a number of places (Fig. 1-1) so as to form lamellae (cristae), which are covered by so-called elementary particles containing components of enzyme systems. Among others, the enzymes of the citric acid cycle, of oxidative phosphorylation and of the chain of respiratory reactions are known to be present in the mitochondria.

Besides specialized organelles for protein synthesis and energy supply, many cells contain *lysosomes* which, among other functions, form the defensive system of the individual cell (Fig. 1-4). These particles play an important part in the intracellular digestion of material originating within the cell and of extraneous material absorbed into the cell; for this purpose they contain hydrolytic enzymes which split various substances into simpler compounds. An example of such cellular defense functions is *phagocytosis*—the action of some cells, notably white blood cells (leukocytes), which engulf and destroy foreign bodies such as bacteria. The enzymes in the lysosomes are so powerful and concentrated that they can completely dissolve and digest their own mother cell itself if they are accidentally released as a result of severe damage to the cell, e.g., by burning or mechanical action. Thus the lysosomes have an active role in the "self-cleaning" of tissue that has been partly destroyed by accidental causes.

The cell's protoplasm occupying the space between the nucleus and organelles within the cell membrane is more specifically termed *cytoplasm*. Its purpose is, however, more than just that of a "filling" and medium of transport between the individual components of the cell. It also contains a number of enzymes, e.g., for the anaerobic breakdown of sugar, synthesis of glycogen and fatty acids, and breakdown of amino acids.

The *cell membrane* forms a protective enclosure around the contents of the cell and serves as its boundary with regard to external influences. In unicellular organisms everything

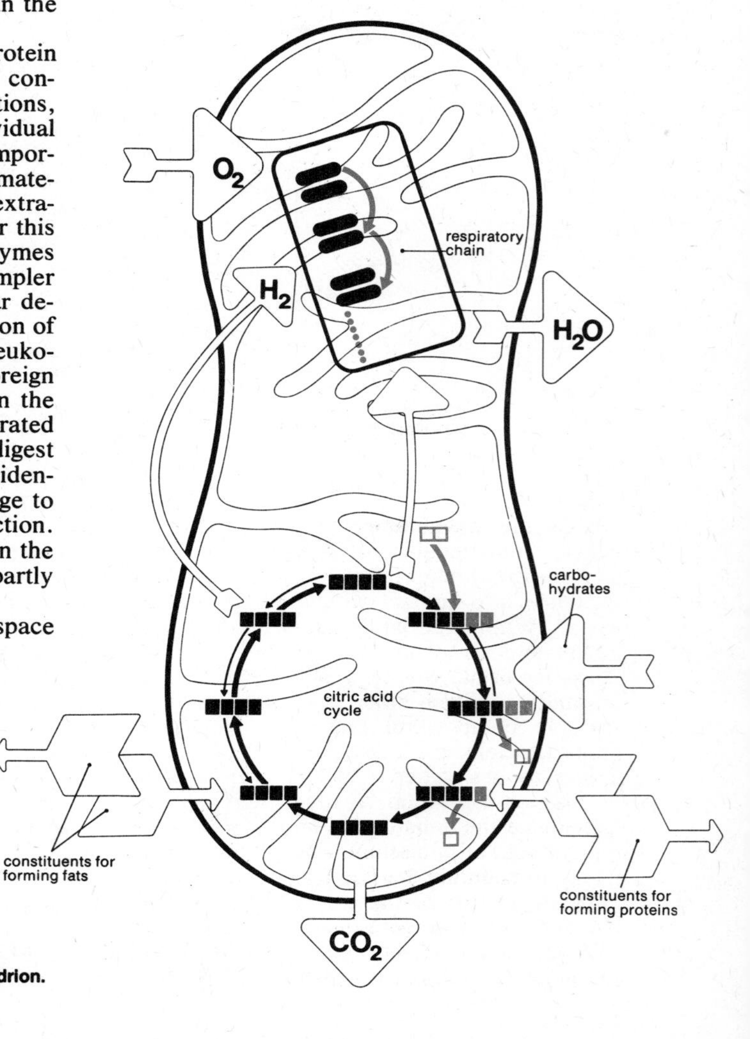

FIG. 1-3. Mitochondrion.

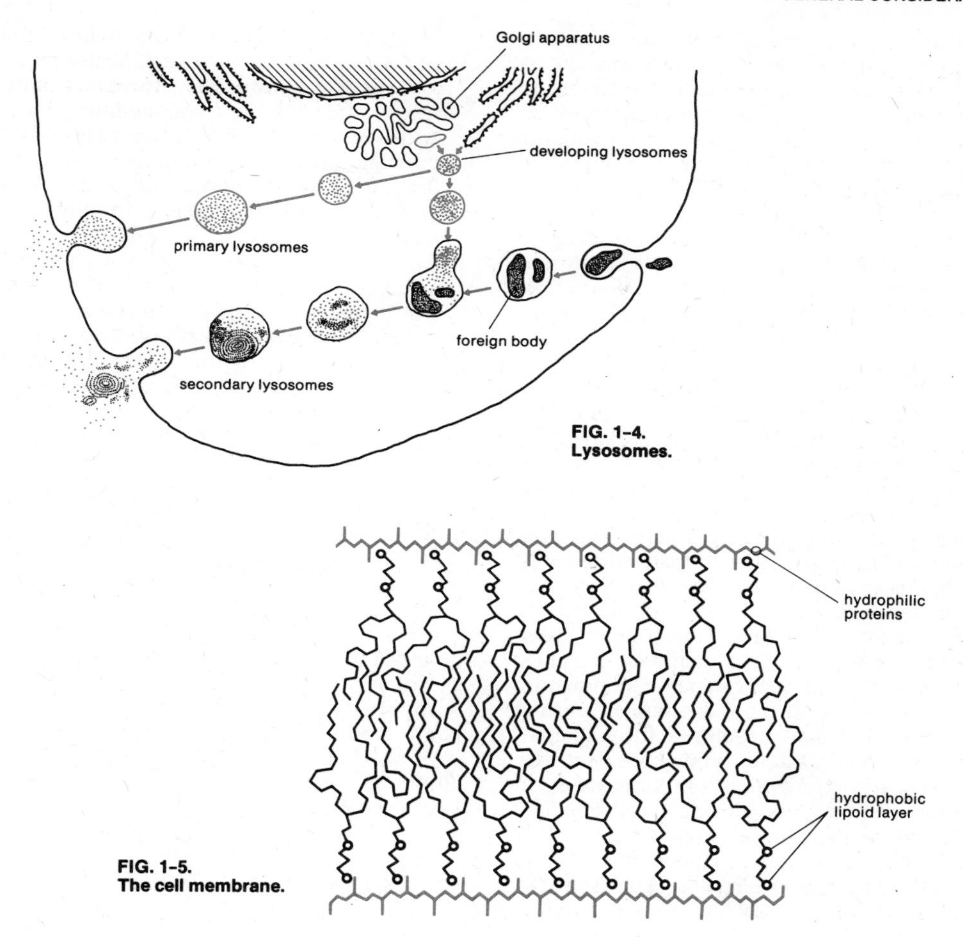

FIG. 1-4.
Lysosomes.

FIG. 1-5.
The cell membrane.

outside the membrane is environment. In higher, multicellular organisms (including man) the great majority of the body's cells do not come into contact with the environment proper. Within the body, however, the cells are surrounded by an "internal environment" consisting of the *intercellular fluid* (or interstitial fluid), of which blood plasma is a component. In composition this fluid medium in which the cells are constantly bathed is very similar to sea water. Thus the intercellular fluid is rich in sodium, as contrasted with the fluid inside the cells themselves, which has a high content of potassium. The cell membrane serves to maintain the cell's chemical individuality in relation to the surrounding medium. This function is rendered more difficult by the fact that the cell is not just a closed system, a mere bag of fluid, but is in constant communication with its environment: the cell membrane has to be permeable to various substances so that the cell is able to absorb nourishment and discharge its waste products. For this reason the membrane is not simply a dead skin, but is in fact a complex structure possessing specific properties and performing functions involving energy expenditure geared to the cell's metabolism (Fig. 1-5).

Only 10 millionths of a millimeter thick, the cell membrane contains enzymes and, in combination with these, so-called *ion pumps* which discharge potassium into, and sodium out of, the interior of the cell and which thereby ensure that the cell preserves its own internal condition in relation to the intercellular medium surrounding it. In addition, the membrane has the property of *selective permeability,* i.e., it allows certain ions to pass through it, while barring the passage of others, in such a manner that the positively charged potassium ions impart a positive electric charge to the outside of the membrane in relation to the inside thereof. This difference in charge amounts to a potential difference of about 80 millivolts between the outside and inside of the membrane. In nerve cells this so-called resting potential breaks down during the passage of an impulse and is momentarily reversed, so that the inside becomes electrically positive in relation to the outside as a result of a selective increase in the sodium permeability of the membrane. This potential reversal is of extremely short duration (1–2 thousandths of a second); it travels as a wave of potential change along the nerve fiber at speeds of up to 150 meters per second and serves to conduct the impulses ("messages") by which the nervous system functions. Muscle cells behave similarly in that a potential change associated with an external stimulus is transmitted into the interior of the muscle fibers and initiates their contraction.

Besides the operation of ion pumps, other active material transfer functions are performed by the cell membrane in connection with the intake of nutrient substances and the elimination of waste products: for example, glucose and amino acids may be selectively concentrated within the cell and thus stored in readiness to provide energy for the cell and sustain its specific synthesis function.

The differentiation and specialization that make some cells diverge so widely from the basic type of cell from which they originated are manifested both in their structure and in their biochemistry. This is the fundamental reason why whole assemblies of similar cells can combine to form special *tissues* performing different specific functions.

Epithelium (epithelial tissue) is the name given to tissue composed of firmly coherent sheets of cells forming a lining or covering for internal or external surfaces of the body. Depending on the shape of the constituent cells, a distinction can be drawn between columnar, cuboidal and squamous epithelium (in order of diminishing relative height); also, between simple and stratified epithelium, according to whether the sheet is one cell thick or many. Epithelium may function more particularly as a protective sheath or covering (as in the outer skin of the body) or it may serve primarily for the interchange of matter, e.g., resorption by the epithelium of the intestine, and secretion in certain parts of the renal tubules (Fig. 1-6a and b).

Connective tissues in general are characterized by having their intercellular space filled with fluid, semifluid or solid substance (intercellular substance), sometimes also containing fibrous elements that enhance a connecting, supporting and strengthening function. This category comprises, in addition to loose and dense connective tissues, also those tissues known as cartilage and bone. Connective tissue proper consists of stationary cells (fibrocytes) and cells that are able to move about (migratory cells: histiocytes and others). More particularly typical of this tissue is the presence of collagenous and elastic fibers in the fluid intercellular substance. Loose connective tissue interconnects cells of many different types to combine them into tissues and organs. Many organs such as the heart, lung, kidney, brain and intestine are moreover enclosed within a layer of connective tissue variously referred to as a "sheath," "membrane," "capsule," etc. Dense connective tissue forms tendons, which connect muscle to muscle or muscle to bone. Connective tissue, together with the small blood vessels embedded in it for supplying blood to the adjacent cells, furthermore plays an important part in various types of inflammation and also, by the formation of new connective tissue and development of new blood vessels, in cicatrization (healing by scar formation) after injury or damage to tissues.

Cartilage consists of cells embedded in an elastic ground substance, or matrix, which is strengthened by connective tissue fibers. Depending on the content of fibrous or reticular elements present in it, a distinction can be drawn between hyaline cartilage (e.g., the cartilage covering the articular surfaces of bones, costal cartilage connecting the ribs to the sternum, and laryngeal cartilage) and fibrous cartilage (e.g., intervertebral cartilage). Hyaline cartilage also occurs as the elastic and as yet uncalcified precursor of bone during skeletal development and also in the healing of bone fractures.

Bone is the material of which the skeleton is composed; it supports the softer parts of the body and also forms a protective sheath or casing (the spine and the skull) for the central nervous system and the highly developed sensory organs. In addition to organic ground substance, bone contains solid inorganic material: calcareous apatite. Bone-forming cells (os-

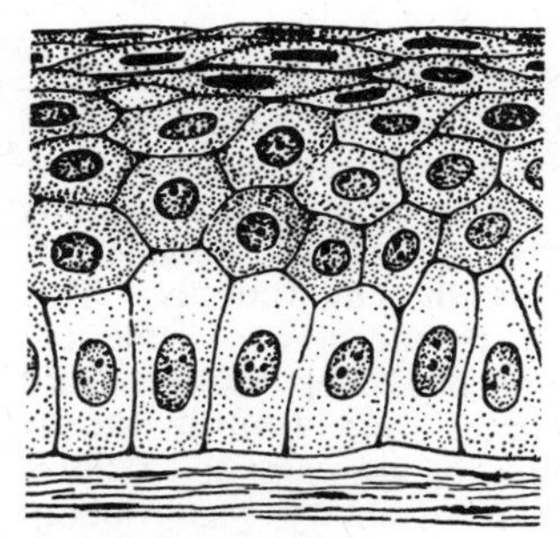
(a) Stratified squamous epithelium.

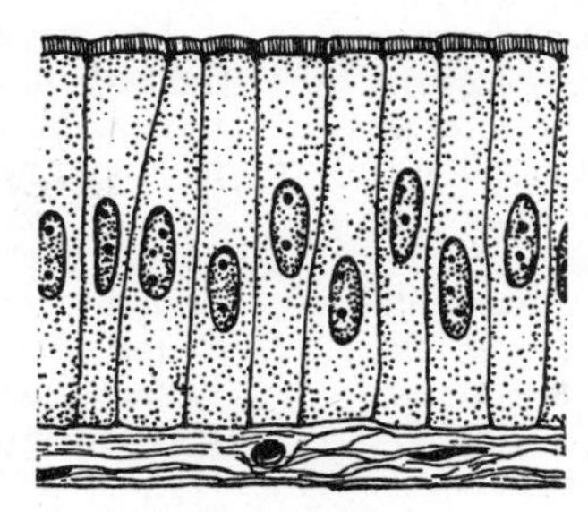
(b) Columnar epithelium.

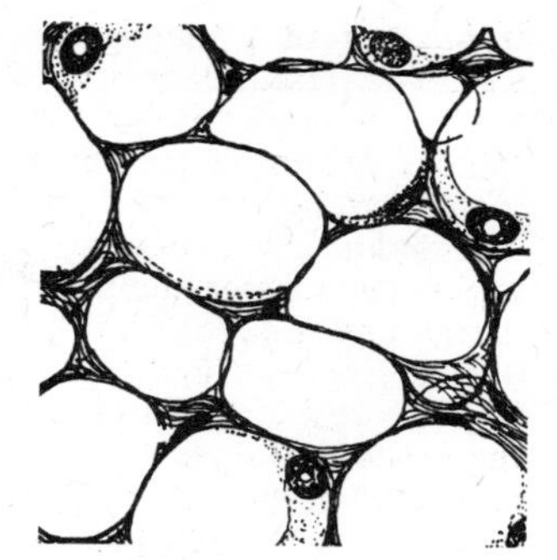
(c) Adipose tissue.

(d) Connective tissue fibers.

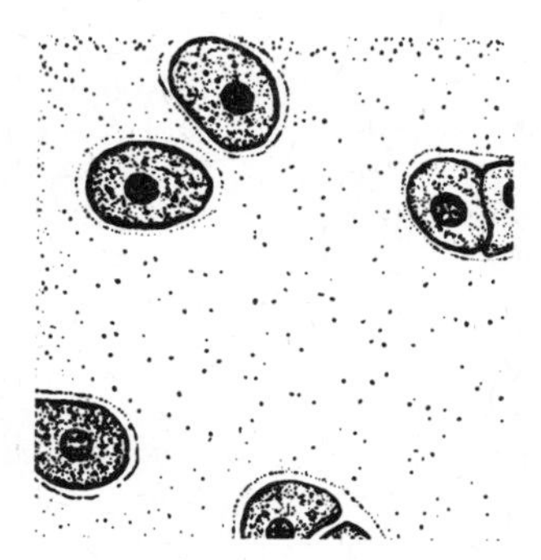
(e) Hyaline cartilage.

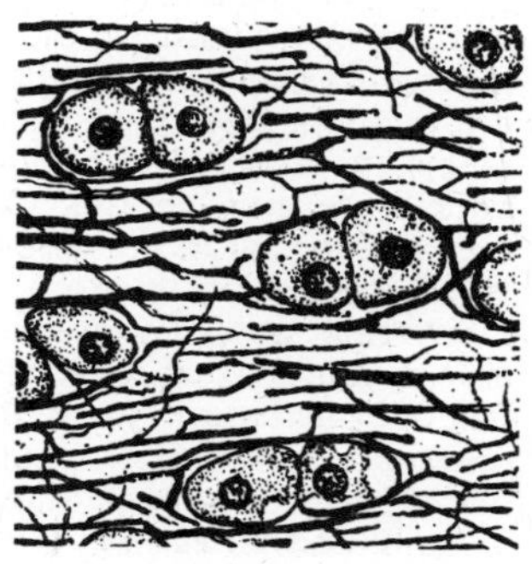
(f) Fibrous cartilage.

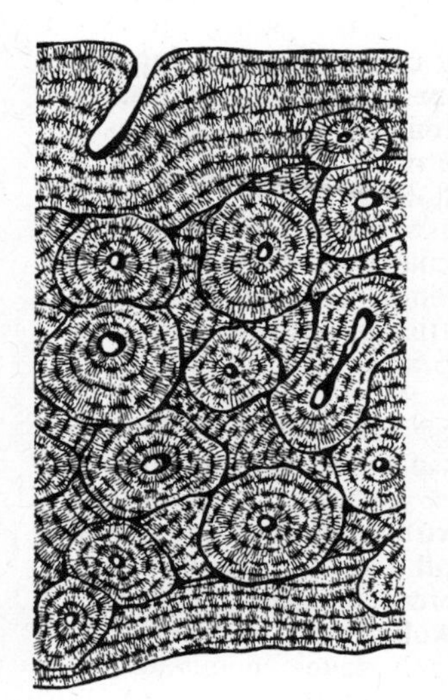
(g) Bone.

(h) Muscular tissue (heart).

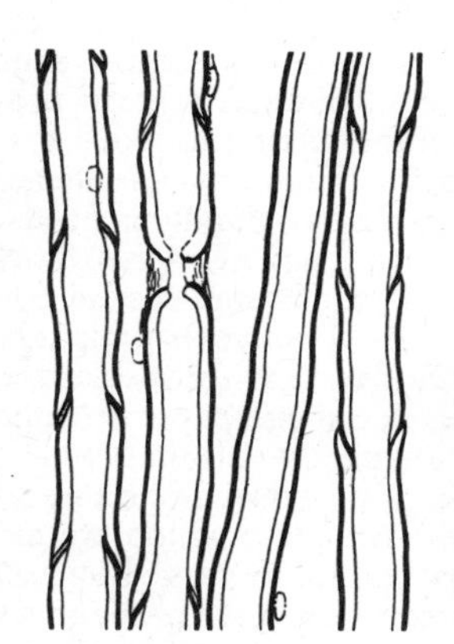
(i) Nerve fibers (medullated).

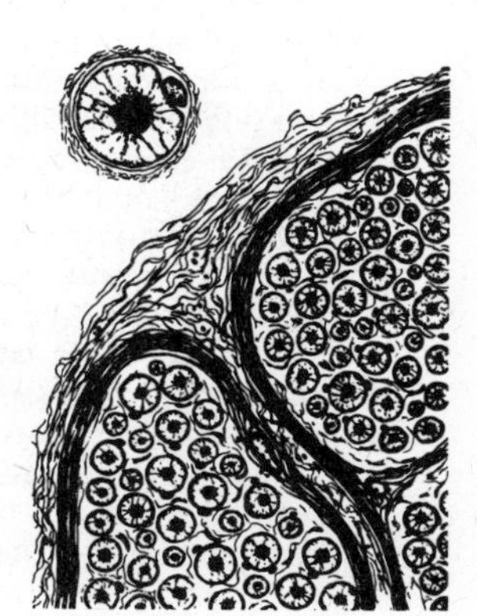
(j) Peripheral nerve (cross section).

FIG. 1-6.

teoblasts) and bone-removing cells (osteoclasts) are continually at work in the formation and breakdown (by resorption) of bone substance. According to whether bone is formed in the enveloping membrane or in the interior of a hyaline cartilaginous precursor (or model), we speak of perichondral (or intramembranous) and enchondral ossification, respectively. The first type of bone formation is exemplified by the ends of long bones, while the second type (i.e., enchondral ossification) occurs in the shafts of such bones. The human body contains 206 bones which, depending on their functions, differ greatly in shape and size and are connected to one another by various types of joints (see pages 374 ff).

Muscular tissue consists of specialized cells which are contractile, i.e., capable of shortening (contraction) in response to stimuli. These cells are elongated in shape—in striped muscle they have evolved into multinuclear "muscle fibers" several centimeters in length—and contain special proteins (actin and myosin) which are responsible for their contractility. A distinction is to be drawn between smooth muscle and striped muscle (see pages 360 ff). An intermediate type is the cardiac muscle (see page 366), which, though striped, consists of fibers that are (as in many smooth muscles) interlinked so that an impulse can spread from one end of a fiber to another as well as along the length of a fiber, thus ensuring that all the fibers will contract exactly in time with one another (they form what is known as a syncytium: see page 367). The muscular contractions are initiated by potential changes in the membranes enclosing the cells and fibers (as already referred to).

The functions of *nervous tissue* comprise the reception of stimuli by the sense organs, processing the stimuli in the central nervous system, and responding to stimuli through the action of muscle cells or glandular cells. Essential to the performance of these functions are, firstly, reliable and rapid conduction of impulses ("messages") between periphery and center and, secondly, a highly specialized system—possessing learning capacity and memory—for coordinating the stimuli. The conducting function of nervous tissue is performed by nerve fibers (axons) which are extensions, sometimes several meters in length, of the nerve cells (neurons). More particularly the medullated or myelinated (well-insulated) nerve fibers attain high conduction speeds for the electric *action potential,* which is the wave of sudden change in the resting potential when an impulse is conducted (see above).

The stimulus coordinating function is performed by the central nervous system, comprising the brain and spinal cord. For this purpose there are millions of elaborately branched nerve cells, all provided with junctions (synapses) whereby they are interconnected to branches of other nerve cells. At a synapse, stimulation of one nerve cell by another takes place. Since most synapses transmit in one direction only, stimulation may fail to occur if the impulse arrives from the wrong direction (see page 339). The nerve cell represents an instance of extreme specialization, a degree of differentiation that is possible only at the expense of other properties. For example, mature nerve cells have lost the ability to propagate by division. Nervous tissue as a whole is moreover characterized by the development of a special tissue, called *neuroglia,* which forms a supporting network for the nerve cells.

INFLAMMATION

Inflammation is the body's defensive response, closely involving the blood vessels, to injury or irritation. The purposes of inflammation are to prevent the injurious or irritant action from spreading, to reduce its intensity, to cleanse the tissue affected, and finally to create the conditions for repairing the harm done.

Inflammations are usually caused by pathogenic (disease-producing) microorganisms such as bacteria, but other causes may also give rise to typical inflammatory reactions, e.g., heat, cold, tissue irritant poisons, foreign bodies that have penetrated into tissues, or even cast-off pieces of tissue which have the same effect as foreign bodies. If pathogenic microorganisms are not involved in causing inflammation, it is referred to as aseptic inflammation. Depending on the duration, a distinction can be drawn between acute and chronic inflammation.

Acute inflammation is directly observable when it occurs at the surface of the body, e.g., in the form of sunburn, boils on the skin, or the common cold affecting the mucous membranes of the nose. In all cases four distinctive signs of inflammation are manifest: redness (rubor), heat (calor), swelling (tumor) and pain (dolor). The development of an inflammation can be observed under the microscope in suitably "transparent" tissues, such as the mesentery of the small intestine (the mesentery is

FIG. 1-7.
Various levels of blood flow through capillaries.

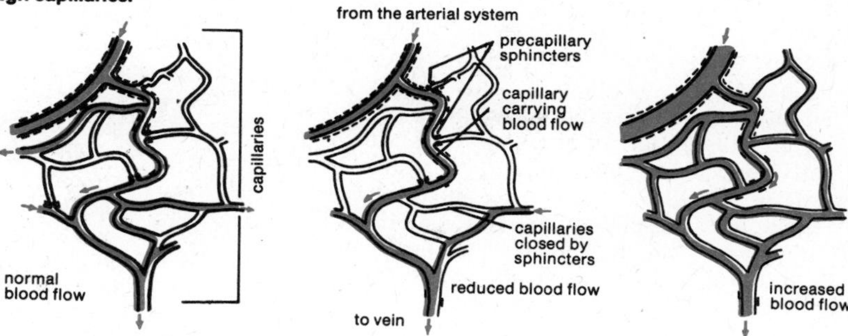

a peritoneal fold containing blood vessels and connecting the intestine to the back of the abdomen). For the same purpose of observation a "transparent chamber" may be inserted into a rabbit's ear, for example. When inflammation-producing substances are applied to the experimental specimen, local changes in the flow of blood are the first signs to appear. The flow through the capillaries, after sometimes briefly decreasing, undergoes a very marked increase (*inflammatory hyperemia*), which is the direct cause of the redness and heat around the center of inflammation and may occur as a result of dilatation of the arterioles (the minute arteries leading to the capillaries). Alternatively, it may be caused by a paralysis of the sphincters (circular constricting muscles) at the entrance to the network of capillaries which forms a sort of bypass to the terminal blood vessels and which can be temporarily disconnected if necessary, as in the event of inflammation (Fig. 1-7).

The third visible symptom of acute inflammation, *swelling* of the tissue, is also due to the locally increased flow of blood and higher internal pressure in the capillaries, as a result of which larger amounts of watery fluid are exuded from the blood into the surrounding tissue. Actually, dynamic equilibrium between the blood and the tissue fluid (more specifically: between the plasma and the interstitial fluid) has to be continually restored. This is achieved through the thin water-permeable capillary wall forming the boundary between the blood vessel and the tissue. The discharge of water from the capillaries is essentially a filtration process in which the blood pressure causes water contained in the blood plasma to be forced through the capillary wall into the tissue. Reentry of water from the tissue into the capillaries occurs farther downstream, at the ends of the capillaries, as a result of the water-attracting effect of the plasma proteins. In this part of the capillaries the blood pressure has diminished to such an extent that it can no longer counteract this reentry effect.

An increase in capillary blood pressure intensifies the discharge of fluid into the tissue; the resulting excess of (interstitial) tissue fluid is called *edema*. In the case of inflammatory hyperemia, the blood pressure in the capillaries is indeed increased because the sphincters in the blood vessels leading to them (the arterioles and precapillaries) are dilated. Besides, the capillaries themselves become more highly permeable as a result of dilatation and the effect of chemical inflammatory substances. The filtering action of the capillary wall becomes coarser, so that, in addition to watery fluid, larger molecules are now also able to pass. For this reason the water forced out of the capillaries into the tissue—the inflammatory edema—often contains a high concentration of proteins which, now outside the capillaries, exert their water-attracting action and cause more water to be discharged into the tissue (Fig. 1-8). This

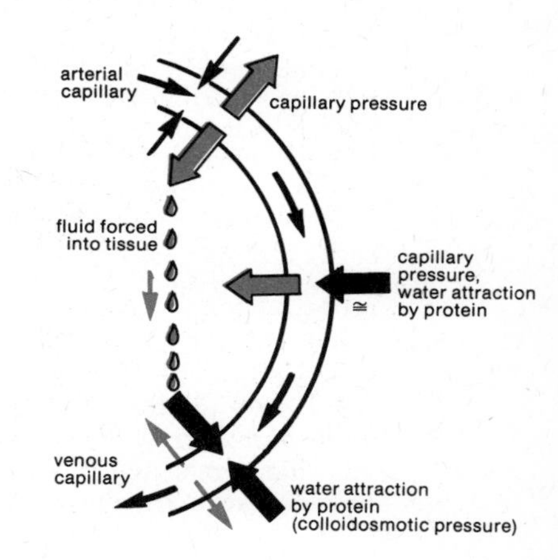

FIG. 1-8
Formation and absorption of edema.

fluid containing blood proteins is called *exudate*. It distinguishes inflammatory edema from ordinary dropsy, which is edema in which the excess water in the tissues does not contain protein substances, but only such small-molecular substances as salt and glucose. If substantial numbers of cells subsequently also enter the exudate, they are referred to as *infiltrate*. Exudate and infiltrate are the underlying phenomena associated with inflammatory swelling and its attendant pain.

In addition to hyperemia and swelling, the phenomenon called *stasis* may be a determining factor in the progress of the inflammation. In general, this term denotes a stagnation of flow of fluids, here more particularly the flow of blood. If too much water is filtered out of the blood, as a result of excessive increase in blood pressure or increase in capillary permeability, the blood remaining in the blood vessels becomes less fluid (because of increased internal friction due to the high concentration of cells in it). As a result, stasis may occur, i.e., the capillaries become blocked, and this in turn may result in local necrosis (death of areas of tissue). This condition may be beneficial in sealing off the center of inflammation together with the bacteria or other irritants present in it, these then being destroyed along with part of the tissue that suffers necrosis.

If the composition of exudate containing proteins corresponds approximately to that of blood serum, it is referred to as *serous exudate,* and the inflammation itself is called a serous inflammation. Serous exudate from inflamed mucous membrane may be produced, for example, in the early stages of the common cold and is then described as catarrh, especially when the mucous membrane of the nose is affected. In the skin, a collection of serous exudate below the epidermis may cause a blister, as may occur as a result of a burn or of chafing by ill-fitting footwear. Serous exudate may also accumulate in internal tissue cavities, e.g., in the pleural cavity, the condition then being known as pleurisy (inflammation of the pleura, the double layer of membrane surrounding the lung). In such cases the exudate, especially when purulence or suppuration (formation of pus) occurs, will have to be removed by draining the cavity (Fig. 1-9).

There the term *fibrinous exudate* describes exudate containing fibrin, the blood protein that causes clotting. The clotted exudate may, as in diphtheria, remain adhering to the mucous membrane and may even form false membranes (pseudomembranes). There is often no

FIG. 1-9
Purulent exudate in the pleural cavity (empyema).

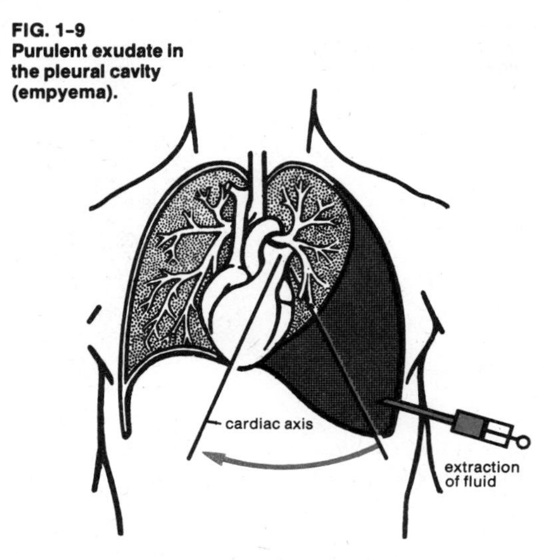

clear-cut borderline between serous and fibrinous inflammation, which may shade off into each other, as in pneumonia (see page 132), where the serous inflammation of the lungs gradually changes into a fibrinous one.

Exudate contains white blood cells (leukocytes) which have leaked from the blood vessels and become yellowish by fatty degeneration; they are the characteristic component of *pus*. Pus produced by inflammation of mucous membrane can flow away directly, as occurs in certain types of catarrh. A localized accumulation of pus resulting from displacement or disintegration of tissue is called an *abscess*. Abscesses may form in virtually any organ of the body. Typical examples are the so-called otogenous brain abscess, which may develop from a chronic inflammation of the middle and internal ear and which involves the bone (Fig. 1-10); the peritonsillar abscess (quinsy), which may follow a purulent inflammation of the tonsils (Fig. 1-11); the abscess around the root of a tooth in consequence of inflammation of the periodontal membrane (Fig. 1-12); abscesses of the breast and the sweat glands (Fig. 1-13); and the whitlow (paronychia), which may start in the subcutaneous tissue of a finger or at the edge of a nail and cause a septic fingertip (Fig. 1-14). Suppuration infiltrating into degenerating tissue produces a condition called a *phlegmon*. When pus collects in existing cavities of the body, such as the pleural cavity, it is referred to as *empyema*, e.g., pleural empyema (Fig. 1-15). If blood is present in the exudate, it is called *hemorrhagic exudate;* a *putrid exudate* is caused by the presence of putrefactive

FIG. 1-10
Brain abscess arising from ear inflammation.

uvula

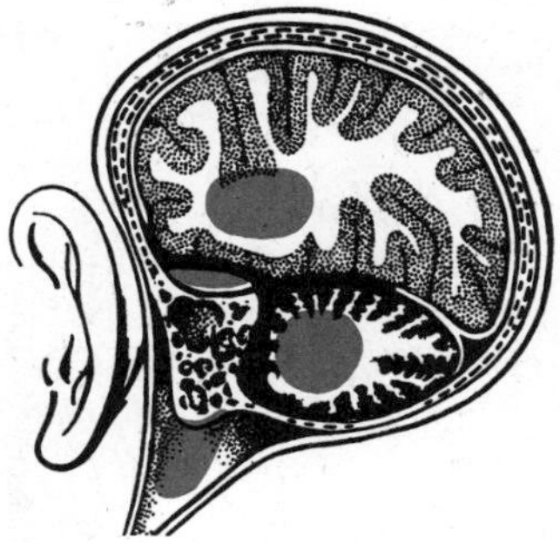

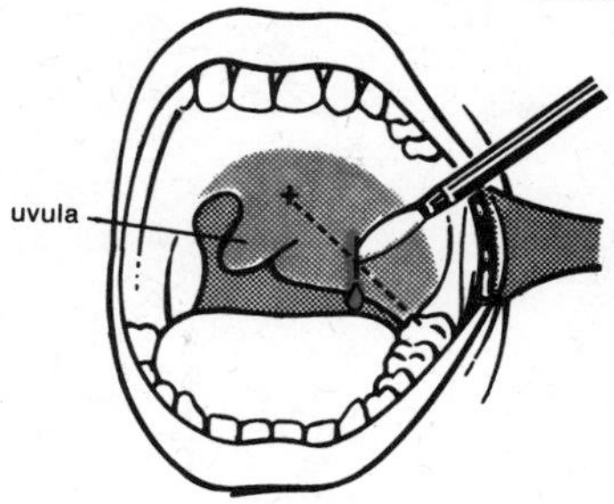

FIG. 1-11.
Lancing a peritonsillar abscess.

FIG. 1-12.
Abscess around root of tooth.

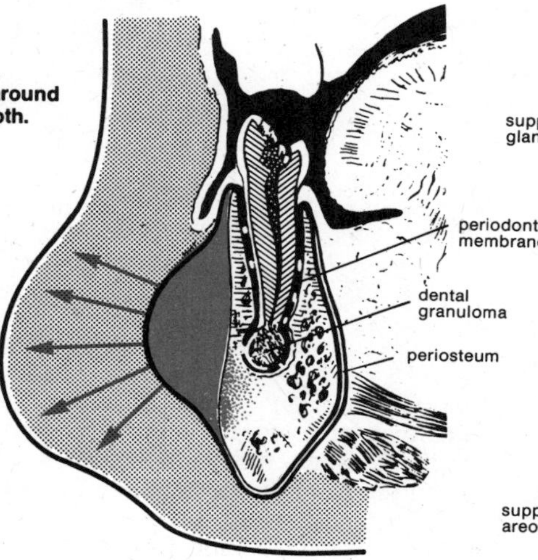

periodontal membrane
dental granuloma
periosteum

suppurating glandular tissue
suppurating areolar tissue

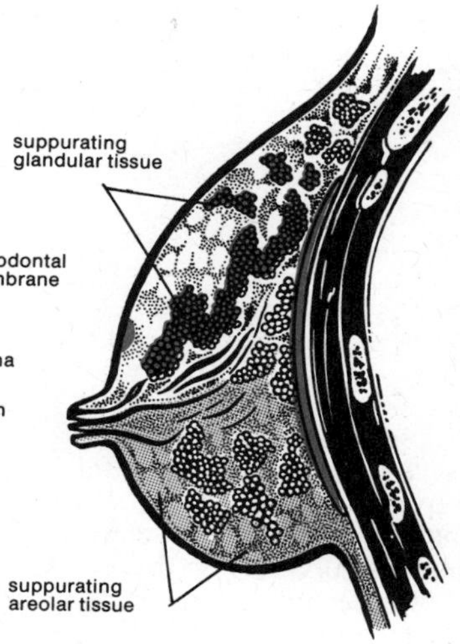

FIG. 1-13.
Abscess of the breast.

FIG. 1-14.
Suppurative inflammation of the fingertip.

in the skin
under the periosteum
in the subcutaneous connective tissue
in the bone
in the tendon sheath
in the joint

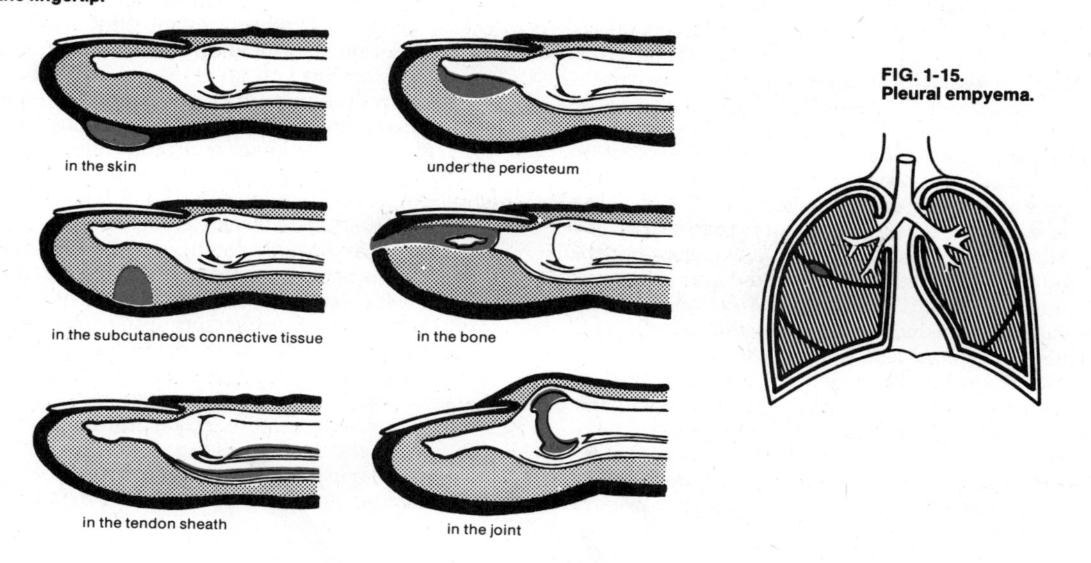

FIG. 1-15.
Pleural empyema.

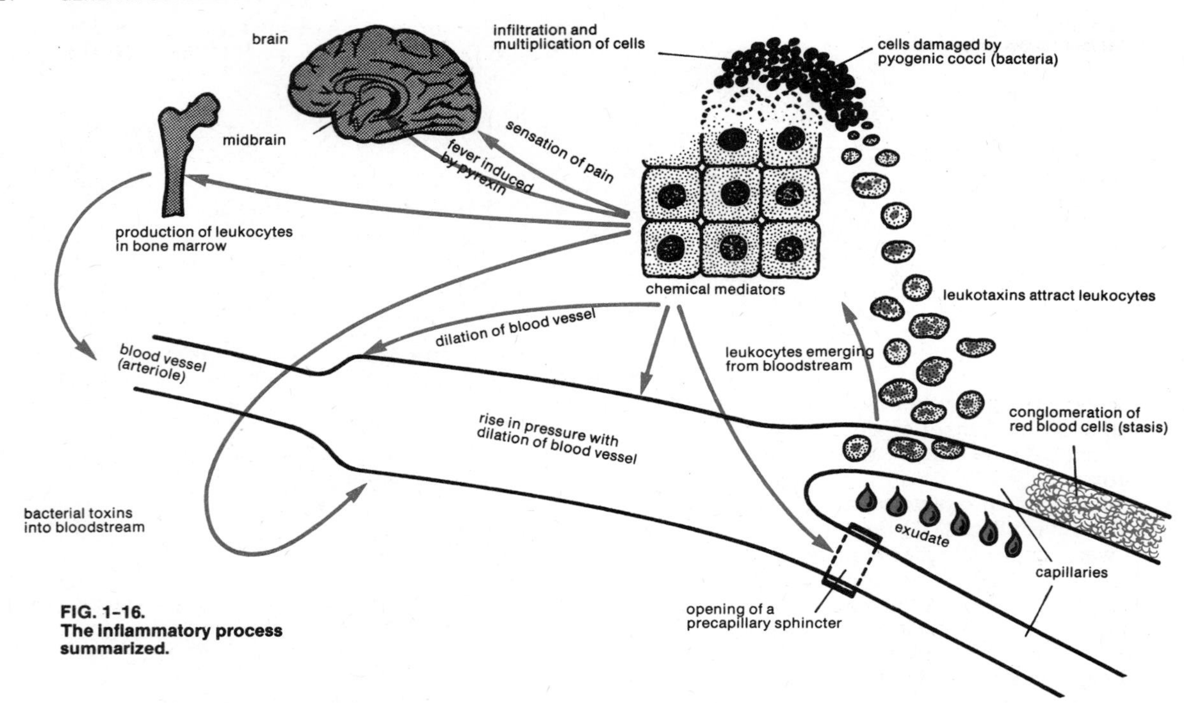

FIG. 1-16.
The inflammatory process summarized.

bacteria. Whereas serous exudate may be returned to the blood (partly via the lymph) and fibrinous exudate may be dissolved by enzymatic processes, pus usually has to be removed from the body for healing to take place. Pus may be discharged spontaneously from an abscess; alternatively, surgery (incision or puncturing) may be necessary.

The inflammatory process can be summarized as follows: Inflammation is a typical reaction of the blood vessels to injury or irritation. It used to be supposed that the irritant agent directly acted upon the smallest terminal blood vessels, but it is now considered that the latter are affected indirectly through the action of substances, chemical mediators, which are liberated from the damaged tissue cells. The release of these substances that affect the blood vessels is the direct consequence of the injury or irritation that started the inflammatory process. They include *histamine*, a degradation product of protein which is released also in allergy reactions, as well as *serotonin* and *heparin*. These chemical mediators are formed in a special sort of cell, the mast cells, which are found in the vicinity of blood vessels. That they do indeed act as "mediators" is borne out, among other things, by the fact that they dilate the capillaries and increase the blood flow rate. Biochemical analysis of inflammatory exudates has revealed the presence of other, mostly proteinlike substances resulting from the inflammatory breakdown of cells, in addition to those already mentioned: *leukotaxin*, which induces white blood cells (leukocytes) to come out of the bloodstream near the center of inflammation and is furthermore a factor promoting leukocytosis (increase in the number of white blood cells); *necrosin*, which destroys tissue; and *pyrexin*, which induces fever (Fig. 1-16).

Leukocytosis and fever are indications that the organism as a whole may become affected as a result of inflammation. The increase in the number of leukocytes is brought about by stimulation of the bone marrow in which they are formed.

Fever (pyrexia) is due to disturbed heat regulation caused by pyrogenic (fever-producing) substances, variously known as pyrexins and pyrotoxins, which are generated at the center of inflammation and act upon nerve centers in the midbrain, where the heat regulation is located. Pyrogenic substances may be of "foreign" origin, e.g., toxins produced by bacteria (so-called endotoxins) or may be generated within the body itself, e.g., when breakdown of tissues or blood effusions occur. The heat-

FIG. 1-17.
How fever arises and recedes.

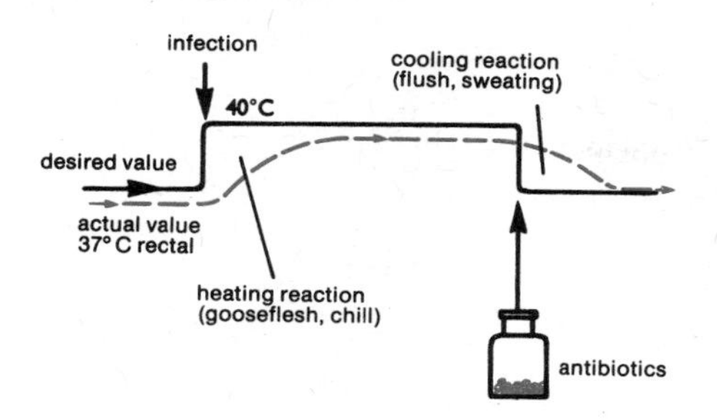

regulating center in the midbrain operates rather in the manner of an ordinary thermostat in maintaining the body temperature at approximately 37°C (98.6°F). It does this by controlling heat production and heat dissipation. With fever (Fig. 1-17) the "thermostat" is, as it were, set to a higher temperature by the action of a pyrogenic substance such as the pyrexins produced in an inflammation. For instance, if the temperature setting is thus raised to 40°C (104°F), the sensitivity of the heat-regulating center is reduced: normal temperature is now measured as 3°C too low and is sensed as "cold." The body counteracts this by increasing its heat production and reducing the dissipation of heat. The patient experiences symptoms such as "gooseflesh" and chill. As a result of the counteraction, the internal temperature of the body rises to 40°C, above which normal regulation is resumed at the higher level. When the pyrexins disappear, the temperature-regulating center is reset to the normal temperature of 37°C. A "hot" sensation is then felt, attended by heat dissipation by conduction, radiation and water evaporation (accompanied by flush and sweating).

The effect of fever on the course of a disease varies and must be judged in each case individually. The rise in body temperature may be beneficial. On the other hand, fever may be harmful or indeed dangerous on account of excessive acceleration of metabolism and breakdown of proteins of vital importance to the body. A temperature of 43°C (109.4°F) is likely to result in death; the highest body temperature known to have been survived by a human being is 43.5°C (109.8°F) (rectal measurement).

A characteristic cell reaction associated with inflammation is multiplication of cells, with cellular infiltration into connective tissue: leukocytes, especially the type known as neutrophil granulocytes, make their way through the walls of the capillaries into the tissue. For this to happen, three conditions must be fulfilled:

1. Attraction of leukocytes (white blood cells) to the center of inflammation by leukotaxis, a particular variety of chemotaxis as found also in unicellular organisms. It is due to the presence of chemical substances (leukotaxins) released from cells affected by inflammation.
2. A certain stickiness of the leukocytes, so that they will adhere to the endothelium cells forming the lining of the capillaries. This condition is fulfilled by the slowing down (prestasis) of the blood flow in the inflamed area.
3. The leukocytes must come into contact with this endothelium. This is by no means self-evident, as leukocytes, being relatively large blood cells, tend to move along centrally, not along the walls, in small blood vessels. This situation is changed, however, when the blood flow slows down as a result of conglomeration of the red cells (erythrocytes) into larger masses which push the leukocytes to the outside of the flow, i.e., to the walls of the capillaries (so-called granular blood flow) (Fig. 1-18). Only when this occurs do the leukocytes actually pass through the walls.

Each leukocyte extends thin protoplasmic protrusions (pseudopodia) which make an entry into the gap between two adjacent endothelium cells of the capillary wall, and the whole leukocyte then squeezes through the gap. Within a few minutes the leukocyte has thus entered the connective tissue and, attracted by chemotaxis, is "creeping" toward the center of inflammation. It travels at a speed of about 5 mm per day (Fig. 1-19).

One of the principal functions of the leukocytes is *phagocytosis,* i.e., the ingestion and digestion of bacteria and other "foreign" particles.

The neutrophil granulocytes are the first white blood cells to make their way out of the bloodstream; they are followed by lymphocytes, another kind of white blood cell, which perform various functions and are more characteristic of chronic inflammation. Lymphocytes remain viable in the tissue for a longer time than the granulocytes. Also, lymphocytes can transform themselves into other types of cells: into monocytes, for example, or into histiocytes, the latter being the resting form of

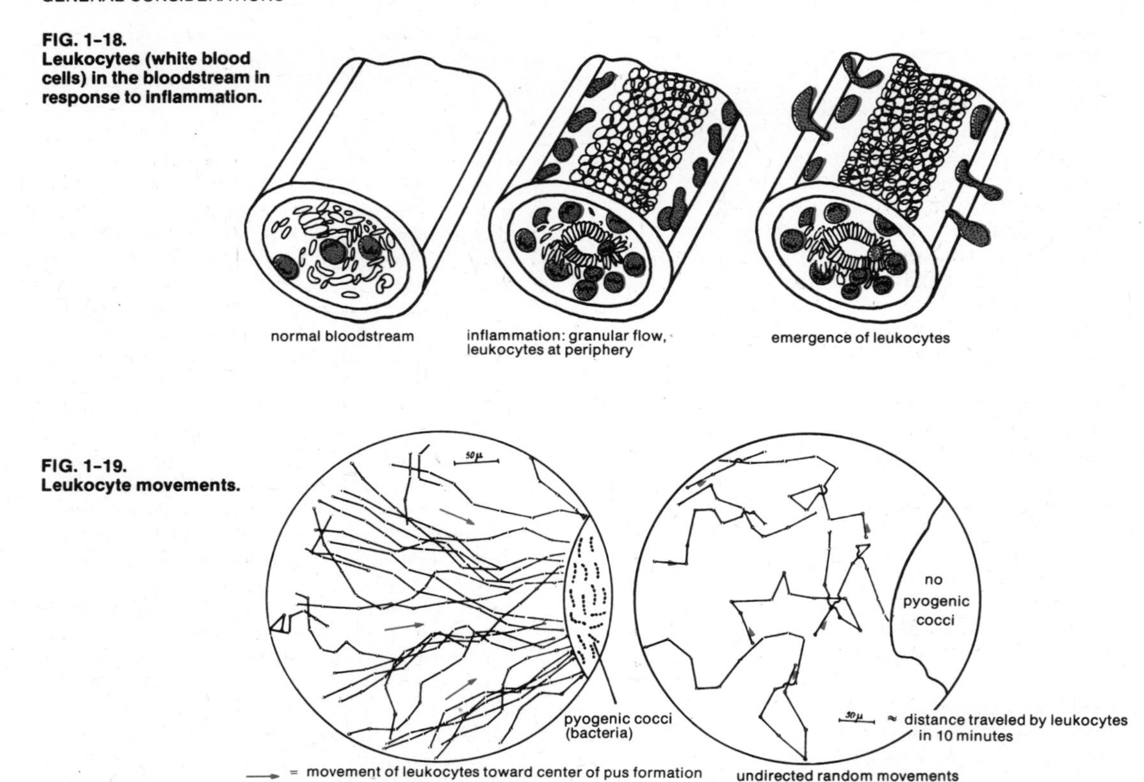

FIG. 1–18. Leukocytes (white blood cells) in the bloodstream in response to inflammation.

FIG. 1–19. Leukocyte movements.

macrophages (phagocytic wandering cells) which are present in connective tissue. Alternatively, lymphocytes can return into the blood with the lymph flow; this is probably involved with their function in detecting the presence of antigens (foreign substances, bacteria, etc.) and raising the alarm. Monocytes are formed from emigrated lymphocytes, and also emerge direct from the blood, of which they are a normal component; in the tissue they are in part transformed into histiocyte-type cells. In a case of acute inflammation that lasts a long time, or indeed in chronic inflammation, transformation of lymphocytes and monocytes (together with the histiocytes) will result in an overall increase in the number of fibroblasts, these being cells that produce collagen and elastic fibers and thus form new connective tissue.

Besides cellular there is humoral (chemical) defense, especially against viruses and bacteria that have entered the body. These humoral defensive processes, involving such conceptions as lysozyme, properdin, immunoglobulin and antigen-antibody reaction, play a prominent part in combating infections and infectious diseases.

Chronic inflammation may occur as such from the outset, or it may develop from an acute inflammation. Whereas the latter is characterized by exudation of fluid, the prominent feature of chronic inflammation is proliferation, i.e., rapid reproduction and formation of tissue. Nevertheless the small blood vessels and surrounding tissue-forming cells play an important part in this type of inflammation, too. The blood vessels put out offshoots which in turn branch out and finally present an overall appearance of numerous reddish granules, described by the term *granulation tissue*. Such tissue is, for example, formed on the bottom of old wounds or ulcers and grows toward the surface. Between the blood vessels are mainly lymphocytes and numerous fibroblasts which ensure that the space between the offshoots is continually filled with new connective tissue.

Granulation tissue is responsible, among other things, for the repair of superficial defects in tissue; it then occurs at the bottom of an ulcer (ulcus) and is often covered by a surface layer of serous or purulent exudate. Abscesses that have persisted for a long time also become surrounded by an abscess membrane of granulation tissue. If the abscess rises slowly to the surface from below, a passage enclosed by granulation tissue may develop and maintain a permanent tubelike connection (fistula) between the old center of inflammation and the skin or the mucous membrane.

As healing of the chronic inflammation advances, the number of cells in the old center of inflammation decreases, while that in the connective tissue fibers increases, until the fibers eventually force back the blood vessels. The *scar* (cicatrix) formed in this healing process tends to shrink as a result of contraction or aging of the blood cells in the connective tissue. Such contraction may be beneficial by giving additional strength to the former center of inflammation. If the cicatrix encloses a hollow organ, shrinkage may, however, have disagreeable effects by constricting the passage or orifice (stenosis: stricture due to contracted cicatrix). This may, for instance, occur at the outlet from the stomach to the intestine after healing and cicatrization of gastric ulcers (pyloric stenosis). Contraction of scar tissue often also plays a part in cardiac valve defects; the contraction of scars formed on extensive burn wounds on the skin can be objectionable from a cosmetic point of view.

WOUNDS AND WOUND TREATMENT

A wound is an injury to tissue due to an external cause. Superficial *scratches* or *abrasions* (grazes) usually require no treatment and will heal of their own accord. However, if they are relatively extensive and occur especially on the hands or feet, it is advisable to inject with tetanus antitoxin (see page 334). Simple smooth incisions (*cuts*) usually cleanse themselves by the discharge of blood and will then readily heal. On the other hand, *lacerations* (tears) or *contusions* (bruises), such as often occur in accidents, frequently are heavily contaminated with dirt. Also, they are usually more extensive than would appear from a superficial inspection. Quite often the surrounding tissue in the edge zone of the wound has been damaged by mechanical action and dies after a time.

"Compound" or "complicated" wounds usually occur as a result of injury also involving organs situated deeper down, such as the intestine or the lung in the case of gunshot or stab wounds. Such wounds are especially prone to infection. In *bites* there is usually only a small external opening, but the teeth that caused the wound infect it more particularly with anaerobic bacteria, which can grow and multiply in the absence of atmospheric oxygen. The canine teeth of wild animals, cats and dogs produce deep puncture wounds. Bites should therefore always be regarded as infected and be treated by a doctor.

The two principal hazards of open wounds are bleeding and infection. In superficial abrasions only minor branches of blood vessels and capillaries are involved. The blood merely oozes out of the wound. Larger wounds may result in damage not only to these minor blood vessels, but also to veins and arteries. When a vein is opened, blood of dark red color flows out in a steady nonpulsating stream. Minor *venous hemorrhages* can usually be stanched by means of a pressure bandage and by elevating the wounded limb or other part.

Arterial hemorrhages are characterized by the pulsating ejaculation of bright red blood from the wounded artery. Frequently the injured walls of small arteries become convoluted, so that in such cases, too, the bleeding can be stanched by elevating the parts affected. Otherwise, as a first-aid measure, the limb should be tied with a tourniquet applied above the wound, i.e., to the part nearer the heart. A tourniquet should, however, never be applied for longer than about 10 to 15 minutes; it should be released at regular intervals for half a minute or so, for otherwise there will be a risk of devitalizing the limb.

The further treatment of *major wounds* should always be left to a doctor. In the case of a deep, gaping flesh wound the edges of the wound are stitched together by means of a simple skin suture. As firmly united edges heal quickly and are cosmetically favorable (less scarring), the aim will always be to achieve a smooth suture. For this purpose, bites or contusions have to be "cleaned up" by cutting away parts of the affected tissue, the edges of the wound then being united by stitching. The contaminated and no longer viable tissue is removed with scissors, knife and forceps (Fig. 1-20). It is essential that the wound receive such treatment within about six hours; otherwise the bacterial spores that have entered the wound will develop into bacteria which can

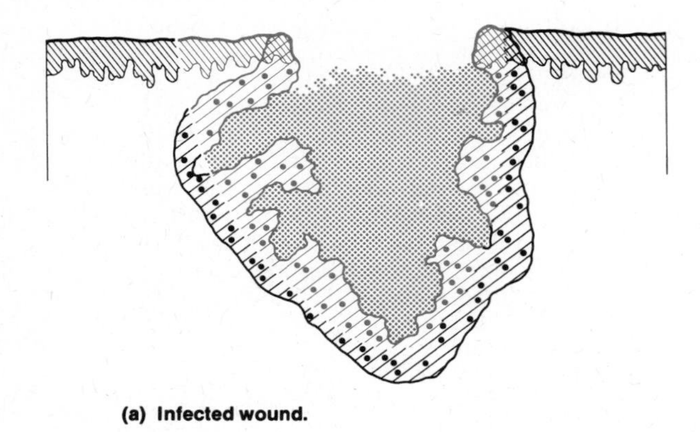

(a) Infected wound.

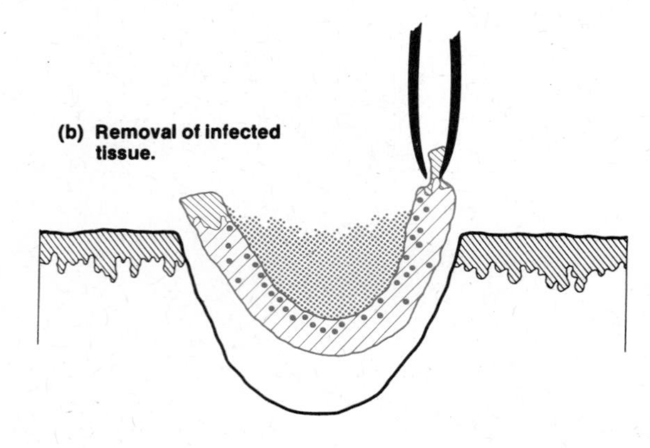

(b) Removal of infected tissue.

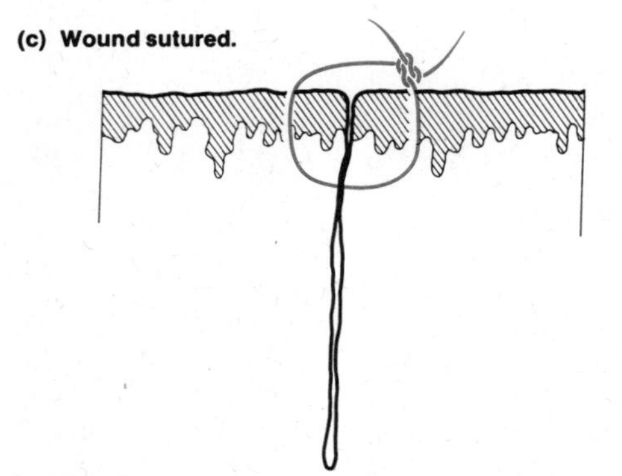

(c) Wound sutured.

FIG. 1-20.

grow and multiply: the infected wound becomes inflamed (Fig. 1-21). Apart from stanching the bleeding, the most important requirement in the treatment of wounds is to rest the wounded part of the body, in order to assist undisturbed healing.

WOUND INFECTION

In healthy people the blood seeping from a wound will clot within minutes to form a crust or scab. The scab acts as a natural protection against the entry of bacteria, against further loss of blood and against drying of the wound surface; it should therefore not be removed. Also, it is better not to swab the wound surface, as this is liable to delay scab formation and to transfer large numbers of germs to the wound.

Nearly every wound is infected from the outset. Without medical attention the number of microbes that have penetrated into it will increase. They may make their way from the edges of the wound into the adjacent skin and cause *erysipelas,* a painful spreading febrile inflammation (Fig. 1-21a). As destroyed tissue material and bacterial toxins are carried along by the lymph, bacteria capable of reproduction may enter the lymphatic system and cause inflammation there. Red streaks on the skin and painful swelling of the lymph nodes are always signs that the local inflammation at a wound is developing into a general infection (blood poisoning) (Fig. 1-21b). If bacteria penetrate into the interstices of the subcutaneous cellular tissue, a so-called cellular tissue inflammation develops (Fig. 1-21c). The skin is reddened, hot and swollen; it "throbs" in the inflamed area. Such symptoms indicate that surgical intervention is necessary.

Not infrequently the surroundings of an infected wound take on a bluish-red color. This is because the bacteria penetrate into the adjacent blood vessels and set up inflammation in them. As a result, blood clots are formed and minor blockages occur, eventually followed by congestion of the blood vessels in the area affected (Fig. 1-21d).

If the escape of pus from an infected wound is prevented, e.g., because the opening is too small as in the case of a stab wound, an accumulation of pus—an abscess—may form (Fig. 1-21e). The action of the bacterial toxins, which attract white blood cells, causes more and more tissue cells to be involved in the abscess, which cells then die and thereby enlarge the cavity. *Pus* consists of bacteria, white blood cells and remnants of damaged tissue. It is a by-product of inflammation—the battle that the body fights against invading bacteria (see also pages 109 ff).

If the body's defensive forces succumb to the destructive action of the bacteria, *blood*

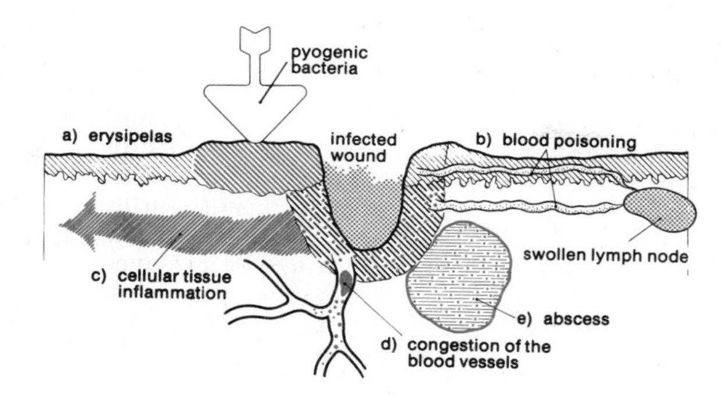

FIG. 1-21.
Forms of wound infection.

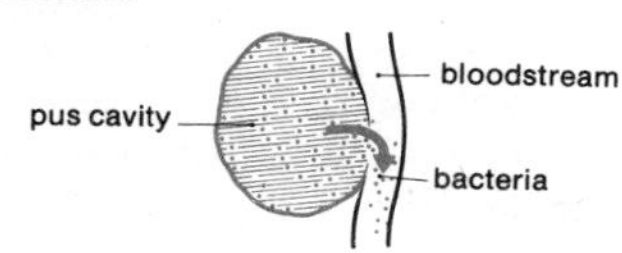

poisoning (sepsis) may develop at any stage of the wound inflammation. When that happens, the pyogenic (pus-producing) microorganisms penetrate into the bloodstream, where they multiply. Blood poisoning starts with a chill, i.e., an attack of shivering, followed by a rapid rise in body temperature to between 39° and 41°C (102.2° and 105.8°F) (Fig. 1-22).

The increased metabolic rate due to the fever causes the heart to beat faster. A rule of thumb states that for each 1°C (1.8°F) rise in temperature the pulse rate increases by 12 beats per minute, the normal rate being 75–80. The patient usually suffers from severe headache, loss of appetite and nausea. In the past, blood poisoning was very dangerous, often fatal. Nowadays it can be successfully treated

FIG. 1-22.
Blood poisoning.

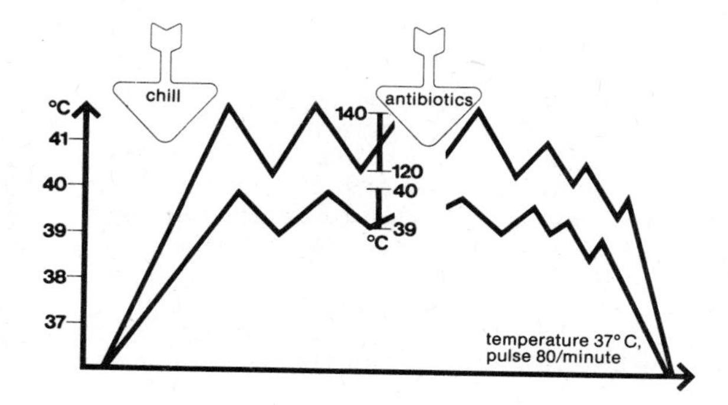

with antibiotics such as penicillin or tetracycline. With their help it is possible even in severe cases to bring the temperature down, reduce the pulse rate, and finally achieve a cure by eradication of the bacteria (Fig. 1-22).

DRUGS

Drugs in a technical sense are chemical substances, or preparations from such substances, whereby the condition or the functions of the human body or mental diseases can be diagnosed (diagnostic drugs) or be remedially influenced (remedial drugs). Drugs can for the present purpose be taken to comprise those substances that can be used to replace the body's own fluids or physiologically active components. These include blood substitutes and hormones. Furthermore, drugs include substances that are able to render harmless or destroy pathogenic (disease-producing) microorganisms, parasites and toxins which enter the body from outside, e.g., antibiotics, serums and antitoxins.

In nontechnical language the term *drugs* is applied also to narcotics or habit-forming substances. These will not be considered in the present discussion, however.

A relatively small number of drugs make up deficiencies of essential substances and can thus be said to be directly beneficial and not produce ill effects. Most drugs, however, act by interfering with particular chemical processes or by disturbing the function of a particular organ. The direct effect is usually harmful to the organ or the tissue concerned, though the result to the body as a whole will, on balance, be beneficial. The action of many drugs depends on their ability to act in the same manner, and produce the same effect, as a natural reagent in the body. An ideal antibacterial drug would kill bacteria but have no harmful effect on human tissue. In reality this can seldom be achieved; it is usually a matter of degree, a compromise. For example, the arsenic compounds formerly used in the treatment of syphilis are poisonous to all living cells, but are more lethal to the parasite than to the patient.

A drug cannot exercise its effect until it gets to the intended place of action. So it must first enter the patient's body, i.e., it has to be injected or be absorbed through the skin or a mucous membrane. When a drug is injected directly into the bloodstream, it does not first have to be absorbed, but it still has to be transported to its place of action.

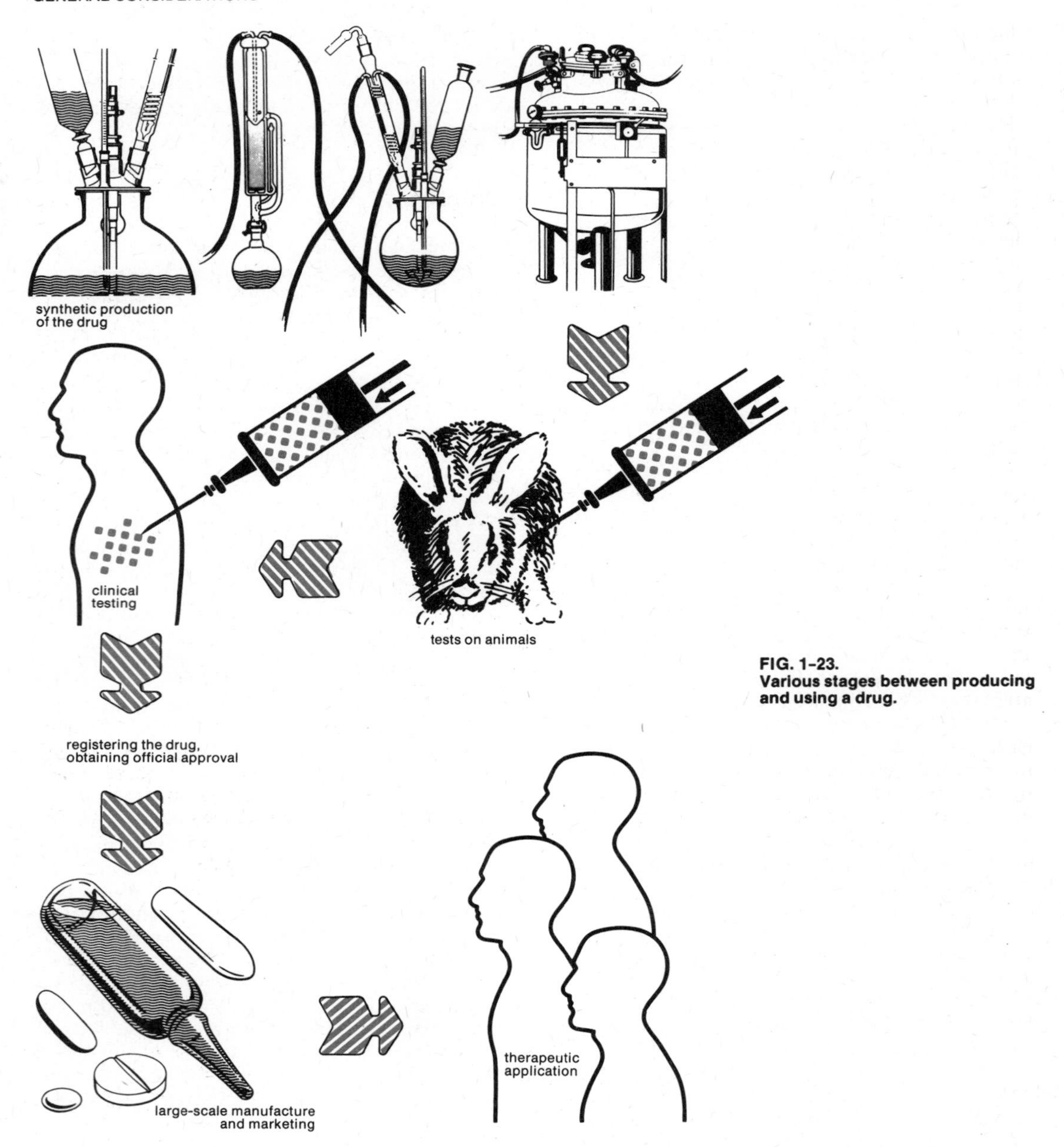

FIG. 1-23.
Various stages between producing and using a drug.

In connection with the *absorption* of drugs from the surface of the skin or through mucous membranes of the stomach and intestines, an important part is played by (passive) diffusion and (sometimes) active transport processes. The rate of absorption will depend on where the point of input is located and on the physiochemical properties of the drug. Absorption of any particular drug is accelerated by small molecular size, low electrochemical charge and good solubility in water and fat; it also depends on the blood flow volume and the local permeability of the drug. Contrary properties, such as large molecular size and low solubility of the drug, will reduce its rate of absorption—a fact that is utilized in the production of slow-acting drugs designed to develop their action over a long period.

For *local treatment* it is important that a drug should confine its action to the point of application and should, after absorption, not significantly affect the rest of the organism. The possibility of this type of action is not confined to the skin and those mucous membranes that are easily accessible. For example, activated charcoal can hold many substances to its surface (by adsorption) and may be administered as an antidote to poisons in the bowel; drugs injected into joints can locally relieve diseased conditions. A drawback of local treatment applied to the skin or mucous membranes is the relatively major hazard of sensitization (see allergy, pages 45 ff).

For general treatment, drugs are usually administered *orally*, i.e., through the mouth. The rate of absorption will depend, among other things, on the form in which the drug is taken (tablet, pill, capsule, solution), the properties of the drug itself, and the functional condition and filling of the stomach and bowel. After being absorbed through the intestinal wall, generally in the small intestine, the drug is carried by the blood to the liver, where it may be intercepted or chemically changed. This journey to the liver can be avoided by administering the drug *sublingually* (under the tongue, in the form of capsules or pastilles) or *rectally* (through the rectum, in the form of suppositories). These methods are used especially in cases where the drug would be broken down quickly in the liver or be destroyed by the gastric or intestinal juices if it were taken orally; also, in this way, the liver is spared the action of the drug.

Injection avoids some of the disadvantages associated with giving drugs orally. It enables the drug to make a rapid and effective entry into the body and in an accurately controlled dose. Against this, it requires special care and skill, so that it is usually not possible or advisable to allow the patient to inject himself. The speediest dispersion of the drug is obtained by *intravenous injection*, i.e., into a vein. Substances injected into a muscle (*intramuscular injection*) or into the peritoneal cavity (*intraperitoneal injection*) are, however, absorbed into the bloodstream almost as quickly. On the other hand, drugs injected *subcutaneously*, i.e., under the skin, are absorbed at an appreciably slower rate because the subcutaneous tissue is less well supplied with blood vessels.

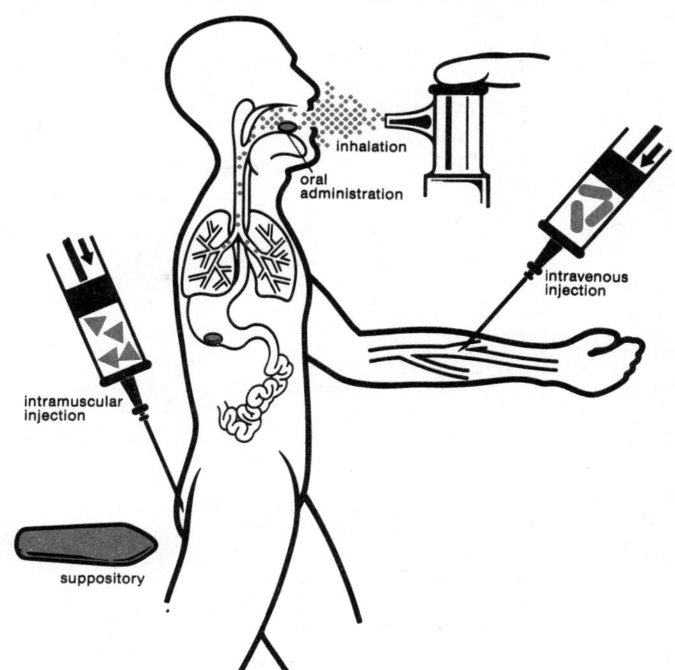

FIG. 1-24.
Various methods of administering a drug.

More or less localized rapid action is obtained when drugs are injected into joints (*intraarticular injection*) or into the cerebrospinal fluid (*intralumbar injection*). Drugs in gaseous or vapor form (anesthetics, for example) can be quickly absorbed by *inhalation* (Fig. 1-24).

After performing their desired actions, drugs must be eliminated from the body and/or be broken down to inert substances. Most drugs are eventually discharged in the urine and feces; some are excreted in perspiration, saliva, milk or exhaled air (gaseous anesthetics, in particular). Of special importance is the ratio between the rate of intake and the rate of discharge of a drug, as this ratio significantly affects the concentration of the drug in the blood and tissues and the duration of its action.

ANESTHESIA

Anesthesia means loss of sensation, with or without loss of consciousness. It refers more specifically to the temporary elimination of the sensation of pain for the purpose of performing surgical operations, usually by means of drugs called anesthetics (pharmacologically induced anesthesia). *General anesthesia* affects the entire body, with loss of consciousness; it is based on controlled functional inhibition or benumbing of various centers in the brain or spinal cord. It must be clearly distinguished from *local anesthesia,* which affects a local area only, the sensation of pain being eliminated in that area by the action of the anesthetic on nerve tracts or nerves leading to the brain, while there is no loss of consciousness (see page 34).

Anesthetics—drugs that produce anesthesia—penetrate not only into the central nervous system but, theoretically, also into every cell of the body. Not surprisingly, the various types of cells and tissues respond in differing degrees to anesthetics. In man and the higher animals the nerve tissue is particularly sensitive to these drugs. Anesthetics thus cause larger or smaller sections of the central nervous system to be completely numbed (anesthetized), while the other tissues and organs remain relatively unaffected. In this condition all sensation of pain, consciousness and muscular movements are abolished, without serious interference with metabolism, blood circulation and other vital functions. Even within the nervous system there are considerable differences in sensitivity to anesthetics. Most sensitive in its response is the cerebrum, the largest part of the brain, consisting of two hemispheres, where sensory perceptions and physical sensations are processed into what we call consciousness and where, in addition, the higher centers for the voluntary muscular movements are located. Least sensitive is the medulla oblongata—the enlarged portion of the spinal cord in the skull—which contains the vital nerve centers for the blood circulation and respiration. This fact accounts for the typical progress of anesthesia in successive stages. However, the anesthetic will also find its way into other tissues and organs; it accumulates particularly in the subcutaneous fatty tissue. Because of this, patients with a high proportion of such tissue require a larger dose of anesthetic and lose consciousness less quickly than thin ones, but regain consciousness more slowly too, as the anesthetic stored in the fatty tissue prolongs its action.

During the course of anesthesia the various centers in the brain and spinal cord are not put out of action all at once, but in a certain sequence (*stages of anesthesia*) (Figs. 1-25, 1-26), whereby a step-by-step abolition of the corresponding physical functions is achieved.

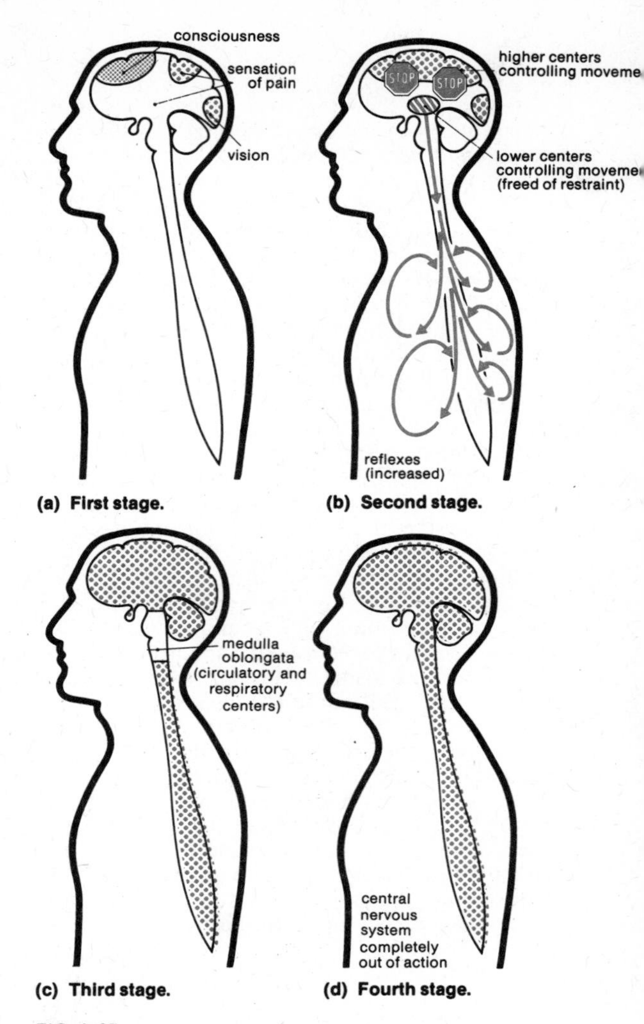

FIG. 1-25.
The four stages of anesthesia: how the central nervous system is progressively put out of action by the anesthetic.

In the first stage of general anesthesia, the *analgesic stage,* the patient first goes into a condition resembling intoxication in which the consciousness is progressively dimmed and is filled with jumbled notions and mental images, often of a dreamlike character. While still partly conscious, the patient loses his sensation of pain, so that in this stage minor surgery can be performed without discomfort to him. The pupils, breathing, muscular reflexes and circulation are still largely unaffected.

The second stage of anesthesia, the *excitation stage,* commences with complete loss of consciousness. As a result of numbing of the voluntary movement centers in the cerebral cortex, which in turn control subsidiary move-

MUSCLES OF THE FRONT OF THE BODY (TOP LAYER)

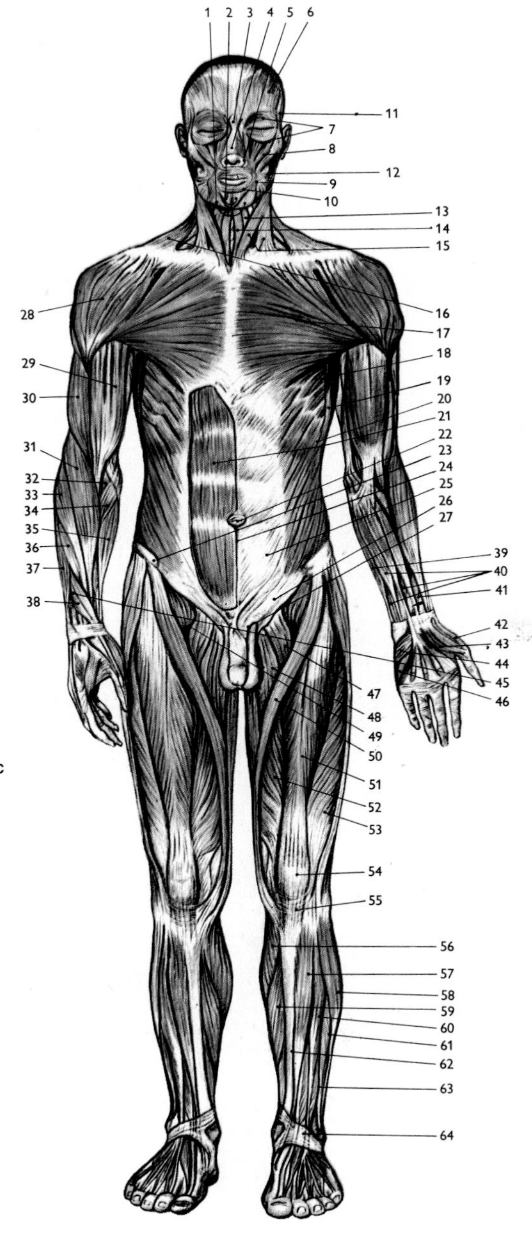

I. Head

Muscles of facial expression

1 Musculus triangularis (depresses angle of mouth)
2 M. mentalis (elevates lower lip)
3 M. depressor glabellae
4 M. nasalis (draws outer wall of nostril downward)
5 M. levator labii superioris (elevates upper lip)
6 M. frontalis (moves scalp, raises eyebrows)
7 M. orbicularis oculi (closes eyelid, wrinkles forehead)
8 Musculi zygomaticus major and minor (draw upper lip up and out)
9 M. risorius (laughing muscle: draws angle of mouth outward)
10 M. orbicularis oris (closes lips)

Muscles of mastication

11 M. temporalis (closes jaws)
12 M. masseter (closes jaws, principal muscle of mastication)

II Neck

13 M. thyrohyoideus
14 M. sternohyoideus
15 M. sternocleidomastoideus (rotates and depresses head)

III. Chest and abdomen

16 M. trapezius (draws head back and to the side)
17 M. pectoralis major (flexes and rotates arm)
18 M. serratus anterior (elevates ribs)
19 M. latissimus dorsi (adducts, extends, and rotates arm)
20 M. obliquus externus abdominis (contracts abdomen)
21 M. rectus abdominis (compresses abdomen)
22 Umbilicus (navel)
23 Spina iliaca ventralis (spine of hipbone)
24 Linea alba (white line)
25 Vagina musculi recti abdominis (sheath of muscle)
26 Ligamentum inguinale (inguinal ligament)
27 Canalis inguinalis (inguinal canal) with funiculus spermaticus (spermatic cord)

IV. Upper arm

28 M. deltoideus (deltoid muscle: raises and rotates arm)
29 M. biceps brachii (biceps: flexes arm)
30 M. brachialis (flexes forearm)
31 M. brachioradialis

V. Forearm

32 M. pronator teres
33 M. extensor carpi radialis longus (extends and abducts wrist)
34 M. flexor carpi radialis (flexes and abducts wrist)
35 M. palmaris longus (tightens palmar fascia, flexes wrist)
36 M. extensor carpi radialis brevis (extends and abducts wrist)
37 M. extensor digitorum communis (extends fingers and wrist)
38 M. extensor pollicis brevis (extends thumb)
39 M. flexor carpi ulnaris (flexes and abducts wrist)
40 M. flexor digitorum superficialis (flexes hand)
41 M. flexor pollicis longus (flexes thumb)

VI. Hand

42 M. abductor pollicis brevis (abducts thumb)
43 M. flexor pollicis brevis (flexes thumb)
44 M. palmaris brevis (wrinkles skin on inside of hand)
45 M. abductor pollicis longus (abducts thumb)
46 M. flexor digiti quinti brevis (flexes little finger)

VII. Thigh

47 M. pectineus (flexes and adducts thigh)
48 M. adductor longus (adducts and flexes thigh)
49 M. iliopsoas (or M. psoas: flexes and adducts thigh)
50 M. sartorius (flexes and rotates thigh and leg)
51 M. rectus femoris (extends leg)
52 M. vastus tibialis
53 M. vastus fibularis
54 Patella (kneecap)
55 Ligamentum patellae (patellar ligament)

VIII. Leg and foot

56 M. gastrocnemius (flexes foot and leg)
57 M. tibialis anterior (elevates and flexes foot)
58 M. peroneus longus (extends foot)
59 M. soleus (extends and rotates foot)
60 M. extensor digitorum longus (extends toes, flexes foot)
61 M. peroneus brevis (extends foot)
62 Tibia (shinbone)
63 M. extensor hallucis longus (extends foot)
64 Ligamentum cruciforme pedis (cruciate ligament)

MUSCLES OF THE BACK OF THE BODY (TOP LAYER)

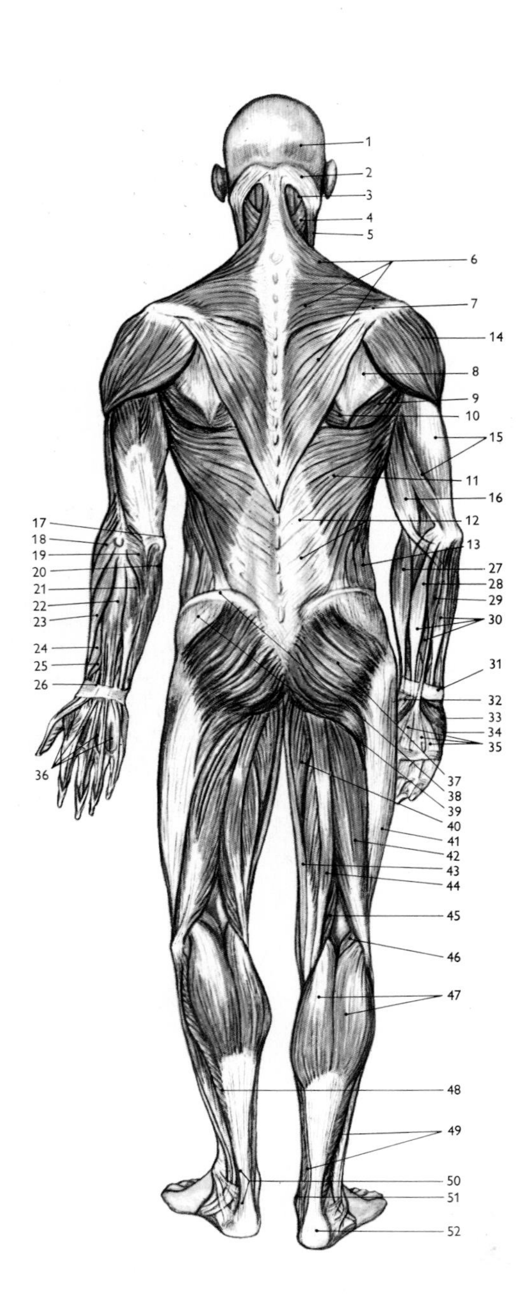

I. Head and neck
1 Musculus occipitalis (moves scalp, raises eyebrows)
2 Septum nuchae
3 M. transversooccipitalis
4 M. splenius capitis
5 M. sternocleidomastoideus (rotates and depresses head)

II. Back and abdomen
6 M. trapezius (draws head back and to the side)
7 Spina scapulae (spine of shoulder blade)
8 M. infraspinatus (rotates arm)
9 M. teres major
10 M. rhomboideus major (elevates shoulder blade)
11 M. latissimus dorsi (extends and rotates arm)
12 Fascia thoracolumbalis
13 M. obliquus externus abdominis (contracts abdomen)

III. Upper arm
14 M. deltoideus (deltoid muscle: raises and rotates arm)
15 M. triceps brachii (triceps: extends arm)
16 M. biceps brachii (biceps: flexes arm)

IV. Forearm
17 Caput ulnare (ulnar head)
18 M. extensor carpi radialis longus (extends and abducts wrist)
19 M. anconeus (extends forearm)
20 M. flexor carpi radialis (flexes and abducts wrist)
21 M. extensor carpi ulnaris (extends and abducts wrist)
22 M. extensor digitorum communis (extends fingers and wrist)
23 M. extensor carpi radialis brevis (extends and abducts wrist)
24 M. abductor pollicis longus (abducts thumb and wrist)
25 M. extensor pollicis brevis (extends thumb)
26 Ligamentum carpi dorsale
27 M. brachioradialis (flexes forearm)
28 M. flexor carpi radialis (flexes and abducts wrist)
29 M. palmaris longus (tightens palmar fascia, flexes wrist)
30 M. flexor digitorum superficialis (flexes hand)
31 Ligamentum carpi volare radiatum

V. Hand
32 M. flexor pollicis brevis (flexes thumb)
33 M. flexor digiti quinti brevis (flexes little finger)
34 M. opponens digiti quinti
35 Aponeurosis palmaris
36 Musculi interossei dorsales (abduct and adduct fingers)

VI. Thigh
37 M. gluteus maximus (extends and rotates thigh)
38 Crista iliaca (iliac crest)
39 M. gluteus medius (abducts and rotates thigh)
40 M. adductor magnus (adducts and rotates thigh)
41 Tractus iliotibialis fasciae latae
42 M. biceps femoris (flexes and rotates knee)
43 M. gracilis (flexes and adducts leg)
44 M. semitendinosus (flexes and rotates leg)
45 M. semimembranosus (flexes and rotates leg)

VII. Leg and foot
46 M. plantaris (extends foot)
47 M. gastrocnemius (flexes foot and leg)
48 M. flexor digitorum longus (flexes phalanges, extends toes)
49 M. soleus (extends and rotates foot)
50 Tendo calcaneous (Achilles tendon)
51 Malleolus tibiae (ankle)
52 Calcaneus (heelbone)

MEDIAN SECTION THROUGH SKULL WITH BRAIN, FACE, THROAT, AND NECK

1 Cutis (skin)
2 Meninges (membranes): dura mater encephali, pia mater encephali, arachnoidea encephali
3 Cerebrum
4 Arteria cerebralis anterior (anterior cerebral artery)
5 Hypothalamus with optic chiasm (crossing of optic fibers)
6 Corpus callosum (transverse band of nerve fibers)
7 Hypophysis (pituitary gland)
8 Diencephalon with thalamus
9 Gyrus precentralis and gyrus postcentralis (cerebral convolutions)
10 Mesencephalon (midbrain) and rhombencephalon (hindbrain)
11 Ventriculus tertius (third ventricle of brain)
12 Corpus pineale or epiphysis (pineal body)
13 Sinus durae matris (drains venous blood from meninges)
14 Sella turcica (pituitary fossa) containing the pituitary gland
15 Pons varolii
16 Cerebellum with arbor vitae
17 Centriculus quartus (fourth ventricle of brain)
18 Os occipitale (occipital bone)
19 Medulla oblongata
20 Cisterna cerebellomedullaris
21 Atlas (first cervical vertebra)
22 Medulla spinalis (spinal cord)
23 Axis (second cervical vertebra)
24 Spinous process of vertebra
25 Body of vertebra
26 Glandula thyroidea (thyroid gland)
27 Cartilago cricoidea (cricoid cartilage)
28 Cartilago thyroidea (thyroid cartilage)
29 Trachea (windpipe)
30 Rima glottidis (glottis)
31 Esophagus (gullet)
32 Epiglottis (shown raised)
33 Mandibula (lower jaw)
34 Tonsilla palatina (tonsil)
35 Uvula
36 Palatum molle (soft palate)
37 Lingua (tongue)
38 Labia inferiora (lower lip)
39 Dentes (teeth)
40 Labia superiora (upper lip)
41 Palatum durum (hard palate)
42 Pharynx (throat)
43 Tonsilla pharyngica (adenoid)
44 Concha nasalis inferior (inferior nasal concha)
45 Concha nasalis media (middle nasal concha)
46 Sinus sphenoideus
47 Os nasale (nasal bone)
48 Lamina cribriformis (cribriform plate of ethmoid bone)
49 Sinus frontalis (frontal sinus)

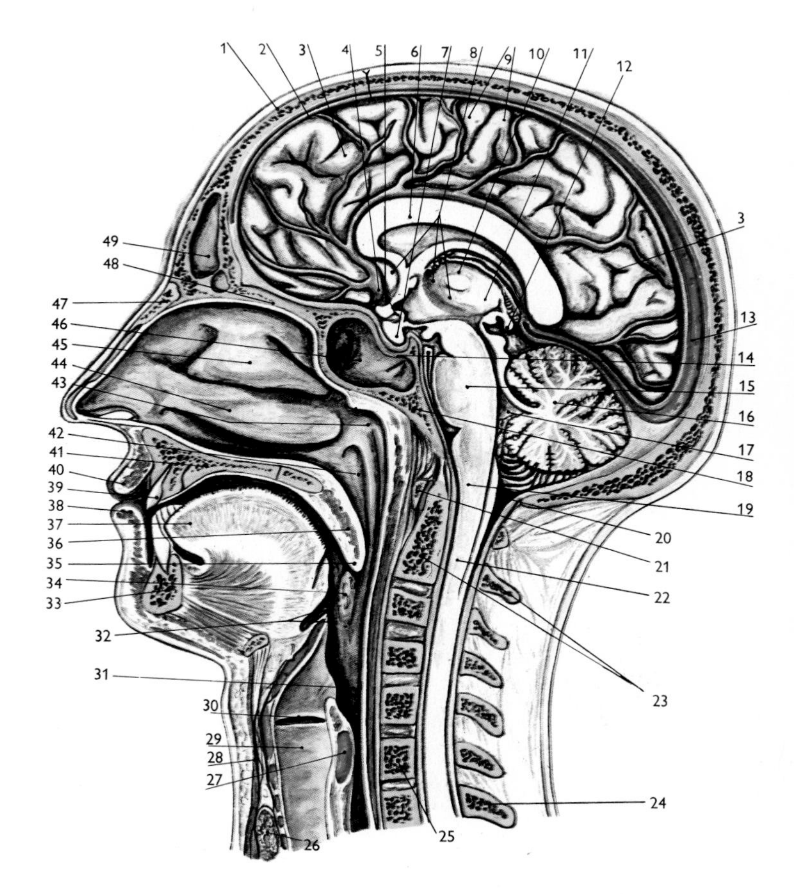

INTERNAL ORGANS OF THORAX AND ABDOMEN (TOP LAYER)

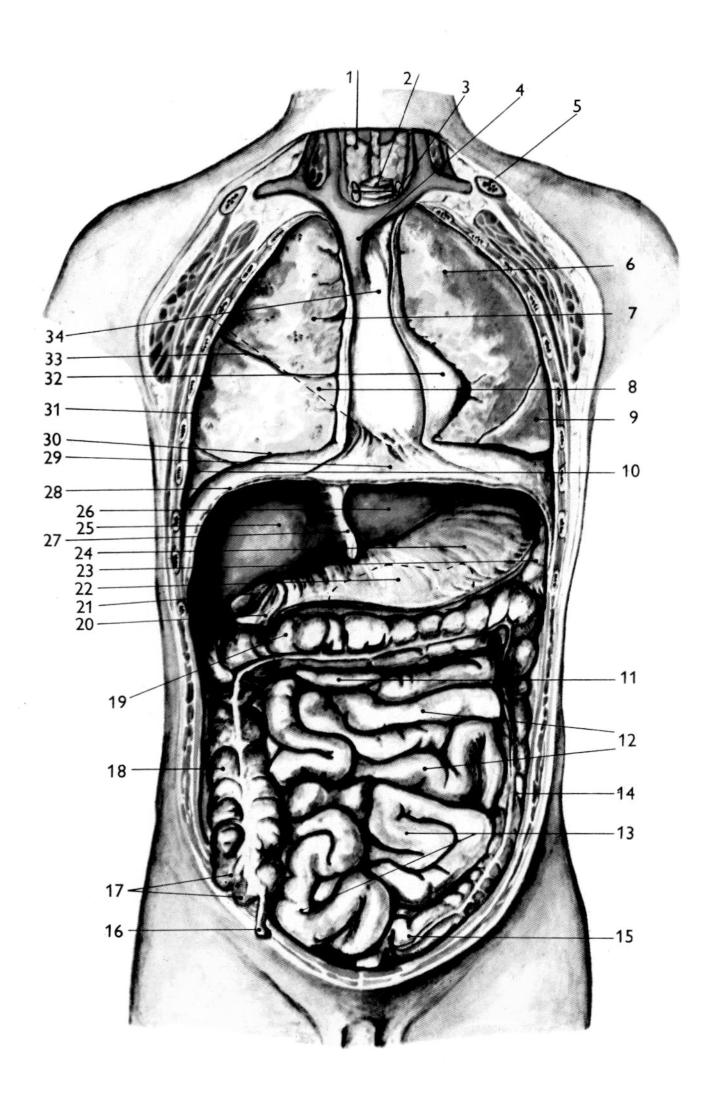

I.

1 Glandula thyroidea (thyroid gland)
2 Trachea (windpipe)
3 Superior vena cava
4 Clavicula (clavicle or collarbone)
5 Left upper lobe of lung
6 Right upper lobe of lung
7 Right middle lobe of lung
8 Left lower lobe of lung
9 Right lower lobe of lung

II. Intestinum tenue (small intestine)

10 Intestinum duodenum (duodenum)
11 Intestinum jejunum (jejunum)
12 Intestinum ileum (ileum)

III. Intestinum crassum (large intestine)

13 Colon descendens (descending colon)
14 Colon sigmoideum (sigmoid flexure)
15 Appendix vermiformis (appendix)
16 Intestinum caecum (cecum)
17 Colon ascendens (ascending colon)
18 Colon transversum (transverse colon)

IV.

19 Pylorus (lower orifice of stomach)
20 Vesica fellea (gallbladder)
21 Pancreas (in dotted outline: behind the stomach)
22 Lien (spleen)
23 Ventriculus (stomach)
24 Lobus hepatis dexter (right lobe of liver)
25 Lobus hepatis sinister (left lobe of liver)
26 Mesohepaticum

V.

27 Peritoneum
28 Diaphragma (diaphragm or midriff)
29 Pleura diaphragmatica (pleura on upper surface of diaphragm)
30 Pleura costovertebralis (parietal pleura)
31 Pericardium
32 Pleura pulmonalis (pleura investing the lungs)
33 Mediastinum (interpleural space)

MEDIAN SECTION THROUGH THE MALE PELVIS

1 Vas deferens
2 Ampulla of the vas deferens
3 Intestinum rectum (rectum)
4 Prostata (prostate gland)
5 Glandula bulbourethralis (Cowper's gland, bulbourethral gland)
6 Anus with sphincter muscles
7 Scrotum
8 Epididymis
9 Vas deferens
10 Testis (testicle)
11 Preputium (prepuce)
12 Glans penis
13 Penis
14 Urethra
15 Corpus cavernosum penis (erectile tissue)
16 Corpus spongiosum penis (erectile tissue)
17 Symphysis
18 Vesica urinalis (bladder)

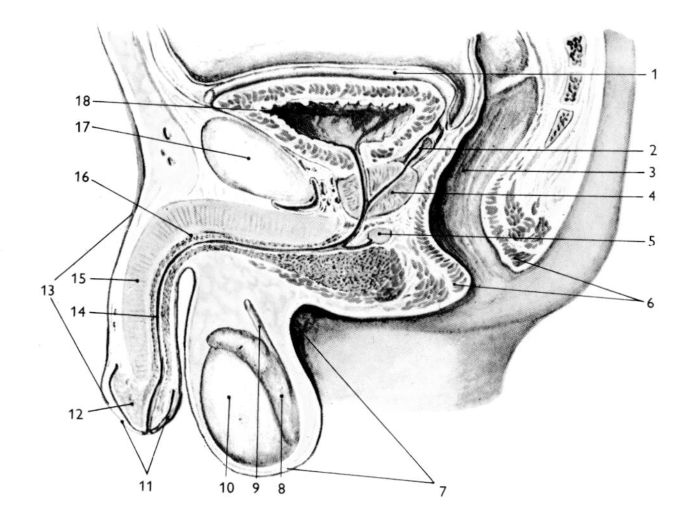

MEDIAN SECTION THROUGH THE FEMALE PELVIS

I.
1 Promontorium (promontory of sacrum)
2 Os sacrum (sacrum)
3 Douglas's pouch
4 Os coccygis (coccyx)
5 Intestinum rectum (rectum)
6 Anus with sphincter muscles
7 Perineum
8 Urethra
9 Symphysis
10 Vesica urinalis (bladder)
11 Ligamentum rotundum

II. Uterus (womb)
12 Fundus uteri (fundus of the uterus)
13 Corpus uteri (body of the uterus)
14 Myometrium (muscular wall of the uterus)
15 Endometrium (lining of the uterus)
16 Cavum corporis uteri (uterine cavity)
17 Isthmus uteri (isthmus of uterine cavity)
18 Cervix uteri (cervix of the uterus)
19 Os uteri externum/internum (external/internal os: external and internal openings of cervical canal, respectively)
20 Portio vaginalis

III.
21 Vagina
22 Hymen

IV. Pudendum or vulva (external female genitals)
23 Labium minus pudendi
24 Labium majus pudendi
25 Clitoris with erectile tissue

V. Uterine appendages
26 Tuba uterina (fallopian tube or oviduct)
27 Ovarium (ovary)

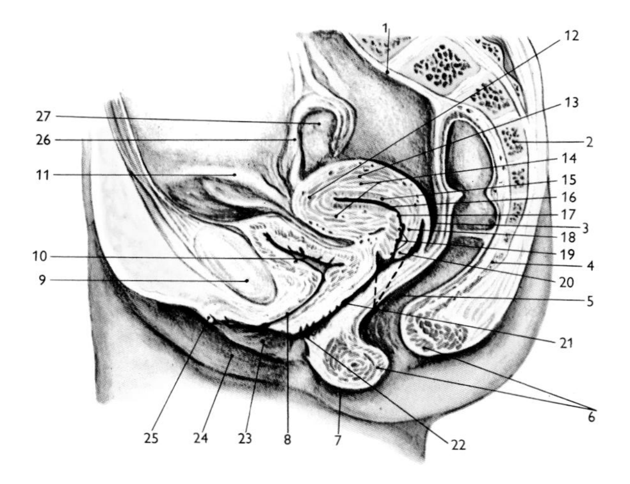

OBLIQUE FRONTAL SECTION THROUGH THORAX

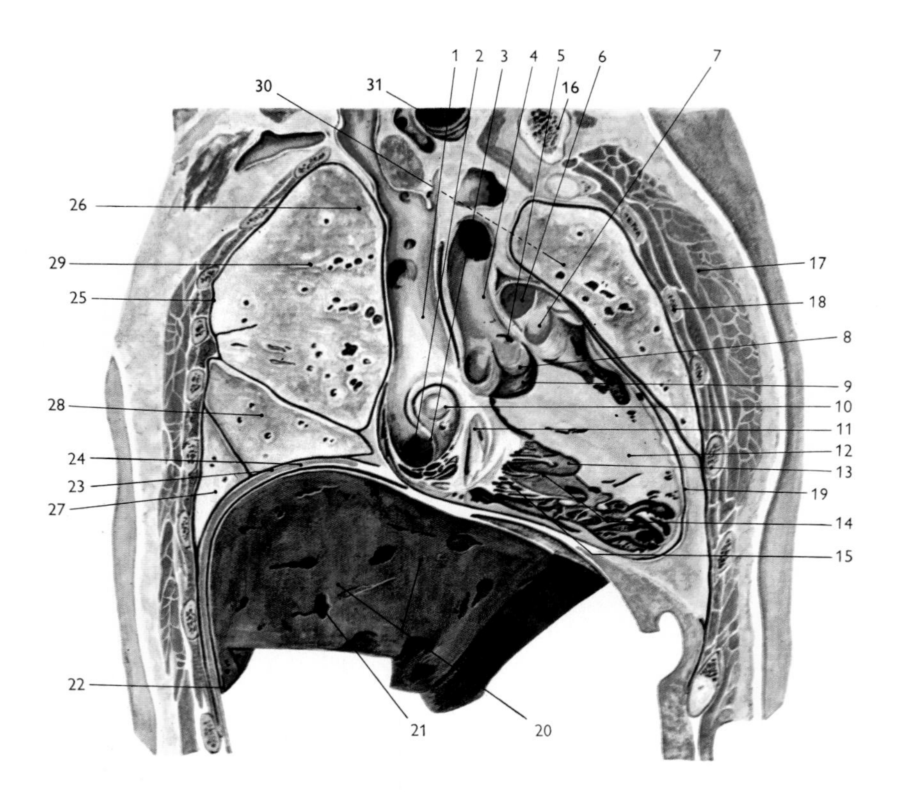

I. Heart

1 Superior vena cava and section through right atrium
2 Atrium cordis dextrum (right atrium of heart)
3 Outlet of vena cava caudalis
4 Aorta (main arterial trunk)
5 Inlet of arteria coronaria sinistra (left coronary artery)
6 Arteria pulmonalis (pulmonary artery)
7 Semilunar valves of pulmonary artery
8 Semilunar valves of aorta
9 Ventriculus sinister cordis (left ventricle of heart)
10 Fossa ovalis
11 Valvula tricuspidalis (tricuspid valve)
12 Septum ventriculorum (interventricular septum)
13 Musculi papillares (papillary muscles)
14 Ventriculus dexter cordis (right ventricle of heart)
15 Chordae tendineae

II.

16 Section through clavicula (clavicle or collarbone)
17 Muscles of wall of thorax (chest)
18 Section through costae (ribs)
19 Pericardium
20 Hepar (liver)
21 Blood vessel in liver
22 Peritoneum
23 Pleura diaphragmatica (pleura on upper surface of diaphragm)
24 Diaphragma (diaphragm or midriff)
25 Pleura costovertebralis (parietal pleura)
26 Pleura pulmonalis (pleura investing the lungs)

III. Lung

27 Right lower lobe
28 Right middle lobe
29 Right upper lobe
30 Left upper lobe
31 Trachea (windpipe)

INTERNAL ORGANS OF THORAX AND ABDOMEN (DEEP LAYER)

I. Blood vessels

1 Arteria carotis communis (common carotid artery)
2 Superior vena cava
3 Arteria subclavia (subclavian artery)
4 Vena subclavia (subclavian vein)
5 Ascending aorta and arch of the aorta
6 Descending thoracic aorta
7 Inferior vena cava
8 Arteria renalis (renal artery)
9 Vena renalis (renal vein)
10 Abdominal aorta
11 Arteria iliaca communis (common iliac artery)
12 Arteria iliaca interna (internal iliac artery)
13 Arteria iliaca externa (external iliac artery)

II. Air passages

14 Trachea (windpipe)
15 Lymph nodes of the hilus of the lung
16 Right upper lobe of lung
17 Bronchus
18 Left upper lobe of lung
19 Right middle lobe of lung
20 Left lower lobe of lung
21 Right lower lobe of lung
22 Diaphragma (diaphragm or midriff)
23 Recessus pleurae

III. Urinary tract

24 Vesica urinalis (bladder)
25 Ureter dexter (right ureter)
26 Ureter sinister (left ureter)
27 Ren sinister (left kidney)
28 Pelvis renalis (renal pelvis)
29 Ren dexter (right kidney)
30 Corpus suprarenale (adrenal or suprarenal gland)

IV. Alimentary tract or canal

31 Intestinum rectum (rectum)
32 Nervus femoralis (femoral nerve)
33 Lymph nodes of the abdominal viscera
34 Cardia (upper orifice of stomach)
35 Esophagus (gullet)
36 Nervus vagus (10th cranial nerve)
37 Plexus brachialis (brachial nerve plexus)
38 Glandula thyroidea (thyroid gland)

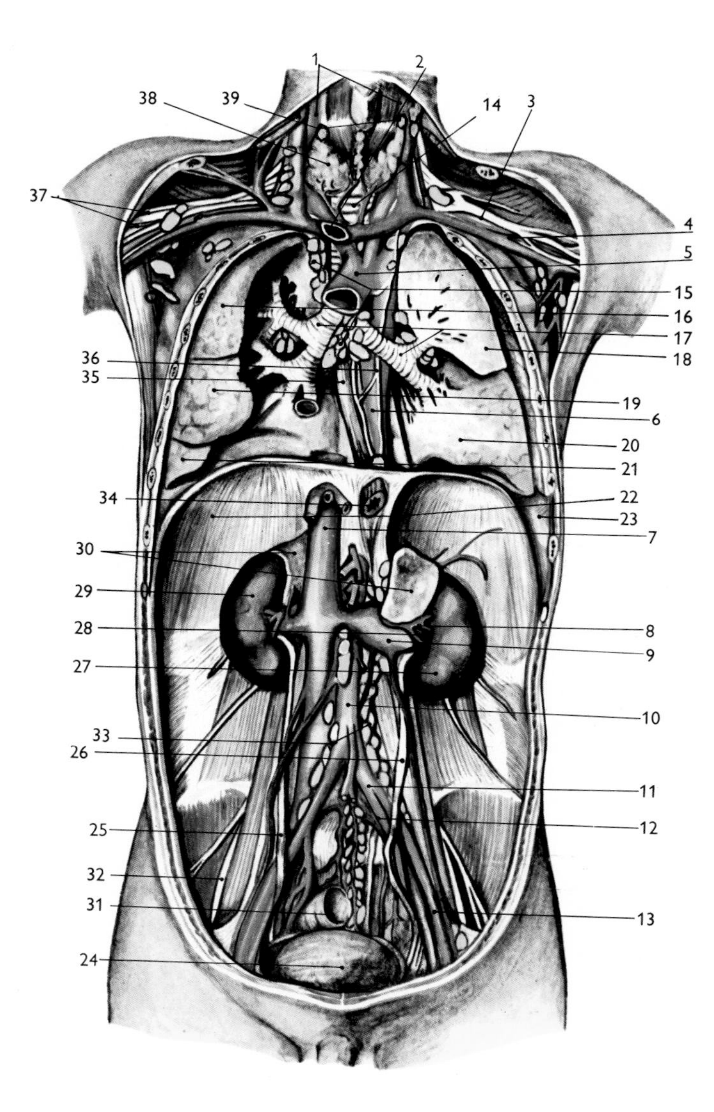

CIRCULATORY SYSTEM (SCHEMATIC, SHOWING MAIN BLOOD VESSELS)

I. Arteries (carry blood from the heart to the tissues)

1 Arteria pulmonalis (pulmonary artery), from right ventricle
2 Ventriculus dexter cordis (right ventricle)
3 Aorta (main arterial trunk), from left ventricle
4 Ventriculus sinister cordis (left ventricle)
5 Arteria carotis communis (common carotid artery), suppies head
6 A. subclavia (subclavian artery), supplies upper limbs and thorax
7 A. gastrica (gastric artery), supplies stomach
8 A. linealis (lineal or splenic artery), supplies spleen
9 A. hepatica communis (common hepatic artery), supplies liver
10 A. renalis (renal artery), supplies kidneys
11 A. mesenterica (mesenteric artery), supplies intestines
12 A. iliaca communis (common iliac artery), supplies pelvic organs
13 A. femoralis (femoral artery), supplies lower limbs

II. Capillaries

14 Blood capillaries are minute blood vessels which convey blood from the arterioles (minute arterial branches) to the venules (tiny veins which come together into larger ones).

III. Veins (carry blood to the heart)

15 Inferior vena cava, drains the abdomen and lower limbs, leads to right atrium
16 Atrium dextrum (right atrium)
17 Vena portae (portal vein), drains blood from abdominal organs and conveys it to the liver
18 V. hepatica (hepatic vein), drains blood from liver into inferior vena cava
19 V. pulmonalis (pulmonary vein) leads to left atrium
20 Atrium sinistrum (left atrium)
21 Superior vena cava, drains head and neck, upper limbs, and thoracic wall, leads to right atrium

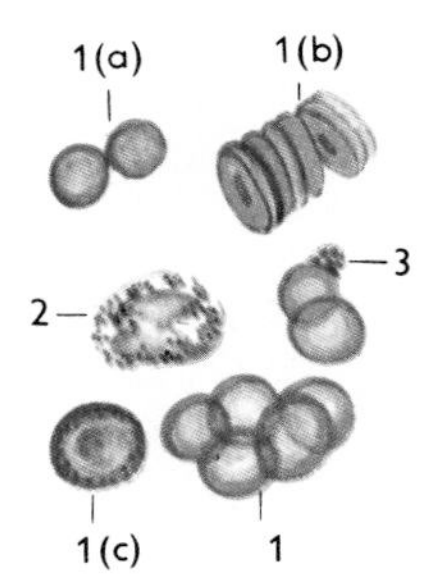

HUMAN BLOOD CELLS

(stained for examination under the microscope)

1 red blood cells (red blood corpuscles, erythrocytes)
 (a) erythrocytes nearing the end of their useful life, affected by pitting
 (b) erythrocytes arranged side by side
 (c) erythrocyte with basophilic material stained as blue spots
2 white blood cells (leukocytes)
3 platelets (thrombocytes)

VARIOUS FORMS OF WHITE BLOOD CELLS (LEUKOCYTES)

1 neutrophil granulocyte with unsegmented nucleus
2 neutrophil granulocyte with segmented (polymorphic) nucleus
3 eosinophil granulocyte
4 basophil granulocyte
5 small lymphocyte
6 large lymphocyte
7 monocyte

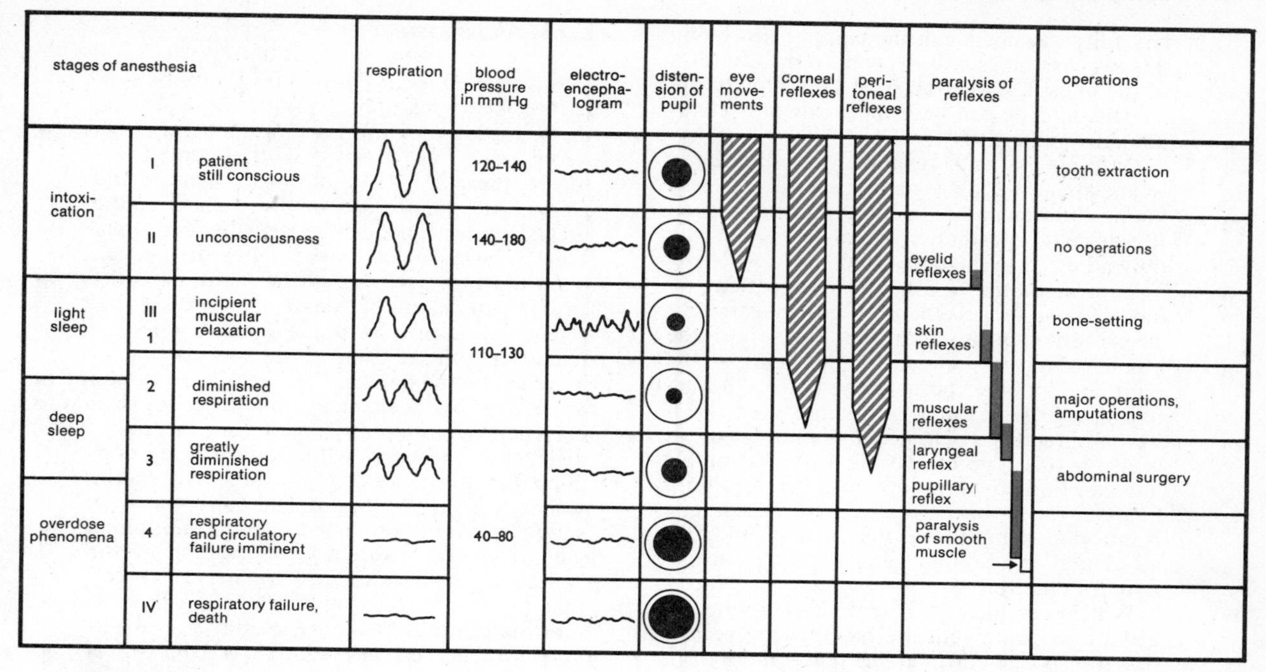

stages of anesthesia			respiration	blood pressure in mm Hg	electro-encephalogram	distension of pupil	eye movements	corneal reflexes	peritoneal reflexes	paralysis of reflexes	operations
intoxication	I	patient still conscious		120–140							tooth extraction
	II	unconsciousness		140–180						eyelid reflexes	no operations
light sleep	III 1	incipient muscular relaxation		110–130						skin reflexes	bone-setting
deep sleep	2	diminished respiration								muscular reflexes	major operations, amputations
	3	greatly diminished respiration		40–80						laryngeal reflex pupillary reflex	abdominal surgery
overdose phenomena	4	respiratory and circulatory failure imminent								paralysis of smooth muscle	
	IV	respiratory failure, death									

FIG. 1-26.
Characteristics of anesthesia in the various stages. (Courtesy of Guedel.)

ment centers in the midbrain, the overall motor control is abolished. This is often manifested in restlessness and uncontrolled movements, which is the reason why the patient has to be strapped down before being anesthetized on the operating table. In this stage of excitation no surgery must be performed, however, as the breathing tends to be irregular and spasmodic, the pulse rapid, and the blood pressure raised; the pupils are enlarged, the eyes held convulsively closed; sometimes there is retching or vomiting. Muscular reflexes still occur and may even be intensified.

In the third stage, the *tolerance stage,* in which the midbrain and spinal cord are increasingly paralyzed, more extensive surgical intervention is possible. The voluntary muscles, except those controlling the respiratory movements, relax, and the reflexes cease. The circulatory functions are adequately maintained; breathing becomes deep, slow and regular. The pupil of the eye contracts again and responds promptly to light. The electroencephalogram, i.e., the curve representing the electrical activity of the brain, undergoes a characteristic change in the tolerance stage of anesthesia: instead of rapid small waves, as are observed in the conscious brain, slow large waves now occur which resemble those produced during ordinary sleep. As the temperature control center is also paralyzed by the anesthetic, the body temperature may go down; the patient must therefore be kept warm if the anesthesia is of relatively long duration.

The fourth stage, the *collapse stage,* will occur only if an overdose of anesthetic is given. Vital centers in the brainstem, particularly the circulatory and the respiratory centers, become paralyzed. After a short transitional phase with shallow irregular breathing, the breathing stops, the blood pressure falls, the pulse becomes irregular and rapid, the pupil is enlarged and no longer responds to light, the electroencephalogram waves become slow and low. At the first sign of this dangerous stage of anesthesia, the further supply of anesthetic to the patient must be cut off at once, and a mixture of oxygen and carbon dioxide should be given to sustain respiration.

On "coming out of the anesthetic" the patient passes through these various stages in reverse order. He must not be left alone until he has fully regained consciousness. How long this process takes will depend on the duration of the anesthesia and also on how rapidly the anesthetic drug can be broken down or eliminated from the body.

Different patients respond very differently to anesthetics. Most sensitive are babies, very young children and old people. For this reason the dose of anesthetic administered must be duly adjusted to the individual.

With *inhalation anesthesia* the patient inhales vapor (e.g., ether) or gas anesthetics. These drugs pass from the lungs into the blood, which is pumped by the heart to the brain and all other parts of the body. Alternatively, anesthetics may be administered by *injection* directly into the bloodstream, the drugs used for this type of anesthesia being soluble in water (mostly barbiturates). There is a third method: *rectal anesthesia*, i.e., the anesthetic is introduced into the rectum in an enema; it is a "gentle" method and is used more particularly for children.

Whichever means of administering the anesthetic is employed, the blood is the vehicle that transports it through the body. It develops its action in the brain and spinal cord. A fat-soluble anesthetic will accumulate preferentially in the nerve centers of the central nervous system, which are rich in fat and well supplied with blood. Anesthetics administered by inhalation include chloroform (now seldom used), ether, halothane and nitrous oxide. The last-mentioned anesthetic, familiarly known as "laughing gas," is not by itself able to produce a sufficiently deep anesthesia for major surgery. Other gaseous anesthetics are ethylene and cyclopropane.

In principle, all anesthetics produce the same result of temporarily paralyzing the central nervous system. Their undesirable side effects vary greatly, however. In order to reduce these effects as much as possible, different anesthetics are often used in combination with one another. As a further means of reducing side effects, one important function of anesthesia, namely, relaxing the muscles, may be performed instead by a different type of drug, such as curare, so that a smaller and safer dose of anesthetic will suffice. Of course, these muscle-relaxing drugs have to be used in conjunction with artificial respiration. Modern mixed anesthesia, using a combination of anesthetics, often begins with a rapidly acting preliminary anesthetic which is injected intravenously and makes the patient lose consciousness gently and smoothly.

Local Anesthesia

A so-called *local anesthetic* is one that affects a local area only, temporarily abolishing the sensation of pain by acting upon nerves or nerve tracts. In contrast with general anesthesia, there is no loss of consciousness, and pain is not suppressed in those parts of the body that are outside the anesthetized area. This is because local anesthetics do not act upon the nerve centers of the brain, as is the case in general anesthesia.

The action of a local anesthetic is due to a temporary interruption of the conduction of impulses by the nerves. Peripheral pain stimuli are caused by mechanical injury affecting specific nerve endings (so-called pain receptors) distributed throughout the body and are conducted by the nerves, for the most part via the spinal cord, to the brain. The conscious sensation of pain is produced in the brain, along with a precise awareness of location of the source of the pain stimuli. If the nerve "cable" that carries the pain impulses to the brain is interrupted, they cannot reach the central nervous system, so that no feeling of pain is produced.

The *conduction of stimuli* in the various nerves of the body is based on a complex electrochemical process. Like other cells in the body, each nerve cell is enclosed in a membrane which separates the interior of the cell from its environment. Inside the nerve cell there is a high concentration of positively charged potassium ions, whereas sodium ions are predominantly present in the fluid medium around the cell. As a result of this particular distribution of electrically charged particles (ions), a difference in electric potential of a few thousandths of a volt—the so-called membrane potential—is produced between the inside and the outside of the membrane. Normally the membrane is more permeable to potassium ions than to sodium ions; but when an impulse is transmitted along a nerve cell, the permeability of its membrane to sodium ions suddenly increases, so that more of these ions can rush into the interior of the cell. During this time the cell membrane itself undergoes a reversal in its electric charge for a fraction of a second. When the impulse has passed, the membrane returns to its normal state of permeability, and the so-called resting potential is restored.

Local anesthetics have the property of being able to lower the permeability of the cell membrane, so that the influx of sodium ions asso-

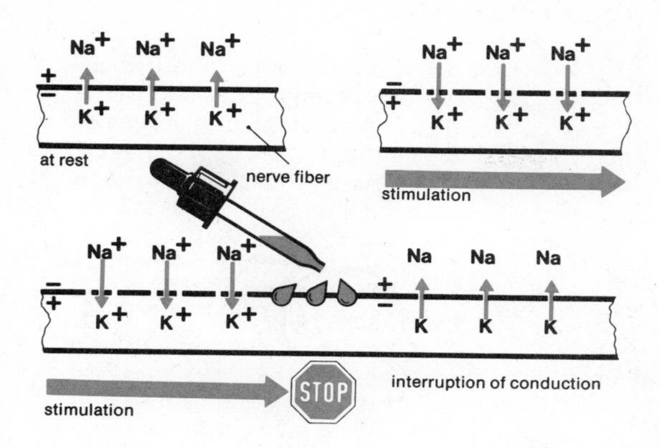

FIG. 1-27.
Interruption of the conduction of impulses in a nerve by local anesthesia.

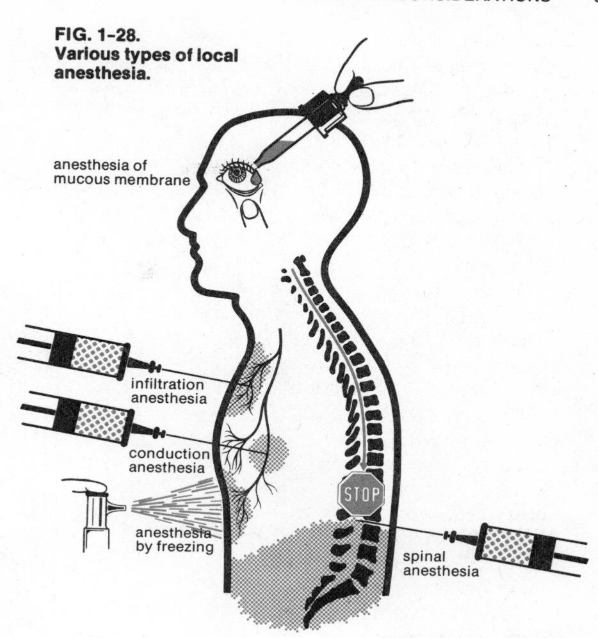

FIG. 1-28.
Various types of local anesthesia.

ciated with the passage of an impulse along a nerve cell is inhibited. As a result, the excitability of the nerve cells is reduced and indeed temporarily abolished altogether, so that no impulses are conducted (Fig. 1-27). In order to develop its action, the local anesthetic has to get to the nerve cell membranes and must therefore first pass through the fatty nerve sheaths. The speed with which the anesthetic takes effect will therefore depend, among other factors, on its fat solubility. As the motor nerves have thicker sheaths than the sensory nerves, they are correspondingly less susceptible to local anesthetics.

With *infiltration anesthesia* (Fig. 1-28), the local anesthetic solution, such as procaine, lidocaine or mepivacaine, is injected directly into the tissues, through which it then spreads, deactivating the local pain receptors and fine nerve branches. With *conduction anesthesia,* the anesthetic is injected close to a nerve trunk in order to suppress the conduction of impulses. *Spinal anesthesia* is produced by injection of the anesthetic into the lower part of the spinal fluid, so that the sensation of pain cannot be transmitted upward to the brain. With appropriate dosage, the entire lower half of the body can be anesthetized in this way. *Superficial anesthesia* can be produced by application of a local anesthetic to mucous membranes (by spraying, for example), from where it makes its way by diffusion to the sensory nerve endings. One way this type of anesthesia is used is in carrying out medical examinations of sensitive organs, especially in ophthalmology and in examination of the nose, throat and ears.

The only suitable anesthetics for superficial application are those which, on account of their solubility in fats, penetrate quickly into the mucous membrane, e.g., such drugs as mepivacaine and dibucaine. Cocaine is a habit-forming drug and is, for this reason, now seldom used as a local anesthetic except occasionally in ophthalmology.

HEART FAILURE, CIRCULATORY FAILURE, RESUSCITATION

In the great majority of cases, death occurs as the result of heart (cardiac) failure, even if originally other organs were diseased. If such "cardiac death" ensues as the final and inescapable consequence of serious illness or injury, then there is nothing that modern medicine can do about it—for instance, in a case of incurable cancer, severe brain damage, or ultimate failure of the diseased heart itself. But if cardiac failure occurs suddenly, as an unexpected event during an illness that the patient could survive with a properly functioning heart, resuscitation can and must be attempted. Examples of such situations (Fig. 1-29) are drowning, certain types of cardiac arrest, asphyxiation, heart failure associated with cardiac infarction, loss of blood, pharmacological anesthesia, and severe electric

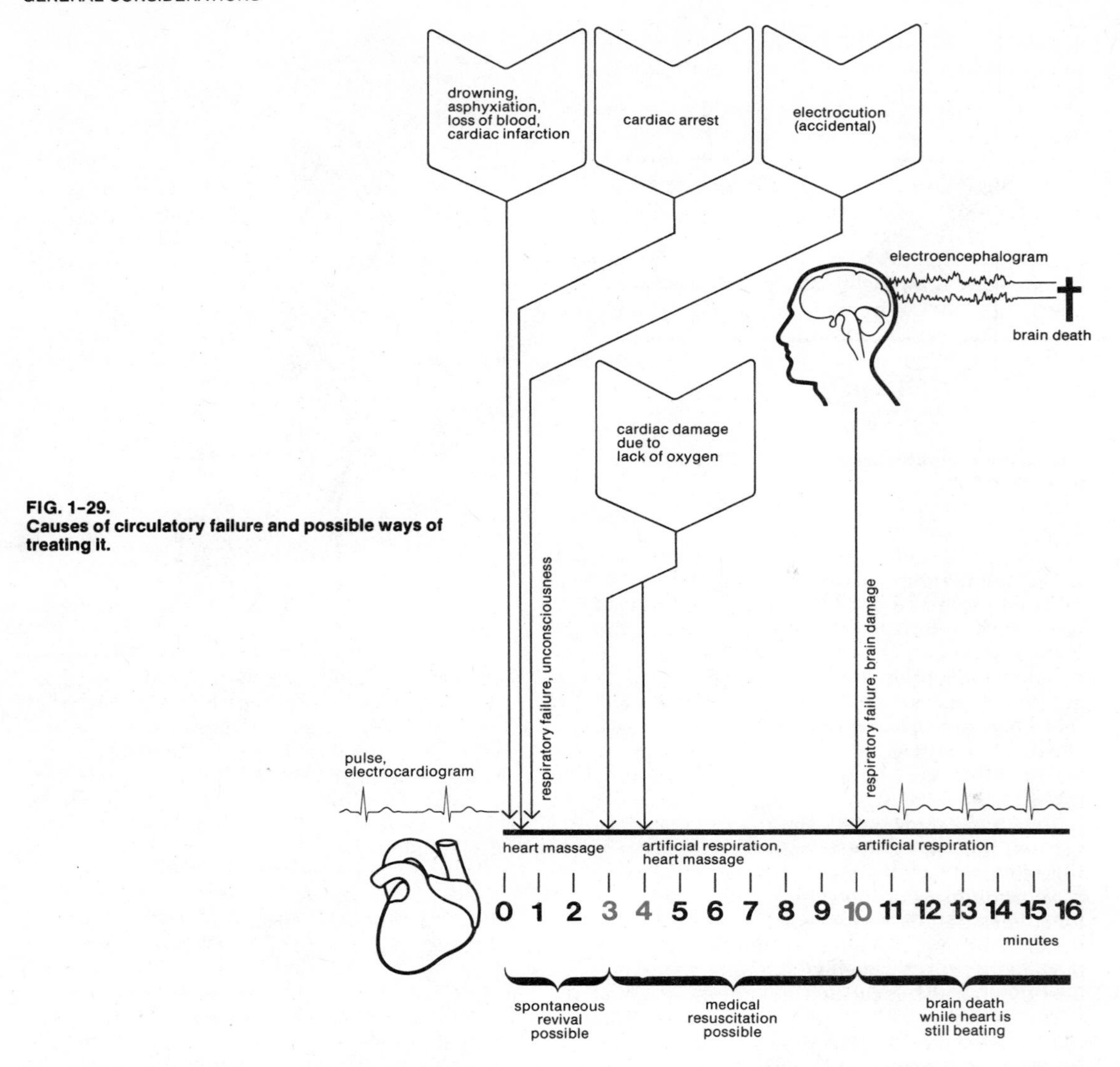

FIG. 1-29.
Causes of circulatory failure and possible ways of treating it.

shock. Only about 10 minutes is available for resuscitation attempts after the heart has stopped beating; within 3 to 4 minutes after circulation has stopped, the heart becomes so damaged by lack of oxygen that it is unable to get the circulation going again without outside help. Although the heart can be restarted even later than this by suitable stimulating action, 10 minutes after the heart has stopped, the brain suffers irreversible damage, and breathing stops. Resuscitation of the heart, if it succeeds, will then, in conjunction with artificial respiration, still only result in a condition called "brain death," i.e., with the brain out of effective action. Ethical, religious and medicolegal problems arise, particularly in cases where organ transplants are sought from a human body that has suffered brain death but whose circulatory functions are still intact. New medical criteria for the reliable diagnosis of brain death will have to be established; one method is the use of the electroencephalogram: record of the electric currents generated by the brain cells (Fig. 1-29).

The short time of only 3 or 4 minutes available to the lifesaver means that, for his efforts to have any hope of succeeding, he should be able at once to recognize a case of cardiac and circulatory failure, or at least to suspect it, and to intervene effectively. Unconsciousness sets in about 6 to 12 seconds after the heart has stopped. There is no pulse; a very faint pulse may, however, remain undetected even by an experienced observer. Then, 15 to 30 seconds after circulation has ceased, breathing stops. The skin now looks grayish; the pupils of the eyes are dilated as a result of central paralysis. Although it requires an electrocardiogram to be absolutely certain that the function of the heart has stopped, resuscitation should, in view of the very short time available for effective intervention, be applied in all circumstances where there is even no more than a suspicion of circulatory failure. Besides, resuscitation can be beneficial if there is serious weakening of circulation without actual cardiac arrest.

Resuscitation may be applied in three stages:

1. Supplying oxygen to the brain by artificial respiration and heart massage
2. Restoring as normal a circulation as possible by injecting drugs that raise the blood pressure and stimulate the heart
3. Stabilizing the condition; removing the causes of cardiac arrest; intensive care

Treatment of circulatory failure must proceed on the basis that two very different conditions may arise: it may be a case of true cardiac arrest, i.e., stoppage of the heart, when the chambers of the heart receive no stimuli nor generate any of their own; alternatively, it may be a case of fibrillation, i.e., quivering or uncoordinated twitching of muscle fibers of the chambers, which can thus no longer contract rhythmically to perform their pumping function.

In a case of heart stoppage it is, first and foremost, necessary to restart circulation and breathing. Fortunately, effective first aid in the form of oral resuscitation and heart massage can be given without special equipment and without the help of a doctor. First, wipe any foreign matter from the unconscious patient's mouth and throat; his head should be in the sideways position while this is being done. Then pull or push the jaw outward, so that the head is tilted back, and move the tongue from the back of the throat. A doctor may, more reliably, insert a special tube into the victim's throat, or insert a tube into the larynx through the glottis (endotracheal intubation). For mouth-to-mouth resuscitation (“kiss of life”), close the patient's nostrils by pinching them together with the fingers, and blow into his mouth (Fig. 1-30). If the mouth has been damaged by injury, hold the mouth and lips shut and blow into the nose. Movements of the patient's chest will indicate whether the efforts are having effect. Blowing should be done vigorously, at a rate of 15–20 breaths a minute. The patient's blood will then soon receive sufficient oxygen. A disadvantage of oral resuscitation is that air may enter the stomach and push the diaphragm upward; this can, however, be counteracted by applying pressure to the pit of the stomach, thus discharging the air in it.

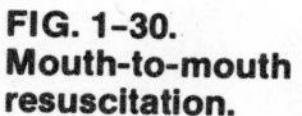
FIG. 1-30. Mouth-to-mouth resuscitation.

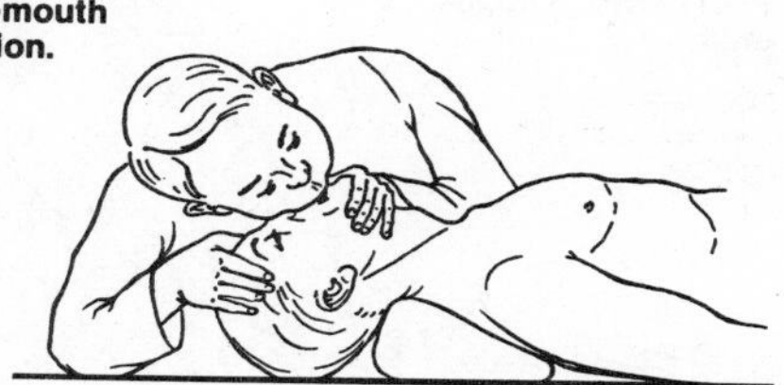

Heart massage (cardiac massage) can be applied in conjunction with resuscitation. If the heart has stopped beating, massage is necessary in order to convey the oxygen, introduced by resuscitation, to the various organs of the body, particularly the brain. Fortunately, heart massage can be applied immediately and can be learned by anyone. Its principle is to replace the spontaneous action of the heart muscle by pressure applied to it externally. In this way the blood in the chambers is pumped through the aorta and pulmonary artery. To apply external heart massage, apply pressure to the thorax at the lower third part of the sternum (breastbone), causing the heart to be compressed between the sternum and vertebral column and thus emptied (Fig. 1-31). On removal of the pressure, the thorax springs back

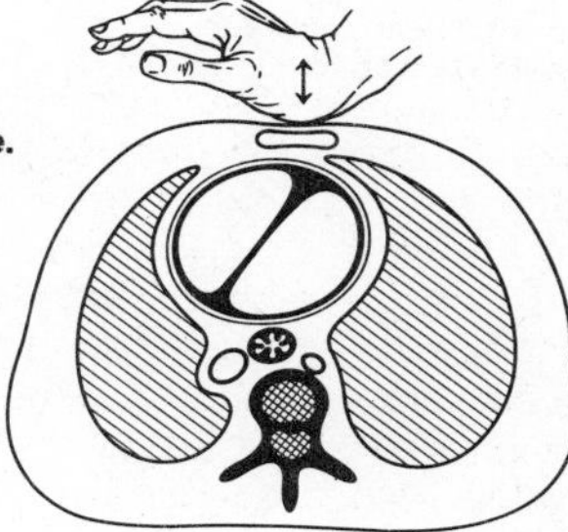
FIG. 1-31. External heart massage.

elastically, so that a suction effect develops in it and draws blood back into the heart. The patient should be lying on a hard base. To depress the thorax about 3–5 cm (1–2 inches) in an adult requires the application of 25–40 kg (55–88 lb) pressure. The operator should therefore place his hands one upon the other and, with arms outstretched, apply the full weight of the upper part of his body (Fig. 1-32). For children and babies, much lower pressures are needed, such as can be applied with one hand or merely with the thumb (Figs. 1-33 and 1-34).

FIG. 1-32. Mouth-to-mouth resuscitation and external heart massage applied to an adult.

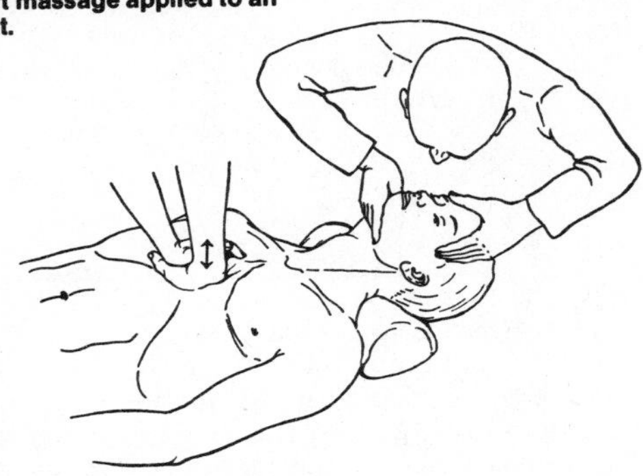

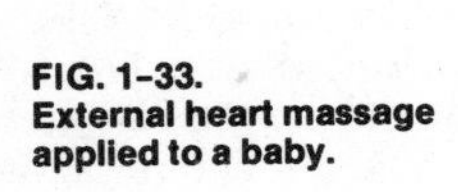

FIG. 1-33. External heart massage applied to a baby.

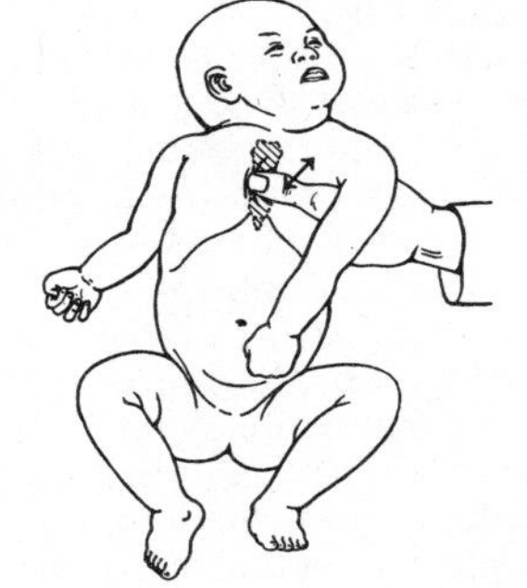

FIG. 1-34. External heart massage applied to a child.

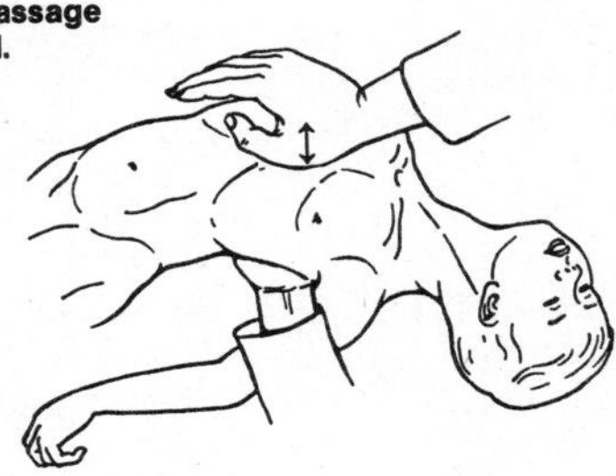

The massage pressures should be applied briefly and vigorously at a rate of 50–70 times a minute. If two operators are available, a ratio of four massage pressures to one mouth-to-mouth blow should be applied. If there is only one operator, it is best to apply a rhythm of 8:2 or 14:3. External heart massage can also be done mechanically with the aid of special equipment. In general, massage will be effective if it achieves an "artificial" pulse. As a visible result, the gray color of the patient's skin disappears, and his dilated pupils contract. In a hospital, the thorax can be opened and the heart massaged by direct manual manipulation. Direct massage may be necessary in the case of rigid thorax, after heart operations and in the event of injury to heart or thorax. It is, however, a complex procedure and is usually not resorted to until the possibilities of external massage have been exhausted.

Resuscitation does not end with restoration of heart activity and breathing. Quite often the heart must also be artificially stimulated with adrenaline; excess acid that has been formed will have to be neutralized, and lost blood or fluid replaced.

The electrocardiogram can be used for distinguishing between two types of circulation failure which, though requiring the same resuscitation treatment, must subsequently be treated differently. First, there is true cardiac arrest, in which the electrical activity of the heart muscle ceases (see page 64). In such cases it can be attempted to revive the heart with an electric pacemaker. The other type of failure is due to fibrillation, i.e., uncoordinated twitching of the heart muscle, so that it no longer functions effectively as a pump, though it is in a state of high excitation, as the electrocardiogram shows. Fibrillation can be stopped and normal functioning of the heart restored by means of special electrical equipment which administers powerful shocks to stimulate the heart as a whole and thus compel all the heart muscle fibers to fall into step, as it were.

FROSTBITE

Freezing of a part of the body, resulting in damage to the skin and sometimes to deeper tissues, is called *frostbite*. It is caused by exposure to cold, usually under conditions of low temperature in combination with wind. Not only the temperature of the skin but especially also the flow of blood in the skin play a part, and the latter, in turn, depends on the general condition of the blood circulation and the width of the local blood vessels. Frostbite occurs when the supply of blood to the part affected is so reduced—during cooling or subsequent warming—that a lack of oxygen develops in the tissues. The feet and legs are especially prone. Standing in melting snow and slush for long periods will increase the risk of frostbite, as will the wearing of tight footwear that cramps the movements of the feet. As with burns, three degrees of frostbite are recognized. With first-degree frostbite the tissue at first looks white and taut, as a result of maximum contraction of the blood vessels; after warming, there is complete recovery. Longer exposure to cold produces second-degree frostbite, characterized by severe reddening and swelling of the area affected. When third-degree frostbite occurs, the skin is white and hard through and through; rewarming it will cause blistering and shedding of frostbitten tissue.

In contrast with burns, damage due to frostbite does not affect the organism as a whole. Local aftereffects such as nerve damage, muscular atrophy and thrombosis may occur, however.

In the treatment of frostbite, care must be taken not to rewarm the affected tissue too quickly, for although tissue metabolism is rapidly restored on warming, the contracted blood vessels are not yet able to supply enough oxygen. For this reason frostbitten parts should be kept at a temperature of about 4° to 5°C (39.2° to 41°F). In cases of light frostbite, coffee and alcohol may be useful in stimulating the flow of blood; in severe cases drugs that stimulate circulation are administered, or artificially induced fever may be employed. To prevent secondary infections, frostbitten parts should be covered with sterilized dressings. Rapid rewarming (e.g., in a warm bath) is advisable only in nonsevere cases where exposure to cold was of relatively short duration.

What are commonly known as *chilblains* are dark red or bluish patches which occur on the fingers, toes, soles of the feet, heels, legs, ears, nose or cheeks. Women and young girls are especially liable to be affected. Chilblains are not really a form of frostbite, but are due to predisposed poor adaptability of the blood circulation to temperature variations. For this reason chilblains usually do not occur under severe winter conditions, but are more likely during intermediate periods of wet and cold weather.

Exposure to cold may also cause bluish-red areas of painful *inflammation,* owing to disturbed adaptation to temperature variations. Such inflammation is most likely to occur on parts of the skin where the subcutaneous tissue is poorly supplied with blood, e.g., on the chin in overweight women, on the legs, on the insides of the knee joints and, though more rarely, on the thighs. Inflammation of this kind may also occur on the chin and cheeks of babies who (though otherwise adequately covered) are wheeled around in carriages in cold weather for relatively long periods. Chilblains and inflammation of the skin can be prevented by stimulation of the blood circulation in the parts likely to be affected, by means of hot baths, sauna baths, rubbing the skin with circulation-promoting agents, or artificial sunlight lamp treatment. It is important to wear warm but not tight-fitting clothing. Hot drinks and hot bathing confined to the parts affected are of no use.

BURNS

A vitally important constituent of the tissues of the human body is protein (a collective name for a whole class of complex organic compounds). If protein is heated above 56°C (132.8°F), it undergoes a change in structure, is coagulated and is irreversibly altered; this process is called *denaturation*. Denatured protein acts as a foreign substance in the organism, and as such, it is combated by inflammation or is rejected.

Burns may be caused by hot water, steam, fire or electric current. Certain chemicals, such as acids or alkalis, also produce changes in protein and thus cause tissue damage similar to that found in heat burns. A burn primarily affects the skin. Only the most serious burns cause charring of the muscles, nerves, blood vessels and bones beneath the skin. Such deep-reaching damage necessitates amputation of limbs so affected; on the head or torso it will, in general, quickly result in death.

Three *degrees of burn* are recognized. First-degree burns are superficial in character. The epidermis is affected by a painful reddening, which disappears in a few days, without scarring. Second-degree burning is characterized by extremely painful blistering. Blisters occur when heat damages the capillaries, so

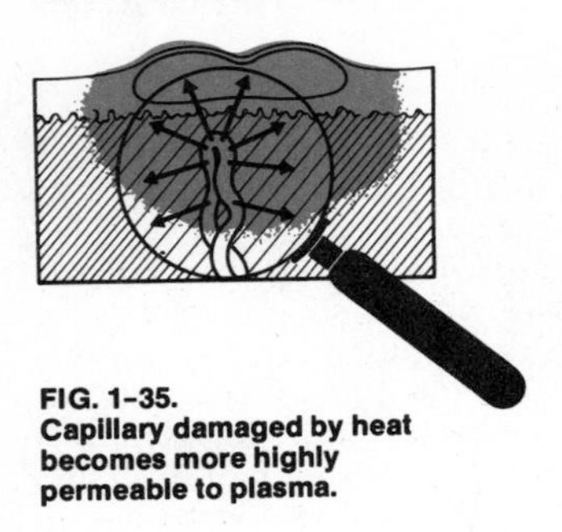

FIG. 1-35.
Capillary damaged by heat becomes more highly permeable to plasma.

	appearance	healing time	scarring	sensation
1st degree		3–6 days	none	very painful
2nd degree	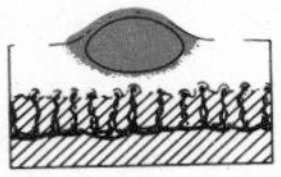	10–14 days	none	very painful
2nd degree	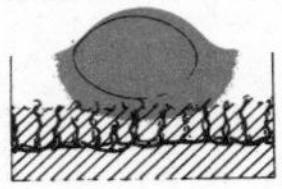	25–35 days	slight scarring	very painful
3rd degree		dead tissue sloughed in 2-3 weeks	scarring (skin grafts often necessary)	no pain

FIG. 1-36.
Severity of burns.

that plasma escapes from these blood vessels and lifts the epidermis (the outer layer of the skin) from the underlying layer (Fig. 1-35). Depending on the depth to which the epidermis has been affected and on whether the capillaries under it have been destroyed or merely damaged, the blisters will appear whitish-gray or red. If the network of capillary blood vessels has not been destroyed, a blister heals in 10 to 14 days, usually leaving no scar. Severer deeper blisters take about a month to heal and leave scars. When third-degree burns occur, the epidermis and also the dermis (the part of the skin containing the blood vessels and nerves) are destroyed, with damage to the underlying other tissues, which may be charred or coagulated, presenting a leathery white appearance. Because nerve fibers have also been destroyed, third-degree burns may be locally insensible and painless. The dead skin and tissues are sloughed (cast off) within two or three weeks and leave a defect whose recovery can be hastened by skin grafting (Fig. 1-36). The patient's fate, however, will depend not just on the depth but also, and not least, on the extent of the damage suffered by the skin. What matters is not so much whether the burned skin can "breathe," but how much loss of fluid occurs.

The severity of burns is roughly estimated as a percentage on the basis of the "9 percent rule" (Fig. 1-37). Burns (even first-degree burns) affecting about two thirds of the total skin area are nearly always fatal. If about one half of the body's surface has suffered burns, there is statistically a more than 50 percent risk of death. Even if only 15 to 20 percent of the surface is burned, serious general symptoms of shock—failure to maintain an adequate circulation of blood, so that oxygen supply to the tissues is disturbed—will result. This shock is caused by the damage to the capillaries already referred to. Blood plasma leaks out from the whole area of the burn and also into the tissues, causing edema (excess of fluid). As a result, the blood thickens and the blood vessels are not properly filled. Extensive burns must therefore, besides being kept sterile, be treated by making up the loss of fluid, this being achieved by injections of blood plasma or

FIG. 1-37.
Nine percent rule for estimating the severity of burns.

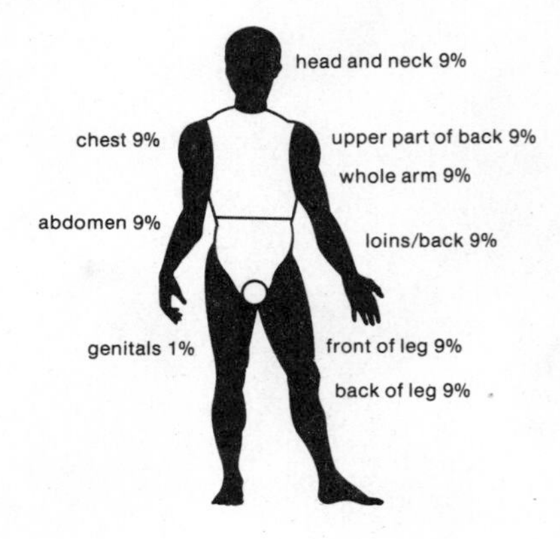

FIG. 1-38.
Contracture of burn scars.

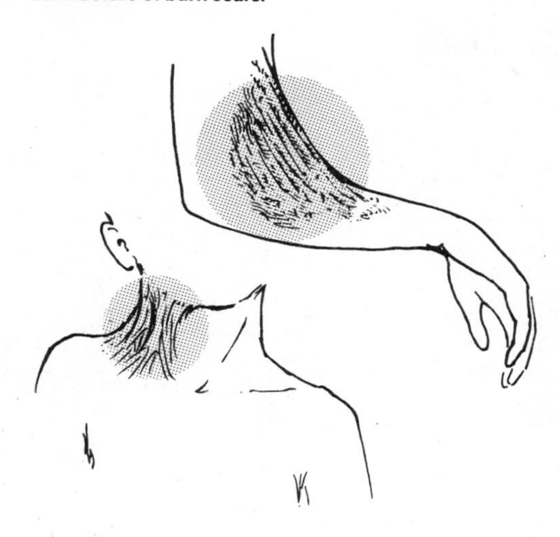

blood substitute (see page 116).

Two or three days after burning, the edema fluid begins to seep back into the bloodstream. Urine production, which was greatly reduced during the first 48 hours, now increases considerably. Large quantities of potassium are excreted in the urine. To protect the heart from overstraining, the supply of fluid to the patient in this stage has to be restricted. The danger of bacterial infection spreading from the remains of destroyed tissue can be combated with antibiotics. Major deep burns require early treatment by skin grafting. Burn scars are particularly liable to undergo contracture (Fig. 1-38).

VACCINES AND IMMUNIZATION

The basis of all vaccination is the *antigen-antibody reaction*. Antigens are substances that induce the formation of antibodies, the latter being protective protein substances produced by the organism in response to the presence of "foreign" substances (antigens), often of bacterial or viral origin. An antibody combines with the antigen that provoked its formation and renders it harmless. Once an antibody has been produced in response to the antigen of a particular disease, the organism can produce it very quickly in response to a subsequent attack of the same disease; this is known as (acquired) *immunity* to that antigen. Antigens are usually proteins, but some fats and polysaccharides may also have antigenic properties. Antibodies are produced in the cells or occur as plasma proteins belonging to the so-called gamma globulins. *Inoculation* (*active immunization*) consists in introducing certain antigens into the organism in order to provoke the formation of defensive antibodies, i.e., it is a deliberate infection with a disease in a mild form in order to confer immunity against a subsequent attack of that disease. Repeated inoculation can thus induce the formation of a relatively large quantity of the antibody. The immunity obtained in this way may remain effective for years and can be restored again and again by booster injections. The drawback of active immunization is that it takes a relatively long time for the immunity to develop, so that it may be too slow-acting in acute cases of disease.

With active immunization a distinction is to be drawn among inoculation with *live microbes*, with *dead microbes* and with *toxoids*. Live microbes used for the purpose are so modified as to be practically harmless, yet still capable of conferring immunity. Dead microbes for inoculation have been killed in ways that likewise preserve their immunity-inducing properties. Toxoids are toxins that have been so treated as to destroy their toxicity but leave them still able to induce the formation of antibodies. The term *vaccination* is sometimes more specifically used to denote the injection of live microbes (attenuated strains of microbes that have lost their virulence). Such vaccines are used against smallpox, yellow fever, tuberculosis and poliomyelitis. The smallpox vaccine contains the live virus of cowpox, which is harmless to man but nevertheless confers immunity to smallpox (Fig. 1-39). For immunization against tuberculosis, live bacteria of bovine tuberculosis are used, which do not produce the disease in man; this treatment must not, however, be given to persons who have already had tuberculosis. Dead microbes are used for immunization against measles, poliomyelitis (Salk vaccine), influenza, rabies, whooping cough, typhoid and cholera (Fig. 1-40). Toxoids are used for conferring immunity against diphtheria and tetanus (Fig. 1-41).

In addition to the classification by immunization with live microbes, dead microbes and toxoids, a distinction can be drawn between vaccines that contain only the antigen in a fluid medium, e.g., a physiological salt solution, and vaccines in which the antigen adheres to the surface of an adsorbent substance, usually aluminum oxide, so that the antigen enters the

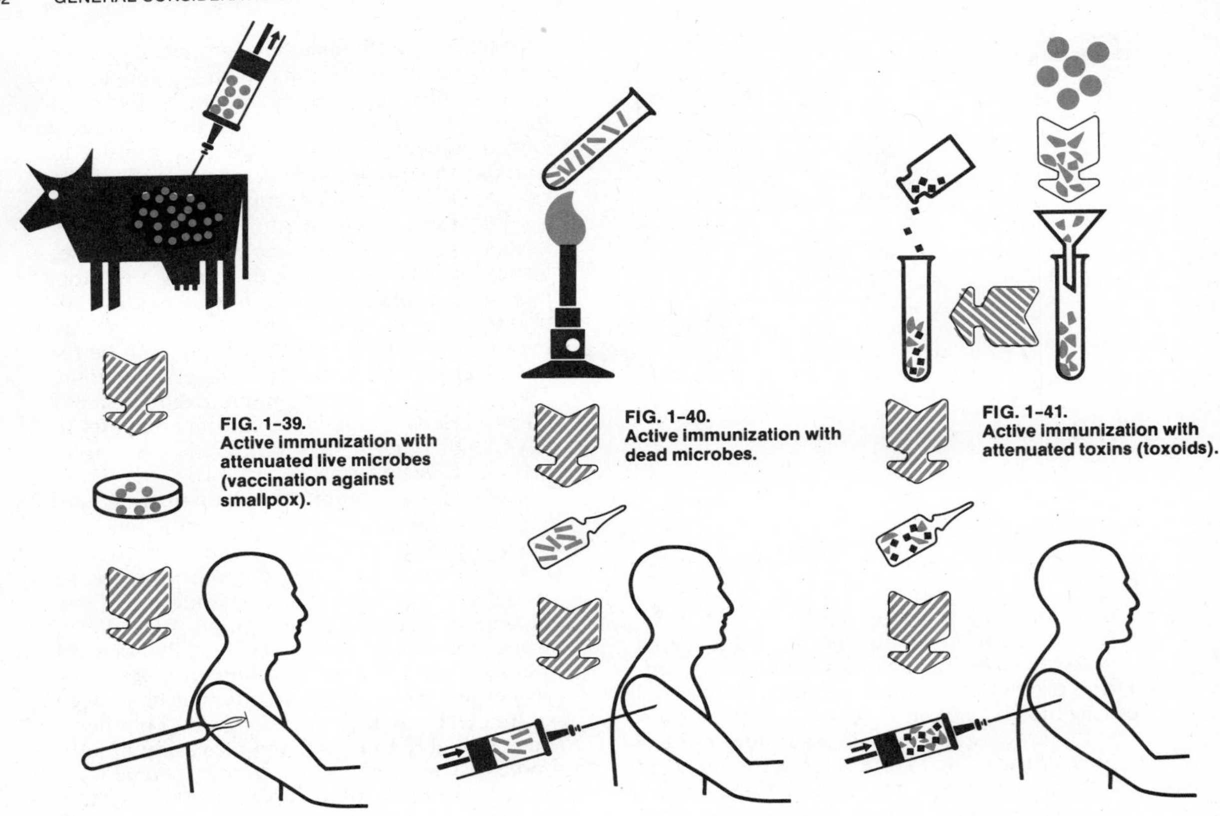

FIG. 1-39. **Active immunization with attenuated live microbes (vaccination against smallpox).**

FIG. 1-40. **Active immunization with dead microbes.**

FIG. 1-41. **Active immunization with attenuated toxins (toxoids).**

bloodstream gradually and the production of antibody is correspondingly increased.

With *passive immunization* (Fig. 1-42) the patient is injected with a serum containing "ready-made" antibodies of animal or human origin—instead of the body having to produce them itself. The advantage of this method is that the treatment acts quickly; its disadvantage is that the immunity is of short duration (only 8–14 days with animal sera). Passive immunization is therefore appropriate only in cases where the time between infection and the anticipated outbreak of the disease is too short to enable the body to build up its own defenses. In most cases a combination of passive and active immunization is applied, in order to obtain a more lasting as well as immediate protection.

Sera for passive immunization are obtained mostly from horses, cows, sheep or pigs. The animal is treated with the antigen of a particular disease until it produces a sufficient amount of antibody. Blood is then taken from the animal and suitably prepared. Untreated blood serum from such an animal is called a *native serum*. Because of their high overall content of foreign proteins native sera are not easily tolerated by the human patient and are not normally employed for immunization. By separating out these proteins and breaking down, by enzyme action, the serum globulin with which the antibodies are associated, a serum is obtained which produces much less of the undesirable and possibly dangerous reaction in the patient to the animal protein and which nevertheless achieves effective immunization. But even with such an improved serum it is always necessary first to test the patient's reaction to it by giving a trial injection. In order to reduce the risk of allergic reactions (due to the formation of antibodies against foreign proteins which are not in themselves harmful) when repeated doses of the same serum are given, the rule is first to administer horse serum, then cow serum and finally sheep serum. Serum treatment is now available for a number of diseases including diphtheria, tetanus, anthrax, meat poisoning and snakebite.

FIG. 1–42.
Passive immunization.

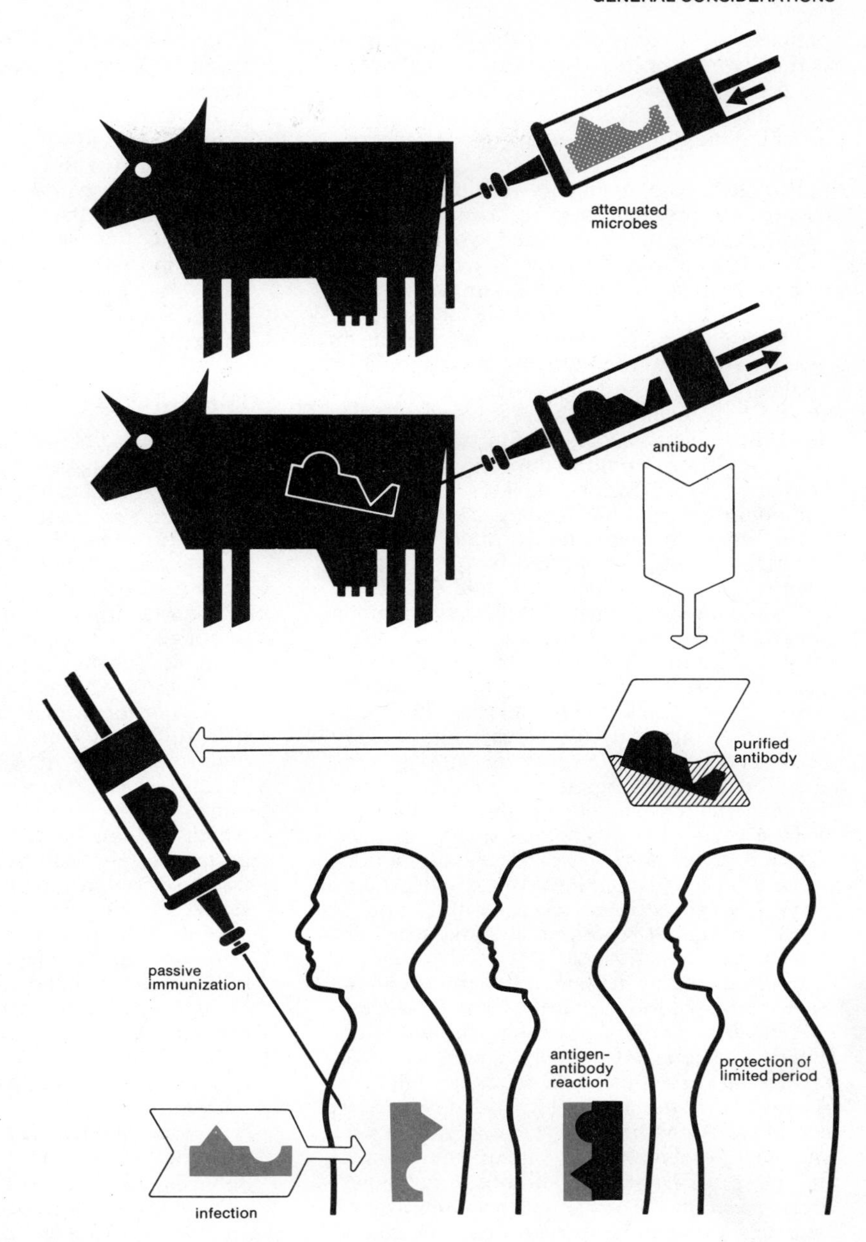

In recent years, besides animal sera, human sera containing antibodies have been used for passive immunization in the form of gamma globulin or hyperimmunoglobulin. *Gamma globulin* is that fraction of the globulin in the blood with which most of the immune antibodies are associated. Immunoglobulins are proteins that are capable of acting as antibodies. *Hyperimmunoglobulins* are associated with specific antibodies to whooping cough, tetanus, mumps, measles and smallpox. By and large, human antibodies have one major advantage over animal antibodies: the patient's defensive mechanism does not respond to them as to foreign proteins, i.e., they do not provoke the production of the human patient's own antibodies to the injected antibodies (which in this particular respect act as antigens), as does occur when the latter are of animal origin. Hence there are no undesirable and potentially dangerous allergic defense reactions.

For greater convenience of immunization against several diseases at once, combined vaccines have been produced which contain the antigens of three, four or five diseases. Of course, it must be ensured that the various antigens do not adversely affect one another.

There still exists divergence of opinion on when, and how often, immunization should be applied. Some experts consider that an infant should be vaccinated against tuberculosis (indeed, as its first vaccination) in the first week of its life. Then, from the third month onward, combined vaccination against diphtheria, whooping cough and tetanus can be given three times, at one-monthly intervals. After the eighth month comes the first vaccination against smallpox. When the child is 2 years old, vaccination against diphtheria, whooping cough and tetanus is repeated, and two doses of oral poliomyelitis vaccine are given. At the age of 6 the child receives further booster doses of diphtheria and tetanus vaccine. If no vaccination against tuberculosis has already been given and if the child has not by then acquired a natural resistance to the disease, TB vaccination should be done now, i.e., when 6 years old. The second smallpox vaccination is given at the age of 12. Finally, a third booster dose against diphtheria and tetanus may be given between the ages of 12 and 14 (with subsequent repetition at three-yearly intervals).

In Germany, as in some other countries, the only immunization required by law is vaccination against *smallpox* (variola). Every child must be vaccinated in the calendar year following the year of its birth. A booster injection is required at the age of 12. The most serious complication that may arise in connection with smallpox vaccination is meningitis, which, though rare, is more liable to occur as the person grows older. To rule out this complication with certainty, older persons are vaccinated with killed smallpox virus.

Poliomyelitis (infantile paralysis) can be caused by three types of virus, and full protection requires vaccination against all of them. In the Salk vaccine the killed virus of all three types is administered in three doses spaced at monthly intervals. In the Sabin oral vaccine the attenuated live virus is given, but not all three types simultaneously: first type I is given, followed by types II and III six to eight weeks later.

The antigen injected against *tetanus* (lockjaw) is a toxoid—a bacterial toxin attenuated by treatment with formaldehyde and adsorbed on aluminum oxide. As untreated tetanus is often fatal and the possibilities of infection are numerous, vaccination against this disease is very important. In the absence of vaccination, temporary (passive) immunity against tetanus after a wound can be given with an injection of antitoxin.

IMMUNITY

Immunity denotes the organism's capacity for successfully resisting the actions of harmful foreign substances or pathogenic (disease-producing) microbes such as bacteria or viruses. It comprises, to begin with, various unspecific reactions which occur even when the body comes into contact with a noxious agent of external origin for the first time. This natural or congenital immunity is characterized by, for example, such actions as phagocytosis (the destruction of pathogenic agents by white blood cells) and local inflammation reactions which help to attract white blood cells to the parts under attack (see pages 108–9). In a more restricted sense, however, immunity refers to more highly specific defensive processes which are not immediately available when a harmful foreign substance enters the organism for the first time, but which are induced as a result of such contact (*specific immunity, acquired immunity*). The "immunological memory," one of the most interesting but not yet fully solved problems surrounding the immunological reaction, ensures that henceforth the specific defensive facilities are immediately available in the event of a second or subsequent contact with the harmful substance in question.

Substances that give rise to an immunological reaction are called *antigens*. The specific defensive substances that the body produces in response to particular antigens are called *antibodies*. Antigens comprise all substances that are "foreign" to the cells responsible for developing and maintaining immunity. Exceptionally, the body's own tissues or substances which normally do not enter the bloodstream may act as antigens. In such cases they may cause what are known as *autoimmune diseases*, in which the body produces a disordered immunological response: failure to distinguish between what is a normal substance and what is foreign, so that antibodies against normal parts of the body are produced. As a rule, however, the immunological response is directed only against pathogenic organisms or their toxic products.

Active immunity is immunity resulting from antibodies formed within the organism's own tissues; it may arise naturally from having the disease, or artificially from vaccination with an attenuated form of the infecting microbe (see page 41). On the other hand, if immunity is produced by injection of a serum already containing the antibody, it is called passive immunity. Most antigens are proteins with a high molecular weight (3800 and upward). In certain circumstances, however, much smaller molecules may act as antigens if they are combined with protein molecules outside and inside the patient's body; such nonprotein substances are called haptenes. Some chemicals, including drugs, may indirectly activate the immunological system and produce a reaction called an allergy (see the next section). Allergy can thus be regarded as another particular instance of disordered immunological response.

How are the immunity-controlling cells, i.e., those responsible for developing and maintaining immunity, able to identify the body's own cells and substances as nonantigenic and to distinguish them from antigenic ("foreign") cells and substances? Actually, at birth a kind of compatibility develops between the human body and its own immunological system; this is known as *immunological tolerance*. One theory is that all those immunity-controlling cells which by their nature are so constituted as to treat the body's own tissues as antigenic are, as a result of coming into contact with these supposedly foreign substances, themselves destroyed at a very early age. If in later life potentially carcinogenic (cancer-producing) cells, possibly with fresh antigenic properties, are formed in the body as a result of mutation, an additional function of the now—i.e., at this more advanced age—no longer adaptable immunological system may consist in detecting such cells as "foreign" and rendering them harmless. There are in fact indications of increased tumor growth after damage to the immunological system.

If immunity-controlling cells come into contact with an antigen, they respond by producing antibodies—very quickly or with a time lag. It is supposed that all these cells responsible for the body's immunity derive from lymphocytes (a particular kind of white blood cell without cytoplasmic granules) which were formed in the marrow and which, under the influence of two central lymphatic organs, become differentiated at a very early age into two specialized families. These directing organs are, respectively, the thymus and certain lymphatic structures in the intestinal region. The thymus is associated with the family of so-called *T lymphocytes*, which, in response to an antigen identified as "foreign" by the immunological system, are transformed into sensitized T lymphocytes specifically directed against that antigen. They contain antibodies which are responsible for the *delayed* immunological reaction or for an allergy of the delayed type. The rejection of homologous transplanted organs is an example of this phenomenon.

The lymphatic structures in the intestinal region induce the formation of *B lymphocytes*, which, under the influence of recognized antigens, undergo differentiation into *plasma cells*. The latter in turn form specific immunoglobulins (proteins capable of acting as antibodies) which circulate freely in the blood plasma and thus bring about *immediate* immunological reaction—namely, precipitation of antigens, agglutination of antigenic bacterial bodies to render them harmless, or dissolving and thus destroying them (the antibodies that perform these functions are called precipitins, agglutinins and lysins, respectively). These free (*humoral*) antibodies can be transmitted to other living creatures by injection with blood serum from a donor and can continue to exert their antigen-destroying action in their new surroundings. This is the principle of passive immunity, so called because immunity is acquired directly as a result of the injection, without demanding any active participation on the patient's part (see pages 42 ff). Certain blood-group properties of some people are also "identified" as antigens against which the immunological system produces antibodies (see pages 46 ff).

The various effects of the repeated encounter of antibodies with the corresponding antigens (antigen-antibody reaction as immediate or delayed hypersensitivity reaction) will be described in connection with allergy.

ALLERGY

Allergy is an immunological reaction to a specific and normally harmless substance (the *allergen*) which in nonsensitive persons will produce no effect. Unpleasant symptoms due to the action of the antibodies produced in response to the allergen (acting as an antigen) are characteristic of the allergic condition. Allergens are not all proteins; they may be drugs or other chemicals. These simpler substances (called haptenes) are thought to combine with proteins in the body to form new "foreign" proteins. The first encounter with an allergen calls forth no adverse reaction, but the body becomes sensitized, i.e., it produces antibodies. On exposure to the allergen a second or any subsequent time, the antibodies will react with it and cause objectionable symptoms. This is the so-called allergic reaction.

Humoral antibodies can produce a very rapid immunological reaction (an allergic reaction of the *immediate type*), as occurs when the antigen is introduced intravenously (e.g., anaphylactic shock) or when there is a repeated or continuous supply of antigen (serum sickness) or when certain kinds of antibodies (reagins) are present which are located mainly in the skin and mucous membranes (allergies of the reaginic type, e.g., hay fever, urticaria, asthma). On the other hand, the antibodies linked to cells (more particularly the white blood cells called lymphocytes) produce only a *delayed* reaction (in particular cases involving an interval ranging from several hours up to about 36 hours) in response to the antigen when it enters the body for the second or a subsequent time. Such a delayed-action allergy is, for example, contact dermatitis, an allergic inflammation of the skin due to repeated contact with the allergen concerned, which may be an apparently innocuous compound such as a perfume or a deodorant. Autoimmune diseases, which are a special type of allergy developed by the immunological system against the body's own substances, have already been referred to.

In the allergic response of the immediate type the consequences of the antigen-antibody reaction are, among others: release of histamine, serotonin and kinin, with local or general expansion of minute blood vessels and escape of fluid into the tissues. Such phenomena give rise to wheals or reddening of the skin and mucous membranes in the case of urticaria (nettle rash); swelling of mucous membranes, sometimes in the region of the larynx, in acute circumscribed edema; lowering of blood pressure in anaphylactic shock. Alternatively, there may be contraction of the smooth muscles in asthma and allergic enterocolitis (inflammation of intestines and colon) and the destruction of blood cells to which haptenes (see above) have attached themselves (as in blood allergies, e.g., agranulocytosis, see page 111).

In the allergic response of the delayed type the immunological reaction of the sensitized lymphocytes plays the decisive part. This may result in tissue damage or destruction (as occurs in the rejection of transplanted organs) or may cause fever and other general symptoms. In such processes the participating lymphocytes release a number of substances which may, among other effects, attract other white blood cells, intensify their phagocytic action, and result in an increased flow of blood through the capillaries.

Anaphylactic shock (serum shock) is due to a massive antigen-antibody reaction in the blood and tissues resulting from the reentry of large quantities of antigen into the blood, e.g., in the case of reinjection of foreign proteins contained in an immunizing serum, when the body has already formed antibodies to those proteins after the first injection of that serum. *Antigen-antibody complexes* which release histamine are formed; these are able to develop their action not only in a localized manner in the tissues but also in the so-called mast cells of the blood in which histamine and histaminelike substances are stored. Besides causing difficulty in breathing, the formation of antigen-antibody complexes in anaphylactic shock results in expansion of the small blood vessels, so that the blood pressure drops. In severe cases death due to circulatory failure may ensue.

The less severe general allergic symptoms associated with *serum sickness* are caused by an antigen-antibody reaction extending over a period of several days. It occurs when the formation of antibodies is started in response to an antigen, e.g., a drug or an immunizing serum, which is being continuously supplied to the patient's body. The antigen-antibody reaction cannot cease until the supply of antigen stops and all the antigen still present in the body has reacted. Thus the allergic symptoms (headache, fever, nausea, nettle rash, edema) may persist for up to a week.

For the diagnosis and treatment of allergic diseases it is very important to know the condition of the allergens and how they enter the body. Macromolecular substances, more particularly foreign proteins, may act directly as allergens. Alternatively, substances consisting of small molecules (so-called haptenes) may attach themselves to large protein molecules. Hence, in principle, almost any substance could potentially act as an antigen (and allergen). Allergens are classified according to the way they enter the body: (1) *inhalation allergens,* which are inhaled and often cause allergic reactions in the respiratory system (e.g., pollen from various plants, especially grasses, and dust, hairs, feathers); (2) *ingestion allergens,* which are taken in with food and which include certain foods themselves (e.g., fish, milk, strawberries; in particular cases: drugs administered as treatment); (3) *injection allergens,* which are introduced by injection into the bloodstream or into other body fluids (e.g., vaccines, blood of the "wrong" blood group, many drugs); (4) *contact allergens,* which cause contact dermatitis as a result of repeated contact with the skin (e.g., soaps, cosmetics, wool, silk, synthetic fibers, various species of plants).

In practice, when an allergy is suspected, it is important to study the past history of the case (possibilities of exposure to the allergen in the home or the place of work, use of drugs and cosmetics, kinds of food consumed, etc.) and to detect the allergen concerned. Various methods of seeking and confirming the presence of allergens exist, and a wide range of extracts of individual allergens and groups of allergens for test purposes are available. They are used especially in the *intracutaneous test* (for the detection of humoral antibodies) and the *epicutaneous test* (for the detection of immune cells; the test preparation is applied generally to the patient's arm or back). Experiments involving exposure of patients to certain suspected agents, or experiments in which patients methodically avoid contact with such agents, can likewise help to identify the allergen.

Treatment of allergic complaints must consist primarily in avoiding future contact with the allergen. In some cases this may necessitate a change of home or job. If contact with the allergen cannot be avoided, it may be possible to *desensitize* the patient by the administration of gradually increasing doses of the allergen. By the end of the course, large doses can be tolerated which would previously have caused severe symptoms. In particular cases the symptoms of the allergy can be relieved or suppressed by drugs such as epinephrine (adrenaline) and related preparations (for treating asthma), antihistamine drugs (for hay fever and urticaria), corticosteroids (which arrest various kinds of inflammation, especially in allergic skin disorders), etc.

ORGAN REPLACEMENT AND TRANSPLANTATION

When vitally important organs fail to function adequately, the survival of the organism as a whole is endangered. There are many diseases that may result in such organic insufficiency or failure. The most important are changes in the blood vessels supplying blood to the organs concerned, chronic inflammation, malignant tumors, and phenomena of wear and exhaustion due to overstrain. In such circumstances important functional tissue of the organs will be destroyed, and the functioning of those organs will be correspondingly diminished.

Quite often the organic insufficiency is progressive and irreversible. If the organs so affected are of vital importance, the organism as a whole can be kept alive only if the defective organ is replaced by a healthy one from a suitable donor or its function performed by a machine.

At present, long-term *replacement* of organs by machines is really possible only for kidneys. Purification of the blood can be done by an "artificial kidney" machine operating on the principle of *dialysis* (selective filtration) if the patient's kidneys have partly or completely stopped functioning. The blood containing waste products for removal is passed through tubes consisting of thin synthetic membranes immersed in a dialyzing solution. The molecules of the waste products make their way through the pores of the membrane and are thus removed from the bloodstream, whereas the vital proteins in the blood serum are unable to penetrate the membrane and are therefore retained in the blood. A similar technique is *peritoneal* dialysis. This is based on the fact that the peritoneum, which lines the abdomen, happens to be a natural membrane of the right type for dialysis. The dialyzing solution is introduced into the abdominal cavity through a hollow needle. Waste products in the bloodstream make their way through the peritoneum into the solution, which is then drained away. By repeated renewals of the solution, fairly effective purification of the blood can be achieved. The frequency of the dialysis treatment will depend on the degree of insufficiency of the patient's kidneys. Drawbacks of dialysis methods are their cost (the artificial kidney is an expensive apparatus) and/or the inconvenience and possible emotional strain on the patient. For these reasons a kidney transplant may be a preferable alternative.

Organ transplantation is a difficult and risky intervention. The main problems are the practical one of obtaining suitable organs for transplanting and the problem of organ compatibility and rejection. The surgical techniques involved have been highly developed and will generally not present major difficulties. The organ to be transplanted from the donor must be suitably viable in the recipient's body. A kidney may be taken from a living donor, in which case the donor's remaining kidney will after a time be able fully to perform the function originally performed by his two kidneys. On the other hand, single organs such as the heart or liver can be obtained only after the donor's death. Ethical problems arise in this context, particularly with regard to the question of when the donor can with certainty be pronounced dead: "brain death" (when the brain has ceased to function for 12 hours) is regarded as a medically reliable criterion. During this period the organ for transplantation must be maintained in a biologically functional condition by artificial respiration. On removal from the donor's body the transplant, i.e., the organ or part thereof to be transplanted, must be kept alive by having a nutrient solution circulated through it.

Efforts are being made to preserve transplants in good condition for prolonged periods and thus establish "organ banks" in which they can be stored in readiness for use when required, on the same principle as the now familiar blood banks for transfusions. Blood is one of the easiest tissues to transplant, but even here it is necessary for the donor's blood to be suitably matched to the recipient's; it must belong to an acceptable blood group. With organ transplants the criteria are much more exacting. An incompatible transplant will be resisted as a "foreign" substance by the recipient's defensive mechanism and eventually rejected. The *rejection reaction* will be more severe to the degree that the transplant is more "foreign" to the recipient. The minimum requirement is that there must be blood

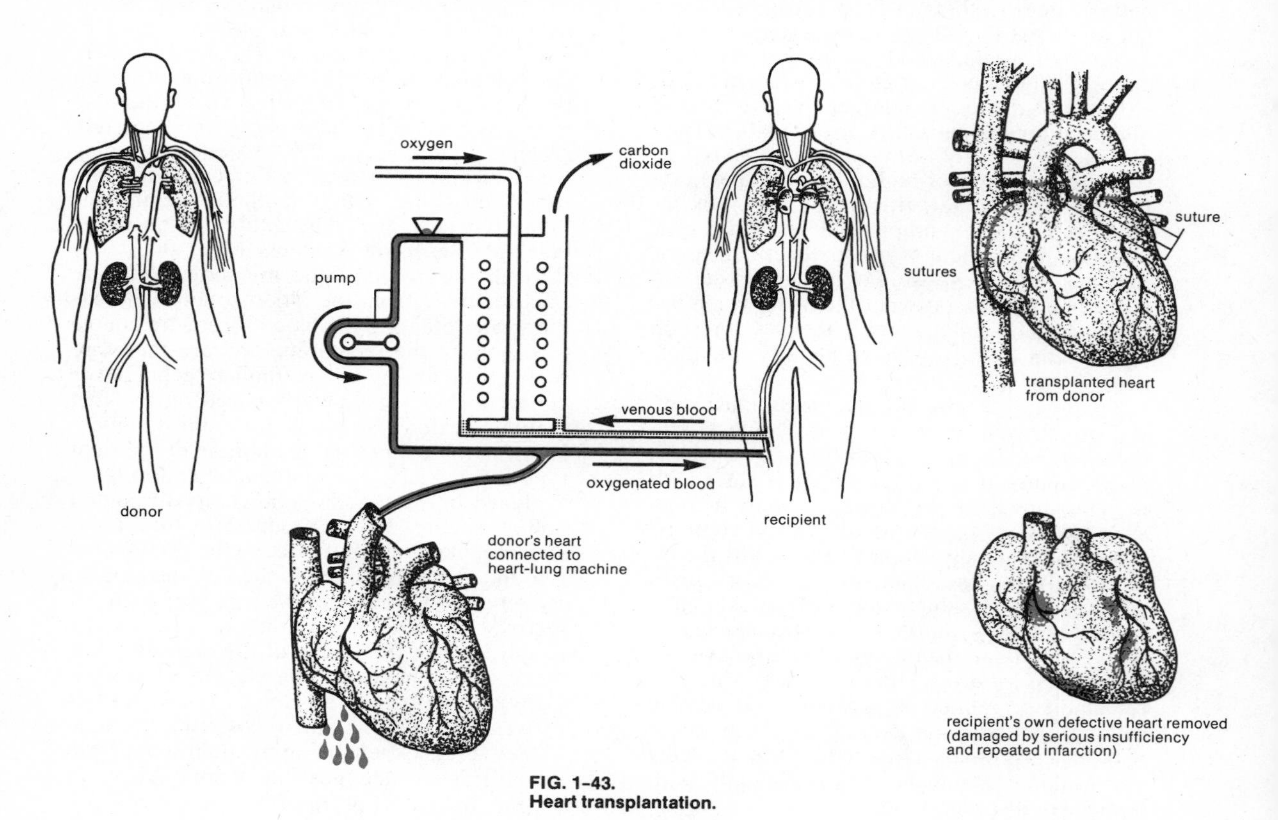

FIG. 1-43.
Heart transplantation.

group compatibility between donor and recipient. Transplants from animals are, for this reason, less likely to be accepted than those from humans; and a transplant from a pig is much less likely to succeed than one from a chimpanzee. Transplants between identical twins produce no rejection reaction.

Rejection is caused by the "foreignness" of the (specific) properties of the transplanted tissue. At present it is not yet possible to determine these properties, and therefore the compatibility of organs for transplantation, with complete certainty. As a precaution against rejection it is therefore endeavored, after transplantation of an organ, to suppress the recipient's defensive reaction by means of drugs, radiation or immunization. The drugs, mainly of the type called cytostatics (alkylating agents and antimetabolites) and corticosteroids, reduce the production of defensive cells, but thereby also weaken the recipient's defensive powers against infection. Irradiation of the transplanted organ, and also of the recipient's blood, can inhibit rejection, but is not without hazards. The recipient's lymphocytes play a major part in rejection. By immunization of animals with such lymphocytes it is possible to obtain an antilymphocyte serum whose antibodies substantially inhibit lymphocyte activity.

Careful attention to the properties of organs, together with suppression of incompatibility phenomena, has made numerous successful transplants possible. For example, thousands of kidneys have been grafted into recipients' bodies, where they have survived for periods of ten years and more. Livers, lungs and pancreases have also been successfully transplanted. Of more than 110 heart transplants carried out, one survived for more than a year and a half.

Heart transplantation comprises three stages: removal of the heart from the donor's body; removal of the recipient's diseased or defective heart; grafting the "new" heart in its place. Recipient and donor are both connected to a heart-lung machine, which enriches the blood with oxygen and removes carbon dioxide as a waste product. The oxygen-laden blood is then pumped by the machine through the circulatory system, but by-passes the heart and lungs (Fig. 1-43). The recipient's chest is opened, the heart is separated from all the blood vessels connecting it to the systemic circulation and pulmonary circulation, and is removed. Parts of the atria are left in situ (Fig.

FIG. 1-44.
Recipient site after removal of diseased heart.

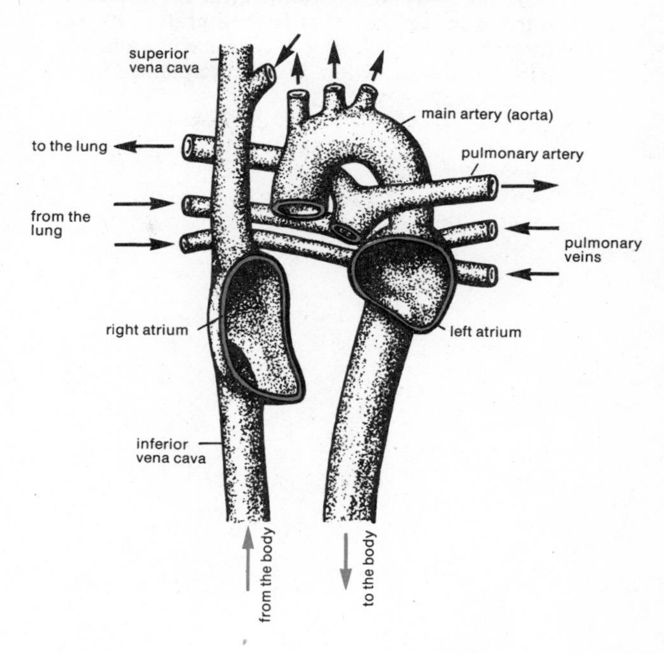

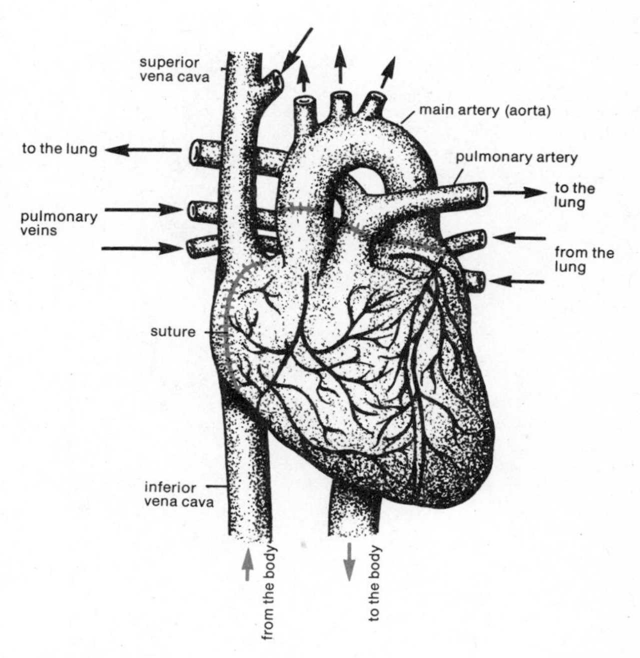

FIG. 1-45.
Donor heart grafted into recipient.

1-44). The donor's heart is similarly removed from the body, but here the atria are not separated from the heart. It is then grafted by sutures between the blood vessel ends and the parts of the atria left in the recipient's body (Fig. 1-45). During this operation the "new" heart is supplied with blood by the heart-lung machine. Finally, it is electrically stimulated to start it beating regularly. When it is functioning properly, the machine is disconnected.

Part Two

THE HEART AND CIRCULATORY SYSTEM

STRUCTURE AND FUNCTION OF THE HEART

The heart, an approximately fist-sized organ of truncated conical shape, is located behind the sternum (breastbone). The base of the heart, to which the blood vessels are connected, is at the top and somewhat to the right, while the apex of the heart is at the bottom and pointing to the left (Fig. 2-1). The so-called apex beat can be felt at the fifth intercostal space (between ribs) (Fig. 2-2).

The heart consists of two halves, right and left, separated by a partition (septum). Each half is subdivided into two chambers: an upper chamber (*atrium*) and a lower chamber (*ventricle*). The ventricle has a thicker muscular wall than the atrium, and the left ventricle is more powerfully muscular than the right ventricle. The *auricle,* a term sometimes used as a synonym of *atrium,* is really a pocketlike protrusion. As the heart occupies a somewhat twisted position in the chest, the right ventricle is mostly visible in the front view (Fig. 2-3). The heart is located in the thoracic mediastinum—the space between the pleural cavities—and is partly covered on both sides by the lungs. By percussion of the thorax (chest) a doctor can determine the position of the heart: light tapping reveals the so-called absolute cardiac dullness, while sharper tapping reveals the larger area of relative cardiac dullness that indicates the true outline and position of the heart (Fig. 2-2). The heart's precise position, and therefore its shape as it appears in an x-ray photograph, depends on particular physical conditions, including the person's constitutional type, the position of the diaphragm, and the respiratory movements. Figure 2-4 shows variants in the shape and outline of the heart. In each drawing the red line indicates the "electrical axis" of the heart, which is of importance in connection with the electrocardiogram (see page 65): normal type, right-hand type, left-hand type.

The *pericardium,* the double membranous bag enclosing the heart, protects it from overstretching. Its inner layer, the visceral pericardium (or *epicardium*), is firmly joined to the surface of the heart. The outer layer (parietal pericardium) consists of firm connective tissue whereby the heart is resiliently suspended from the spinal column, the thorax and the windpipe. The space between the two layers constitutes the pericardial cavity, which is filled with serous fluid acting as a lubricant that enables the layers to slide in relation to each other. In the condition known as pericarditis

FIG. 2-1.
Base and apex of heart.

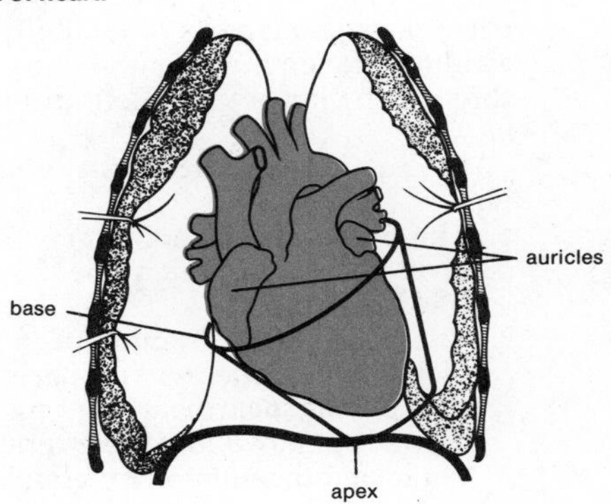

FIG. 2-2.
Site of apex beat.

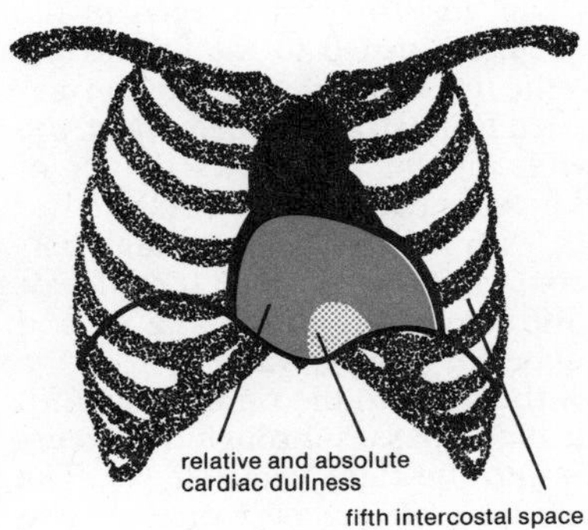

FIG. 2-3.
Front view of heart.

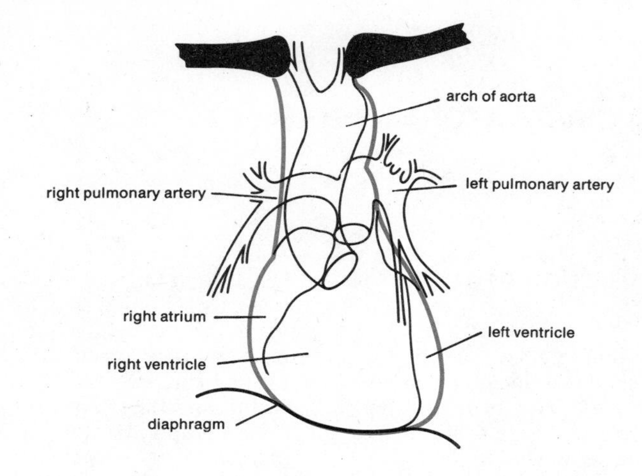

FIG. 2-4.
Various shapes of heart. Red line indicates electrical axis.

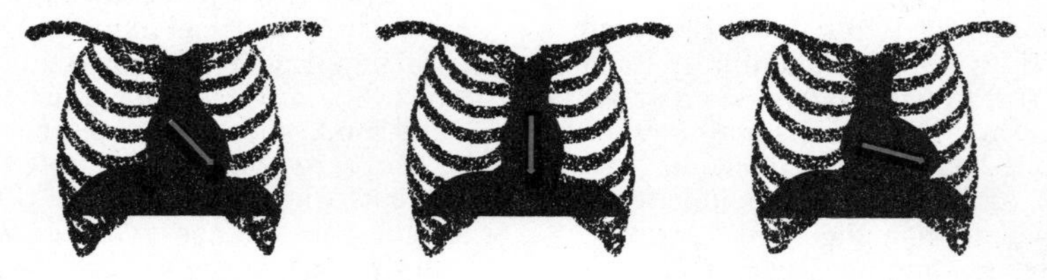

(inflammation of the pericardium) the quantity of this fluid is increased. Within the epicardium is the actual heart muscle (*myocardium*), the inner surface of which is lined with a thin membrane (*endocardium*) from which the heart valves arise.

The left ventricle pumps oxygenated blood through the systemic circulation. After parting with its oxygen, the blood returns to the right atrium, enters the right ventricle and is then pumped through the pulmonary circulation to give off CO_2 and absorb fresh oxygen in the lungs before being returned to the left atrium and thence to the left ventricle. This path traveled by the blood is illustrated schematically in Figure 2-5 while Figure 2-6 shows details of the heart and its connected blood vessels: The venous blood, with low oxygen and high carbon dioxide content, flows through the inferior vena cava (10) from the lower parts and through the superior vena cava (11) from the upper parts of the body to the right atrium (3). From here the blood passes through the tricuspid valve (16) into the right ventricle (1). The latter pumps it through the pulmonary valve (17) into the pulmonary artery (5) and thus to the lungs, where the blood is enriched with oxygen (arterialized) and parts with carbon dioxide. It then flows through the pulmonary veins (7) to the left atrium (15), through the bicuspid (or mitral) valve into the left ventricle (2), from where it is pumped through the aortic valve (18) into the aorta (12, 13), which is the principal artery, and through the systemic circulation of the body. *Systemic* designates the circuit that the blood makes from the left ventricle through all parts of the body to the right atrium; the circuit from the right ventricle through the lungs to the left atrium is the *pulmonary circulation*.

The action of the heart comprises two phases:

1. The phase of relaxation (*diastole*), the passive phase, in which the heart becomes filled with blood
2. The phase of contraction of the heart muscle (*systole*), the active phase constituting the actual heartbeat, during which the blood is pumped into the arteries of the systemic and the pulmonary circulation

FIG. 2-5.
Path of blood.

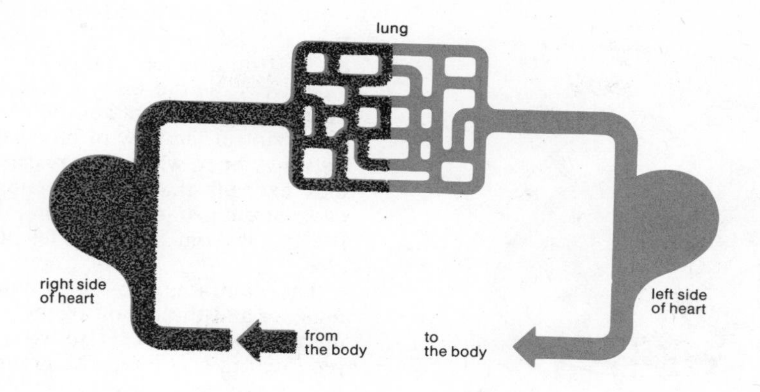

FIG. 2-6.
The heart and its vessels. (Numbers are explained in text.)

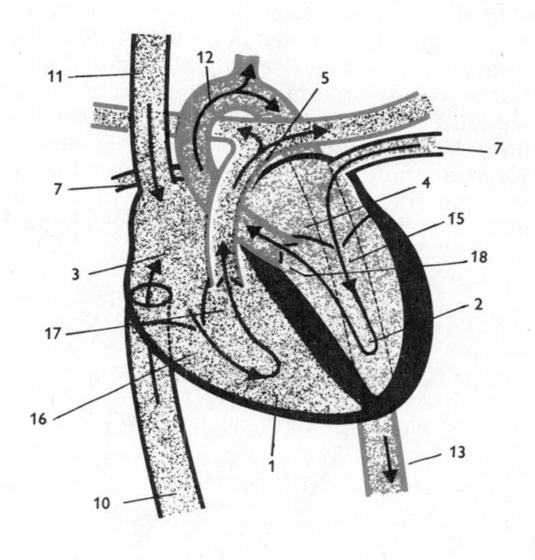

FIG. 2-7.
Path of blood through right atrium and ventricle.

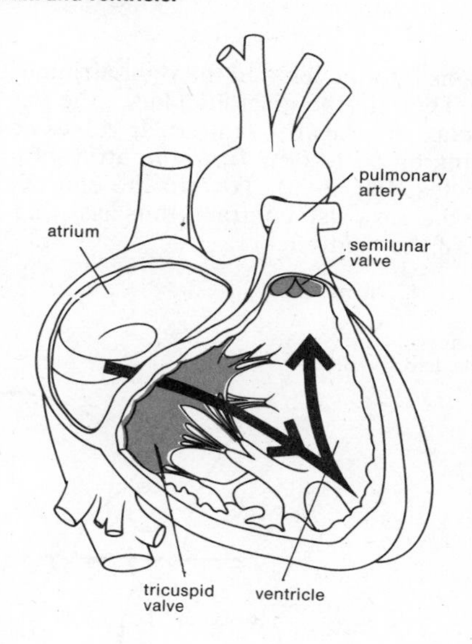

The cyclic succession of diastole and systole—filling and discharging—and directed circulation of the blood is made possible by a system of special valves. These are of two types differing in their construction and function. They have already been mentioned, but will now be considered in more detail (Figs. 2-7 and 2-8).

The *mitral valve* (bicuspid valve) guards the opening between the left atrium and left ventricle; the *tricuspid valve* similarly guards the opening between right atrium and right ventricle. The mitral and tricuspid are the so-called *atrioventricular valves*. During the systolic phase, i.e., when the ventricles contract, these valves are closed, and blood is sucked from

FIG. 2-8.
The valves of the heart.

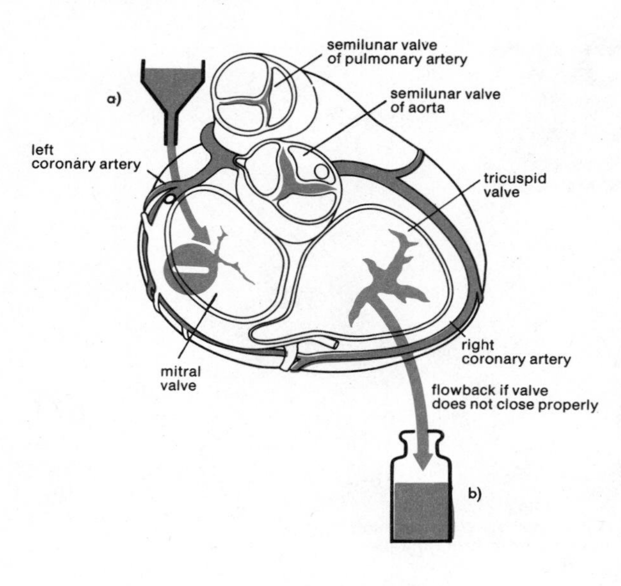

the veins into the left and the right atrium (Fig. 2-9b). Then, in the diastolic phase, the ventricles relax and the atrioventricular valves open, allowing blood to flow from the atria into the ventricles (Fig. 2-9a). Toward the end of this phase the atria also contract, thus assisting the filling of the ventricles.

The *semilunar valves* guard the orifices between the ventricles and the respective arteries the ventricles discharge the blood into: they are the *aortic valve* at the entrance to the aorta (from the left ventricle) and the *pulmonary valve* at the entrance to the pulmonary artery (from the right ventricle). Their function is to control the flow of blood from the heart, i.e., they open when the pressure in the ventricles exceeds that in the major arteries, and close at the instant of diastole, thus preventing the blood from flowing back into the ventricles.

The contraction of the myocardium (heart muscle) and the action of the valves (producing the so-called *heart sounds*) can be visibly recorded in a phonocardiogram. The contraction of the ventricles, the sudden rise in pressure at the beginning of the systole, and the closure of the atrioventricular valves cause the first heart sound (systolic heart sound), which is prolonged and dull; the second sound, short and high-pitched, is caused by the closure of the semilunar valves at the beginning of the diastole (diastolic heart sound). If the valves of the heart are diseased or defective, the heart sounds are characteristically altered and additional sounds (murmurs) may be heard. When a doctor listens to the heart sounds (auscultation), he hears the sound of the aortic valve in the second intercostal space to the right of the sternum (breastbone), the sound of the pulmonary valve in the second intercostal space

FIG. 2-9.
Diastolic and systolic phases.

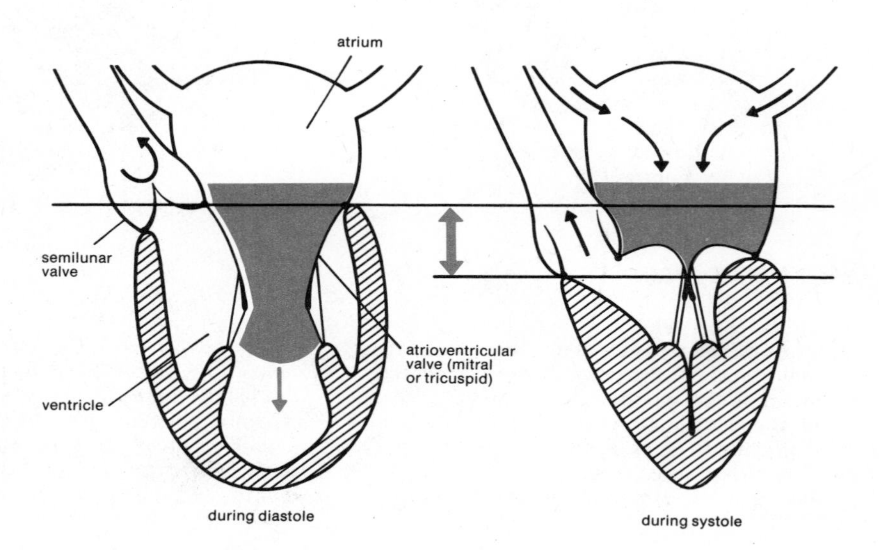

to the left of the sternum, the sound of the tricuspid valve in the fourth intercostal space to the right of the sternum, and the sound of the mitral valve in the region of the heart apex.

The two main phases of the *cardiac cycle* (heart cycle)—systole and diastole—are divided into two subphases, according to the position and controlling function of the heart valves. Thus systole comprises the period of contraction and the period of blood expulsion; diastole comprises the period of relaxation and the period of filling with blood.

The start of the cardiac cycle is marked by contraction of the heart muscle fiber, at first without expulsion of blood from the heart, i.e., without doing any pumping work: the volume in the ventricles does not change in this subphase, but there is a steady rise in pressure. When the pressure built up in the ventricle exceeds the pressure in the artery, the blood is expelled into the artery, while the ventricular volume rapidly decreases. This contraction (systole) does not expel all the blood; a certain residual volume remains in the ventricles. Finally, when the flow from the ventricles has stopped, the semilunar valves (aortic valve and pulmonary valve) close. In the following diastolic period the ventricles relax, so that the pressure in them falls below the pressure in the atria. The atrioventricular valves now open, admitting blood from the atria into the ventricles, so that the latter fill up. The whole cycle then repeats itself. Filling the ventricles is assisted by a small residual pressure that remains in the arteries, which acts through the capillaries and veins and maintains a certain pressure in the atria. The return flow of blood through the veins from the systemic circulation to the right side of the heart is, however, substantially assisted by the movements of the diaphragm (midriff) and the other respiratory muscles, by suction of the valve mechanism, and possibly by contraction of skeletal muscles.

HEART VALVE DISORDERS

In the case of *stenosis* (constriction or narrowing) of the heart valves the blood accumulates upstream of the atrioventricular valve orifices or can be pumped only with difficulty through the semilunar valve orifices. With the condition called *valvular insufficiency* there is imperfect closure of the heart valves, permitting leakage, so that the ventricle has to cope with an extra quantity of blood that is constantly leaking back. Rheumatism (see page 380) affecting the heart usually attacks the valves on the left side, namely, the mitral valve and/or the aortic valve. Rheumatic inflammation of the valve covering produces wartlike excrescences consisting of tiny blobs of coagulated blood which tend to cause adhesions. In addition, the valves become inflamed and swollen and may subsequently suffer distortion by scarring. Rheumatic heart disease thus causes valvular stenosis, e.g., affecting the mitral valve (mitral stenosis), and may also cause other valvular disorders (insufficiency of the aortic valve). The rheumatic growths on the valves may subsequently be attacked by bacterial infection. Before the discovery of penicillin, bacterial inflammation of the endocardium, i.e., the lining membrane of the heart cavities (endocarditis), was usually fatal. Syphilis may cause weakening or local abnormal dilatation of the aorta (aortic aneurysm), which in turn may result in aortic insufficiency. Rheumatic valvular disorders are most likely to affect children and young people, but may also occur in middle age. In many cases the young heart is able to cope with its extra burden of work by developing greater muscular strength. The patient feels no ill effects, and the disorder may be discovered by mere chance, e.g., in the course of a routine medical examination at school or for military service or for life insurance. Cardiac failure will, in such cases, occur only under conditions of overstrain.

Another consequence of valvular disorders particularly associated with bacterial inflammation is that accretionary matter formed on the valves may be carried along with the bloodstream and cause obstruction of blood vessels (embolism) in other parts of the body.

The valves of the left side of the heart are more prone to disease than those on the right side. Constriction of the left atrioventricular orifice (*mitral stenosis*) is four times as common in women as in men. Even if the heart valves suffer rheumatic damage at an early age, the patient usually does not develop trouble until the age of about 30. First symptoms are shortness of breath (dyspnea), tendency to bronchial catarrh, and irregular pulse. Five to ten years later the heart's performance seriously weakens, the left atrium is dilated, and some of the blood pumped by the left ventricle is forced back into the pulmonary veins, so that the blood is no longer adequately aerated.

FIG. 2-10.
X-rays of heart with mitral stenosis.

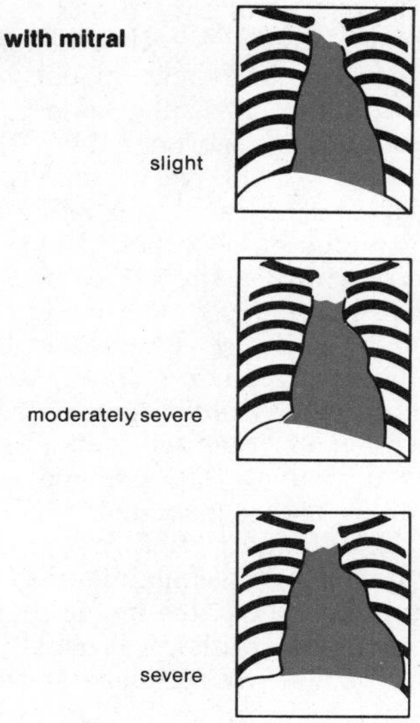

slight

moderately severe

severe

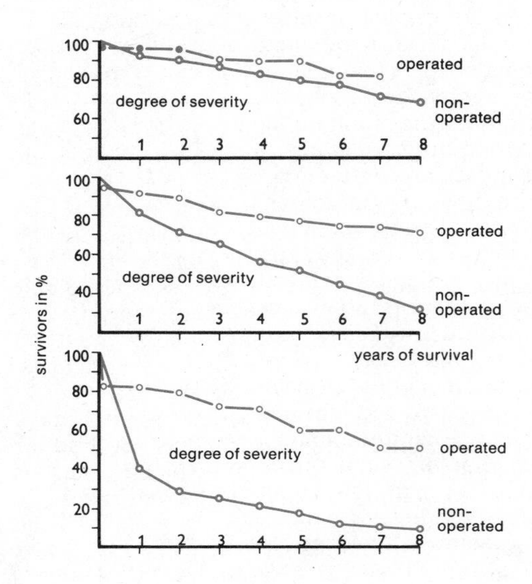

FIG. 2-11.
Survival rates of patients with mitral stenosis, with and without corrective operation.

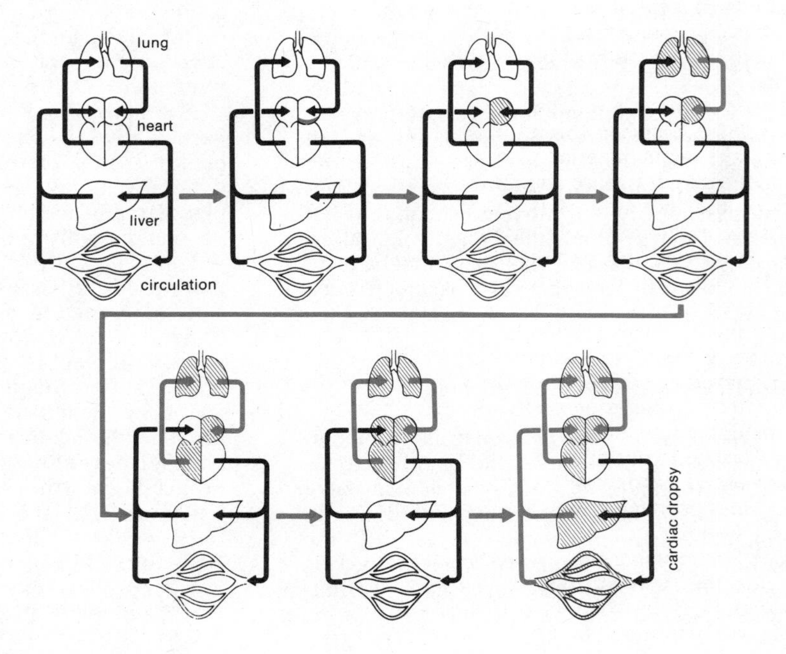

FIG. 2-12.
Increasingly severe congestive phenomena due to mitral insufficiency.

The patient has blue lips, cheeks, ears and tip of the nose. Finally, the right ventricle is unable to develop the force needed to pump the blood efficiently through the congested pulmonary circulation to the left atrium. This condition of cardiac insufficiency manifests itself in cold bluish hands and progressively weakening pulse. Treatment in the past consisted in the avoidance of physical exertion, so as to minimize the strain on the heart, and in administering drugs (digitalis) to stimulate the force of muscular contraction of the heart. Modern surgery to correct mitral stenosis is successful in about two thirds of the cases. It is especially appropriate in the treatment of relatively young patients having difficulty in breathing and can substantially prolong life (Fig. 2-11).

Mitral insufficiency, i.e., imperfect closure of the mitral valve, is encountered somewhat more frequently in men than in women. If there is only a slight defect, the left ventricle can cope with the extra burden of discharging the relatively small quantity of blood that leaks back through the imperfectly closed valve. When the heart's contractile force weakens in due course, however, congestion of blood in the lungs will occur because the left ventricle now cannot pump the full volume of blood. Spitting of blood may occur, and congestion is liable to spread also to the systemic circulation (see Fig. 2-12). Cardiac insufficiency of this kind is also treated with digitalis drugs; in certain cases, replacement of defective valves by artificial valves made of plastics may be appropriate.

Aortic stenosis, i.e., constriction of the semilunar valve between the left ventricle and the aorta, the main arterial trunk, can often be compensated by more vigorous muscular action of the left side of the heart. When the ventricle eventually fails to maintain its function, however, the patient becomes short of breath and suffers attacks of dizziness and fainting as a result of diminished blood supply to the brain. Overexertion of the heart in this stage may result in sudden death. The heart must therefore be protected from strain and be assisted by stimulation with digitalis to combat its insufficiency. Surgery to enlarge the aortic valve orifice is possible, but is difficult if there is advanced hardening (sclerosis) and is generally less promising than similar treatment for mitral stenosis.

The aortic valve may, like the mitral valve, develop the fault of imperfect closure (*aortic insufficiency*), in which case the left side of the heart will respond at first by some enlargement (hypertrophy) of the heart muscle to cope with the extra burden of work. First signs of the disease are fatigue, shortness of breath, and palpitation. As a result of hypertrophy of the left ventricle the heart can cope with these circumstances for quite a long time. In the end, however, cardiac pain and dyspnea (difficulty in breathing) indicate that the heart muscle is overstrained and has developed insufficiency.

Heart valve faults can often be diagnosed by auscultation (listening) with the aid of a stethoscope. Such faults manifest themselves as so-called *heart murmurs,* which are adventitious sounds (e.g., blowing sounds if a valve does not close tightly), which are in a certain regular relationship (in terms of time and location) to the normal sounds of the heart's action. Murmurs are caused by vibrations due to movement of blood within the heart, particularly when the blood is forced through constricted valve orifices or flows back through imperfectly closing valves. Like the normal sounds, murmurs can be heard particularly clearly in certain positions on the chest. X-rays and electrocardiograms reveal that the size, shape and position of the heart are characteristically altered as a result of some types of valvular defect (Fig. 2-10).

CARDIAC INSUFFICIENCY

As a general term, *cardiac insufficiency* denotes the inability of the heart to function normally. In its function as the body's circulating pump the heart is responsible for supplying blood and oxygen to all organs. It delivers blood at a rate appropriate to the demand that the organs, through the central ganglia (relay stations in the nervous system) and cardiac nerves, make upon it. The major oxygen consumers are the skeletal muscles, whose oxygen demand increases very greatly upon physical exertion. Under certain diseased conditions the performance of the heart will diminish instead of responding to the demands for increased action—i.e., insufficiency occurs (Fig. 2-13). This is first likely to be experienced at times of physical exertion, e.g., when mounting stairs or indeed merely when walking. In more serious cases the heart becomes unable to meet the normal demand for blood even in the absence of exertion. Heart failure may occur as the direct consequence of disease of the heart muscle, such as inflamma-

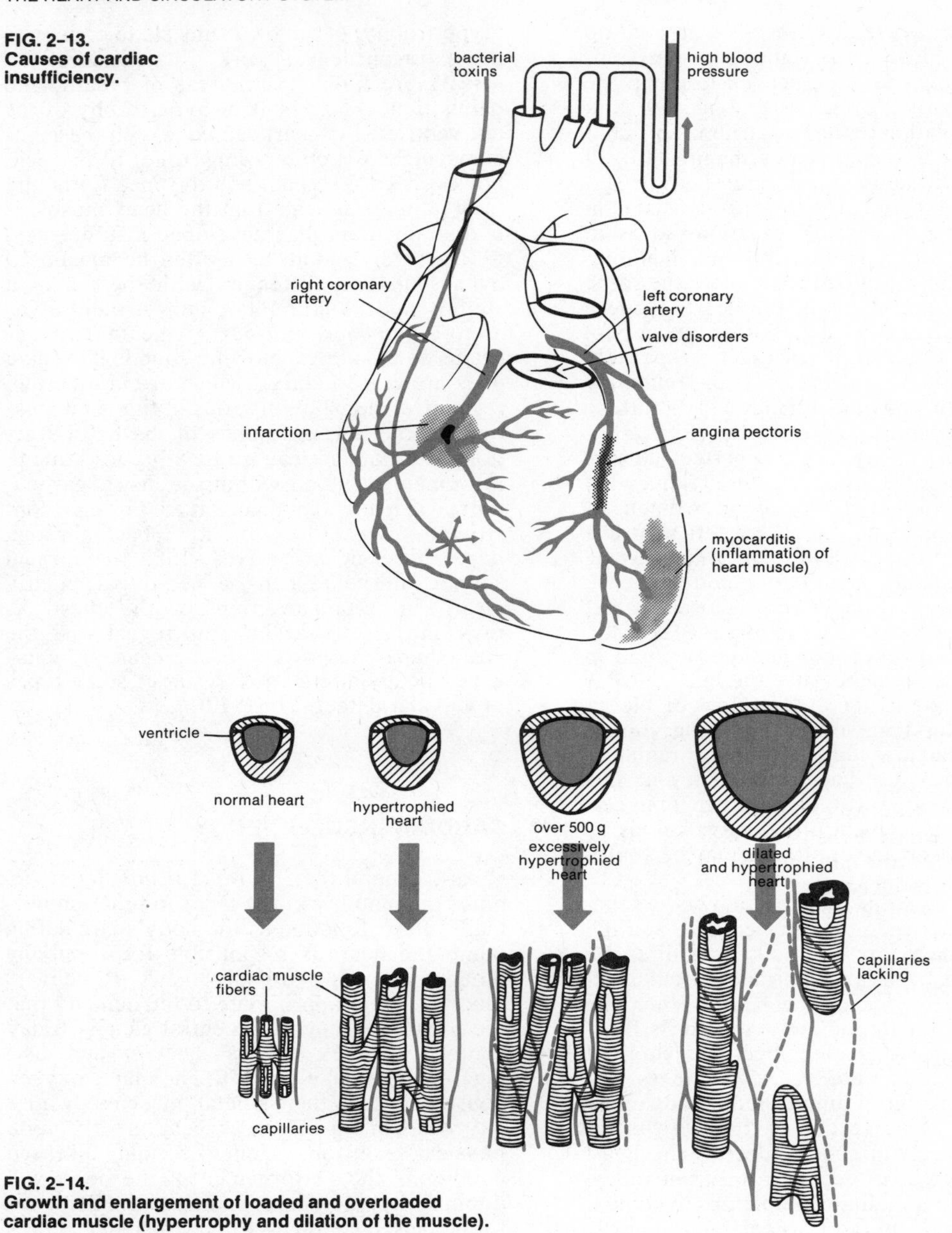

FIG. 2-13. Causes of cardiac insufficiency.

FIG. 2-14. Growth and enlargement of loaded and overloaded cardiac muscle (hypertrophy and dilation of the muscle).

tion. More often, however, the muscle is damaged, not put out of action, by bacterial toxins that invade it. Quite frequently, too, heart failure results from disease or damage of the coronary arteries, which are two branches of the aorta supplying blood to the heart muscle. Disorders in this flow of blood, causing shortage of oxygen in the heart muscle, may manifest themselves in paroxysmal pain: angina pectoris (see page 71); similar pains may be caused by congestion of blood vessels resulting in cardiac infarction (see page 73).

Cardiac insufficiency is most likely to develop as a result of overstraining the heart muscle. If some of the muscle fibers are put out of action, as in cardiac infarction, even the normal work load on the heart may overburden it. Arteriosclerotic disorders of the blood flow leave numerous small scars on the heart muscle, so that the muscle fibers not thus affected are liable to do most of the work and become overstrained.

Healthy heart muscle may also suffer overstrain, however—for example, in consequence of high blood pressure, which makes increased demands in terms of pumping exertion, or in consequence of valvular stenosis, in which case the blood has to be forced through a narrower passage. Valvular insufficiency (imperfect closure) also places an additional burden on the heart, as already explained.

Healthy heart muscle fibers always respond in the same manner to any increase in physical exertion, no matter whether this is due to work, sport or disease. With regular demand for higher performance the fibers become stronger and thicker (hypertrophy), just as the fibers of any skeletal muscles do when regularly exercised (Fig. 2-14). This adaptation of the heart muscle to higher demands is a necessary and meaningful response to such conditions as valvular disorders or abnormally high blood pressure, which the heart can thus cope with and which may therefore remain undetected for a long time. However, as a general rule, the reserve capacity of any hypertrophied organ is reduced, so that it is more likely to fail in the event of a certain degree of overstrain than the same organ which has not developed hypertrophy. For instance, with excessive cardiac hypertrophy the formation of new capillaries is diminished (Fig. 2-14), so that the enlarged muscle cells are less well supplied with blood than in the normal heart. With further overstrain of the muscle, and correspondingly increased metabolism, the muscle will suffer a deficiency of oxygen (anoxia), there will be a certain amount of degeneration accompanied by the formation of tiny scars, the muscle fibers will become flabbier and, as a result of this, the heart will eventually become enlarged (Fig. 2-14).

Depending on whether the left or the right wall of the heart becomes thickened, a characteristic pattern is revealed by percussion and in x-rays. Also, the symptoms will be different according to whether the left or the right side of the heart is affected.

Insufficiency of the Left and the Right Side, General Cardiac Insufficiency

Insufficiency, and possible failure, may affect either side of the heart. The diminished pumping performance that this involves will in either case have its own typical consequences to the blood circulation.

With *insufficiency of the left side* of the heart—caused mainly by high blood pressure, valvular defects of the aorta, and coronary disease (affecting the coronary blood vessels which supply blood to the heart muscle)—the rate at which the left ventricle pumps blood into the aorta is diminished. On the other hand, the right ventricle continues pumping the normal quantity of blood into the pulmonary circulation, which thus becomes overfilled and congested, causing dyspnea (breathlessness) at the slightest exertion and eventually even when the patient is at rest. This *cardiac asthma* is characterized by attacks of breathlessness especially at night. This air hunger becomes gradually worse, and the patient takes on a bluish complexion, when the pumping pressure developed by the (unaffected) right ventricle forces fluid from the blood into the alveoli of the lungs. The condition is called *edema of the lung;* it impairs the normal exchange of gases in the lung (cf. pneumonia, page 132). It can be obviated if the normal pressure in the pulmonary circulation is restored. This can be achieved either by stimulation of the left ventricle, so that it will then more effectively pump away the blood that has accumulated in the pulmonary circulation, or by lowering the excess pressure developed by the right ventricle. This latter effect used to be attempted by opening a vein (by a surgical operation called venesection or phlebotomy); nowadays it is possible to relieve pressure by the controlled removal of fluid containing salt from the blood circulation through the kidneys with the aid of drugs (diuretics, saluretics) (Fig. 2–15). Digitalis compounds are used for stimulating the left ventricle. Sudden attacks of breathlessness are rather dramatic, but not the only signs of insufficiency of the left side of the heart. In the long run, as a result of the excess pressure in the pulmonary circulation, a condition of pulmonary congestion and congestive bronchitis will develop, with characteristic expectoration which sometimes contains blood from ruptured blood vessels in the lungs. Pulmonary congestion is the reason why patients with heart disease cough and clear their throats frequently.

In the case of *insufficiency of the right side* of the heart, the (normal) left ventricle discharges more blood into the systemic circulation than the defective right ventricle can cope with. The systemic blood vessels thus become congested. At first the surplus blood can be accommodated in the veins and capillaries. If the condition of the heart worsens, however, blood congestion in the liver will cause jaundice and dropsy of the belly (ascites); digestion is impaired; the kidneys cease to eliminate salt and water effectively. Finally, fluid from the congested capillaries of the systemic circulation is forced into the tissue cavities. This edema, due to insufficiency of the right side of the heart, first manifests itself in the lower parts of the body—especially around the ankles when the person affected stands for comparatively long periods; later also around the shin and calf of the leg; and when the patient rests in the reclining position, edema will develop at the back of the thighs and on the buttocks. When edematous tissue is pressed with a finger, a shallow depression (pit) is formed because the fluid is temporarily forced into the interstices of adjacent tissue. Escape of fluid from blood vessels occurs not only in the abdominal cavity but also in the thoracic cavity and cardiac cavity. As a result of diminished blood circulation the patient's lips are bluish; he is short of breath and tires easily. Usually the pulse is accelerated, even when at rest, and is often also irregular.

Treatment of cardiac insufficiency will consist, first of all, in stimulating the flagging power of the chambers of the heart. This can be done with the aid of digitalis-based drugs, which develop a long-term action, or with strophanthin, which acts much more quickly after injection, but only for a short time. Such drugs not only strengthen the heart's performance, but also act on the stimulation and transmission of impulses and on the heart frequency. They must therefore be administered only under strict medical supervision, especially as digitalis drugs are liable to accumulate in the heart if given in incorrectly large doses. The accumulation of salt and water in the tissues, causing edema, can be eliminated through the kidneys. For this purpose, in addition to digitalis, drugs called diuretics or saluretics are used, which compel the kidneys to excrete salt. In some cases, as in cardiac asthma, morphine or a similar drug may be administered to calm the patient's anxiety and thus relieve the

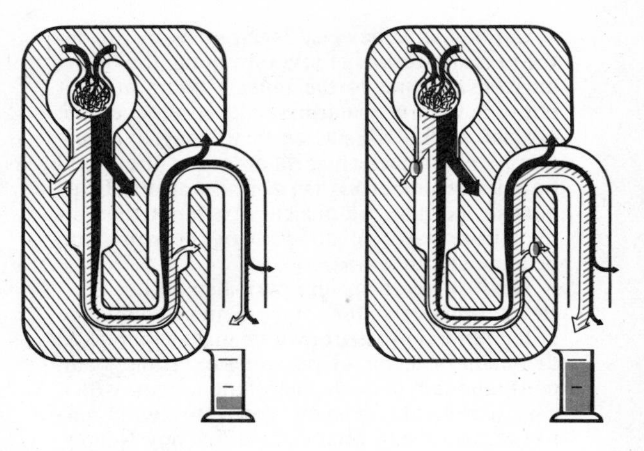

FIG. 2-15.
Suppression of salt absorption and therefore increase in salt excretion by means of a saluretic.

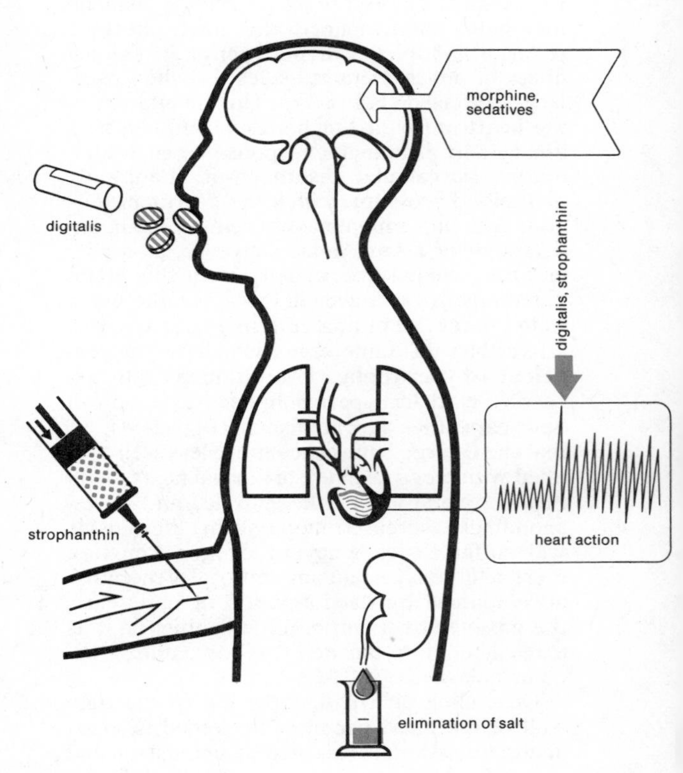

FIG. 2-16.
Treatment of cardiac dropsy.

strain on the heart (Fig. 2-16). In serious cases the patient will have to be kept in bed. A sitting or semisitting posture may be helpful in relieving pulmonary congestion; special beds for cardiac patients facilitate this. In general, the limits imposed on physical activity will depend on the seriousness of the condition. Excessive caution in avoiding all activity can be as unhelpful as excessive activity, unnecessarily prolonged confinement to bed may be harmful in that it increases the tendency to develop thrombosis.

It is especially important that the cardiac patient should keep to a suitable *diet*. The primary aim will be to limit the intake of salt, for the reason that of every gram of unexcreted salt the body will retain about 100 ml of water. A low-salt diet, corresponding to a salt intake of about 3 grams a day, will usually suffice to get rid of edema. Sometimes even this small amount of salt is too much. Formerly, an almost saltless diet, with about 1 gram of salt a day, was prescribed in such cases; nowadays it is more usual to use the low-salt diet (see pages 141 ff) and combine it with doses of a diuretic, which helps the body eliminate salt. Low-salt foods, which are in other respects also suitable for cardiac patients, include white bread, rusks, rice, noodles, boiled potatoes, honey, marmalade, vegetable oil, soft-boiled and scrambled eggs, curds, soft cheese, carrots, lettuce, endive, asparagus, apples, oranges, berry fruits, bananas and stewed fruit. Constant weight checks will reveal, by rapid variations in body weight, whether (and also roughly how much) salt and water have been retained or eliminated. As digestion is impaired by the congestion of blood, the diet should be not only low in salt content but also easily digestible and, with overweight patients, low in calories. Sometimes, in addition to dietetic measures (see page 141), it may be necessary to give a laxative to prevent constipation and flatulence. Treatment often starts with a short hunger cure during which the patient takes no food, only enough fruit juice—generally rather less than 1 liter (about 1 quart) per day—being given to maintain the fluid balance of the body and prevent potassium losses. The hunger relieves the heart of all "digestive work" and completely interrupts the intake of salt, thus enabling the salt-eliminating treatment to take full effect.

POWER EXERTED BY THE HEART, ENERGY SUPPLY, CARDIAC NERVES

The work done by the heart in performing its function comprises: (1) expelling the blood against the pressure existing in the great arteries (aorta and pulmonary artery), and (2) the much smaller amount of work done in accelerating the blood, i.e., imparting a certain velocity to it. The power (the rate of doing work) exerted by the left side of the heart is approximately five times that exerted by the right side, because the pressure to be overcome in the aorta is about five times as high as that in the pulmonary artery. The average pulse rate in middle life is about 75 beats per minute. At every stroke about 80 ml of blood is pumped into the body, or about 8600 liters in a 24-hour period. The work done by the adult heart at each stroke is estimated at 0.15–0.20 meter-kilograms (1.08–1.44 ft-lb), corresponding to 16,000–21,500 mkg (115,200–154,800 ft-lb) per 24 hours, which is equivalent to raising a weight of 1 ton (1000 kg) to a height of between 16 and 21.5 meters every 24 hours. These are, of course, only very approximate figures, but they do give a fair idea of the physical effort developed by the human heart.

A healthy heart adapts its effort within a wide range to the varying demands of the body's functions. Thus, the heart can pump blood at the required rate even against the resistance of raised arterial pressure; also, it can greatly increase the actual rate of flow. This adaptation of the cardiac function is achieved very efficiently through largely automatic mechanisms which are brought into action in response to an increase in the volume of filling of the ventricle. They primarily bring about an increase in the quantity of blood pumped by the ventricle at each stroke (the stroke volume). In addition, the rate of the heartbeat (cardiac frequency) is changed under "remote control" by the cardiac nerves.

It would be incorrect to suppose that the innervation (nerve supply) of the heart serves the same purpose as that of the skeletal muscles. Nervous impulses do not produce the contraction of the heart muscle, but serve only to regulate its action. Depending on requirements, the cardiac nerves control: (1) the rate of the heartbeat (chronotropic action); (2) the strength of the heartbeat (inotropic action); (3)

the conduction of stimuli (dromotropic action); (4) the excitability (sensitiveness to stimuli) of the heart (bathmotropic action).

The heart obtains the energy needed to sustain its high output of work from the nutrient media supplied to it, more particularly the blood flowing through the coronary arteries. Although its weight is only about 0.5 percent of the total weight of the body, the heart requires approximately 10 percent and at times as much as 20 percent of the oxygen absorbed into the bloodstream. A heart weighing 300 grams will, even under resting conditions, consume about 25 liters of oxygen per 24 hours, a quantity that has to be supplied by about 360 liters of blood during that period.

Although the heart muscle is, in principle, similar in construction to a skeletal muscle, there are nevertheless significant differences. Thus the heart muscle contains considerably more mitochondria in its cells, and its rate of combustion metabolism is correspondingly higher. Fats are the principal source of energy for the heart muscle; carbohydrates rank second in this respect. Among the latter, besides glucose, up to 50 percent of the body's lactic acid is consumed in supplying energy to the heart. This is important because in performing physical work the skeletal muscles give off lactic acid to the blood. The heart muscle absorbs less glucose when the concentration of lactic acid in the blood is higher. It can draw upon various nutrient resources for its energy supply, which is thus ensured under almost any conditions.

On the other hand, the heart is very sensitive to lack of oxygen (see pages 69 ff). This is important because the oxygen content of the coronary arterial blood conveyed to the heart muscle is almost entirely consumed even when the body as a whole is at rest, so that there is little room for more intensive extraction of oxygen from the blood. To increase the oxygen supply to the heart, the actual flow of blood itself has to be increased. The critical requirement in maintaining the overall energy performance of the heart is not so much the oxygen demand for the combustion of fats and carbohydrates as the need to keep the heart muscle adequately supplied with oxygen. To get more oxygen, it must be supplied with more blood, and this can be achieved only by dilation (widening) of the blood vessels, more particularly the coronary arteries (see page 68).

The action of the heart and its blood supply are controlled by two sets of antagonistic nerves, i.e., nerves acting in opposition to each other in the effects they produce. The sympathetic nerves act in a positive sense, tending to speed up the function of the heart, and for this reason they are called accelerator nerves. The vagus nerve, belonging to the parasympathetic system, acts in a negative sense, tending to slow down the heart rate (Fig. 2-17).

The *sympathetic* trunk comes from the lateral horn cells of the spinal cord and communicates with the musculature of the atria and ventricles of the heart. These nerves make the heart beat faster, increase the strength of the beat, promote the conduction of stimuli and increase the sensitiveness to stimuli. An especially important function of the sympathetic nerves consists in increasing the contractile force, as a result of which the heart discharges a larger volume of blood by contracting to a smaller volume. This extra blood is derived from the normal residual volume that would otherwise remain in the ventricle at the end of the systole. The blood output from the heart can thus in certain circumstances (e.g., in the body of an athlete performing some strenuous feat) be increased to a multiple of the normal stroke volume. This means that under the action of the sympathetic nerves the reserve capacity of the heart is mobilized. In general, any physical effort will cause these nerves to increase the heart's performance. In a case of cardiac insufficiency, different conditions prevail; although the heart may then have an abnormally large residual volume of blood (due to dilation of the heart), this blood cannot be effectively mobilized as reserve capacity because the contractile force that the heart can develop is too weak.

The *vagus* nerve acts in opposition to the sympathetic nerves, i.e., it produces the corresponding negative effects: slower beat, reduced strength of beat, etc. The left vagus nerve fibers act mainly upon the sinoatrial node, while the right vagus nerve fibers act mainly upon the atrioventricular node. Overstimulation of these nerve fibers may cause "vagal arrest" of the heart. They feed a continuous stream of impulses from the cardiac center in the medulla oblongata (the lower portion of the brainstem) to the nodes. The vagus nerve fibers, however, do not communicate directly with the working musculature of the heart in the way that the sympathetic nerves do.

Neither the sympathetic nor the vagus nerve fibers act directly upon the heart muscle fibers,

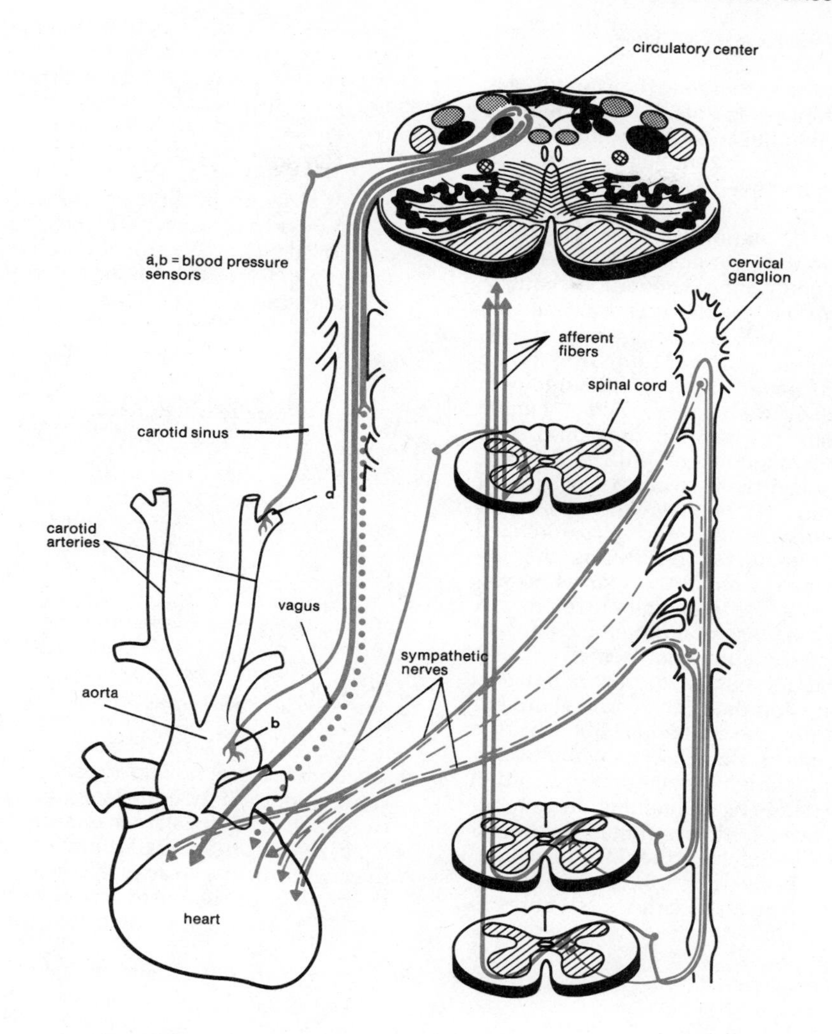

FIG. 2-17.
Innervation of the heart.

but through the medium of special chemical transmitters: acetylcholine for the vagus, noradrenaline (norepinephrine) for the sympathetic nerves. Simultaneous stimulation of the vagus and sympathetic nerves will not simply cancel out their respective actions, because in fact these two sets of nerves act upon different areas of the heart. As a result of this, ventricular fibrillation (see later) may occur, especially if the heart is already deficient. Figure 2-17 shows the cardiac nerves and also illustrates a reflex (which, in general, is an involuntary response to a stimulus) involving the heart and the circulatory system, in this case initiated by so-called pressoreceptors, i.e., sensory nerve endings which are stimulated by changes in blood pressure.

HEART ACTION, HEART BLOCK

The heart itself generates the stimuli that cause contraction of its muscle (autorhythmic action). Under suitable experimental conditions it can therefore be kept "alive" and beating even when removed from the body (Fig. 2-18).

Cardiac self-stimulation is centered in the *sinoatrial node* in the right atrium (Fig. 2-19). In its role as the controlling automation center it is the "transmitter" of the impulses that are passed to the secondary stimulating centers. The first "receiver" of these impulses is the *atrioventricular node*. From here the impulses are conducted by the atrioventricular bundle (bundle of His), which divides into two branches that communicate with the Purkinje fibers and, through these, with the muscles of the heart chambers. The atrioventricular bundle and the Purkinje fibers are two modified forms of muscle fibers constituting the impulse-conducting system of the heart. It is notable that each of the secondary centers can itself take over the task of automatic control in the event of failure of the main center.

The automatic action of the heart is based on the phenomenon that the electric charge of the membrane of the individual fibers in the automation centers decreases spontaneously and produces a stimulus (the action potential) which is transmitted as an impulse. In general, the heart conforms to the rhythm of the impulses issued by the sinoatrial node, the main controlling center, because it is here that the action potential associated with the stimulus is developed more rapidly (Fig. 2-20, top). The stimulation rhythm of the sinoatrial node is between 60 and 80 beats per minute. If this node is put out of action, the secondary centers take over with their own respective rhythms. Thus the rhythm of the atrioventricular node is be-

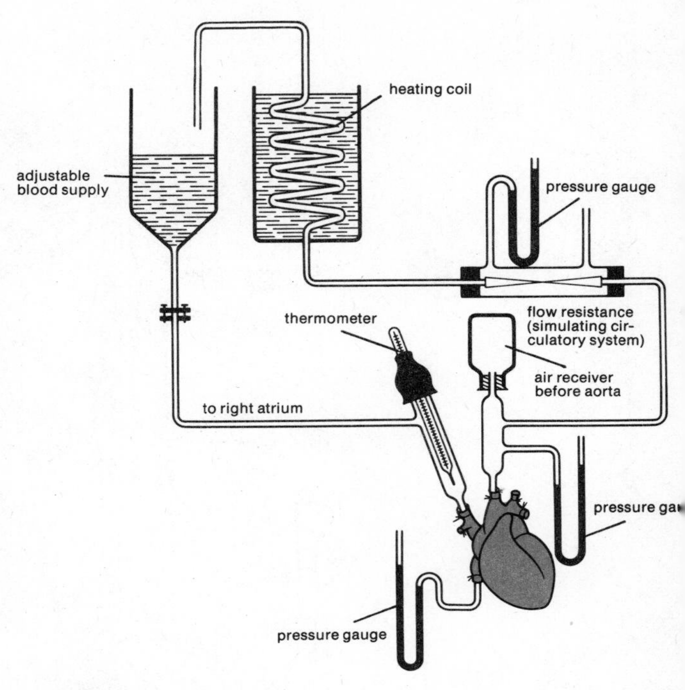

FIG. 2-18.
Heart kept "alive" when removed from body.

FIG. 2-19.
Conductive anatomy of the heart.

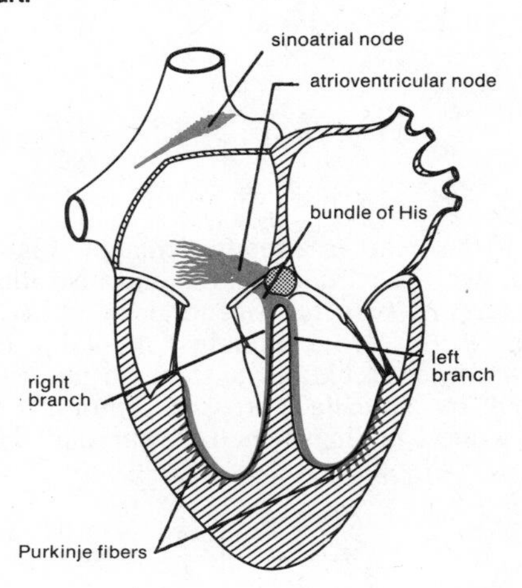

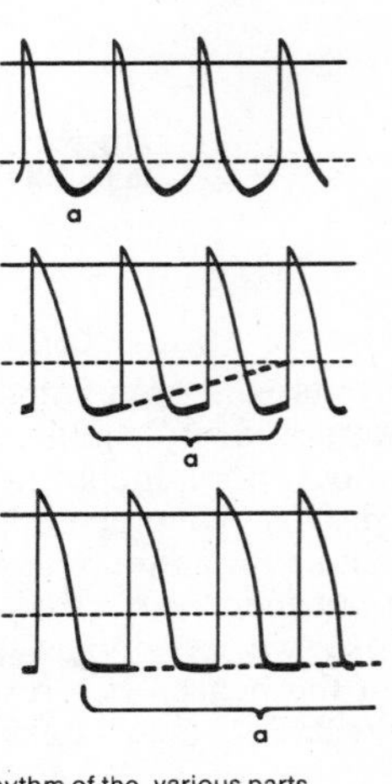

FIG. 2-20.

a = rhythm of the various parts

tween 40 and 60 (Fig. 2-20, middle), whereas that of the ventricles is between 20 and 30 beats per minute (Fig. 2-20, bottom). Because of its rhythm-generating function the sinoatrial node is sometimes called the pacemaker of the heart.

The *electrocardiogram* (ECG) is the record of the electrical activity of the heart. More particularly, it is the record of the sum of all the action potentials generated by the heart muscle and picked up by means of electrodes applied externally to the patient's body. Each stimulated heart muscle cell constitutes an elementary source of electric potential (dipole). Together the cells produce the overall potential, which varies in magnitude and direction during the heartbeat. Greatly simplified, these variations in potential recorded by electrodes applied, for instance, to the limbs can be represented in a triangular diagram (Fig. 2-21). An electrocardiogram is recorded by means of an instrument called an electrocardiograph. The

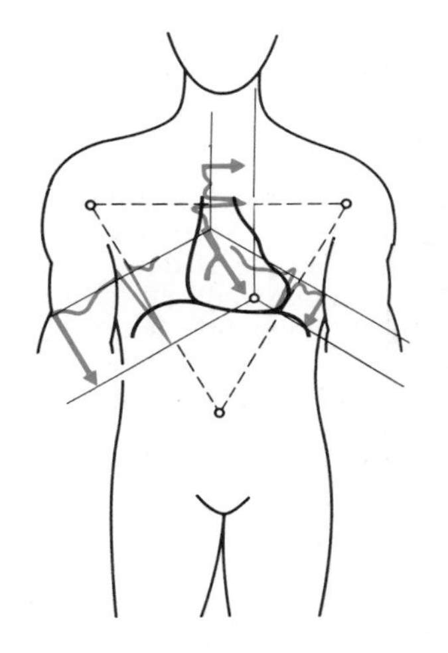

FIG. 2-21.
Electrodes applied to limbs.

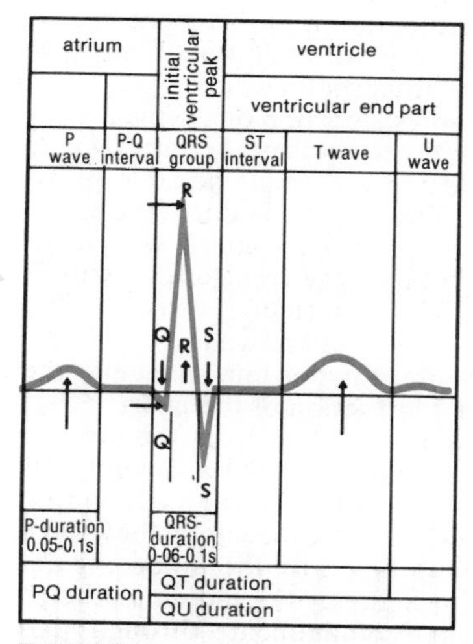

FIG. 2-22.
Electrocardiogram of the extremities.

FIG. 2-23.
Relation between cardiac stimuli and the various portions of the electrocardiogram.

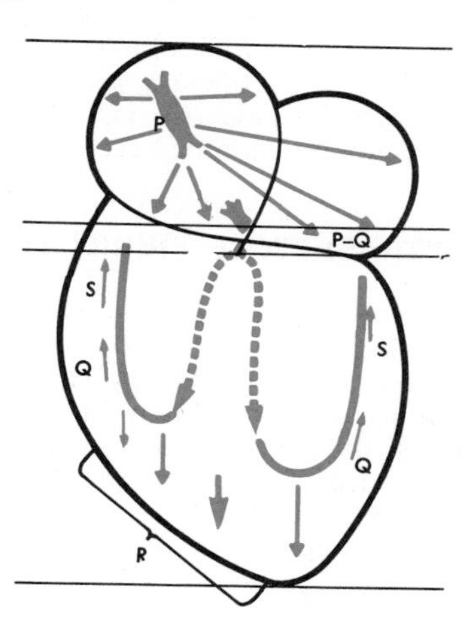

FIG. 2-24.
Heart block.

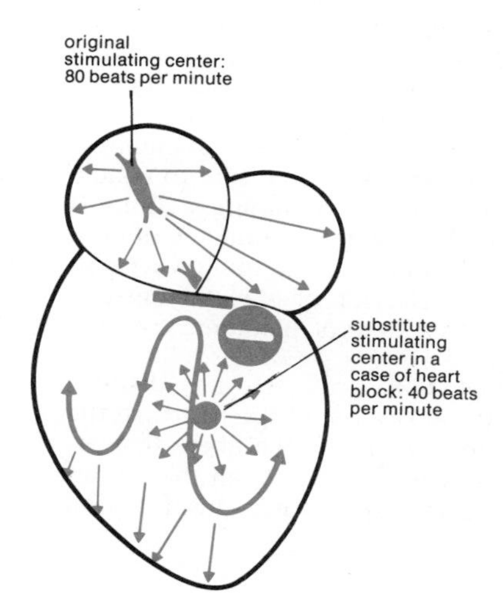

individual records obtained from electrodes applied to certain parts of the body are called leads. The three standard leads are right arm to left arm (lead I), right arm to left leg (lead II), left arm to left leg (lead III). Additional electrodes may be applied to the chest wall to detect minor defects not picked up by the standard leads. Although the electrocardiogram gives information on the electrical activity associated with the contractions of the heart muscle, it does not provide a direct indication of the intensity of these contractions. In order to draw reliable conclusions from irregularities in the electrocardiogram it is therefore necessary to supplement it with additional information obtained by other methods.

Figure 2-22 shows a portion of a typical electrocardiogram. It comprises certain *waves* (variations in electric potential) called P, Q, R, S and T waves. The P wave is due to contraction of the atria; the Q, R, S and T waves are due to contraction of the ventricles, with T marking the end of ventricular activity. The cause of the U wave is unknown.

Three things are of major importance to the stimulation and contraction of the heart. First, the transmission of impulses from the atria to the ventricles is retarded in the atrioventricular node, i.e., a time lag is introduced, so that the ventricles do not contract until they have become filled with blood by the preceding action of the atria. Second, the individual heart muscle fibers intercommunicate through special points of contact (the heart muscle is what is known as a syncytium, i.e., a group of cells with interconnected protoplasm). Because of this the atria and ventricles always contract as a coordinated whole to develop their pumping action. Third, the (electrical) stimulation of the heart muscle fibers is of very long duration in comparison with that of skeletal muscles. For this reason the so-called refractory period (the period of relaxation of the muscle during which excitability is suppressed so that it will not respond to stimuli) is correspondingly long. This is an important condition to enable the heart to develop its pumping action. It makes possible the rhythmic sequence of contraction and relaxation—something very different from the behavior of skeletal muscle, which can go into a state of sustained contraction (tetanus).

The term *arrhythmia* refers to any disturbance of the natural rhythm of the heart. In the condition called *heart block* the fibers that ordinarily conduct the impulses from the atrium to the ventricle fail to perform this function (Fig. 2-24). Another type of arrhythmia is *extrasystole* (see next section), caused by a contraction wave initiated in some part of the heart muscle other than the sinoatrial node. Normally such contraction waves arising from secondary centers are suppressed by the more rapid rhythm controlled by the sinoatrial node.

If conduction of impulses between the atria and the ventricles is completely cut off, the sinoatrial node cannot transmit impulses to the ventricles. In the normal heart, the contraction wave spreads through the atria and stimulates the conducting fibers (bundle of His) which convey the impulse to the ventricles, so that the latter are ready to contract just when they have been filled with blood from the atria. If this conduction of impulses between the atria and ventricles is seriously impaired or fails entirely, heart block occurs. This may vary in degree. With "incomplete" heart block the conduction time of the impulses is prolonged. If conduction fails entirely, the condition is called "complete" heart block; the atrium then beats at its normal rate of 70–80/min, whereas the ventricle independently takes up its own rhythm of 30–40/min, which is not enough to maintain efficient circulation. Heart block may have various causes, such as inflammation of the heart muscle due to diphtheria or acute rheumatism, sclerosis of the coronary arteries, or an overdose of digitalis. Especially in coronary sclerosis, the conduction of impulses may be only temporarily inhibited. It may then be several seconds before the ventricle starts pumping of its own accord (and at its own rhythm); during this interval the blood pressure falls rapidly and the patient may suffer fits of unconsciousness (Stokes-Adams attacks) due to insufficient oxygen supply to the brain (Fig. 2-25). Treatment of this dangerous condition consists in stimulating the ventricle to beat at a faster rate, especially by means of an artificial pacemaker implanted in it which gives electric impulses at the rate of a normally functioning heart.

EXTRASYSTOLE, FIBRILLATION

Extrasystole is caused by a premature contraction wave due to a stimulus from some source other than the sinoatrial node (the natural pacemaker). When the normal impulse arrives, the heart muscle is still recovering from the extrasystole, i.e., the interposed premature beat, and does not respond effectively. One category of extrasystole is due to

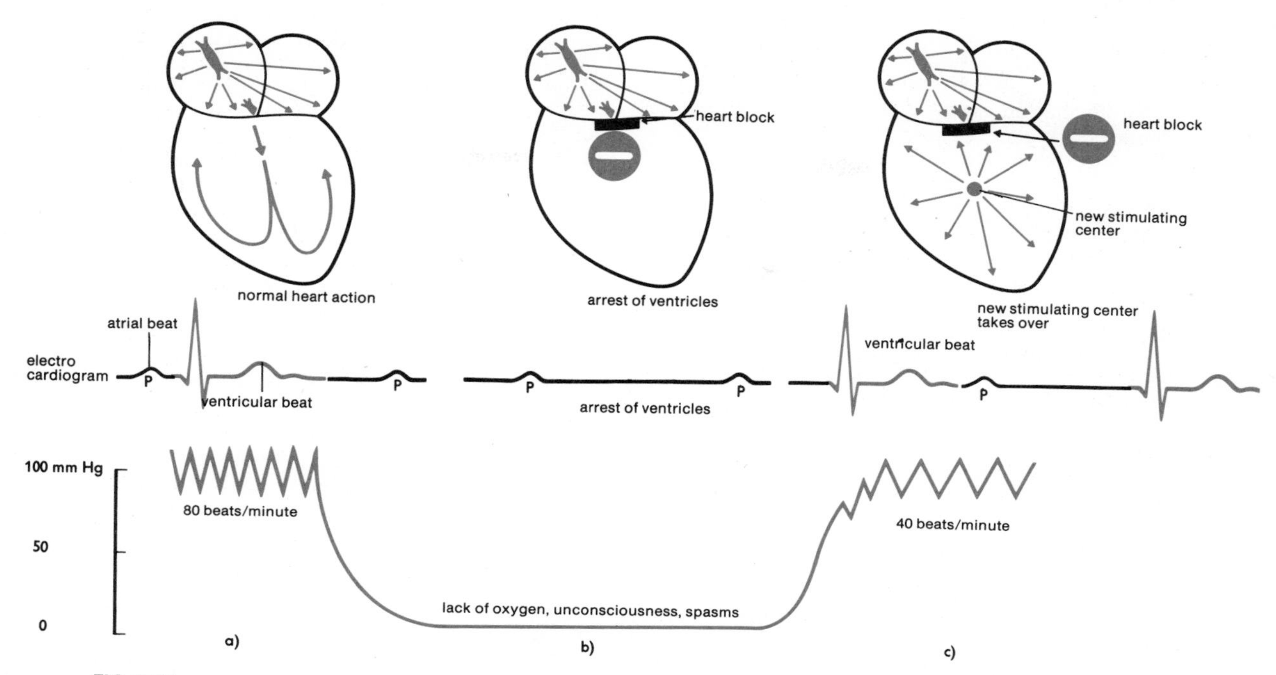

FIG. 2-25.
Heart block and Stokes-Adams attack.

impulses emitted by areas within the atrium or by the atrioventricular node, but extrasystole may also occur as a reflex initiated by stimuli from any other part of the body. The underlying causes may be of various kinds, e.g., an infection, enlargement of the atria (due to vascular defects), coronary sclerosis, or overactivity of the thyroid gland. If the extrasystole arises from within the ventricle itself, it is called ventricular extrasystole.

Besides manifesting itself in certain changes in the action of the ventricle as recorded in an electrocardiogram, the extrasystole is characteristically followed by a long interval until the next normal impulse for the ventricle arrives from the sinoatrial node. The patient may feel extra beats as palpitation (throbbing pulsation) or have an awareness of the unusually long intervals (as though the heart "stops" over and over again). Ventricular extrasystole is often relatively harmless, however. Although it may be an attendant symptom of coronary sclerosis and possibly other defects, it may also occur in the entirely normal heart. A slow pulse rate such as that of a highly trained athlete, and stimuli affecting the sympathetic nerves of the heart, may be conducive to causing extrasystole, as also the action of fear, psychological stimuli, nicotine or caffeine. Treatment is, generally speaking, necessary only if the heart rate becomes excessively high or a major "pulse deficit" is caused by premature extra beats.

A particular form of arrhythmia may occur in a heart affected by insufficiency and having overexpanded atria. An accumulation of extra atrial beats or a rapid succession of stimuli to the atrium may then cause a state of uncoordinated quivering of the atrium at a rate of 500–600/min, so that effective pumping action ceases. This condition is called *atrial fibrillation*. In comparison with the more serious ventricular fibrillation, it is in itself relatively harmless because normally the pumping action of the atria is responsible only to the extent of 10–20 percent for filling the ventricles. The harmful effects are of a different kind: under normal conditions the ventricles are stimulated by impulses at a rate of 70–80/min; with atrial fibrillation the ventricles constantly receive several hundred impulses, of irregular intensity, per minute. The ventricles cannot respond to all these impulses, but only to the effective stronger ones, which are, however, largely haphazard and uncoordinated, so that ventricular contraction is irregular and inefficient. This may precipitate heart failure. If atrial fibrillation is stable, however, its consequences are generally not very serious unless the contraction rate of the ventricles increases excessively or a considerable pulse deficit develops which uselessly absorbs energy. ("Pulse deficit" denotes that the pulse rate counted at the wrist is lower than the actual contraction rate of the heart.) The drug digitalis can be helpful in slowing the ventricle, enabling it to fill between beats.

Atrial fibrillation may turn into atrial flutter, involving a somewhat slower rate of contraction (200–400/min), with the attendant risk that this (relatively reduced) rate of impulses transmitted to the ventricles will cause ventricular flutter and fibrillation, which is rapidly fatal unless corrective action is speedily taken (cardiac massage, electrical defibrillation).

CORONARY ARTERIES

Like any other organ, the heart is supplied with blood by its own blood vessels: the coronary arteries (Fig. 2-26), which are two

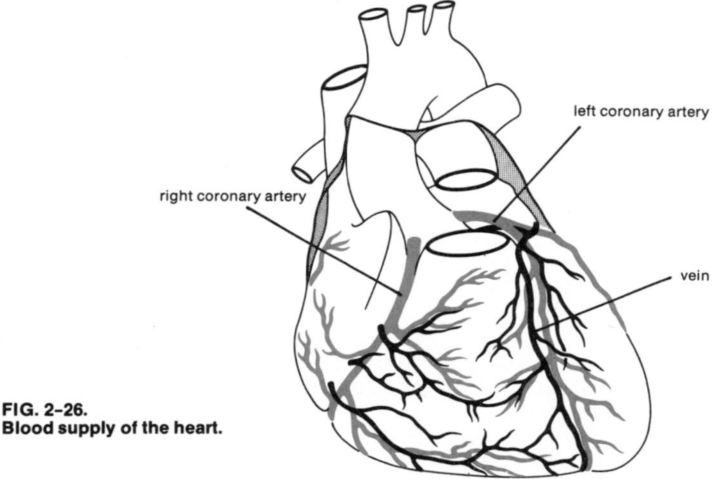

FIG. 2-26.
Blood supply of the heart.

FIG. 2-27.
Factors influencing coronary blood flow.

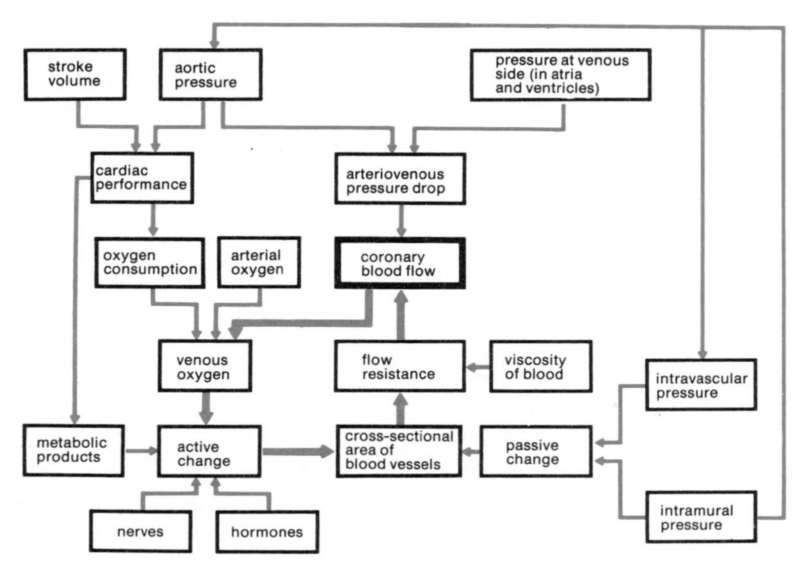

Courtesy Schödel and Grosse-Brockhoff

branches of the aorta that encircle the heart and supply the heart muscle (the myocardium) with blood. The right coronary artery supplies the greater part of the right side of the heart, the rear of the septum, part of the rear wall of the left ventricle, and its rear papillary muscle. The left coronary artery supplies blood to the other parts of the left ventricle, the front half of the septum, and the large papillary muscle of the right ventricle. After parting with its oxygen, the blood is conveyed through the cardiac veins to the coronary sinus in the right atrium.

The coronary arteries arise within the right and left aortic sinuses, at the base of the aorta, just above the rear semilunar valve. The flow through these arteries therefore depends, among other factors, on the pressure in the aorta. An increase in pressure not only propels the blood with greater force through the arteries, but also causes them to dilate. In addition, the flow of blood through the coronary arteries depends to a great extent on the "tissue pressure" in the wall of the heart: During systole, contraction of the heart muscle fibers prevents the blood from flowing into the coronary arteries, whereas the flow of blood in the cardiac veins is increased during systole. It used to be thought that the rhythmic functioning of the heart muscle promoted its own flow of blood by a kind of massage action, but it is now known that the inhibitory effect of the systolic contraction is far more important than such "massage." Furthermore, the internal width of the coronary arteries is determined more particularly by the muscular tone (the state of contraction) of the smooth muscle fibers of these blood vessels. In this connection the metabolism of the heart is especially important. When the heart works harder, an accumulation of waste products is formed in the tissue, which in turn brings about dilation of the coronary arteries. The cardiac nerves also play a role in controlling the dilation of these arteries. Finally, the condition of the walls of the arteries is important. Hardening of their walls due to arteriosclerosis causes them to lose their capacity to dilate, so that under certain circumstances the heart does not get a sufficient supply of blood. A severe pain develops due to shortage of oxygen in the heart muscle (angina pectoris, discussed in the next section). The various factors that together determine the flow of blood through the coronary arteries are summarized in Figure 2-27. In addition to those already mentioned, hormones are also involved.

Of especial clinical importance is the fact that the coronary arteries are functionally so-called end arteries, i.e., arteries whose branches do not communicate with those of other arteries. Only as a result of very gradually progressing coronary blockage can connections (anastomoses) between the coronary arteries develop, so that one of these two arteries can then take over the work of the other. In a case of sudden obstruction of a coronary artery this possibility does not exist, however, and the resulting shortage of blood supply to the heart can cause cardiac infarction (see later).

CARDIAC PAIN

As already mentioned, the heart muscle is supplied with blood by the *coronary arteries*. These differ from the arteries of other organs in the body in that they supply only so much blood as to meet the oxygen requirement of the heart muscle at any given time. If the requirement increases in order to cope with a rise in blood pressure or increased rate of flow to other parts of the body, the heart muscle will—in contrast with skeletal muscles, which in such cases can extract more oxygen from the blood they receive—always have to rely on dilation of its arteries to increase its oxygen supply. This dilation is initiated by waste products which are formed in greater quantity when the heart muscle has to work harder. An increase in arterial blood pressure will also increase the flow of blood to the heart muscle, but will at the same time impose a heavier pumping strain on the heart. Finally, the rate of flow in the coronary arteries can also be varied through the action of the sympathetic cardiac nerves: nervous stimulation makes the heart work harder, so that it uses up more oxygen, the extra oxygen normally being supplied by an increased flow through the coronary arteries. The interaction of the factors that regulate the blood flow to suit the oxygen requirement of the heart muscle functions very efficiently and imposes the least possible strain on the heart when it has to step up its pumping performance to supply more blood to skeletal muscles at times of physical exertion.

On the other hand, if the stimulus to the coronary arteries to supply more blood to the heart muscle originates solely in the sympathetic nervous system (Fig. 2-28), e.g., in consequence of emotional excitation (in the absence of physical effort), the flow through

FIG. 2-28.
Control of blood flow through the heart.

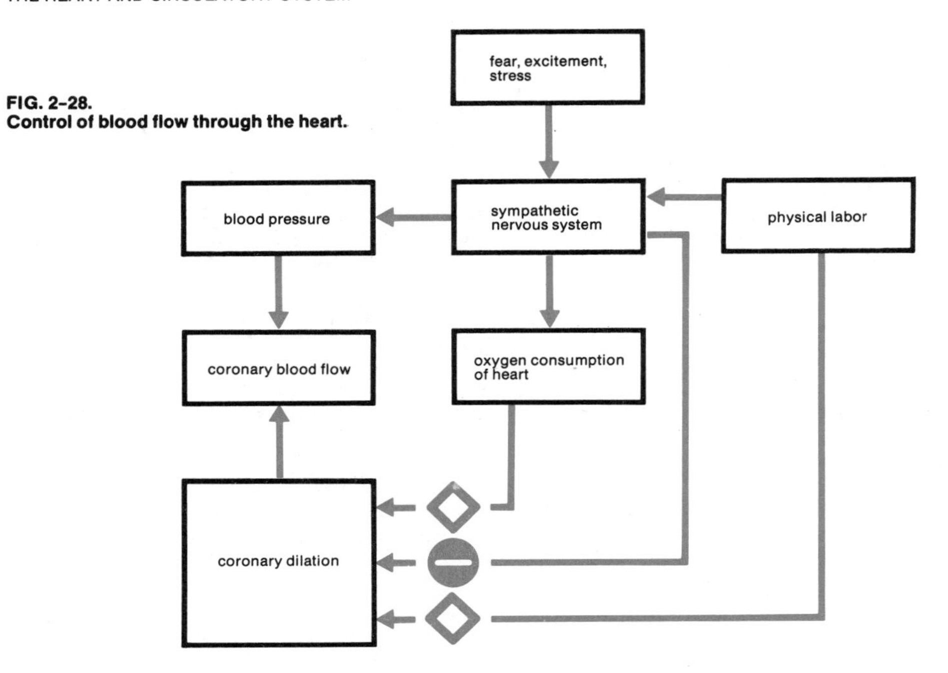

FIG. 2-29.
Causes of oxygen deficiency and cardiac pain.

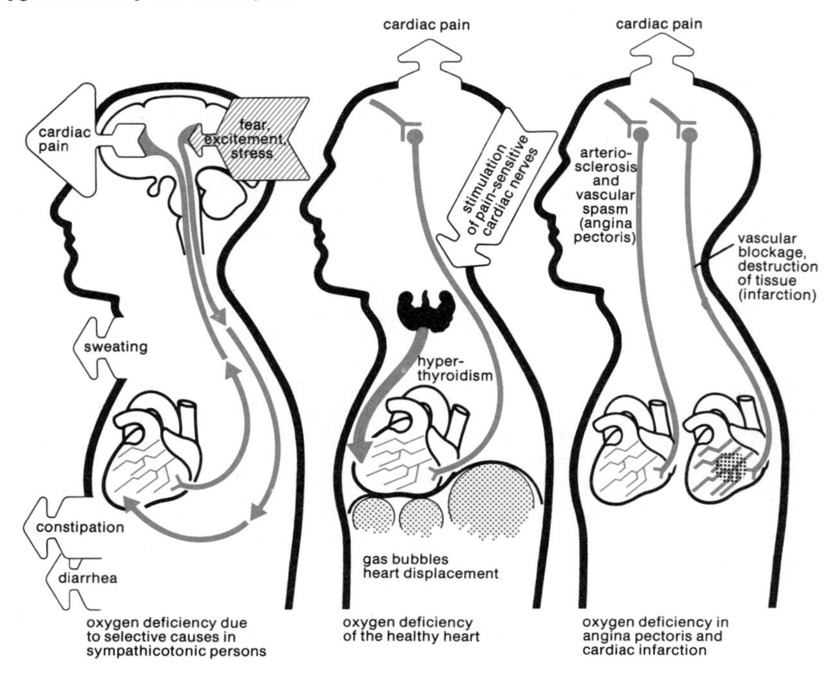

these arteries is not so reliably increased in relation to the heart's rate of oxygen consumption.

This explains how emotional conditions (fear, excitement, tension) may disturb the balance between oxygen requirement and coronary arterial flow (by nervous excitation transmitted through the sympathetic nerves). Such cardiac disorders, which are liable to occur under conditions of complete organic health, belong to the class of *functional disorders,* which occur in the absence of any changes in the organs involved. The indications may be excessively rapid pulse, palpitation, a sense of discomfort in the region of the heart and possibly even quite severe cardiac pain due to lack of oxygen (Fig. 2-29). Most prone to be affected by functional disorders are persons of sensitive disposition, so-called sympathicotonics (see page 330). The symptoms are most likely to manifest themselves at times of physical rest, may continue for hours and are—in marked contrast with those of angina pectoris (see later)—likely to get better with physical activity. Besides these cardiac disorders there are generally also other indications of vegetative lability: tendency to break out in sweat, hot flushes, gastric and intestinal complaints, nervousness, lack of concentration. (The term *vegetative* in this context means functioning involuntarily; thus the sympathetic nervous system is known also as the vegetative nervous system.)

Cardiac pain due to deficient oxygen supply to the healthy heart muscle may occur not only as a result of psychological causes, but also from such causes as lack of oxygen at high altitudes, anemia, or intensified metabolism due to overactivity of the thyroid gland. Excessive gas in the stomach and intestines (flatulence) may exert pressure on the heart and thus cause deficient blood flow with its resultant shortage of oxygen in the heart muscle, manifesting itself as pain. Such gas accumulations may be caused by bolting one's food, consuming effervescent drinks (fizzy lemonade, etc.), or nervous gulping of air into the stomach (see page 154). In the colon, gas may be formed by fermentation or decomposition processes, and lack of exercise promotes flatulence. Obesity tends to increase elevation of the diaphragm (midriff) and in general aggravates existing cardiac disorders.

Cardiac pain may occur even if there is no deficiency of blood and no shortage of oxygen supply to the heart muscle and may even have its origin outside the region of the heart itself, if the pain-sensitive cardiac nerves are stimulated at some intermediate point on their way to the brain centers, e.g., by pressure from bone at their point of entry into the vertebral canal.

Functional heart disorders are largely harmless and, if reliably diagnosed, often require no treatment. If they are due to anemia or thyroid overactivity, treatment will be directed at the basic disease. Sympathicotonic persons may, after having the nature and causes of their trouble explained to them, try to cope with distressing or emotionally straining situations in a more relaxed manner, so as to avoid cardiac involvement as much as possible. In many cases a suitably graded course of physical exercises under medical supervision may be beneficial.

Coronary reserve means the maximum possible increase in blood supply to the heart muscle; in a healthy person it is about 300 percent of the normal at-rest rate of flow. This reserve capacity to supply extra blood depends on how dilatable the coronary arteries are; it may be reduced by degenerative thickening of the walls of the arteries, with loss of elasticity, especially as a result of arteriosclerosis affecting the coronary arteries. Inflammatory diseases, including those of an allergic type, affecting these arteries may have a similar effect. Deficient coronary reserve may not manifest itself under conditions of physical rest, as the coronary flow of blood may be sufficient for basic requirements. But if the heart has to work harder in order to cope with the needs of physical activity, the coronary arteries with their limited capacity for dilation are unable to maintain an adequate flow: the shortage of oxygen supplied to the heart muscle is felt as a severe pain (*angina pectoris,* Fig. 2-29c). It will not, as a rule, occur when the body is at rest.

Like most pains arising in internal organs—and in contrast with pains originating at the surface of the body, which can usually be pinpointed to a certain well-defined area—angina is difficult to localize precisely. The pain is most severe in the upper left part of the chest and usually radiates into the left arm, more rarely to the right shoulder, the neck and the abdominal region (Fig. 2-30). The pain is described as burning or stabbing; it is accompanied by feelings of anxiety, fear of impending death, and breathlessness. The attack is paroxysmal, occurring under conditions of physical exertion (running, climbing stairs, etc.) or sudden cooling of the chest, e.g., by exposure to a draft of air. When the immediate cause, i.e., the exertion or strain, ceases and the patient stops to rest, the pain subsides fairly quickly, usually after a minute or two and at most after about 10–20 minutes. On the other hand, cardiac pains due to nervous or emotional causes tend to be relieved by physical activity.

If the attacks of angina pectoris are of relatively long duration, necrosis (see next section) of the heart muscle fiber due to lack of oxygen may occur. With repetition of the attacks the muscle becomes affected by small infarcts (areas of tissue that have undergone necrosis) and callosities. In contrast with a skeletal muscle which can be rested when oxygen shortage causes pain, the heart muscle must go on working. Further increase in load on the heart will therefore cause a dangerous situation which must, with continuing inadequacy of blood supply, eventually result in heart failure.

Long-term treatment of angina pectoris generally aims, among other measures, at changing the patient's habits and pattern of life. Physical exertion tending to cause spasms of pain should be avoided, because repetition of oxygen shortage can cause further deterioration of the heart muscle, and it is necessary also to avoid emotional stress and excitement liable to bring on an attack. Medically prescribed sedatives may be beneficial. Patients disposed to obesity should reduce weight by suitably dieting, the more so as a heavy meal can cause an attack. Drugs that expand the coronary arteries or reduce the oxygen demand can be helpful in relieving an actual attack of angina pectoris. Besides various new and promising drugs, the familiar older ones, especially nitroglycerin (taken by mouth) or amyl nitrite (inhaled as vapor), are still used.

over 50%
10–30%
10%
5% of persons affected have pain in the regions indicated

FIG. 2-30.
Pain radiating from the heart due to deficient supply of blood to the heart muscle (angina pectoris, infarction).

FIG. 2-31.

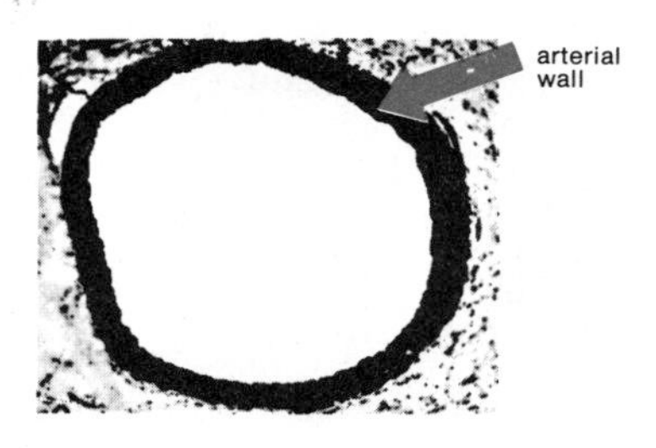

(a) Normal coronary artery.

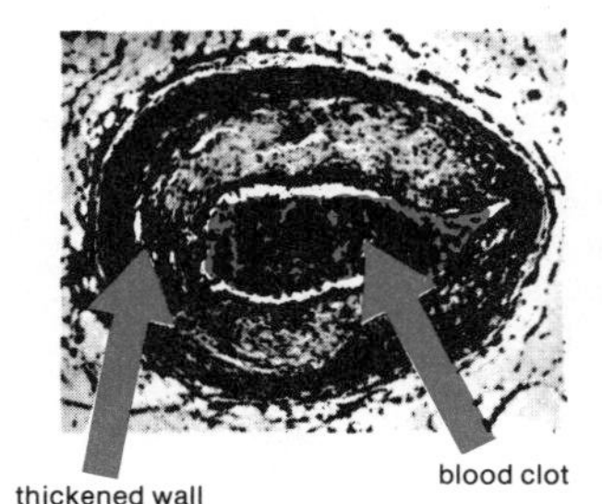

(b) Coronary artery with thickened wall and blocked by a blood clot.

CARDIAC INFARCTION

FIG. 2-32.
Obstruction of upstream arterial branch by thrombosis.

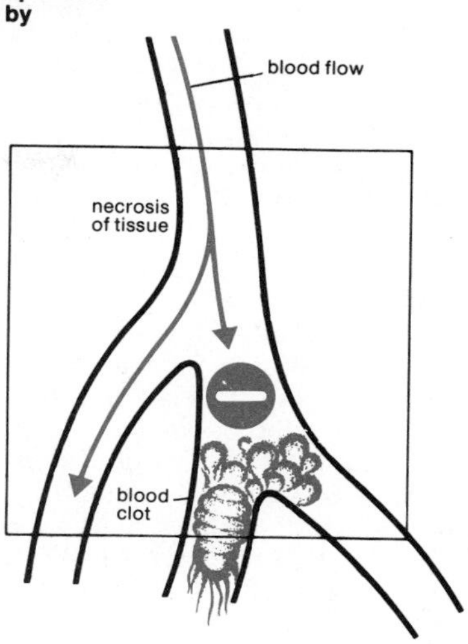

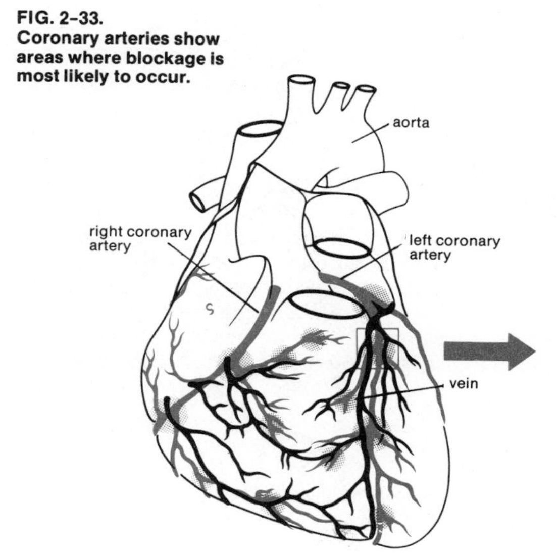

FIG. 2-33.
Coronary arteries show areas where blockage is most likely to occur.

Cessation of blood supply to tissues will cause necrosis, i.e., death of areas of tissue. An area of dead tissue due to this cause is called an *infarct*. In most cases the lack of blood is due to occlusion or stenosis of the supply arteries, e.g., as a result of arteriosclerosis. Cardiac infarction involves damage to muscle fiber; it is thus known also as *myocardial infarction* in that it affects the heart muscle (myocardium). In the great majority of cases the underlying cause is degenerative change in the walls of the arteries due to sclerosis.

Cardiac infarction (myocardial infarction) is the result of a kind of chain reaction whose successive stages have been studied and described in detail by microscopic examination, but the causal relationships have as yet by no means been fully clarified. Microscopically and functionally it is possible to distinguish between sclerosis due to normal aging and sclerosis due to arterial disease. In the former case there is a gradual hardening (loss of elasticity) of the arteries.

Much more serious in its consequences to circulation is *arteriosclerosis* occurring as a result of vascular disease. It primarily affects the lining of the arteries. The layers of tissue of which the lining consists become increasingly permeable and first absorb additional fluid and eventually also foreign substances. Small lacerations form the starting points of minute blood clots which progressively merge, become superimposed upon one another, and may finally block the passage of the artery completely (Fig. 2-31). Such clotting is called *thrombosis*, and the obstructing clot itself is called a *thrombus*. The latter can spread in the now stagnant column of blood in the blocked artery and also obstruct the inlet to the next upstream arterial branch (Fig. 2-32). At the same time the parts of the heart muscle that rely on those branches for their blood supply will suffer infarction due to lack of oxygen.

The extent and dangerousness of an infarction will depend on the diameter and location of the coronary arterial branch concerned (Fig. 2-33). If the area affected by necrosis is very large, the heart as a whole may fail. Extensive infarcts sometimes cause a rent in the heart muscle, resulting in internal bleeding due to cardiac rupture. Even very minor individual infarcts which reduce the heart performance only to an insignificant extent may be harmful by inducing extrasystole or, depending on where they are located, by obstructing the normal conduction of stimuli in the conducting system of the heart muscle.

The cause and development of infarction are, however, determined not only by sclerosis, thrombosis and blood supply by the coronary arteries, but also by the heart's oxygen consumption rate. In the normal heart a shortage of oxygen, such as may occur when the heart has to work harder, will cause the coronary arteries to dilate. Arteries that have abnormally thickened walls and/or are partly blocked have largely lost this capacity for dilation. Therefore any heavier load of work on the heart will further worsen the oxygen shortage of the heart muscle. Stimulation of the cardiac nerves also increases the oxygen demand. That is why exertion, agitation, nervousness and emotional stress may bring on cardiac infarction or adversely affect an already existing infarct.

A number of factors contribute to causing coronary sclerosis. They are therefore contributory factors in the risk of cardiac infarction. One of these factors is overweight, especially, it would appear, if it is due to excessive consumption of saturated (animal) fats, mostly associated with a high standard of living in affluent societies. In the Second World War, when food was generally scarce, there was a decline in the incidence of cardiac infarction (Fig. 2-34). The chief suspect is an excess of beta-lipoprotein in the blood; this is a protein that occurs combined with cholesterol and other fatty compounds found chiefly in animal fats. Other contributory factors are emotional stress, which acts through the cardiac nerves, high blood pressure, metabolic diseases such as diabetes, smoking, probably also excessive consumption of alcohol, and possibly caffeine. Lack of exercise is of particular significance, because physical activity not only helps to keep the skeletal muscles fit but also "trains" the coronary arteries by causing them to dilate from time to time. As a general rule: the more physically strenuous a person's normal occupation is, the less likely is he to develop cardiac infarction. Finally, the person's age is an

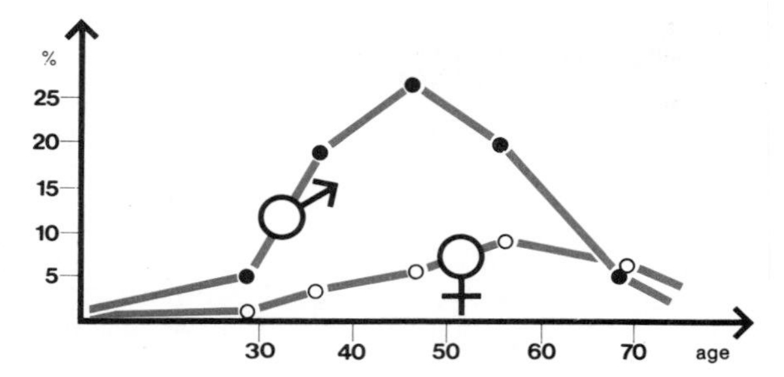

FIG. 2-35.
Frequency of coronary disease at various ages in men and women.

important factor (Figs. 2-35, 2-36, and 2-37).

The overall infarction risk increases with advancing age. In western Europe about one man in every five who survives beyond middle age dies of cardiac infarction, the peak frequency being around the age of 55, whereas for women, who are protected by their reproductive hormones up to the menopause, this peak comes much later, around the age of 70. The negative effect of civilization is manifested not only in the greatly increasing frequency of cardiac infarction, but also in the fact that under modern conditions of life there is a shift to-

FIG. 2-34.
Incidence of cardiac infarction to 1960s (in West Germany).

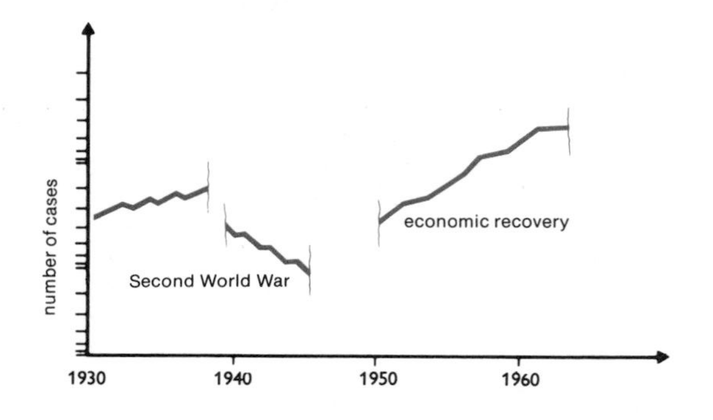

FIG. 2-36.
Incidence of cardiac infarction in both sexes.

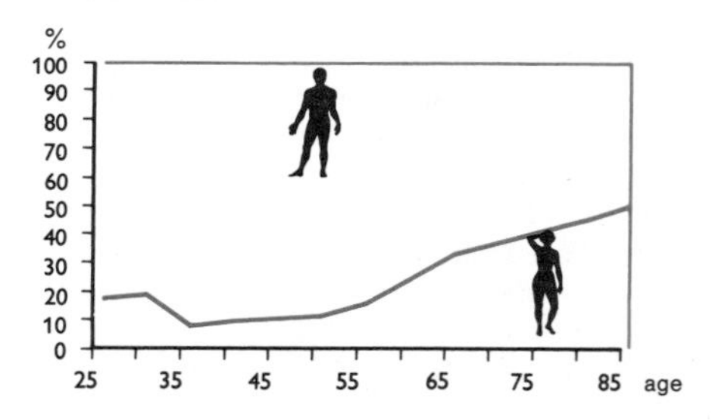

FIG. 2-37.
Incidence of cardiac infarction in men and in women.

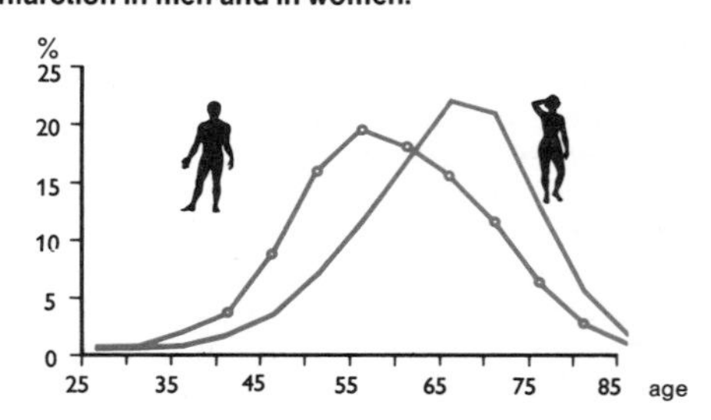

FIG. 2-38.
Changes in electrocardiograms as a result of progressively deficient blood supply with necrosis of tissue.

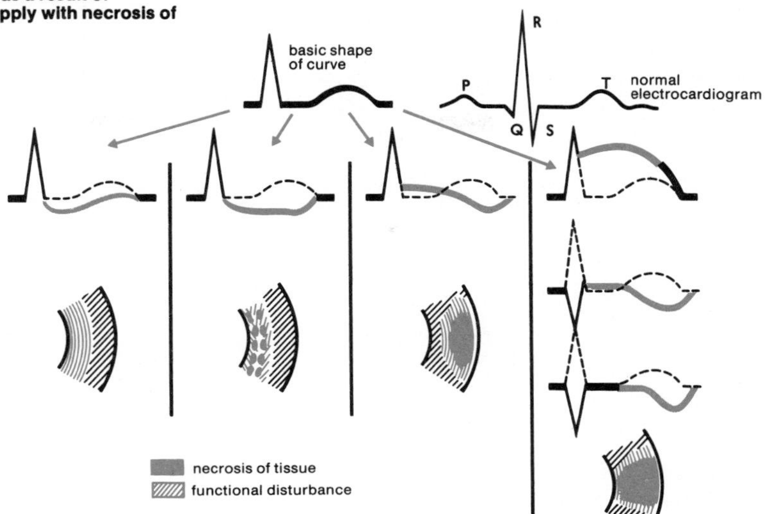

ward younger ages. Thus infarction as a cause of death in men in their 30s and even in their 20s is now by no means rare. Early diagnosis and, especially, the prevention of this potential killer are therefore all the more important.

Any shortage of oxygen supply to the heart muscle is felt as pain, and this applies also in infarction. The pain may be so severe as to produce fear and a sense of impending death. Like the pain of angina pectoris, the pain associated with cardiac infarction is diffuse and difficult to localize. It appears to radiate from the region of the heart and behind the breastbone into the left shoulder and the neck, more rarely to the right shoulder and the head; sometimes stabbing pains shoot into the back and abdomen. There is a possibility of confusing these symptoms with pains originating in the gallbladder or the stomach, the more so as infarcts frequently occur in combination with gallstones or gastric ulcers. In contrast with angina pectoris, where the pain is started by physical exertion and rapidly subsides when the patient rests, the pain associated with infarction may occur also under resting conditions, often at night, and may continue for much longer periods.

In addition to the known risk factors that promote the development of cardiac infarction in the long run, it is often possible to pinpoint certain causal factors that are of a more immediate effect. These include emotional stress, physical overexertion, lack of sleep with excessive indulgence in coffee or tea, heavy meals, alcoholic intoxication, and rapid changes of weather associated with the passage of weather fronts.

As contrasted with angina pectoris, with cardiac infarction a sudden general circulatory disturbance usually occurs, with a fall in blood pressure, feeble and irregular pulse, and pallor, while the skin feels cold and is covered with sweat. This alarming condition is called circulatory shock (pages 86 ff). Conspicuous though these symptoms are, there also exist so-called silent infarctions which run their course unnoticed until detected in a routine medical checkup. Depending on individual sensitivity, infarction pains may be experienced and rated differently by different patients. As a rule, in a case of acute symptoms it is essential to send for a doctor at once, who will then often take steps to have the patient clinically examined even if there is no more than a suspected infarction.

In the diagnosis of cardiac infarction, the electrocardiogram is a very important aid (Fig. 2-38). Typical deviations from the normal pattern indicate not only the presence of the infarct (the area of affected tissue) but also its situation in the heart muscle. Laboratory tests reveal an increase in the number of white blood cells, a rise in the sedimentation rate of the erythrocytes (red blood cells) or increased

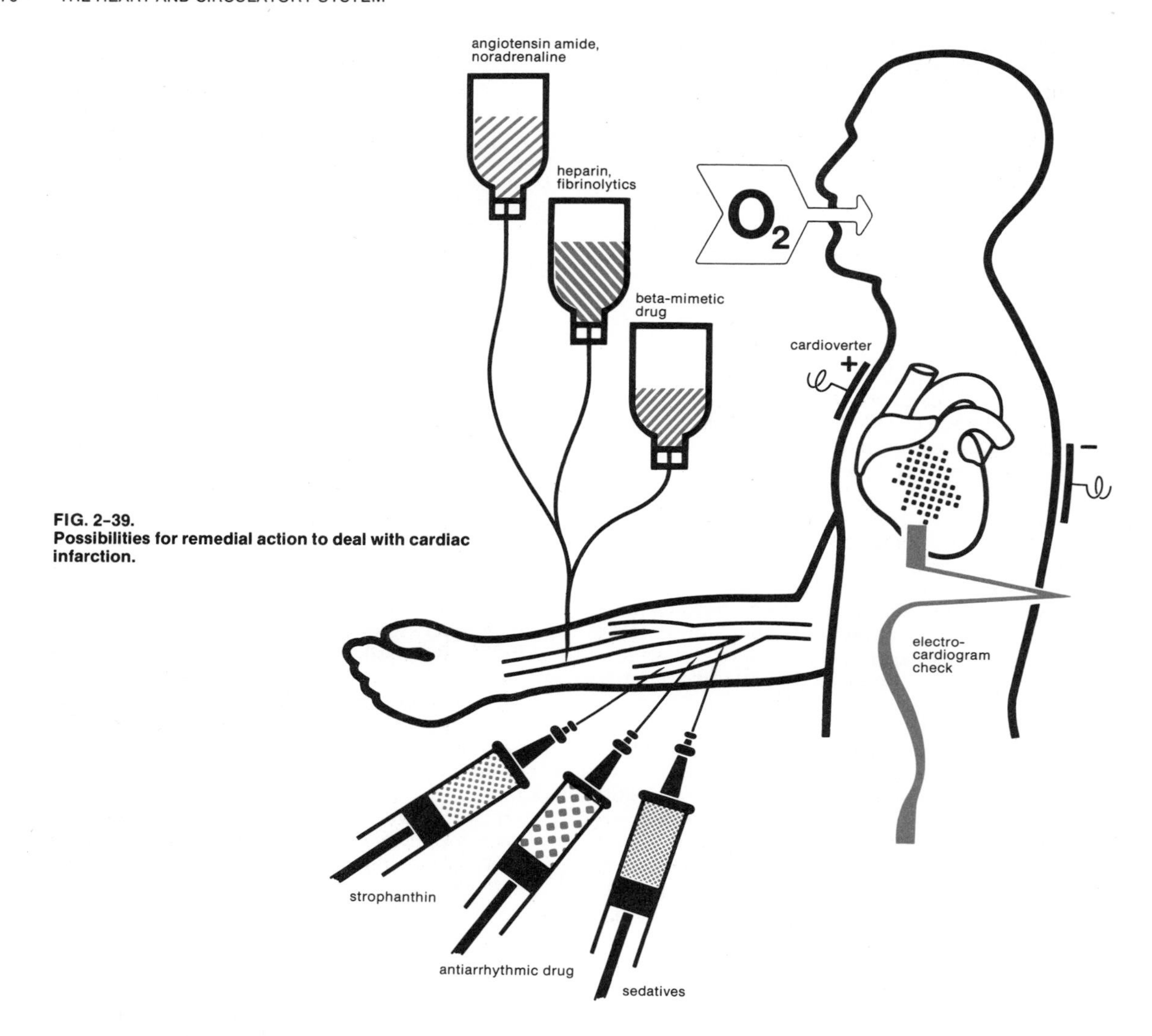

FIG. 2-39.
Possibilities for remedial action to deal with cardiac infarction.

blood sugar values. Enzyme diagnosis is an important modern development. The cells of the body produce various proteinlike organic catalysts called *enzymes* which control certain processes in the metabolism of the cells. If a cell is damaged by contusion or by lack of oxygen, the cell membrane will release enzymes through it into the bloodstream. In a blood sample they can be detected by their characteristic metabolic action in laboratory tests. In a case of cardiac infarction three enzymes are present in the blood for several days: creatine phosphokinase, transaminase and lactic dehydrogenase. With the aid of enzyme diagnosis it is in certain circumstances possible to obtain an early warning of infarction and to take action to prevent progressive obstruction of the blood vessels involved.

In the treatment of infarction the first requirement is strict confinement to bed in order to relieve the oxygen demand and allow healing by cicatrization (scar formation) of the heart. Morphine or one of its derivatives is given to calm the patient and relieve pain. Dangerously low blood pressures are combated by drugs that raise the pressure. Measures liable to cause dangerous disturbances in the heart rhythm can be applied only under suitable clinical conditions. In the hospital, treatment often starts with the administration of oxygen, while the behavior of the patient's heart is constantly monitored by an electrocardiograph. Low blood pressure is treated with noradrenaline or angiotensin amide. The weakened heart muscle is assisted with digitalis drugs or strophanthin. Irregular heartbeat (ventricular extrasystole) is controlled with infusions of antiarrhythmic drugs while electrocardiographic monitoring of the heart is maintained; and if the heart rate is too slow, it can be speeded up with so-called beta-mimetics. If these measures prove inadequate and ventricular fibrillation or indeed cardiac arrest occurs, a cardioverter (an electrical defibrillation device) or an artificial pacemaker can be used.

At the stage of cicatrization of the infarct it is important to prevent new blood clots from forming and being carried along with the blood, and, of course, to prevent a recurrence of infarction. For this purpose, anticoagulant drugs (heparin, which is rapid-acting, or coumarin, which develops its action more slowly, over a longer period of time) and sometimes also fibrinolytic drugs which dissolve clotted blood may be given (Fig. 2-39).

If the attack runs a favorable course, the acutely dangerous stage of infarction will pass in three or four days. Unless complications set in, strict confinement to bed may cease after three or four weeks. Suitable physical exercise is advised transitional to the aftertreatment stage of convalescence. To enable the patient to resume normal activities may, among other precautions, necessitate changing his habits and manner of life (weight-watching, moderation in eating and drinking, no smoking, properly regulated exercise, etc.). Strenuous physical effort will henceforth have to be avoided.

BLOOD CIRCULATION, BLOOD PRESSURE, BLOOD VESSELS

The circulatory system is a transport system that performs the function, among others, of regulating the supply of blood, and therefore of nutritive substances, to suit the needs of the various organs of the body.

The circulatory system (cardiovascular system) includes two pumps, namely, the left and the right side of the heart. The right side circulates the blood through the *pulmonary* circuit, i.e., through the lungs, while the left side circulates it through the *systemic* circuit, comprising all the other parts of the body. In the arteries of the systemic circulation the pressures are substantially higher than anywhere else in the vascular system. For this reason the circulation as a whole may alternatively be subdivided into a *low-pressure* and a *high-pressure* system. The low-pressure system comprises the pulmonary circulation and the venous part of the systemic circulation and is known also as the capacity system; it extends from the capillaries of the systemic circulation through the right side of the heart and the pulmonary circulation to the left ventricle. The high-pressure system (arterial system, resistance system) comprises only the aorta and the arteries of the systemic circulation up to the capillaries thereof. The left ventricle of the heart "oscillates" with alternate systole and diastole between the high-pressure and the low-pressure system (Fig. 2-40).

The supply of blood to the various organs of the body is accomplished by the high-pressure system through the *arteries* and *arterioles* (smallest branches of the arteries) which finally branch into the *capillaries*, these being the very fine blood vessels in which the actual "work" of the blood is done—i.e., it is here that the exchange of materials between blood and tissue takes place, involving the supply of nutritive substances to, and the removal of waste products from, the tissues. The capillaries communicate with a dual return flow system. In the first place, this comprises the *veins*, which carry the major part of the blood and, with it, the readily diffusable substances; secondly, there are the *lymphatic vessels*

FIG. 2-40. Circulatory system.

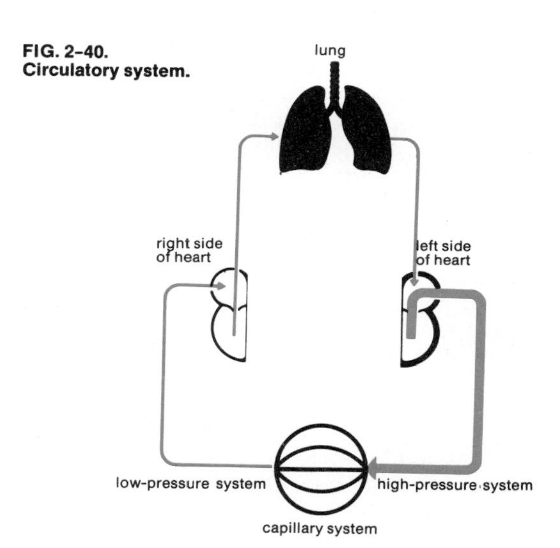

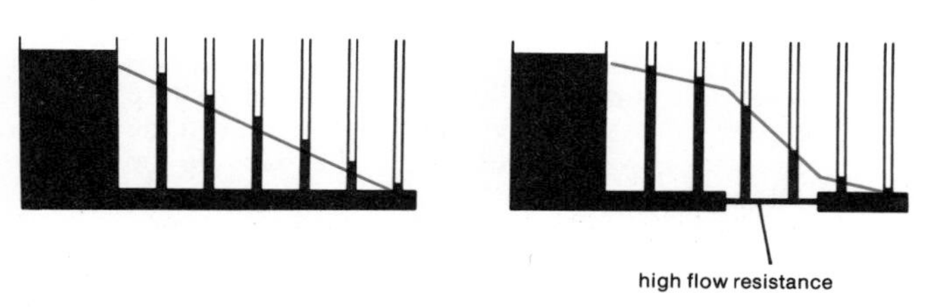

FIG. 2-41.
Principle of frictional resistance.

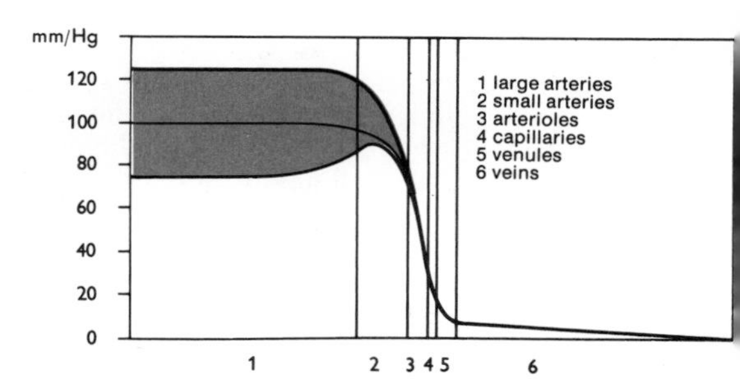

FIG. 2-42.
Pressure conditions in systemic circulation.

through which, among other things, the larger molecular substances (in particular, proteins and also, from the intestines, certain solids) are transported.

A fluid can flow in a tube only if there is a difference in pressure between the beginning and the end of the tube. The pressure needed to propel the blood through the arteries is generated by the heart acting as a forcing pump. As a result of frictional flow resistance, the pressure diminishes toward the periphery of the circulatory system. The flow resistance depends mainly on the length and internal diameter of the blood vessels: it is proportional to the length and inversely proportional to the square of the internal cross-sectional area thereof. The longer and, especially, the narrower a tube is, the greater will be the fall in pressure due to frictional resistance of the fluid along the wall of the tube (Fig. 2-41), and the less will be the rate of flow through it. Even quite minor changes in the internal diameter of a blood vessel will therefore produce considerable changes in the rate of blood flow. Figure 2-42 represents the pressure conditions existing in the various parts of the systemic circulation. It shows that the greatest fall in pressure occurs in the arterioles, which is due to the fact that the internal diameter of these blood vessels decreases rapidly with repeated branching.

By *blood pressure* is generally understood the pressure existing in the large arteries at the height of the pulse wave (systolic pressure). Blood pressure is usually measured indirectly with a device called a *sphygmomanometer*, invented by Riva-Rocci at the end of the last century (Fig. 2-43). It comprises an inflatable cuff connected to a manometer tube in which the height of a column of mercury indicates the pressure in the cuff. The latter is wrapped round a limb, usually the upper arm, and inflated until the pulse in the artery is suppressed because the artery is squeezed shut. The cuff is then gradually deflated, so that blood begins to flow through the artery again: turbulent flow (a) will cause a characteristic sound, audible in a stethoscope. The pressure indicated by the manometer at that instant corresponds to the systolic blood pressure. With further deflation of the cuff the turbulent flow in the now fully open artery becomes so-called laminar ("smooth") flow (b), and the sound ceases. The pressure measured at that instant corresponds to the diastolic blood pressure. The systolic and the diastolic pressure measured (in millimeters of mercury) by this method are usually expressed as two values preceded by the letters RR (denoting Riva-Rocci). For instance, RR/140/80 signifies a systolic pressure of 140 and a diastolic pressure of 80 mm of mercury. In the normal adult when at rest the

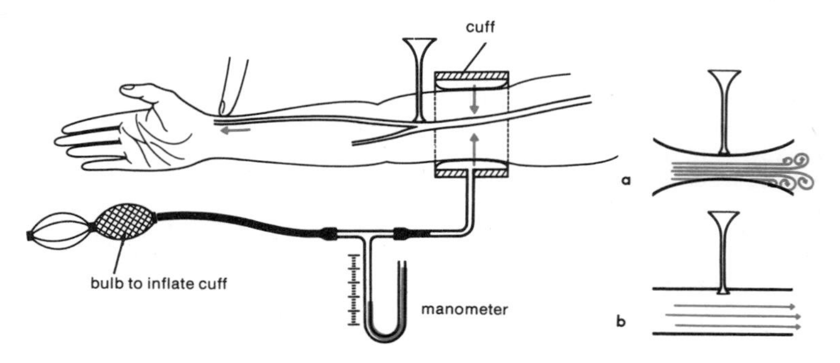

FIG. 2-43.
Sphygmomanometer.

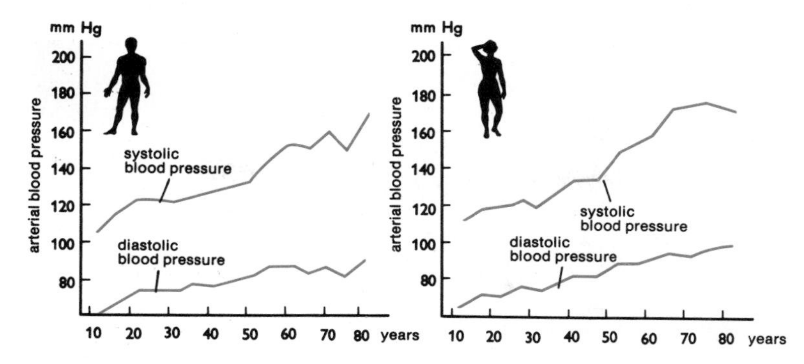

FIG. 2-44.
Variations in blood pressure by age in males (left) and females (right).

systolic pressure averages 120, but may be as high as 140 or as low as 100. Blood pressure varies with age and sex (Fig. 2-44), also with altitude, body weight, muscular development, and physical (fatigue) or mental (worry) states. The "rule" that the normal systolic blood pressure is equal to 100 plus the person's age is not reliable; there are wide individual variations.

With its systolic contractions the left ventricle of the heart pumps blood in rhythmic spurts into the aorta and thence into the major arteries. If the arteries were rigid, like metal tubes, the blood flow would momentarily stagnate in the intervals between the contractions, i.e., during the diastolic periods. Actually, the major arteries, especially the aorta, are elastic and therefore expand at each systole. The energy stored up in the artery walls each time they expand is largely recovered during the subsequent diastole: the arteries then contract by their own elastic action and impel the blood in the peripheral direction also while the aortic valves are closed (Fig. 2-45). As a result, the bloodstream, though generated as a series of separate spurts from the ventricle, is smoothed to a continuous (though still perceptibly pulsating) flow which is maintained during the diastolic periods.

With regard to the blood flow the major arteries, especially the aorta, function in much the same manner as an air receiver does in equalizing the flow of air from a compressor, though in an air receiver it is the elasticity of the compressed air itself that damps the pulsations due to the pumping strokes. This function is reflected in the anatomic structure of the major arteries. There is a predominance of elastic fibers which ensure the expandability of these blood vessels, which are for this reason called elastic arteries. Toward the periphery of the body the elasticity of the arteries diminishes; instead, they take on a muscular character, capable of contraction and relaxation so as to regulate the flow of blood to the tissues. Relatively the largest variations in flow cross-section occur, under the influence of vascular nerves and products of metabolism, in the *arterioles*. These are in fact the blood vessels in which most of the flow resistance is developed and which ultimately determine how much blood any particular organ receives.

The *pulse* is the rhythmic throbbing caused by the regular contraction and dilation of the wall of an artery due to the systolic wave. Propagation of the wave with its relatively high velocity is due to transmission of energy rather than mass (the velocity of the blood flow is considerably lower than that of the pulse wave). The velocity is measured by taking the pulse at two different points along an artery

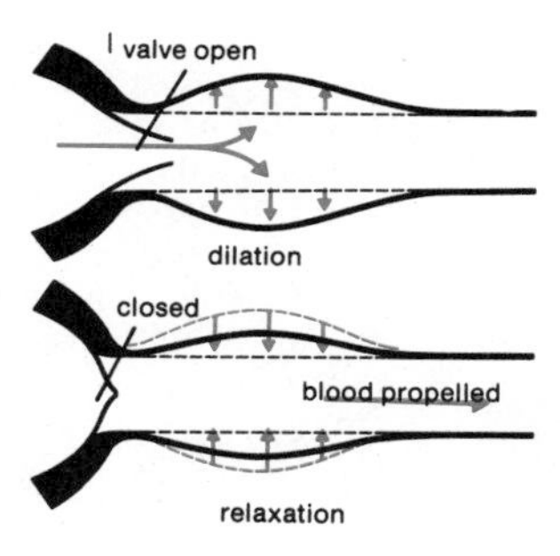

FIG. 2-45.
Propulsion of blood through artery.

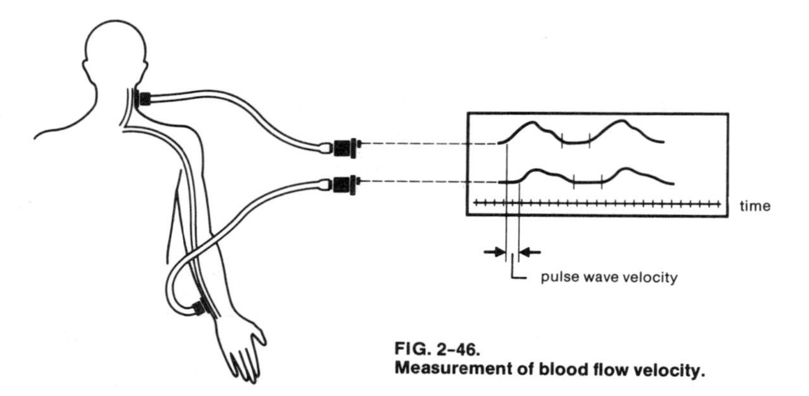

FIG. 2-46.
Measurement of blood flow velocity.

and finding the time it takes to travel a given distance (Fig. 2-46). The pulse can be felt wherever an artery can be compressed against bone; usually it is felt in the radial artery at the wrist. The pulse provides important information on the activity of the heart (especially the heart rhythm) and the condition of the cardiovascular system. A count of the pulse rate gives the heartbeat frequency.

Whereas the blood pressure in the arteries oscillates between an upper (systolic) and lower (diastolic) value, the flow in the *capillaries* takes place with hardly any variations in pressure. Only in circumstances of greatly lowered peripheral resistance is a capillary pulse detectable, which is always an indication of a preceding marked dilation of blood vessels. Capillaries are thin-walled minute vessels which connect the smallest arteries (arterioles) with the smallest veins (venules). Exchange of materials between the blood and the surrounding tissues takes place through their walls, more particularly the passage of nutritive substances and oxygen from the blood to the tissues and of waste products from the tissues to the blood.

Capillaries are noncontractile, and their width (internal diameter) is controlled by changes in blood pressure due to regulatory adjustments of the flow resistance in the arterioles, which are contractile, as already stated. The width of the arterioles, in turn, is mainly controlled locally by products of metabolism and adapts itself to requirements.

Circulatory reflexes acting through the sympathetic vascular nerves (vasomotor nerves) ensure that the arterial flow pressure is maintained. The vasomotor nerves are of two types: vasodilator nerves, which conduct impulses that cause a blood vessel to dilate, and vasoconstrictor nerves, which similarly bring about constriction of a blood vessel.

The low velocity of flow in the capillaries and the large overall surface area of their walls due to the elaborate branching of these tiny blood vessels promote the exchange of materials between the blood inside them and the tissues surrounding them. In a person of medium build the total length of the capillaries is about 100,000 kilometers (62,000 miles), with a wall surface area of 6000–7000 m^2. The capillary walls are semipermeable, i.e., they allow water and small molecules of dissolved substances (salts, glucose, amino acids, etc.) to pass, but not proteins.

The exterior of the capillary wall is not in direct contact with the tissue. There are, instead, cavities containing interstitial fluid which acts as the go-between in effecting the exchange of materials. This exchange conforms to the physical laws of filtration and diffusion. In the arterial capillaries the blood pressure exceeds the forces opposing it; as a result, plasma is filtered from the capillaries into the surrounding interstitial space. In the venous capillaries the situation is reversed (Fig. 2-47); here plasma flows back into the capillaries. In connection with these phenomena the colloid osmotic pressure, i.e., the water-combining power of the plasma proteins, plays an important part. Figure 2-48 illustrates the permeability of the capillary wall with regard to various substances contained in the blood.

The parts of the circulatory system which comprise the arterioles, the capillaries and the venules are collectively called the terminal blood vessels. It is in these that the so-called *microcirculation* takes place.

The capillaries discharge into the *venules*

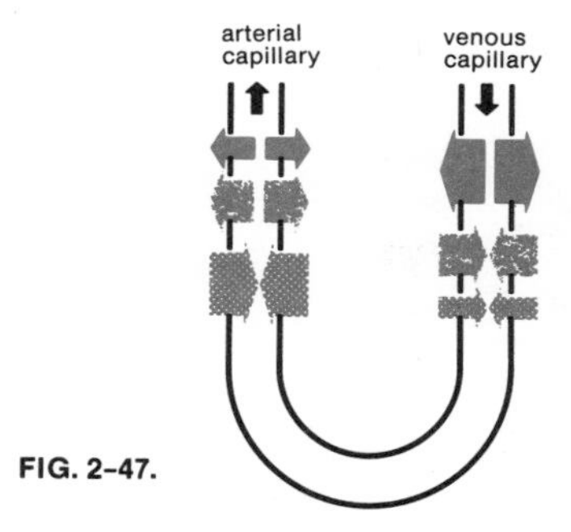

FIG. 2-47.

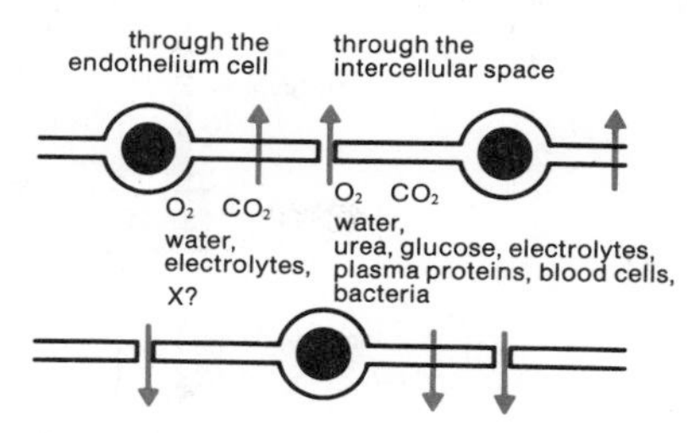

FIG. 2-48.
Capillary exchange.

(tiny veins) which in turn come together into progressively larger and larger *veins*, which finally return the blood to the left atrium of the heart (*venous return*). The small- and medium-sized veins, which have very thin walls, can vary their internal diameter as a result even of quite small variations in blood pressure, so that they can either discharge the blood or provide storage capacity for it. The veins in the limbs have thicker walls because in man, with his upright gait, these veins have to resist larger pressures than do the veins in the neck, for example. However, the effective hydrostatic pressure is reduced by the presence of valves in the veins. These valves, which prevent flowback, break up the column of blood in a vein into short segments (Fig. 2-49). Contractions of muscle fibers in the walls of the veins help to keep the blood flowing, but the main factor in this respect is the general activity of the muscles in various parts of the body: they produce a squeezing action which moves the blood in the veins along from valve to valve. In this way physical exercise involving muscular exertion helps the blood circulation. The pulsation of the arteries augments this effect by compressing the adjacent veins (Fig. 2-49).

To enable the suction described on page 82 to develop, which assists the return flow of the blood to the heart (as do also stretching movements of the body or rising from a doubled-up attitude, Figs. 2-51 and 2-52), the veins must not collapse upon themselves, but must remain open. In the thorax this is achieved by the elastic pull of the lungs which keeps the veins expanded; in other parts of the body a similar effect is obtained by the way in which the veins are associated with the adjacent tissues (Fig. 2-49). When veins situated near the heart

FIG. 2-49.
Venous blood flow.

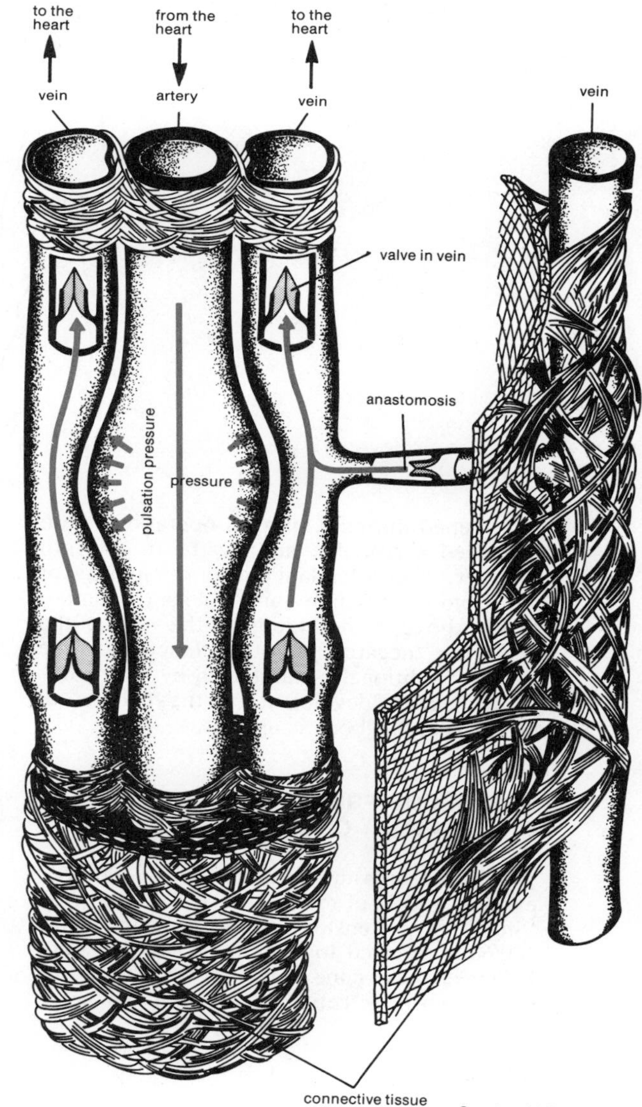

Courtesy M. Ratschow

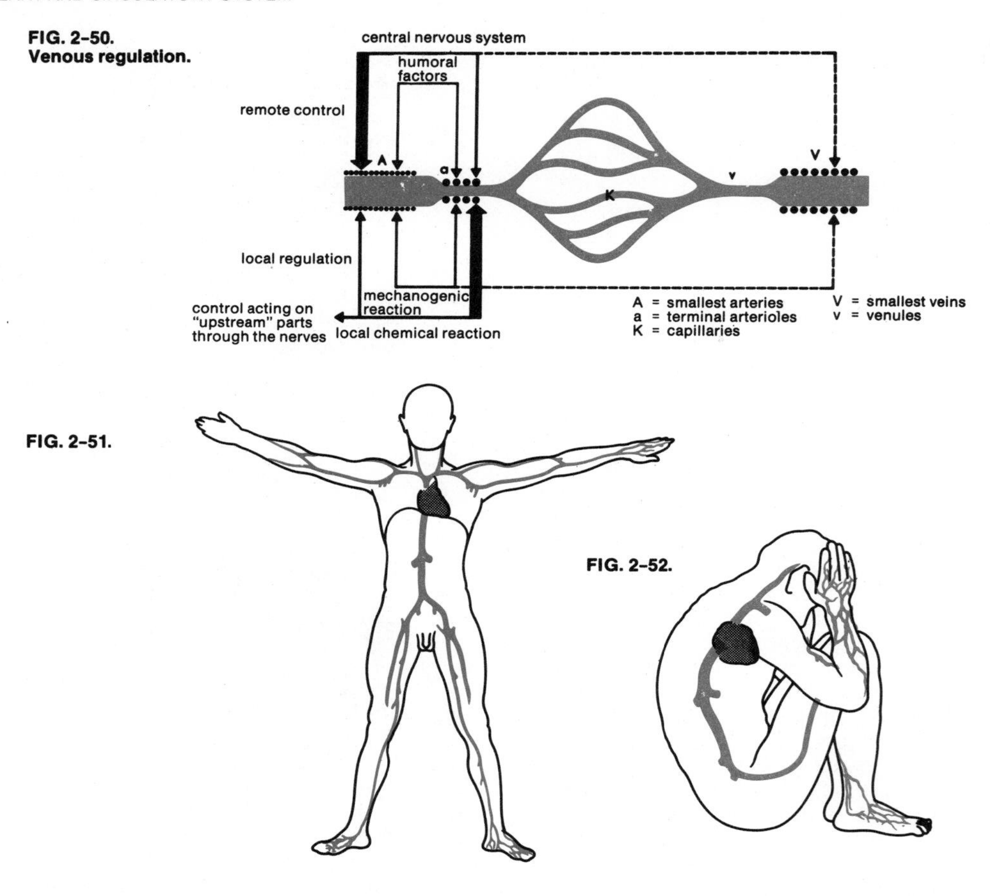

FIG. 2-50. Venous regulation.

FIG. 2-51.

FIG. 2-52.

are opened during a surgical operation, there is indeed a risk that air may be drawn into them, with the attendant danger of blockage of small blood vessels by air bubbles (air embolism). Above the heart, e.g., in the veins of the neck, the "negative pressure" (suction) in the veins is additionally maintained by the hydrostatic suction developed by the column of blood when the body is upright.

LOW BLOOD PRESSURE, FAINTING, CIRCULATORY COLLAPSE

If blood pressure under resting conditions exceeds 150/90 of mercury, it is referred to as "high" (hypertension); blood pressure below 100/60 is referred to as "low" (hypotension). Either of these conditions may be a symptom of long-term defective control, in the sense that in the main blood pressure controlling centers in the brain the "desired value" to be attained by the pressure has been shifted (up or down) in relation to the normal value, whereas the subordinate centers controlling the circulatory system perform their function correctly and thus duly produce the desired pressure in accordance with the (wrong) instructions they receive from the brain.

Thus there is a form of low blood pressure called *primary* or *essential hypotension*. It is constitutional, i.e., it is determined by individual predisposition and is characterized by constant pressures of around 100/60 mm. Figure 2-53 illustrates how such pressures are produced by the circulatory organs. The basic cause of the low pressure is that the rate of blood discharge from the heart is too low in relation to the width (internal diameter) of the blood vessels. Such blood pressure values are

FIG. 2-53.
Production of hypotension.

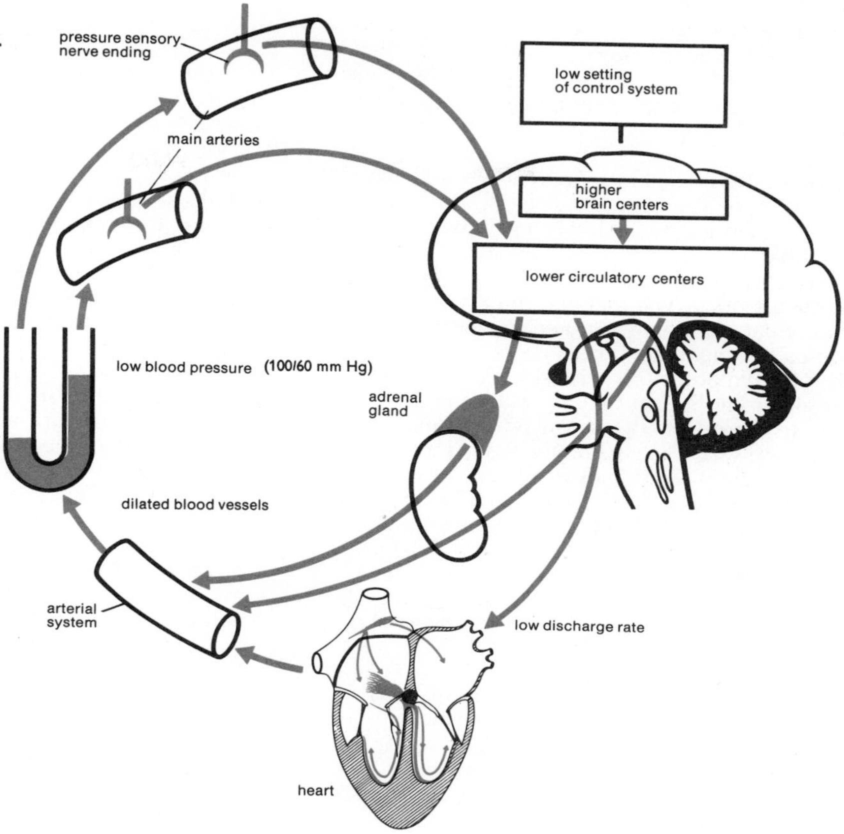

pressure sensory nerve ending
low setting of control system
main arteries
higher brain centers
lower circulatory centers
low blood pressure (100/60 mm Hg)
adrenal gland
dilated blood vessels
arterial system
low discharge rate
heart

FIG. 2-54.
Symptoms of hypotension.

low setting of control system ("reference value" too low)
low blood pressure
organs poorly supplied with blood
dilated blood vessels
low blood discharge rate from heart
lack of energy, poor concentration, dizziness, headache
flickering before the eyes
pressure or twinges around heart
100/60 mm Hg
spasms of internal organs
cold limbs, pallor

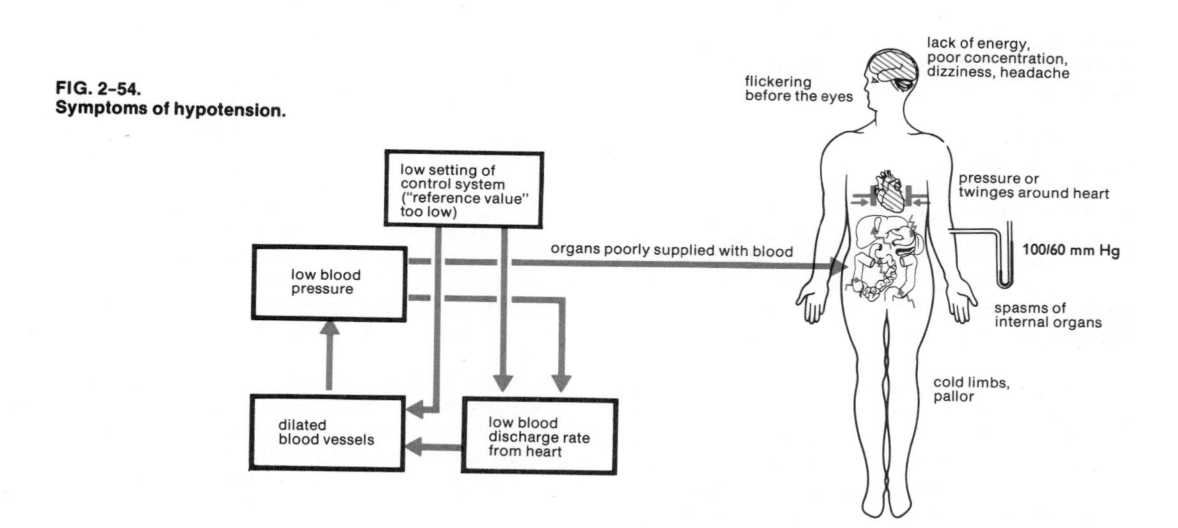

by no means rare. As they sometimes tend to "run in families," they can be regarded as statistical variants at the extreme lower borderline of what is conventionally rated as the "normal" range, just like bodily height or weight. Psychological factors, strain, eating habits and, in general, the person's manner of life also play a part. Low blood pressure is found more particularly in "overgrown" youths who, in popular speech, are said to have "outgrown their strength," in persons of asthenic body type (tall and thin, with inferior muscular development), and in persons of the so-called vagotonic type.

The symptoms associated with hypotension result from the fact that the rate of blood supply to the organs is too low on account of the low pressure. The patient's face is pale, his limbs feel cold, he finds it difficult to concentrate, he is listless and disinclined to exert himself, he wakes up in the morning feeling tired, needs more sleep, suffers from dizziness, headache, a sense of flickering before the eyes, sometimes also cardiac complaints, a feeling of pressure or twinges around the heart (Fig. 2-54). Other complaints that may be associated with low blood pressure are stomach and bladder cramps and menstrual irregularities. Most of these symptoms become worse when the patient stands, because standing brings about a further lowering of blood pressure. On suddenly rising to his feet, or as a result of standing for a long time, the patient may faint.

Though disagreeable, the symptoms of essential hypotension are generally harmless. Treatment will usually be aimed at "training" the blood vessels by suitable exercises, sports activities, massage, alternate hot and cold baths or showers, and adequate adjustment of the patient's manner of life. If the symptoms nevertheless persist, the patient should avoid standing for long periods and should certainly persevere with his exercises, etc., under medical supervision. The strenuousness of these activities should gradually be increased. If general remedial measures fail to restore the blood pressure to more normal values, drugs may be given to raise the pressure. They compensate for the underactivity of the sympathetic nervous system and cause the peripheral blood vessels to contract.

A more cautious approach is advisable in a case of *secondary hypotension,* which may be caused by, for example, infectious diseases, certain forms of heart disease, cardiac infarction, anemia, tuberculosis, loss of blood, adrenal failure, defective pituitary gland, or defective functioning of the thyroid gland. All these possible types of secondary hypotension will have to be judged and treated individually, according to the underlying diseased condition.

A particular form of "acquired" hypotension occurs in sportsmen and athletes whose circulatory system is, as a result of intensive physical training, geared to high performance and is characterized by relatively low blood pressure during normal physical activity. Besides low pressure, there are also other signs to indicate that this is a physical condition capable of coping with heavy extra loads when called upon—for example, a slow pulse which can be more than doubled under load, and a large athlete's heart which can pump out a great quantity of blood at each beat.

In some persons the circulatory control, which should come into action on rising from the reclining to the upright position, operates with a delay. Also, its functioning may become defective when the person has to stand for long periods. When suddenly getting up from a chair or getting out of bed, or after standing for some time, the patient complains of a feeling of emptiness in the head, flickering before the eyes, dizziness and nausea; he may break out in sweat and even have a fainting fit. These symptoms, due to a fall in blood pressure and a deficient supply of blood to the brain, are brought about by the following set of conditions: The quantity of blood in the body (about 5 liters) is insufficient to keep the entire vascular system completely filled at all times. Particularly in the highly dilatable veins, where the pressure is very low in comparison with that in the arteries, there is always space available to accommodate blood which normally is in circulation or is "parked" in other dilatable sections of the system, e.g., in the blood vessels of the lungs. While the patient is lying down, for example, the pulmonary circulation does indeed contain substantial reserves of blood. When he stands up, the force of gravity acts upon the distribution of the blood in the body. The blood will now tend to collect in the veins of the limbs, especially the legs. Thus something like 0.5–1 liter of blood may remain standing in these newly available "parking zones," in particular if the veins are flabby and if large accelerative forces act on the body (e.g., in an aircraft flying on a sharply curved course or in a rocket during launching). If the veins in the legs do not contract and if blood also remains in the old "parking zone" in the lungs, too little blood will be available to sus-

FIG. 2-55.
Compensatory control of circulation.

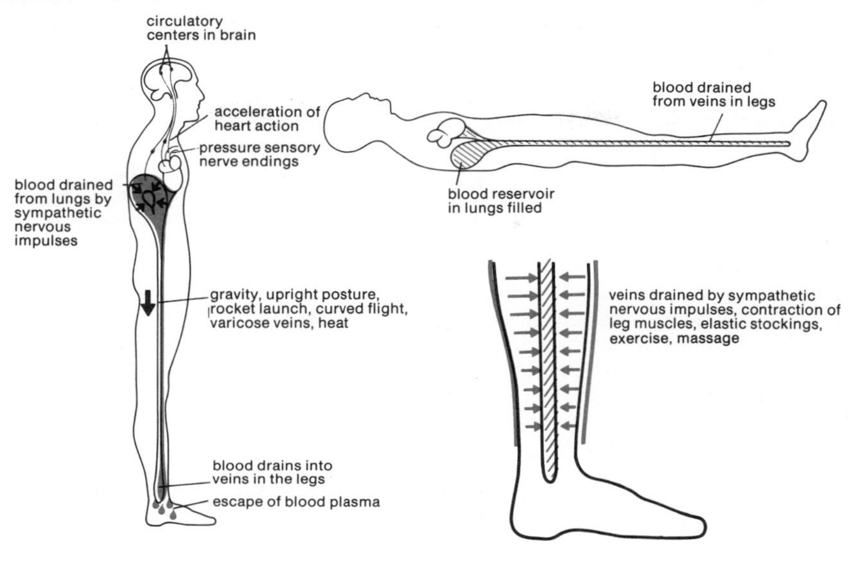

FIG. 2-56.

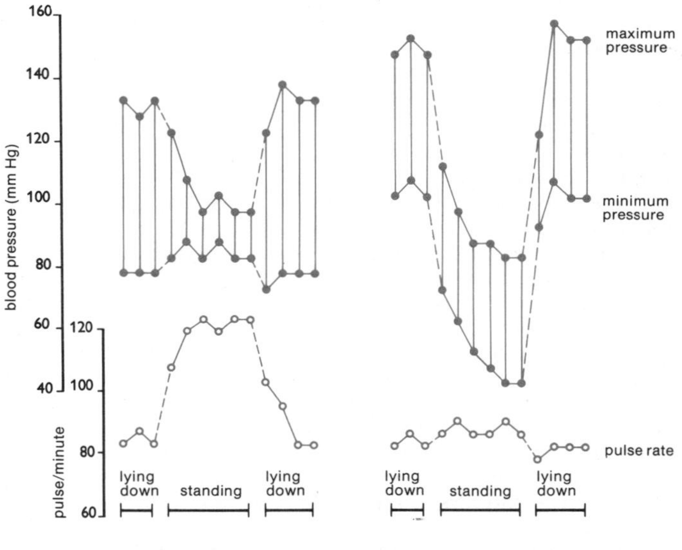

(a) Blood pressure falls as a result of blood draining into the veins of the legs; increased heart action prevents fainting.

(b) Blood pressure falls as a result of blood draining into the veins of the legs; the heart is not stimulated to extra action by sympathetic nervous impulses, the supply of blood to the brain diminishes, fainting occurs.

tain proper circulation, and pressure will drop, resulting perhaps in an insufficient supply to the brain. This is, of course, more likely to happen if the blood pressure is already low from the outset and if, to make matters worse, the patient has extensive varicose veins (see page 95). In a normal person, rising from a reclining to an erect position will not be attended by circulatory deficiency; instead, a compensatory control mechanism comes into action: the veins in the leg tighten and the blood vessels of the lungs disgorge their extra contents into the circulation. Both these effects are brought about by means of pressure-sensitive nerve endings which act through circulatory nerve centers and sympathetic nervous impulses transmitted to the heart and blood vessels (Fig. 2-55). In some persons and in certain situations these regulating processes, which compensate for the effect of gravity, are or become inadequate (Figs. 2-55 and 2-56). Tall people are, understandably, more likely to be affected in this way—especially if they are of the asthenic body type or are vegetatively labile.

For *prevention*, it is most important to avoid situations liable to upset the circulation and cause fainting by displacement of blood within the system. Standing for fairly long periods, especially in warm surroundings, makes the condition worse because the blood vessels are then wider open and respond less readily to nervous impulses striving to constrict them; in addition, blood plasma is exuded through the walls of the capillaries into the interstices of tissues or fluid is lost by sweating. Standing completely still or being raised to the upright position with outside help (i.e., not by one's own effort) is particularly unfavorable in this respect because then the leg muscles, which would otherwise by their rhythmic action help to empty the blood-filled veins of the legs, remain inactive and thus do nothing to assist circulation. For this reason the legs should be moved a little before standing up, in order to stimulate the circulatory control center. Muscular contractions cannot empty the veins, however, if the valves in the veins fail to close properly. In such cases, elastic stockings or supporting bandages can help to reduce the blood storage capacity of the distended veins (varicose veins). In general, measures aimed at stimulating and improving the circulation, as already mentioned, can be helpful (massage, suitable exercises, etc.). Resting in the reclining position with legs raised prevents fainting. In some cases drugs that cause constriction of the blood vessels may be given.

Fainting (syncope) is loss of consciousness due to insufficient blood supply to the brain. This may occur, for example, if severe emotional stress (such as excessive joy, fright or horror) or intestinal pain causes a slowing of the heart rate and general dilation of the blood vessels, so that the blood pressure is lowered. In some susceptible persons faintness may occur as a result of standing for long periods, e.g., a soldier standing at attention, and is due to pooling of blood in the legs (Fig. 2-56b). As a rule, fainting occurs suddenly, but sometimes it is preceded by warning signs such as pallor, dizziness, shivering, sweating and nausea. Faintness at the sight of blood or injury is a familiar event. Common fainting is harmless and usually passes when the patient lies down or lowers his head, whereby the flow of blood to the brain is fully restored. Cold stimuli and the proverbial bottle of smelling salts can help to speed the regaining of consciousness.

Collapse is a sudden failure of vital power which may be the result of disturbed blood circulation due to heart failure, for example, in which case the blood vessels are contracted in striving to compensate for the fall in blood pressure, or may be due to a lowering of pressure by general dilation of the vessels. The common faint is an example of the latter type of collapse and is generally harmless. Collapse of long duration may turn into circulatory shock, however.

HIGH BLOOD PRESSURE

In young adults the average (normal) blood pressure is 120/80 mm of mercury, representing the systolic and the diastolic pressure respectively (see page 78). In a person suffering from high blood pressure (*hypertension*), particularly the small peripheral branches of the arteries are narrowed, so that the flow resistance is increased. The diastolic pressure is especially increased in consequence of this. Blood pressure generally increases with age, the systolic value averaging about 130 at 40 years of age and about 145 at 60. The increase in diastolic pressure is relatively less with advancing age and seldom exceeds 90 mm in a normal elderly person. As a rough general rule, blood pressures above 150/90 are regarded as "high."

Statistics indicate that at present some 43 percent of people die of cardiovascular disease, of whom about 60 percent succumb to the consequences of high blood pressure. In about 20 percent of the cases of high blood pressure the condition is attributable to particular organic diseases, including 14 percent in which there is kidney disease, e.g., chronic nephritis or a constriction of the renal artery. In such *renal hypertension* there is at an early stage a marked narrowing of the peripheral blood vessels. It is believed that the kidney, with its constricted blood vessels, forms renin, a proteinlike substance, which in turn reacts with a serum globulin in the blood to produce angiotensin, a so-called pressor substance, i.e., its action is to raise the blood pressure. When the constriction of renal blood vessels is relieved or the diseased kidney is itself removed, the blood pressure goes down to normal values. In a further 3 percent of cases the hypertension is of hormonal origin, due to excess activity of the pituitary gland (hypophysis cerebri) or of the adrenal glands. Finally, just under 1 percent of all hypertensive conditions are of nervous origin, e.g., in consequence of infantile paralysis, disseminated neuritis or brain tumors.

In something like 80 percent of cases of high blood pressure there is no clearly assignable cause. Such cases are referred to as *primary* or *essential hypertension*. A number of contributory factors within this type of high blood pressure has been identified, e.g., age, familial predisposition, obesity, excessive intake of salt in food, lack of exercise, emotional conflict. Essential hypertension develops usually in middle or late middle age. It is much commoner in women than in men. Before the blood

FIG. 2-57.
Symptoms of advanced hypertension.

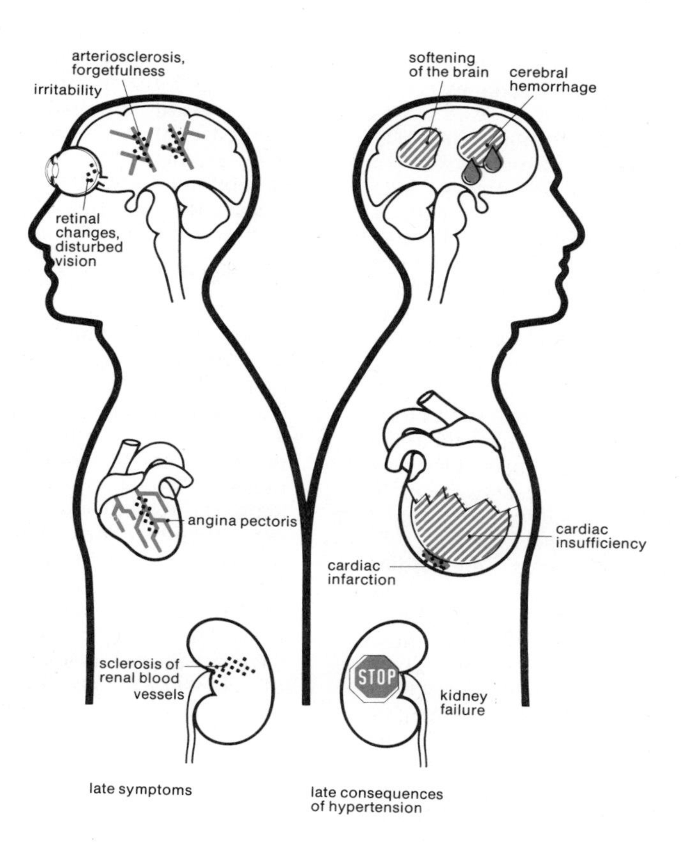

pressure settles to a steady high value there may be a transition period of 5–20 years in which the pressure varies (labile hypertension).

High blood pressure does not at first produce specific symptoms. Headache, dizziness, buzzing in the ear, fatigue, insomnia, palpitation and irritability may be indications, but usually do not appear serious enough for the patient to seek medical advice. For this reason the detection of high blood pressure is often the result of chance, in the course of a routine checkup or of an examination required for some other purpose. Yet it is important to detect the condition at an early stage, because timely treatment can arrest its further development and prevent harmful consequences. High blood pressure eventually overloads the blood vessels and brings on premature sclerosis. As a result, the blood pressure is further increased and serious vascular damage in various organs may be caused.

The symptoms occurring at an *advanced* stage of *hypertension* (Fig. 2-57) are characterized particularly by sclerotic changes in the coronary arteries and in the blood vessels of the brain and kidneys. Sclerosis of the arteries that supply blood to the *brain* produces symptoms such as forgetfulness, diminished concentration and increased irritability. Subsequent effects may include thrombosis of an artery of the brain, cerebral hemorrhage, and "softening of the brain" (a slow apoplexy, or stroke, due to progressive rupture of small blood vessels: see page 344). Sclerosis of the blood vessels that supply blood to the fundus (posterior inner part) of the *eye* produces changes in the retina resulting in impaired vision. As it can be directly observed (with an instrument called an ophthalmofundoscope), examination of the fundus is, along with blood pressure measurement, an important aid in the detection of hypertension. Sclerosis of the *coronary arteries* reduces their reserve capacity to increase the supply of blood to the heart and is thus liable to cause angina pectoris and, subsequently, coronary thrombosis and cardiac infarction. Sclerosis of the blood vessels of the *kidneys* has particularly unfavorable effects because the deficient supply of blood to those organs will result in the release of renin, which in turn will further raise the blood pressure.

Once essential hypertension has developed, it may be years or indeed tens of years before death ensues in consequence of heart failure (in about 50 percent of cases) or of complications in the cerebral blood vessels (in about 25 percent of cases). In order to prevent the harmful effects and aftereffects of high blood pressure, the object of any treatment will be to lower the pressure to something near the normal average for the patient's age. This cannot be achieved by means of drugs alone; even more important is an appropriately adjusted manner of life. Since overweight and excessive intake of salt are additional risk factors, a change of diet will be necessary. Therefore, along with measurement of the blood pressure at regular intervals, weight-watching is an important means of monitoring the progress of the treatment. If necessary, the patient will have to take a slimming course. Intake of salt should not exceed 3–6 g per day (low-salt diet, see page 141). Coffee and tea will do no harm if not taken in excessive quantities. The same is true of alcoholic drinks. Too much smoking is harmful.

Treatment of chronic hypertension by means of drugs has made great advances in recent years. There is, however, as yet no ideal drug that is fully effective and without undesirable side effects. Long-term treatment of high blood pressure is therefore usually based on a combined therapy. In each individual drugs must usually first be tested to find which one or combination is most effective. In mild cases it will usually suffice to give doses of reserpine (a blood-pressure-lowering drug) in combination with a diuretic (a drug that increases the secretion of urine) (see Fig. 2-58). If these measures are unsuccessful, increasing amounts of guanethidine may additionally be given.

Treatment for high blood pressure is usually a long-term (in many cases a lifelong) course. Its sudden interruption can be just as harmful to the patient as the withholding of insulin from a diabetic. It is necessary to be on the lookout for side effects of drugs. With drug therapy there is always a risk that the blood pressure is lowered too much. When that happens, the patient will, especially when standing, experience the characteristic symptoms of hypotension (low blood pressure) such as singing or buzzing sounds in the ears, flickering before the eyes, dizziness and perhaps faintness. The patient is mostly able to cope with such situations himself by sitting or lying down (if necessary, raising the legs and lowering the head). It may be helpful for one to be instructed in how to measure one's own blood pressure at intermediate times between seeing the doctor.

FIG. 2-58.
Treatment of hypertension.

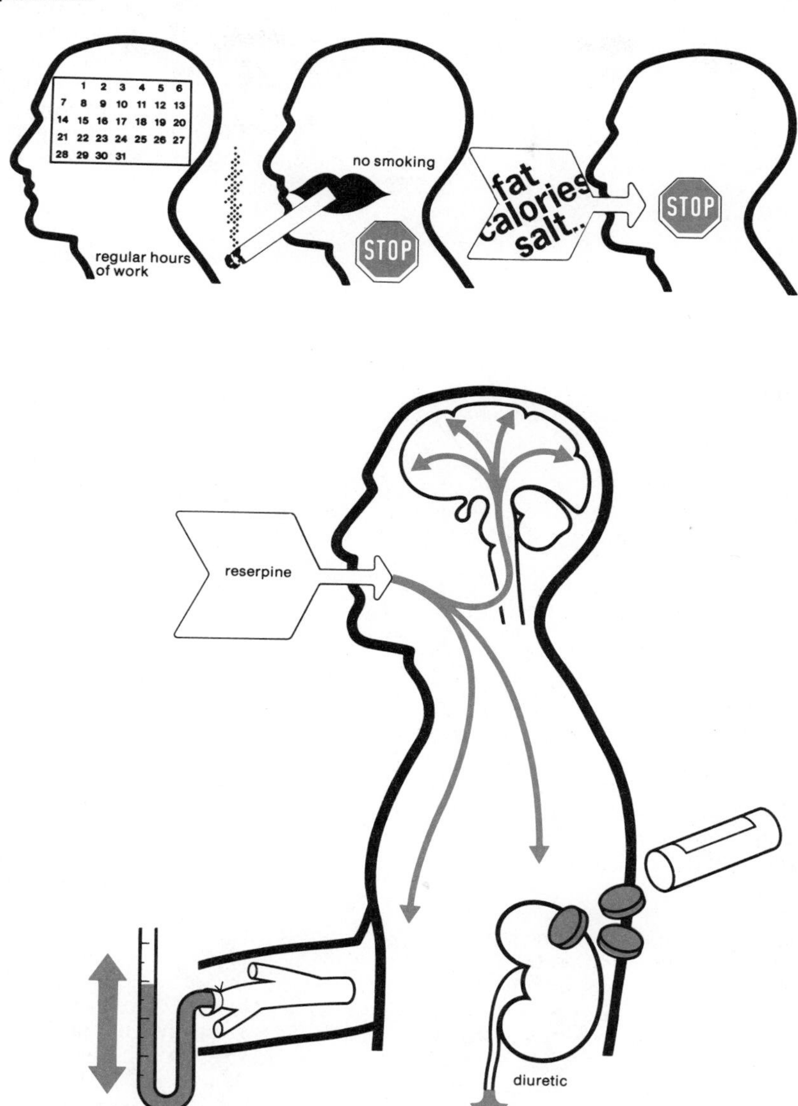

ARTERIAL STENOSIS, ARTERIAL OBSTRUCTION

Arteries are blood vessels that carry blood away from the heart. This function is of vital importance in supplying oxygen to all parts of the body; proper flow of blood in the arteries is therefore a primary requirement. The walls of these blood vessels are reinforced with elastic fibers and muscle. Healthy arteries can be contracted or dilated by their muscle fibers to adjust the rate of flow to suit the requirements of the parts of the body served. If arteries harden and lose their elasticity, as occurs in

FIG. 2-59.
Collateral arteries.

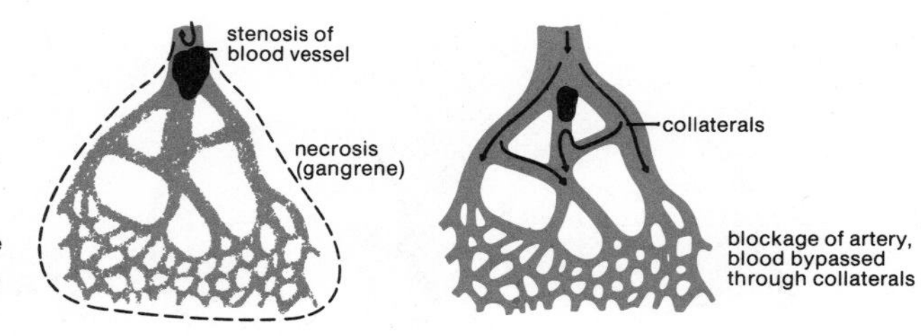

FIG. 2-60.
Bypass circulation when artery becomes obstructed.

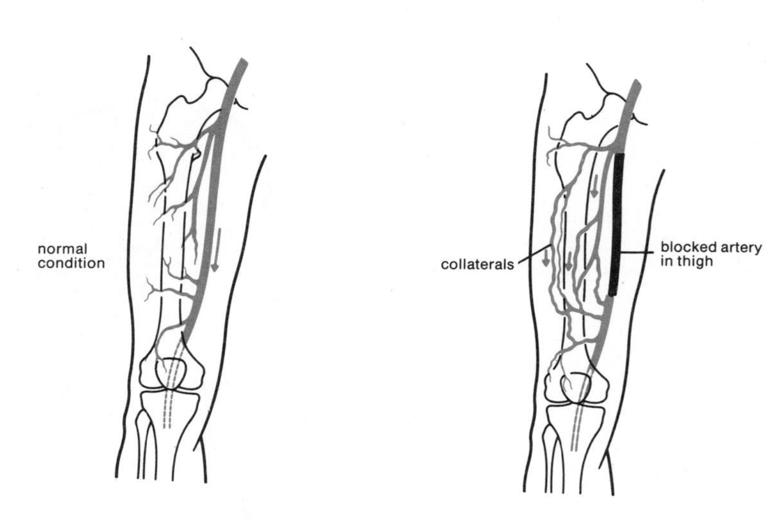

FIG. 2-61.
Normal arterial system.

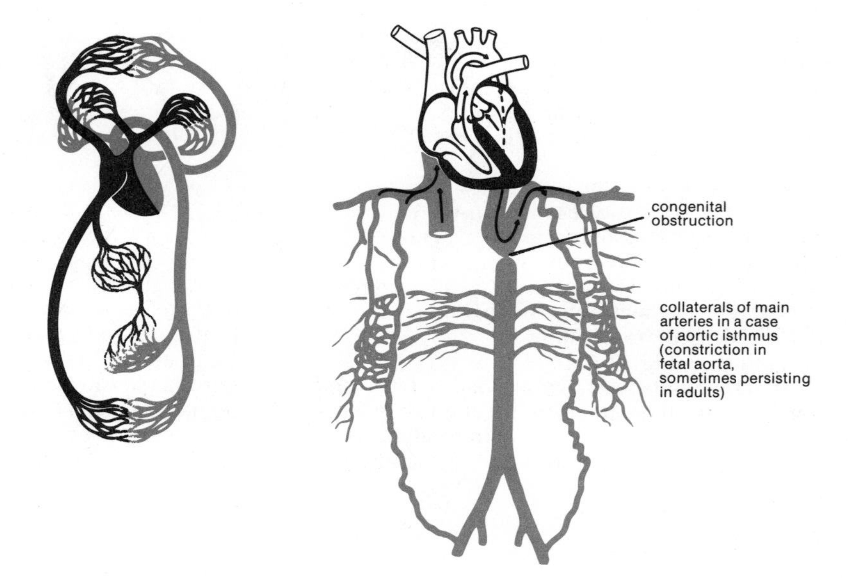

arteriosclerosis, they become incapable of adequate dilation to meet an increased demand for blood. For instance, with arteriosclerosis of the legs, walking or other exertion of the legs will cause pain in the calves. This pain will disappear quite soon with rest. Such temporary discomfort is characteristic of a condition in which the sclerotic arteries are still just wide enough to provide an adequate supply of blood when the limb or other part concerned is at rest. However, if the flow passage of the artery is reduced to substantially less than half its original size as a result of thickening of the lining, difficulties will arise even in the resting condition. If complete blockage occurs, the tissues supplied with blood by the artery will be starved of oxygen and undergo necrosis. The consequences of *stenosis* (narrowing) of an artery may vary greatly, depending on the branching pattern of the artery, the rate at which stenosis develops, and the tissues or organs supplied by the artery. Necrosis will not necessarily occur, however, unless the tissue has to rely for its blood supply on only that one artery affected.

Quite frequently a particular area is supplied by two or more arteries which communicate with one another through cross-connections. When one of the arteries is threatened by obstruction, the cross-connecting secondary blood vessels will be dilated by the stimulus arising from a shortage of oxygen. Eventually, bypass blood vessels (so-called collaterals) will be formed, which can sometimes provide an alternative route and thus take over the duties of the obstructed blood vessels (Fig. 2-59). It may, however, be several weeks before the collaterals become properly functional (Fig. 2-60), and they can therefore not be relied on in a case where arterial obstruction develops rapidly. In the developing fetus the blood vessels possess an extremely high degree of adaptability and versatility in compensating for possible hereditary vascular defects by forming collaterals (Fig. 2-61).

Arterial obstruction may develop gradually or may occur as the result of a sudden blockage (embolism). Some vascular diseases are associated with ulcerous deterioration of the walls of the blood vessels affected. Blood clots will establish themselves in such diseased arteries, despite rapid flow of blood (arterial thrombosis, Fig. 2-62a). These degenerative changes usually develop gradually. On the other hand, the effects of embolism due to blood clots carried along in the bloodstream and becoming wedged in an artery are likely to be sudden and therefore more acutely dangerous (Fig. 2-62b).

Less vitally important arteries may develop a "silent" diseased condition, i.e., without obvious symptoms or signs, in the event of disturbed blood supply or obstruction. Lack of oxygen will, however, usually produce deficiency symptoms in the areas dependent on the arteries affected. Cramplike pains are felt in the limbs, especially as a result of muscular activity; obstruction of the blood flow to extensive areas will cause *gangrene* (necrosis, or death, of tissue due to deficient blood supply) of the limbs. This may take the form of dry gangrene, when the part that dies has little blood and remains aseptic, or moist gangrene, when the part is infected with putrefactive bacteria. In such cases, amputation of the limb may be the only possible course of action.

Reduced flow of blood through the coronary arteries also manifests itself in pain due to shortage of oxygen (angina pectoris, see page 71), and blood clots may cause necrosis of the heart muscle (cardiac infarction, see pages 73 ff).

Obstruction of the renal arteries, i.e., those which supply blood to the kidneys, causes the release of substances that raise the blood pressure. Eventually, failure of the excretory function for the removal of toxic minerals and waste products will occur as a result, or a renal infarct (necrosis of tissue in the kidney) may develop.

Stenosis of cerebral arteries causes dizziness, mental confusion and loss of memory. Finally, "softening of the brain" is liable to occur, which may be attended by the paralysis phenomena characteristic of a stroke (apoplexy, see page 344).

Among the diseases tending to obstruct or narrow the arteries there are several distinct syndromes, each with a different cause, extent and course (Fig. 2-63). (*Syndrome* denotes a group of signs or symptoms that collectively characterize a disease.)

Buerger's disease occurs chiefly in men under 45 years of age. It is aggravated by smoking and it particularly affects the legs, the disease then being typified by intermittent limping. Progressive narrowing of the inflamed and structurally changed arteries in the legs sometimes causes ulceration and gangrene, so that amputation may be necessary. Treatment

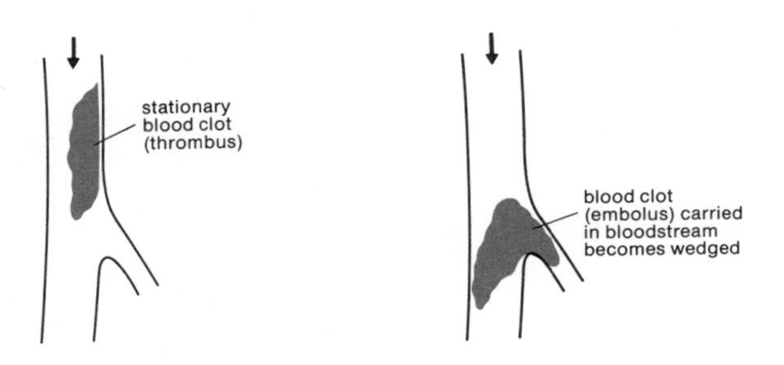

FIG. 2-62.
Arterial embolism.

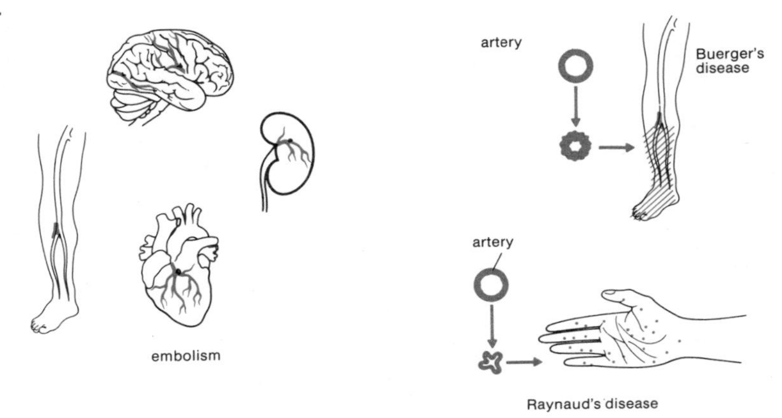

FIG. 2-63.
Stenotic arterial diseases.

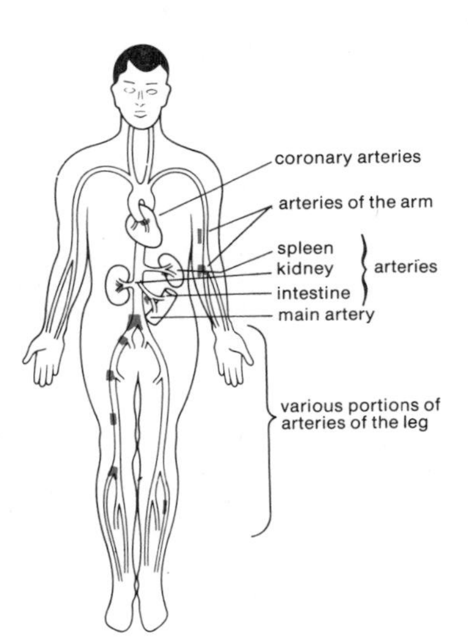

FIG. 2-64.
Common sites of embolism.

will consist in, first of all, giving up smoking completely. Drugs to dilate the arteries can sometimes improve the blood flow and also assist the formation of collaterals. Suitable exercise may also be beneficial.

Raynaud's disease is usually found in women under 40 years of age and manifests itself in the extremities (feet and hands, especially the latter). As distinct from Buerger's disease, the deficient flow of blood is here due to exaggerated contraction of the blood vessels rather than thickening of their walls. The contraction occurs as a spasm in response to emotional stress or to slight cold which would not affect a normal person. The symptoms are most likely to arise in cold and wet weather, particularly in the autumn. The attack is accompanied by pallor and by coldness of the fingers or toes; it may last from a few minutes to an hour or more. There is generally no danger to life, though in comparatively rare instances gangrene may develop. Treatment consists in protecting the patient from cold, abstention from smoking, massage and suitable exercises. In severe cases of Raynaud's disease and Buerger's disease, sympathectomy—i.e., cutting the sympathetic nerves to the affected limb—may relieve the condition by preventing reflex contraction of the arteries. Blocking of the sympathetic nervous system by injection of drugs can produce the same result.

Acrocyanosis is the name given to a blotchy reddish-brown discoloration of the skin sometimes found in young women with low blood pressure. These external symptoms appear on the limbs and are especially noticeable on exposure to cold. The cause is a constitutional and, in itself, harmless deficiency in the blood supply to the skin, attended by a sensation of cold in the areas affected. In young girls similar bluish blotchy discolorations may occur more particularly on the legs. The condition does not call for any special treatment. Exercise and sport can improve the regulation of the blood supply to the skin.

An embolus is anything causing a blockage of an artery: a blood clot, an air bubble, etc. The condition itself is called *embolism*. If the embolus is a blood clot, it is called a thrombus, with the term *thrombosis* to denote the condition where partial or complete arterial blockage is caused by a fragment of clotted blood carried in the bloodstream. Such clots are usually formed in the veins, from where they are carried through the right atrium and ventricle of the heart and into the lung blood vessels (pulmonary embolism). Venous blood clots will enter the arteries of the systemic circulation only if there are "short circuits" between the arterial and the venous system. This may, for example, exist in the case of a congenital defect in the septum of the heart, so that there is not complete separation between the left and right sides. Otherwise thrombosis of the systemic arterial system arises mainly in the left side of the heart itself. With stenosis of the mitral valve, clotting may occur in the overdilated and fibrillating atrium. In a case of infarction, clots may form on the inside of the wall of the ventricle, and fragments of these may be carried along in the blood. Inflammation of the heart valves may also be responsible for the formation of clots. Detached fragments of clots tend to become lodged in particular parts of the arterial system, e.g., in the arteries of the leg, kidneys or brain (Fig. 2-64). The resulting sudden blockage of the blood vessel affected will cause immediate symptoms, which in other respects are similar to those which appear when a blockage develops more gradually as a result of local formation of an arterial thrombus. Treatment may consist in the injection of drugs which dilate the arteries; in the event of an actual blockage, drugs to dissolve the clot may be administered or the clot may have to be removed by surgery.

ARTERIOSCLEROSIS

Arteriosclerosis is a term applied to various pathological conditions characterized by hardening, thickening and loss of elasticity of the walls of blood vessels, especially arteries. Fats, proteins and mineral substances are deposited in the walls. Eventually the internal membrane (lining) of the artery becomes ulcerous, and blood clots attach themselves to it, thus gradually reducing the channel of the artery until there is complete blockage. Such changes may occur simultaneously in many arteries or may affect only individual arteries or limited parts of the arterial system. For example, if the condition is mainly localized in the arteries of the heart, brain or kidneys, it is called coronary sclerosis; cerebral sclerosis or nephrosclerosis, respectively. The term *atherosclerosis* is more appropriately applied to the common form of arteriosclerosis associated with atheroma—i.e., degeneration of the wall of the artery characterized by deposits of fatty material in the lining.

The precise cause of arteriosclerosis is not known. It would appear to be due to a whole complex of factors, one or another of which may predominate in any particular case. The principal ones involved are as follows:

1. Age. Sclerotic changes in the coronary arteries are found in 30–40 percent of men between 20 and 30 years of age, and in 70–80 percent of men of 70 and over. Nevertheless, arteriosclerosis is something more than just a normal process of deterioration with age.
2. Hereditary factors.
3. Metabolic diseases, including diabetes and gout.
4. High blood pressure. It intensifies the arteriosclerosis, while the latter in turn raises the blood pressure (see pages 86 ff).
5. Fat content of the blood. In some ethnic groups there appears to be a close link between the frequency of arteriosclerosis and the amount of fat consumed in food. The view has been put forward that consumption of saturated fats (animal fats) is especially harmful in this respect; the fatlike substance called cholesterol, found in artery linings affected by atheroma, is regarded (but not conclusively) as the chief suspect.
6. Hormone balance. Like cardiac infarction, arteriosclerosis is rare in women below the age of 40; after the hormonal readjustments of the menopause, however, such differences between the sexes eventually disappear.
7. Lack of regular exercise (due to a sedentary manner of living), exposure to occupational stress and strain, and the effects of nicotine (smoking) are furthermore considered to be risk factors in the development of arteriosclerosis.

Of particular importance to the syndrome (the characteristic group of signs and symptoms) of arteriosclerosis is the decrease in the internal diameter of the arteries affected, so that the blood supply to the organs concerned is reduced. Quite often the supply is adequate for the at-rest condition; but since the sclerotic arteries can hardly undergo any further dilation, with physical exertion the increased demand for blood cannot be satisfied. Blood clots may eventually attach themselves to the ulcerated lining of the artery, with further reduction of the flow channel. As a result, the tissues deprived of a proper oxygen supply may become gangrenous.

Characteristic symptoms occur especially when sclerosis affects the arteries supplying the heart, the brain or the legs. If the coronary arteries are affected, angina pectoris will develop (see page 71), and with further advance of the sclerotic changes accompanied by the formation of blood clots, cardiac infarction may ensue (see pages 73 ff). Sclerosis of the arteries of the brain will, particularly as a result of clotting, cause "softening of the brain" and perhaps apoplexy (stroke) (see page 344). Sclerotic constriction of the arteries of the legs causes difficulty in walking, especially when climbing stairs or going uphill. Deficient blood supply to the legs manifests itself in severe pain, just as angina pectoris does in the heart. The patient limps intermittently and has to rest frequently. As the disease progresses, the distances that can be walked without acute discomfort become shorter and shorter, until the legs ache even when resting. Eventually, gangrenous areas may develop on the feet. Complete obstruction of arteries by clots is especially liable to cause gangrene.

HEMORRHOIDS

Hemorrhoids (piles) are vascular tumors due to distended veins, similar to varicose veins, which occur in the anal canal a short distance above the anus. They may be located above or below the junction of anal canal and rectum, and are called internal or external hemorrhoids, respectively. Like varicose veins, hemorrhoids are usually associated with a constitutional weakness of the local connective tissue. Straining because of constipation, coughing and nose blowing increase the pressure in the abdominal cavity and may be contributory to hemorrhoids. Certain anal veins (the upper hemorrhoidal veins) discharge into the portal vein; because of this, congestion of the portal vein due to liver disease (cirrhosis) may cause hemorrhoids (see pages 94 ff). It is estimated that at least a third of all people have hemorrhoids.

External hemorrhoids are visible externally, whereas internal ones can be detected only by probing with a finger. As the name implies, the characteristic symptom of hemorrhoids is bleeding. This is particularly true of small internal ones, especially when the patient is constipated. The actual loss of blood is generally not serious, and the fresh blood on the feces is not likely to be mistaken for blood discharged from a cancer of the rectum (see page 496). Although there may be discomfort, hemorrhoids are generally not painful except when they become inflamed. Pain in the anal region may be due also to other causes, such as inflammation of mucous membranes, anal carcinoma (cancer), anal fissure and abscesses.

Inflamed hemorrhoids discharge blood until the blood in them eventually clots, and the hemorrhoids themselves waste away and sometimes even fall off. Quite often so-called anal fissures may develop; these are cracklike ulcers which are painful and very slow to heal. Frequent recurrence of hemorrhoids may lead to anal eczema, abscesses or ulcers. A burning sensation and itching, especially at night, are signs of such developments. Anal fistulas may also form. Sometimes thickened hemorrhoids heal and turn into inconspicuous skin flaps.

If symptoms are not severe, treatment consists in preventing constipation and keeping the stools soft and regular. Usually this can be achieved by a suitable diet with plenty of indigestible roughage, such as vegetable fibers and cereals, which acts as a stimulant to intestinal action. Sometimes purgatives are necessary. To train the anal sphincter (the circular muscle constricting the orifice), it has been suggested that the patient should contract this muscle 30, 40 and eventually 80 times for periods of 2 seconds, followed by relaxation for 3 seconds. These exercises should be repeated two or three times a day. It is claimed that they help to relieve congestion in the veins in the anal region and thereby reduce the tendency for hemorrhoids to develop. Suppositories may be prescribed for local pain relief. In some cases severe pain is caused when the anal sphincter grips very tightly around small hemorrhoids or anal fissures. In dealing with such cases the use of suppositories containing a spasmolytic drug to relax the sphincter muscle is indicated. Eczema and inflamed small hemorrhoids can be treated with ointments that relieve inflammation. In the absence of inflammation, hemorrhoids may eventually heal of their own accord. If they do not respond to treatment, they may have to be removed by a surgical operation, especially if they occur frequently, become inflamed, are painful and are attended by anal fissures. The operation consists in ligaturing the blood vessels supplying the hemorrhoids and then removing the latter. As hemorrhoids are often due to constitutional weakness of connective tissue, their subsequent recurrence cannot be ruled out.

VARICOSE VEINS

The term *varicose* refers to abnormally enlarged and twisted veins, generally in the legs. The function of the veins is to return the spent blood—with low oxygen content and carrying waste products—from the various organs to the heart. In these blood vessels the pressure is relatively low, a few millimeters of mercury, and for this reason the veins need have only thin walls to withstand the pressure. When a person is standing upright, the static pressure due to the weight of the blood in the body additionally exerts a pressure of 75–100 mm of mercury on the walls of the veins in the legs. This extra pressure should cause the veins to distend, but normally this does not happen, because there are two factors that relieve the walls of the veins. For one thing, the veins contain one-way valves which subdivide the column of blood into smaller sections and thus reduce the pressure in each section between valves. Secondly, contraction of the muscles in the leg exerts a squeezing action on the veins and helps the blood along, this effect being further strengthened by the pulsation of the adjacent arteries (Fig. 2-65a–c). Hence it is evident that standing still, with the leg muscles

FIG. 2-65.
Factors influencing venous pressure.

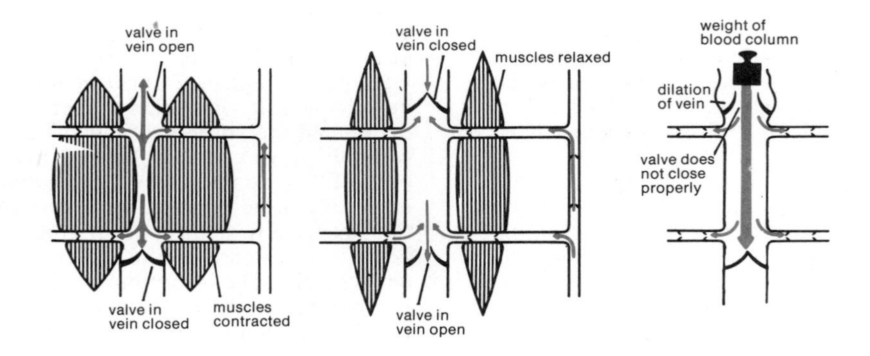

FIG. 2-66.
Venous valves.

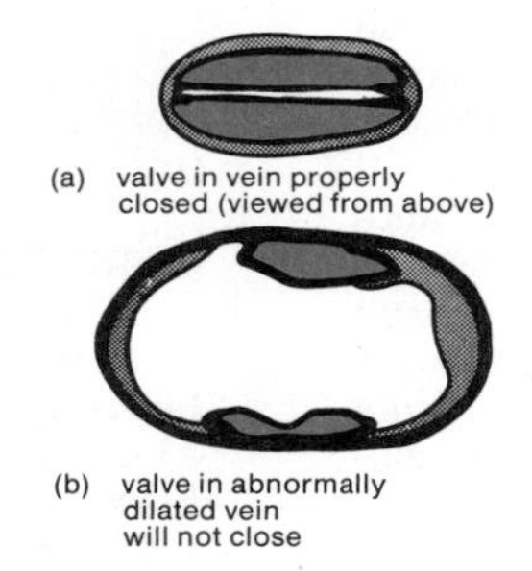

at rest, constitutes an additional pressure load on the veins in the legs. In particular, the weight of the blood column in the body as a whole acts on those sections of the veins that have not been emptied by muscular action and are therefore not relaxed.

If the veins, because of a constitutional predisposition, are too easily distendable, they can be so overdistended that the valves fail to close properly (Fig. 2-66a and b). The column of blood standing in a vein thus becomes longer, and its weight forces open other valves lower down, so that progressively greater lengths of the vein become overdistended and more valves fail to close. At first such veins are emptied by the squeezing action of the muscles when walking, which causes the blood pressure in the veins to fall from 100–120 mm to 30–40 mm of mercury (Fig. 2-67). In a later stage of the disorder, however, this is no longer possible, the walls of the veins having become so flabby that they remain permanently distended and the veins themselves become swollen, twisted and knotted. The congested muscles cause cramplike pains; blood plasma discharged from the swellings causes hardening of the subcutaneous connective tissue; the skin thickens and presents a brownish discolored appearance; nagging pains are felt also at night.

When the patient is standing, the varicose veins may contain as much as 1 liter of blood. They can be emptied by placing the legs in a raised position or by the application of external pressure. When the patient rises to his feet, however, the blood flows back into the reservoir formed by the varicose veins. The effects of this sudden movement of blood are similar to those of loss of blood: fall in blood pressure, dizziness, a feeling of weakness, perhaps even

FIG. 2-67.
Fall in blood pressure in veins during walking.

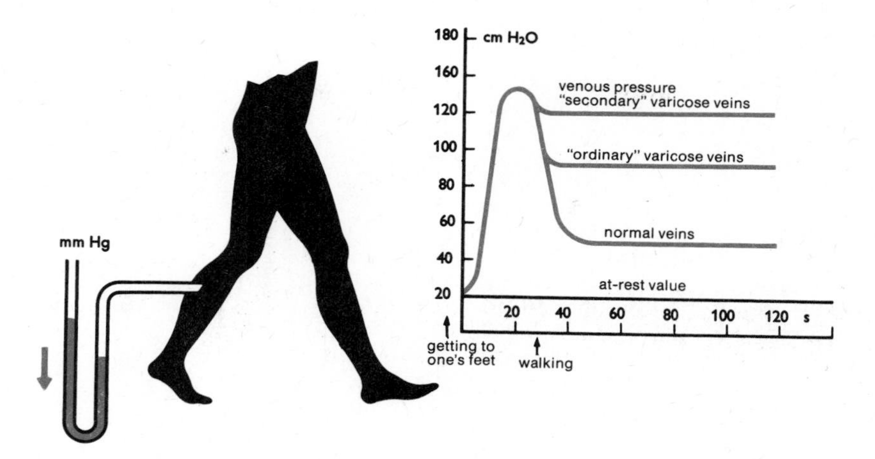

heart trouble or fainting. In the long run there is a risk that clotting will occur in the sluggishly flowing blood (thrombosis) and that the walls of veins will become inflamed (thrombophlebitis).

Varicose veins are often a symptom of serious weakness of connective tissue which may manifest itself also in such disorders as flat feet, hemorrhoids (piles), and abnormalities of posture and gait. Occupations that involve much standing (waiters, policemen, housewives) are a contributory factor. Pregnancy, obesity and certain items of clothing (e.g., garters) also contribute to causing varicose veins. The disorder is found more frequently in women than in men (Fig. 2-68). Prevention and nonoperative treatment are similar to the methods for low blood pressure as regards exercise and manner of life to be adopted by the patient, the more so as varicose veins contribute to low blood pressure (see pages 82 ff). Gymnastic exercises, sport, hot and cold baths or spongings can increase the tension of the walls of the veins. Relief is obtained by frequently resting the legs in a raised position and by the wearing of medically approved elastic stockings, whereby the accumulation of blood in the varicose veins can be reduced.

Fully developed varicosities will not disappear as a result of such remedial measures, however. They can only be dealt with by sealing off the vein with ligatures or by surgical removal. Before any such action is taken it must be ascertained whether it is indeed a case

FIG. 2-68.
Incidence of varicose veins in men and women at various ages.

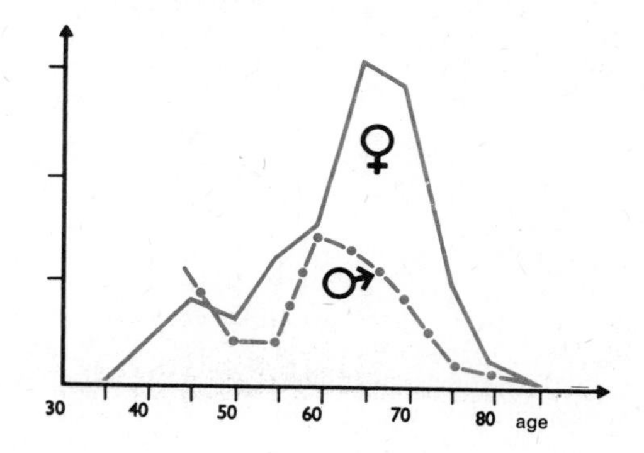

of varicosities affecting the saphenous (superficial) veins of the leg, as envisaged here, which are caused by increased blood pressure in the legs in combination with a constitutional predisposition. After these veins have been sealed off, their function can be taken over by veins located deeper in the leg and thus better supported by surrounding muscles. But if the deep-lying veins have also deteriorated as a result of particular inflammatory processes, the veins at the surface will be urgently needed to help carry the blood normally carried by the deep veins.

DEEP VENOUS DISORDERS

Disorders in blood flow through the deep veins have more disagreeable consequences than the more common varicosities encountered in the superficial veins. Normally, the deep veins have to convey the major part of the blood from the leg (Fig. 2-69), and they can usually cope with the relatively small extra quantity that would otherwise be carried by the superficial veins. But the reverse is not possible: if the deep veins cease to function efficiently, secondary varicosities will develop in the superficial veins as a result of overloading; furthermore, congestion phenomena will occur in the parts from which the discharge of blood through the veins is deficient.

Congenital weakness of connective tissue is not so prominent a contributory factor in causing this condition as in the case of ordinary varicose veins. Deficient flow in the deep veins is, instead, usually caused by inflammatory damage to the walls of these veins and their valves, possibly in conjunction with blood clot formation. If the inflamed valves of the deep veins fail to close efficiently, the pressure developed by the column of blood will, especially when the patient stands, cause fluid from the blood to be forced into the adjacent tissues (edema of the leg, Fig. 2-70).

If the deep veins are partly or wholly obstructed by clots, the resulting back-pressure and congestion will not disappear even when the patient lies down. Such blood clots (thrombi) may be the result but also the cause of inflammation of a vein (*phlebitis*), more particularly the condition called *thrombophlebitis*—i.e., inflammation of a vein associated with thrombosis. It occurs not infrequently as an aftereffect of a surgical operation; overweight and advanced age are contributory factors, as is also impaired circulation due to varicose veins, cardiac insufficiency or arteriosclerosis. Local signs of inflammation, such as reddening, tenderness (sensitivity to pain upon pressure) and swelling are liable to occur. More serious is the risk that a fragment of blood clot may be dislodged and travel to the lungs, where it may obstruct an artery (pulmonary embolism, see page 100).

If clotting in the deep veins becomes chronic or if new flow channels are established after those veins have become obstructed, a state of congestion and poor circulation will prevail, so that the blood tends to stagnate (Fig. 2-70). Swelling develops, and the skin takes on a shiny and brownish appearance, becoming thickened. Pressure exerted with a finger will leave a depression (pit) that remains visible for a time. Such edema is a sign that local escape of fluid from the bloodstream has occurred. Eventually, eczema and persistent *leg ulcers* (Fig. 2-71) may develop. There is poor circu-

FIG. 2-69. Superficial and deep veins in the leg.

FIG. 2-70. Secondary varicose veins as a result of thrombosis in the deep veins of the leg.

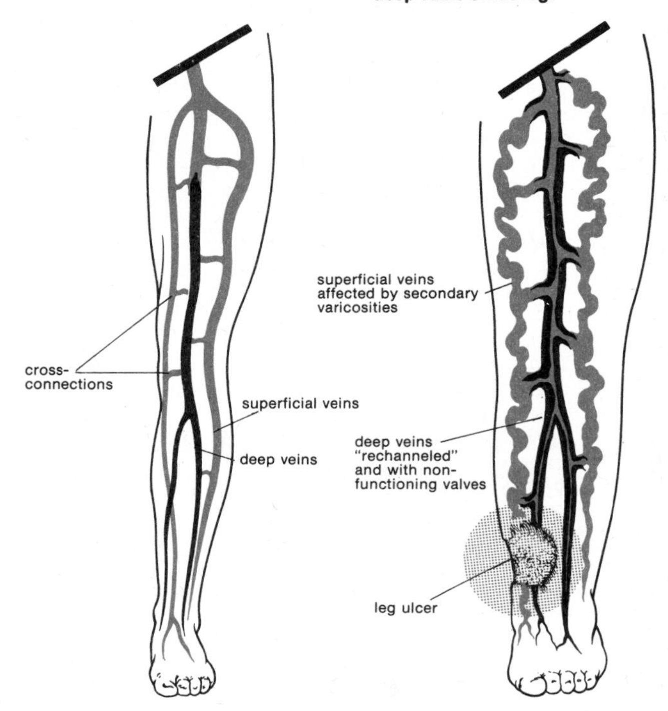

FIG. 2-71. Leg ulcers.

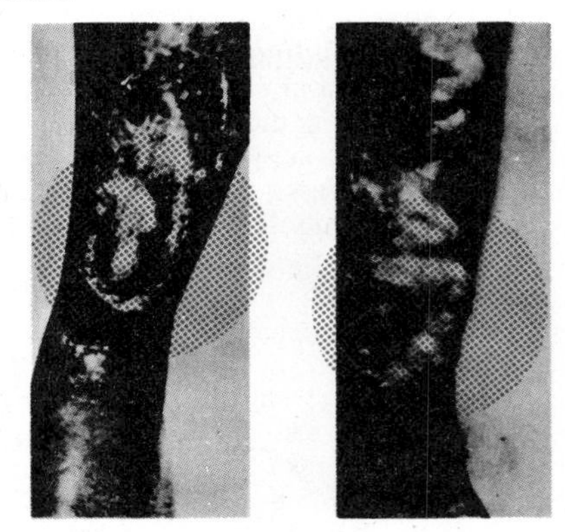

lation in such ulcerated areas, which are pale, have raised edges and are very slow to heal. The overall syndrome is that of *chronic venous insufficiency*.

Treatment should aim at preventing such conditions from arising or at halting their progress. It will include, for example, preventive action against further clotting and will comprise measures to relieve pressure in the deep veins of the legs. Standing and sitting should be avoided as much as possible, and the legs should be kept in an elevated position when resting. Unfavorable factors are overweight, the wearing of tight garters, and lifting or carrying heavy loads involving abdominal muscular strain, which raises the blood pressure in the veins. Walking and swimming assist and train the squeezing action of the muscles and thus help to discharge blood from the veins and improve circulation in them. Pressure bandages can, as a result of narrowing the veins by external compression, help to restore the valves of the veins to something like normal functioning. However, bandages that are too tight, especially those which are tighter at the top than at the bottom, are liable to act like a tourniquet and may worsen the chronic venous insufficiency.

VENOUS THROMBOSIS

The term *thrombosis* denotes the more or less complete obstruction of a blood vessel or heart cavity by a clot of blood (thrombus). The formation of clots is most likely to occur in veins (venothrombosis or phlebothrombosis), and fragments of clots may travel in the bloodstream via the right side of the heart into the pulmonary circulation and cause obstruction there (pulmonary embolism, see next section).

The reason clots develop mostly in veins is that clotting is promoted by a slowing of the rate of flow, as is bound to occur in parts of the extensive venous system anyway. Additional congestion of veins, as in varicose veins or during postoperative rest, promotes thrombotic developments. Furthermore, damage affecting the walls of the blood vessels often also plays a part, e.g., arteriosclerosis of the coronary arteries in connection with cardiac infarction. The damaged wall of the artery is then the site and starting point of a blood clot which at first adheres firmly to it.

Such clotting has much in common with the normal coagulation of blood and can therefore be described as "coagulation in the bloodstream." First, the blood platelets, which move along mainly at the periphery of the slowed-down flow, attach themselves to the damaged wall and become covered with fibrin filaments (so-called plate thrombus, Figs. 2-72 and 2-73a). Then coagulation takes place, involving platelets and the combined action of several substances (see page 118). In addition to the "white" plate thrombus, a "red" coagulation thrombus is formed (Fig. 2-73c). In between them is a transition zone in which layers of platelets alternate with layers of white and red cells (stratified thrombus, Fig. 2-73b).

The clotting tendency of blood increases under certain conditions—e.g., after an operation, in pregnancy and childbirth—and generally whenever coagulation processes are occurring anywhere in the body in response to wounds or inflammation. Clots would form even more frequently but for the fact that the coagulative factors are continually being neutralized and any clotted blood formed is dissolved (cf. fibrinolysis, page 118). Blood clots form mostly in the lower limbs. Superficial venous thrombosis, as may occur in varicose

FIG. 2-72.
CROSS SECTION THROUGH VEIN.

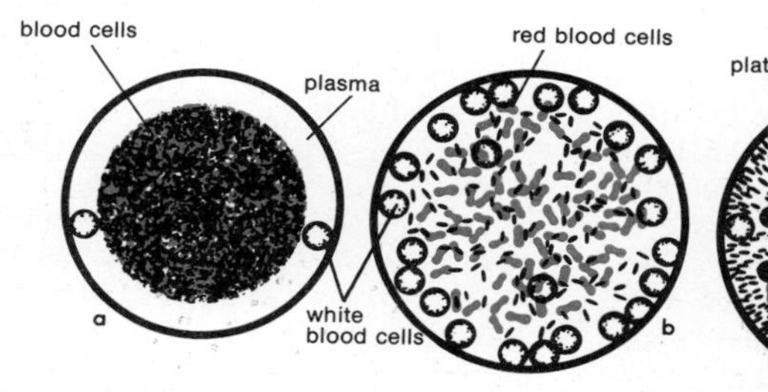

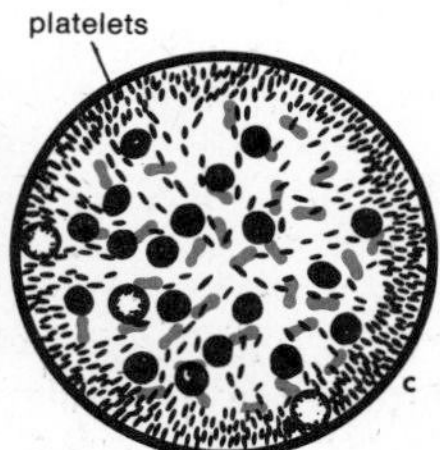

(a) Normal rapid blood flow. **(b) Moderately retarded flow.** **(c) Greatly reduced flow.**

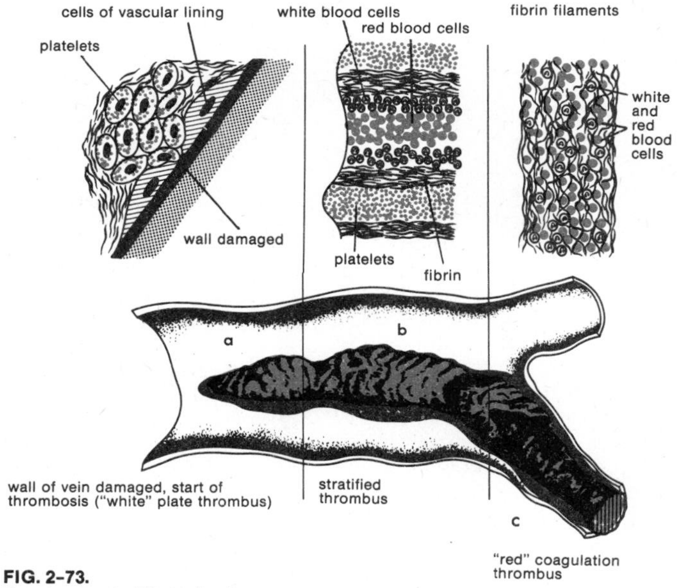

FIG. 2-73.
Development of thrombosis.

veins, is usually of only local significance. Deep-lying clots in the main blood vessels, on the other hand, first cause tension and pain on the sole of the foot and in the calf, attended by reddening, followed by bluish discoloration and edema of the leg. Particularly the large deep-lying clots are liable to be carried along in the bloodstream and may travel to the lungs, there causing extensive embolic obstruction of blood vessels (pulmonary embolism). To prevent this it is therefore necessary to prevent or at least speedily detect thrombosis of the veins in the legs and pelvic region. In addition to the aforementioned symptoms, minor attacks of fever and quickened pulse may be signs of local inflammation of veins (phlebitis) or first minor movements of clotted blood. The first few weeks after an operation, injury, infection or childbirth are regarded as a critical period for the development of thrombosis. After about four weeks this hazard will generally have passed. Thrombosis in the deep veins of the leg is dangerous and is held to result in death due to pulmonary embolism in something like 10 percent of cases. Dangerous, too, is the progress of thrombosis in the large venous trunks, possibly resulting in serious circulatory disorders or gangrene of the leg affected.

PULMONARY EMBOLISM

Embolism is the obstruction of a blood vessel by a blood clot, air bubble or globules of fat. The obstructing matter (embolus) is carried in the bloodstream and, depending on its size, eventually becomes lodged in the branches of the arterial system. Blood clots are formed mostly in the veins of the systemic circulation (see page 99)—e.g., thrombosis of the leg—and an embolus, after passing through the right atrium and ventricle of the heart, is most likely to become lodged in a pulmonary artery (Fig. 2-74). Large clots formed in the deep veins of the leg may thus block pulmonary arteries. With the resulting decrease in flow of blood through the lungs, the systemic circulation is also impaired. Collapse occurs, with rapid fall in blood pressure, shock, pallor, abnormally rapid heart action, sweating, dyspnea and cyanosis (bluish complexion due to oxygen deficiency in the blood) (see page 37). Patients in this dangerous condition often succumb to the pulmonary embolism; or otherwise serious degenerative changes occur in those areas of the lung that have had their blood supply cut off, so that a pulmonary infarct (necrosis of lung tissue) may occur, which may involve the pleura and cause

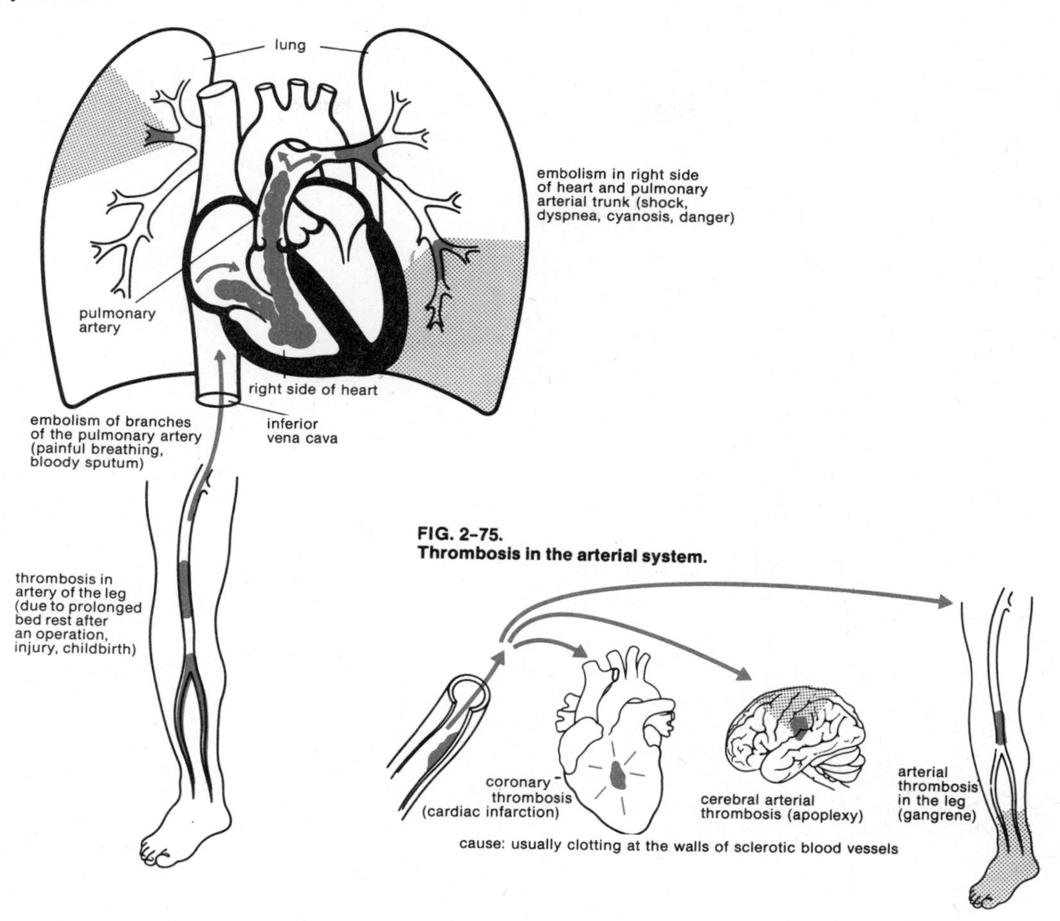

FIG. 2–74.
Pulmonary embolism.

FIG. 2–75.
Thrombosis in the arterial system.

painful breathing. The patient usually also coughs up blood and has fever. The right side of the heart is overloaded and overdilated.

In the comparatively rare cases where suitable operating facilities are immediately available, the treatment of massive pulmonary embolism will consist in very speedily opening the patient's thorax and removing the blood clot. Otherwise there is a high risk of death. Since about 10 percent of all cases of thrombosis in the deep veins of the leg end fatally by pulmonary embolism, the first aim should be to prevent thrombosis from developing after operations or childbirth and also to take precautions to prevent clots from traveling in the bloodstream. Suitable movement therapy can counteract sluggish blood flow and thereby reduce the risk of thrombosis (medical gymnastics, elevation of the legs, getting the patient up out of bed as soon as possible after an operation or childbirth). Anticoagulant drugs also reduce the risk (heparin, coumarin derivatives, see page 119). Such drugs may, for example, be given in the period of increased thrombosis hazard during the first six or eight weeks after a major operation of the abdominal cavity or pelvic region. With reduced risk of thrombosis in the leg arteries, the risk of pulmonary embolism is correspondingly reduced. Recently formed clots can be dissolved by injection of a solvent agent (such as the enzyme called streptokinase) into the affected vein.

Though more rarely, clotting may also occur in arteries (*arterial thrombosis*). It mostly originates with serious sclerotic degeneration of the arterial wall. Since the arteries, in contrast with veins, hardly ever occur in pairs to provide reserve capacity, even quite minor arterial obstructions can be very dangerous. An arterial thrombosis may result in, for example, cardiac infarction, apoplexy, or gangrene of the extremities (Fig. 2-75). Clots traveling in the bloodstream can cause arterial embolism. Such clots can, generally speaking, originate only in the left side of the heart or in the arterial system itself, as clots of venous origin normally end up in the pulmonary circulation. An important exception is formed by cross-connections of the veins and the arteries of the systemic circulation, thus bypassing the lungs. These are abnormalities that may occur as a result of malformations of the heart, e.g., when there is a hole in the septum between the left and right atrium. Clots that have developed in the veins may thereby find their way into the left side of the heart instead of into the lung and then be carried in the bloodstream as though they had been formed in the heart itself. As a result, obstruction of arteries is liable to occur, which can have serious consequences in the brain, the kidneys or the limbs (apoplexy, renal infarction, gangrene of the extremities, Fig. 2-76).

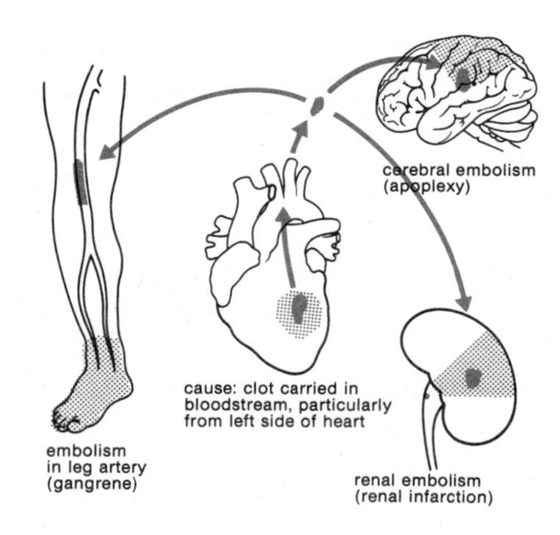

FIG. 2-76.
Embolism in the arterial system.

Part Three

THE BLOOD

Because of the many duties that the blood has to perform within the organism, it claims a special place among the body's vitally important organs. The principal functions of the blood are the transport of oxygen from the lungs to the tissues (respiratory function), transport of carbon dioxide from the tissues to the lungs and of waste matter to the kidneys (excretory function), transport of nutritive substances from the intestines and liver to the tissues (nutritive function), and transport of vitamins and hormones. Furthermore, the blood forms a system of defense against infection, which it does through the action of antibodies and phagocytes, and it plays an important part in thermal regulation by the removal of excess heat from the interior of the body to the surface.

Like other tissues, blood is composed of cells and an interstitial fluid (plasma). Blood *plasma* is a pale yellowish liquid containing inorganic salts, carbohydrates (especially glucose, the so-called blood sugar), fats, vitamins, waste matter and plasma proteins (albumins, globulins and the fibrinogen that plays an important part in the clotting of blood). Of the blood *cells,* by far the most numerous (99 percent) are the red cells (erythrocytes). The rest is made up of white blood cells (leukocytes) and so-called platelets. The platelets, of which there are 250,000–400,000 per mm^3 of blood, are formed by extrusion from giant cells in the bone marrow and are important in clotting and in the arrest of bleeding; for this reason they are sometimes known also by the name of thrombocytes. They live for periods ranging from 2 to 10 days, after which they are destroyed in the spleen.

The *volume of blood* in the human body corresponds to about 7 to 8 percent of the body weight; a person weighing 70 kg (165 lb) has about 5 to 5.5 liters (4½ to 5 quarts) of blood. The blood cells account for approximately 45 percent of this volume, the remainder being plasma. This percentage of blood cells (for practical purposes: the percentage of red cells, these being in the vast majority) in the total blood volume is called the hematocrit, and 45 percent is an overall average; in men the average is rather higher than in women.

The composition of human blood can be summarized as follows:

blood plasma
(1) proteins
(2) minerals
(3) substances transported: nutritive substances, hormones, enzymes, vitamins

blood cells
(1) red cells (erythrocytes)
(2) white cells (leukocytes)
(3) blood platelets (thrombocytes)

RED BLOOD CELLS (ERYTHROCYTES)

The red cells (erythrocytes) give blood its color. In man and other mammals these cells have no nucleus and therefore cannot increase their numbers by division. Each cell is a biconcave disc with a diameter of about 7.5 microns (1 micron = $\frac{1}{1000}$ millimeter) and a thickness of 2 microns. Because of this shape the red blood cells have a very high ratio of surface area to volume, which facilitates efficient exchange of gases. The body of the cell consists of a spongelike mass containing the oxygen-carrying red pigment *hemoglobin* and is enclosed in a membrane consisting of proteins in combination with lipoid substances. The characteristics of the blood groups (blood group antigens) are contained in the membranes of the red cells.

In the capillaries the red blood cells are subject to conditions of relatively severe mechanical wear and tear. The life-span of a red cell is about 120 days. The number of red cells averages about 5 million per mm^3 (the number is higher in men than in women). The total number in a person of average size and weight is about 25×10^{12} (25 million million). Each day about 2×10^{11} new red cells have to be produced, i.e., a production rate of 2.4 million per second. The normal number of red blood cells is kept constant within fairly narrow limits. If their number is increased (e.g., due to transfusion) or decreased (e.g., due to loss of blood), a compensatory adjustment takes place. To make up for losses and normal destruction of red cells, new ones are formed in the red bone marrow; the process of formation is called *erythropoiesis*. This process is intensified or diminished to compensate for a temporary decrease or increase in the number of red cells; the actual life-span of the individual cells remains unchanged. The cells are continuously dying and disintegrating; the final breakdown of the red cells is accomplished in the reticuloendothelial system, especially the spleen and liver. The reticuloendothelial system comprises phagocytic cells in various parts of the body which are able to engulf and destroy foreign matter (bacteria, worn-out tissue, blood cells that have reached the end of their useful life-span).

The red cells are heavier than the blood plasma and will settle if the blood is prevented from clotting. In the laboratory test for *sedimentation* the speed at which the red cells settle in blood to which an anticoagulant has been added is observed. The test is performed in a long thin tube. The actual technique employed may vary. In one method the time required for the red blood cells to settle a certain distance is determined (sedimentation time); another method determines the distance the cells settle in a given time (sedimentation rate). Normal sedimentation rates in a one-hour period are 3–8 mm in men, 6–12 mm in women. The sedimentation rate can provide an indication of the presence of disease. The rate increases in various infections, in cancer, and in pregnancy.

ANEMIA

Literally meaning "lack of blood," *anemia* is actually characterized by a shortage of hemoglobin, usually associated with a decrease in the number of red blood cells (erythrocytes). The effects of anemia are due to deficient supply of oxygen to all parts of the body.

Anemia may be caused by abnormal loss of blood, defective production of red cells, or excessive destruction of red cells (Fig. 3-1). So-called *aplastic anemia* is the result of aplasia (i.e., failure to develop normally) affecting the bone marrow. This defect may be congenital or be caused by toxic chemical agents or by x-rays or other ionizing radiation. In severe cases there is suppression of all blood cell formation; the patient can then be kept alive only with transfusions. Certain drugs may cause anemia as an allergic reaction of the marrow (e.g., to chloramphenicol or sulfonamides). *Deficiency anemia* results from a lack of an essential ingredient in the diet or from the inability of the intestine to absorb it. Thus there may be a deficiency of iron (causing iron-deficiency anemia, page 107) or of certain substances essential to hemoglobin formation (pernicious anemia, page 107, or anemia due to folic acid deficiency). Excessive destruction of red blood cells (*hemolytic anemia*) may be due to any one of a number of causes, e.g., a congenital abnormal fragility of these cells, malfunction of the spleen, or the effects of toxic agents.

The circumstances attending a major loss of blood due to hemorrhage are exceptional in that there arises not only a deficiency of blood cells but also of blood fluid. Symptoms associated with defective filling of the blood vessels therefore dominate the patient's condition: fall in blood pressure, abnormally rapid heart action, dyspnea, pallor, sweating, restlessness, confusion and thirst. A state of shock may ensue, i.e., a condition characterized by disturbed oxygen supply to the tissues and impaired return of blood to the heart; severe shock is fatal, particularly in consequence of inadequate flow of blood to the brain. The rapid loss of more than 0.5 liter of blood may already be dangerous. A loss of 1.5 liters is likely to cause disturbed vision, spasm, and loss of consciousness and may even result in death. Treatment consists in quickly stanching the bleeding and making good the loss of blood by means of transfusions.

In a case of continuous minor "chronic" hemorrhage the blood fluid can be replaced more quickly than the hemoglobin. The symptoms are then generally similar to those of other forms of chronic anemia. The significant central features of such conditions are deficiencies in the oxygen-carrying function of the blood, accompanied by symptoms such as pallor, dyspnea (shortness of breath), quickened pulse, dizziness and tinnitus (buzzing or other sounds in the ear). Loss of blood may be due to oozing—e.g., from hemorrhoids (piles), the uterus, a chronic gastric ulcer or intestinal cancer. Chronic hemorrhage is accompanied by depletion of the iron stored in the body (iron-deficiency anemia).

Abnormally increased rates of destruction of

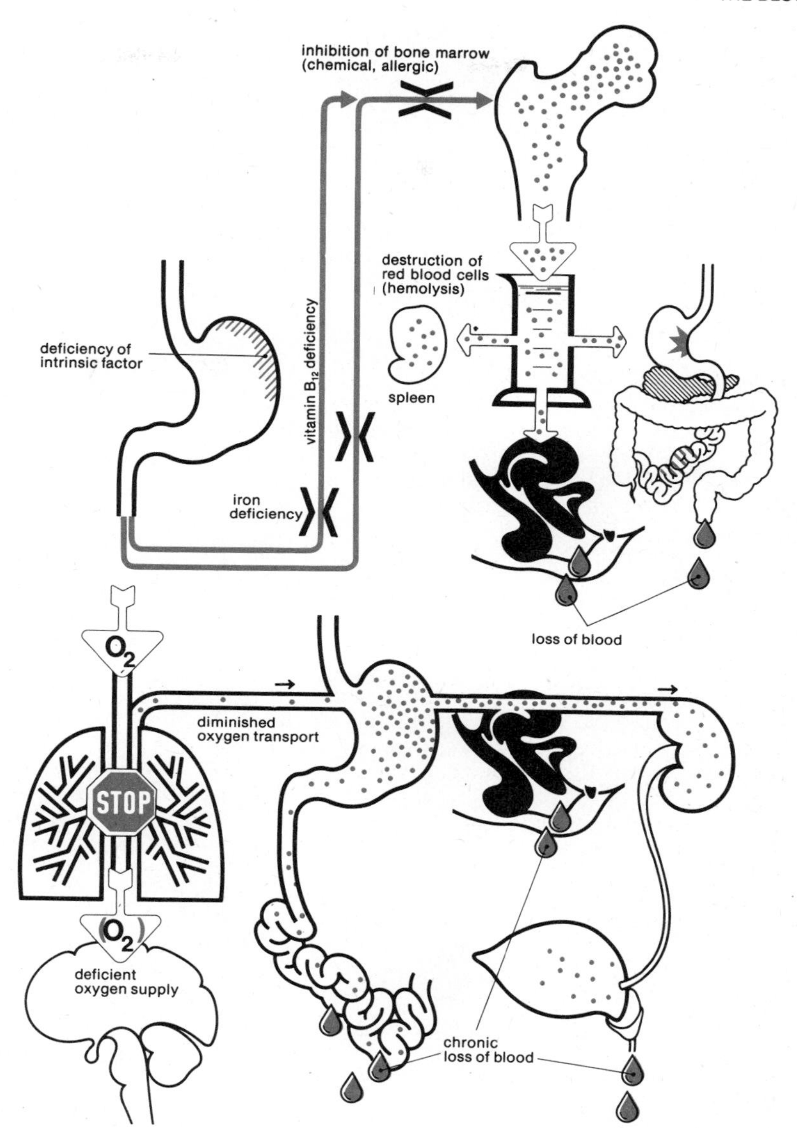

FIG. 3-1.
Causes of anemia.

FIG. 3-2.
Consequences of anemia.

red blood cells (*hemolysis*) have effects similar to those of chronic hemorrhage, except that the loss of hemoglobin is internal instead of external. There may, moreover, be no depletion of iron, which is a favorable aspect of the condition; on the other hand, there is the disadvantage that more decomposition products arise from the breakdown of the red cells. These products have to be eliminated with the bile (gall) from the liver, more particularly in the form of bilirubin, the orange or yellow pigment formed from the decomposition of the hemoglobin of the red cells. If the liver is diseased or overburdened, bilirubin will accumulate in the body, causing hemolytic jaundice; products of the breakdown of bilirubin then also appear in the feces and urine.

Hemolytic anemias may be due to congenital fragility of the red blood cells, which are thus more liable to be destroyed. They comprise some hereditary diseases: notably, sickle-cell anemia, which is characterized by the presence of large numbers of crescent or sickle-shaped red cells in the blood and occurs almost exclusively in African populations; Cooley's anemia (thalassemia), found mainly around the Mediterranean and resulting from a recessive trait responsible for interference with hemoglobin synthesis.

Iron-Deficiency Anemia

The red blood cells carry oxygen to the various organs of the body and also play a part in the removal of waste products from them. They perform these functions with the help of the red pigment *hemoglobin*. This substance consists of a protein (globin) and an iron-containing pigment (heme), the latter being the actual oxygen carrier. Hemoglobin is produced in the red marrow of the flat bones of the skull, in the breastbone (sternum), in the vertebrae, and in the end parts of the long bones. If insufficient iron is available to these hemoglobin-producing centers, the deficiency will result in anemia—the commonest form, namely, iron-deficiency anemia.

The human body contains about 4 to 5 g of iron, of which about 3 g is present in hemoglobin. The rest is in the liver, spleen and bone marrow, where the iron is obtained from the breakdown of red blood cells which have reached the end of their life-span and is stored as an iron-protein compound; some iron is also stored in the red pigment of the muscles and, to a less extent, in certain metabolic enzymes. In blood plasma, iron is combined with a carrier protein called *transferrin* and is withdrawn from the reserves and transported to the blood-producing centers as and when it is required there.

The daily *iron requirement* is 0.5–1 mg for men, 2–3 mg for women of menstruating age. Normal iron losses occur as a result of exfoliation (scaling-off) of cells of the epithelium of the skin and the intestinal canal. In women there is loss due to menstruation. These losses are compensated by absorption of iron from the small intestine.

The normal level of iron in the body (Fig. 3-3) is regulated, not by excretion, but by absorption of iron. The deciding factors with regard to the rate at which iron absorption into the body occurs are the rate at which new red blood cells are formed and the degree of saturation of the iron storage centers (spleen, liver,

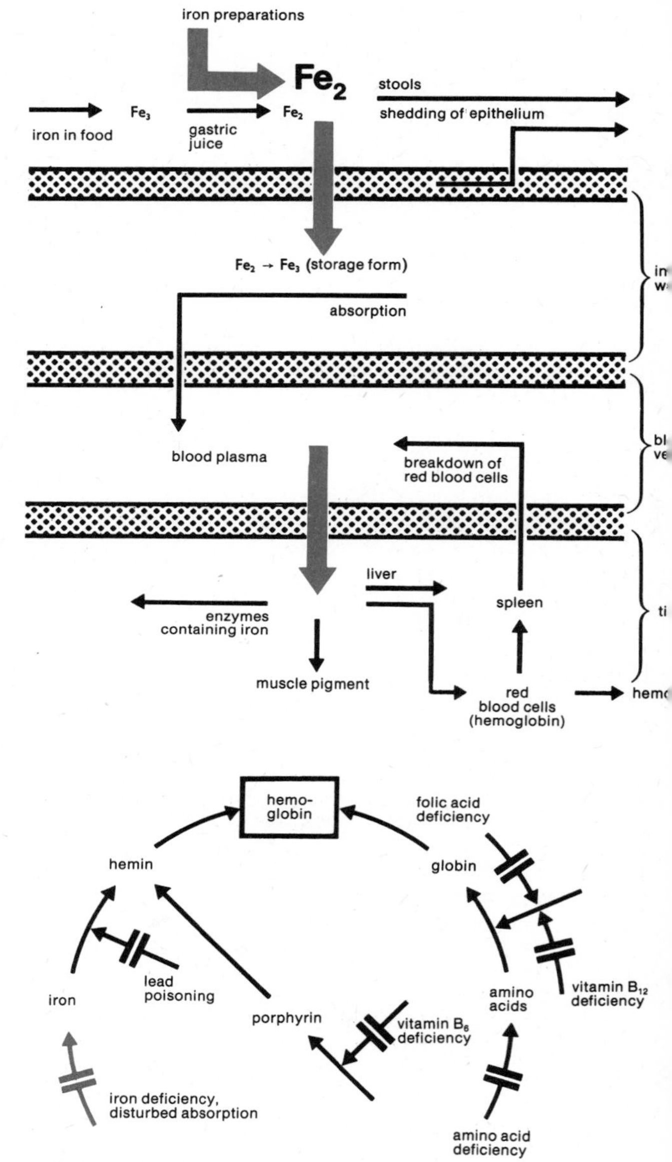

FIG. 3-3.
The body's iron balance.

bone marrow, intestinal cells) in which the iron is found combined with a specific protein (*apoferritin*). Iron deficiency is seldom caused by inadequate iron content of the diet: only in incorrectly nourished infants, in victims of starvation or in persons who have lived on a very one-sided diet. Poor utilization of available iron supplies due to degeneration of the intestinal villi or to lack of hydrochloric acid in the stomach (e.g., after surgical removal of the stomach) may also occur. Most frequently, however, iron deficiency is caused by loss of iron due to chronic slow bleeding. Such bleeding may take place from gastric ulcers, bleeding polyps, cancer of the intestine and, in women, from the genital organs.

Iron-deficiency anemia, like any other anemia, reduces the capacity of the blood to carry oxygen. The consequences of the diminished oxygen supply to the organs are similar to those of other types of anemia, the main symptoms being pallor, weakness, dizziness, breathlessness, accelerated pulse, headache and tiredness. Characteristic of iron deficiency in particular are certain changes in the epithelium (surface layer of cells) of the skin and mucous membranes, with loss of hair, dry skin, split and deformed fingernails, painful and smooth tongue, and difficulty in swallowing. In the blood the hemoglobin goes down to 60 percent of the normal value, the red cells are small and of low hemoglobin content, and the iron content of the blood plasma is also reduced.

Treatment aims primarily at removing the underlying cause. In addition, iron is given in order to compensate for the deficiency. The simplest and most reliable method is to give the patient tablets containing stabilized divalent iron (e.g., ferrous sulfate). Iron compounds color the feces black and may make the patient unwell or cause diarrhea.

Pernicious Anemia

Hemoglobin is produced in the red marrow of the bones from amino acids and iron, a process that requires the assistance of certain auxiliary substances (biocatalysts), particularly vitamin B_{12}. This vitamin is found in foods of animal origin such as meat, eggs, milk, cheese and especially liver. It is absorbed in the small intestine with the aid of a protein (called *intrinsic factor*) formed in the lining of the stomach. There are thus two essential functions involved: adequate supply of vitamin B_{12} and adequate secretion of intrinsic factor. The absorbed vitamin B_{12} is stored in the liver and is fed into the blood-forming centers as and when it is required (Fig. 3-4). Vitamin B_{12} is essential not only to the formation of red blood cells but also to the healthy development of the cells of other tissues. This is so because the vitamin plays a part in the formation of nucleic acid and is therefore needed for nuclear growth and division. Disturbed development of cells is a consequence of vitamin B_{12} deficiency and therefore especially likely to occur in tissues with a rapid turnover of cells. Such tissues in-

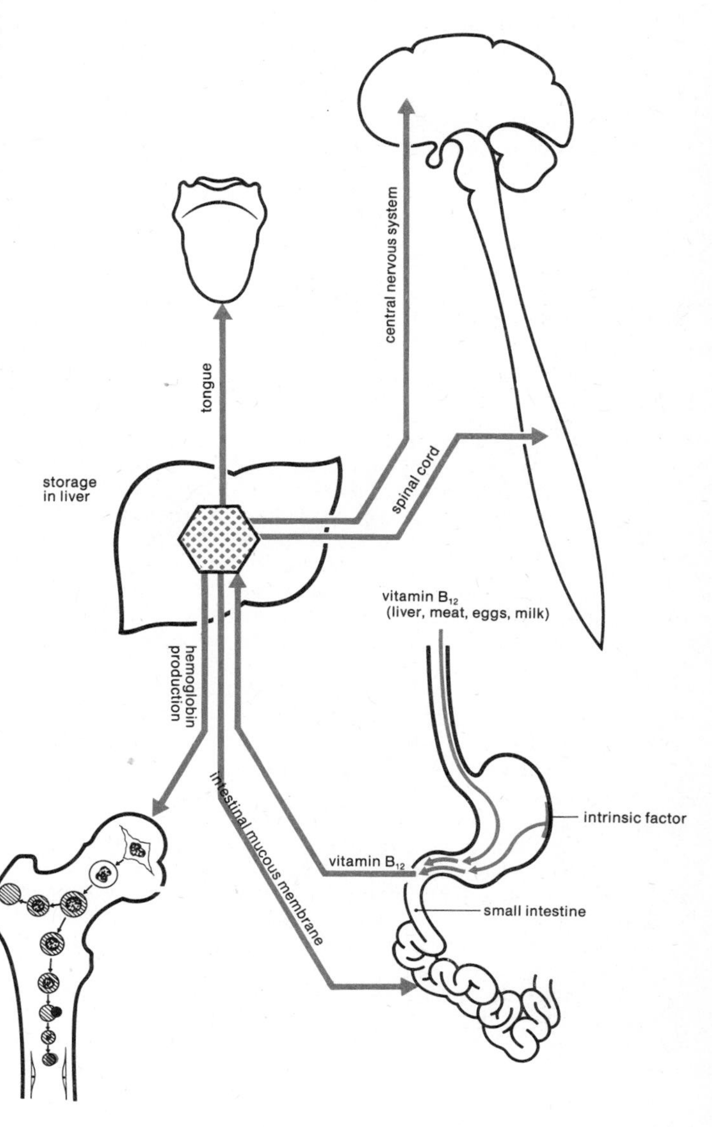

FIG. 3-4.
Absorption and action of vitamin B_{12}.

clude the blood-forming centers, e.g., the gastrointestinal mucous membranes (Fig. 3-4). Without this vitamin the bone marrow is unable to produce enough red blood cells, and those that are formed are immature, misshapen and fragile.

Vitamin B_{12} deficiency is caused by (1) inadequate vitamin supply, as may occur in strict vegetarians; (2) inadequate absorption of the vitamin because of deficient production of intrinsic factor or because of functional deficiency of the intestinal lining; (3) breakdown of the vitamin in the intestine by certain parasitic worms or, more rarely, by bacterial action. Which of these causes is responsible can be detected by giving the patient vitamin B_{12} in combination with radioactive cobalt, so that its progress in the body can be observed. In practice, failure to absorb the vitamin is by far the most frequent cause of its deficiency in the body.

Folic acid is another vitamin, unrelated to B_{12}, whose deficiency causes anemia by arresting the development of red blood cells. The effects of folic acid deficiency are confined to the anemia itself and do not involve other tissues in the way that vitamin B_{12} does.

The most important form of anemia associated with vitamin B_{12} deficiency is *pernicious anemia*. An incurable disease until the 1920s, it develops chiefly in persons over 45 years of age, as a result of gradual atrophy of the parts of the stomach lining that produce intrinsic factor, the essential protein that makes absorption of the vitamin possible. Usually this condition is associated with failure to produce hydrochloric acid; it may "run in families," and the incidence of stomach cancer in pernicious anemia patients is unusually high.

As contrasted with the disturbed nuclear metabolism, the formation of hemoglobin is not affected in pernicious anemia. The red blood cells become overcharged with hemoglobin, and the anemic condition is characterized by the presence of numerous abnormally large blood cells (macrocytes). However, because of the deficient production of red cells, the overall hemoglobin content of the blood is seriously reduced, and there are all the usual signs of anemia, such as weakness, tiredness, pallor, breathlessness and palpitation. Frequently there is also jaundice, with yellowish skin and puffy face. Since the metabolism of the nervous system is also dependent on vitamin B_{12}, pernicious anemia may—as distinct from other types of anemia—be accompanied by sensations of tingling of the hands and feet ("crawling ants"). Less frequently there are muscular weakness, paralysis of muscles, and defective coordination of muscular movements. There may also be signs of mental disturbance. Disturbed development of the cells, moreover, manifests itself in deterioration of the mucous membrane of the tongue, which eventually takes on a smooth red appearance and feels sore. Pernicious anemia is treated by giving the patient regular doses of vitamin B_{12} as intramuscular injections. This treatment controls the disease quite effectively, but in severe cases it cannot entirely rectify defects that have occurred in the nervous system.

WHITE BLOOD CELLS (LEUKOCYTES)

In contrast with red blood cells, white blood cells (leukocytes) are nucleate cells of various kinds and sizes. They tend to move along at the outside of the bloodstream, i.e., close to the walls of the blood vessels. These cells can change their shape, can move spontaneously (whereas red cells are carried passively by the blood) and can in certain circumstances make their way out of the bloodstream. Normally there are about 5000–10,000 leukocytes per cubic millimeter of blood. Three types of leukocytes are to be distinguished with regard to shape, function and origin (Fig. 3-5 illustrates the various types and indicates their relative numbers). *Granulocytes* (about 70 percent of the total number of white blood cells) are so called because their cytoplasm contains granules; the nucleus is of somewhat elaborate segmented appearance, and for this reason these cells are known also as polymorphonuclear leukocytes. According to the staining properties of the granules by certain dyes, granulocytes are classified into three subtypes: neutrophil, eosinophil, and basophil granulocytes. *Lymphocytes* (about 25 percent of the white blood cells) are smaller, about the same size as red blood cells, and have a large round nucleus. *Monocytes* (about 5 percent) are much larger, about three times the size of red cells, and have a lobed nucleus.

The relative numbers of the different types and subtypes of leukocytes can provide indications of diseased conditions and of the various stages of a disease. Monocytes and neutrophil granulocytes are able to make their way out of the vascular system and to destroy bac-

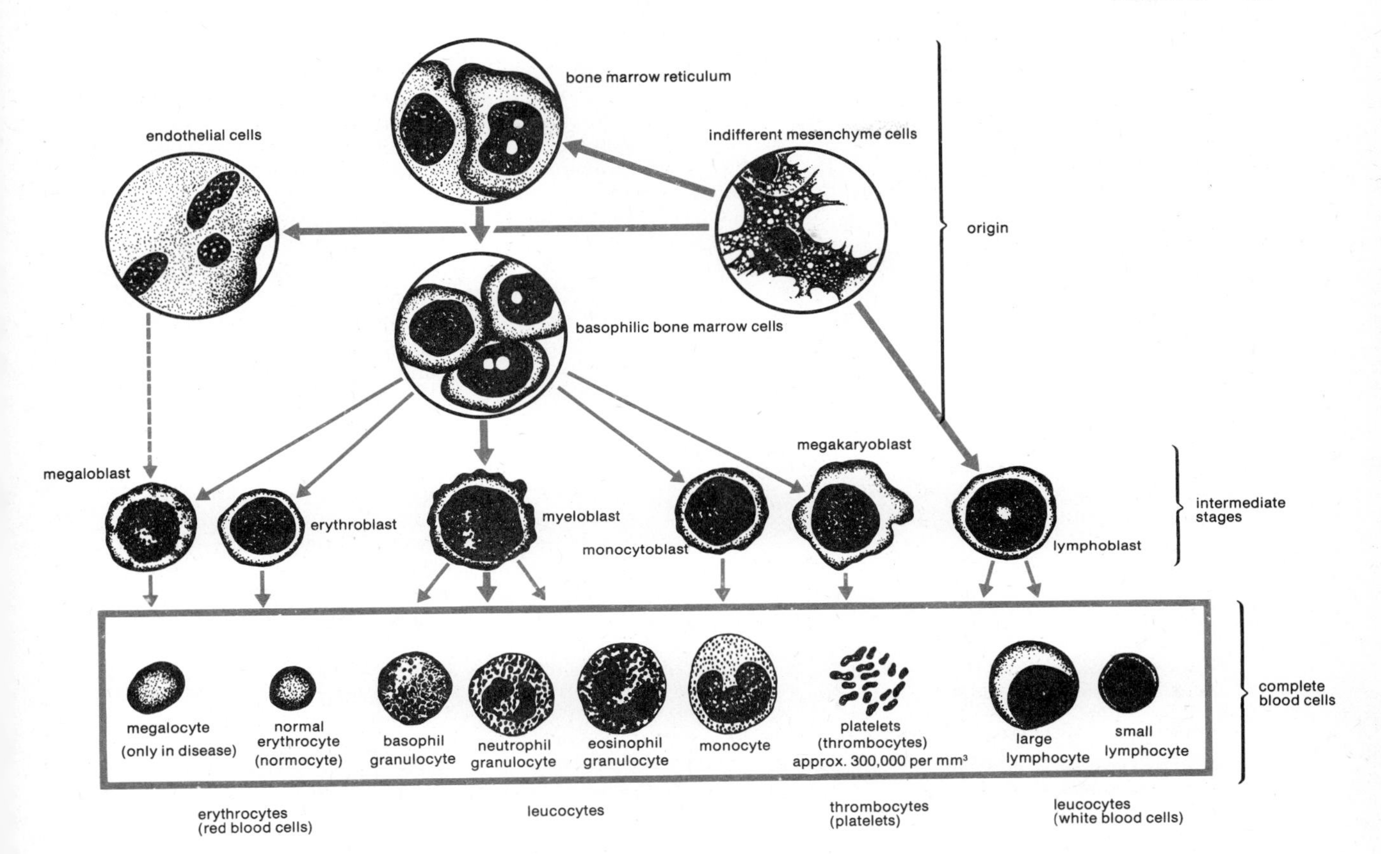

FIG. 3-5.
Development of blood cells.

teria by engulfing them (phagocytosis). These leukocytes are attracted to an inflammation by chemical stimuli (chemotaxis); dead leukocytes and their remains form pus in the inflamed tissue. Eosinophil granulocytes occur in increased numbers in the blood particularly in a case of allergic disease. On the other hand, in the early stages of a feverish condition there is usually an overall increase in the number of leukocytes, but the number of eosinophil granulocytes decreases. The basophil granulocytes contain heparin, which is a substance that inhibits coagulation, and histamine, which lowers blood pressure. The lymphocytes occur in increased numbers particularly in response to infections. They are associated with specific defense and immune reactions (e.g., rejection reaction after transplantation, delayed-action allergy, see page 46). The granulocytes are formed in the bone marrow, the lymphocytes in the lymphatic tissues. The origin of the monocytes has not yet been fully clarified.

LEUKOCYTOSIS, LEUKEMIA

In infections there is increased production of granulocytes, of which immature forms (with rod-shaped nuclei) then occur in the blood. The presence of above-normal numbers of granulocytes is called *granulocytosis*. During the course of an infection there will also be an increase in the number of monocytes *(monocytosis)*. The lymphocytes, which play a part especially in connection with immunological processes, increase in number usually only in a later stage of the disease *(lymphocytosis)*. The general term for the increase in the num-

FIG. 3-6.
COURSE OF AN INFECTIOUS DISEASE OF SHORT DURATION WITH LEUKOCYTOSIS.

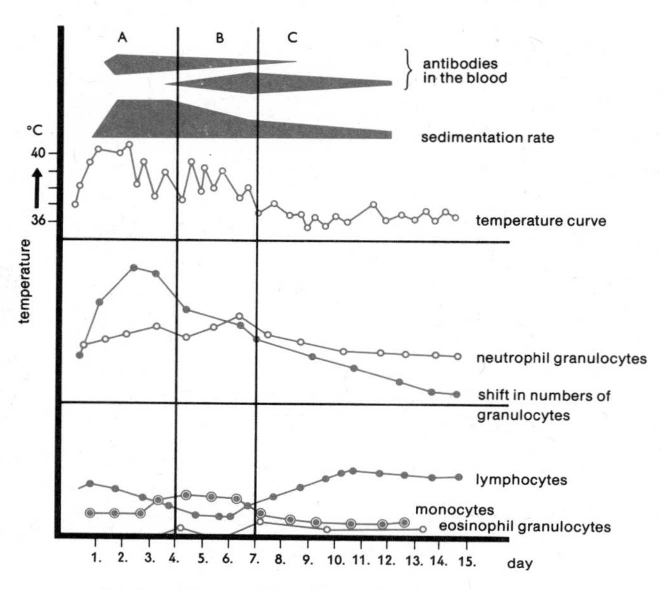

(A) Stage in which granulocytes increase in number.

(B) Stage in which monocytes increase in number.

(C) Stage in which lymphocytes and eosinophil leukocytes increase in number.

FIG. 3-7.
Frequency of types of leukemia by age.

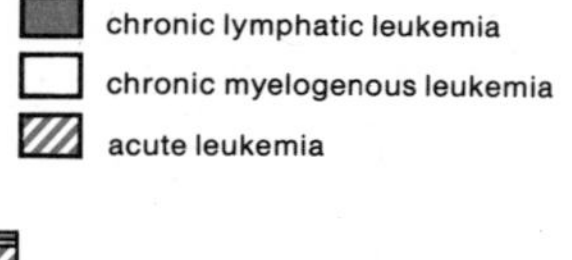

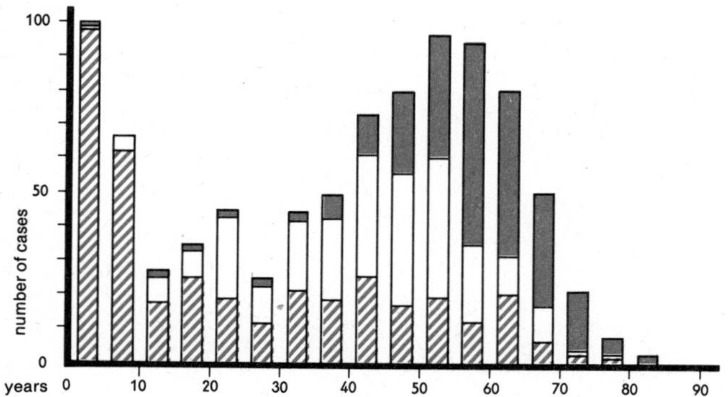

ber of white blood cells (leukocytes) in response to an infection is called *leukocytosis* (Fig. 3-6). At such times the number of white cells per mm³ of blood, normally between 5000 and 10,000, may become as high as 20,000–30,000. This is a natural defense reaction; when the infection subsides, the number of white blood cells returns to normal.

In the disease called *leukemia* there is an abnormal increase in the number of leukocytes. Although the cause is unknown, it can be described as cancer of the white blood cells. Viruses possibly play a part in causing the disease. More particularly, leukemia is a cancer-like proliferation affecting the centers where white blood cells are formed in the bone marrow, spleen and lymph nodes, causing leukocytes in excessive numbers to be released into the blood. In various forms of leukemia there is an abnormal increase in the granulocytes, the lymphocytes or the monocytes. At the same time, large numbers of immature white cells in various stages of development are formed in the bone marrow or the lymphatic tissue. For diagnosis, a specimen of marrow from the breastbone or tissue from lymph nodes may be examined. *Myelogenous leukemia* involves the blood-forming bone marrow; *lymphatic leukemia* is characterized by a marked increase in the size of the spleen and lymph nodes. Both types of the disease may be acute or chronic; the acute forms are commoner in children than in adults (Fig. 3-7), and in such cases immature leukocytes are often present in the blood. Leukemia characterized by large numbers of mature leukocytes occurs in patients of relatively advanced age, including especially the chronic form of lymphatic leukemia, in which the patient's life expectation is usually 3 to 5 years, sometimes as much as 10 years. On the other hand, untreated acute leukemia may be fatal within a few weeks.

The main sign of leukemia is an increase in the number of white blood cells to 100,000–200,000 per mm³ (counts of up to 300,000 per mm³ have been reported). Sometimes, however, there is no very marked increase of white cells in the blood; but in such cases the bone marrow contains abnormally large numbers of immature white cells. These "clog" the marrow and obstruct the formation of red blood cells and platelets, so that the patient becomes anemic and develops symptoms of platelet deficiency (blood will not clot properly; purpura, i.e., bleeding under the skin, occurs). As the abnormally numerous white blood cells are immature or incapable of performing their normal functions, especially in defending the body against infection, the disease is often complicated by infections against which the white cells provide little or no protection and which may themselves therefore prove fatal. In myelogenous leukemia particularly the spleen and in lymphatic leukemia particularly the lymph nodes are greatly enlarged. In monocytic leukemia, a rarer form of the disease, there is especially an abnormal increase in monocytes; symptoms are severe hemorrhages and swelling of the gums.

There is no known cure for leukemia. Treatment consists in radiation (x-rays, radioactive phosphorus) of the spleen and the lymph nodes, or sometimes of the whole body, in order to arrest the formation of abnormal cells. Cytostatic drugs such as aminopterin, which suppress the growth and proliferation of cells, are also given. Blood transfusions and anti-infective precautions help to combat the anemia and provide a defense against infectious disease.

AGRANULOCYTOSIS

The number of white cells (leukocytes) in the blood may be temporarily increased as a normal response to infection or be abnormally and excessively increased as in leukemia. Conversely, there may be an abnormal decrease in the number of white cells, a condition called *leukopenia*. The number of granulocytes among the white blood cells (see page 108) may be reduced as a result of shortened lifespan or obstruction of the producing centers in the marrow. A mild deficiency may be harmless, but in an acute form the disease, characterized by absence or severe shortage of granulocytes and then known as *agranulocytosis* (or *granulocytopenia*), is very dangerous. If leukopenia consists more particularly in a deficiency in the number of lymphocytes, it is called *lymphopenia* and frequently occurs in the early stages of infections, together with a decrease in the eosinophil granulocytes; it is found also in cases of overactivity of the adrenal cortex.

Agranulocytosis due to suppression of the normal activity of the bone marrow is a dangerous disease, often of sudden occurrence, which can, however, usually be treated with some success if detected in time. The white blood cell count, normally 5000–10,000 per mm³, drops to 800–1000 per mm³, so that defense against infection is greatly weakened. As a result, influenzalike infections, sore throat and tonsillitis (agranulocytic tonsillitis) are liable to develop. At a later period in the disease the mouth, throat, rectum and genitals may become ulcerous, while the condition is further attended by accelerated pulse rate and weakness (Fig. 3-8).

Suppression of the activity of the marrow may be due to oversensitivity to viral or bac-

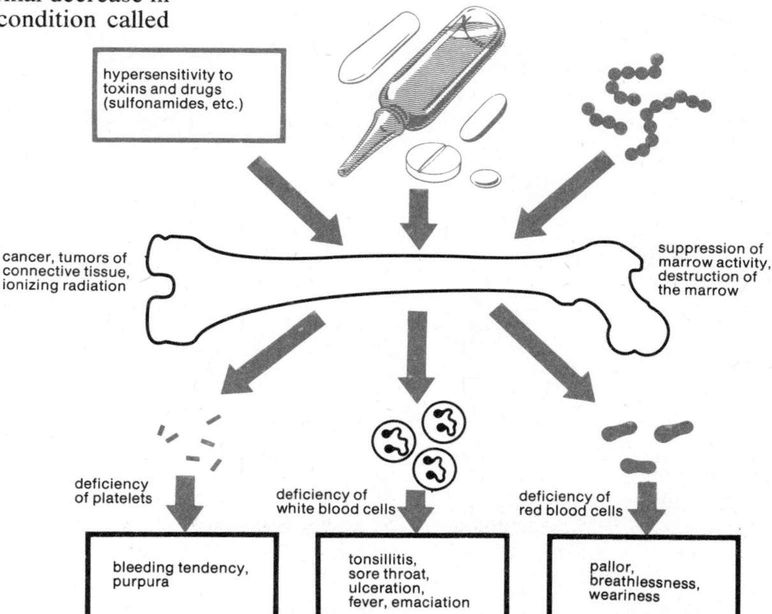

FIG. 3-8.
Characteristics of agranulocytosis.

terial toxins or may be brought about by radiation or, in certain patients, by the action of particular drugs. Thus the occurrence of severe sore throat during a course of treatment with sulfonamides, thyroid inhibitors, antispasmodic agents, and other drugs must always be regarded with suspicion.

Treatment of agranulocytosis consists first in stopping the intake of harmful drugs, so as to give the marrow an opportunity to recover. Infections can be treated or prevented with penicillin or other antibiotics and with blood proteins which strengthen the defenses. Sometimes hormones isolated from the cortex of the adrenal gland are given, and blood transfusions to replace the shortage of white cells may be necessary.

Sometimes the suppression of marrow activity includes the formation not only of white blood cells but also of red blood cells and platelets. This condition involving a reduction in all cellular elements of the blood is known as *pancytopenia*. The symptoms are then those of anemia (tiredness, pallor, breathlessness) and of platelet deficiency (bleeding tendency, especially under the skin: purpura). The treatment of this condition is generally similar to that of agranulocytosis.

Besides such temporary suppression of activity of the bone marrow in producing blood cells, there are instances of an entirely different kind where actual destruction or deterioration of the marrow occurs. This is sometimes due to cancerlike proliferations of the marrow, which may become atrophied (wasted) or calcified (osteopetrosis: "marble bones"). In advanced cases it may be attempted temporarily to make up for the deficiency of blood cells by means of transfusions. For the rest, treatment must, as in acute agranulocytosis, consist in giving protection against infection.

LYMPHATIC SYSTEM, HODGKIN'S DISEASE

Chyle is the milklike fluid in the lacteals and lymphatic vessels of the intestine (see page 166) and contains the products of digestion, especially absorbed fats. After the chyle enters the lacteals it is known as *lymph*. Fluid exuded from the blood vessels into the tissues is collected and returned to the bloodstream by the *lymphatic vessels* (also known as *lymphatics* or *lymph vessels*). These convey the lymph toward the heart and contain valves like those in

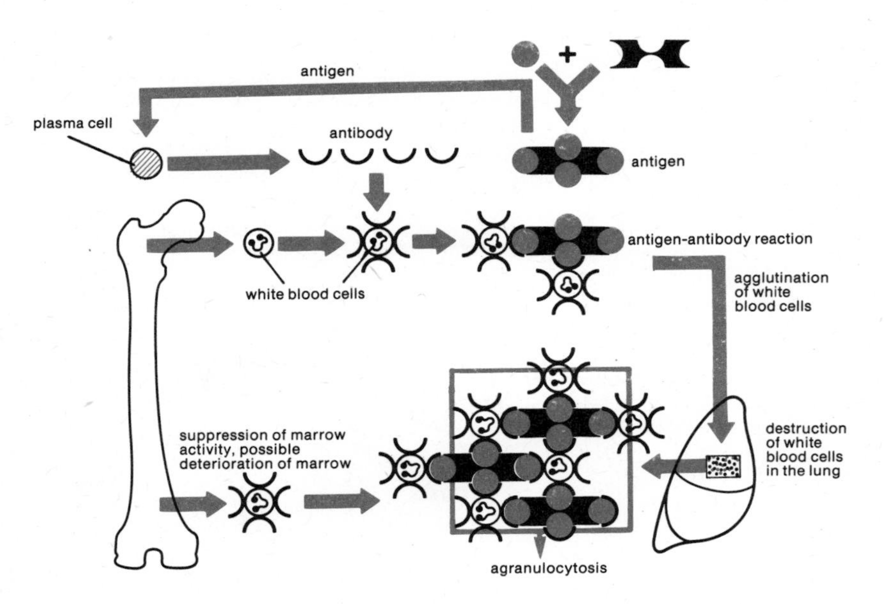

FIG. 3-9.
Development of allergic agranulocytosis.

the veins. They thus form a second pathway for fluid returning from the tissues, the veins being the first pathway. Unlike blood, however, lymph flows in one direction only. The lymphatic vessels unite to form larger vessels, which in turn form two main ducts (thoracic duct and right lymphatic duct) which discharge the lymph into the great veins at the root of the neck, so that the lymph becomes part of the bloodstream. Before entering the blood, the lymph is filtered through one or more *lymph nodes* (lymph glands).

Lymph is generally a clear colorless fluid, except in the lacteals draining from the intestine, where the presence of absorbed fats gives it a milky appearance. Its composition may vary considerably in different parts of the body. In peripheral lymph vessels it is similar to blood plasma, though with a lower protein content. The normal components of lymph are proteins, salts, organic substances, and water, as well as cells (notably lymphocytes) formed in the lymph nodes and other lymphatic organs. Lymph is formed in tissue spaces all over the body. It indeed resembles the tissue fluid that bathes all the cells, but is inside the closed system of lymphatic vessels and lymphatic capillaries. Tissue fluid drains into the latter by filtration and diffusion.

Lymphoid (or lymphatic) tissue produces certain cells (mainly lymphocytes, also monocytes and plasma cells) and acts as a filter for lymph and blood. The lymph nodes are composed of lymphoid tissue, which is found also in the tonsils and adenoids, along the digestive tract, and in the spleen. The nodes act as filters which stop bacteria and other foreign particles, including cancer cells; in addition, they play an important part in the immune response, along with other lymphoid tissue, notably in the spleen. Acting as barriers, the lymph nodes prevent the spread of bacterial infection from the tissue spaces into the bloodstream. The nodes occur in groups, particularly in the armpits and groin, at the root of the lungs, at the sides and back of the neck, and around the large veins in the abdomen. The *thymus* is a glandular organ belonging to the lymphatic system. It lies behind the upper part of the breastbone. Its cortex consists of dense lymphoid tissue. In young children the thymus is large and consists mainly of thymocytes (developing lymphocytes) and is concerned with the development of immunity. After puberty, the thymus gradually diminishes in size and eventually consists merely of fatty tissue. The *spleen* is an important organ not only of the lymphatic but also of the *reticuloendothelial system,* which comprises phagocytic cells in the various parts of the body—i.e., cells that engulf and destroy foreign matter (bacteria, worn-out tissue, spent blood cells). Broadly speaking, the spleen consists of blood-filled lymphoid tissue. In the embryo the blood cells are formed in the spleen, thymus and liver. Subsequently the spleen becomes concerned mainly with the destruction of spent or defective red blood cells; it also forms antibodies to maintain immunity against disease.

Hodgkin's disease is an uncommon disorder of the lymphoid tissue, especially the lymph nodes and spleen. The cells of the tissue proliferate, the lymph nodes become enlarged, and the patient's immunity against infection is impaired. The disease is not attributable to any detectable microorganism and is regarded as a particular form of cancer. Swelling of the lymph nodes of the neck is usually the first symptom that appears. The disease, which is malignant and progressive, causes anemia; the patient grows weaker and weaker, with death sometimes occurring after only a few years, though with modern treatment many patients survive for very much longer periods. Although incurable, it is treatable by radiotherapy (irradiation with x-rays, radium, etc.) and to some extent also by drugs such as nitrogen mustard, used also in the treatment of leukemia, with which Hodgkin's disease has certain features in common.

BLOOD GROUPS, RHESUS FACTOR

When blood from one person is injected into the bloodstream of another, *agglutination* (clumping) and destruction of the injected "foreign" blood cells may occur. This agglutination is due to defense reactions in the recipient's blood in response to the intrusion of the donor's blood. The underlying cause is incompatibility of the properties of different blood groups (blood types); these properties relate to hereditarily determined antigens which are contained especially in the enveloping membranes of the red blood cells. In certain cases the antigens are attacked by specifically incompatible antibodies of the other

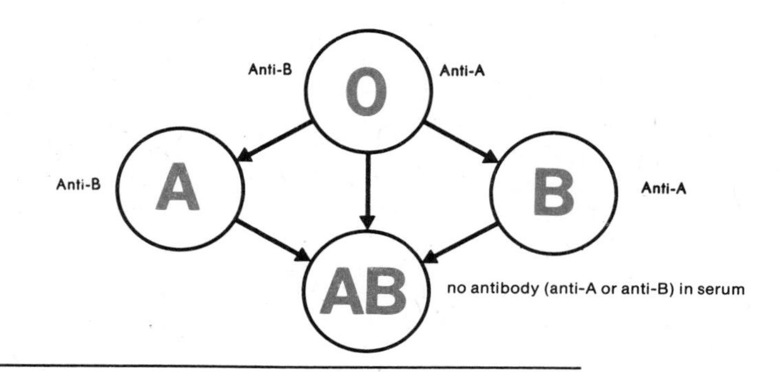

FIG. 3-10. ABO system.

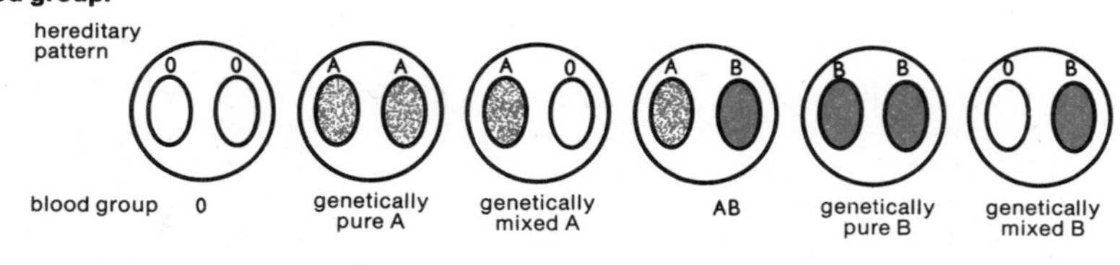

FIG. 3-11. Inherited combinations of blood group.

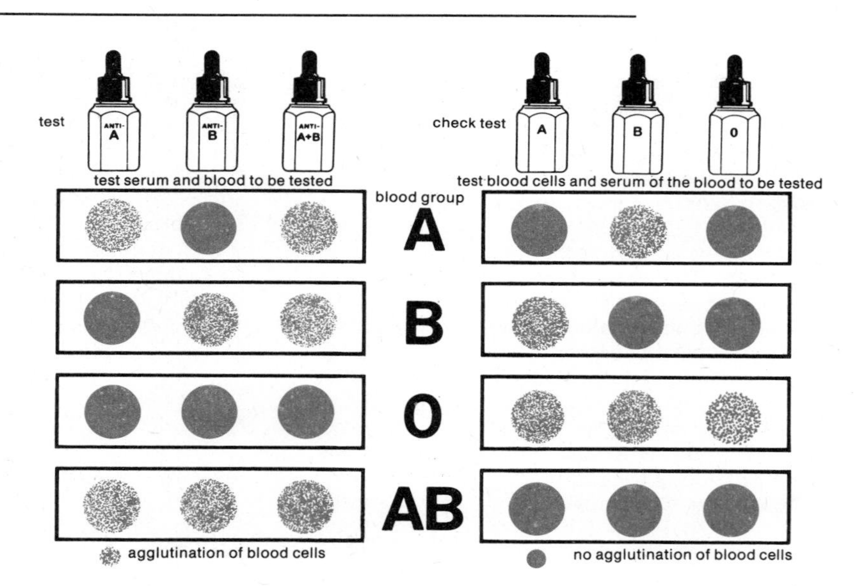

FIG. 3-12. Testing for blood group.

person's blood. The antigenic properties associated with a particular blood group are not, in fact, confined to the cellular elements of the blood, but extend to almost every cell in the body. It would thus be more correct to speak in general of cell group properties rather than blood group properties. Indeed, in certain persons these specific properties are detectable even in secretions such as saliva or sweat.

Of most practical importance is the so-called *ABO system* comprising blood groups A, B, AB and O. The two antigens associated with this system are designated by the letters A and B. If the red blood cells (and other cells) of a person contain the hereditary antigen A, his blood is said to belong to group A. If the antigen B is present instead, the person has blood of group B. If no antigen A or B is present, the blood belongs to group O. In each of these groups the blood plasma contains certain antibodies which characteristically attack only the blood cells of "foreign" blood groups. For example, the plasma of the blood group A contains the antibody "anti-B," which agglutinates blood cells of group B. Similarly, the plasma of blood group B contains "anti-A." In blood group O the plasma contains both antibodies—i.e., "anti-A" and "anti-B" are present. In blood group AB, on the other hand, the plasma contains no antibodies. The antibodies of the ABO system are so-called preformed antibodies, which develop in the very young infant in the first ten days after birth and increase up to the age of about 10 years.

Inheritance of the ABO system conforms to Mendel's laws. The A and B properties are dominant over O, which in this respect behaves genetically as a blank. For example, a person who inherits A from each parent, and a person who inherits A from one and O from the other parent, both belong to blood group A (combinations AA and AO, respectively). In all, there are six combinations: AA, AO, BB, BO, AB and OO, corresponding to the four groups A, B, AB and O. In a case of disputed paternity the hereditary patterns of the ABO system can sometimes provide definite proof that a person belonging to a certain blood group cannot be the father. The four groups of the ABO system are present in varying proportions. In Britain, for instance, 46 percent of the population belongs to group O, 42 percent to group A, 9 percent to group B and 3 percent to group AB. In Central Europe the corresponding proportions are 40, 40, 13 and 7 percent.

Determining the blood group (blood typing) is done by placing a drop of test serum A (containing anti-B) and a drop of test serum B (containing anti-A) on a slide; a drop of the blood to be tested is then added to each test serum drop (Fig. 3-12). According to whether or not agglutination of the red blood cells occurs, the blood group is determined as follows:

Test Serum A	*Test Serum B*	*Blood Group (or Type)*
–	–	O
–	+	A
+	–	B
+	+	AB

(+ denotes agglutination, – denotes absence of it)

Blood typing is especially important in connection with *transfusion*. If the patient (the recipient) is given blood against which his own blood has antibodies, the transfused blood will be agglutinated. The destroyed red cells set up a dangerous reaction, possibly with fatal consequences. In order to rule out possible mistakes in blood typing, the tests are duplicated and, in addition, a cross-test between the donor's and the recipient's blood is carried out before transfusion. Recipients belonging to group AB have been called "universal" recipients because their serum will not agglutinate cells of any other blood group; similarly, donors of group O have been called "universal donors" because in principle their blood can be transfused without danger of harmful reactions into persons belonging to any other group. These designations are now regarded as misleading. There are many blood-type factors in addition to those of the ABO system, and it must therefore not be assumed that group O blood can be given to persons of a different blood group unless further tests of compatibility are made. In modern practice it is usual to transfuse blood only between donor and recipient both belonging to the same group. Besides the ABO system of blood groups, the rhesus system involving the so-called Rh factor (rhesus factor) is important (see below). There are also other systems, such as the P system, the MN system, and the systems associated with the leukocytes and thrombocytes (platelets). They are, however, of practical significance only in special cases, e.g., where multiple transfusions are given.

The *rhesus system* is so named because of a factor (antigen) discovered in the red blood cells of the rhesus monkey and present also in about 85 percent of human beings. Such persons are said to be Rh-positive. In the other 15 percent the Rh factor is absent, such persons being Rh-negative. In contrast with the antibodies anti-A and anti-B which are preformed, i.e., present almost from birth, the antibodies against rhesus antigen are formed in response to sensitization. Thus, if Rh-positive red blood cells are injected into an Rh-negative person, the Rh factor acts as a "foreign" substance and calls forth a defense reaction; antibodies (anti-Rh) are formed to destroy the injected red cells. In a first transfusion, this process of antibody formation is too slow to cause serious trouble. But if the recipient subsequently receives another transfusion of Rh-positive blood, the antibodies which he has meanwhile formed will destroy the red cells in the transfused blood. Sensitization that produces the antibody in an Rh-negative person may also occur as a result of pregnancy. The Rh factor is inherited, so that in a case where an Rh-negative woman is fertilized by an Rh-positive man, she may become sensitized by the blood of the Rh-positive fetus at the time of birth, so that she will then produce antibodies against the Rh factor. In a subsequent pregnancy, if the fetus is again Rh-positive, these antibodies in the mother's blood may filter across the placenta and destroy the red blood cells in the unborn infant.

LOSS OF BLOOD, BLOOD REPLACEMENT

The normal volume of blood in a human being corresponds to 7 to 8 percent of the body weight, the average quantity in an adult being 5 to 5.5 liters (4½ to 5 quarts). The body is able to compensate fairly easily for a loss of up to 20 percent of this volume. The peripheral blood vessels, especially in the muscles, are constricted so that the remaining blood is so distributed that the vital organs (brain, heart, kidneys) are still adequately supplied. If the loss exceeds about one third of the total volume of blood, however, the vascular system can no longer be kept properly filled. The blood pressure falls, and the pulse rate increases from normally about 80/min to 110–120/min. The patient feels weak and cold and breaks out in a sweat. Blood pressures around 70 mm of mercury and a pulse of 140/min usually are signs of acute danger to life. They indicate that about half the total volume of blood has been lost from the system and that a state of shock is imminent.

External indications of circulatory shock due to loss of blood are bluish pallor, cold skin (especially on the nose and fingers), cold sweat, a burning thirst. Dyspnea (air hunger) is increasingly severe as more red blood cells, the oxygen carriers of the blood, are lost from the system. The victim's breathing is usually increased; nausea and vomiting frequently occur; there is retention of urine. The victim usually remains conscious for a long time, feels anxious and afraid, loses interest in his surroundings, and finally lapses into unconsciousness.

Treatment of the shocked condition aims at making up for the loss of fluid volume in the circulatory system and replacing the lost red blood cells. Both these objects are fully achieved by *blood transfusion,* for which blood of the same group as the patient's own is normally used (see pages 115 ff). As an interim emergency measure a transfusion of *plasma,* containing no blood cells, may be given instead of whole blood. Freeze-dried plasma in powder form can be stored for months without refrigeration. A so-called *blood substitute* may alternatively be used; this is a synthetic solution (so-called plasma expander) which increases the plasma volume in the recipient's blood vessels and enables his remaining red cells to be utilized as efficiently as possible. These are substances with molecules corresponding in size to those of certain plasma proteins and capable of remaining in circulation for a sufficient length of time. At present, three such substitutes are widely used: a high-molecular synthetic-resin-type product, a biosynthetically produced polysaccharide broken down to the desired molecular size, and chemically converted gelatin.

HEMORRHAGE (BLEEDING)

Loss of blood due to hemorrhage occurs as a result of an injury or of pathological changes affecting the walls of blood vessels. Depending on the type of blood vessel from which the blood is discharged, a distinction is drawn between arterial, venous and capillary hemorrhages. If injury affects only the capillaries, blood oozes in a blob from the wound. When a vein is ruptured, a steady flow of dark red blood is discharged. Blood from a ruptured artery comes out in rhythmic spurts and is of a bright red color. Furthermore, internal and external hemorrhages are to be distinguished. Whereas an external hemorrhage is immediately detected, an internal one may result in substantial losses of blood before the consequences become manifest.

Minor hemorrhages are arrested within a few minutes by the natural process of coagulation (clotting, see next section). This process cannot, however, cope unaided with more serious cases of bleeding, especially from ruptured arteries, because the blood is discharged from the wound at such high pressure that clotting is prevented. In such circumstances it is necessary to take appropriate action to stop the bleeding. Sometimes it will suffice merely to elevate the injured part of the body and apply a *pressure bandage* (Fig. 3-13). For this purpose a dressing pad or, in an emergency, a folded handkerchief should be pressed against the wound and be held in position with a firmly wrapped bandage. To deal with a major arterial hemorrhage it will usually be necessary to

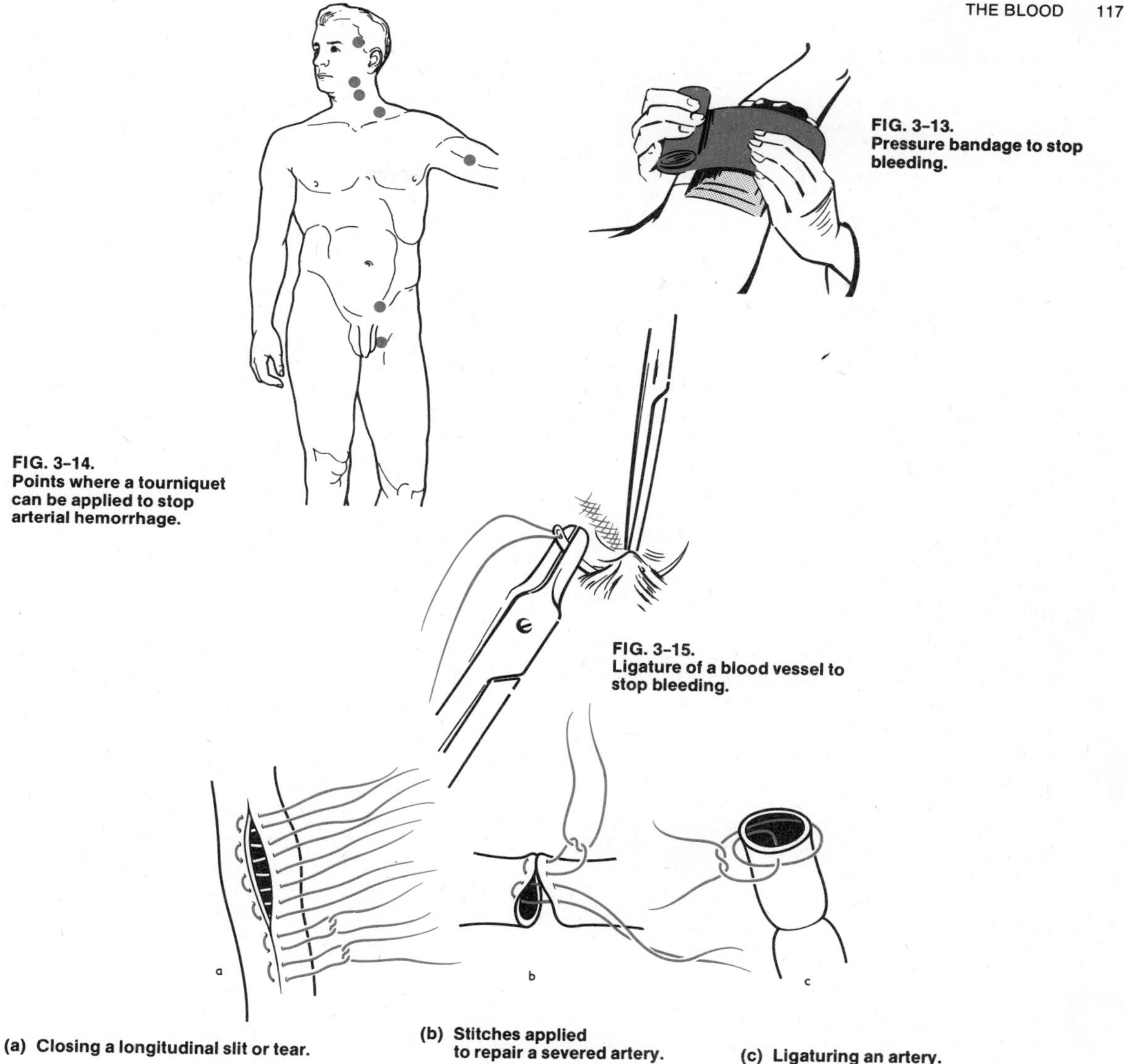

FIG. 3-13.
Pressure bandage to stop bleeding.

FIG. 3-14.
Points where a tourniquet can be applied to stop arterial hemorrhage.

FIG. 3-15.
Ligature of a blood vessel to stop bleeding.

(a) Closing a longitudinal slit or tear.

(b) Stitches applied to repair a severed artery.

(c) Ligaturing an artery.

FIG. 3-16.
SUTURES IN DAMAGED ARTERIES.

apply a *tourniquet* to the artery at a point between the heart and the wound. This can be done particularly at certain points as shown in Figure 3-14. Makeshift tourniquets consisting of a handkerchief, scarf, garter, stocking or belt must not be too tight, as this could cause permanent damage to adjacent nerves. Also, interruption of the blood flow for too long periods is liable to cause lack of oxygen below the tourniquet, which should therefore be released after 12–18 minutes, retightening if necessary.

A patient who has suffered a major loss of blood should be laid with his head and upper part of the body in a low position and with the limbs elevated, in order to utilize all remaining blood to supply the brain and heart. As a general rule, the victim of a serious hemorrhage should be removed to a hospital as speedily as possible. In a case of even relatively minor hemorrhage associated with injuries, it is generally advisable to call a doctor. If the location of the bleeding is internal, e.g., in a miscarriage, it can be reduced by application of an ice bag.

For definitive treatment to stop hemorrhage, large blood vessels have to be secured with special clamps and be ligated (Fig. 3-15). Smaller ones may be closed by compression or sealed by electrocautery with an electric knife. Major blood vessels have to be sutured (stitched) (Fig. 3-16); a damaged portion may be replaced by a length of plastic tube. Hemorrhage involving a large area, as in a ruptured spleen or liver, is first covered with gauze dressings saturated in a salt solution; as such wounds are difficult to close by suturing, they may be closed by having portions of the peritoneum stitched onto them. Hollow organs, e.g., the uterus or the nasal cavity, are packed with strips of sterile gauze to stop bleeding.

Chemical agents that arrest the flow of blood (hemostats) are used externally and locally (on wounds) or internally (in cases of disturbed clotting). Material based on gelatin or foamed fibrin is applied externally for the purpose, as are also the coagulation-promoting substances thrombin and fibrin. Small superficial wounds may be treated with an astringent (styptic), a substance that contracts the blood vessels, usually a salt of a metal such as lead, iron or zinc. Bleeding disorders—characterized by wounds that bleed for too long, spontaneous bleeding, e.g., blood discharged under the skin (purpura)—are treated with vitamin K or the deficient blood-clotting substances may be given.

CLOTTING OF BLOOD

Clotting (coagulation) serves the very important purpose of protecting the body against loss of blood. It is a highly complex enzymatic process which takes place in several stages (Fig. 3-17). The clotting mechanism is initiated as follows: When blood is shed from a wound, the substance called *thromboplastin* is liberated from the injured tissues and from degenerating platelets. Thromboplastin converts *prothrombin* (formed in the liver) to *thrombin*. This conversion constitutes the first stage of the actual clotting process, thrombin being the clotting enzyme. In the second stage the thrombin converts *fibrinogen*, which is a soluble substance, to insoluble *fibrin*. Fibrin forms a meshwork of fibers in which red blood cells are enmeshed, thus forming a clot. Then, in the third stage, the fibrin meshwork slowly contracts and squeezes out a clear yellow fluid called *serum* (which is blood plasma minus the fibrinogen) so that a firm clot is left. The conversion of fibrinogen to fibrin is governed by a number of factors. Although most of these are present in circulating blood, they are ordinarily prevented from acting by counterbalancing influences exercised by a fibrinolytic (i.e., having a dissolving or splitting-up action on fibrin) system. This latter action is developed also after temporary clotting in a blood vessel and thus causes the clot to dissolve with the aid of an enzyme called *plasmin*.

An important criterion in the diagnosis of clotting disorders, or bleeding disorders, is the *clotting time*. This is the time required for a small amount of blood to coagulate. In a test tube this time is normally 5–8 minutes. Natural clotting of blood from injured tissue takes place more quickly because here the blood platelets (thrombocytes) with their capacity to agglomerate (agglutinate) by sticking to one another and to the edges of the wound also play a part, as do moreover certain vascular reactions. Thus the actual *bleeding time*, i.e., the time required for blood to stop flowing from a small wound, is shorter than the clotting time.

Disorders of clotting—hemorrhagic or *bleeding disorders*—result from a deficiency of substances that promote clotting or from an excess of those that inhibit it. Also, the fibrinolytic system may be overactive. The symptoms of such disorders are that wounds continue to bleed for too long and that hemorrhage can occur spontaneously under the mucous membranes or the skin (purpura) or in other parts of the body.

Besides hereditary bleeding disorders (hemophilia, see next section; lack of fibrinogen) there are also acquired defects of clotting. Of most frequent occurrence among these is lack of prothrombin. Normally this substance is produced in the liver with the aid of vitamin K. Disturbance of fat absorption by disease of the intestinal mucous membrane or by obstruction of the flow of bile, which is essential for the absorption of fats (see page 189), may disturb the absorption of vitamin K (dissolved in fat) from the intestine. In another type of spontaneous bleeding disorder, too large amounts of the clotting substances are lost by slow internal bleeding or are consumed in the vascular system (e.g., in serious infections).

Clotting may be deliberately inhibited by drugs in order to prevent thrombosis and embolism or in a case of cardiac infarction. For instance, the coumarin drugs displace vitamin K from the liver and thus bring about a deficiency of prothrombin. Other substances, such as heparin, neutralize thrombin when it is formed. Such drugs obviously involve an increased risk of hemorrhage as a side effect.

Treatment of bleeding disorders consists in making up for a deficiency in any particular coagulative factor, in giving the patient doses of vitamin K, in using drugs to counteract the fibrinolytic system, and in avoiding trauma (injury or wound) as much as possible.

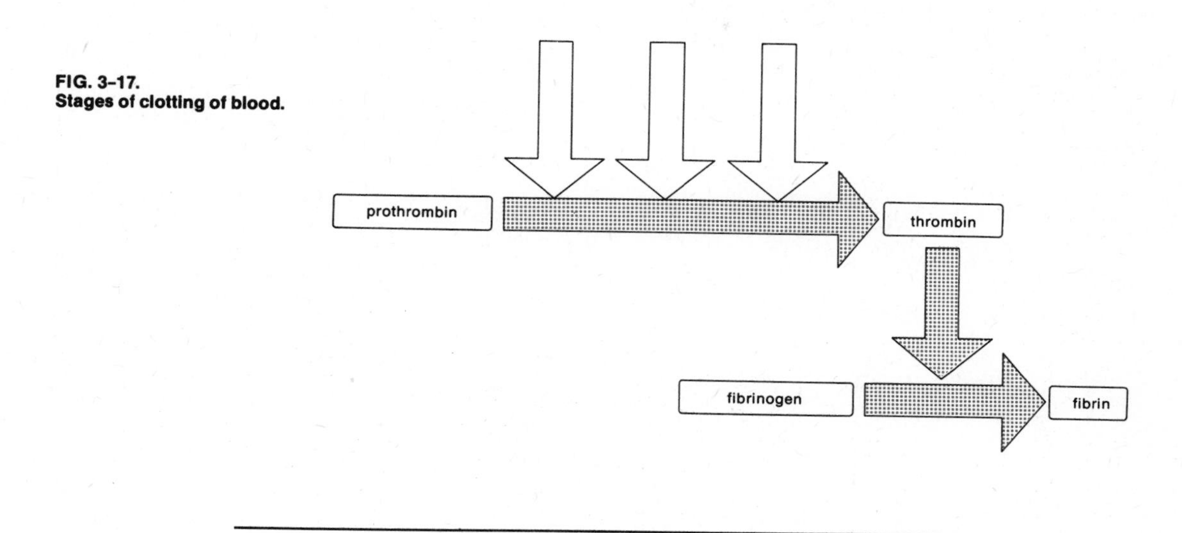

FIG. 3-17.
Stages of clotting of blood.

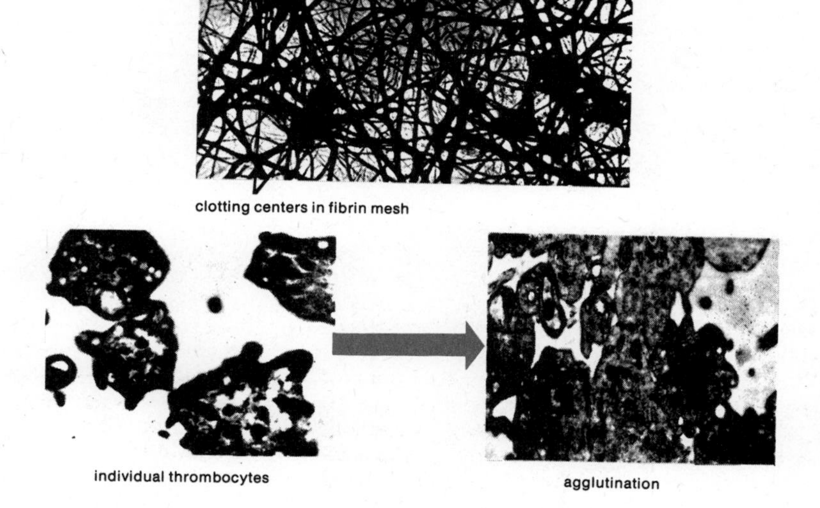

FIG. 3-18.
Clot formation.

HEMOPHILIA

Hereditary bleeding disorders are due to congenital deficiency or absence of a coagulative factor (antihemophilic factor). Most important of these disorders, besides some rarer forms of disease, is *hemophilia*. It occurs in two forms, which differ only in that in one of them the factor VIII is absent (hemophilia A), while in the other the missing factor is IX (hemophilia B). The missing factor in each form of the disease is a protein which the hemophiliac is unable to synthesize. Hemophilia is a rare disease: form A occurs in about 5 people per 100,000, while form B occurs in only 1 per 100,000. The disease is inherited as a sex-linked recessive character which manifests itself only in the male sex. Women, though they may be "carriers" of the disease, are not themselves affected by it. The sons of a hemophiliac will all be normal, his daughters will all be carriers. The sons of a carrier will have a 50 percent probability of being hemophiliacs; her daughters will have a 50 percent probability of being carriers. In the rare instance where a hemophiliac marries a carrier their children may include a female hemophiliac (also a carrier of the disease, Fig. 3-19).

Though in other respects apparently healthy, the hemophiliac is always in danger of developing a serious hemorrhage, because even quite minor wounds, or the bleeding that follows a tooth extraction or an operation, or even a knock or bump, may result in a dangerous loss of blood. There is a tendency to bleed internally without apparent cause. Characteristic of hemophilia is bleeding in the joints, especially at the knee, which occurs suddenly and in consequence of the least injury. Repeated hemorrhages in early childhood may result in a stiffening of the joint. Carriers, i.e., women in whom the disease is present as a recessive hereditary character, may also tend to have spontaneous hemorrhages and increased loss of blood during menstruation, because in some of these women the blood clotting time is longer than normal. On the other hand, not all hereditary hemophiliacs are so seriously affected by the disease: in some hemophilic families, factor VIII is not entirely absent, but is reduced to 15 percent of the normal amount; this is enough to prevent dangerous bleeding, which occurs only if factor VIII is reduced to below 5 percent.

There is no known cure for hemophilia. The hemophiliac must live very carefully so as to minimize the risk of external or internal bleeding. In an emergency, a transfusion of blood or plasma can temporarily supply the missing factor VIII or IX. Factor VIII does not endure in blood stored for some length of time in a blood bank, so that transfusions for hemophilia A require fresh blood. Higher levels of the missing factor in the hemophilic patient's blood can be attained by the transfusion of dried normal plasma or the purified hemophilic factor itself. Factor VIII survives in the recipient's blood for 4–8 hours, factor IX for about 20 hours.

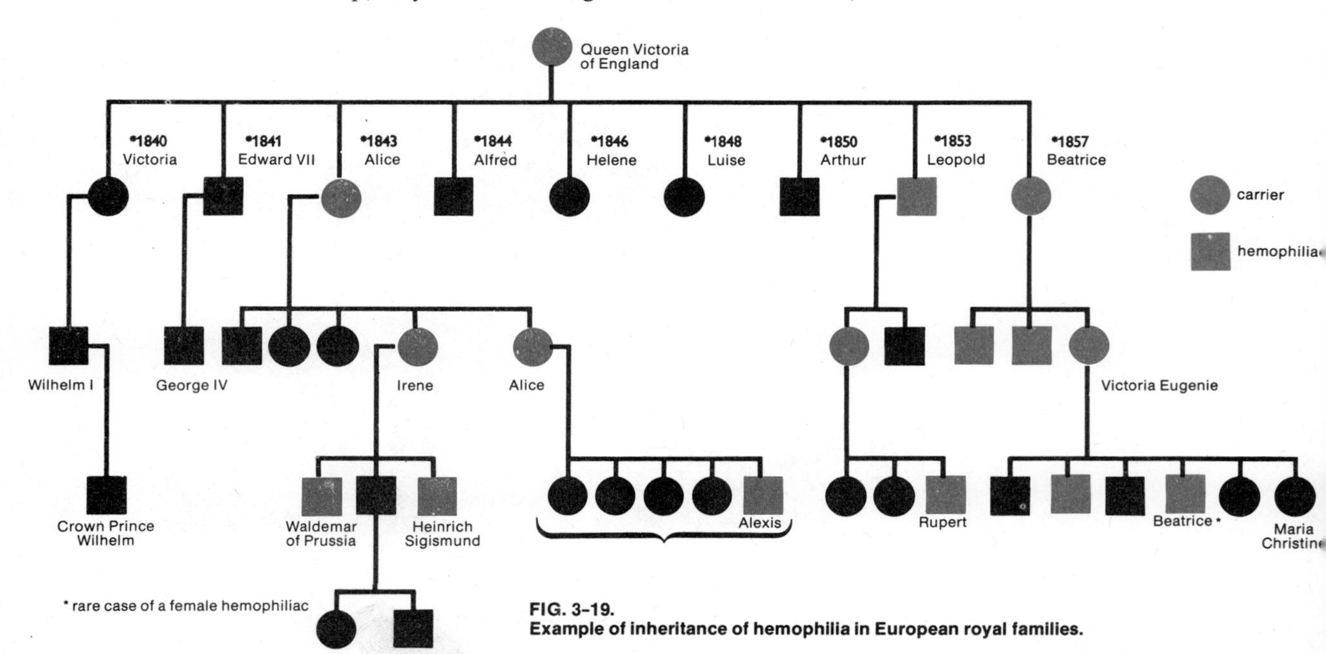

FIG. 3-19.
Example of inheritance of hemophilia in European royal families.

Part Four

THE RESPIRATORY ORGANS

NOSE, THROAT, WINDPIPE

Respiration comprises a complex series of interlinking functions which all make their contributions to the oxidative metabolism of the cells and thus to the energy balance of the body (Figs. 4-1 and 4-2). These include the exchange of gases between the blood and the external air (external respiration, pulmonary respiration), the transport of oxygen and carbon dioxide in the blood, the exchange of gases between the blood and the tissue cells, and finally the oxidation processes in the metabolism of the cells (internal respiration).

The organs that play an important part in external respiration are those whereby atmospheric air is introduced into the bronchial system and alveoli of the lungs: the nose with its accessory cavities, larynx, windpipe (trachea), bronchial system (air passages in the lungs), muscles of respiration, and thorax, together with the respiratory center (a region in the medulla oblongata which regulates the movements of respiration). The breathing of the skin is part of the external respiration, but accounts for no more than 1 percent of the oxygen intake.

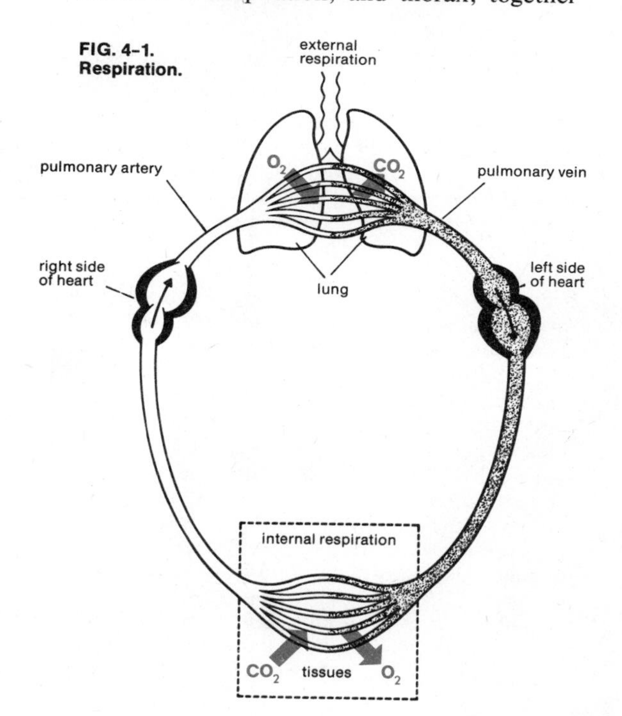

FIG. 4-1. Respiration.

The *nose,* as an auxiliary organ of respiration, tests the air for respiration and prepares it by warming, moistening and filtering (Fig. 4-3). The nasal cavity is divided into two not completely symmetrical chambers by the nasal *septum.* Each chamber is further subdivided into three passages separated by the nasal *conchae* (turbinate bones), which increase the surface area of the cavity and serve also to protect the olfactory epithelium and inner air passages. The conchae are so positioned that the inhaled air comes into contact with all parts of the internal surface of the nasal cavity. The bones surrounding the nose contain cavities (sinuses) which communicate with the nasal cavity by small openings in the side wall.

The interior of the nose is lined with *mucous membrane,* whose surface layer consists of ciliated respiratory epithelium, i.e., covered with microscopic hairlike structures (cilia) which perform undulatory motions that propel the mucus together with dust and any other foreign matter toward the passages at the back of the nose (choanae) which communicate with the pharynx, or forward toward the nostrils (Fig. 4-4). The epithelium contains goblet cells and tiny glands which secrete the mucus that serves to moisten the respiratory epithelium and the inhaled air. The lacrimal glands, which secrete tears, are also involved in this process; they communicate with the nasal cavity by the nasolacrimal duct. The nerve endings for the sense of smell are located only in a small area at the top of the nasal cavity. These so-called olfactory receptors form part of the olfactory membrane.

The mucous membrane of the nose contains a network (plexus) of veins which can swell or contract under the control of small muscles which can throttle down the flow of blood. This plexus serves to warm the air flowing past it; its swelling and contraction are reflex move-

FIG. 4-2.
Lung function in respiration.

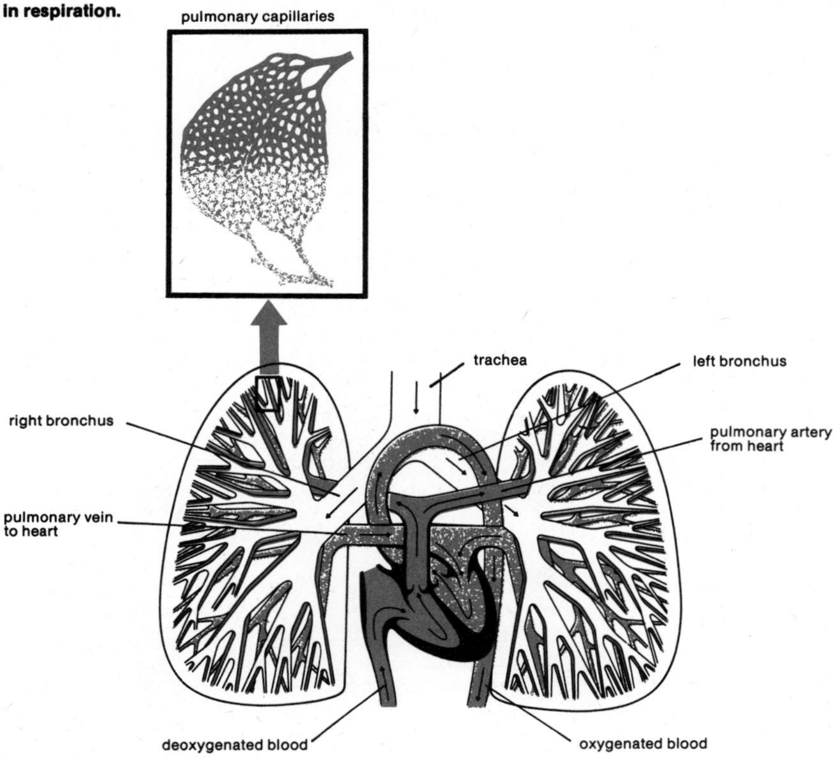

ments. Cold feet, for example, may cause contraction of the plexus, so that the mucous membrane can then no longer properly perform its warming function; this in turn may result in catching cold. Other stimuli cause distension of the blood vessels, accompanied by the secretion of watery fluid (running nose, see next section).

Closely associated with the nasal cavity are the paranasal *sinuses,* i.e., cavities in the adjacent bones (frontal, maxillary, ethmoidal and sphenoidal). They are lined with the same kind of epithelium as the nasal cavity itself and appear to perform the same functions: warming, moistening and filtering the air.

The *pharynx* is the space behind the nasal cavity, oral cavity and larynx. It thus comprises three sectors, one above the other: nasopharynx, oropharynx and laryngopharynx. The nasopharynx is the upper sector of the pharynx continuous with the nasal passages and is for breathing only; during swallowing, it is closed by the soft palate. The bottom sector is the laryngopharynx (laryngeal pharynx), which is concerned only with swallowing. The oropharynx, the intermediate sector, is a passage for air as well as food. The pharynx is surrounded by lymphoid tissue, more particularly the *adenoids* at the back of the nasopharynx and the *tonsils* between the pillars of the arch of the palate. The tonsils and adenoids are believed to have a role in filtering microbes from the inhaled air and aiding the formation of antibodies or white blood cells, but this function can hardly be an important one.

In the pharynx, more particularly in the oropharynx, the air passage and the food passage intersect. The *epiglottis* is a flap of cartilage located behind the root of the tongue; normally during breathing, the air passes behind the epiglottis and between the vocal cords in the larynx; during swallowing, the epiglottis is lowered from the upright position to cover the vocal cords and prevent food from entering the trachea.

The *trachea* (windpipe) extends from the larynx to the lungs. It is lined with mucous membrane whose epithelium (covering layer)

FIG. 4-3.
Nasal anatomy.

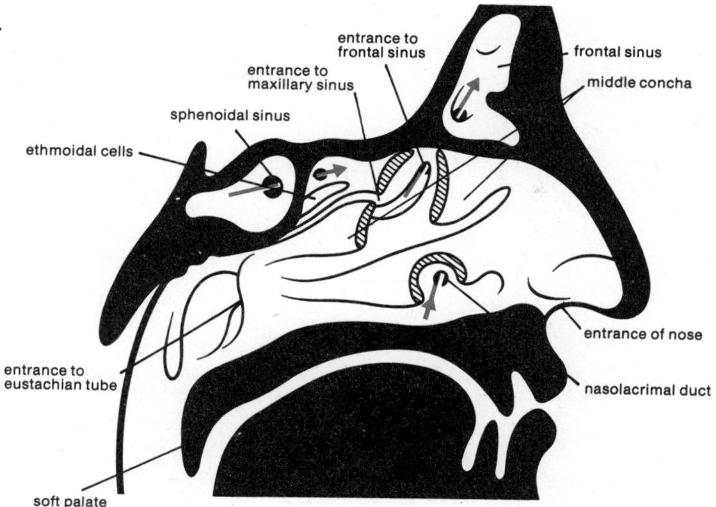

FIG. 4-4.
Propulsion of matter by nasal cilia.

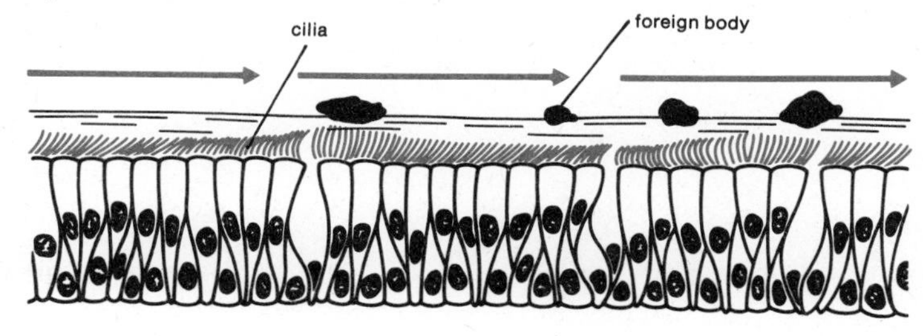

FIG. 4-5.
Foreign body in bronchus.

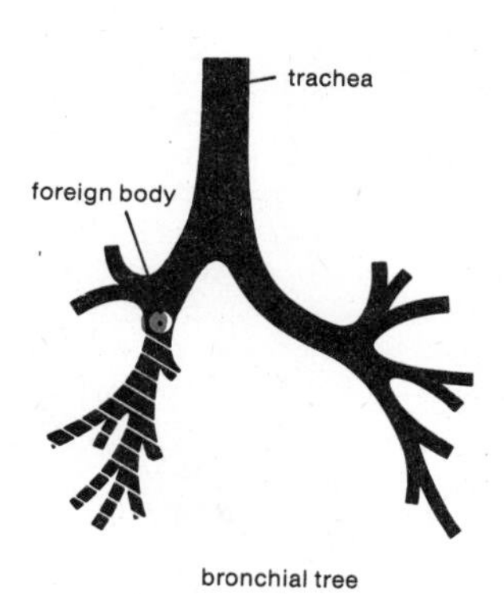

secretes mucus and is covered with cilia. The undulations of these fine hairlike structures maintain an upward movement of mucus. At its lower end the trachea divides into two main branches (bronchi), one for each lung. It is surrounded by cartilaginous rings which keep it permanently open, i.e., it does not become flattened when not in use, as the esophagus does. Normally the mucus secreted in the trachea is just enough to keep the mucous membrane moist. Irritation or inflammation of the membrane increases the secretion in the trachea and bronchi; on auscultation the presence of this mucus produces an abnormal sound (rale). Excess mucus may be expelled by expectoration. The right bronchus at the base of the trachea is somewhat less curved than the left, so that foreign bodies which accidentally find their way into the trachea are more likely to end up in the right bronchus (Fig. 4-5).

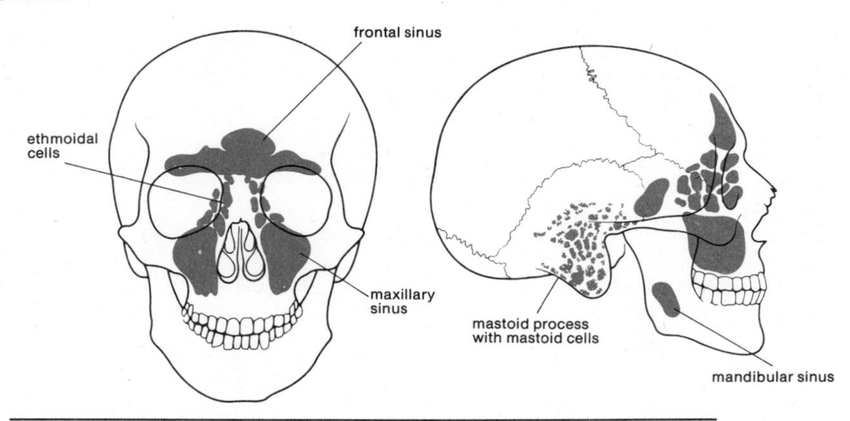

FIG. 4-6.
Paranasal sinuses.

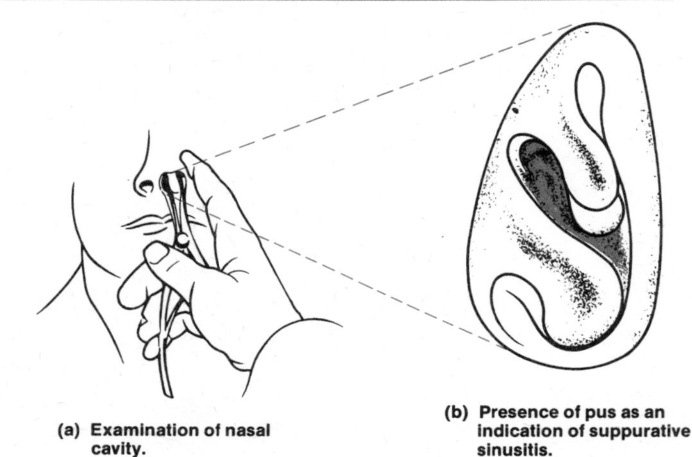

FIG. 4-7.

(a) Examination of nasal cavity.

(b) Presence of pus as an indication of suppurative sinusitis.

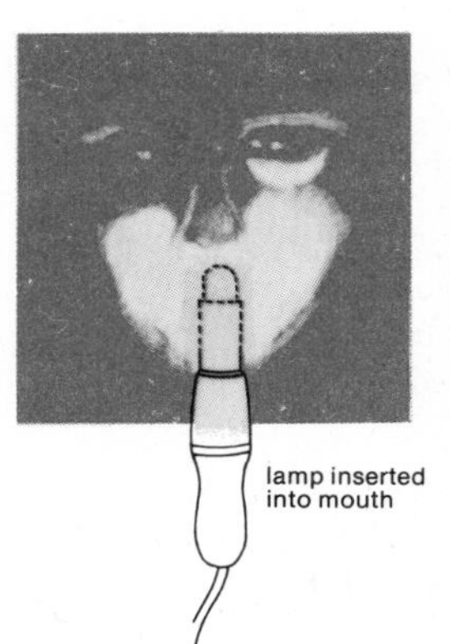

FIG. 4-8.
Reduced translucence due to pus in maxillary sinus.

COMMON COLD, SINUSITIS

Rhinitis is inflammation of the mucous membrane of the nose. Though often caused by an infection, a similar condition can alternatively be caused by irritants such as fumes, smoke or heat. So-called vasomotor rhinitis may be caused by neurovascular imbalance, especially by nervous impulses arising from emotional excitement. Hay fever results from an allergic reaction to a pollen antigen. However, rhinitis is generally caused by pathogenic organisms (microbes), e.g., as an attendant condition of influenza, measles or diphtheria, but more often—in the form known as the common cold—as a result of viral infection.

Over thirty types of virus (so-called *rhinoviruses*) are known to be responsible for the *common cold*. The infection is spread by droplets expelled when the patient sneezes or coughs. The viruses tend to vary in potency and may be more or less dormant on the mucous membrane of the nasopharyngeal cavity, where they can be activated by exposure of the patient to cold or wet conditions or sudden changes in temperature or even by the mere suggestion of "catching cold." More aggressive strains of virus will penetrate into the mucous membrane without waiting to be activated, where they cause reddening and swelling, while a thinly fluid mucus is secreted ("running nose"). If, in addition to viruses, bacteria also establish themselves in the mucous membrane, the secretion will contain pus and take on a greenish-yellow color. The patient's temperature usually does not rise above 38°C (100.4°F), but this may sometimes be a transitional condition to bronchitis, laryngitis and pharyngitis (see page 126) and to influenza. There exists as yet no specific treatment to neutralize the rhinoviruses. Fever can be kept down by antipyretics; and sweating induced by hot packs can prevent chill. Inhalants and various drugs which shrink the mucous membrane of the nose and thus make breathing easier can control only individual symptoms. Unfortunately, the immunity resulting from the common cold is feeble or short-lived, and the production of vaccines against it is made difficult by the fact that so many viruses can cause it.

Acute rhinitis may spread to the middle ear or the sinuses. A person may be repeatedly afflicted with rhinitis, which can thus become chronic in character. This sometimes leads to congestion of the mucous membrane on the conchae, to difficulty in breathing through the nose and finally to inflammation of the sinuses *(sinusitis)* (Fig. 4-6). Sinusitis may also, though more seldom, occur when the root of a tooth becomes septic or as the result of injuries or contact with chlorinated water. The most frequent inflammation of this type is in the maxillary sinuses, i.e., in the jawbone, but the others may also be affected. The symptoms of sinusitis are usually not very characteristic: fever, pallor, headache (especially when stooping). Maxillary sinusitis often causes frontal headache; but the pain may alternatively be centered in the upper jaw and radiate to the temples. In frontal sinusistis the frontal region is painful to external pressure. Suppuration in the sinuses causes typical streaks of pus (Fig. 4-7); it also shows up on x-rays and reduces translucence (Fig. 4-8). Acute sinusitis may lead to complications by spreading to adjacent bones and to the orbits (eye sockets); there is also a risk that it may spread inward and cause meningitis.

Sinusitis is treated with cold compresses, with nasal drops which shrink the mucous membrane, and by drainage of the sinuses. In persistently severe cases it may be necessary to use antibiotics to control the pyogenic (pus-forming) microorganisms. Infected sinuses may have to be rinsed out, with a sodium bicarbonate solution, for example. Chronic sinusitis, which is often due to the presence of a polyp, may require surgical treatment to remove the polyp; the sinus is cleaned out, and an enlarged opening into the nasal cavity (for ventilation and drainage) is formed. A polyp is a usually harmless tumor which is attached to a mucous membrane by a short stalk (pedicle) and looks like a small mushroom.

TONSILLITIS

Tonsillitis is inflammation of the tonsils. Children are sometimes affected by a catarrhal inflammation involving the entire lymphatic tissue of the pharynx, including especially the adenoids. Mostly this condition is due to an acute viral infection attended by strong fever, headache and vomiting. Streaks of mucus and pus on the reddened mucous membrane indicate that the adenoids are affected. There is usually also rhinitis, sometimes bronchitis. Suppuration of the lymph nodes on the rear wall of the pharynx may cause a so-called retropharyngeal abscess in young children, characterized by difficulty in breathing and swallowing and possibly becoming dangerous as a result of constriction of the respiratory passages. The abscess can be lanced or punctured to relieve the pus.

Tonsillitis is common especially in children and young persons, though no age is exempt. Exceptionally, tonsillitis may be caused by a viral infection or by spirillum bacteria (Fig. 4-9) or be associated with diphtheria. The microbes most frequently responsible for tonsillitis, however, are pyogenic streptococci. Hemolytic streptococci may cause scarlet fever,

FIG. 4-9.
Ulcerative tonsillitis (spirillum infection).

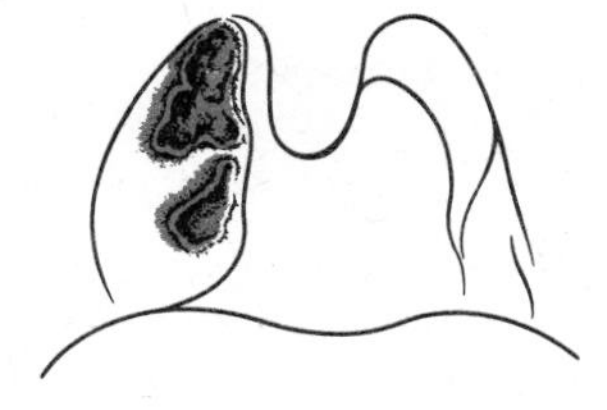

FIG. 4-10.

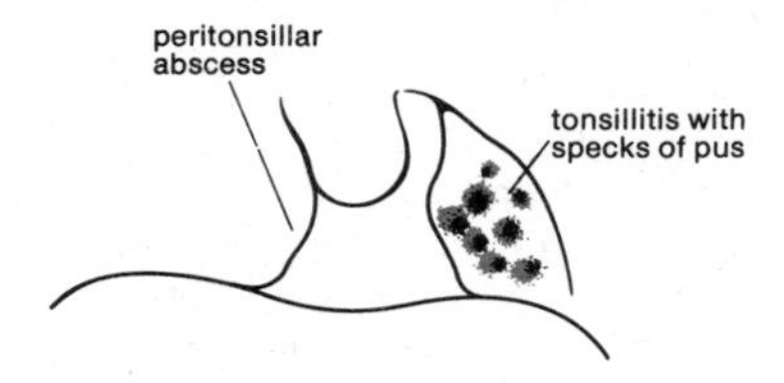

which is an infection of the pharynx accompanied by a rash (see page 465). *Pharyngitis* (sore throat) is one of the commonest of all infections. When it spreads and involves particularly the tonsils, it becomes tonsillitis. Streptococcal infection may spread and cause rheumatic disorders, cardiac inflammation affecting the heart valves, and acute nephritis (inflammation of the kidneys). Other possible complications associated with streptococcal and also with staphylococcal tonsillitis are suppurative otitis media (inflammation of the middle ear) and the formation of an abscess in the region of the tonsil (peritonsillar abscess). This latter condition is called *quinsy* and may be accompanied by one-sided inflammation of the soft palate, with difficulty in swallowing and in moving the jaw, so that the speech is slurred. If untreated, it can even lead to blood poisoning. The abscess (Fig. 4-10) can be treated with antibiotics; surgical incision may be required to release the pus.

There are good reasons not to treat acute tonsillitis lightly, even though it most frequently heals completely without any special medical treatment. The onset of tonsillitis is usually quite sudden, with painful swallowing, headache, fever, aching limbs, depression, and swelling of the lymph nodes of the jaw and neck. In young children there is strong fever, drowsiness, vomiting, sometimes stomachache. After two or three days the tonsils become covered with a whitish coating. In about four or five days the condition begins to improve. Besides antipyretics, antibiotics (e.g., penicillin) are given in the treatment of streptococcal tonsillitis. Nephritis occurring as a possible complication can be detected by blood pressure checks and urine tests. Peritonsillar abscesses may be lanced or punctured to extract the pus (Fig. 4-11).

Recurrent suppurative inflammation of the tonsils can turn into chronic tonsillitis. In that condition the tonsils may be enlarged but also scarred and may contain pus in deep crypts (cavities). In adults such chronically inflamed tonsils are suspect as centers of infection that may affect rheumatic joints and inflamed nerves. The tonsils are sometimes surgically removed (the operation being called tonsillectomy) as a precaution against scarlet fever, rheumatism, nephritis, quinsy and blood poisoning.

FIG. 4-11.
(a) Extraction of pus.

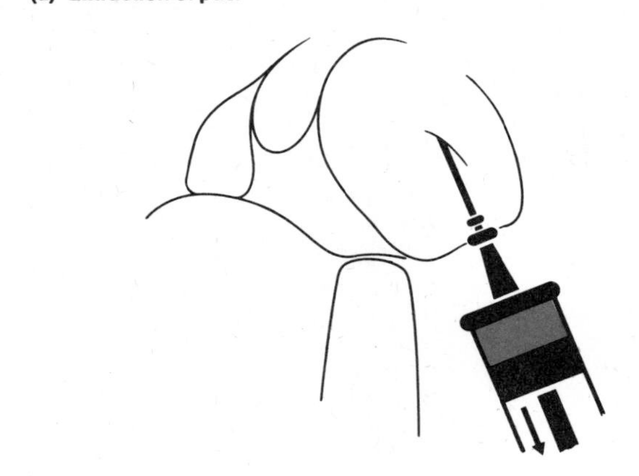

(b) Incision of peritonsillar abscess.

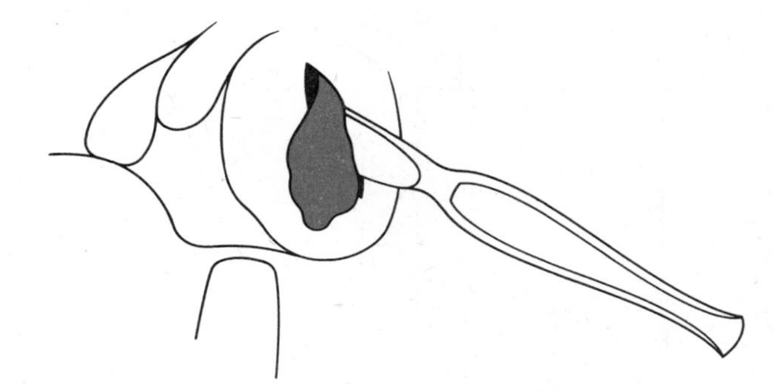

LUNGS

The lungs are a pair of cone-shaped organs contained within the pleural cavity and enclosed by a membrane called the *pleura*. They fill most of the thorax and are interconnected by the *trachea* (windpipe) and the two main branches thereof *(bronchi)*. The right lung has three *lobes*, the left lung has two. The fissures

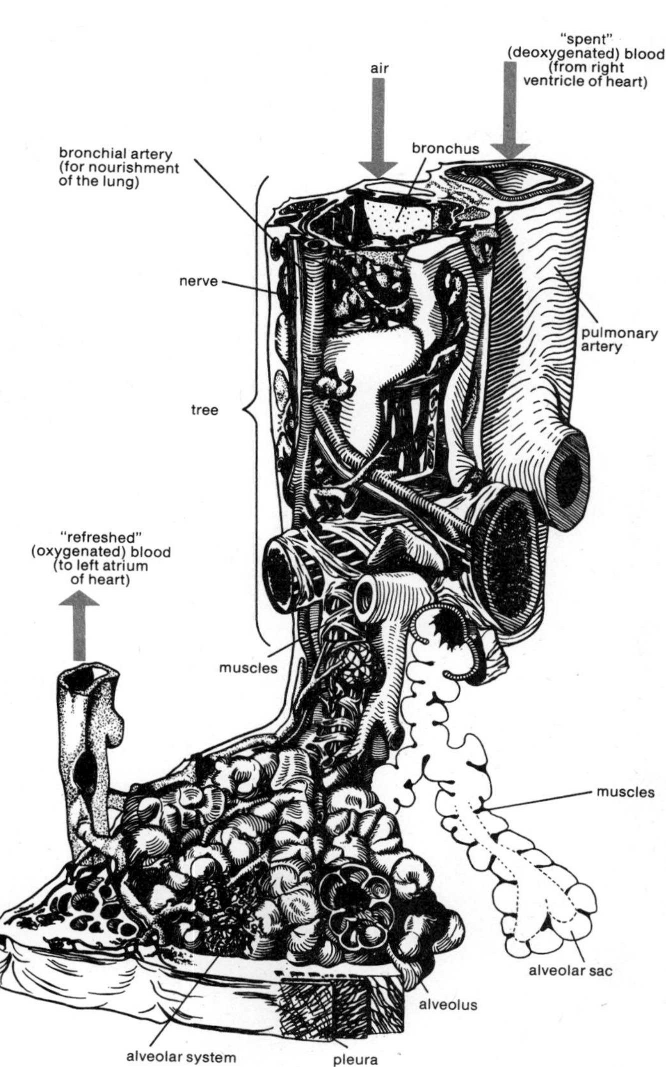

FIG. 4-12.
Bronchial tree and alveolar system.

separating the lobes extend to the hilum (root of the lung). The pleura forms a closed sac; one part of the pleura (visceral pleura) covers the surface of the lung; the other part (parietal pleura) lines the wall of the thoracic cavity. These two layers of the pleura are normally in contact and can slide over each other during the breathing movements. They are lubricated by a film of tissue fluid between them, so that there is no friction between the lung and the ribs. In a healthy person the pleural cavity is therefore really almost nonexistent; as a result of disease or injury the visceral and the parietal layer may become separated, however. If air gets into the pleural cavity, the lung collapses by its own elastic tension (Fig. 4-26).

The main bronchus, which connects the trachea to each lung, divides into a branch to each lobe. These branches divide again and again into progressively smaller bronchi and finally into thin tubes called *bronchioles*. The whole system of air passages resembles the branches of a tree and is therefore sometimes referred to as the bronchial tree.

The bronchioles terminate in small cavities (alveolar sacs), each of which has several compartments called *alveoli*. The walls of the alveoli are very thin and contain capillaries, i.e., very fine blood vessels; it is here that the blood takes up oxygen from the air in the alveoli and discharges carbon dioxide into this air, which is then exhaled. This is the so-called external or pulmonary respiration, as distinct from the internal respiration, which denotes the exchange of oxygen and carbon dioxide between the blood and the cells of the body. There are between 300 and 450 million alveoli, with a total respiratory surface of about 80–120 m^2. The capillaries in the walls of the alveoli form a network in the meshes of which the alveolar epithelial cells are located. Some of these cells have phagocytic properties, i.e., they can ingest bacteria, dust and other foreign bodies. The actual *respiratory epithelial cells* are extremely thin and thus highly permeable to the gases of respiration (Fig. 4-13).

The primary function of pulmonary respiration is to remove spent air with a high content of carbon dioxide and to replace it with fresh air with a high content of oxygen. The difference in pressure needed for this is produced by rhythmic enlargement and contraction of the thoracic cavity, whereby atmospheric pressure forces air into the lungs (inspiration, inhalation) and air is subsequently expelled (expiration, exhalation). Inspiration takes place when the diaphragm (midriff) and the intercostal muscles (between the ribs) enlarge the thoracic cavity. When the intercostal muscles contract, they lift the ribs, making the rib cage and therefore the thoracic cavity deeper from front to back and wider from side to side

FIG. 4-13.
Alveoli of lung.

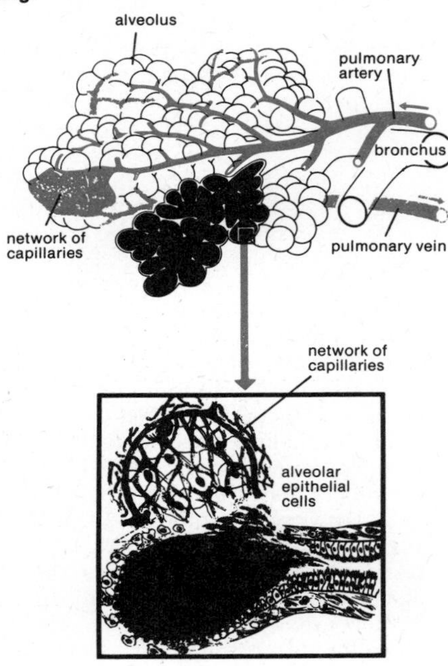

FIG. 4-14.

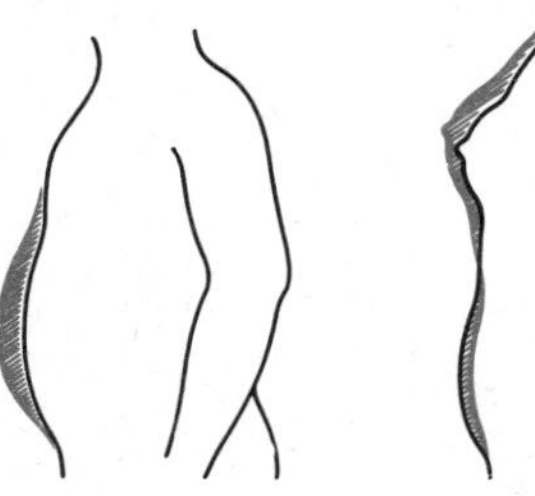

(a) Abdominal respiration. **(b) Costal respiration.**

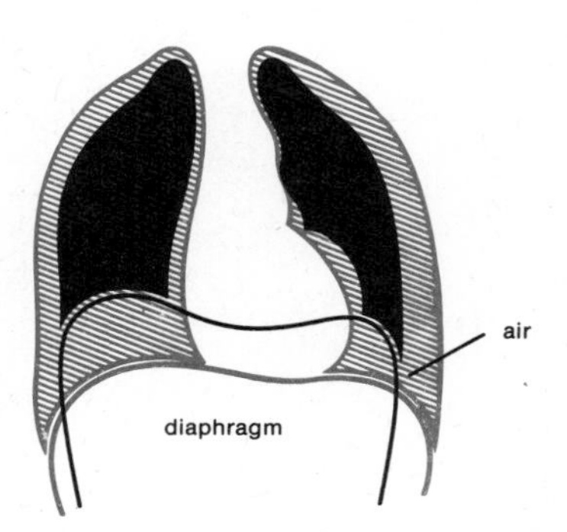

FIG. 4-15.
Diaphragmatic movement in respiration.

FIG. 4-16.
Lung volumes.

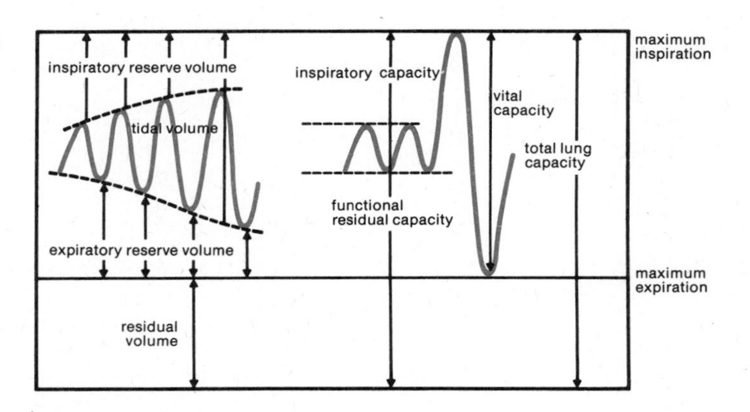

(costal respiration, Fig. 4-14). Contraction of the diaphragm causes it to flatten its domed shape, so that the height of the thoracic cavity is increased, while the muscles of the front wall of the abdomen relax to accommodate the movement (abdominal respiration, Fig. 4-14a).

Expiration is largely passive, involving hardly any muscular contraction. Broadly speaking, the chest and lungs spring back elastically, of their own accord, to their original position. The elastic contraction expels the air. In normal quiet breathing, costal respiration and abdominal respiration contribute about equally to the overall movement of air. With advancing age the thorax becomes stiffer, and abdominal respiration takes on a larger share. In pregnancy, when the abdomen has less "give," costal respiration is relatively more important.

The *rate of respiration* in an adult is 10–15 breaths per minute. At each breath about 0.5 liter of air is breathed in and out (the so-called *tidal volume*). Deep breathing can increase this volume to about 2.5 liters. With quiet breathing, about 2.5–3 liters of air remains in the lungs after expiration. By a violent effort it is possible to blow out up to half this volume; the air then still remaining in the lungs is called the *residual volume*. The maximum volume of air that a person can breathe out after taking the deepest possible breath is about 4 liters (*vital capacity*). These are, of course, only average figures, and there are considerable individual variations.

BRONCHITIS

Bronchitis is inflammation of the mucous membrane of the air passages (bronchi) of the lungs. It may be either acute or chronic. *Acute bronchitis* is sometimes an accompanying phenomenon of an infectious disease such as influenza, measles, whooping cough or malaria. It may also be caused by inhalation of dust or chemical irritants (noxious fumes, etc.). Mostly, however, bronchitis occurs as a more or less independent disease caused by an infectious agent, particularly a virus, usually followed by bacterial infection. Infection is often preceded by the common cold. Predisposing factors are exposure to wet or cold conditions, which reduce the local resistance, so that pyogenic organisms such as streptococcus, staphylococcus or pneumococcus can establish themselves in the bronchial mucous membrane (Fig. 4-17a and b). These organisms first set up a "dry," subsequently a catarrhal (with plentiful secretion of mucus) and finally a purulent inflammation (i.e., with formation of pus). In adults the bronchitic inflammation is usually confined to the mucous membrane of the large bronchi, but in children, especially babies, and in elderly people the inflammation not infrequently affects also the terminal bronchioles and alveoli (bronchopneumonia).

The first symptoms of acute bronchitis are a sensation of soreness behind the breastbone and a dry cough. Coughing is attended by pain in the chest. In this stage of "dry" inflammation, expectoration is viscid, "glassy" in appearance, and not plentiful. In a later stage it becomes fluid and then slimy due to the presence of pus and white blood cells. Eventually, large amounts may be coughed up. There is fever, accompanied by headache, malaise and loss of appetite; the temperature rises above 38°C (100.4°F). The general symptoms are, however, usually rather mild, and circulatory disorders occur only in severe cases or if pneumonia develops. In general, an attack of bronchitis lasts only a few days, but it may be repeated and eventually turn into chronic bronchitis. Acute bronchitis is treated mainly with inhalants. If the sputum is viscid and tenacious, expectorants may be given to loosen the mucus and facilitate coughing. Persistent dry coughing, i.e., unaccompanied by expulsion of moisture, can be harmful because the effort of coughing imposes a strain on the pulmonary tissue and the circulation. In such cases a cough medicine to suppress the cough may be helpful. Such medicines are especially useful in dealing with bronchitis in the aged and in patients with circulatory disorders. If spasm of bronchial muscles (spastic bronchitis) additionally occurs, a transitional condition to bronchial asthma may develop.

Chronic bronchitis may develop from repeated attacks of acute bronchitis. The term *chronic* becomes applicable when the disease is recurrent year after year and continues for months. Predisposing or contributory factors are cold weather, damp foggy weather, allergic reactions, and regular and prolonged exposure to dust, fumes and chemical irritants (Fig. 4-17c). Smoking, an important predisposing factor, comes within the last-mentioned category. The main symptom of chronic bronchitis is a persistent cough with sputum. The amount of sputum (mucus, phlegm) brought up is generally not more than about 20 ml per day. Bronchitic coughing is, in the long run, liable to cause distension and weakening of the alveoli. This condition, called *emphysema,* is attended by shortness of breath; subsequently there may also be constriction of pulmonary blood vessels and possible failure of the right side of the heart (see also page 60). Chronic bronchitis may be treated with antibiotics, expectorants or cough-suppressing medicines. Irritants such as smoking or exposure to wet and cold conditions should be avoided. Sometimes a change of climate or environment is helpful.

FIG. 4-17.

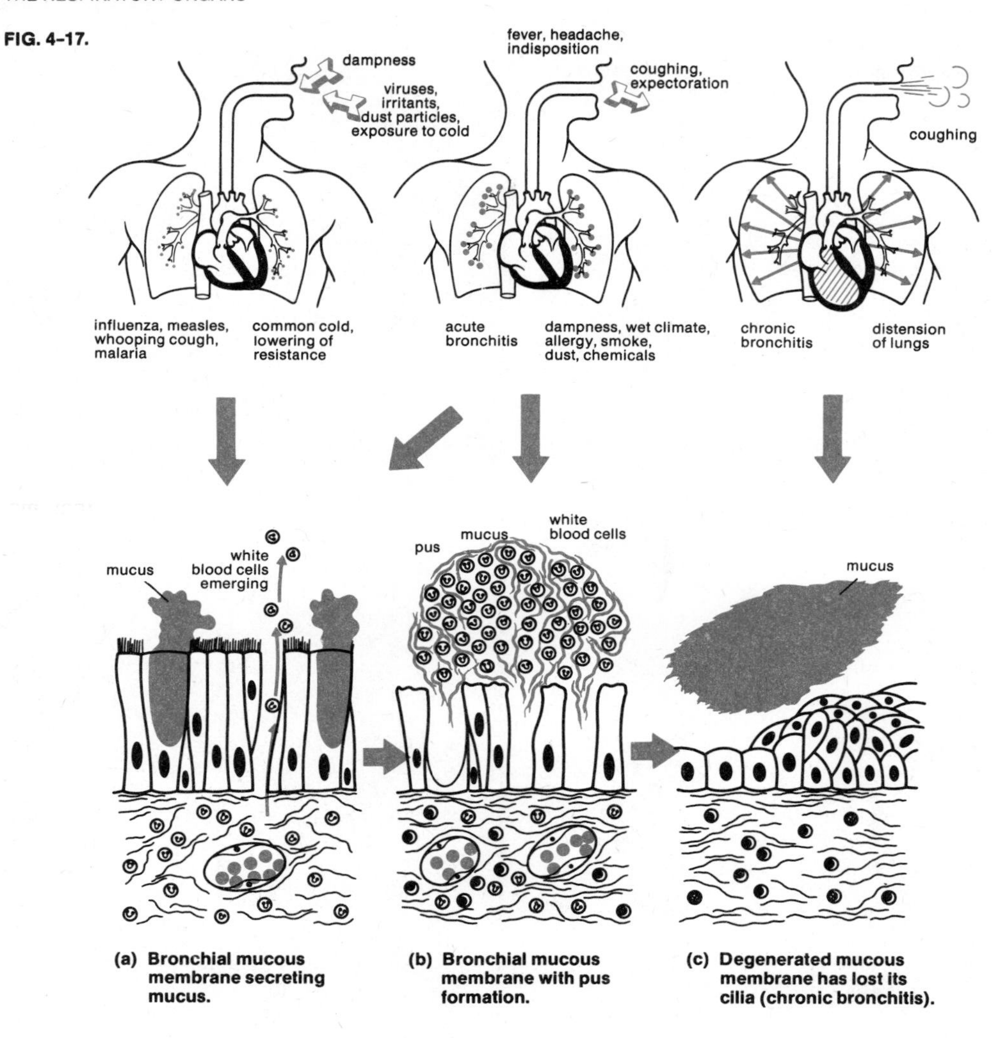

(a) Bronchial mucous membrane secreting mucus.

(b) Bronchial mucous membrane with pus formation.

(c) Degenerated mucous membrane has lost its cilia (chronic bronchitis).

BRONCHIAL ASTHMA

Asthma is paroxysmal dyspnea, i.e., difficult breathing, the common form being bronchial asthma (as distinct from cardiac asthma, which is unrelated to it but has similar symptoms). It is caused by spasm and tightening of muscles around the finer branches of the bronchi in the lungs and is aggravated by congestion of the air passages due to inflammatory swelling of the bronchial mucous membrane and increased secretion of viscid mucus in all parts of the bronchial system. The spasmodic contraction of the bronchial muscles can be partly counteracted by breathing in, as this helps to widen the passages, but the patient has difficulty in breathing out, and this is accompanied by a wheezing sound. In a severe attack the patient gasps for breath, and his face turns purple from lack of oxygen. An attack of asthma may vary greatly in length and intensity. It may last a few minutes or several hours; exceptionally, it may continue for weeks. A severe and prolonged attack is called *status asthmaticus,* and the usual drugs for relieving spasm are unable to control it because the small bronchi are obstructed by dried mucus. It is a dangerous condition which may overstrain the right side of the heart and cause circulatory failure.

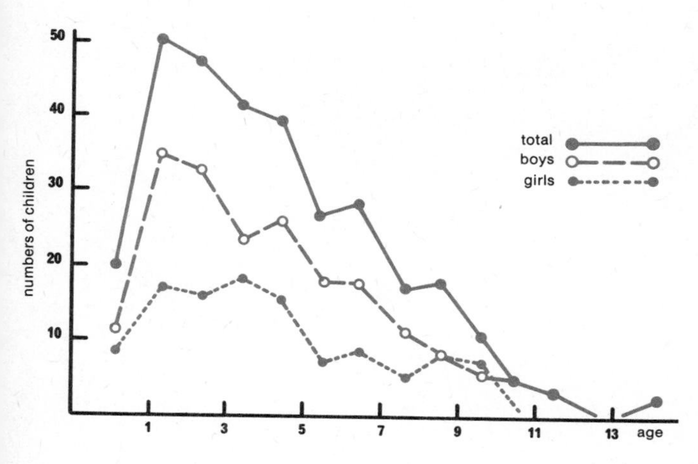

FIG. 4-18.
First occurrence of bronchial asthma in children of both sexes

A number of factors are involved in causing bronchial asthma: allergic reactions, emotional stress, recurrent infections of the upper air passages. The allergic symptoms arise from an antigen-antibody reaction (see page 46); the allergic sensitivity often "runs in families." Also, bronchial asthma may occur in combination with other allergic diseases (urticaria, hay fever, eczema). Asthma due to psychological factors is frequently encountered in sensitive, timid children. Also, asthma may develop as a result of infections affecting the air passages at an early age and be dependent to some extent on weather and climatic conditions. These various causes also play a part in bringing on an attack. Thus, allergic attacks follow exposure to the allergen concerned, which may be a particular species of pollen, dust, a particular food, etc. Other factors that may cause an attack are emotional stress, anxiety, conflicts, anticipation of some important event, etc. In the long run, a kind of reflex pattern may develop, in which an attack is triggered off more or less automatically in a particular situation.

Treatment of bronchial asthma (Fig. 4-19) aims at stopping an attack and reducing the tendency to have an attack. Attacks can be relieved by various drugs which may be injected (e.g., adrenaline), swallowed (e.g., ephedrine), inhaled as a spray (e.g., isoprenaline) or taken as a suppository (e.g., theophylline). If asthma occurs in response to an allergic condition, calcium or antihistamine preparations may be useful. It is sometimes possible to make a precise identification of the allergen and to cure or relieve the asthma by desensitization (see page 47). In nervous or irritable asthmatics a marked reduction in the frequency of attacks may be achieved by sedatives and tranquilizers. Severe chronic asthma, especially the persistent condition called status asthmaticus, can be treated with corticosteroids (adrenal cortical hormones). In dealing with status asthmaticus, oxygen may have to be given as an emergency measure. If the asthmatic condition is governed mainly by psychological factors which reinforce the physical causes, treatment with drugs may be ineffective. In such cases a change of climate or environment or emotional factors may improve the patient's condition; psychotherapeutic treatment is sometimes helpful.

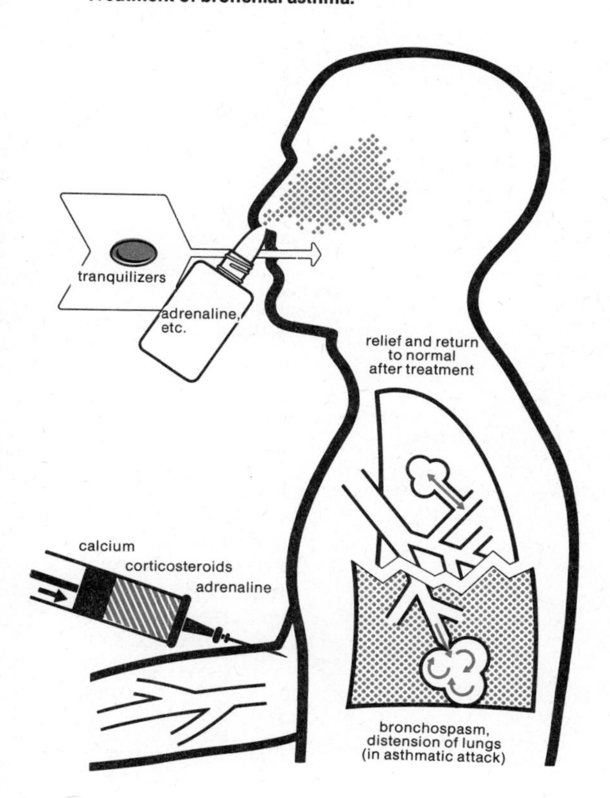

FIG. 4-19.
Treatment of bronchial asthma.

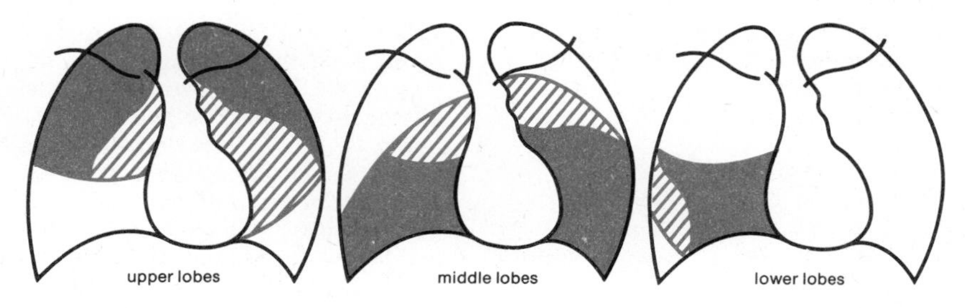

FIG. 4-20.
Lobar pneumonia as revealed by x-ray examination.

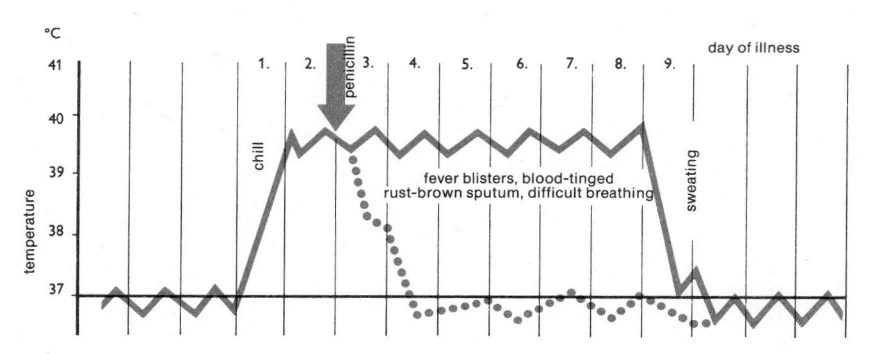

FIG. 4-21.
Course of pneumonia with and without treatment with drugs.

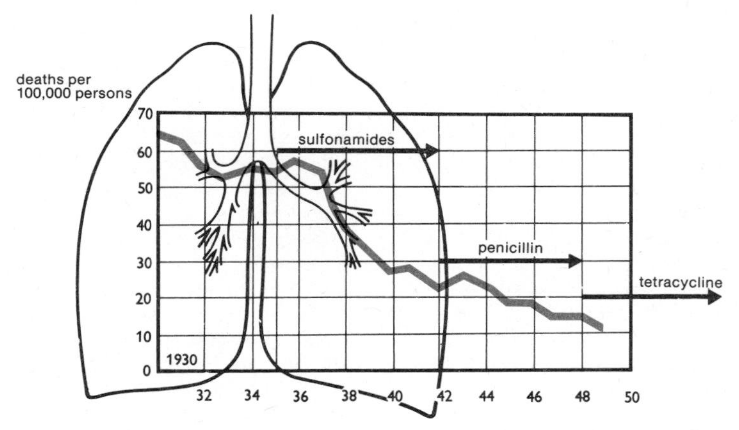

FIG. 4-22.
Drug treatment and mortality in pneumonia.

PNEUMONIA

Pneumonia is inflammation of the lungs caused by infection (viruses or bacteria), chemical irritants or other agents. Common bacterial pneumonia (*lobar pneumonia*) is usually brought about by pneumococci and has the characteristics of an infectious disease. It is not contagious, however, in that pneumonia is rarely transmitted from person to person. Besides, these microorganisms are often also found in the oral cavity in healthy persons.

Factors as yet not definitely known, perhaps allergic reactions, minor injuries, or disorders of the type popularly referred to as colds and chills, can suddenly bring on an attack of pneumonia.

The disease starts abruptly and with acute symptoms: coughing, chill, quickened breathing, numbness, temperature around 40°C (104°F), fever blisters, and a pulse rate of about 120/minute. There is pain in the chest due to inflammation of the pleura; pain is felt especially when the patient coughs. On the second day a viscid rust-brown sputum tinged with blood is brought up. In severe cases the respiratory function may be seriously impaired.

Percussion, auscultation, or x-ray examination (Fig. 4-20) shows one or more lobes of the lungs to contain more fluid and less air than they normally do. Indeed, the presence of excess fluid in the alveoli is a distinctive feature of the disease. A closer examination of the tissues would reveal that the blood vessels of the lung are numbed by bacterial toxins on the first and second day of the disease and become congested with blood; plasma, but also red cells and coagulating substances from the blood, escape into the alveoli and obstruct them, which explains the rust-brown sputum and dyspnea. Subsequently, white blood cells also escape; the red cells are dissolved and are, from the tenth day onward, absorbed into the bloodstream together with the fluid contents of the alveoli, and the patient recovers. The fever in lobar pneumonia may disappear suddenly after about a week (crisis) or gradually (lysis). A complication may arise in that the infection can spread outside the lung, and pus may collect in the pleural cavity, i.e., between the lung and the outer wall of the chest (empyema). With antibiotics this condition can be prevented, and in otherwise healthy patients the mortality from penumonia is very low. Without treatment, however, one in three cases ends in death. In general, pneumonia can be effectively treated with penicillin and other antibiotics and with sulfonamides (Figs. 4-21 and 4-22), which shorten the course of the disease and reduce its severity. Modern treatment has thus largely eliminated the danger of pneumonia except in babies and old people.

In the form of the disease known as *bronchopneumonia* there is no sudden attack of entire pulmonary lobes; instead, groups of alveoli close to the larger bronchi become affected, so that small inflamed areas occur scattered throughout the lungs. The onset of the disease is therefore less sudden, and there is no pain due to involvement of the pleura. The fever fluctuates; sputum is purulent and slimy. Bronchopneumonia is most likely to develop as a complication of other diseases, especially bronchitis, or as an attendant illness of measles, whooping cough, etc. A frequent causal factor is obstruction of small bronchi, so that a group of alveoli becomcs deprived of air and thus forms a starting point for inflammation. This form of penumonia is most likely to occur in babies and in elderly people. In the very old, it may be most serious because of the additional strain imposed on the heart (particularly the right side of the heart).

Unlike bronchopneumonia, *viral pneumonia* occurs suddenly and in this respect resembles lobar pneumonia. It usually lasts for five to eight days. It does not respond to treatment with antibiotics. In general, however, it is a relatively mild disease.

PLEURISY

Pleurisy (pleuritis) is inflammation of the pleura due to infection by bacteria or viruses. The pleura is a double membrane that enfolds the lungs. It comprises an outer and an inner layer enclosing a cavity which is normally of almost zero width because the layers are separated by only a very thin film of lubricating fluid. Pleurisy usually develops as a result of pneumonia or pulmonary infarction. In old people, pleurisy may also be an attendant symptom of a malignant tumor (in the lung, the stomach or the thyroid gland). More rarely, inflammatory involvement of the pleura may occur in allergic or rheumatic diseases, in inflammatory epigastric disorders, or in consequence of cardiac infarction.

In *dry pleurisy* the pleural membrane is covered with a fibrinous exudate, which roughens its surface, so that the two layers do not slide smoothly over each other. As a result, there is pain during coughing and respiration. To ease the pain the patient hunches his back and takes shallow breaths. Through a stethoscope a rubbing sound, somewhat like the creaking of a shoe, is heard as the patient breathes. Treatment of dry pleurisy will depend on the nature of the underlying disorder. Healing can be speeded by applying hot swathes to the chest and by rubbing with embrocations (liniments) that help blood circulation.

Not infrequently, dry pleurisy occurs as a

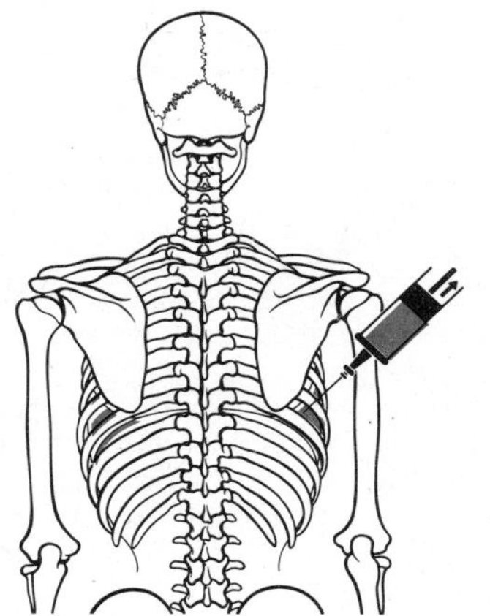

FIG. 4-23.
Extraction of fluid from the pleural cavity for examination in a case of pleurisy.

transitional stage to *pleurisy with effusion,* characterized by accumulation of fluid in the pleural cavity, i.e., between the lung and the wall of the chest. This fluid acts as a lubricant to the inflamed membrane, so that breathing now causes little or no pain. Excessive fluid can displace the lung, also the heart in a severe case. Symptoms are coughing, difficulty in breathing; bacterial infection is accompanied by fever. The presence of fluid in the pleural cavity can be detected by percussion and auscultation, more reliably by x-ray examination. Depending on the nature of the effusion, a distinction can be drawn between *serous pleurisy,* which is characterized by serous effusion (having the nature of serum) and occurs chiefly in association with allergic, rheumatic or tubercular conditions, and *purulent* (or *suppurative*) *pleurisy,* which occurs most frequently as a complication of pneumonia and in which the effusion contains pus. If a large quantity of pus thus accumulates in the pleural cavity, the condition is called *empyema.* The various types of pleurisy can be more precisely diagnosed by extraction and examination of a specimen of the effusion. This is done by means of a syringe and hollow needle inserted between two ribs into the pleural cavity (Fig. 4-23). The fluid thus extracted is examined for its content of proteins and cellular matter and the presence of microorganisms. The extraction of fluid also constitutes a treatment of the pleurisy, as it reduces the fluid in the cavity and relieves the pressure on the lung. Purulent pleurisy is treated with antibiotics (general and local). Severe or persistent cases involving empyema or excessive adhesions of the pleura may require surgery.

PNEUMOTHORAX

Pneumothorax is a general term designating the presence of air in the pleural cavity. Normally this cavity is of almost zero thickness, containing merely a thin film of fluid which lubricates the sliding contact of the inner and outer layer of the pleura and also helps to keep these layers in contact with each other by its adhesive action. The contact is maintained largely by a negative pressure in the pleural cavity, so that the air in the lung keeps the lung expanded and pressed against the wall of the chest. This pressure in the pleural cavity fluctuates from −8 cm to −4 cm water gauge with alternate inspiration and expiration (Fig. 4-24). If air (or any fluid) gets into the pleural cavity, the normal expansion of the lung is restricted and respiration therefore impaired. A larger amount of air in the cavity will cause collapse of the lung.

Pneumothorax may have many different causes. It can be the result of an injury of the chest or of lung disease, so that air enters the pleural cavity from the exterior of the chest or from within the affected lung. Perforation of the lung by a fractured rib may cause pneumothorax, as may also the rupture of a superficial lung abscess such as a tuberculous abscess. *Artificial pneumothorax* is induced intentionally by injecting a measured amount of air into the pleural cavity (Fig. 4-25). It used to be employed in the treatment of tuberculosis, the object being to give the diseased lung a temporary rest by preventing its expansion. With the discovery of streptomycin and other modern drugs this method is now much less frequently applied. With a "closed" pneumothorax (Fig. 4-25), once it has been induced, there is no further exit or entry of air. If the lung is not fully collapsed, it will still participate to some extent in the respiratory movements of the chest. The closed penumothorax of one lung causes little discomfort to the patient, as the other lung can quite adequately meet the body's normal respiratory requirements.

An *open pneumothorax* (Fig. 4-26) causes complete collapse and immobility of the lung; the respiratory movements of the chest cause air to flow alternately in and out of the opening formed in the chest wall. If the opening in the chest wall or in the lung is a slit which acts as a one-way valve, so that air can enter but not easily escape from the pleural cavity, the condition is called a *valvular* (or *tension*) *pneu-*

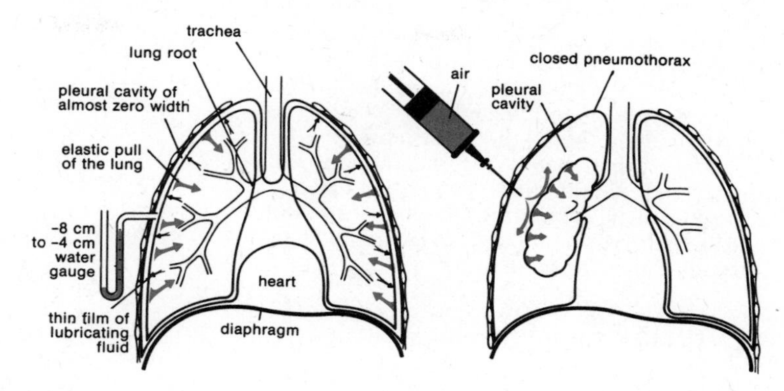

FIG. 4-24.
Lung is expanded in the chest.

FIG. 4-25.
Closed pneumothorax.

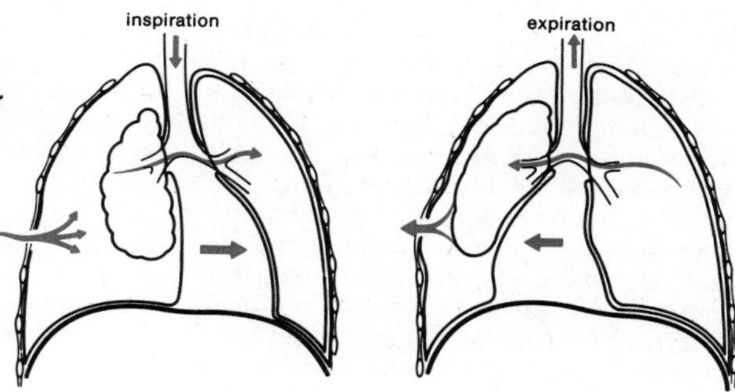

FIG. 4-26.
Open pneumothorax: air flows alternately in and out of the opening in the wall of the chest during breathing; air also passes to and fro between the two lungs (the diaphragm and heart are displaced sideways in rhythm with the breathing).

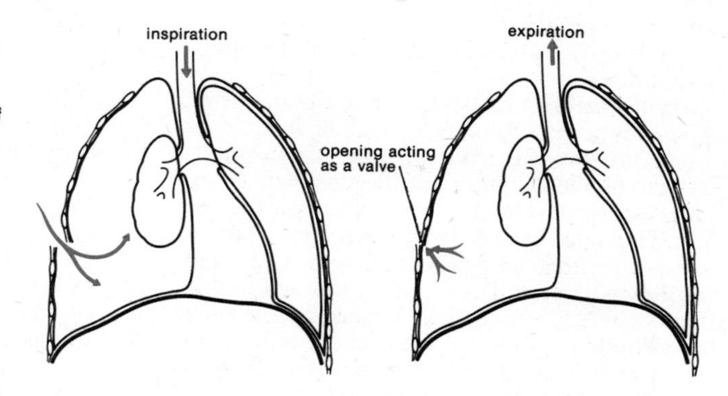

FIG. 4-27.
Valvular (tension) pneumothorax with buildup of pressure in the pleural cavity.

mothorax. The pressure that then builds up in the pleural cavity will displace the diaphragm and the heart toward the healthy side of the chest (Fig. 4-27). Such displacement of the diaphragm may in turn cause obstruction of major veins and thus reduce the respiratory efficiency of the sound lung. In such cases the patient suffers dyspnea and cyanosis (face and extremities take on a blue appearance due to lack of oxygen) and is in great distress. If the hole in the chest wall in open pneumothorax is a large one, breathing trouble arises from so-called diaphragmatic flutter, and the venous return of blood is impaired, resulting in falling blood pressure, oxygen deficiency and cyanosis.

The air that has penetrated into the pleural cavity in closed pneumothorax is gradually absorbed by the bloodstream. A valvular pneumothorax may have to be treated surgically to get rid of the valve action after relieving the pressure that has built up in the pleural cavity. In the event of major chest injury resulting in an open pneumothorax the latter must first, by means of airtight bandaging, be transformed into a closed pneumothorax. Next, the damage to the chest wall is repaired by surgery and the aperture sutured. Suction drainage can help to reexpand the collapsed lung.

PULMONARY EMPHYSEMA

This condition is characterized by overinflation and distension of the alveoli of the lungs. The walls dividing the alveoli deteriorate and eventually disappear (Figs. 4-28 and 4-29). As a result, the available internal surface area for the exchange of gases, especially the absorption of oxygen into the blood, is decreased and there is moreover destruction of capillaries, so that the pulmonary circulation is impaired. Pulmonary emphysema is a fairly common disease and is the cause of between 20 and 30 percent of all certified cases of invalidism.

Above the age of 40, aging of the elastic lung tissues, which is associated with the gradual disappearance of their fibrous structures, inevitably results in some distension of the lungs, a condition referred to as *atrophic* (or *senile*) *emphysema*. Another form of emphysema, unrelated to age, develops as a result of convulsive closure of bronchioles and small bronchi, as may occur in bronchial asthma and (acute or chronic) bronchitis. In such cases the action of the inspiratory muscles, which are more powerful than the expiratory muscles, causes acute distension of the lungs, which in its earlier stages is still completely reversible. Prolonged overinflation will, however, eventually give rise to the aforementioned degenerative changes of the pulmonary tissue. Emphysema also occurs when part of the lung becomes scarred and contracts (as in tuberculosis) or is removed (as in lung cancer). The remaining pulmonary tissue then expands to fill the extra space in the chest.

The first symptoms of emphysema are breathlessness due to physical exertion, a persistent dry cough, and dizziness when coughing or stooping. Less characteristic are general symptoms such as headache, weakness, insomnia, lack of appetite, and constipation. In the course of time the breathlessness (dyspnea) becomes more pronounced. It is not always due solely to the reduction in gas-exchange surface area but may also be due to progressive failure of the right side of the heart, which becomes overburdened as a result of the constriction of the air flow passages. Besides, the overinflated chest, which remains immovable in the inspiratory position ("barrel chest"), allows only a limited and inadequate amount of respiratory air movement. The diaphragm strives to make up for this deficiency

FIG. 4-28.
Alveoli in emphysema.

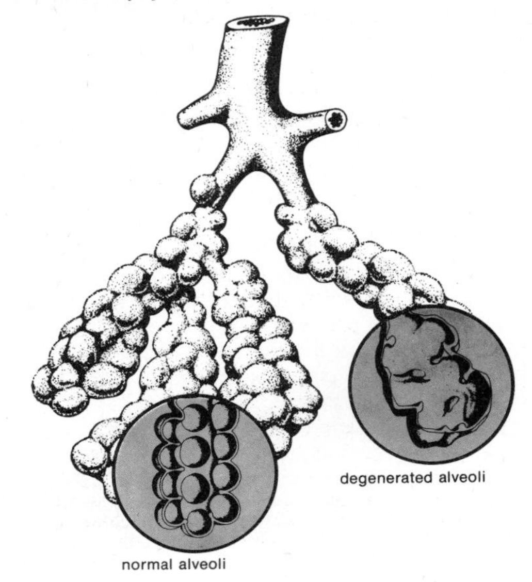

FIG. 4-29.

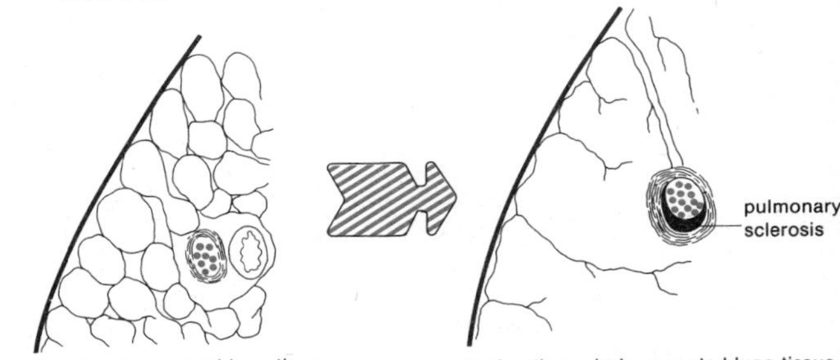

by extra exertion (increased abdominal respiration). The abnormally high air content of the lungs is revealed by a loud and deep sound on percussion, and also by the high penetrability of the thorax to x-rays.

Treatment of pulmonary emphysema will be directed toward dealing with the basic cause, such as bronchitis or dusty environment, with a view to preventing further deterioration of the condition. Lost lung tissue cannot be replaced, and treatment will therefore also aim at giving the patient relief and enabling him to make the best possible use of his remaining respiratory capacity. Breathing exercises and precautions to prevent bronchial disorders are most important. Allergically conditioned (asthmatic) bronchial spasms can be treated with corticosteroids. Since the patient's breathing mechanism is deficient, he must on no account be given—for whatever reason—any centrally acting drugs that tend to suppress respiratory activity (morphine, large doses of soporifics). He should spend as little time as possible resting in bed. Deficiency affecting the right side of the heart may require treatment with digitalis drugs.

PNEUMOCONIOSIS

Pneumoconiosis is a condition of the respiratory tract due to inhaled dust of various kinds. Prolonged inhalation of fine dust particles of a particular composition causes characteristic irritation phenomena affecting the lung tissue. To this irritant action the lung responds, as it does to other harmful actions (pneumonia, tuberculosis), by fibrosis, i.e., the formation of fibrous scar tissue in the connective tissue framework. This transformation of respiratory tissue to more or less inactive fibrous tissue reduces the efficiency of the respiration and of the pulmonary circulation. This, in turn, places an extra strain on the right side of the heart.

Pneumoconiosis may occur after years of exposure to certain industrial dusts, e.g., in coal mines (*anthracosis*), in ironworks (*siderosis*) or in asbestos factories (*asbestosis*). The most frequently encountered type of pneumoconiosis, however, is *silicosis*, a chronic debilitating illness which is caused by inhalation of silica (quartz) dust. It occurs in quarrymen, miners, sandblasters, stonemasons, ceramic workers, etc. Damage to the lungs is permanent, so that prevention of silicosis is more important than treatment. Preventive measures include suppression of dust (e.g., by wet drilling), dust extraction, adequate ventilation, wearing of protective masks in dusty surroundings, etc. Proper instruction of employees in the hazards of silicosis, and regular medical examination to detect early signs of the trouble, are also essential. Any assessment of pneumoconiosis and the invalidism or disablement resulting from it must be based on the consideration that the condition is not only irreversible but, if anything, tends to worsen.

Silicosis is caused by the inhalation of very fine particles of silica dust (below 0.001 mm size) which find their way into the alveoli, where they are ingested into the epithelial cells. The silica is slightly soluble; it thereby damages the cells and eventually kills them, so that some of them are cast off and are brought up by coughing. Some remain in the lung tissue, while others are carried by the lymph flow to the pulmonary lymph nodes. The silicate crystals are released from the dying cells and can then be ingested into other phagocytic cells and cause harm to them also (Fig. 4-30a). In the course of time a nucleus of tissue surrounded by dying phagocytic cells is formed. The process is continued and repeated until in the end a small fibrous nodule is formed. When the nodule grows larger, the tissue cells at its center may degenerate and decompose (Fig. 4-30b). Subsequently a number of nodules merge or coalesce to form large nodules the size of a plum, sometimes even considerably larger. These nodules develop especially in the top part of the lung, where most of the dust is trapped (Fig. 4-30c). The consequences of silicosis are (1) increased infection hazard (bronchial catarrh, pneumonia, tuberculosis, Fig. 4-30e); (2) impaired blood circulation through the lungs, with increased strain and possible failure of the right side of the heart (Fig. 4-30f); (3) impaired respiration (Fig. 4-30d). This last-mentioned effect is especially significant and is due to loss of respiratory tissue in the lungs, to the formation of fibrous scar tissue with loss of elasticity, to attendant phenomena, and to local emphysema (overdistension) resulting from shrinkage of adjacent fibrous tissue.

Depending on the extent of the fibrous degeneration and the very characteristic x-ray image with shadows of varying number and size, three stages of silicosis can be distinguished. In a relatively early stage of the dis-

FIG. 4-30.
Features of silicosis.

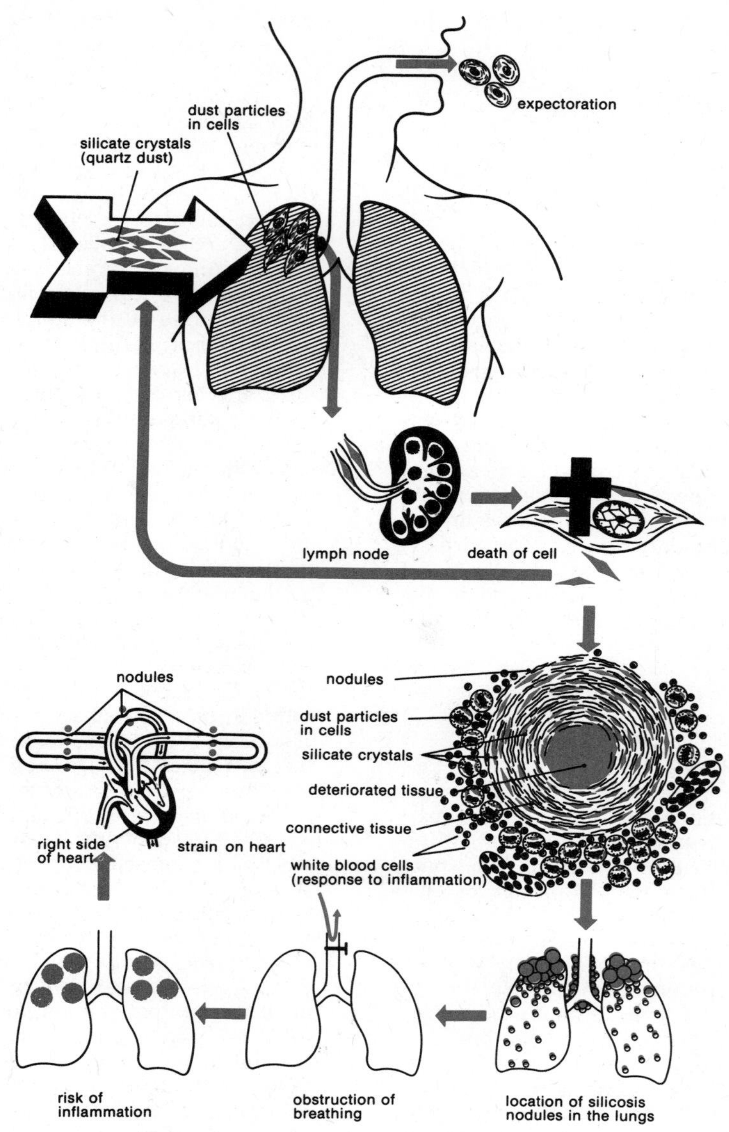

ease the symptoms are generally confined to dyspnea upon physical exertion, sometimes with signs of cyanosis. There is a persistent dry cough. In the more advanced stage, bronchitis or pulmonary emphysema occurs. Gray or black sputum is a sign of deterioration of tissue in the large nodules in which blood supply is poor. Finally, the strain on the heart increases, and symptoms of insufficiency of the right side of the heart appear.

Part Five

THE DIGESTIVE SYSTEM

NUTRITION

A properly balanced diet is necessary for the growth and maintenance of the body. It should be composed of the basic nutrients—protein, carbohydrate and fat—in suitable proportions and should furthermore contain sufficient minerals, vitamins, trace elements and roughage. The food should also be tasty and be so prepared that it is properly digestible and thus utilizable by the body. Most suitable for human beings is a mixed diet of animal and vegetable foodstuffs.

In the alimentary canal (or tract) the nutrients are converted to a soluble and therefore absorbable form, carried by the blood to the various tissues in the body, and there oxidized in the cells with the aid of a number of enzymes. This process is comparable to combustion which produces heat and energy. The waste products of combustion are eliminated from the body, particularly in the feces, the urine and the exhaled air.

The value of food as a source of energy for the body's needs is expressed in *Calories*. In physics the calorie is the amount of heat required to raise the temperature of 1 gram (g) of water by 1°C. The amount of heat required to raise the temperature of 1 kilogram (kg) of water by 1°C is therefore 1000 calories, a quantity that is known as the *kilogram calorie,* the *kilocalorie* or the *large calorie*. With reference to human nutrition this latter unit is simply called the Calorie, spelled with capital C to distinguish it from the small or gram calorie. The caloric value of food is the number of Calories it would yield on complete combustion. The caloric value of foods differs greatly; for instance, 1 g of carbohydrate or 1 g of protein has a caloric value of 4.1 Calories, whereas the caloric value of 1 g of fat is 9.3 Calories.

Carbohydrates and fats serve mainly as providers of energy, whereas proteins (and amino acids) are needed mainly for the building and replacement of the body's cells and for the production of hormones and enzymes. In a correctly balanced diet about 55–60 percent of the caloric requirement is supplied by carbohydrates, 25–30 percent by fats, and 10–15 percent by proteins, with the approximate rule that the daily intake of proteins should be 1 g per kg (2.2 lb) of body weight. In young persons, in pregnant women and also during the lactation period the protein requirement is about 1.5 g per kg of body weight.

Not all types of dietary protein are equally suitable for building up the body's own proteins. For example, the German Society for Nutrition recommends that the daily protein intake of an adult should comprise at least 0.4 g of animal protein per kg of body weight and not less than 30 g of such protein in all. Amino acids are the building blocks of which proteins are constructed, and they are the end products of protein digestion or hydrolysis. The biological value of proteins depends on their content of so-called essential amino acids, i.e., those which the human body cannot synthesize for itself but has to obtain from the food consumed. Some proteins lack one or more of the essential amino acids, and in this respect animal foods are better than vegetable. If proteins of animal and vegetable origin are mixed together, the biological value of the mixture can be higher than would correspond to the individual components.

The most important carbohydrate in human food is starch, which is found in many foods, especially cereals, potatoes and rice. In the alimentary canal, starch is converted to glucose which, in turn, is converted to glycogen (animal starch) and stored in the liver. To meet the body's requirements, glycogen can at any time be converted back to glucose and burned in that form.

When the intake of food, especially in terms of Calories, is in excess of immediate requirements, the surplus is stored in the form of fat, whatever the nature of the basic nutrient from which the Calories have been derived. In this sense all foods are fattening, if they have any caloric value. When the need for extra fuel arises, fat can be broken down and burned. Because of its high caloric value, fat is the body's most important reserve of energy. Chemically, fats are so-called *triglycerides,* i.e., esters of glycerol and fatty acids. Natural fats and oils usually are mixtures of triglycerides. There is no difference in principle between fats and oils, except in their consistency at ordinary environmental temperatures. The fatty acids of the fats in the human diet are mainly stearic, palmitic and oleic acids. Fats are also important for the absorption of the fat-soluble vitamins A, D and K, which can pass the wall of the intestine only in combination with fats. So-called lipoids (lipids) are fat-like substances which contain other groups than the glycerol and fatty acids that make up the true fats; the most important of them is *cholesterol.* Most fats and lipoids can be synthesized in the body and are therefore not essential to the diet. However, the body is unable to synthesize some vitally important fatty acids, such as linoleic acid and linolenic acid (vitamin F). The daily intake of these essential fatty acids should be about 4–6 g; this quantity is present, for example in two tablespoonfuls of sunflower seed oil, in 45 g of margarine or in 150 g of butter (100 g equals approximately 3½ oz).

DIET

Diet, in the general sense of the word, means food consumed in the normal process of living, but more specifically it refers to a prescribed allowance of food suited to a particular state of health or disease (special diet). The science of the use of foods in health and illness is called *dietetics.* For example, the diet appropriate to *acute gastritis* (inflammation of the mucous membrane of the stomach, see page 160) should conform to the following rules: 1–2 days' fasting, during which the patient takes only weak tea without milk or sugar (ordinary tea, camomile tea or fennel tea), followed by 1–3 days' sloppy diet (gruel of oatflakes, rice or linseed). If the patient suffers from loss of appetite or tends to vomit, the gruel should be boiled in water only, but with sufficient salt to make up for the losses of salt due to vomiting. When the condition improves, the gruel may be boiled in veal broth or in diluted milk, and an egg may be beaten into it. Lightly toasted white bread or rusks, salted tomato or carrot juice, and mineral water are also allowed. In a case of *acute gastroenteritis* (inflammation of stomach and intestinal tract) the patient will often tolerate raw grated apple or mashed banana. This should be followed in due course by a *bland diet* designed to avoid gastrointestinal irritation. The transition to a normal diet will depend on how well the patient feels. In general, any particular diet should be properly tolerated before proceeding to the next diet.

A bland diet may be long-term or indeed permanent in cases of gastrointestinal disorders such as chronic gastritis, peptic ulcers or stomach reduced by partial surgical removal. Such a diet should therefore be well balanced, varied, appetizing, ensuring an adequate intake of calories, minerals and vitamins. As a rule, the patient should take four or five small meals distributed through the day, so as not to overload the stomach and also not to leave it empty for any great length of time. If the stomach produces too little acid and digestive enzymes, the necessary acids and enzymes must be taken with the food. Eating slowly and chewing the food well will also relieve the digestive load on the stomach and bowels. Food and drink should not be consumed too hot, nor should they be icy cold. Coffee, sweets, smoked meats, rich (especially fried) foods, alcohol and nicotine must be avoided.

In a case of *chronic constipation* (see page 173) the first aim should be to stimulate the action of the bowels (intestines). The physiological stimulus for bowel movement is the pressure exerted on the intestinal wall by the contents. In seeking to improve sluggish bowel action, the diet should therefore strive to ensure adequate filling of the intestinal canal; this can be achieved by consumption of bulky food, i.e., food having a high content of roughage, particularly cellulose fibers, which are indigestible and stimulate intestinal peristalsis (vegetables, fruit, salads, cereals). Being indigestible, the cellulose remains present in the bowels all the way, keeps them expanded and, by absorption of water, prevents the feces from becoming too hard and dry. Linseed, in particular, absorbs water and swells, thereby producing bulky and soft stools. Cold drinks taken on an empty stomach stimulate the bowels by reflex action. Bowel movement is stimulated also by lactic acid as contained in yogurt, buttermilk and sauerkraut, as well as by fruit juices and effervescent drinks. Nuts and dried figs likewise help to reduce intestinal sluggishness. Dietetic measures should, however, be backed up by consistent self-imposed discipline on the patient's part in striving to regularize the motions.

To combat *flatulence* (see page 153), foods with a high cellulose content, indigestible foods and effervescent drinks should be avoided. Uncooked food, provided it is free from cellulose, and fresh vegetable or fruit juices are permissible.

In dealing with a sudden onset of *acute diarrhea* due to, or associated with, increased fermentation or putrefactive bacterial action in the large intestine, the main dietetic principle is first to relieve the overburdened intestine by fasting for 1–3 days until it has emptied. Drinks containing tannic acid, such as tea, are suitable for quenching thirst. A raw apple diet is also beneficial in such circumstance: five meals a day, each consisting of 250–300 g of grated apples, including the skins, to which a little lemon juice and sugar are added. Mashed bananas, or carrots boiled in salt water and then mashed, are also suitable. As in a case of gastritis, this diet should be followed by a few days on slops, with a gradual transition to a fuller bland diet.

In a case of *chronic diarrhea* the first step must be to seek the causes—e.g., disease of the liver or pancreas, overactivity of the thyroid gland, ulcerative colitis, excessive nervous tension—and try to remove them. A bland diet of easily decomposable carbohydrates and proteins with relatively little fat, but with sufficient fresh vegetable and fruit juices, is usually tolerated quite well.

A special diet for persons afflicted with disorders of the *liver,* such as cirrhosis, will be rich in carbohydrates and high-grade (i.e., animal) protein and will contain sufficient vitamins. Fats should include only those that are easily digestible and those with a high content of unsaturated (essential) fatty acids. Liver trouble is often associated with loss of appetite. For this reason the diet should strive to be as appetizing as possible. For example, vegetable and fruit juices are to be recommended; these may be enriched with carbohydrates in the form of glucose or fructose. Carbohydrates are particularly important in a diet intended for a patient with liver trouble, because if the glycogen content of the liver falls too low, the resistance and the detoxifying function of that organ will suffer. In addition to the readily absorbable simple sugars, other suitable carbohydrates are the easily decomposable ones found in white bread (non-new), rusks, crackers, maize starch, rice, semolina, oat flakes, sago, noodles, groats and floury potatoes. The diet should furthermore include fine vegetables, particularly on account of the minerals and vitamins in them. Besides carbohydrates, high-grade proteins are also beneficial to the liver cells. Certain amino acids, in particular, fortify the defensive power and regenerative capacity of those cells. Depending on the degree of damage suffered by the liver, its ability to transform and synthesize proteins may be impaired, and for this reason the appropriate protein intake can best be individually determined. On average, the daily intake of protein should be 1 g per kg of body weight, two thirds of which should be of animal origin. In particular, milk products are to be recommended—especially skim-milk curds. The so-called egg-and-potato diet (1 egg + 500 g of potato) has also been found beneficial. Fish and boiled lean meat are usually well tolerated by sufferers from liver disorders. Easily digestible fats such as butter, dietary margarine, maize germ and sunflower seed oil may be used in the diet, whereas fat meat and rich fried foods should be avoided. Some liver diseases are associated with impaired excretion of salt and water and therefore involve a tendency to edema, especially in cirrhosis of the liver (see pages 193 ff). In such cases the intake of salt should be restricted. Since persons afflicted with liver trouble usually have poor appetite, it is generally better not to apply a strictly low-salt diet and, instead, merely to reduce the salt intake to 3 g per day.

Impaired excretion of salt and water, which manifests itself more particularly in *edema* (abnormal accumulation of fluid in the tissues), may have widely varying causes. It may occur as a result of, among other causes, kidney diseases, cardiac insufficiency, cirrhosis of the liver, and hypertension (high blood pressure). The simplest and sometimes indeed the only effective treatment consists in restricting the patient's intake of salt. A healthy person needs a few grams of salt daily, this being necessary to make up for the salt excreted in urine and sweat; the actual quantities vary and will depend on such factors as climatic conditions and the person's activities and manner of life.

The *low-salt diet* can be subdivided into three stages or degrees of strictness: a salt-reduced diet with comparatively slight restriction of salt intake (2.4–2.5 g of NaCl per day), moderate restriction (1 g of NaCl per day), severe restriction (0.5 g or less NaCl per day). In the first of these three stages the food can be cooked with a little salt; in the second stage the total daily amount of salt allowed is only about a quarter of a teaspoonful; and in the third stage with severe restriction of intake no salt at all should be added to the food. Since sodium is contained in virtually all foods anyway, the choice of food should moreover aim at minimum sodium content. In order to make the changeover to saltless food easier, spices and herbs having a low salt content may be used in cooking and can to some extent compensate for loss of flavor. Persons afflicted with kidney and liver diseases often have poor appetite, and it is therefore especially important to provide them with a varied and appetizing diet which is moreover rich in carbohydrates and has a relatively high content of fat as well as containing the necessary vitamins. Intake of liquids will be regulated according to the preceding 24 hours' urine quantity. Tea and fruit juices, sweetened with glucose, are suitable breakfast drinks, as also tomato juice. Furthermore there are such products as strictly low-sodium toast or rusks, which may be taken with unsalted butter, honey or jam. Since ordinary full-cream milk has a relatively high sodium content; a strictly low-sodium dietary milk obtained in powder form should be used instead. The following vegetables have a very low sodium content: asparagus, salsify, lettuce, red cabbage, aubergine (eggplant), kohlrabi, cucumbers, tomatoes and peppers. These can be suitably accompanied by rice, sago, noodles, semolina and potatoes, which should be prepared with butter or fine oil to improve flavor and enhance their caloric value. The ordinary varieties of cheese, except curds and special dietary cheese, contain too much salt and are therefore unsuitable; the same applies to sausages and smoked meats. Meat, fish and eggs may, if due allowance is made for their sodium content, be included in the diet, provided that excretion of the protein decomposition products from the blood plasma through the kidneys is still functioning normally.

Persons with *essential hypertension* (see pages 87 ff) not due to a kidney disease are often overweight so that, besides restriction of salt intake, it is especially necessary to reduce weight in order to relieve the strain on the cardiovascular system. In terms of diet this can be achieved by means of very low-calorie "juice and fruit days." On such days the patient is given 200 ml (6–7 fluid oz) of fresh fruit juice five times a day, or 1 kg (2.2 lb) of fruit spread out over the day. On the other days a low-salt diet with limited-calorie content should be given; in particular, the intake of fat must be restricted because of its high caloric value.

In a case of *cardiac dropsy* (edema due to cardiac disease) the low-salt diet serves to support the treatment with drugs. To begin with, the patient can advantageously be put on a fruit juice diet, under medical supervision: he is given 200 ml of fresh fruit juice five times a day, and the amount of urine excreted is monitored. Fruit juices contain little sodium, but quite a lot of potassium, which promotes the elimination of the edematous fluid. Besides, fruit juices impose less burden on the digestive organs and thus make less severe demands on the heart and circulatory system than an ordinary diet does. It is usually on such fasting days, which should be spent resting in bed, that the edema begins to decrease.

VITAMINS AND ENZYMES

Vitamins are vitally essential organic compounds which the human body is unable to synthesize for itself and which must therefore be absorbed in the food. In contrast with the other essential components of diet, vitamins are needed only in small amounts and are therefore of no importance as fuel to provide energy. Their importance consists in their function in the metabolism of the cells. Although vitamins are complex chemical substances, the nature, chemical structure and composition of most of them are known. The majority have been isolated and some have been synthetically produced. Vitamins are unstable. Oxidation, heat, strong acids, light and aging destroy them.

The intracellular metabolic processes are promoted by *enzymes,* which are organic catalysts produced by the living cells; they are complex proteins which can induce chemical changes in other substances without being changed themselves. Enzymes are specific in their action, i.e., they will act only upon a certain substance or group of chemically related substances. There are many hundreds of known enzymes. An enzyme consists of a specific component, called the *apoenzyme,* which is able to detect and recognize its particular substrate (the substance acted upon by the enzyme) but which is unable to develop its catalytic function until it has been activated by another component, the *coenzyme.* The coenzyme is a substance of low molecular weight and is not produced in the body. Vitamins are, or help to form, coenzymes. For example, vitamin B_1 linked to phosphoric acid is the coenzyme of carboxylase, an enzyme that brings about the removal of the carboxyl group (COOH) from amino acids.

Complete or partial absence of a vitamin from a person's diet results in a deficiency disease known as *avitaminosis*. With present-day normal diet an actual vitamin deficiency is rare. Under particular conditions, however, the need for certain vitamins may increase. For instance, women in pregnancy and during lactation, and also young children in the stage of rapid growth, require more vitamin A and vitamin D. Deficiency can in such circumstances result in the disease called *rickets*, characterized by deficient deposition of lime salts in newly formed bones, causing abnormalities in their shape and structure (see page 147).

Vitamin B_{12} can be absorbed by the small intestine only with the aid of a specific substance (intrinsic factor) produced in the mucous membrane of the stomach. In the event of atrophy of this mucous membrane, production of intrinsic factor is impaired or ceases. As a result, vitamin B_{12} deficiency occurs, irrespective of the actual intake of this vitamin in the food, and pernicious anemia develops (see page 107).

Of considerable practical importance is the fact that some vitamins are soluble in water, whereas others (vitamins A, D, E and K) are soluble in fat. Fat-soluble vitamins are absorbed from the intestine into the blood together with fat. Deficiency may arise in consequence of diseases that interfere with the digestion and absorption of fat. For instance, vitamin K can be absorbed only in the presence of bile. If the flow of bile is obstructed and functioning of the liver is impaired (as in jaundice), a deficiency of vitamin K occurs, which in turn results in a lack of prothrombin and thus impairs the clotting of blood.

An excess of the coumarin derivative that counteracts vitamin K may also cause deficiency of this vitamin. Injections of vitamin K are given to compensate for this. This vitamin and some others can be synthesized by certain microorganisms in the intestine, so that an additional supply of this vitamin is thereby provided. A broad-spectrum antibiotic, which is effective against a wide variety of microorganisms, may destroy these useful ones and therefore, as a side effect, cause vitamin deficiency.

OBESITY

Obesity (corpulence, adiposity) is excess fat on the body. A slight amount of overweight in athletes and in persons regularly engaged in strenuous physical labor may arise in consequence of muscular development. In persons afflicted with dropsy there is overweight due to abnormal accumulation of fluid in the tissues. Obesity is not just an aesthetic problem, but can also fairly be described as a disease, as it appears from the statistics of life insurance companies (and from such statements as "suicide with fork and knife" and "man digs his grave with his teeth").

The cause of obesity is excessive intake of food and/or too low a level of energy consumption: in relation to his physical activity the obese person always eats too much. Such people sometimes assert (and believe) that they "extract more nourishment from their food" than thinner people do. But this can hardly be true, since the feces of the latter are not found to contain more undigested or unabsorbed nutritive matter than those of fat people. Obesity can, in other words, be said to result from an imbalance between food eaten and energy expended. The underlying causes are usually quite complex and difficult to ascertain precisely. In fewer than 10 percent of cases of obesity is there any detectable specific abnormal condition or disease to account for it. Such abnormalities include, for example, damage to certain centers in the midbrain which in normal persons produce a sensation of repletion. These appetite-controlling centers are called *appestats*, in analogy with the thermostat as a controller of temperature. Disturbed glandular function, e.g., overactivity of the pancreas or the adrenal cortex (Cushing's syndrome, see pages 358 ff), may also—though rarely—cause obesity.

If there are, as in over 90 percent of obesity cases, no detectable physical causes, it may readily be supposed that psychosomatic factors are involved, i.e., emotional factors which can cause physical changes in the body. Sometimes a person's environment, a bad example, a desire to be sociable or convivial, or mere habitual overeating (cooks, butchers, overfed children, elderly people who ignore the fact that their caloric requirements are diminishing) is the deciding condition. In some instances sheer self-indulgence is the key factor; more frequently it is a desire to seek solace for loneliness, disappointment, grief or frustration that drives people to pamper themselves with too much food. Women in the menopause often indulge in sweets and rich foods as a kind of compensation before they come to terms with their "change of life." Boys approaching puberty sometimes put on weight, perhaps encouraged by a too-indulgent mother. Such children are fat simply from overeating and not because of any glandular disorder, even though in such cases puberty may be late in coming. Sporting activities and a reduced intake of food will generally help to achieve readjustment to the normal environment and life pattern of boys of their own age.

The distribution of excess fat on the body differs between the sexes. In men, the stomach, back and neck are the areas where fat more particularly accumulates; in women it is in the hips, buttocks, thighs and upper arms that are most affected. With increasing obesity these differences tend to disappear because of the general accumulation of fat on all parts of the body in men as well as women. Fatness in women sometimes occurs only below the navel and especially in the thighs and legs.

To determine whether someone is overweight it may be necessary to refer to statistical tables, which sometimes also take account of build and bone structure. A typical (very approximate) "formula" is that a person's weight in kg should be equal to his height in cm minus 100; for example, a person 1.75 m (175 cm, or 5′9″) in height should normally weigh 75 kg (165 lb). Another formula, somewhat more refined, is: weight in kg is height in cm multipled by chest circumference in cm, divided by 240.

An experienced observer can judge obesity by the thickness of the folds that can be felt at the back of the upper arm, at the side of the waist or on the back below the shoulder blades.

In view of the reduced life expectancy associated with obesity, this condition can, as already remarked, justifiably be regarded as a disease (Fig. 5-1). Although fat people at first feel quite well, shortness of breath and reduced capacity for physical exertion are signs that they have excess weight to carry and move about. Breathing difficulties may occur also because the fat abdomen impedes the respiratory movements of the diaphragm. The life-shortening effect of obesity is more particularly bound up with the complications it is liable to involve. Arteriosclerosis and high blood pressure are likely to develop more quickly and in a more serious form in the obese; gout and renal calculus (stone in the kidney) are more frequent. Accumulation of fat, moreover, increases the body's consumption of insulin, and as a result of premature exhaustion of the pancreas, there is a higher risk of diabetes. Other disorders that are more likely to occur in overweight persons include flatfoot, trouble due to premature wear of joints (knee, hip, vertebral joints), varicose veins and thrombosis (Fig. 5-2).

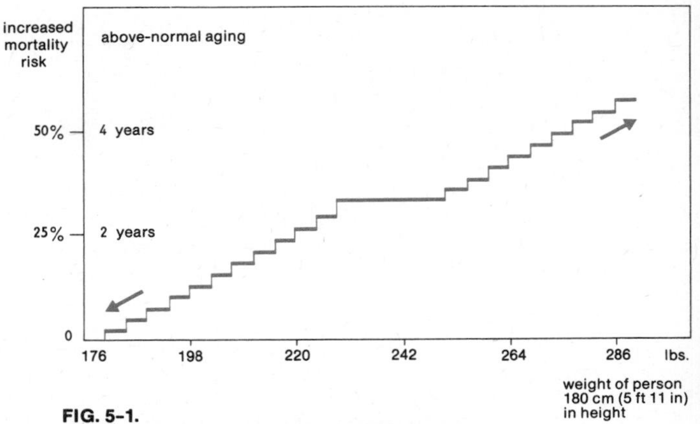

FIG. 5-1.
Premature aging and mortality of obese persons.

The treatment of obesity should consist first and foremost in effectively reducing the consumption of food. The patient should be made to realize that he eats too much in relation to his physical activity. Many fat people more or less consciously conceal or gloss over this fact, are "secret eaters," and have to be educated to control their appetite and keep down their caloric intake to a precisely prescribed amount. A person doing strenuous physical labor requires about 4000 Calories per day, a housewife requires 2600, a man in a sedentary occupation requires 2500, and a woman in a sedentary occupation requires 2200. The intake of obese persons who wish to reduce weight should be kept 1000 Calories below these values. Since 1 kg of adipose tissue ("fat") with low water content has an energy equivalent of above 6000 Calories, a daily "deficit" of 1000 Calories will cause the patient to lose weight at a rate of about 1 kg (2.2 lb) per week. At first, as a result of water and salt losses, the rate of weight reduction is more rapid.

Caloric intake with 1 g of fat consumed is about 9, with 1 g of carbohydrate or protein about 4, and with 1 g of alcohol about 7 Calories. In a long-term dieting course it does not greatly matter what particular combination of

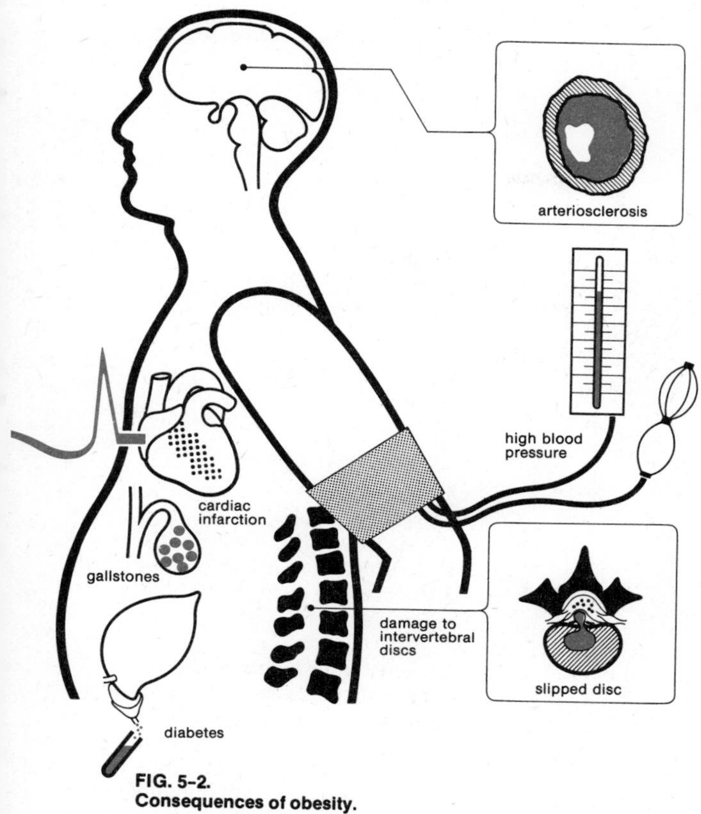

FIG. 5-2.
Consequences of obesity.

foods is consumed, provided that its caloric value is within the prescribed limits and that it provides at least 65–80 g of protein daily, together with adequate potassium and vitamins. People with a tendency to put on weight should live by the principle: "Stop eating just when you are enjoying it most." By the same token, special weight-reducing "formula" diets based on water-soluble nutrients in powder form, with no specially attractive flavor, are often very effective in achieving their object. Some patients cannot endure such diets in the long run and need something more appetizing. For a long time the so-called Hollywood diet was popular, based on lean meat (beef or veal steaks) and fish.

Physical activity is beneficial for various reasons, but will help to reduce weight only if the higher rate of energy consumption is not canceled or indeed exceeded by increased intake of food. On the other hand, when someone gives up walking to work every day, for example, or even when he moves into a ground-floor apartment instead of having to climb stairs several times a day, he may noticeably put on weight. Consuming 60 Calories less energy per day corresponds to a daily weight increase of 10 g. Over a whole year this could mean an increase of about 3.5 kg (7½ lb).

The effectiveness of massage and sauna baths as a means of losing weight is often overrated. In the sauna, more water and salt than fat are shed, while massage is more likely to have a slimming effect on the masseur than on the massaged. Obesity due to psychological, more particularly emotional, factors can sometimes be cured by psychiatry or even by an act of will on the patient's own part, once he has become consciously aware of the true nature of his weight problem. Drugs and other preparations designed to curb or reduce appetite should be used only under proper medical supervision; they can be useful mainly in conditioning the patient to the right state of mind, but cannot relieve him of the need to change his eating habits.

EMACIATION

Emaciation denotes the consequences of malnutrition. It is therefore an abnormal condition, quite distinct from congenital leanness which exists despite adequate nutrition and which normally involves no hazard to health. Emaciation may occur as a result of increased caloric losses (e.g., in diabetes associated with excessive excretion of sugar in the urine, see page 179, or in hyperthyroidism, i.e., overactivity of the thyroid gland, see page 352). Much more often it is brought about by an insufficient supply of calories. Hunger is today still the cause of millions of deaths in underprivileged parts of the world. Deficient nutritional intake may, however, be caused by obstruction of the pylorus (the orifice through which food passes from the stomach to the duodenum) or by severe and prolonged diarrhea. A particular form of emaciation in Western societies occurs from loss of appetite (anorexia). This may take the morbid form of a neurosis called anorexia nervosa, which occurs chiefly in girls and young women (see later). As a result of the rejection of food the patient becomes undernourished, so that the body's resistance to infection is lowered. Physical and mental illness of a more general kind may also cause loss of appetite. In chronic infectious diseases, such as tuberculosis or cancer, toxic substances or decomposition products may

FIG. 5-3.
Causes of emaciation.

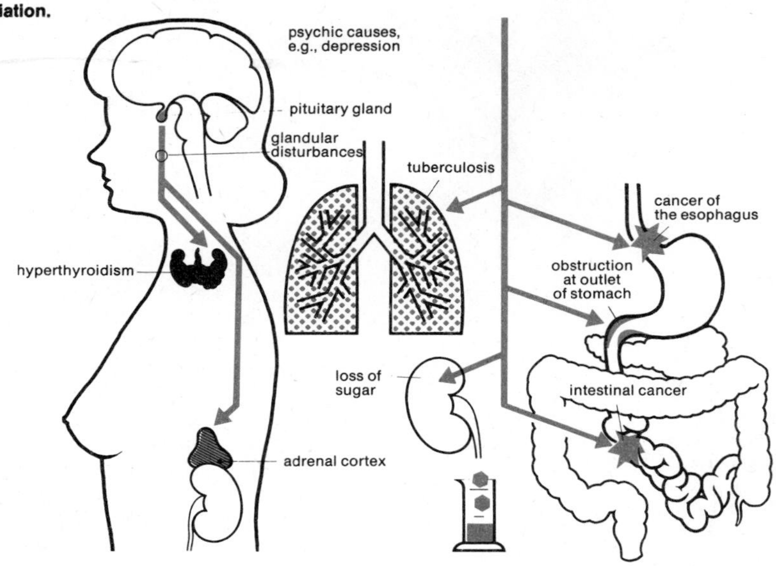

FIG. 5-4.
Consequences of emaciation.

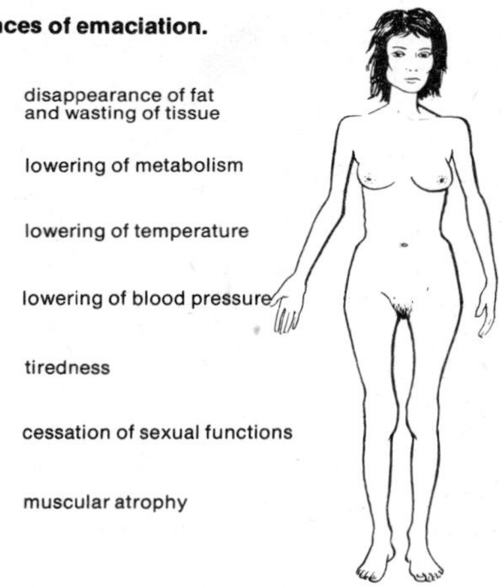

impair the patient's ability to derive nourishment from the food consumed. Rapid loss of weight under unchanged conditions of diet and manner of life must therefore be viewed with suspicion, and medical advice should be sought.

Besides underweight, symptoms associated with emaciation (Fig. 5-4) include lack of interest, listlessness, lack of energy, tiredness and diminished sexual urge. The body's other functions, too, perform in a low key, as it were; blood pressure, blood sugar content, basal metabolism and body temperature are all lowered. The skin is yellowish and pale, the bones protrude, and deep hollows appear under the collarbones. Often there are diarrhea and copious discharge of urine. Eventually the mind is affected; there is breakdown of cerebral matter, resulting in apathy, confusion and semiconsciousness. Death often occurs in consequence of circulatory failure, usually brought on by a supervening infection.

In many cases of emaciation, treatment will consist in treating the basic disease, if any, such as tuberculosis, cancer or glandular disorders. Indifference to food due to neurosis may be combated by tempting the patient with an appetizing high-calorie diet, taking the best possible advantage of such residual appetite as the patient still possesses. Cool and soft foods such as ice cream may help to stimulate the appetite. As a last resort, feeding through the nose may be utilized, with liquid nutrients given five or six times a day. In extreme cases, nutrients such as amino acids, glucose, salts and vitamins can be infused intravenously (intravenous feeding).

Artificial feeding may be necessary in dealing with *anorexia nervosa,* a neurotic condition chiefly affecting girls and young women. It is accompanied by absence or suppression of menstruation (amenorrhea). The patient is usually ill-adjusted to her environment and finds it difficult to make the transition to adult womanhood. Her neurosis may be a kind of protest against growing up, in the sense that she rejects adult sexuality, the prospect of childbearing, etc., this state of mind being reflected in her rejection of food. At first she may be motivated by an exaggerated desire to reduce weight and be slim or has an unfounded fear of the supposed harmful effects of certain foods. Such a patient is often active and energetic, even overactive, until emaciation and sheer physical weakness prevent this. Some cases of anorexia nervosa are not detected until they are well advanced, because the patient appears to eat enough, but afterward secretly gets rid of her food by vomiting. This is also likely to happen if food is forced on her. The condition is potentially dangerous because of the risk that an infection may supervene in the patient's weakened state. Nearly one in three cases of anorexia nervosa ends in death despite artificial feeding and psychiatric treatment.

VITAMIN D, RICKETS, OSTEOMALACIA

Vitamin D is a fat-soluble vitamin which may be absorbed with the food or be formed in the skin by the action of sunlight. The general designation comprises at least two substances which are closely related chemically and are almost identical in their action—namely, vitamin D_2 (*calciferol*) and vitamin D_3 (*cholecalciferol*). Vitamin D_3 is found in cod liver oil and in various other foods of animal origin, including egg yolk. Vitamin D_2 is of vegetable origin and is found especially in yeast and bread. The substance contained in yeast is *ergosterol,* which, on exposure to ultraviolet radiation, is transformed into vitamin D_2; ergosterol present in the skin as an inert substance is thus activated by sunlight.

Vitamin D is essential in *calcium* and *phosphorus* metabolism; it is therefore necessary to the formation of healthy bone. The human body contains between 1.2 and 1.5 kg of calcium; 99 percent of this quantity is present as calcium phosphate in the bones. The rest of the body's calcium is essential to a number of vital processes. Calcium ions play a part in the clotting of blood, the transmission of nervous impulses to muscles, and the contraction of muscles. They are also important with regard to the (normally low) permeability of the cell membranes and the walls of the capillaries. A constant level of calcium of about 0.1 per liter of blood is maintained by the interaction of vitamin D with the hormone secreted by the parathyroid gland (parathormone). Vitamin D promotes the absorption of calcium and phosphate from the intestine, assists the deposition of calcium salts in the bones, and reduces the excretion of calcium. When the level of cal-

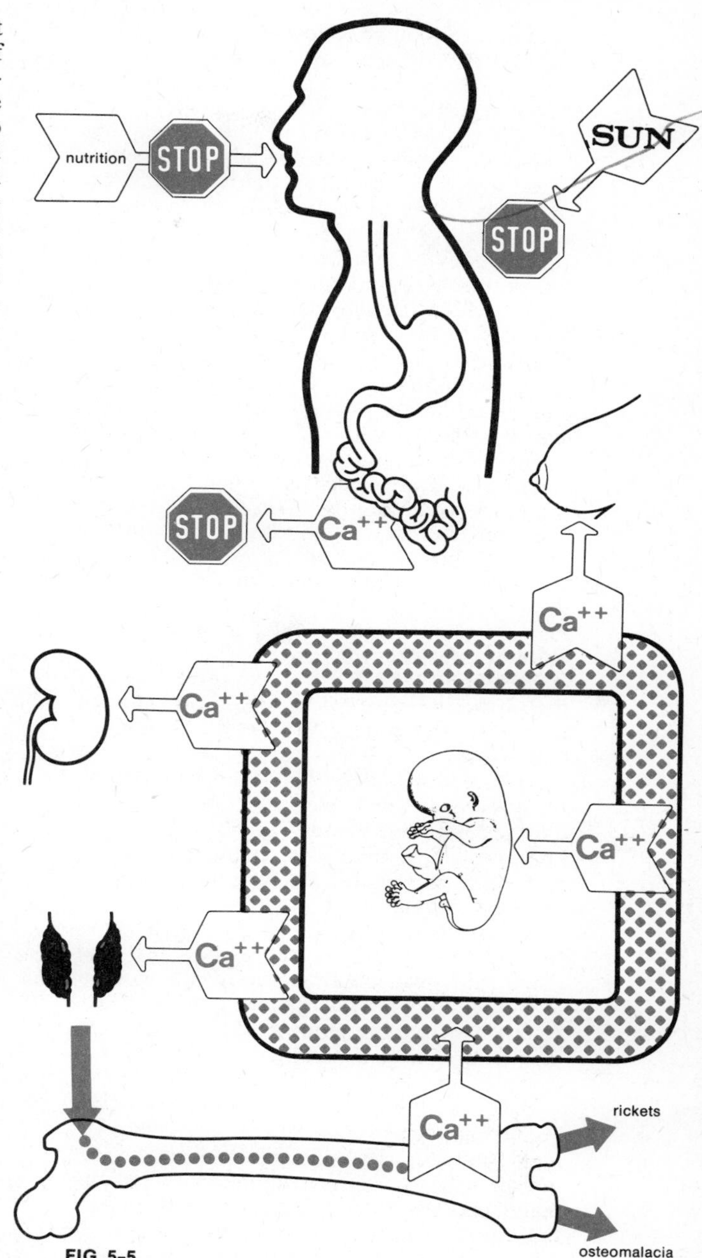

FIG. 5-5.
Causes and consequences of vitamin D deficiency.

cium in the blood goes down, parathormone is released from the parathyroid gland and dissolves calcium from the bones, thus making it available to the blood. In this way, under the protection of vitamin D, calcium can be taken from older and more stable bones and be transferred by the blood to zones where bone growth is still in its early and as yet cartilaginous stage.

Deficiency of vitamin D causes *osteomalacia,* a disease marked by softening of the bones, which become flexible and brittle, so that deformities arise. It occurs in adults, chiefly in women and especially when the calcium requirement is increased during pregnancy and in the lactation period. Children may be afflicted with a form of osteomalacia called *rickets* (rachitis). It may occur in young children who are deprived of sunlight (which forms the vitamin in the skin) and whose diet does not contain enough vitamin D (which is present especially in fish, eggs and milk). Despite adequate intake of calcium, absorption of this mineral from the intestine is so diminished that not enough is available for normal bone building. Besides, the low blood calcium level causes increased release of parathormone with the further adverse effect of withdrawing calcium from the bones. As a result, not enough calcium salts are deposited to give rigidity to newly formed bones. The condition is most likely to occur in babies and infants, also in children between 5 and 7 years of age, when bone growth is rapid and requires much calcium and phosphate. Rachitic bones remain soft and become deformed by the action of body weight and pull exerted by muscles. Bowlegs and deformations of the chest (flattened sides, so-called rachitic rosary: nodules at sternal ends of ribs), pelvis and skull (head squarish in outline, thin skull bones) occur. The teeth are also affected by deficiency of calcium salts.

Babies and infants should regularly be given vitamin D to prevent any risk of rickets. If a child is already afflicted with the disease, treatment will consist in providing a carefully regulated diet with adequate calcium and substantial amounts of the vitamin. This should be done under regular medical supervision; otherwise there may be a risk that, under the calcium-depositing action of vitamin D, the calcium level in the blood is lowered too much, so that *spasmophilia* (a tendency to tetany: spasm of muscles) may develop, with convulsive muscular twitching affecting, among other parts, the hands and mouth. Possible spasm of the glottis in such cases involves a risk of choking.

Excessive use of vitamin D in adults and in children must be avoided, as it can lead to *hypervitaminosis,* with toxic effects and risk of death due to excessively high blood calcium levels, resulting in deposition of calcium in what normally are soft tissues, especially in the kidneys.

GOUT

Gout is a congenital disease which occurs predominantly in men after the age of 40. It is due to hyperuricemia, i.e., abnormally high uric acid content of the blood caused by disturbed excretion of this acid. As a result, uric acid is precipitated in tissues, especially at joints and in tendon sheaths. The symptoms of gout are more particularly caused by crystals of uric acid salts in and around certain joints.

Progressive chronic gout usually begins with acute arthritis (inflammation of a joint), mainly in the lower limbs, the great toe being nearly always involved. Gout affecting the joints of the foot and great toe is called *podagra.* An acute attack of gout begins suddenly, with extremely painful inflammation and swelling of the joints affected. Sometimes it is preceded by rheumatic trouble, nervousness, nausea and increased discharge of urine. Without treatment an attack may last for days or weeks, then subsides, leaving the patient free from discomfort until the next attack. In the further course of the disease the articular cartilage is destroyed by the uric acid crystals precipitated in the joints. The joints may become permanently deformed and stiff, with large deposits of uric acid (which are called *tophi*) that are externally visible as swellings. Sometimes the acid is also deposited in the kidneys and may damage them.

Acute gout is treated with colchicine, an alkaloid drug derived from the autumn crocus (*Colchicum*), which relieves the pain fairly quickly. Other drugs may also be used, such as glucocorticoids or phenylbutazone. The treatment of chronic gout strives to lower the uric acid level in the blood by increasing the excretion of the acid. As the acid is filtered in the kidneys and is then largely reabsorbed in the renal tubules, excretion of uric acid can be greatly increased by administration of a drug, such as probenecid or sulfinpyrazole, which inhibits this reabsorption. Long-term treatment, consistently applied, will gradually re-

FIG. 5-6.
Treatment of gout.

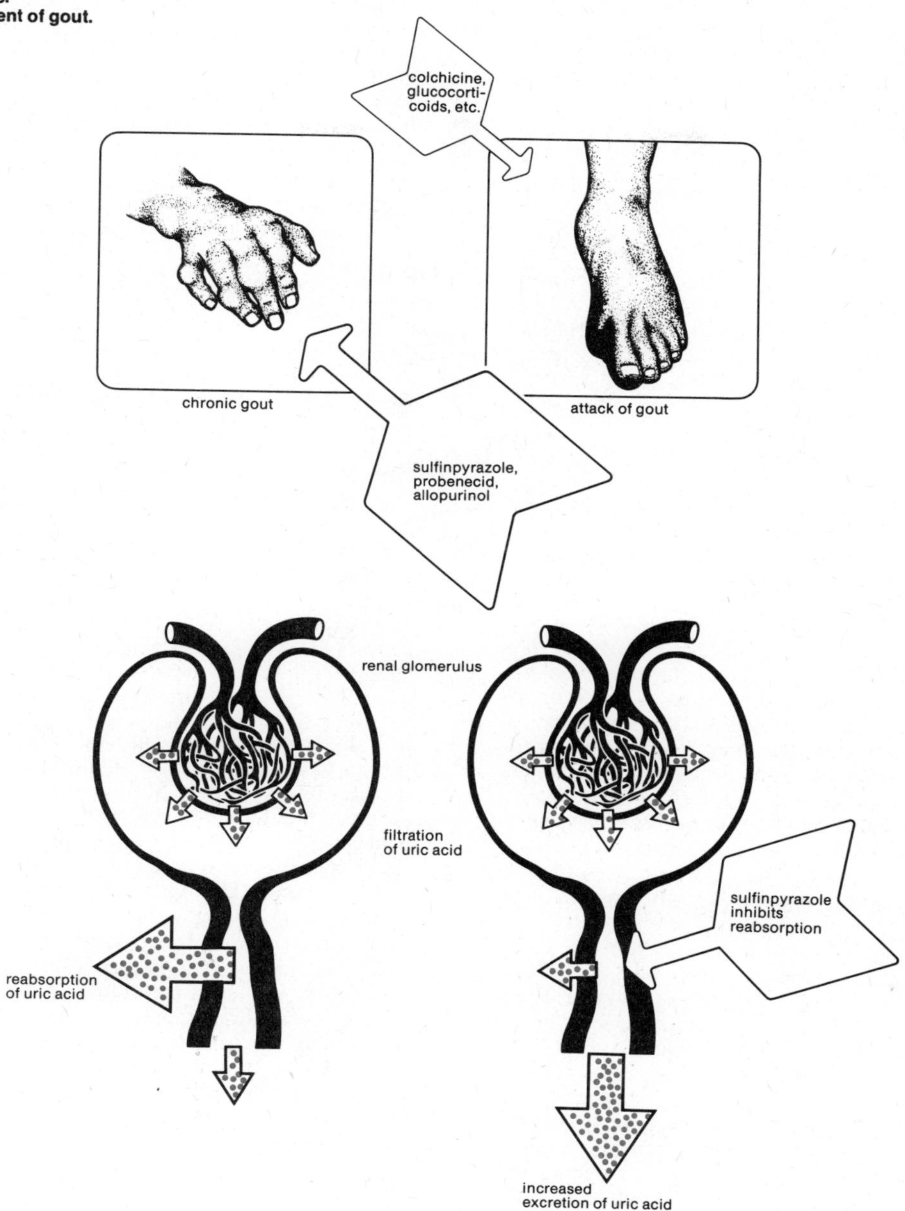

duce the gouty swellings, and the deposits of uric acid in the joints are also reduced. Drug therapy should be supported by a suitable diet devoid of foods rich in purines (a group of compounds that includes uric acid). Thus meat, liver, kidney, anchovies, peas and beans should be avoided. Another type of treatment for gout aims at inhibiting the formation of uric acid in the body and thereby lowering the uric acid level in the blood. Thus the drug allopurinol diverts the enzyme that promotes the formation of the acid.

FIG. 5-7.
Anatomy of the mouth.

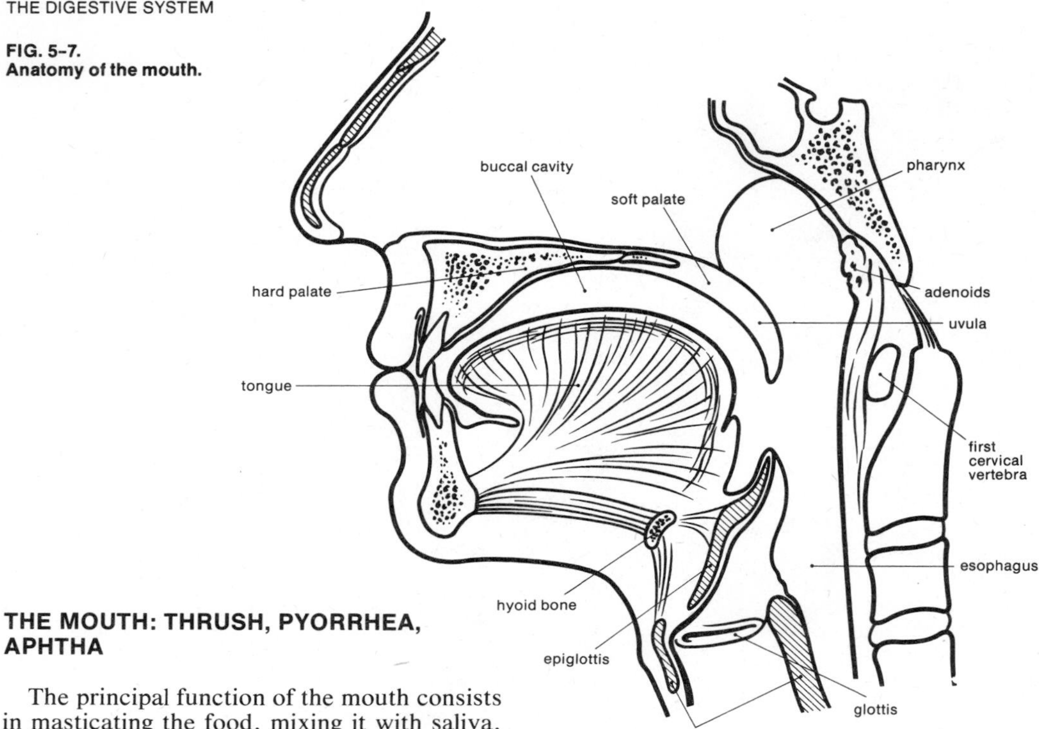

THE MOUTH: THRUSH, PYORRHEA, APHTHA

The principal function of the mouth consists in masticating the food, mixing it with saliva, and passing suitable quantities to the esophagus (gullet) for swallowing. The mouth cavity is lined with mucous membrane amply provided with tactile sensory nerve endings as well as taste-sensitive and temperature-sensitive ones. Thus the mouth ensures a last check on the food before it is swallowed. In addition, in man, the mouth has an important function to perform in connection with speech (see pages 431 ff).

The mouth comprises two parts: the vestibule and the buccal cavity. The *vestibule* is bounded by the lips and cheeks anteriorly and by the gums and teeth posteriorly. The *buccal cavity* is bounded by the teeth and jaws in front and at the sides; to the rear it communicates with the oral part of the pharynx; the roof is formed by the *palate,* and the floor is formed by muscles and the attachments of the tongue. The front two thirds of the palate is bony (hard palate); behind it is the soft palate, whose movements are important in swallowing and in speech; the uvula hangs down from it. The floor of the mouth is covered by the *tongue,* a highly mobile muscular organ covered with mucous membrane containing large numbers of glands and sensory cells. The posterior part of the top of the tongue is called the *dorsum* and carries tiny projections, the papillae; taste buds are scattered in the mucous membrane of this part. The membrane of the undersurface of the tongue forms a fold, the *frenulum,* which is attached to the floor of the mouth. On each side of the frenulum are the openings of the ducts of the submandibular and sublingual salivary glands. There is a third pair of salivary glands, the parotid glands, on each side of the face below the ear. *Saliva* is secreted as a reflex action in response to mechanical or chemical stimuli acting on taste buds. Saliva begins the process of food digestion by moistening the contents of the mouth, acting as a lubricant and solvent. It contains enzymes, including *ptyalin,* which splits starch into smaller molecules as the first stage of its subsequent conversion to simple sugars.

Large numbers of bacteria are generally present in the mouth. Nevertheless, injuries to the mucous membrane seldom become infected: local resistance to harmful bacteria appears to be very high. There are, however, certain diseases, mostly of an inflammatory character, that do affect the mouth. *Thrush* is an infection caused by a fungus (*Candida albicans*) and is characterized by the formation of white patches on the mucous membrane.

The fungus is normally suppressed by bacteria, but if the latter are destroyed by antibiotics or if the patient's general resistance is lowered, the fungus may develop and spread, eventually forming a white coating to the whole of the inside of the mouth cavity. The disease can be very painful. It can usually be cured by fungicide mouthwashes; in some cases, however, antibiotics may be necessary.

Gingivitis is inflammation of the gums from bacterial infection. It may be due to improper dental hygiene or to other causes. *Pyorrhea* is caused by an infection that produces a discharge of pus from the gums, which become swollen and subsequently recede, so that the teeth look long and become loose. The breath has an offensive smell (halitosis).

Aphtha is a term sometimes used as a synonym for thrush, but it also denotes a small ulcer on the mucous membrane of the mouth, usually in the vestibule. Such ulcers are often a symptom of viral infection (herpes simplex) or, less frequently, are an accompanying phenomenon of infectious diseases such as measles, scarlet fever or whooping cough. Treatment is by means of mouthwashes.

HEARTBURN, ERUCTATION

Acid liquid rising from the stomach causes a burning sensation in the esophagus (gullet). Called *heartburn,* it is an accompanying symptom of indigestion (dyspepsia) and occurs when the cardia—i.e., the stomach inlet orifice from the esophagus—is open. Heartburn is usually due to a temporary excess of hydrochloric acid secreted by the stomach lining, but may also be caused by other acids formed in fermentation processes in the stomach. These are mostly acetic acid, lactic acid or propionic acid, which are formed especially after a heavy meal of rich or sweet foods—particularly if the stomach itself produces relatively little hydrochloric acid. These fermentation products may also cause halitosis (offensive breath). Heartburn is usually felt to be located at the lower end of the sternum; sometimes it is felt all along the esophagus up to the root of the tongue. As a rule, an antacid, i.e., a drug that counteracts acid, brings relief. Sodium bicarbonate is a familiar antacid; there are many others, e.g., aluminum hydroxide or magnesium oxide.

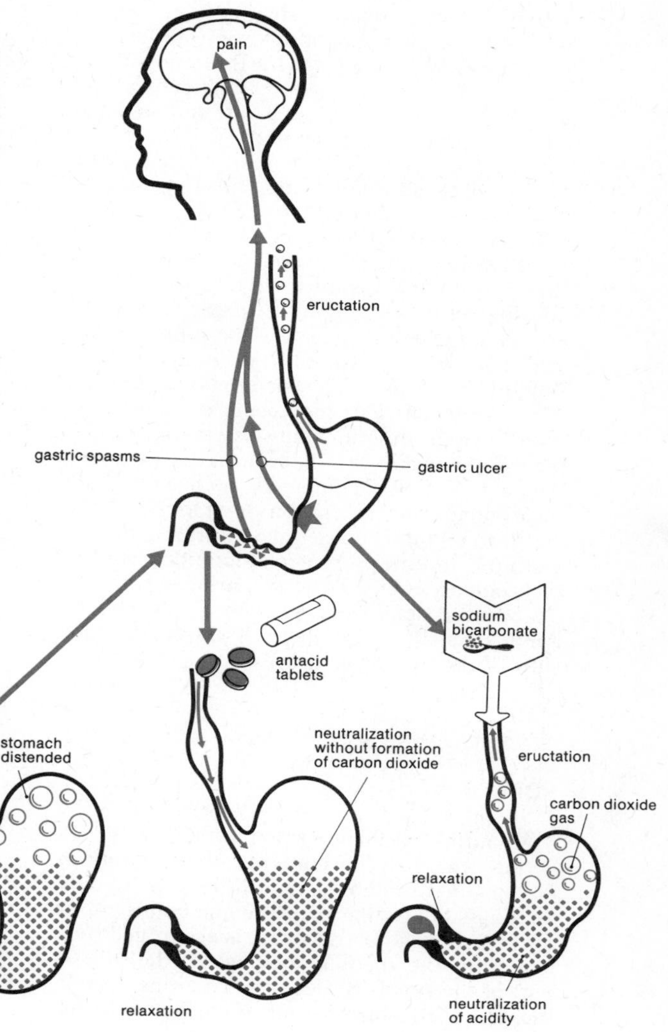

FIG. 5-8. Increased eructation due to disturbed gastric function, and relaxation of gastric spasm due to antacid.

Heartburn due to contact of acid gastric juice with the lining of the esophagus often occurs in connection with so-called *hiatus hernia,* a hernia of the diaphragm (midriff) in which part of the stomach bulges up through the opening (hiatus) in the diaphragm through which the esophagus passes. This type of hernia is very common, especially in fat and elderly people, but usually does not cause symptoms other than discomfort due to indigestion and heartburn. In most cases these can be relieved by adjustment of the diet.

With *eructation* (*belching*) the glottis is closed and gas is expelled from the stomach upward through the cardia and esophagus, accompanied by the characteristic belching sound. There is usually some gas present in the stomach, as is apparent from the level of liquid revealed on x-ray. Furthermore, air is swallowed in the processes of eating and drinking. It is this gas and air that are discharged in belching, accompanied by a mild sensation of relief. In babies, belching may be necessary to get rid of air that fills the stomach and prevents further intake of food.

Some nervous people and people who are in the habit of eating their food hastily may swallow considerable quantities of air along with it, causing eructation. More often, excessive eructation is due to spasm of the esophagus or stomach wall associated with involuntary swallowing of air and a sensation of distension of the stomach. This may occur in connection with various functional disorders of the alimentary canal, e.g., heartburn, gastritis or peptic ulcer. Some antacids such as sodium bicarbonate release carbon dioxide gas in the stomach and thus cause uncomfortable distension; this in turn promotes eructation. Sodium bicarbonate is not entirely harmless; it may disturb the acid/alkali balance of the blood. Antacids such as aluminum hydroxide do not suffer from this drawback.

VOMITING

Vomiting is an important protective reflex action whereby the contents of the stomach are ejected through the mouth. It is assisted by abdominal muscular pressure and contractions of the esophagus. The reflex is controlled by a center in the medulla oblongata, the lowest part of the brain. Besides the vomiting center, the so-called trigger zone is responsible for vomiting. The trigger zone is an area in the cerebral cortex which responds particularly to chemical stimuli (bacterial toxins, tartar emetic, pregnancy toxins, decomposition products formed as result of exposure to x-rays) and which in turn transmits stimuli to the vomiting center, which then causes the vomiting reflex (Fig. 5-9) to take place. The vomiting center itself is closely linked to other vegetative centers, particularly the respiratory center in the medulla oblongata. This is evident from the fact that vomiting is usually preceded by a sensation of nausea with increased salivation and slower breathing, together with retching, uncoordinated respiratory movements and slackening of the stomach muscles. Deep breathing can sometimes suppress the urge to vomit. A distinction can be drawn between *cerebral* vomiting, due to direct stimulation of the vomiting center, and *peripheral* vomiting, which is brought about by indirect stimulation of that center.

Vomiting is initiated by processes in the brain itself (e.g., increase in cerebral pressure due to certain diseases), emotional factors (disgust, excitement, etc.), and vasomotor processes (e.g., in migraine). In addition, smells and mechanical stimuli, especially in the throat, may bring about retching and vomiting. Vomit is discharged from the stomach when the pressure in the stomach exceeds about 25 cm water column or when chemical stimuli act upon the mucous membrane of the stomach or intestine. In seasickness or travel sickness, motion disturbs the balance organ in the inner ear, as a result of which the vomiting center is stimulated.

Impulses from the vomiting center travel along the cerebral nerves to the throat and palate, along the vagus nerve to the stomach, along the phrenic nerve to the diaphragm, and along spinal nerves to the respiratory and abdominal muscles. Impulses to the vomiting center travel along sensory fibers coming from the region of the pylorus and duodenum and along the sympathicus (sympathetic nervous system), while those from the gastric region travel along the vagus nerve and the relevant cerebral nerves. The trigger zone receives stimuli through the blood. Vomiting begins with a deep breathing movement, while the glottis and the nasopharyngeal cavity are closed (the latter by elevation of the soft palate). The esophagus is greatly distended as a result. The cardia and the stomach itself relax, and the contents of the stomach are ejected by powerful contractions of the diaphragm and

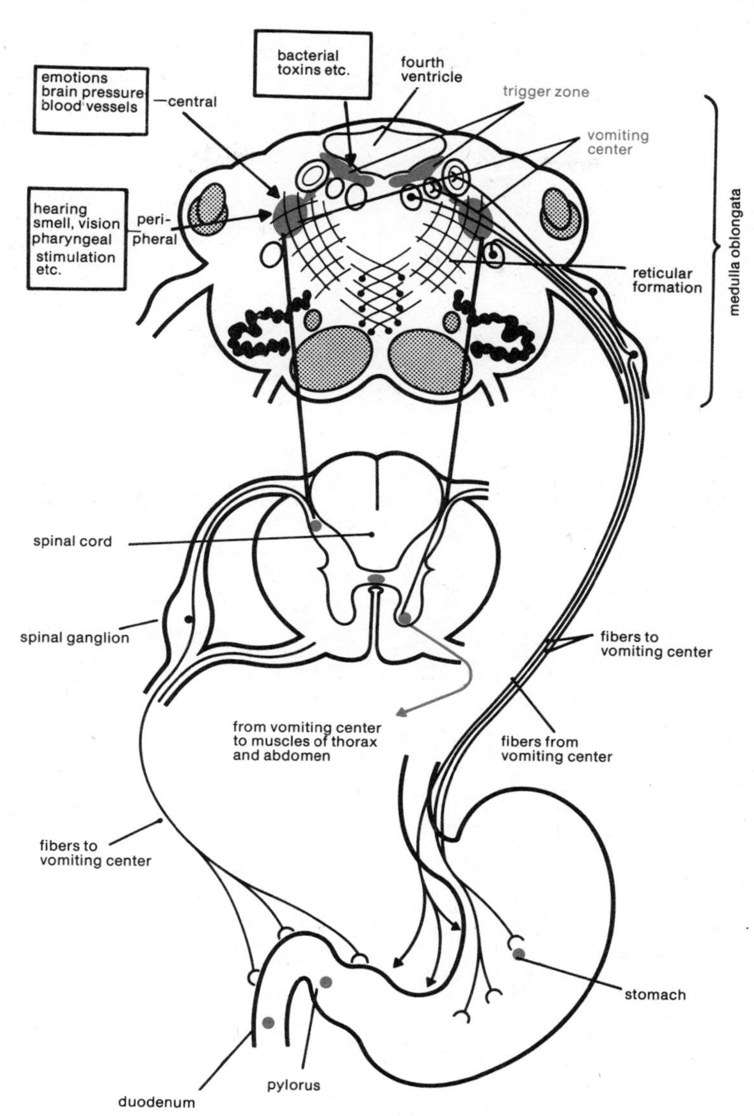

FIG. 5-9.
Vomiting reflex.

stomach muscles. Sustained or repeated vomiting can lead to excessive loss of water, salt and acid. The acid/alkali balance of the blood may be disturbed in consequence.

FLATULENCE

Flatulence is excessive gas in the stomach and intestine, which causes discomfort. The person affected tries to relieve this by eructation (belching) and/or by the expulsion of flatus. The accumulation of gas causes a sensation of repletion and distension, sometimes accompanied by painful intestinal spasms. If the distended stomach and large intestine exert pressure on the diaphragm and thus tend to thrust it and the heart upward, the symptoms of angina pectoris may ensue (see page 71). Sometimes there is difficulty in breathing and a tendency to have nightmares.

FIG. 5-10.
Causes and effects of flatulence.

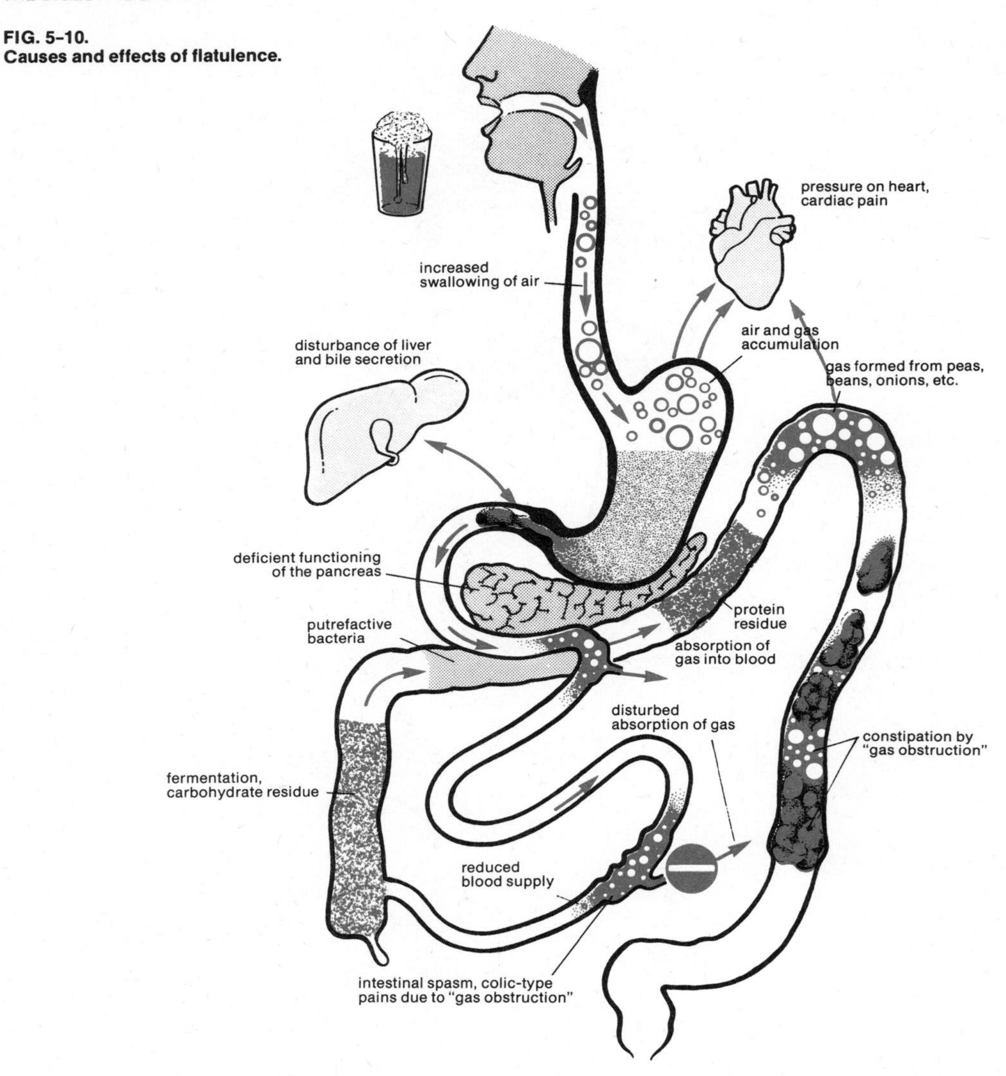

Most cases of flatulence are due to excessive swallowing of air, which may occur as the result of gulping one's food or of a nervous habit. Self-control or a more relaxed approach to eating and drinking can remedy this. Also, the rate at which the intestinal gases are absorbed into the blood or are expelled as flatus plays a part. Another cause of flatulence may be a deficient supply of blood to the intestine; this may be a sign of cardiac insufficiency. Furthermore, the tendency to flatulence increases when insufficiently digested food reaches the lower part of the alimentary canal and there gives rise to fermentation and putrefaction. This is liable to occur not only when the gastrointestinal system is overburdened with excessively large or heavy meals, but also if the digestive capacity of the juices secreted by the system is diminished. In such cases, enzyme preparations may be helpful in combating the condition. Foods with a high content of cellulose—e.g., cabbage, onions, radishes, peas and beans—promote flatulence. Overactivity of the pancreas, diseases of the liver or gallbladder, chronic diseases of the alimentary canal, and diarrhea may also be contributory factors. Flatulence is frequently an attendant phenomenon of constipation.

If flatulence is due not merely to improper eating and drinking habits or to nervous swallowing of air, but to such causes as disease of the pancreas, liver or alimentary canal, or to diarrhea or constipation, it will be necessary to treat the basic condition or disease. The same applies to cardiac or circulatory disorders insofar as they result in impaired blood supply to the intestine and thus contribute to flatulence. An obvious precaution is to avoid all food that particularly promotes the formation of gas. Thorough mastication and adequate mixing of each mouthful with saliva will improve digestion. Moderate physical activity of any kind stimulates bowel action and reduces the tendency for gas to be trapped and cause painful spasms. In severe cases of flatulence causing distension and discomfort, drugs to relieve spasm can help. Carminatives are preparations that are claimed to relieve flatulence; they may be given in an enema. Purgatives should not be taken for long periods without medical advice.

STOMACHACHE, PAIN IN THE BOWELS

Stomachache, i.e., pain in the stomach, may be due to many causes (Fig. 5-11)—for example, peritonitis, intestinal diseases or disorders, flatulence. Pain radiating from small areas actually affected by a disease may involve the entire abdomen and may even extend into it from the thorax, e.g., in cardiac infarction. Other pains may be more precisely localized; for instance, pain in the stomach is often confined to the epigastric region (upper part of the abdomen, over the stomach), pain in the small intestine is located in the middle part of the abdomen, and pain in the large intestine is centered in the hypogastric region (below the navel). Smaller areas where disease is concentrated may be associated with even more closely circumscribed areas of pain.

Disorders affecting individual organs are sometimes recognizable not only by their location but also by the nature of the attendant pain. Thus, a gastric ulcer or a duodenal ulcer causes a piercing pain; there are painless intervals, and the pain may recur after a certain lapse of time after a meal. Intestinal pains are frequently of the colic type, i.e., occurring in acute spasms. The pain associated with colic of the bile ducts (biliary colic) likewise manifests itself in spasmodic attacks; it is centered in the right-hand part of the upper abdomen and radiates along the lower edge of the ribs into the right shoulder. Renal colic causes pain that radiates from the kidney region along the ureter in the forward and rearward directions. Biliary colic and renal colic make the patient restless, as opposed to pain originating in the peritoneum, e.g., appendicitis.

Even greater differences are found to exist on comparison of pain originating internally with pain due to a superficial cause. Superficial pain proceeds only from the skin and the adjacent regions of mucous membrane at the surface of the body. A burn or a prick with a needle, for example, causes a clearly marked and closely circumscribed pain which produces a defensive flight or retreat reaction as a reflex. Such reactions are essentially meaningful in that they cause the organism to retreat

FIG. 5-11.
Causes of stomachache.

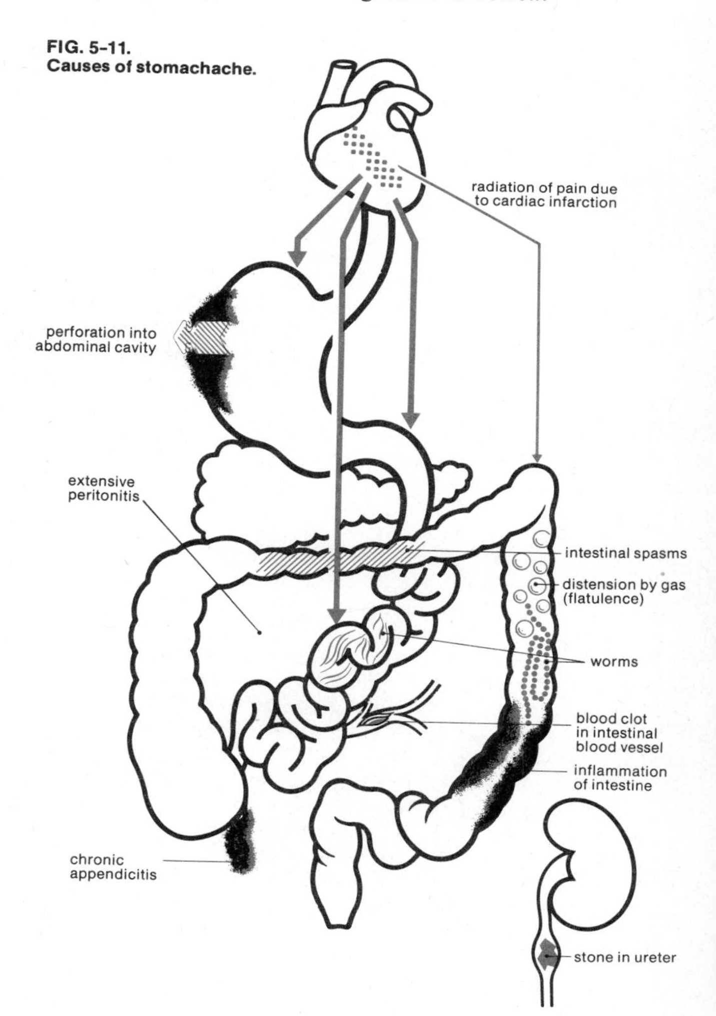

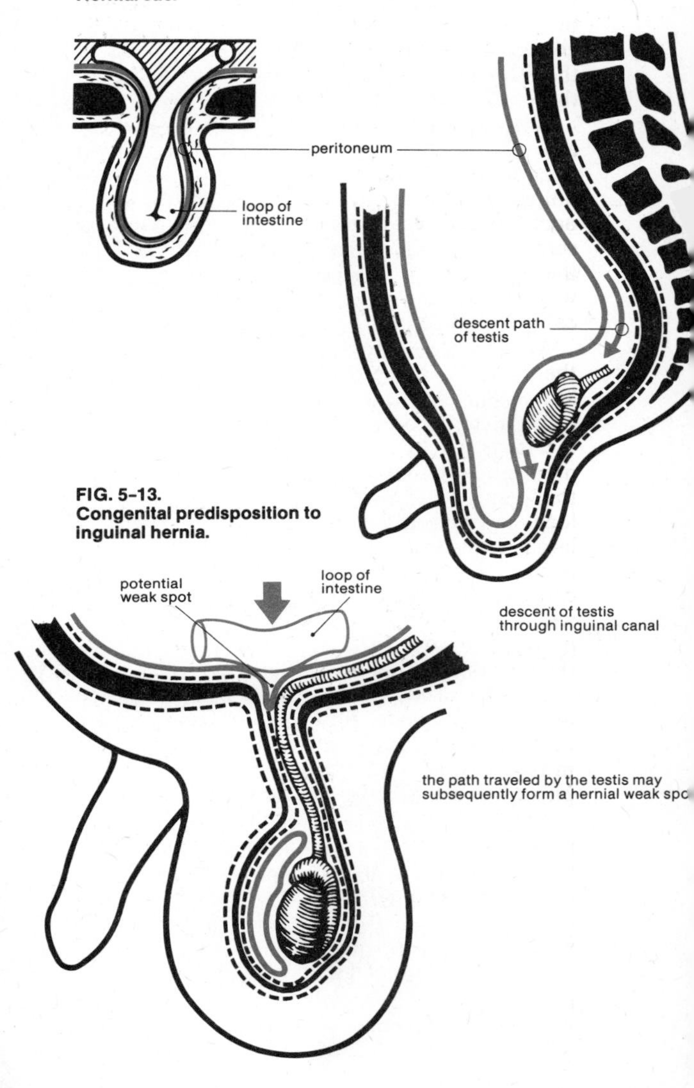

FIG. 5-12. Hernial sac.

FIG. 5-13. Congenital predisposition to inguinal hernia.

and thus withdraw itself from the cause of the pain. On the other hand, the deep-seated pain associated with disorders or diseases affecting internal organs is not so definite and clearly marked in character, but duller and yet agonizing and often of a more distressing kind. It is more difficult to locate in spatial terms and radiates to its surroundings. Internal pain of this kind does not give rise to defensive reactions of retreat or flight; instead, it induces the body to adopt an attitude of rest that will give relief.

Cuts or lesions of the heart or intestines do not in themselves cause pain, as distinct from the pain that such wounds cause in connective tissue, muscles and the edges of bones. Intestinal pain, when it occurs, is caused by inflammation, distension or deficient supply of oxygen. In many cases intestinal pain is probably due primarily to spasm of smooth muscle, which may locally cause distension, ischemia (local and temporary deficiency of blood supply) and a resulting shortage of oxygen.

Radiation of pain from a focal point occurs because different organs and parts of the body are interlinked by their pain-transmitting nerve fibers which lead to the same or interconnected nerve centers. For instance, pain originating in the gallbladder may be transmitted to the area of skin between the shoulder blades, and pain from the kidneys may be transmitted to the groin; in each case the common nerve center is located in the same section of the spinal cord.

HERNIA

Hernia (rupture) is the protrusion of an organ through the wall of the cavity in which it is normally contained. The abdominal organs (viscera) are enclosed within a cavity formed partly by bone and partly by muscle. This cavity is lined with a thin membrane, the *peritoneum,* of which a second layer moreover covers the abdominal organs. If there is a weak spot in the wall of the abdominal cavity, the organs (especially the small intestine) may protrude through a gap formed at such a spot, causing the peritoneum enfolding the organs and lining the cavity to bulge out like a pouch. This is called a hernial sac (Fig. 5-12). Such hernias of the viscera are most liable to occur in places where the abdominal wall is penetrated, e.g., by the esophagus or some other canal. *Hiatus hernia* (diaphragmatic hernia) has already been described.

The commonest type of hernia, particularly in men, is *inguinal hernia,* which occurs in the region of the groin. The hernial sac with the intestine protrudes into the inguinal canal, which contains the spermatic cord in the male. It is the canal through which the testis descended into the scrotum. In its descent the testis carries a pocket of peritoneum with it; this subsequently undergoes atrophy, but it may persist and thus constitute, as it were, a preformed space into which the hernial sac can subsequently bulge (Fig. 5-13).

In another form of hernia, which occurs in both sexes but is commoner in women, a loop of intestine can protrude through the canal containing the femoral artery and vein where they leave the abdomen and pass into the thigh; this is called *femoral hernia*.

Umbilical hernia occurs at the navel and is also more frequent in women than in men, especially in fat middle-aged women whose abdominal muscles are flabby. In babies it may occur as a result of weakness of the umbilical scar. If the hernia protrusion extends through the front wall of the abdomen, e.g., where the wall is stretched or is weakened by a scar, it is called *ventral hernia*.

Many people with hernia are unaware of their condition because the hernia, such as a femoral or an inguinal hernia, may cause little or no actual discomfort. If the condition is detected at all, it is usually in the course of a medical examination carried out for some entirely different reason. Quite often, though, a local swelling is noticed, which may become larger when the patient strains or coughs. The hernia may then become troublesome and cause discomfort when the patient is standing, walking, at work, etc. At first it is sometimes possible to push the intestinal protrusion back through the entrance of the hernial sac into the abdominal cavity. However, when pressure and friction cause inflammation and coalescence, the hernia can no longer be "reduced" in this way. It then often gives rise to nagging pains, nausea and constipation.

Further complications may arise. Thus, the protruding loop of intestine may become obstructed with gas and fecal matter, and there may be strangulation, i.e., the loop is pinched at the entrance to the hernial sac *(strangulated hernia)*. As a result, the supply of blood to the hernia may be cut off, so that gangrene may develop. This is a dangerous condition, rectifiable only by surgery. Strangulation may occur after a severe strain, such as lifting a heavy weight or severe coughing, when the inflamed or distended abdominal muscles around the entrance to the hernia may suddenly contract. The usual symptoms are sudden pain, followed by vomiting and obstruction of the intestine. If peritonitis develops, there are also fever, pallor, rapid pulse, and the face is characteristically sunken and hollow. In order to avoid complications, the best treatment for hernia is surgery. A truss, which is a device comprising a supporting pad for holding a hernia in place, is used only in cases where an operation is impractical or ill-advised.

STOMACH

The stomach is a baglike enlargement of the alimentary canal. It serves as a receptacle for the storage of food, which it prepares with the aid of a digestive fluid (gastric juice) secreted from its lining. After its sojourn in the stomach the food mass, now referred to as *chyme*, is discharged through the pylorus into the duodenum, the first part of the small intestine.

The principal parts of the stomach are the *cardia* (cardiac orifice) through which the food enters the stomach, the *fundus* which rises domelike to the left of the esophagus, the *body* of the stomach, and the pyloric *antrum* which leads to the *pylorus* (pyloric orifice). The stomach is bounded on the right by the short and concave lesser curvature, and on the left by the longer and convex greater curvature (Fig. 5-14).

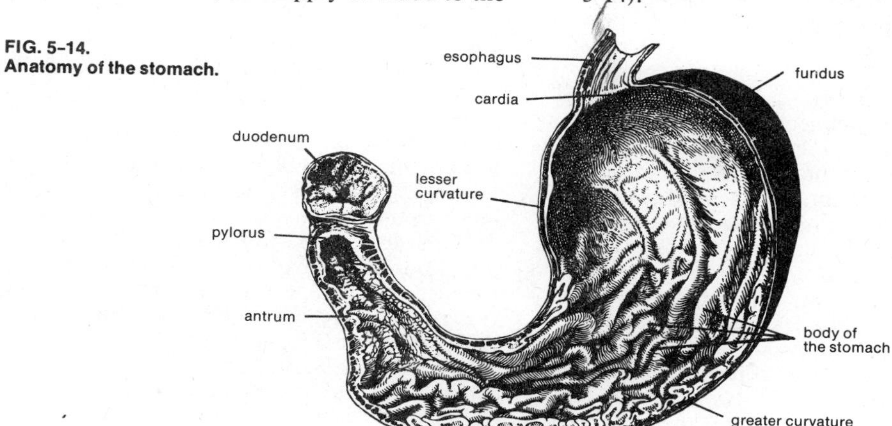

FIG. 5-14.
Anatomy of the stomach.

The *wall* of the stomach is 2–3 mm thick and consists of four layers (from the outside to the inside): serous coat, muscle coat, submucous coat, mucous membrane. The mucous membrane, which forms the inner lining of the stomach, contains three kinds of *glands:* (1) mucus-secreting glands at the cardia; (2) glands in the fundus and body comprising three types of cells: chief cells secreting enzymes (particularly pepsin), parietal cells secreting hydrochloric acid, and mucous cells secreting mucus; (3) mucus-secreting glands at the pylorus which also produce gastrin, a hormone that stimulates the secretion of the glands in the body of the stomach (see also Fig. 5-19).

The *mucus* secreted in the stomach can combine with hydrochloric acid and thus perform the important task of protecting the mucous membrane from being digested by its own juices. It also protects this membrane from damage due to mechanical causes, heat and enzyme action.

The principal function of the *hydrochloric acid* formed by the stomach lining is to bring about initial conversion of proteins and to create a chemically suitable environment for the action of pepsin. Also, the acid kills bacteria that enter the stomach with the food. Acid passing into the intestine with the chyme stimulates secretion of the pancreatic juice. If the stomach produces insufficient hydrochloric acid (anacidity), the digestion of proteins is particularly disturbed. Excessive acid causes heartburn and is a contributory factor in the development of peptic ulcers.

The chief cells of the stomach lining secrete pepsinogen, the precursor of *pepsin,* the most important enzyme of gastric juice. This enzyme, which is formed in the stomach in the presence of hydrochloric acid, converts proteins by hydrolysis into proteoses and peptones. A peptone is a secondary protein formed by enzyme action; a proteose is an intermediate product between protein and peptone.

No enzymes for the digestion of carbohydrates are formed in the stomach. However, the enzymes present in the saliva which break down carbohydrates, notably the eyzyme ptyalin, continue to perform their action in the stomach until the contents are mixed with hydrochloric acid. Small quantities of *lipase,* a fat-splitting enzyme, are secreted in the stomach, but fats for the most part pass undigested through the stomach. The gastric mucous membrane also produces a substance called *intrinsic factor* which is essential to the absorption of vitamin B$_{12}$.

FIG. 5-15.

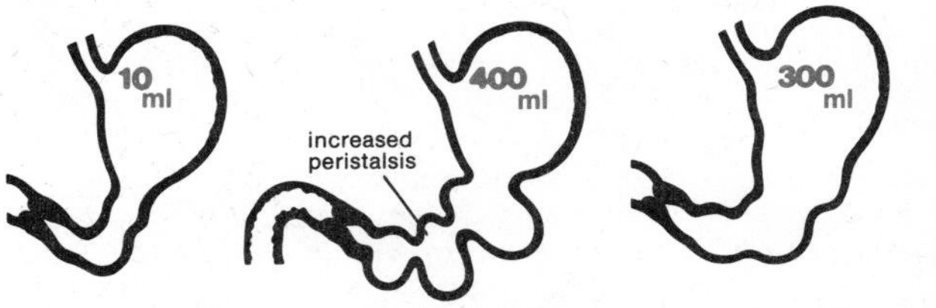

(a) Secretion by stomach at rest. **(b) Emotional stimulation by excitement.** **(c) Stimulation by sight or smell of food.**

Even when at rest the stomach secretes small quantities of digestive fluids at a rate of about 10 ml per hour (Fig. 5-15a). After intake of food this rate of secretion may increase a hundredfold. Gastric secretion can be stimulated merely by sight or smell of food (Fig. 5-15c), also by emotional stimuli such as excitement (Fig. 5-15b). As soon as food enters the mouth and is chewed, gastric secretion is further increased. The fact that secretion of digestive fluids is stimulated even without direct contact with food was demonstrated by Pavlov's famous experiments on dogs around 1900. He showed that salivation and gastric secretion were induced not only by stimuli connected with food, but also by other stimuli (e.g., a ringing bell) that dogs had learned to associate with food. Pavlov called these acquired responses "conditioned reflexes." In general, thorough mastication, as well as pleasant taste and appetizing appearance of the food, are important factors in getting the digestive process going.

After mastication and mixing with saliva in the mouth, the food descends through the esophagus and enters the stomach. The cardiac orifice closes by reflex action, thus preventing the contents of the stomach from flowing back into the esophagus. The orifice also opens by reflex action, brought about by stimulation of the mucous membrane of the esophagus by mouthfuls of food descending through it. Hot or pungent food causes a convulsive contraction (cardiospasm).

With regard to gastric activity, two functionally different parts of the stomach may be distinguished: the digestive part and the eliminatory part. Only the latter performs definite peristaltic movements—movements associated with *peristalsis,* which is the process by which the contents are propelled along the ali-

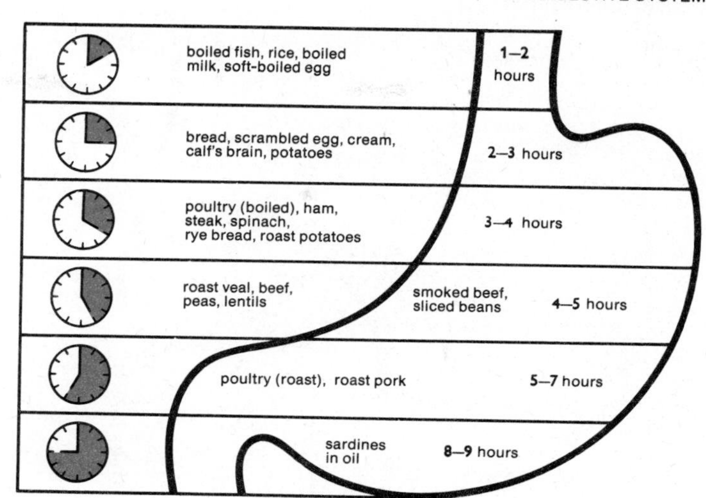

FIG. 5-16.
Retention time of various foods in stomach.

mentary canal. Alternate waves of contraction and relaxation pass along the wall of the canal. Peristalsis is brought about by the longitudinal and circular muscle fibers in the wall; it is a reflex action, induced by distension of the wall and regulated by autonomic nerves. The normal direction of peristalsis is from the mouth toward the anus. Gastric movements are to some extent determined also by the composition of the food. Fats damp down peristalsis, carbohydrates intensify it. The retention time of food in the stomach thus varies widely (Fig. 5-16).

The *gastric phase* of digestion (Fig. 5-17, see also Fig. 5-19) is initiated by chemical stimulation of the mucous membrane of the stomach and also by distension of the muscles in the pyloric region. Foods that are to some extent already predigested (e.g., meat extract) are especially effective in so stimulating the membrane, whereas carbohydrates, for example, are ineffective. Chemicals and digestive products mostly do not exercise a direct action on the gastric glands; instead, they stimulate the pyloric mucous membrane to form gastrin, a hormone that acts through the blood and causes secretion of more gastric juice. The secretion of pepsin in the chief cells of the glands of the fundus and body of the stomach is possibly also promoted by the action of another "digestive hormone." The hydrochloric acid produced by the parietal cells itself appears to act as a regulator, in that strongly acid contents will, on descending farther, depress hydrochloric acid secretion and reduce gastrointestinal motility. As all these reactions are controlled mainly by the pyloric glands, the digestive activity associated with them is sometimes referred to as the pyloric phase of digestion. Local chemical mediators participating in this activity include histamine and acetylcholine (more particularly concerned with the vagus nerve, which increases gastric activity).

The gastric phase is followed by the *intestinal phase* of digestion (Fig. 5-18, also Fig. 5-19). It commences as soon as the food has been liquefied in the stomach. Important factors governing the discharge of the chyme (liquefied food mass) into the duodenum are the volume of chyme, i.e., the pressure it exerts inside the stomach, and its composition. So

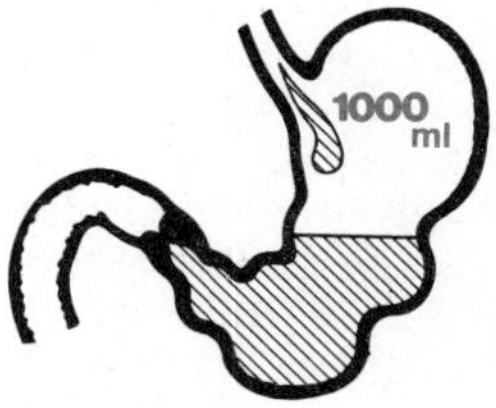

FIG. 5-17.
Gastric phase of digestion.

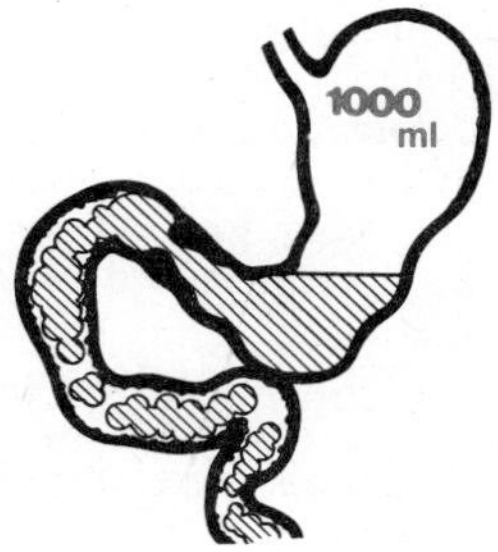

FIG. 5-18.
Intestinal phase of digestion.

FIG. 5-19.
Digestive action of the stomach.

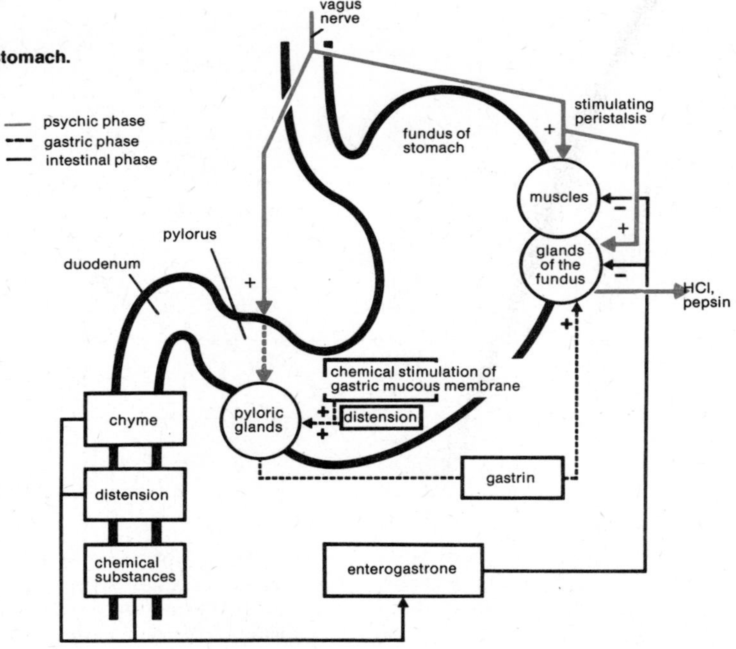

long as the stomach is not much filled and its peristaltic activity is relatively slight, so that the pressure of the contents of the stomach does not exceed the pressure in the duodenum, no discharge through the pylorus takes place. This discharge is controlled by, among other factors, *enterogastrone,* a hormone secreted by the mucous membrane of the duodenum. By depressing gastric motility and secretion, this hormone regulates the release of food from the stomach into the intestine. Whereas fats promote the secretion of enterogastrone and thus retard the emptying of the stomach, carbohydrates and proteins are less active in this respect. This is why fats are retained for longer periods in the stomach (Fig. 5-16).

GASTRITIS

Gastritis is inflammation of the mucous membrane of the stomach. It is not a very precise term and can comprise conditions ranging from mild to dangerous. Acute gastritis is a very common ailment caused by irritation of the stomach lining by excessive eating and drinking, insufficient mastication of food, food poisoning, alcohol (heavy drinking), food or drink that is too hot or too cold, acids, and some drugs (e.g., aspirin). It may, however, also occur as an attendant symptom of acute infectious diseases such as influenza, typhus, pneumonia or diphtheria.

As a result of gastritis, the movements of the stomach are disturbed, with a sensation of repletion, loss of appetite, nausea, eructation, vomiting, dry mouth. Even small quantities of food and drink are rejected by the stomach until it has had time to settle.

In cases of gastritis where the cause is known to be harmless, no specific treatment is necessary. If the condition is due to an acid irritant, it can be counteracted by a dilute alkaline solution; similarly, the effects of an alkaline irritant can be combated by giving the patient diluted vinegar or lemon juice to drink. In doubtful cases, milk may be given. Severe irritation of the stomach by caustic liquids requires urgent medical treatment; otherwise the stomach wall may be perforated. In a case of poisoning with vomiting the treatment may include gastric lavage, i.e., washing out the stomach by means of liquid introduced through an inserted tube. If the inflammation of the mucous membrane is merely an attendant phenomenon of some basic disorder, treatment should primarily aim at remedying the latter.

Chronic gastritis may develop from a recurrence of the acute form. Quite often, too, it develops in consequence of gastric ulcers. Chronically irritated and inflamed mucous

FIG. 5-20.
Gastroscopy.

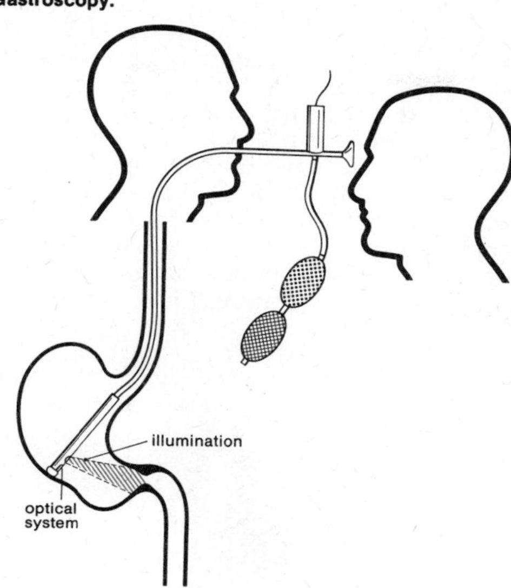

FIG. 5-21.
Obtaining a specimen of gastric mucous membrane by suction biopsy.

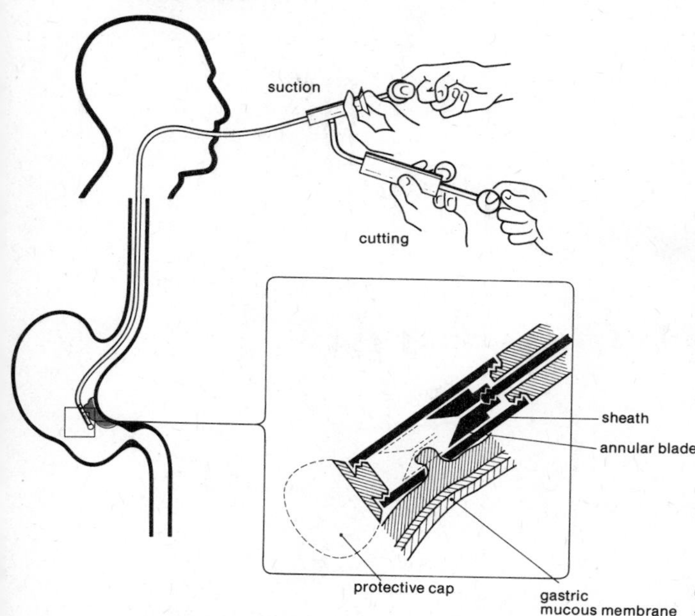

membrane of the stomach, causing no appreciable discomfort to the patient, is a fairly common condition, however, even without acute gastritis as a precursor and without being caused by ulcers. The condition is believed to be induced, at least partly, by the stresses and pressures of modern life and by irregular eating habits. Alcohol, highly spiced foods, foods consumed too hot or too cold, and insufficient mastication are among factors that are liable to cause stomach trouble, especially in persons susceptible to such disorders. In a number of cases there is infiltration of the mucous membrane, as can be detected by x-ray and by visual inspection through a gastroscope, which is a lighted tube equipped with an optical system (Fig. 5-20). There is usually also hyperacidity, i.e., an excess of acid in the stomach contents. In other cases atrophy of the mucous membrane may occur; in about one third of patients this is associated with a deficiency of acid (cf. pernicious anemia, page 107).

Gastritis characterized by bleeding and involving more or less extensive affected areas of the stomach lining is of infrequent occurrence. As chronic gastritis is a somewhat indefinite condition, a reliable diagnosis of the basic cause is possible only by microscopic examination of a specimen of tissue from the lining. The specimen may be obtained by the suction biopsy technique (Fig. 5-21). The symptoms associated with chronic gastritis are not very characteristic and are often difficult to distinguish from those of other disorders of the gastrointestinal canal. No pain is felt in the mucous membrane itself. If there is a deficiency of acid, the patient may have a sensation of repletion, accompanied by flatulence, be unable to tolerate heavy foods, and be afflicted with painful muscular spasms of the stomach wall. An excess of acid may cause heartburn. Depending on the acidity of the gastric contents, antacids or dilute hydrochloric acid may be administered. Also, there are numerous antiphlogistics (agents that relieve inflammation) available. It is especially important to avoid known irritants and to put the patient on a suitably bland diet.

PEPTIC ULCER

An ulcer is an open sore or lesion of the skin or a mucous membrane. A peptic ulcer is an ulcer of the lining of the gastrointestinal canal and is usually either a *gastric ulcer* (in the stomach) or a *duodenal ulcer* (in the duodenum, the first part of the small intestine). Such ulcers, which occur most commonly in middle age, are due to the action of the gastric juice, notably hydrochloric acid and the enzyme pepsin. They are therefore a result of a kind of self-digestion, causing local erosion of the intestinal or stomach wall. Why this should happen is not quite clear, since acid and pepsin are normal digestive secretions. The mucous membrane is, as a rule, protected by mucus. Furthermore, the acid in the contents of the stomach is quickly neutralized in the duodenum, and the mucous membrane possesses considerable regenerative powers with speedy replacement of damaged cells. It is supposed that the ulcer results from a temporary or local failure of these protective influences.

The normal equilibrium between the two sets of opposing influences—i.e., those which protect and those which tend to attack the mucous membrane—may be disturbed by external or internal factors. Chief among external factors is the diet. In principle, any type of damage that causes inflammation of the mucous membrane (see preceding section) may also cause ulcers; indeed, gastric ulcers and gastritis are frequently present in one and the same patient. In this context the action of certain drugs suspected of promoting or contributing to the development of peptic ulcers must be mentioned; they include salicylates (notably aspirin), corticosteroids, phenylbutazone and others. The internal factors that may disturb the equilibrium between protective and destructive influences to which the mucous membrane is exposed include particularly the gastric nerves, whose activity is controlled by the central nervous system, and endocrine (hormone-secreting) glands. The nerves may intensify the secretion of the aggressive gastric juice and stimulate the stomach movements to convulsive contractions. Such local muscular spasms of the stomach wall not only are responsible for the pain felt by the patient, but also cause a harmful ischemic condition (insufficient supply of blood) in the area affected. Strictly localized ischemia may also be caused by vascular spasms due to nervous stimulation. Such areas of insufficient blood supply damage the stomach lining and cause ulcers in the form of round eroded patches.

It is believed that hereditary factors are involved in about 40 percent of all cases of gastric ulcer. Such ulcers are often encountered in persons having a labile vegetative (autonomic) nervous system who tend to develop vascular diseases, migraine and high blood pressure.

Ulcer patients are often introverted characters with unfulfilled ambition in whom various emotional factors (worry, sorrow, agitation, pressure of work) upset the vegetative nervous system and hormone balance. Perhaps also associated with emotional factors is the fact that younger men more frequently develop gastric ulcers than older men and that ulcers are two and a half times more frequent in men than in women. The seasonal concentration of cases of ulcer trouble in the spring and in the autumn is also notable (Fig. 5-22). In certain circumstances gastric ulcers may, instead of developing slowly, form very quickly, e.g., as a result of extensive burns, shock, and large variations in temperature. Liver diseases pro-

FIG. 5-22.
Symptoms due to gastric and duodenal ulcers.

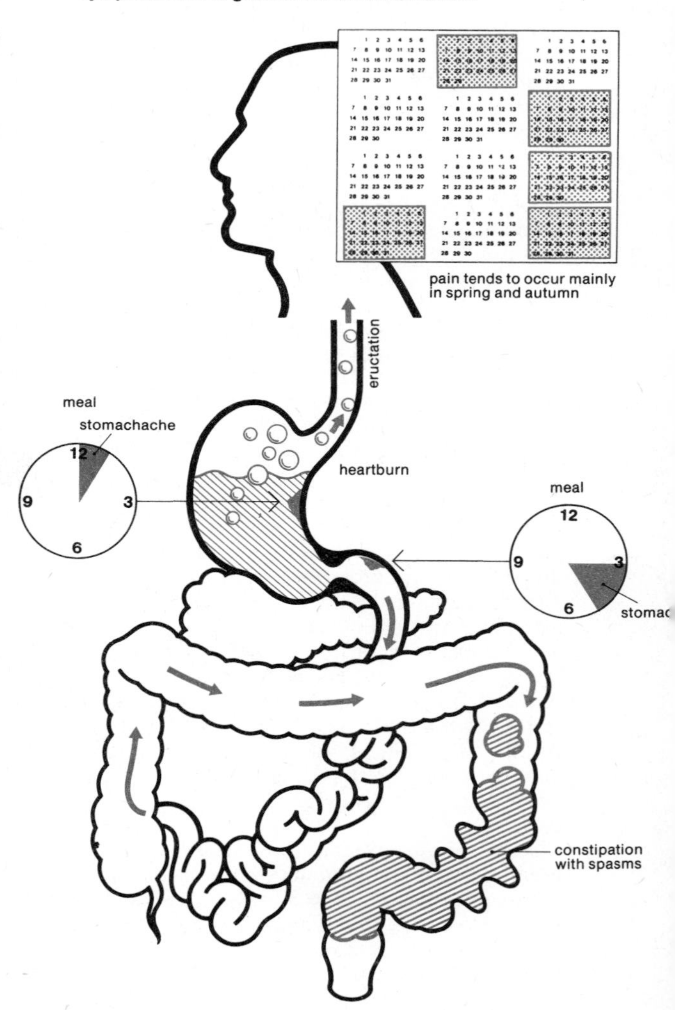

mote the occurrence of ulcers; the reason for this is unknown. The fact that something like 80 percent of all gastric ulcer patients are heavy smokers may mean that nicotine plays a part in causing ulcers. On the other hand, it may mean that heavy smoking and ulcers are both to some extent attributable to common underlying (emotional) factors. Peptic ulcers are a relatively common disorder and account for 2 to 4 percent of all internal disease. Pathologists assert that about 7 percent of all people have a gastric or a duodenal ulcer at some time in their lives.

Gastric ulcers occur mainly in the bottom third of the stomach, in the region of the greater curvature (Fig. 5-23). Duodenal ulcers usually develop in the first 5 cm beyond the pylorus. Both types give rise to "gnawing" pains in the epigastric region. With a *duodenal ulcer* this pain is most likely to occur one to three hours after meals. It often disappears when the patient takes a little food ("hunger pains"). There are also hyperacidity of the stomach contents and constipation. Quite often, too, the patient feels ravenously hungry. Sweets and foods that stimulate the secretion of acid in the stomach are likely to start or intensify the pain. A *gastric ulcer* may produce similar symptoms, but the pain occurs during or soon after meals. Alternatively to hyperacidity, a condition of lowered hydrochloric acid secretion (hypoacidity) may be associated with a gastric ulcer. Other troubles that may accompany such ulcers are eructation, salivation, faintness and vomiting; with hyperacidity there may also be heartburn.

Various transitions to the symptoms of gastritis may occur, which is not surprising, since gastritis is not infrequently an attendant condition of gastric ulcer (see preceding section). Alcohol, coffee, highly spiced foods, etc., aggravate the disorder. Seasonal variations, with a tendency for gastric trouble to occur in spring and autumn, are observed. Peptic ulcer patients are often thin, with typically gaunt faces and hollow cheeks. More than 90 percent of all gastric and duodenal ulcers are detectable by x-ray. Gastroscopic examination may be necessary to distinguish the condition from cancer.

With suitable precautions and treatment, the great majority of peptic ulcers heal and require no surgical intervention. A complication that may arise is *bleeding*, which occurs if the ulcer affects the blood vessels in the wall of the stomach or intestine. Such internal hemorrhages may, in severe cases, cause the patient to vomit blood, with shock, possible collapse and risk of death. Sometimes the hemorrhage develops more surreptitiously: the patient gradually weakens, but does not realize that anything is seriously the matter with him until he notices that his feces are very dark-colored (tarry stools) as a result of the presence of

FIG. 5-23.
Section through a gastric ulcer.

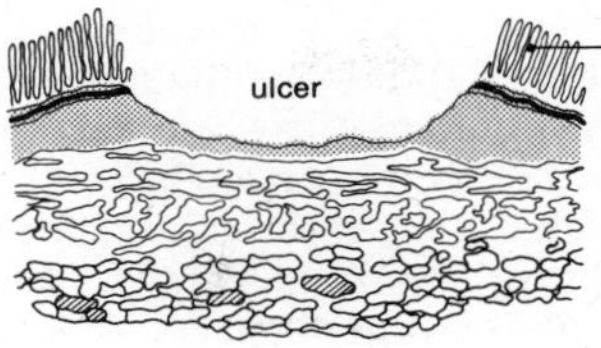

normal mucous membrane
ulcer

FIG. 5-24.
Hourglass stomach.

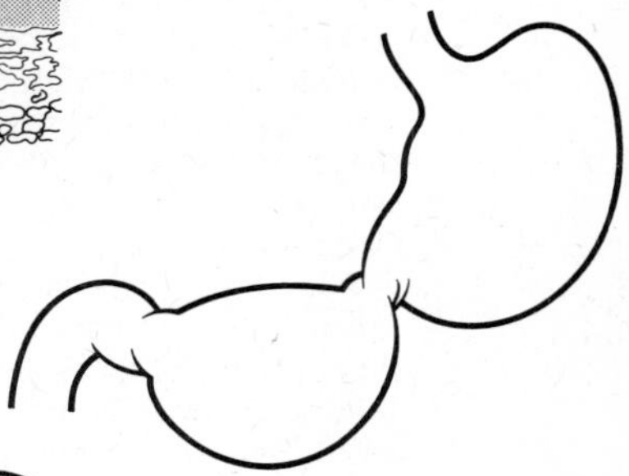

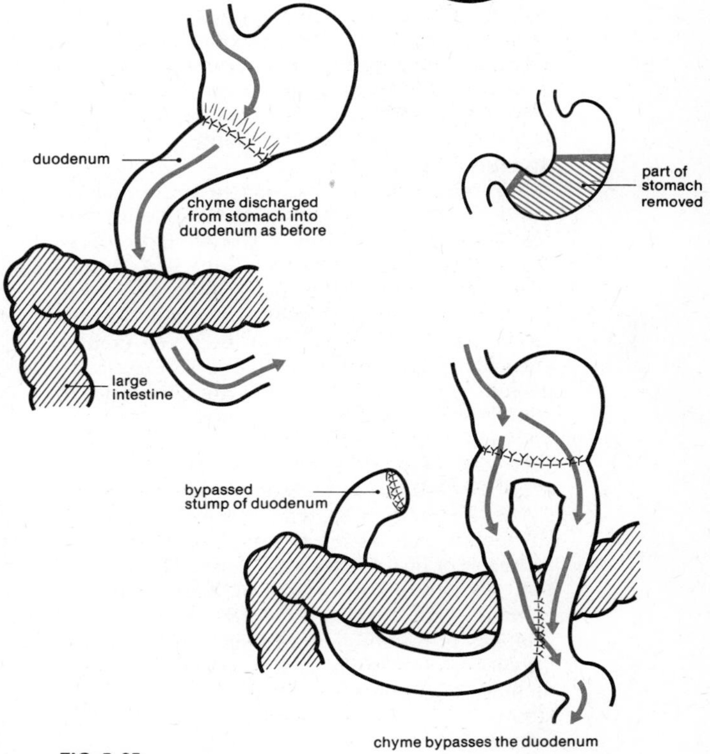

duodenum
chyme discharged from stomach into duodenum as before
large intestine
part of stomach removed
bypassed stump of duodenum
chyme bypasses the duodenum

FIG. 5-25.
Operations for partial removal of the stomach.

breakdown products of blood discharged into the alimentary canal.

Another possible complication is *perforation* of the stomach or duodenum wall, with leakage of the contents into the abdominal cavity. The onset is accompanied by acute pain spreading all over the abdomen, which becomes rigid, while the pulse is rapid and feeble, and blood pressure abnormally low; often there is vomiting. This is a dangerous condition, and peritonitis will develop unless an operation is performed in time.

Ulcers that heal of their own accord usually form harmless scars. Sometimes, however, this process of cicatrization is accompanied by severe contraction of tissue, which, depending on the location of the scar, may cause narrowing of the stomach outlet (pyloric stenosis) or produce a so-called hourglass stomach (Fig. 5-24), divided into two parts by a waistlike constriction. These complications make the emptying of the stomach slower and more difficult, resulting in loss of appetite, a feeling of repletion, and eructation.

An important feature in the treatment of peptic ulcers is to give the stomach sufficient rest. The diet should be suitably bland so as to avoid gastrointestinal irritation and reduce stomach activity (see page 140). Frequent small meals are preferable to widely spread larger ones. Drugs that may be given include antacids, which neutralize acidity (e.g., magnesium compounds), and anticholinergics, which suppress the parasympathetic nerves and thus damp down activity, especially the secretion of acid in the stomach. Sedatives may be helpful if the patient is under emotional stress. In patients in whom peptic ulcers are suspected to have been brought on by stress, a change of life style or situation may be essential. Psychotherapy may be necessary to achieve such a change. If there are frequent relapses or if complications set in, surgical removal of a portion of the stomach may be indicated, e.g., removal of the lower part of the stomach, where most acid is secreted, or an operation to bypass the duodenum (Fig. 5-25).

SMALL INTESTINE

The *intestine* or *bowel,* as a comprehensive term, refers to the whole of the alimentary canal beyond the stomach, i.e., from the pylorus to the anus. It comprises the small intestine, the large intestine and the rectum. With the aid of the secretion of the intestinal glands (intestinal juice), pancreas and liver (bile), the digestion of the food is completed and the nutritive constituents are absorbed in the intestine.

As a result of rhythmic contraction and relaxation of the longitudinal muscle of the intestine, its contents (the chyme from the stomach) are mixed with the digestive juices and, to assist absorption, moved along and thus brought into contact with fresh areas of the mucous membrane with which the intestine is lined. These rhythmic movements (*peristalsis*) are initiated as a reflex by internal pressure due to the presence of food in the intestine. Peristalsis is of two types: a slow wave over a short length of the intestine moves the contents onward about 12 cm; from time to time there is, in addition, a so-called peristaltic rush, which is a rapid movement that may involve the whole intestine, moving all the contents before it. These movements are autonomous, i.e., they are initiated by the muscles of the intestine themselves. The movements are coordinated by Auerbach's plexus, which is a plexus of sympathetic nerves situated between the longitudinal and the circular fibers of the muscular coat of the stomach and intestine. Another nerve plexus, called Meissner's plexus, is situated in the mucous membrane of the small intestine and controls the movements of the villi.

The small intestine, in which most of the absorption of digested food takes place, consists of the *duodenum* (separated from the stomach by the pylorus), the *jejunum* and the *ileum.* Apart from the duodenum, which is secured to the back of the abdomen by a layer of peritoneum, the small intestine hangs in the mesentery, a loose fold of the peritoneum.

The wall of the intestine is composed of the mucous membrane (the inner lining), the submucous coat, the muscle coat and the serous coat. The mucous membrane of the small intestine (Fig. 5-26) is covered with a large number of tiny projections, called *villi,* which serve to greatly increase the area of the epithelium available for absorption. Each villus is a fingerlike projection covered by epithelium and containing at its center a lymph capillary and an arteriole which feeds the capillary plexus of the villus. The epithelium is composed of two types of cells: tall columnar cells (Fig. 5-27a) and, scattered among these, goblet cells which secrete mucus. Nutritive substances such as amino acids, sugar and salts are absorbed from

FIG. 5-26.

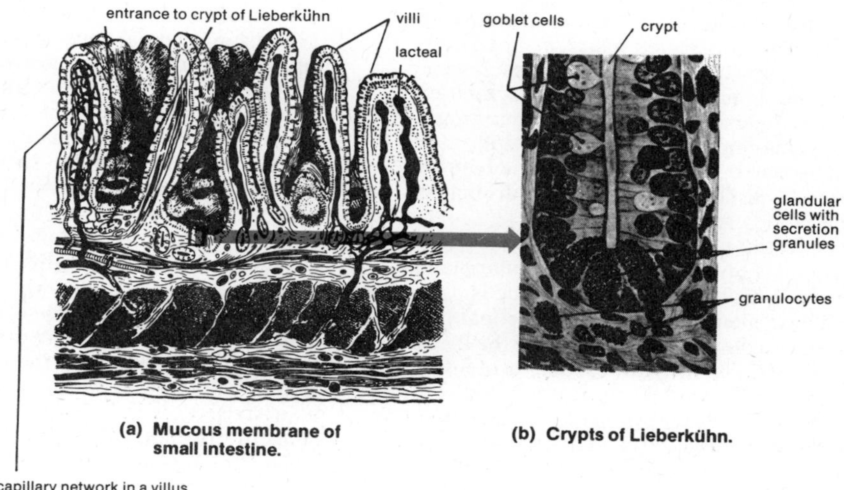

(a) Mucous membrane of small intestine.

(b) Crypts of Lieberkühn.

FIG. 5-27.

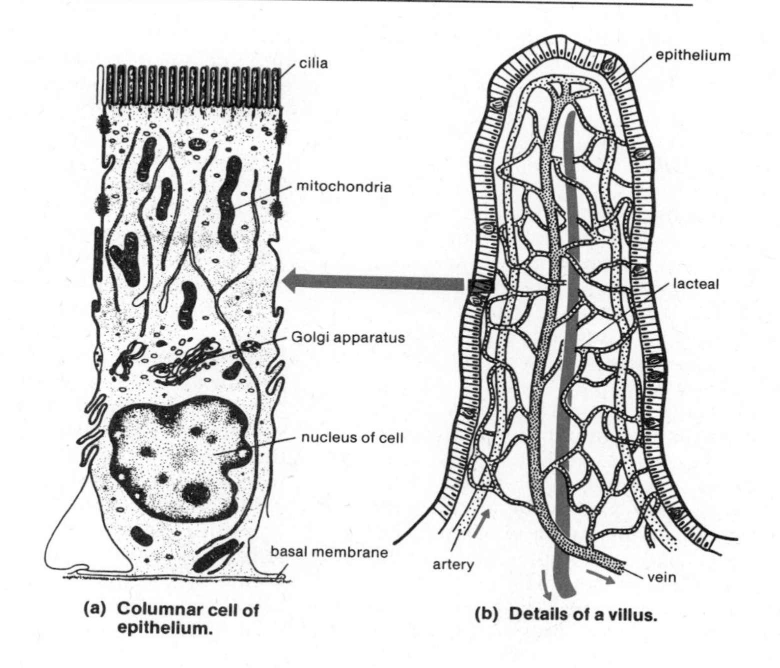

(a) Columnar cell of epithelium.

(b) Details of a villus.

the contents of the intestine through the epithelial cells and pass into the blood capillaries, which drain into a venule of the portal system through which the blood is conveyed to the liver. The villus also contains muscle fibers, which cause it to perform rhythmic contractions that discharge the blood from the capillaries in the villus (Fig. 5-27b). The action of the villi is initiated by mechanical and chemical stimuli (digestive products, spices, coffee, etc.); they are also stimulated by villikinin, a substance that is secreted under the influence of digestive products and hydrochloric acid and is conveyed in the bloodstream to the villi.

Fats are absorbed into the central lymph capillary, which is called a *lacteal,* a term denoting an intestinal lymphatic vessel which takes up chyle and passes it to the lymph circulation. Chyle is a milky fluid consisting of lymph containing products of digestion, especially absorbed fats, which is carried by the lymphatic vessels and eventually discharged into the bloodstream.

Between the villi are depressions (crypts of Lieberkühn, Fig. 5-26b) forming tubular glands which contain large secretory cells (cells of Paneth). These glands secrete the intestinal juice containing enzymes which complete the breakdown of fats, carbohydrates and proteins to simple molecules which can be absorbed.

The ileum, the end portion of the small intestine, enters the large intestine at a T-junction. The short blind portion of the large intestine is the cecum. The outlet of the ileum is guarded by the ileocecal valve, formed of folds of mucous membrane. This valve prevents the contents of the large intestine from flowing back into the ileum.

LARGE INTESTINE

The large intestine comprises the *cecum,* the *colon* and the *rectum.* Digestion is completed in the large intestine and water is absorbed from the contents, as a result of which the chyme gradually acquires the character and consistency of feces. Numerous bacteria, particularly the coli bacterium, flourish in the large intestine, where they assist in further breaking down the intestinal contents. Some of the bacteria play an important part in synthesizing the vitamins B and K.

The *cecum* is only about 7 cm long and terminates in a tail-like narrow blind tube, the *appendix,* which may vary greatly in length between one individual and another, but is about 10 cm long on average. Appendicitis is inflammation of the appendix (see page 172). The *colon* comprises the *ascending* colon, on the right side of the abdominal cavity, the *transverse* colon, which stretches across the abdominal cavity and hangs in a fold of the peritoneum, and the *descending* colon, which runs down the left side into the pelvis (Fig. 5-28). The end portion of the descending colon, before it joins the rectum, is called the *pelvic* colon. Intestinal gases tend to collect at the bend at the left side where the transverse colon joins the descending colon, causing flatulence and "stitch in the side."

The longitudinal muscle fibers of the large intestine occur in three narrow bands (*taeniae coli*) between which the wall of the intestine forms bulges (sacculations). The circular muscle performs contractions which cause, alternately, the folds and the sacculations to contract, thus exercising a kneading action on the contents and also helping to propel them along in the manner of a peristaltic movement. More important for propelling the contents of the large intestine are the major colon movements which occur every four to six hours (more particularly when eating and during defecation). In addition to peristaltic movements there are also movements in the opposite direction (antiperistalsis), which help to consolidate the feces to the appropriate consistency.

The mucous membrane of the large intestine has columnar epithelium and mucus-secreting goblet cells, also crypts of Lieberkühn, but villi are not present. The quantities of mucus secreted here are larger than in the small intestine, the mucus being necessary for lubricating the contents as they become increasingly solid.

The colon continues in the pelvis as the *rectum,* which begins at the middle of the sacrum (the central bone of the pelvis) and follows its curve downward; it finally bends backward at a right angle and continues into the *anal canal.* The rectum is covered only in front by the peritoneum. Because of its location, the rectum is accessible for surgery without involving the necessity to open the abdominal cavity. The anal canal is about 4 cm long; it is directed downward and backward, at right angles to the rectum. The outlet (*anus*) is closed by two sphincters (rings of muscle fibers)—namely, the internal sphincter, which is under the control of the autonomic nervous system, and the external sphincter, which is under voluntary control. Defecation (evacuation of the bowels) is resisted by the external sphincter until the moment is opportune, when it is allowed to relax.

INTESTINAL OBSTRUCTION

The intestinal canal may become constricted or indeed completely obstructed in many ways (Fig. 5-29), including (a) inflammatory adhesion and coalescence of tissues; (b) intestinal tumors or foreign bodies which become enlarged with accretions of fecal matter; (c) gall-

FIG. 5-28.
Features of the large intestine.

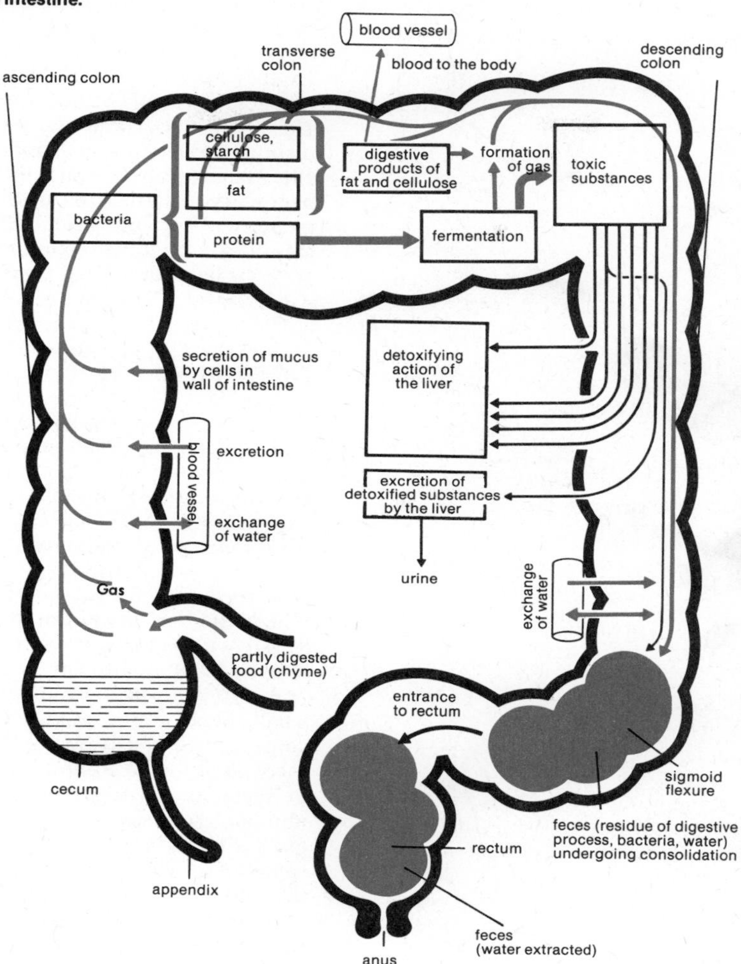

stones which become immovably wedged; (d) tangled roundworms (see later); (e) twisting (volvulus) of a loop of intestine; (f) intussusception, whereby one part of the intestine slips or is drawn onto another part just below it (occurs chiefly in children); (g) congenital defects; (h) hernia, the commonest cause of obstruction.

The symptoms associated with intestinal obstruction vary; they depend on where the obstruction is situated, whether it is a constriction or an actual blockage, whether only the passage of feces or also the blood supply to the intestine is disturbed. If the intestine is not paralyzed, it will strive by increased activity to overcome the obstruction. These efforts are accompanied by sounds, and painful spasms occur, which are detectable as externally visible movements. If the intestinal canal is constricted, but not completely obstructed, the increased activity may be unable to expel the feces, or they may emerge in a flattened or pencil-thin strand. The trapped intestinal gases cause distension of the abdomen. If there is complete obstruction, no defecation is possible. Nausea, vomiting and eructation occur. If the obstruction persists, the intestinal activity will cause fecal matter to be regurgitated into the stomach and subsequently to be expelled by vomiting. Inflammation causes

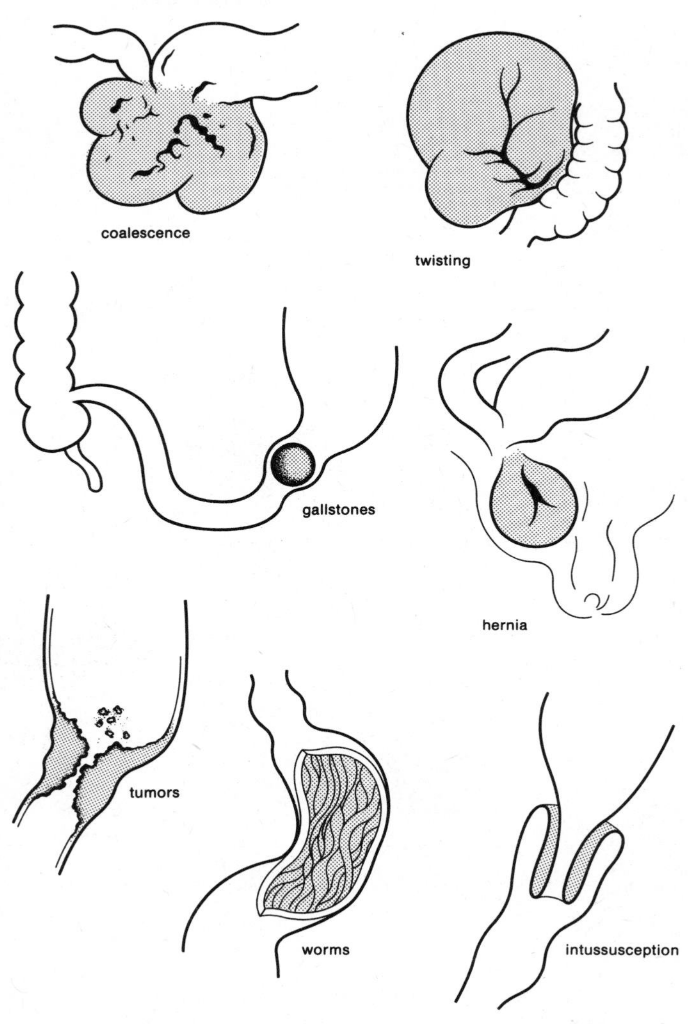

FIG. 5-29.
Causes of intestinal obstruction.

fever and accelerated pulse. Eventually, the exhausted intestine becomes paralyzed, the temperature goes down, the patient weakens and is then in a dangerous condition. Serious losses of salt and water from the body into the distended intestine occur, and infections which may spread from the intestine to the peritoneum constitute a further hazard. All these phenomena develop more quickly the nearer the obstruction is to the stomach. If the obstruction is farther down toward the anus, it may be several days before the whole intestine becomes filled with fecal matter.

The syndrome can develop very suddenly and progress rapidly if the obstruction of the intestine involves obstruction of blood supply. Abdominal rigidity, acute pain and vomiting are the early symptoms. These are particularly characteristic of strangulated hernia (page 157), volvulus or intussusception. Since an intestinal obstruction can seldom be cleared by external action, surgical intervention is usually necessary. Pending the operation, the patient must not be given food, only sips of liquid if this does not induce vomiting, and must on no account be treated with purgatives.

TAPEWORMS

Tapeworms are parasitic worms of the class *Cestoda*, phylum Platyhelminthes. Like other intestinal worms found in man, they are parasites in the sense that they live at their host's expense, claiming a share of the nutritive substances from the contents of the intestine. The two commonest tapeworm species are the beef tapeworm (*Taenia saginata*), a parasite of bovine cattle, and the pork tapeworm (*Taenia solium*), which is parasitic in pigs. Tapeworms live in the small intestine.

A typical worm of this type comprises a small head (scolex) with suckers—and, in some species, also hooks—for attachment to the wall of the intestine, and a ribbonlike string of segments (proglottids), which may vary in number up to several thousand. The worm attains a length that depends on the number of segments it has at any given time (4–10 m for the beef, 2–3 m for the pork tapeworm). The segments increase in size toward the tail end of the worm; the oldest segments are those farthest from the head. The worm consists of a string of immature, mature and ripe segments.

These worms are bisexual and self-fertilizing. The terminal ripe segments containing the eggs break off and are discharged with the host's feces. Outside the host's intestinal canal, but sometimes already inside it, the detached segments disintegrate, releasing the eggs, which then develop into tiny embryonic tapeworms (oncospheres) provided with six hooks. When swallowed by the appropriate intermediate host (cow or pig), the oncospheres penetrate the wall of this host's intestine and enter the bloodstream, which distributes them through the animal's body. They establish themselves in muscle tissue, in the lungs, the

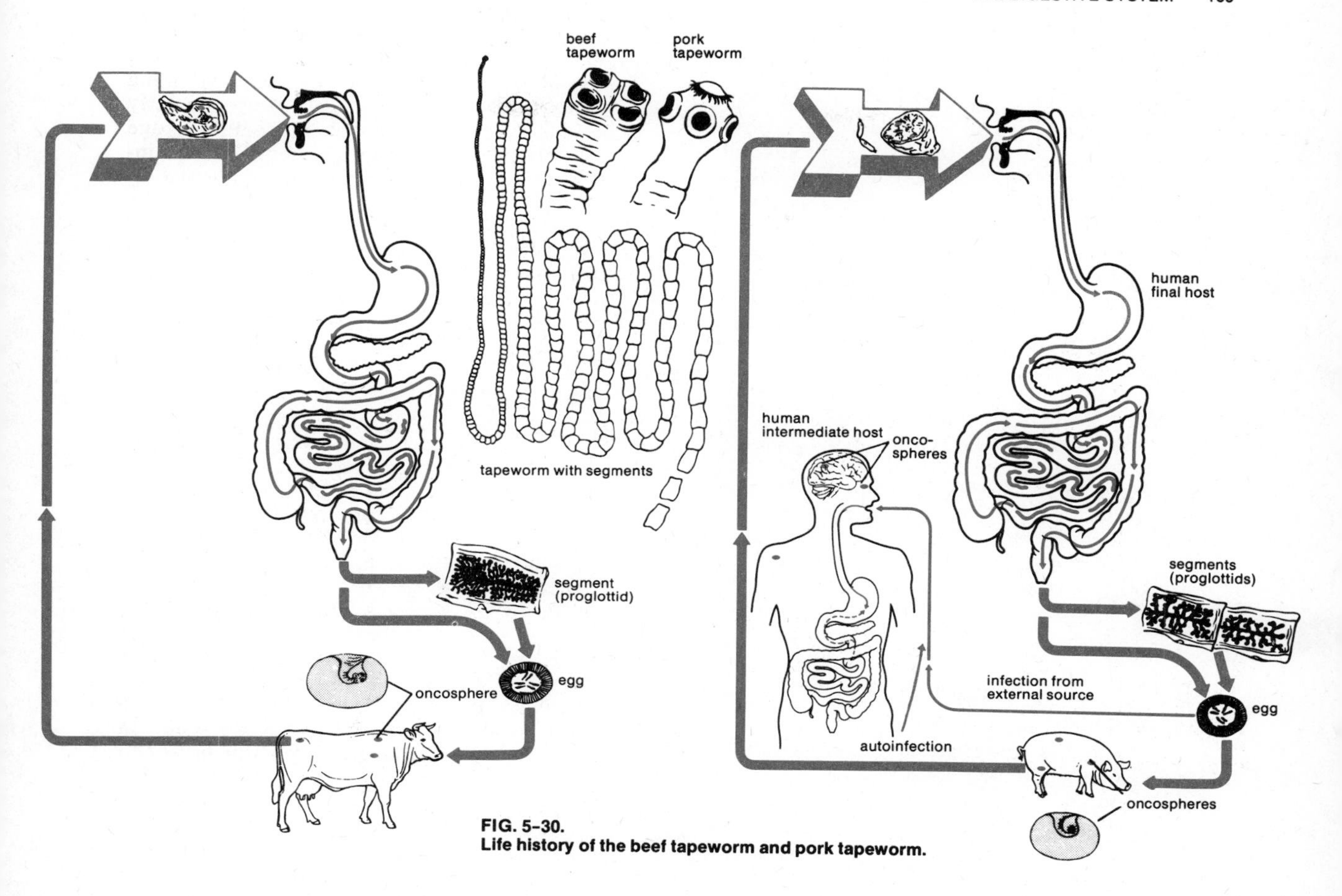

FIG. 5-30.
Life history of the beef tapeworm and pork tapeworm.

heart or other organs. Here each oncosphere develops into an encysted larva known as a bladder worm (cysticercus), about 5–10 mm in size, which remains alive and infectious for about one year, after which it calcifies. Infestation of a human host occurs as a result of eating uncooked beef or pork containing bladder worms. On entering the human intestine the bladder worms become active, attach themselves to the intestinal wall and, in a few months, grow into mature tapeworms.

In the final host, i.e., man, tapeworm infestation generally causes few and uncharacteristic symptoms. Allergic phenomena, digestive disorders, stomach ache, ravenous appetite or loss of appetite, nausea, diarrhea or constipation, headache, irritability and other nervous complaints have been ascribed to tapeworms. The parasites may also, though rarely, cause appendicitis and, exceptionally, intestinal obstruction. The presence of eggs and segments in the feces provide a definite diagnosis; examination of the segments enables the species of tapeworm to be determined. Tapeworms are effectively expelled by treatment with worm-killing drugs. Usually there is only one tapeworm present in the intestine of the host. Without treatment the worm may live for years, producing ten or more new segments a day. The host will not necessarily suffer discomfort or significant harmful consequences from its presence.

For the pork tapeworm, man is not only the final host but sometimes also the intermediate host. For the latter situation to arise, the worm's eggs, not the bladder worms, must be carried into the human intestine with food. This may occur as a result of self-infection or contact with feces of another person with pork tapeworm infestation. The oncospheres that develop from the eggs make their way into the human host's blood circulation (just as they otherwise do in the pig's) and develop into bladder worms in various organs of the body. In connective tissue and under the skin they become encapsulated and subsequently calcified. Their presence in the eye or the brain can be serious. In the brain they may cause symptoms similar to those of a tumor or of meningitis and may also simulate various nervous disorders.

The *fish tapeworm* (*Diphyllobothrium latum*) needs two intermediate hosts, namely, a crab and a certain freshwater fish. The parasite is transmitted to man by the eating of raw fish. If the worm is lodged high up in the intestine, it extracts vitamin B_{12} from the intestinal contents and may thus produce the symptoms of pernicious anemia.

The *dog tapeworm* lives in dogs, cats and foxes as the final host species, with man as its intermediate host, in whom the parasite forms bladder worms in the liver, where they may proliferate and cause a cancerlike condition.

FIG. 5-31.
Life history of the roundworm.

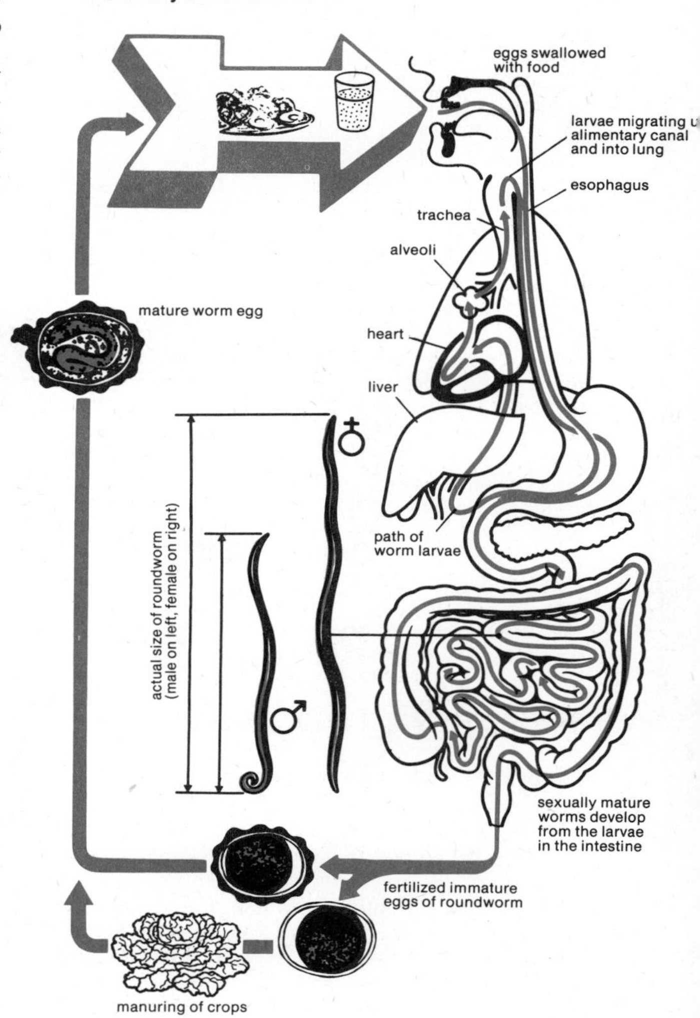

ROUNDWORMS

Roundworms are parasitic worms of the class *Nematoda,* phylum Nemathelminthes. The name *roundworm* is more particularly applied to worms of the genus *Ascaris,* especially *Ascaris lumbricoides,* while those of the genus *Enterobius* are called threadworms or pinworms (see next section). Roundworms are passed from person to person; there is no intermediate host. The ordinary roundworm is a yellowish creature, not unlike an earthworm. The males are up to about 20 cm, the females up to about 40 cm, in length. The adult worms live in the human small intestine, where they consume some of the host's semidigested food. The females lay large numbers of eggs which are expelled with the feces. Even the fertilized eggs are not infectious until they have undergone a maturing process outside the host's body, more particularly on the surface of damp soil, in the shade, with plentiful oxygen supply and a temperature of about 25°C (77°F). Under these conditions tiny larvae develop inside the eggs within two to three weeks. The mature eggs then enter a new host (another person) by ingestion with contaminated food (especially raw vegetables grown with human feces for manure) or water or by hand. On arrival in the intestine, the larvae emerge, penetrate the intestinal wall and enter the bloodstream; via the liver and the left side of the heart they eventually reach the lungs, where they become lodged in the capillaries. The larvae then make their way out of these blood vessels and into the alveoli of the lungs. This stage in the development of the parasite sometimes causes symptoms such as coughing, slight fever and some increase in the number of eosinophil leukocytes in the blood. In rare cases the symptoms may be those of more or less extensive pneumonia. From the lungs the larvae migrate up the respiratory passages to the throat and are swallowed, so that they arrive in the small intestine, where they develop into adult roundworms in about two or three months after passing through the lungs. A new cycle is then started.

In many cases the presence of the adult roundworms in the small intestine does not cause any appreciable symptoms; but massive infestation sometimes constitutes a mechanical obstruction, and the waste products of the parasites' metabolism may have toxic or allergic effects, including stomach trouble or stomach ache, loss of appetite or ravenous appetite, nausea, vomiting, loss of weight, diarrhea or constipation. These are uncharacteristic symptoms and require confirmation, which can be provided by detection of worms or their eggs in the feces. In rare cases tangled worms may cause intestinal obstruction, especially in children. Individual worms may migrate upward or downward in the intestinal canal and plug up the bile duct, the appendix or, very rarely, even make their way from the throat into the Eustachian tube and thus into the middle ear. Roundworms can be effectively treated with drugs.

THREADWORMS, HOOKWORMS

Threadworms (also known as pinworms) are parasitic nematodes of the genus *Enterobius*, especially *Enterobius vermicularis*, which are common in temperate climates, children being especially likely to be affected. Like the roundworm, the threadworm is passed from person to person. There is risk of self-infection, i.e., the patient's hands may be contaminated with threadworm eggs discharged in his own feces, so that he may swallow some of the eggs. The male adult worm is 2–5 mm long, the female 9–11 mm.

The worms attach themselves to the mucous membrane of the cecum, the appendix and the adjacent parts of the large and the small intestines. When the females are laden with fertilized eggs, they detach themselves from the intestinal wall and migrate down the large intestine. They escape from the anus at night, when the sphincter is relaxed. Each female then lays thousands of tiny eggs on the skin around the anus, after which she makes her way back into the intestine. These migrations of the worms cause itching in the anal region. The patient (usually a child) scratches and reinfects himself when he sucks his fingers. The eggs, thus swallowed, duly arrive in the intestine, where they hatch. The larvae that emerge from them develop into adult threadworms.

Person-to-person transmission of threadworm eggs may take place by contact with bedclothes, underclothing, direct mouth-to-mouth contact, etc. Some of the egg-laden

FIG. 5–32.
Life history of the threadworm.

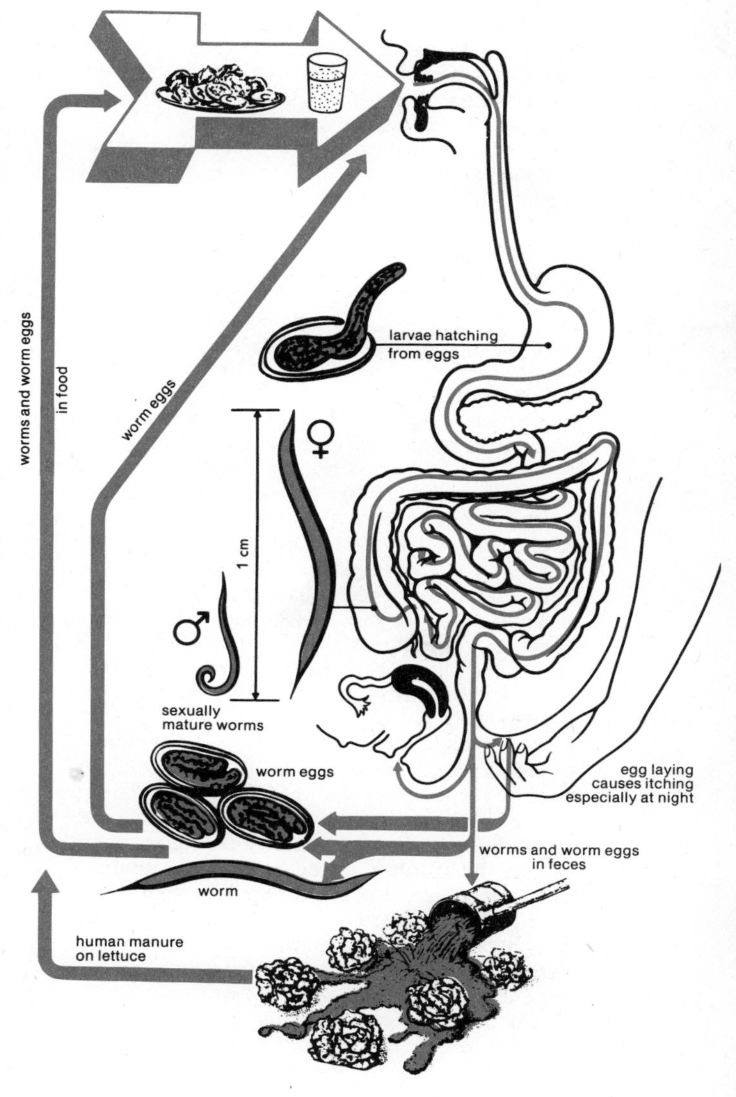

worms are discharged with the feces, so that, as with roundworms, there is a risk of contamination of food and drink, especially vegetables grown with human feces as manure.

As a rule, threadworm infestation causes no appreciable symptoms and has no significantly harmful effects. Sometimes, however, the worms attached to the wall of the intestine may set up irritation or even cause symptoms resembling those of appendicitis. Also, waste products discharged by the worms may give rise to allergic reactions. Exceptionally, worms that migrate out of the intestine may penetrate into other cavities, such as the vagina, and cause inflammation there. The irritation and itching due to the egg-laying activities of the female worms can be objectionable, and a secondary infection of the area around the anus may occur as a result of scratching. In children this may cause loss of sleep, irritability, nervousness, loss of appetite and loss of weight. Treatment with a worm-killing drug can be effective only if the self-infecting cycle is interrupted. Antiseptic and soothing ointments are applied to the rectal area to reduce itching. Cleanliness is all-important.

Hookworms are nematode parasitic worms of two species (*Ancylostoma duodenale* and *Necator americanus*) which are common in warm countries. Infecting larvae live in soil contaminated by feces and enter through the skin of the bare feet. They make their way into the bloodstream, which carries them to the lungs; they then travel up the bronchial tubes to the throat and are swallowed. Thus they eventually reach the duodenum, where the worms attach themselves to the lining and cause persistent bleeding and anemia. Larvae passing through the lungs may cause pneumonialike symptoms.

APPENDICITIS

Appendicitis is inflammation of the appendix, which is the thin tail-like portion at the end of the cecum (the first part of the large intestine). The appendix varies greatly in length, but in the majority of people is about 8–10 cm long. It is a retrogressive organ which in a remote ancestor of man presumably played an important part in the digestion of cellulose, but which has retained no useful function.

The causes of appendicitis are not precisely known. In general terms, it is caused by infection, possibly brought about by obstruction of the appendix by a small particle of hard matter. Appendicitis is commonest in young people, especially those in their late teens. It is very rare in children under the age of 5 or in adults over 50. Broadly speaking, appendicitis can be said to occur at an age when the lymphatic tissue that is present in the appendix (as in the tonsils) reaches its highest level of development and activity. It is also suspected that the appendix is especially susceptible to inflammation because it is only a thin tube with a not very plentiful supply of blood. Even a minor inflammation could cause blockage; the resulting pressure that built up in the appendix would reduce the flow of blood, thus making the lining more vulnerable to bacterial attack. Intestinal worms and grape pips are sometimes suspected of causing appendicitis by obstructing the appendix, but in fact this rarely, if ever, occurs.

In acute appendicitis the inflammation may spread in a matter of hours and perforate the wall of the appendix, so that its infected contents enter the abdominal cavity and set up peritonitis. After a few hours an abscess may then develop in the region of the appendix. Not all cases of appendicitis run this severe course, however; in mild ones the inflammation may subside of its own accord.

The diagnosis of appendicitis is not always easy, because pain in the right side, one of the common symptoms, may have other causes. Also, the pain from a diseased appendix may be reflected in other parts of the body. First signs of appendicitis are fever and chill, loss of appetite, nausea and vomiting, accompanied by pain, usually on the right side below the level of the navel. The pulse is accelerated, the tongue becomes furred (coated), and there is tenderness and rigidity over McBurney's point (which is located halfway between the navel and the crest of the hipbone).

If appendicitis is suspected, the patient should be kept in bed and should eat nothing, pending the arrival of a doctor. Purgatives or enemas must on no account be given. Pain-relieving drugs may, by suppressing the pain, mask the seriousness of the condition. The standard treatment for appendicitis is appendectomy, i.e., surgical removal of the appendix. The operation is a safe one, and with modern drugs and antibiotics the risk of peritonitis has been greatly reduced.

Chronic appendicitis ("grumbling appendix") is a somewhat vague disease. It is supposed to cause symptoms which, if any, may simulate those of peptic ulcer or gallbladder disease.

CONSTIPATION

Constipation denotes difficult or infrequent defecation (evacuation of the bowels) due to sluggishness of the bowels or some other cause. Although one bowel movement per day is popularly regarded as "normal" or "ideal" or indeed essential to the preservation of health, there is really no hard-and-fast rule about this, and habits may vary greatly from one person to another. A change in the quantity or type of food consumed may alter the pattern. Some healthy persons have two or three movements a day, whereas others suffer no ill effects from having only two or three a week. It is justifiable to speak of constipation only if the stools, whether regular or not, are inadequate or of hard and dry consistency.

In addition to the subjective feeling of "being constipated," which is often exaggerated, the diagnosis of constipation must be guided by externally detectable changes of the gastrointestinal canal and, more particularly, by the attendant and consequent symptoms of retarded evacuation of the bowels. In constipation the descending colon (at the left side) is usually overfilled and distended, the abdomen is taut and tender (sensitive to pain on pressure). Often the tongue is furred (coated), there is loss of appetite, bad taste in the mouth, offensive breath, headache and weariness. These symptoms, if they are present at all, are rather vague and may indeed falsely be attributed to constipation. Typically, they disappear after an evacuation.

Not infrequently, the real or supposed symptoms of constipation reflect the patient's exaggerated and possibly hypochondriacal preoccupation with his gastrointestinal functions and physical well-being in general. Underlying it may be the baseless fear that he will somehow be "poisoned" by his own intestinal wastes (i.e., his feces) if he retains them in his body for more than a certain length of time. Evacuation of the bowels thus appears all-important to him, and purgatives become his indispensable standby.

The habitual and ill-considered use of purgatives (or laxatives) is, unfortunately, so widespread that the doctor more often has to advise patients to stop taking them than to prescribe them. Prolonged and excessive self-treatment with these drugs may cause serious functional disorders of the bowels and other organs (Fig. 5-33). Chronic diarrhea, whether due to disease or drugs, may result in significant losses of sodium, potassium, and calcium and thereby weaken the intestinal muscles; it may also cause losses of protein, which are reflected in a lowering of blood protein values.

Primary or "true" constipation, as described previously, is not usually due to any discernible structural change in the intestinal canal; as a rule, it is no more than a functional disorder. Faulty diet is often to blame. If the foods ingested are deficient in roughage, they are already completely digested in the small intestine and become too greatly consolidated in the large intestine. The mechanical stimuli that induce the activity of the intestinal muscles, and the distension of the rectum that normally gives rise to the defecation urge, are then deficient or absent. Some foods and certain drugs tend to harden the stools and thus additionally impede bowel evacuation. On the other hand, roughage—indigestible fibers of vegetables, fruit, cereals and whole-meal bread—adds bulk to the contents of the large intestine and helps to keep it active.

Besides unsuitable diet, disturbed evacuation habits (repression of stools, confinement to bed, unfamiliar surroundings, etc.) are a frequent cause of primary constipation. As a result, there is a damping down and eventually a complete absence of the reflex actions associated with activity of the large intestine. The feces retained in the intestine become progressively more consolidated and more difficult to evacuate. Some people suffer hardly any disagreeable effects from this, whereas others are afflicted with a feeling of repletion, loss of appetite, eructation, painful flatulence, weariness, listlessness and headache.

A particular type of damping down of bowel action occurs when the filling of the rectum with fecal matter no longer acts as a stimulus to evacuation (rectal constipation). This sluggishness of the bowel is especially common in older people afflicted with chronic constipation or in bedridden patients. Sometimes the consolidated fecal matter that accumulates in the rectum becomes so hard that it cannot be dislodged even by a lubricating enema (consisting of oil). In such cases it may be necessary to break up the solid mass of feces by mechanical means.

Some foods and drugs, especially antacids

FIG. 5–33.
Consequences of ill-considered habitual use of purgatives and laxatives.

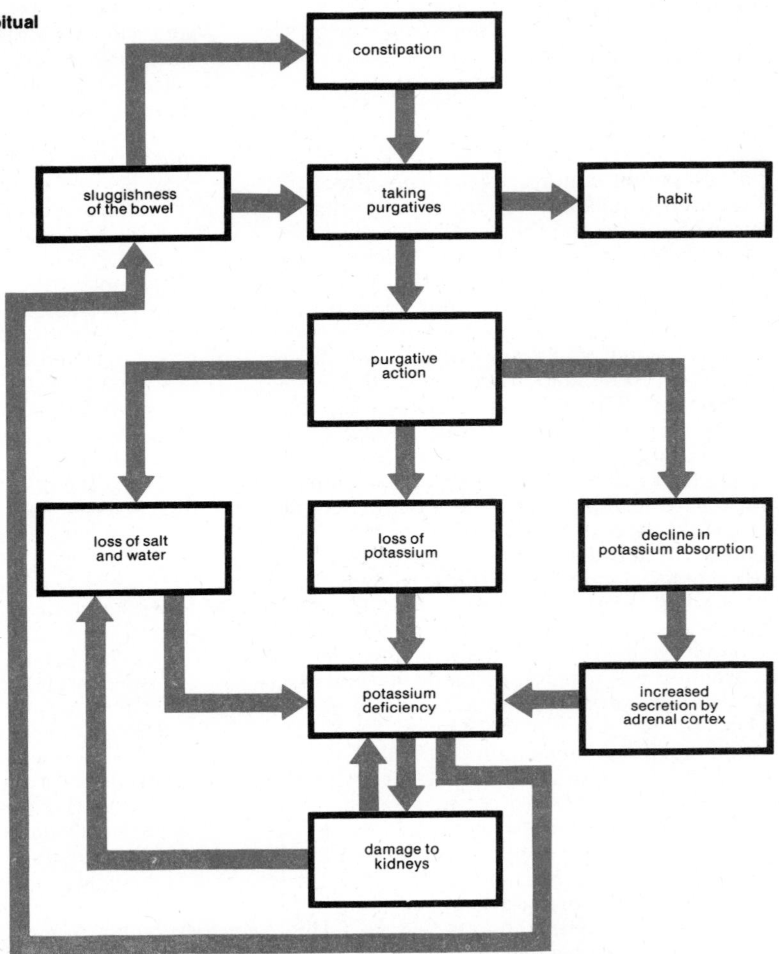

such as aluminum hydroxide and calcium carbonate, tend to harden the stools. Other drugs (e.g., atropine and opiates) are constipating because they paralyze the intestinal movements.

Apart from primary constipation due to bowel sluggishness there is also constipation due to overstimulation of the bowel. This disorder occurs sometimes in conjunction with peptic ulcers, especially duodenal ulcers. In this type of constipation the overstimulated intestine locally undergoes convulsive contractions. As a result, emptying the intestine is delayed, and the feces are discharged as pellets. If these contractions occur in the lower part of the large intestine, the portion of the intestine above it will press against the osbtruction and become overdistended. This causes a disagreeable feeling of repletion, nagging abdominal pains, sometimes colic-type symptoms. The presence of mucus in the stools indicates that the intestinal secretion is also responding excessively.

To be distinguished from primary (or functional) constipation is secondary (or "attendant") constipation, which is associated with detectable changes in the gastrointestinal canal. Thus, gallbladder trouble (see pages 199 ff), appendicitis (page 172), painful disorders of the anal region such as hemorrhoids (page 95), twisting of the intestine upon itself (volvulus) or other obstructions (page 167), and indeed any disease accompanied by fever, may cause constipation. Sudden constipation in elderly people can be a sign of cancer of the intestine (page 495).

DIARRHEA

Diarrhea is characterized by frequent and watery evacuations. It is primarily due to increased peristalsis and is a frequent symptom of gastrointestinal disturbances. Normally the digestion of food and absorption of nutritive substances are completed by the time the chyme has reached the end of the small intestine. The contents of this intestine, reduced in volume but in a liquid condition, then enter the large intestine. Here water is extracted from the contents which thus become gradually consolidated into feces consisting largely of indigestible residual matter such as cellulose. This process of consolidation may be disturbed in various ways (Fig. 5-34).

Since the movements of the intestine are controlled by the nervous system, emotional factors such as fear, stress or fright may accelerate the evacuation of feces. In such circumstances there is no time for the feces to consolidate, and diarrhea is the result. In susceptible persons nervous diarrhea may become more or less habitual and occur even on the least provocation. In such cases there may be an alternation of diarrhea and constipation until the bowel regains its normal condition of filling.

Diarrhea may in general occur if the contents of the alimentary canal pass too quickly

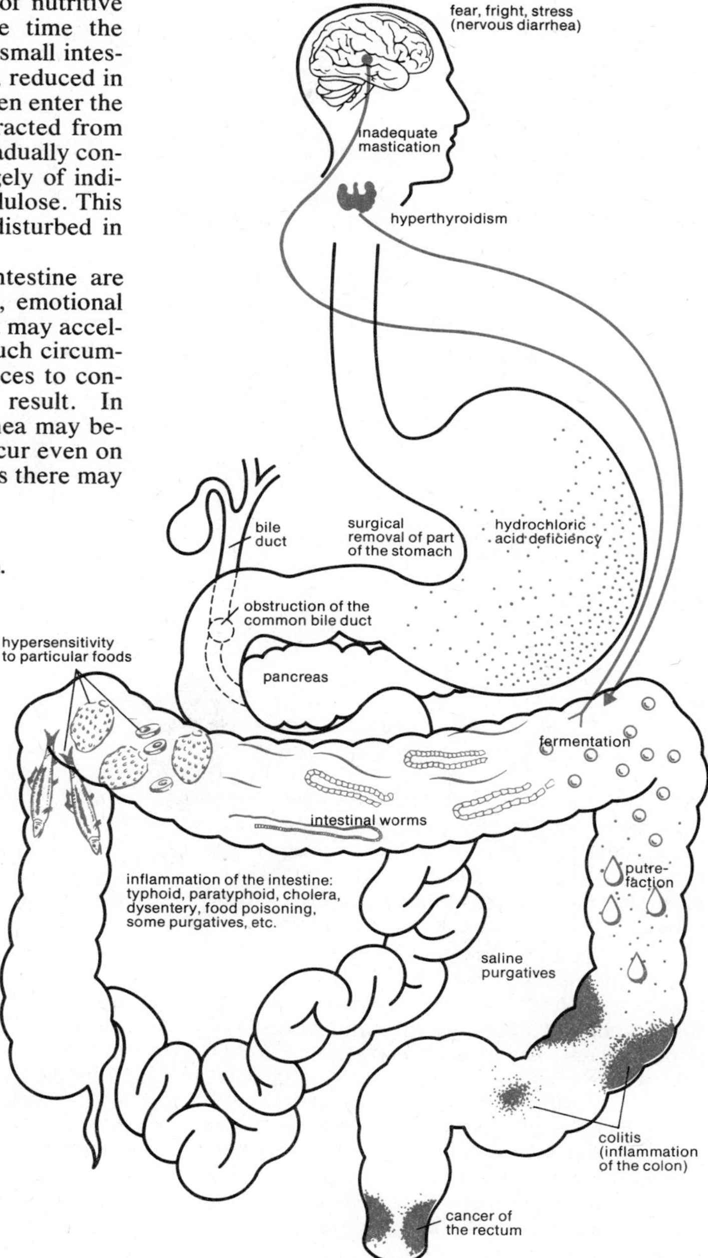

FIG. 5-34
Causes of diarrhea.

through the stomach and small intestine. As a result, the large intestine becomes overdistended and therefore strongly stimulated to action. Furthermore, bacterial activity in the large intestine causes decomposition of the remains of the food, involving putrefaction of proteins and fermentation of carbohydrates. Putrefactive processes give the stools a liquid consistency, with a dark brown color and offensive smell. Fermentation produces large light-yellow stools of thicker consistency, with a pungently acid smell, the evacuation of which is accompanied by much flatus.

Insufficiently digested chyme may reach the large intestine in consequence of insufficient mastication (possibly due to defective teeth), hasty eating, deficiency of hydrochloric acid in the stomach, deficiency of other digestive secretions of the stomach and pancreas, deficiency of bile, or surgical removal of part of the stomach. Excessively large meals may cause diarrhea in a perfectly healthy person because of a temporary relative deficiency of digestive secretions, which the body is simply unable to produce quickly enough in sufficient quantity to cope with the abundance of food to be digested.

Diarrhea may also occur as a result of inflammatory irritation of the intestinal mucous membrane: diseases that may cause this include cholera, dysentery (page 456), typhoid and paratyphoid fever (page 453), worm infestation, irritant drugs and poisons, bacterial food poisoning, etc. The effects of inflammation of the small intestine are particularly pronounced, resulting in stools that contain blood and are colored greenish by bile. Inflammation of the large intestine can be very obstinate. It may, as a result of increased stimulation of intestinal movement, increased secretion of mucus and reduced absorption of water from the contents of the intestine, cause acute diarrhea, sometimes developing into chronic diarrhea. In certain cases of ulcerous inflammation of the large intestine there is a discharge of blood and pus in the stools.

Even relatively minor attacks of diarrhea may be symptomatic of serious diseases, e.g., cancer of the rectum (see page 496), if blood and mucus are present in the stools. Disturbed activity of endocrine glands, e.g., hyperthyroidism (see page 352), sometimes causes diarrhea, as do also allergic reactions to certain foods, drugs or chemicals. Sustained diarrhea is attended by loss of water and salts, which can, if left untreated, bring about collapse, renal failure and shock (see page 208). In serious cases these losses must first be corrected before any further treatment is undertaken. Home remedies such as strong tea, arrowroot, cornflour, adsorbents such as aluminum hydroxide, etc., can be helpful in soothing the irritated intestinal mucous membrane. Severe cases may require additional therapy. Drugs that reduce intestinal activity, e.g., antispasmodics, may bring relief.

PANCREAS

The pancreas is a large elongated gland, 15–22 cm long, situated in the upper part of the abdominal cavity, behind the stomach; it discharges in common with the bile duct into the duodenum (Fig. 5-35). It is really a combination of two organs—an exocrine gland (producing an "external" secretion of digestive enzymes) and an endocrine gland (producing an "internal" secretion of hormones into the blood). The endocrine cells are arranged in groups called islets of Langerhans and secrete the hormones insulin and glucagon. The exocrine cells secrete pancreatic juice containing enzymes to assist digestion in the small intestine. The two organs are intimately associated with each other, and disorders of one are liable to affect the other. Because of their common outlet, the pancreatic duct and the bile duct also often share the same diseased condition.

Under the microscope the *islets of Langerhans* (Fig. 5-36) appear as lighter-colored roundish groups of smaller cells among the exocrine cells and the ducts that drain the pancreatic juice. The endocrine islet cells are not all of one and the same kind. There are alpha cells which secrete *glucagon,* a hormone that raises the blood sugar level by stimulating the breakdown of glycogen and the release of glucose by the liver, and beta cells which secrete *insulin,* a hormone that promotes glycogen formation and thus lowers the blood sugar level (i.e., the content of glucose in the blood). Deficient functioning of the endocrine cells of the pancreas causes diabetes (see pages 179 ff). Whereas this disease is due to a (relative) deficiency of insulin, the opposite condition—i.e., an excess of insulin (hyperinsulinism), resulting in a deficiency of sugar in the blood (hypoglycemia)—is also harmful and may, in a severe case, cause loss of consciousness.

Pancreatic juice is secreted (Fig. 5-37) into the tributaries of the pancreatic duct, which

FIG. 5-35. Pancreas.

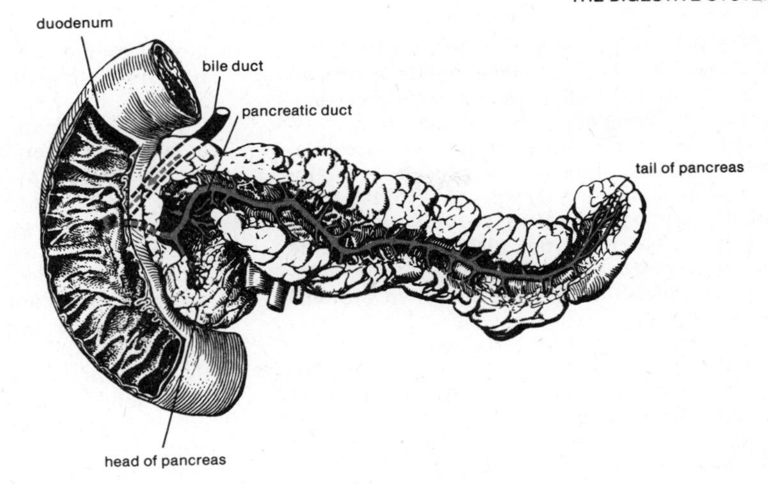

FIG. 5-36. Glandular tissue as seen under the microscope.

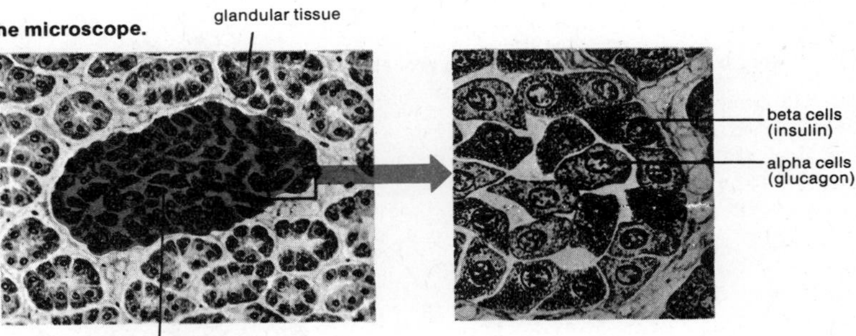

FIG. 5-37.

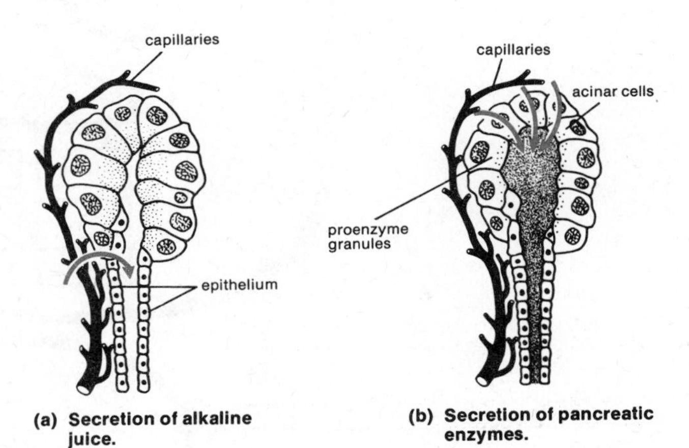

(a) Secretion of alkaline juice. **(b) Secretion of pancreatic enzymes.**

runs along the length of the pancreas and has its outlet into the duodenum with the bile duct. Secretion of pancreatic juice occurs reflexly through the vagus nerve in response to the taste, smell or sight of food and when certain hormones (secretin and pancreozymin), released by the mucous membrane of the duodenum in response to chyme from the stomach, act on the exocrine cells. Pancreatic juice contains enzymes for the digestion of proteins, carbohydrates and fats: namely, trypsinogen, chymotrypsinogen, amylase, maltase and lipase. The juice is alkaline and, with bile, neutralizes the acid contained in the chyme discharged from the stomach. In the intestine, under action of enterokinase present in the intestinal juice, trypsinogen is converted to trypsin, the active form of the enzyme, which hydrolyzes proteins and also promotes the conversion of chymotrypsinogen to chymotrypsin, which likewise breaks down proteins by hydrolysis. In the presence of bile, pancreatic lipase (steapsin) hydrolyzes fats to glycerine and fatty acids. Bile has an emulsifying action which facilitates the digestion of fats by lipase; also, the bile salts form compounds with the fatty acids and are necessary for their absorption. Carbohydrates are digested by amylase and maltase, which hydrolyze starch to maltose and then to glucose (Fig. 5-38).

As the secretion of pancreatic juice is initiated by reflex action as soon as food enters the mouth, the chyme that is passed through the pylorus after digestion in the stomach finds pancreatic juice ready and waiting for it in the intestine. There are actually two types of pancreatic juice. First, under the influence of hydrochloric acid and also of fats, secretin is released from the duodenum and brings about an abundant alkaline secretion from the pancreas; this is the first type of juice. Then, more particularly under the influence of partly digested proteins, pancreozymin is released, which brings about—as does also a stimulus transmitted through the vagus—the secretion of the

FIG. 5-38.
Action of pancreatic enzymes in the digestive process.

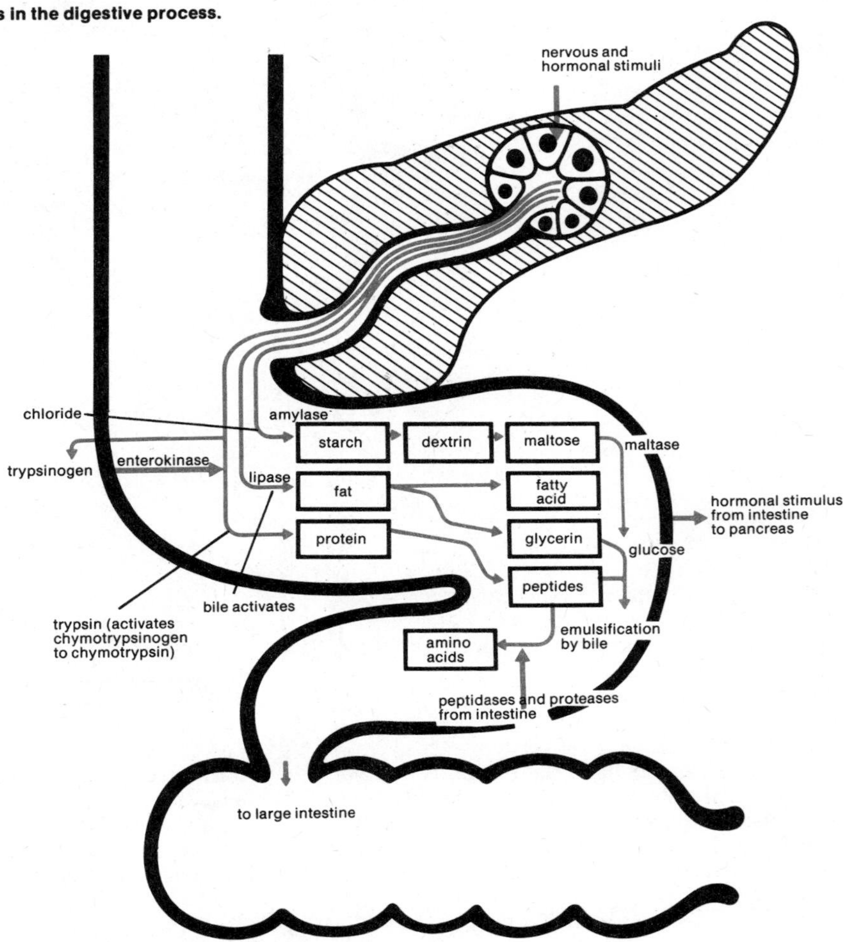

second type of juice containing the pancreatic enzymes (Fig. 5-37b).

The pancreas may become inflamed or may be affected by tumors, notably cancer. Acute inflammation of the pancreas (*pancreatitis*) may develop very suddenly, mainly in middle-aged women and often after a heavy meal. Symptoms are sudden and intense pain in the epigastric region, eructation, vomiting and collapse. As the region around the pancreas is also likely to be inflamed, the pain may radiate to other parts of the abdomen and thus make diagnosis more difficult, with risk of confusion with other disorders (possibly even cardiac infarction). The condition is mostly caused by obstruction of the pancreatic duct at or near its outlet into the duodenum. The severity of the symptoms is due to the destructive action of the highly active pancreatic juice which, instead of being discharged into the intestine, now penetrates into its own glandular tissue, where it destroys cells and causes bleeding. In severe cases digestive juices may leak into the abdominal cavity, where they cause peritonitis and dropsy and destroy the surrounding adipose tissue (fat). Obstruction of the pancreatic duct, as the cause of acute pancreatitis, may be due to a gallstone, disease of the bile duct, heavy drinking (which may cause swelling of the mucous membrane), or injuries. Treatment will consist of giving the pancreas rest by suitable restriction of the patient's diet, and losses of liquid may have to be replaced by intravenous infusions.

Chronic pancreatitis may develop from the acute form, but frequently also occurs as an independent disease. It is characterized by painful attacks, with pain radiating from the epigastric region, especially to the back. The further symptoms include fatty stools, loss of weight, and sometimes presence of sugar in the urine if the islets of Langerhans are affected. Treatment will consist in giving the pancreas as much rest as possible: small low-fat meals, no alcohol, no coffee, no cold drinks. If large areas of the pancreatic glandular tissue are put out of action, the loss of pancreatic enzymes will cause malabsorption, i.e., inadequate absorption of nutrients from the intestine, leading to symptoms of malnutrition. The deficient enzymes can, however, be replaced by enzymes taken in tablet form.

Tumors of the pancreas originate in either the endocrine or the exocrine part of the organ. Proliferation of the beta cells of the islets of Langerhans may cause overproduction of insulin. Growths affecting the exocrine part may cause harmful displacement of tissue or, if they are malignant, are liable to spread to adjacent organs. Cancer of the head of the pancreas causes characteristic symptoms by blocking the bile duct; jaundice develops in consequence. Treatment of cancer may require surgical removal of the pancreas. The pancreatic secretions will then have to be replaced by enzyme tablets and insulin injections.

DIABETES

Diabetes mellitus, commonly referred to simply as *diabetes,* is a serious chronic disorder of carbohydrate metabolism resulting from a deficiency of the hormone insulin, which is secreted by the pancreas (see preceding section), and is characterized by increased blood sugar content (hyperglycemia) and the presence of sugar (glucose) in the urine (glycosuria). There is also disturbance of fat and protein metabolism. Symptoms furthermore include increased urine production, thirst, and increased intake of food. Severe thirst may be the first symptom of which the diabetic patient becomes aware.

The less familiar disease called *diabetes insipidus* is caused by inadequate secretion of vasopressin, the antidiuretic hormone secreted via the pituitary gland. This disease is characterized by thirst and excessive production of urine. It can be controlled by vasopressin replacement therapy.

The precise basic cause of diabetes mellitus is not known, but a large proportion of cases are attributable to genetically determined premature exhaustion of the ability of the pancreas to produce insulin. This is sometimes referred to as primary diabetes.

In a minority of cases, the absence or deficiency of insulin may be the result of inflammation of the pancreas, malignant tumors, or overactivity of those glands which secrete hormones that counteract insulin production—e.g., the pituitary gland, the adrenal cortex and the thyroid gland. This is secondary diabetes.

It is reckoned that between 1 and 4 percent of all people are diabetic (Fig. 5-39), though about half of them are unaware of it, while the number of so-called latent diabetics in the population as a whole is considered to be even

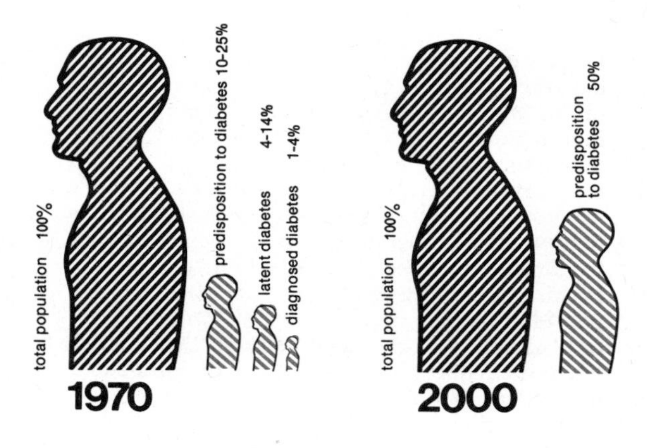

FIG. 5-39.
Incidence of diabetes.

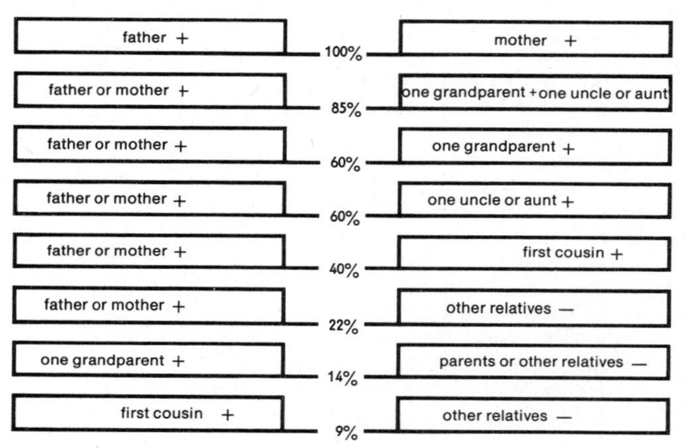

FIG. 5-40.
Probability of hereditary predisposition, depending on occurrence of diabetes among relatives.

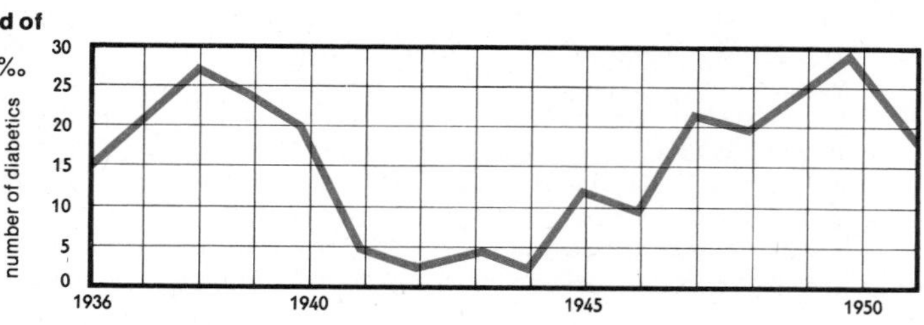

FIG. 5-41.
Decrease in the incidence of diabetes during period of food shortage in Second World War.

substantially larger (4–14 percent). The latter category comprises people whose urine and blood sugar content, though normal in ordinary circumstances, changes abnormally as a result of increased sugar intake. Depending on the person's way of life and habits, the latent condition may develop into overt diabetes. The hereditary predisposition to diabetes in the general population is estimated at 10–25 percent, and actual diabetes is liable to develop in something like 10–20 percent of those in whom this predisposition exists. About 75 percent of all cases of the disease occur in persons over 50 years of age.

Up to the early 1920s more than 40 percent of those afflicted with the disease died in an acute diabetic coma, but with modern treatment this proportion of deaths has been dramatically reduced. Nowadays the most frequent causes of death in diabetics are supervening diseases which affect the vascular system. The life expectancy of persons in whom diabetes develops in middle age is statistically 4–10 years shorter than in the population as a whole; even in cases of juvenile diabetes, which is more serious than diabetes of the middle-aged and in former times almost invariably caused the patients to die young, the statistical life expectancy is now shortened by no more than 10 years.

Encouraging though these figures are in the fight against an incurable disease, there is hope for further progress. The hereditary diabetic predisposition is nowadays more frequently passed on to the next generation. In the past, only about 2–5 percent of all diabetic women became pregnant, whereas now the pregnancy of correctly treated female diabetics is scarcely lower than that of healthy women. For this reason, within a few decades, about 50 percent of all people may be carriers of the genetic factor that predisposes to diabetes (Figs. 5-39 and 5-40). Besides further improvement in the treatment of the disease, and especially of vascular disorders associated with it, increasing importance will attach to prevention, or at least early detection and treatment, by systematic examination of large sections of the population. For every known diabetic there is, at present, believed to be at least one undetected diabetic.

Furthermore, efforts should be made to prevent latent diabetes from becoming overt diabetes. For this, cooperation on the part of the latent diabetic is essential. To a great extent he has his fate, medically speaking, in his own hands. Diabetes statistics for the war years (Fig. 5-41), when food was scarce and rationed, confirm what doctors had long suspected on the evidence of their experience—namely, that among other undesirable effects of overeating and obesity, there is an increased tendency to diabetes. Additionally, muscular activity and exercise not only are an important factor in the treatment of the disease, but are of considerable preventive value, as is also a low-calorie and low-fat diet. Factors that sometimes tend to bring on diabetes are pressure of work, nervous stress, pregnancy, menopause, severe infectious diseases, and certain drugs, e.g., corticosteroids used in the treatment of chronic rheumatism and asthma.

Sugars and starches are the principal carbohydrates in diet. There are many kinds of sugars. They can be subdivided into *monosaccharides* (simple sugars) e.g., glucose and fructose, and *disaccharides* (complex sugars which can be split into two molecules of monosaccharides), e.g., maltose, sucrose (cane sugar) and lactose (milk sugar). Glucose is the most important carbohydrate in body metabolism and is formed during the digestion of all carbohydrates. It is a normal constituent of blood (''blood sugar''), and in diabetes it is excreted in the urine. In a healthy person the blood sugar level is between 0.6 and 1.1 g per liter. More than 99 percent of the blood sugar filtered through the kidneys is returned to the blood, so that urine normally contains only small traces of sugar. Only when the blood sugar level exceeds about 1.7 g per liter (the renal threshold) is there an overflow of sugar (glucose) through the kidneys into the urine.

The regulation of the blood sugar content at a constant level is illustrated schematically in Figure 5-42. The regulating system comprises a control center in the brain, various hormones which are carried by the blood and transmit instructions to the liver and muscles (which carry out these instructions by making the necessary corrections or adjustments), and sensory apparatus for continuously monitoring the blood sugar content and reporting it back to the control center. A large intake of sugar, e.g., when a person eats a handful of sweets, produces a ''sugar load,'' which can, however, speedily be compensated and evened out by the regulating system. If the sugar intake is very large, the blood sugar level will, even in a healthy person, briefly rise above 1.7 g per liter, and sugar will for a time be discharged into the urine.

Figure 5-42 shows that various hormones raise the blood sugar level and that only insulin is able to lower it. Hence a deficiency of insulin causes the blood sugar level to rise, resulting in an overflow of sugar into the urine.

Failure of the pancreas to produce a normal amount of insulin causes an *absolute* deficiency of this hormone. However, the symptoms of diabetes may also be produced by a *relative* deficiency of insulin, which may be due, for example, to the action of antagonistic secretions or antibodies which neutralize it or to a diseased condition of the cells of the liver, so that they do not respond normally to insulin. In either case, i.e., absolute or relative insulin deficiency, the consequences to the carbohydrate metabolism are the same.

Although increased blood sugar is one of the symptoms of diabetes, this excess of glucose in the blood is not in itself harmful. A brief increase in the blood sugar level as a result of, for example, eating a quantity of sweets will merely constitute a transient disturbance in the regulating system of a healthy person and will soon be compensated by increased secretion of insulin (Fig. 5-42). It cannot be ruled out, however, that constant overloading with sugar may in the long run exhaust the capacity of the pancreas to produce the hormone, at any rate in persons with latent diabetes or a predisposition to the disease. This explains why a low-calorie diet can have a preventive effect with regard to diabetes and also why the hormones that act in opposition to insulin can produce secondary diabetes (growth hormone, hormones of the adrenal cortex, thyroid hormone). The rise in blood sugar and the presence of sugar in the urine, while not in themselves harmful, are important as symptoms of disturbed carbohydrate metabolism. Although there is excessive urine production, accompanied by severe thirst, and a loss of

FIG. 5-42.
Regulation of blood sugar content.

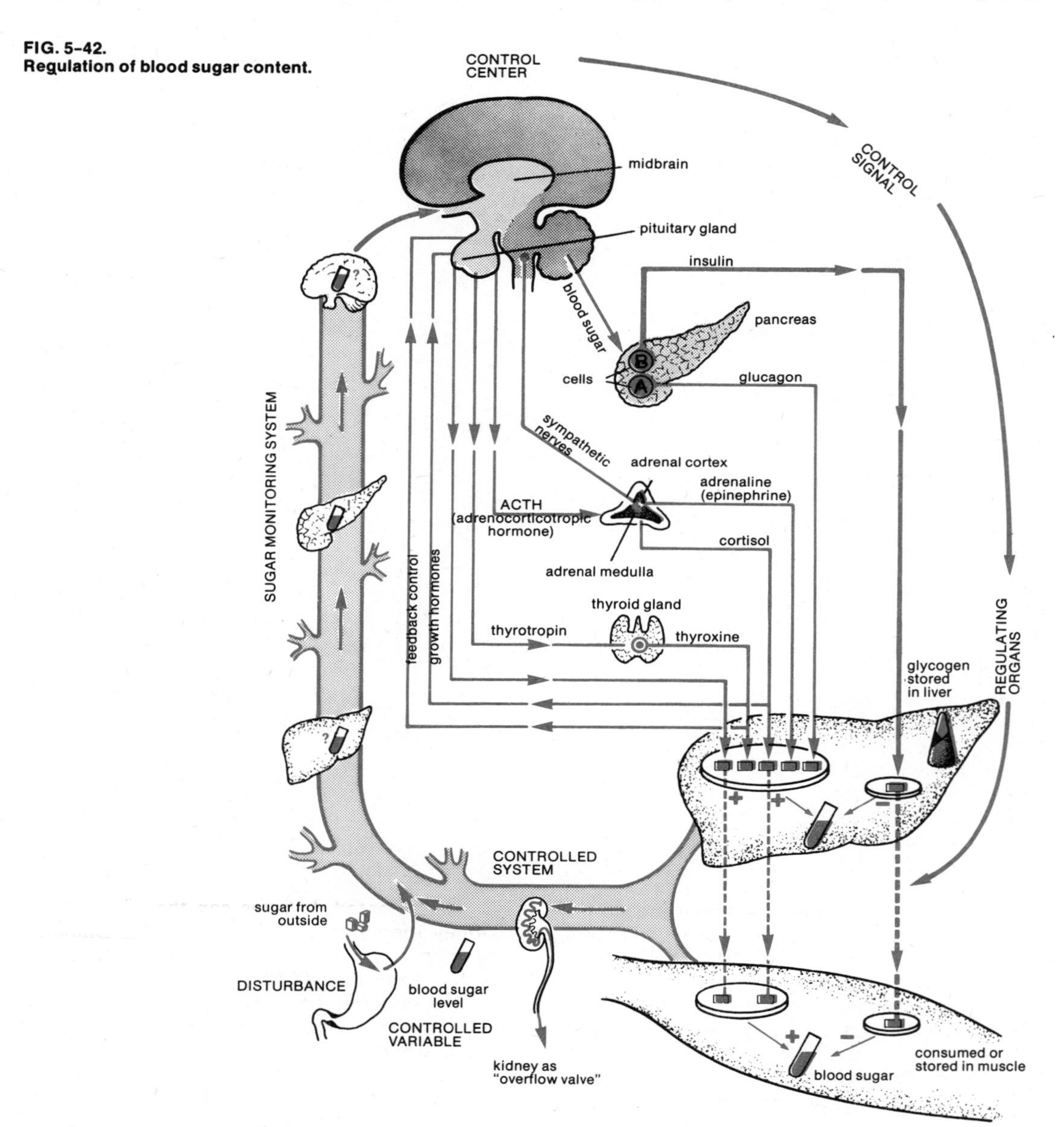

FIG. 5-43.
How insulin works.

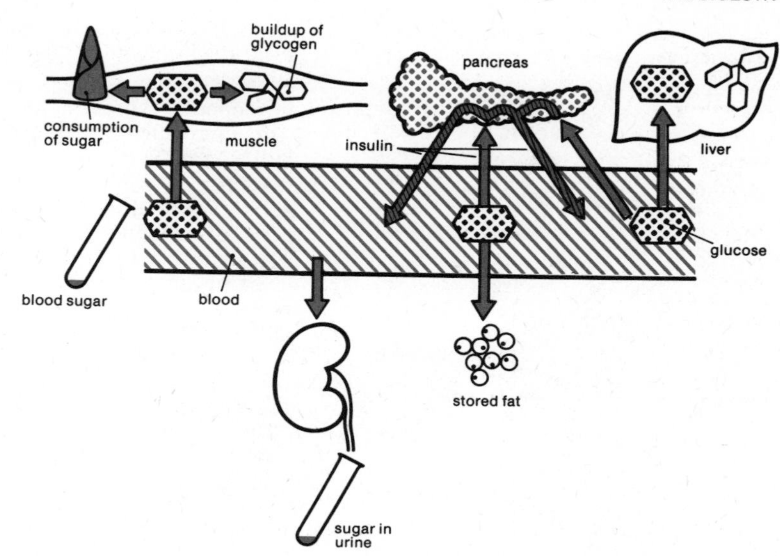

about 200 g of sugar per day, these losses can be made good and are not in themselves serious.

In a normal person the pancreatic hormones insulin and glucagon regulate the deposition of glycogen. Also known as animal starch, glycogen is the form in which carbohydrate, broken down to glucose by digestion, is stored in the body for future conversion back into glucose and for use in performing muscular work and producing heat. It is formed from glucose and some of the fat and protein in the blood. In general, the glycogen stored in the body (particularly in the liver and muscles) is broken down to glucose at the same rate as glucose is burned to provide energy. As a result, the concentration of glucose in the blood (the blood sugar level) remains approximately constant. Insulin promotes and glucagon inhibits the conversion of glucose into glycogen and the storage of the latter in the liver and muscle tissue. Other hormones also play a part in the process. Insulin also promotes the uptake of glucose by the muscle cells (where it is burned) and other cells and ensures that glucose can also be stored as fat (Fig. 5-43). Without insulin, glucose is not burned as fuel nor stored, but simply accumulates in the blood and spills over into the urine.

The main danger of insulin deficiency (Fig. 5-44) lies in the fact that, along with the faulty utilization of glucose, the body's fat metabolism is also defective. The chemical building-up of fat is disturbed. As sugar cannot be burned, fat and protein are burned instead. Complete combustion of fat, however, is possible only in the presence of certain substances formed in the combustion of glucose. Since no glucose combustion takes place, the combustion of fat cannot be completed, but stops at intermediate products called *ketone bodies* or *acetone bodies,* e.g., acetoacetic acid. Accumulation of these toxic acid substances cause *acidosis,* i.e., disturbance of the acid/alkali balance of the blood toward acidity. The patient's breath has a sweet fruity odor (smell of acetone). Unless treated in time, he lapses into a state of unconsciousness (diabetic *coma*), which terminates in death. With continuing lack of insulin the coma deepens, the blood pressure goes down and there is circulatory failure. The high blood sugar level can, in such circumstances, become directly dangerous by its water-binding and water-expelling. Correct doses of insulin can prevent diabetic coma, but an overdose of insulin may induce hypoglycemic coma and can thus also be dangerous.

Other symptoms associated with diabetes are caused by degeneration of small blood vessels, particularly in the kidneys and in the eyes. Untreated diabetes can therefore cause kidney disease and defective vision. Diabetic neuritis sometimes affects the nerves of the limbs. Atheroma (fatty degenerative arteriosclerosis) of the large arteries is liable to occur in combination with diabetes. Also, the patient has low resistance to bacterial infection.

Diabetes may remain undetected for a long time, and in the earlier stages of such mild cases the patient's health is not significantly

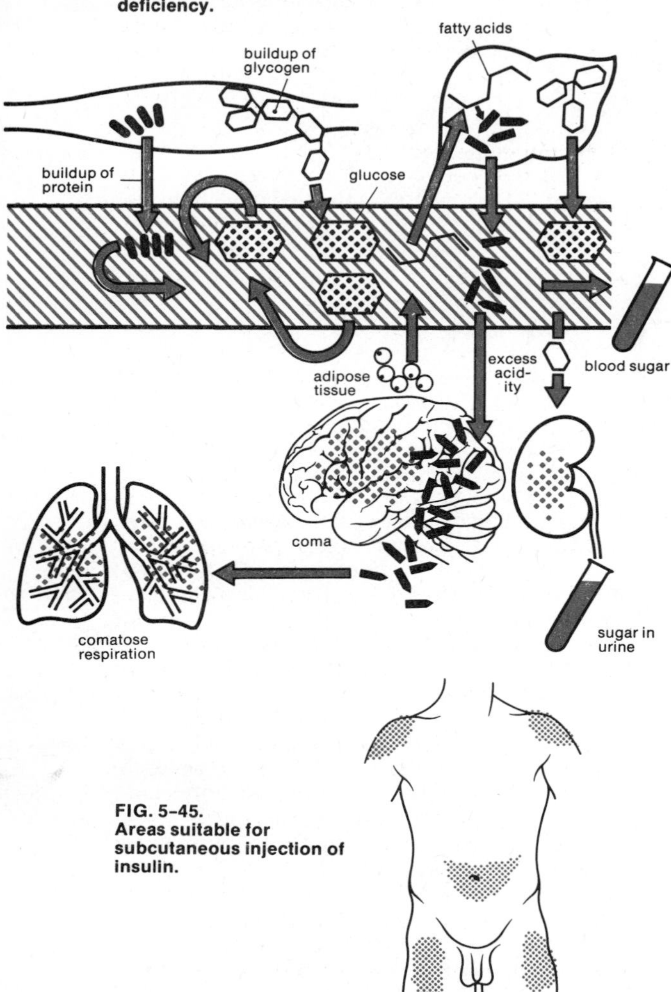

FIG. 5-44. Consequences of insulin deficiency.

FIG. 5-45. Areas suitable for subcutaneous injection of insulin.

impaired. It is nevertheless important to detect the disease as early as possible—by systematically conducted mass examination of sections of the population and/or routine tests on urine—because appropriate measures, taken in time, can have a favorable effect on the further course of the disease.

Many early symptoms of diabetes are directly attributable to disturbance of the body metabolism, whereas others are more particularly due to loss of sugar and water (increased thirst, increased and more frequent urination, also at night, weariness, itching, tendency to skin infections). In about 50 percent of all cases, diabetes is preceded by obesity, and not infrequently there is at first temporarily a sugar deficiency due to increased utilization of sugar in the body (symptoms: outbreaks of sweating, ravenous hunger, weakness, trembling, headache, faintness, diminished concentration and generally lowered performance.

Laboratory tests show the blood sugar level to be not infrequently above the normal range of 0.6–1.1 g per liter; frequently it is at or above the renal threshold of 1.7–1.8 g per liter after meals. The patient's insulin level in the blood is, at first, not lowered and may indeed even be somewhat raised. In a healthy person only very small amounts of sugar (under 0.15 g per day) are excreted in the urine. The urine of a diabetic may contain up to 200 g of glucose per day. This amount of sugar is so great that the urine tastes sweet: hence the name "diabetes mellitus," in that "mellitus" means "honeyed" or "sweet as honey." Such urine leaves a white stain on drying.

The presence of sugar in the urine may, alternatively, be due, not to an increase in blood sugar level, but to faulty kidney function, so that the renal threshold for sugar is abnormally lowered (renal diabetes, see Fig. 5-42).

Physical changes that manifest themselves in the course of the disease and are symptoms of disturbed metabolism, hyperglycemia and accumulation of waste products include changes affecting the *skin* (red face in consequence of distended capillaries, yellowish color of the palms of the hands and the soles of the feet, boils, and fungal infections between the fingers and toes, in the pubic region and under the breasts), changes affecting the *eyes* (conjunctivitis, deposition of fat in the eyelids, cataract, and retinal disorders, especially baglike enlargements of the blood vessels of the retina), and *cardiovascular changes* (early development of arteriosclerosis, risk of cardiac infarction increased 2- to 10-fold, risk of circulatory disturbance of the lower limbs, with vascular obstruction and possible gangrene, increased 20-fold). Angina pectoris is encountered in more than 50 percent and high blood pressure in more than 60 percent of all diabetics. The capillaries of the skin and kidneys, in particular, are affected by premature thickening of the walls. In addition to changes affecting the kidneys, there is often excretion of protein and the occurrence of pyelitis (inflammation of the renal pelvis), which occurs four to five times more frequently in diabetics than in nondiabetics. Diabetics are also more prone to be affected by cirrhosis of the liver, gallstones, obesity, nervous disorders, and cerebral sclerosis with cerebral hemorrhage and softening of the brain.

With proper treatment and care these complications can largely be controlled or prevented. Diabetes is also a major cause of impotence in men and sterility in women. Not infrequently, the disease starts during pregnancy. Also, the probability of miscarriage and premature birth is increased in diabetic women, with a relatively high stillbirth rate. Because of the increased supply of nutrients to the fetus, diabetic mothers often give birth to very large and yet immature babies (weighing 5–7 kg, or 11 to 15 lb), who are particularly vulnerable and are three to five times more likely to be congenitally deformed than children of healthy mothers. Maternal mortality has been greatly reduced with modern care and treatment of diabetic women; infant mortality is reduced if birth can be induced in time or the baby be delivered by Cesarean section.

Treatment of diabetes is, broadly speaking, based on diet, insulin and exercise. In about a third of all cases the disease is treated with diet alone, another third are moreover given drugs by mouth (tablets), and the final third receive insulin injections. Diet and a suitably adjusted way of life are important in all cases, including those where the disease is present merely in latent form.

A well-regulated life implies regular physical activity and regular hours of work, without hurry, without pressure and with the least possible emotional stress. Muscular action (in one's work, in gardening, in walking, in nonstrenuous sports) lowers the blood sugar level and reduces the demand for insulin because glucose can then more readily penetrate into the muscle fibers and is more efficiently burned there. Besides, physical activity is relaxing and helps to control harmful overweight. Obesity appreciably worsens the metabolic situation; insulin demand is increased in proportion to the increase in adipose tissue, and free fatty acids inhibit the action of insulin on the muscles.

The diabetic's *diet* should be a low-calorie and low-fat regime. The caloric value required is determined from tables that indicate the patient's desired body weight for his age and build. Per kilogram (2.2 lb) of desired body weight the daily caloric requirement is 25 Calories in the absence of physical work, 50 Calories if the patient does fairly strenuous work. After the age of 50 the caloric requirement is less, so that for each additional 10 years of age the caloric intake should be reduced by 10 percent. In a case of overweight of more than 10 percent above the desired body weight, it will be necessary to put the patient on a reducing diet containing only 15–20 Calories instead of 25 Calories per day per kg of desired body weight. Of course, all such important details of diet and of treatment in general should be decided in consultation with a doctor, who will also advise on how much physical activity is suitable for the patient.

One gram of carbohydrate or protein produces about 4 Calories, 1 g of fat produces about 9 Calories. Standardized diets have been worked out in which the necessary proportions of carbohydrates, proteins and fats are outlined. The diet may be based on a certain number of units of bread, fruit, vegetables and milk, each unit of these foods being equivalent to 10 g of carbohydrates. Carbohydrate units are interchangeable in principle, but 25–50 percent of the total carbohydrate intake should be in the form of fruit and vegetables, while the remainder should consist of bread and milk units. For diabetics of normal body weight the subdivision of the total permissible intake of calories in terms of carbohydrates, proteins and fats should be so arranged that 40–45 percent of the daily food requirement is supplied by carbohydrates. However, the daily carbohydrate intake should preferably be not more than 250–300 g and not less than 120 g, because a certain minimum of carbohydrates is essential for the proper combustion of fat. Carbohydrate tolerance is checked on the basis of blood sugar values in the process of "standardizing" the patient under medical supervision. The rest of the food requirement is supplied by 20–25 percent proteins and 25–35 percent fats. On an average, it is necessary to allow about 10.5 g of carbohydrates, 5.5 g of proteins and 3.5 g of fats per 100 Calories intake. For overweight patients and for those of advanced age, the proportion of fat in the diet should be reduced.

To avoid sudden heavy loads on the metabolism and blood sugar control system, meals should be well spread out, e.g., six meals a day. Foods with a high content of roughage (indigestible cellulose fibers), such as fruit and vegetables, are especially beneficial, as they release their carbohydrate content only at a

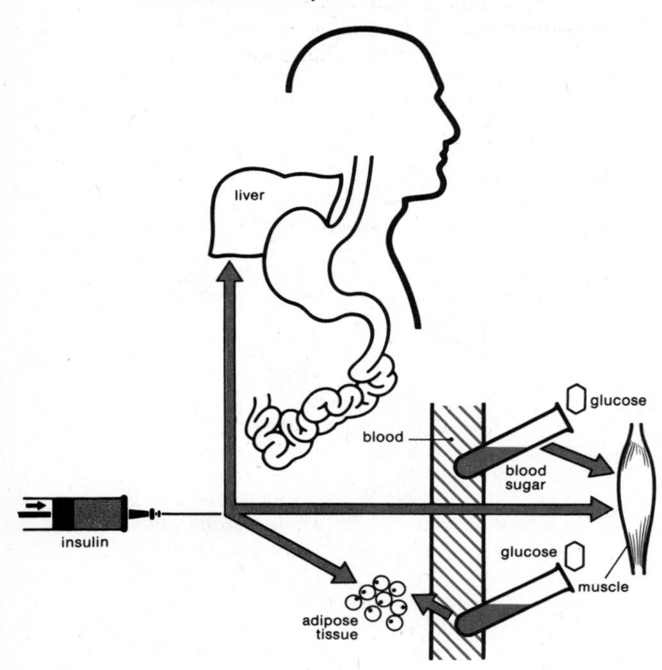

FIG. 5-46.
Action of insulin injection.

relatively slow rate. Besides, they are better able to satisfy the feeling of hunger than more easily digestible food. Pure sugar and highly sweetened foods and drinks, sweet wines, spirits and beer should be avoided by diabetics. On the other hand, synthetic sweeteners are acceptable in terms of caloric intake, though some are now suspected of producing harmful side effects.

Experience shows that, unfortunately, only about a third of all diabetics are able to exercise the sustained self-control needed to take full advantage of therapy and live a reasonably normal and healthy life.

The antidiabetic hormone *insulin* is chemically a protein. Its composition and structure are known, so that its synthetic production is possible, in principle at least. Insulin for the treatment of diabetes is, however, still obtained from animal pancreas, mostly from cows and pigs. To be effective, insulin has to be injected into the body (Figs. 5-45 and 5-46). Conventional insulin is used up within a few hours. To achieve a longer and more gradual action in the body, chemically modified types are available. Thus there is a choice of fast-acting, intermediate-acting and long-acting insulin preparations. There is no such thing as an average dose of insulin for diabetics; dosage varies greatly from one patient to another, and each case must be studied individually. Where possible, the patient is taught to give himself insulin. Injections are given subcutaneously, i.e., under the skin. As treatment continues for years and indeed for decades, needles and syringes must be carefully cleaned and sterilized as a regular routine, and the areas of the skin where the needle is inserted should be varied from day to day.

In about 10–15 percent of cases insulin produces allergic reactions. The most serious side effect, however, is *hypoglycemia,* i.e., excessive lowering of the blood sugar in consequence of an overdose of insulin, which removes glucose from the blood by increased combustion and also by increased storage of glycogen. As the brain is dependent on a constant supply of glucose from the blood, its efficiency is impaired as soon as the blood sugar level falls below a certain value, usually about 0.4 g per liter. The symptoms are headache, a feeling of weakness, sweating, ravenous appetite, shivering, lowered mental concentration, drowsiness and, finally, loss of consciousness (coma). If the diabetic recognizes the symptoms in time, the attack can be stopped by eating sugar. In more serious cases, intravenous injection of glucose may be necessary. Hypoglycemic coma comes on much more rapidly than diabetic coma (the latter is caused by insulin deficiency) and may therefore be es-

FIG. 5-47.
Action of oral antidiabetic drugs.

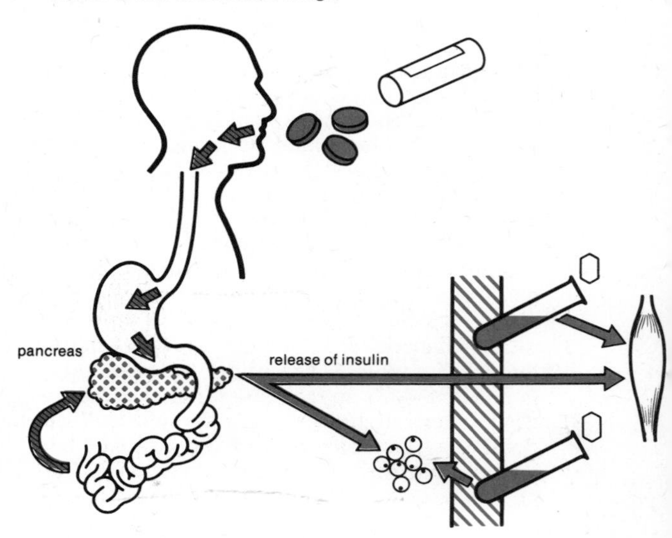

pecially dangerous if the patient is, for example, driving a car or operating a machine. Factors that, in addition to excess insulin, may contribute to hypoglycemia are faulty diet and infections.

Acidosis and diabetic coma can also be prevented by recognition of the early symptoms and prompt treatment. Symptoms of acidosis are pain in the abdomen, nausea, vomiting, drowsiness and difficult breathing. The sweet fruity odor of the patient's breath and his heavy breathing are characteristic of approaching coma. There is a gradual lapse into unconsciousness. Treatment consists in injecting with insulin and replacing the losses of water and salt.

Relatively mild cases of diabetes which cannot be controlled by diet alone can be treated with *oral antidiabetic drugs* taken in tablet form. These drugs stimulate the beta cells of the islets of Langerhans to secrete more insulin (Fig. 5-47). They can, of course, be helpful only if such cells still exist in the pancreas and are capable of functioning. If the islets have atrophied or if the pancreas has been destroyed by disease or surgically removed, antidiabetic tablets are of no use. Older diabetics, in whom the disturbance of metabolism by the disease is not yet of long duration, often respond well to these drugs. In such patients the tablets can replace about 20 units of injected insulin per day. This amount is not quite half the natural insulin production of a healthy person.

LIVER

The liver is the largest gland in the human body. It plays an important part in the metabolism of carbohydrates, fats and proteins. It also performs a number of other functions: storage of glycogen, fat, iron and certain vitamins; secretion (production) of bile; synthesis of proteins (plasma proteins, fibrinogen, prothrombin, etc.) and other substances; detoxification, i.e., neutralization or destruction of harmful or unwanted substances; excretion of waste substances in the bile (including bile pigments). The adjective *hepatic* pertains to the liver.

The liver lies in the upper right quarter of the abdominal cavity, just under the right dome of the diaphragm and protected by the ribs. It functions as an exocrine gland in secreting bile and as an endocrine gland in synthesizing various substances and secreting them into the blood. Apart from the lungs, the liver is the only other organ that receives both arterial and venous blood (Fig. 5-48). The *hepatic artery* supplies the liver with oxygen. All the venous blood from the digestive organs,

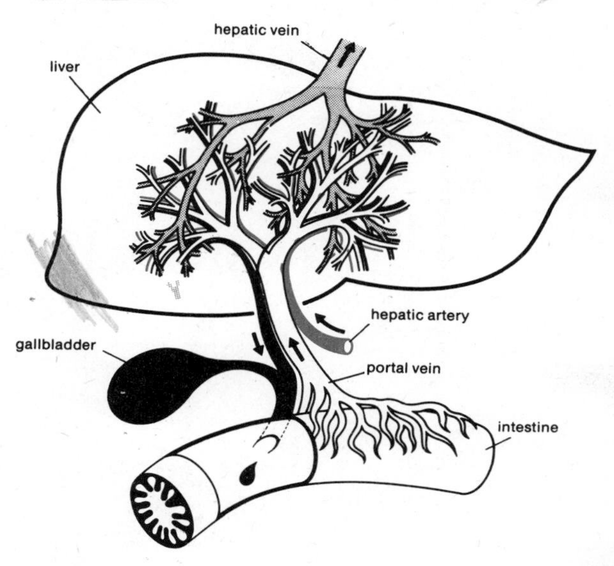

FIG. 5-48.
Blood supply of the liver.

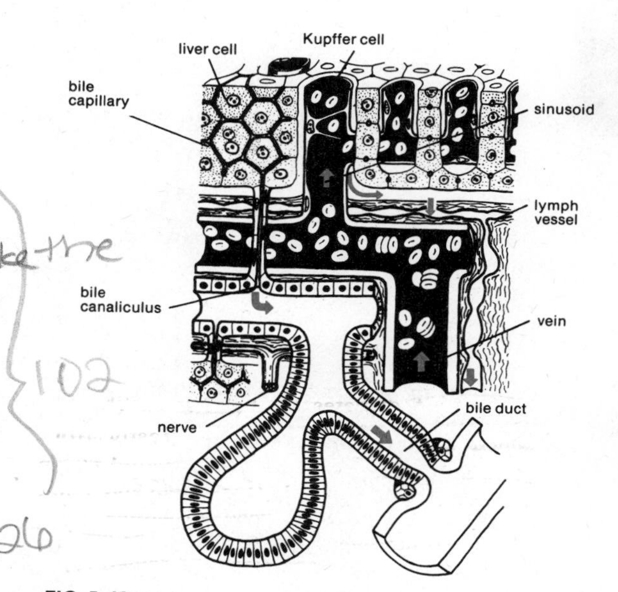

FIG. 5-49.
Detail of structure of the liver.

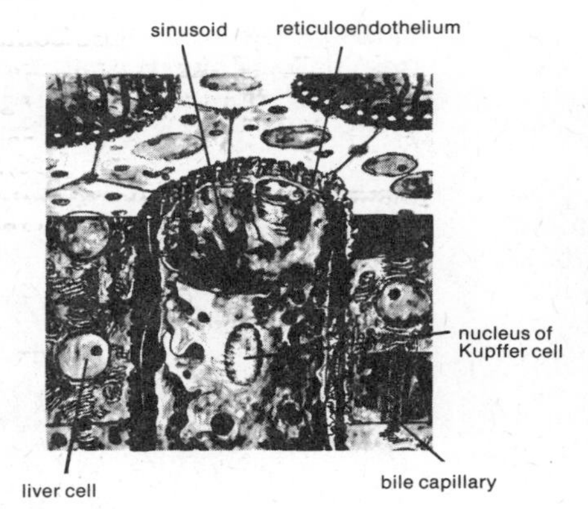

FIG. 5-50. Detail of liver tissue.

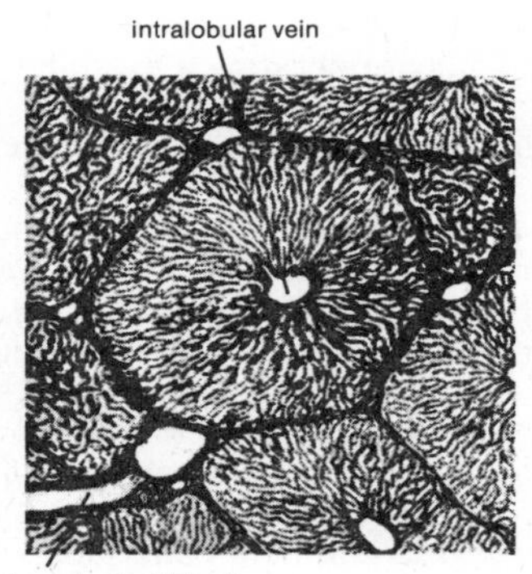

FIG. 5-51. Liver lobule in cross section.

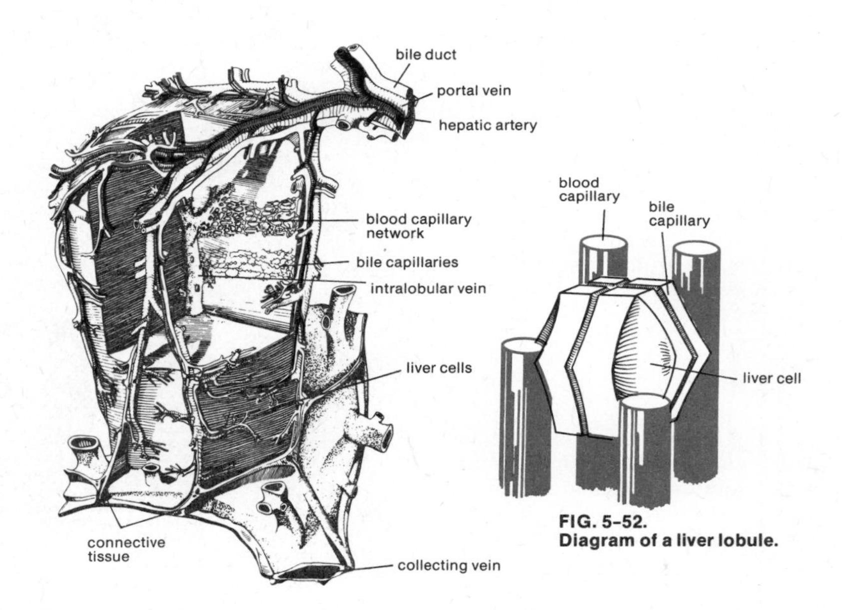

FIG. 5-52. Diagram of a liver lobule.

with the nutritive substances absorbed from the intestines, is carried by a system of veins into the *portal vein,* which also receives venous blood from the pancreas, spleen and gallbladder. The blood in the portal vein thus contains nutrients from the intestine, insulin from the pancreas, iron and bile pigments from the breaking down of red blood cells in the spleen. All these substances are carried by the portal vein direct to the liver, where this vein subdivides into a system of capillaries in which most of the blood drained from the organs of the abdominal cavity is processed and monitored before being passed to the systemic circulation.

The microscopic structure of the liver (Fig. 5-49) is uniform throughout. The liver cells are all very similar to one another. They are ar-

FIG. 5-53.
Formation and excretion of bile pigments.

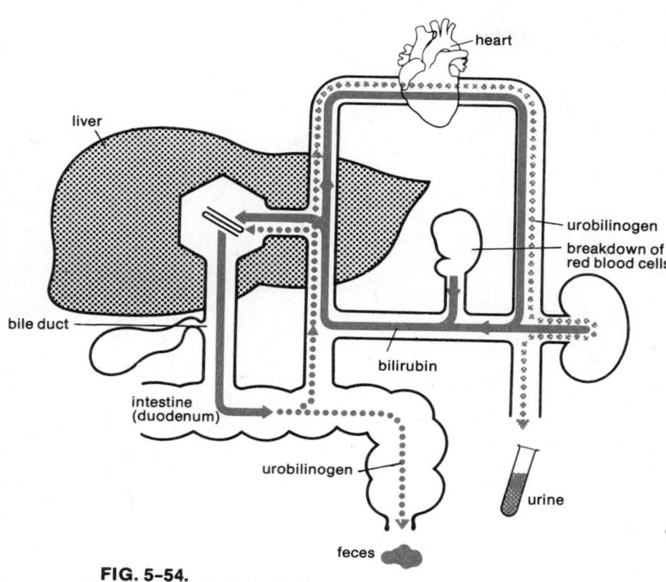

FIG. 5-54.
Retention of bile pigments due to obstruction of the common bile duct.

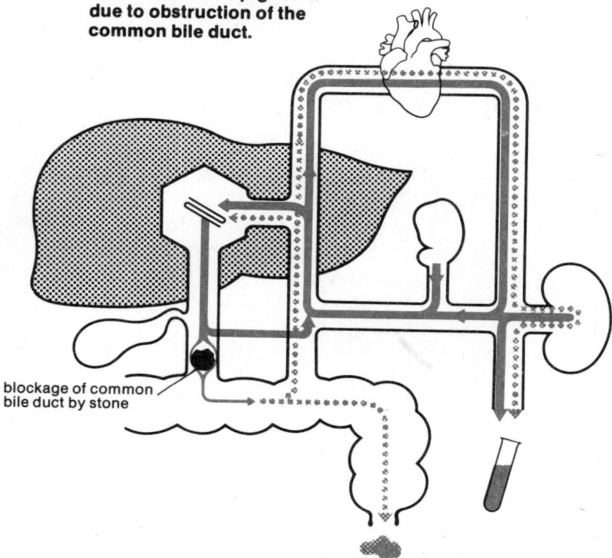

ranged in roughly six-sided *lobules*. In a microscope section the cells are revealed to be in columns radiating from the center of each lobule and are separated from one another by fine blood vessels (*sinusoids*). The walls of the sinusoids are lined with endothelium cells (as in all blood vessels) but also contain large phagocytic cells of the reticuloendothelial system (Kupffer cells), which can engulf and destroy foreign bodies such as bacteria and remains of broken-down cells (Fig. 5-50). The sinusoids communicate with the hepatic artery and with the portal vein. The blood in the sinusoids is in direct contact with the liver cells. The surface of each cell faces a sinusoid and is bathed with blood containing oxygen (from the hepatic artery) and nutrients and insulin (from the portal vein). From this blood the cell extracts what it requires and discharges into the blood the substances that it has synthesized or stored (e.g., glucose). The opposite surface of the cell faces a bile capillary into which it secretes bile. At the center of each lobule is the intralobular vein (Fig. 5-51).

The bile secreted from the liver cells is collected in intercellular capillaries which drain into larger passages (canaliculi) and ducts which finally merge into the common hepatic duct. The latter unites with the cystic duct from the gallbladder to form the common bile duct which discharges into the duodenum. The *gallbladder* is a pear-shaped bag in which the bile (gall) is stored and concentrated. When food enters the duodenum from the stomach, the gallbladder ejects bile into the duodenum. *Bile* is a thick yellow-brown fluid containing bile pigments (Fig. 5-53) derived from the hemoglobin of red blood cells that have reached the end of their life-span (the main pigments being bilirubin and biliverdin), bile salts (formed in the liver from cholesterol), cholesterol and other substances. The bile salts are essential for the digestion and absorption of fats in the small intestine.

Glucose brought to the liver from the intestine is converted to glycogen and stored. As the need arises, glucose is re-formed from glycogen and released into the bloodstream. Glycogen is stored by deposition in the central areas of the lobules, while the precursors of bile are secreted as granules at the periphery. Glycogen deposition and bile secretion appear to take place in alternate phases. After a meal the liver enters the *assimilatory phase*, in which glycogen is formed and deposited in the cells. Deposition of glycogen begins in the central areas of the lobules and progresses toward the periphery; as the glycogen content of the liver increases, the whole organ increases in size and weight. At the same time, the secretion of bile, which was active at the beginning of this phase, steadily decreases. When the digestion of food has been completed, bile secretion is at a minimum and the liver cells are full of glycogen. Now the *secretory phase* begins. Granules of secretion accumulate along the bile capillaries, as well as water, proteins and products of metabolism. Secretion progresses from the peripheral to the central areas of the lobules. As more and more bile is secreted, the glycogen content of the cells is reduced to a small amount. The whole cycle is then repeated.

JAUNDICE

Jaundice (icterus) is characterized by yellow discoloration of the skin and mucous membranes and is due to deposition of excessive bile pigment from the blood. It is not a disease in its own right, but a symptom of various possible diseases or disorders of the liver and bile ducts. Bile pigments are produced by the breakdown of worn-out red blood cells and are normal excretions of the liver into the bile. An excess of bile pigment in the blood may be due to excessive pigment formation, to obstruction of the bile ducts (Fig. 5-54) or to faulty functioning of the liver cells.

The red blood cells have an average lifespan of about 120 days, at the end of which they are destroyed, mainly in the *spleen,* which consists of a spongelike mass of lymphatic tissue and is situated on the left, between the stomach and the left kidney. It is composed of reticuloendothelial cells. In the embryo the spleen forms all types of blood cells, but in the adult only lymphocytes and monocytes. It also acts as a filter and storage reservoir for blood, and it plays a part in the formation of antibodies on which immunity depends.

The hemoglobin of the blood is thus broken down into bile pigments, of which bilirubin is the most important. Bilirubin undergoes a minor chemical change in the liver, is excreted in the bile and enters the duodenum. Here it is converted by bacterial enzymes into *urobilinogen,* a brown pigment. Some of this substance is reabsorbed into the blood (and eventually reexcreted in the bile), while the remainder is discharged with the feces. Since urobilinogen is responsible for the color of feces, stools become light gray when bile production or bile excretion is disturbed. A small proportion of the urobilinogen is normally excreted in the urine (Fig. 5-53). If excretion of bile from the liver is obstructed or impaired, there is an increase in the amount of pigment in the urine, which takes on a dark color. At the same time, the increased concentration of bilirubin in the blood gives the skin and especially also the white of the eyes a yellow color. With larger amounts of pigment the color may be darker, ranging to dark brown or indeed green.

As already stated, jaundice may be due to disturbed function of the liver cells so that they are unable to dispose of bile, e.g., as a result of hepatitis, cirrhosis or certain poisons (Fig. 5-55). Or there may be obstruction of the bile canaliculi, with reabsorption of the pigment into the blood (e.g., in response to certain drugs or in hepatitis), or obstruction of larger bile ducts (e.g., by a tumor or by a gallstone blocking the common duct). In all these cases jaundice occurs because bile fails to reach the duodenum (*cholestatic jaundice*). Alternatively, jaundice may be due to the breakdown of excessive numbers of red blood cells (Fig. 5-56), so that the liver is unable to cope with the amount of pigment formed, as occurs in certain types of anemia (page 104), malaria (page 475), blood group incompatibility between mother and child (page 282), transfusion of blood of an incompatible group (page 115), and the action of certain poisons.

In addition to the yellow color of the skin and mucous membranes, other symptoms associated particularly with cholestatic jaundice are itching (due to accumulation of bile acids in the body), nausea and an inability to tolerate fatty foods. The stools are light-colored and fatty because the digestion of fats in the intestine is disturbed on account of the absence of bile. On the other hand, in a case of jaundice due to excessive pigment formation from the breakdown of red blood cells there is no itching, the digestion of fats is not impaired, and the stools are dark-colored.

HEPATITIS

Hepatitis is inflammation of the liver, usually as a result of viral infection. There are two types of acute hepatitis: *epidemic hepatitis* (epidemic jaundice), which is transmitted from person to person by direct contact or through contaminated food or water, and *serum hepatitis,* the virus of which is transmitted by blood or plasma that may be present as traces on hypodermic needles used for injections. There

FIG. 5-55.
Causes of jaundice due to the liver or the bile ducts.

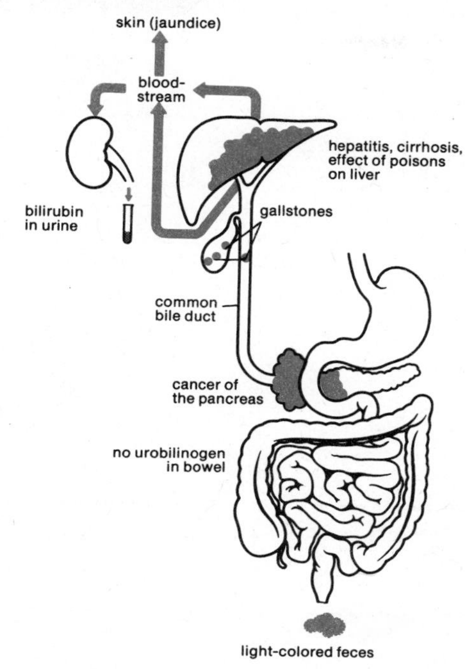

FIG. 5-56.
Jaundice due to excessive production of bile pigments resulting from excessive breakdown of red blood cells.

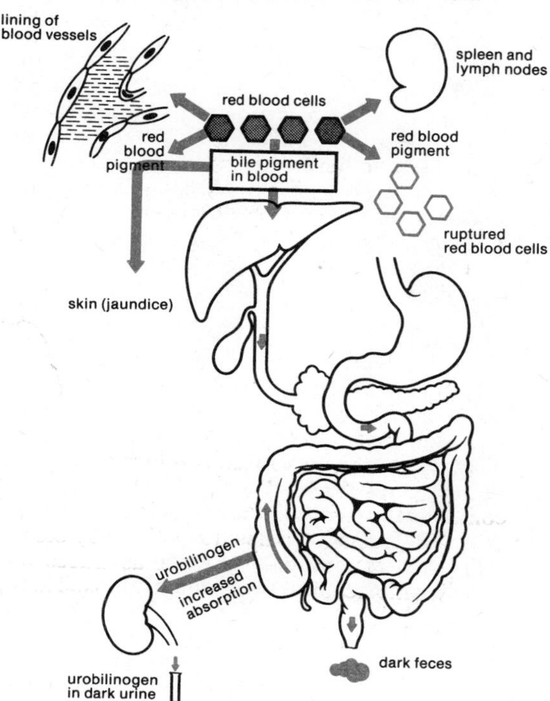

are other forms of hepatitis, but they are relatively rare and occur only as possible consequences of other infectious diseases such as yellow fever, poliomyelitis, typhoid fever, tuberculosis and syphilis.

The incubation period of the virus of serum hepatitis may be as long as six months. Symptoms of epidemic hepatitis may not appear until six weeks after infection. Because of their long incubation periods, it was not until the Second World War that these diseases were clearly recognized. Epidemic hepatitis occurred among troops and especially in prisoner-of-war camps. General weakness, undernourishment and infectious diseases lower the resistance to hepatitis. The epidemic type is transmitted by direct contact with an infected person or contaminated food or water. The virus may be carried by milk or by mussels which live in polluted water. It is excreted in the patient's urine and feces, so that unhygienic conditions evidently promote the transmission of epidemic hepatitis. The traces of contamination that are responsible for transmitting serum hepatitis are not always destroyed by ordinary sterilization. However, special precautions in the sterilization of surgical instruments, and the introduction of throwaway hypodermic needles which are discarded after only one use, have greatly reduced the risk of infection under modern hygienic conditions. Drug addicts, who are often not particular about hygiene, are most likely to catch the disease. Unfortunately, the virus of serum hepatitis is present in about 1 percent of all donors who give blood for transfusion. These carriers of the virus show no symptoms, so that the presence of the virus in their blood remains unsuspected. Prevention of so-called transfusion hepatitis (serum hepatitis transmitted by blood transfusion) therefore continues to be a problem.

Acute hepatitis (Fig. 5-57) initially affects the connective tissue that forms the walls of the lobules in the liver. Soon the liver cells themselves are attacked, as a result of which the secretion of bile ceases. The liver becomes enlarged and tender (sensitive to pain upon pressure), causing a sensation of dull pressure in the upper part of the abdomen, on the right; jaundice develops (see page 190), and there is itching. Bile pigments are excreted in increased amounts in the urine, which appears dark, but are absent in the stools, which are therefore of a light gray color. These characteristic symptoms of hepatitis are often pre-

FIG. 5-57.
Course run by acute hepatitis due to virus A (epidemic) and virus B (serum hepatitis).

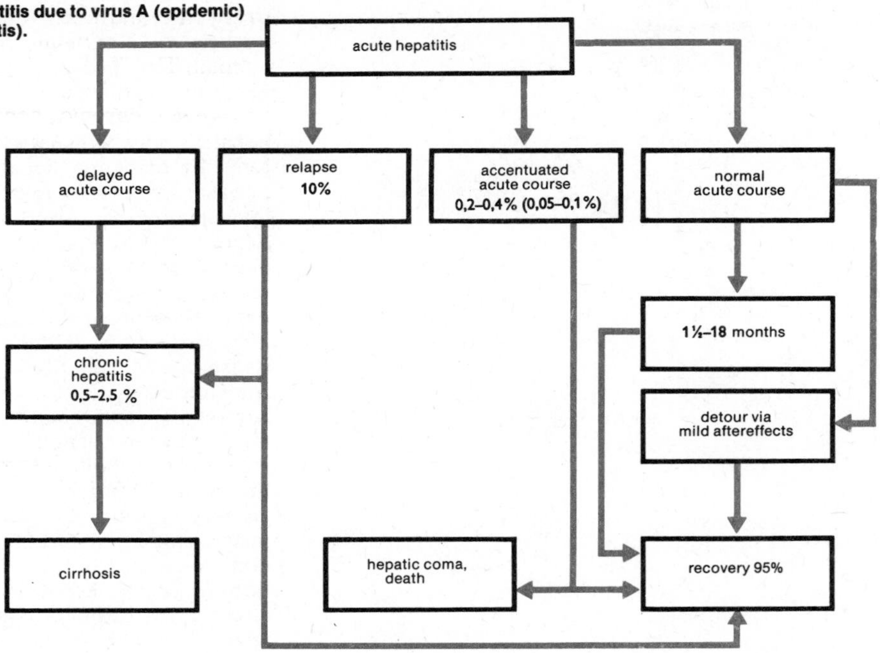

FIG. 5-58.
(a) The normal liver quickly excretes the injected pigment with bile; the concentration in the blood rapidly decreases.
(b) The diseased liver cells are unable to absorb the pigment from the blood; the concentration in the blood remains high.

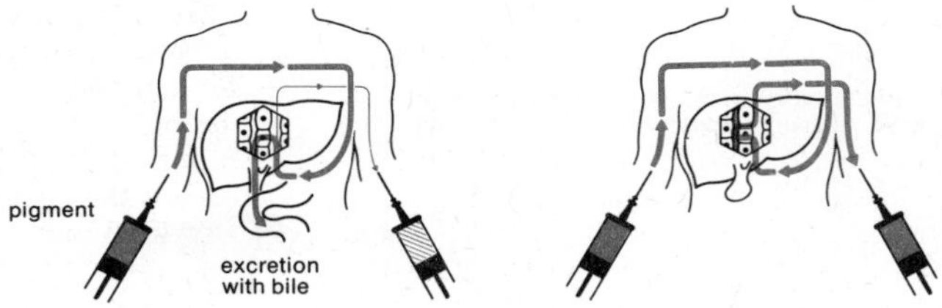

FIG. 5-59.
Needle biopsy of the liver.

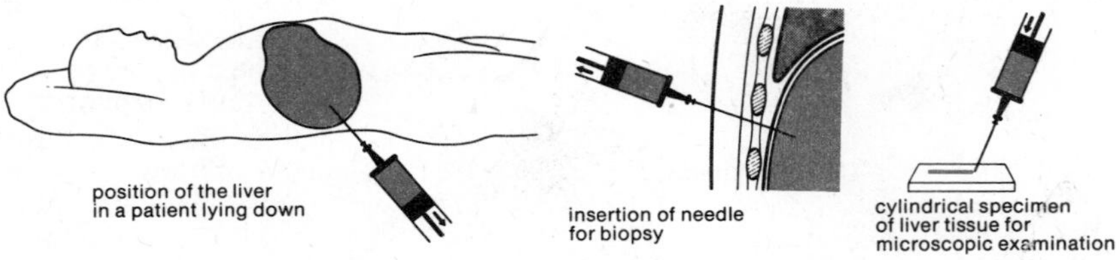

ceded by nausea, vomiting, loss of appetite, constipation or diarrhea, weariness, pains in the muscles and joints, fever, and an inability to tolerate fatty foods, alcohol and tobacco. In a severe case the patient becomes irritable, depressed and listless. Unconsciousness may occur in very severe cases. This *hepatic coma* is a sign that, after extensive destruction of liver cells, substances that impair the function of the brain have accumulated in the body.

Fortunately the disease is seldom fatal (in only 0.2–0.4 percent of all cases, with a mortality rate of only 0.05–0.1 percent in young and healthy persons who catch the disease).

Though there is complete recovery in the great majority of cases, chronic hepatitis sometimes develops. With proper treatment of acute hepatitis the first signs of improvement appear two to three weeks after the onset of the disease; complete recovery takes at least six to eight weeks, occasionally much longer.

The process of recovery can be monitored by laboratory tests. These include transaminase determination. Transaminase is an enzyme of protein metabolism and is found chiefly in the liver and the cardiac muscle. Injury of either of these tissues liberates the enzyme into the bloodstream, where its presence indicates severe damage to cells. By means of electrophoresis it can furthermore be ascertained whether the process of protein synthesis of the liver has been impaired. Injection of pigments which combine with bile and whose concentration in the blood can be monitored provides an indication of whether the hepatic secretory function has been fully restored (Fig. 5-58). In doubtful cases confirmation can be obtained by endoscopy (inspection of cavities by means of a device comprising a tube and optical system) and by biopsy (microscopic examination of a small specimen of liver tissue extracted with the aid of a needle, Fig. 5-59).

Treatment of hepatitis consists mainly in keeping the patient strictly in bed and prescribing a suitable diet (see page 142). There is no specific treatment for either type of acute hepatitis. Sometimes the damaged liver cells can be fortified by cortisone derivatives. In cases where particularly susceptible persons have been in contact with the infection, or if such persons have to be given a blood transfusion, gamma globulin may be injected as a precaution; this protective protein reduces the virulence of the hepatitis, should it develop.

CIRRHOSIS

Cirrhosis is a chronic disease of the liver, in which damaged liver tissue is replaced by dense connective (fibrous) tissue, a kind of scar tissue. Contraction of this fibrous tissue may impair the circulation of blood through the liver, causing congestion in the portal vein system and generally affecting the function of this organ, accompanied by degenerative changes of its cells. The precise cause of cirrhosis is often not ascertainable. Nutritional deficiency (e.g., lack of proteins) or the action of certain poisons can cause it. So can heavy drinking over a long period of time, usually many years, but it is possible that cirrhosis is then due to a combination of alcohol and malnutrition. The disease may also occur as a consequence of various chronic disorders of the liver—e.g., chronic hepatitis, which, in a small minority of cases, may develop from acute hepatitis (see page preceding section).

The symptoms of cirrhosis are a sensation of pressure in the upper part of the abdomen, repletion, flatulence, weakness, depression, loss of weight, nausea and inability to tolerate fatty foods. The liver is enlarged. The connective tissue formed by the disease tends to displace the liver cells and to cut off their blood supply.

Why acute hepatitis should develop into chronic hepatitis in some cases is uncertain. It may be because the patient leaves his bed too soon, or because of unregulated diet, indulgence in alcohol, inflammation of the bile ducts, or diabetes. Some cases of chronic hepatitis are cured, however, whereas others may turn into cirrhosis.

The effect of alcohol is to cause fat globules to accumulate in the liver cells. This condition remains reversible, i.e., curable, for years, but eventually an increase in connective tissue takes place and cirrhosis sets in. The consumption of alcohol, in conjunction with the malnutrition and protein deficiency of chronic alcoholics, appears to be particularly harmful in this respect (see page 141). In alcoholics, malnutrition may occur because of impaired digestion due to damage to the stomach lining; also the body uses alcohol as a source of energy and thus derives little benefit from the food consumed. Excessive intake of fats, obesity, and diabetes are believed to be contributory factors to this condition.

Symptoms at first are few and not severe. There may be a feeling of repletion and pressure in the upper part of the abdomen, a lowering of physical and mental performance, sometimes swelling and tenderness of the liver. High-protein diet, vitamins, abstention from alcohol, and the reduction of overweight can, as a rule, correct this condition. Untreated, it may lead to cirrhosis.

Cirrhosis may also be caused by chronic congestion of bile or by prolonged congestion of blood in the liver, e.g., in a case of failure of the right side of the heart. Chronic inflammation of the bile ducts may likewise cause it. Quite often, cirrhosis is due, not to one particular cause, but to a whole set of factors. It

inflammation of bile passages
chronic hepatitis
"ill-treatment" of the liver by bad habits, etc.
chronic congestion of blood in the liver
chronic congestion of bile
diabetes, protein deficiency
alcohol, excessive fat intake, hunger
acute hepatitis due to viral infection

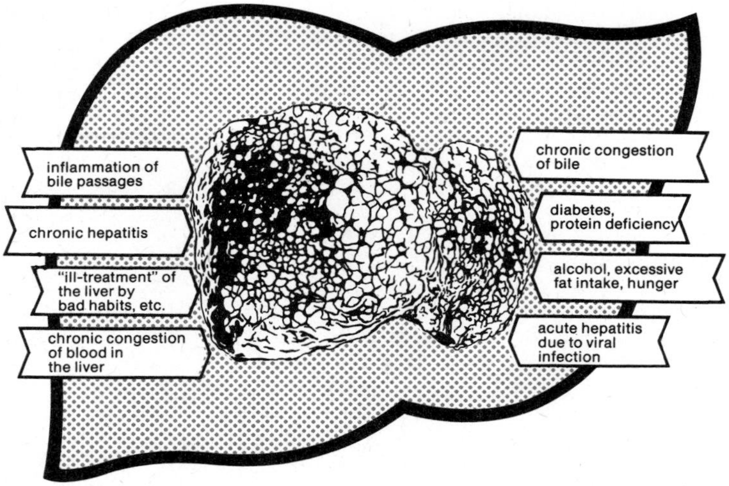

FIG. 5-60.
Causes of cirrhosis of the liver.

usually begins with the formation of dense connective tissue, which is chronically irritated and eventually contracts, so that the liver becomes small and hard. As a result, the liver cells are cut off from their blood supply, so that their function is impaired and they degenerate. More and more fibrous connective tissue is formed until the lobular structure of the liver is destroyed and its surface presents an irregular and knotted appearance. In the earlier stages of the disease there may be few if any symptoms. In an advanced case the symptoms are due to the effect of tissue contraction upon blood vessels, bile ducts and liver cells.

Increased flow resistance caused by constriction of the blood vessels in the liver raises the pressure in the portal system (*portal hypertension*). As a result, fluid is expelled into the abdominal cavity, where it produces *ascites* (dropsy of the belly). There are certain minor blood vessels which connect the portal vein to the veins around the esophagus and the rectum (Fig. 5-61). Severe cirrhosis causes these communicating blood vessels to become distended, with the result that bleeding occurs in the esophagus and at the rectum, where a condition resembling hemorrhoids (piles) arises. Esophageal bleeding may become dangerous, with vomiting of blood, if rupture of blood vessels occurs.

Congestion of bile ducts prevents the discharge of bile. Damage to the liver cells by pressure eventually results in failure of liver function.

Beginning gradually, with rather unspecific

hepatic vein
esophageal vein
portal vein
superficial abdominal vein
hemorrhoidal veins

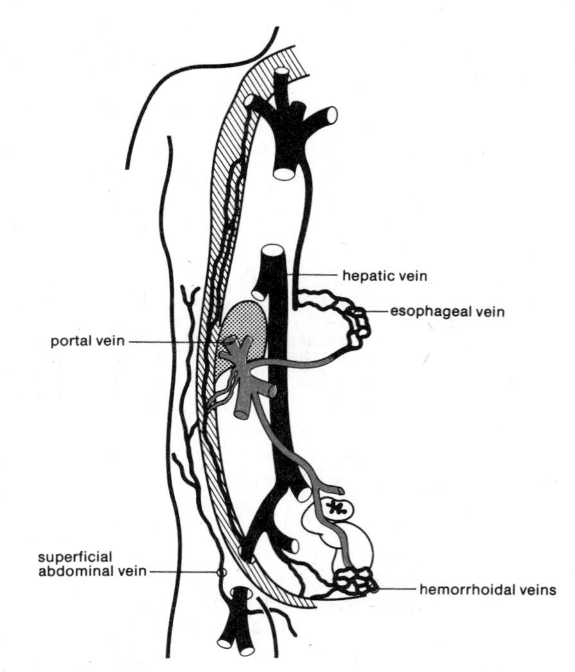

FIG. 5-61.
Short circuit between portal vein and the vein system.

symptoms (weakness, depression, gastrointestinal trouble, loss of appetite, aversion to fat and meat), cirrhosis causes a general lowering of activity and reduced enjoyment of life. There are various external signs, such as dilated capillaries of the skin (sometimes displaying a radiating "spider" pattern), hour-

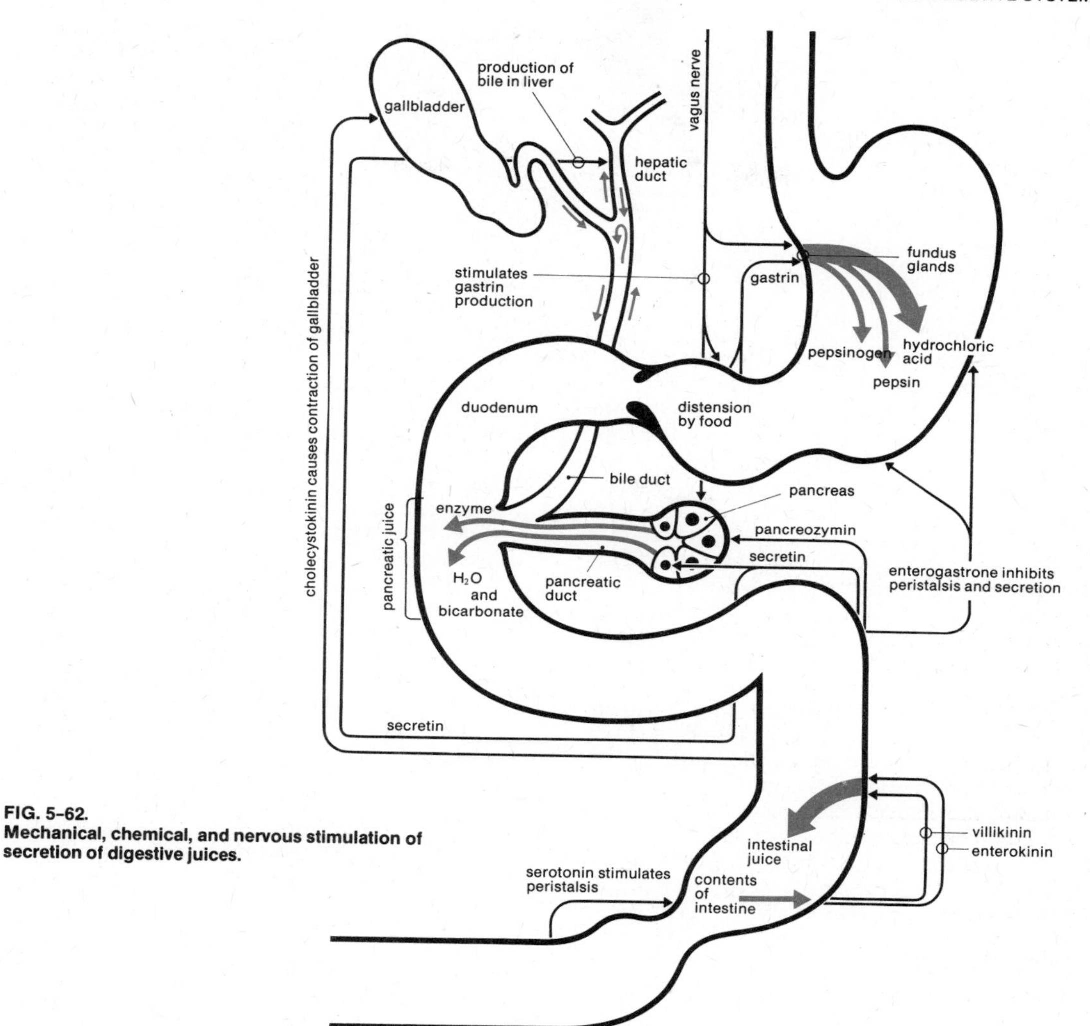

FIG. 5-62.
Mechanical, chemical, and nervous stimulation of secretion of digestive juices.

glass-shaped nails, reduction of body hair, and bleeding tendency. In doubtful cases the diagnosis can be confirmed by endoscopy and biopsy.

Physical decline, emaciation and mental changes may initiate the final stage of the disease, terminating in coma. This state of unconsciousness is attributable to paralysis of the brain by ammonia produced by intestinal bacteria. In a healthy person the ammonia is absorbed into the blood and carried to the liver, where it is metabolized. In advanced cirrhosis the ammonia accumulates in the blood (ammonia toxicity). Coma can sometimes be prevented by killing the ammonia-producing bacteria.

Treatment of esophageal hemorrhage may require surgery. Efforts must be directed at reducing portal hypertension by means of an operation whereby blood is bypassed into the inferior vena cava, this being the vein through which blood from the abdomen and pelvis reaches the heart. This operation may also relieve ascites.

In general, alcoholic cirrhosis can, at least in its earlier stages, be greatly improved by abstinence from alcohol. Suitable diet and a well-regulated way of life can often also arrest the progress of the disease, even though the proliferations of connective tissue cannot be made to disappear. The diet should aim at relieving the liver cells; it should be a high-calorie diet rich in proteins (see pages 141 ff). If the liver cells have become affected by fatty infiltration, fat intake should be reduced to under 20 g per day. Treatment of ascites requires a low-salt diet.

GALLSTONES

As explained earlier, bile (gall) is formed in the liver and stored in the gallbladder. Bile contains excreted waste products (including bile pigments) and is essential for the digestion and absorption of fats in the small intestine. This digestive function is performed more particularly by the bile salts, i.e., salts of bile acids. These acids are reabsorbed from the intestine to be used again by the liver.

The flow of bile (Fig. 5-63) depends on the activity of the liver cells in secreting it, on the freedom from obstruction of the bile ducts, and on the vigor with which the gallbladder contracts. The cystic duct from the gallbladder unites with the hepatic duct from the liver to form the *common bile duct*. The outlet of the common bile duct into the duodenum is surrounded by a sphincter (circular muscle), which can close it. When the sphincter contracts and thus closes the outlet, bile flows into the gallbladder. When the pressure of the bile in the gallbladder and duct exceeds a certain value, the sphincter opens and admits bile into the duodenum. When food, especially fat, enters the duodenum, its mucous membrane secretes a hormone that is carried by the blood to the gallbladder, which contracts in response to this hormone, while the sphincter relaxes (Fig. 5-64). This sequence may be disturbed by various causes, e.g., emotional factors and faulty nervous control, which may bring about simultaneous contraction of the gallbladder and closure of the sphincter. When this happens, an abnormal rise in pressure develops in the gallbladder and ducts, accompanied by a sensation of pain (biliary colic).

The secretion of bile is stimulated by the bile acids returned to the liver and by the hormone called secretin formed in the mucous membrane of the duodenum (this hormone also stimulates the secretion of pancreatic juice). Certain foods, such as meat (especially liver) and rhubarb, also particularly stimulate bile secretion. Overflow of bile into the pancreas, which could have harmful consequences to the latter, is normally prevented by the presence of a higher pressure in the pancreatic duct than in the common bile duct.

In general, *colic* is a spasm in an internal organ, accompanied by pain and caused by powerful contraction of involuntary muscle. *Biliary colic* (gallbladder colic) is characterized

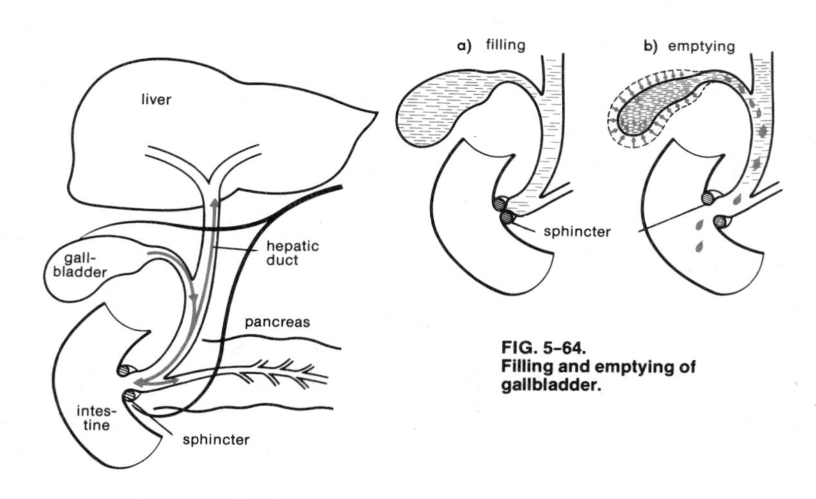

FIG. 5-63.
Bile flow paths.

FIG. 5-64.
Filling and emptying of gallbladder.

by acute intermittent pain originating in the gallbladder and/or bile ducts. It may be due to inflammation of these organs or, though rarely, to faulty nervous control which causes an excessive rise in pressure in the gallbladder (as already mentioned). Most frequently, however, colic is caused by a gallstone becoming wedged in a bile duct, where it irritates the mucous membrane of the lining and repeatedly stimulates the muscles of the duct to eject the stone. This is what causes the recurrent spasms. Colic begins suddenly, usually after a heavy meal, the attendant symptoms being restlessness, eructation, sometimes vomiting. In many cases these symptoms increase in intensity for half an hour or so. The pain is localized in the region of the liver, on the right, under the costal arch, and radiates into the right shoulder. Without treatment, the colic pains may continue for hours, leaving the patient with a feeling of soreness in the upper part of the abdomen, on the right. When biliary colic is caused by a gallstone wedged in a bile duct (Fig. 5-65), the symptoms and consequences vary according to the location of the stone. A stone wedged in the neck of the gallbladder sometimes falls back into the bladder: the pain then suddenly ceases. If the stone is small enough to pass out of the bladder, it is quite likely to be discharged without further trouble into the duodenum and thus be eliminated with the feces. If it becomes temporarily wedged in the cystic duct, the flow of bile from the gallbladder is obstructed and jaundice develops as a result. The stone may, alternatively, be too large to pass through the sphincter at the outlet of the common bile duct and may become impacted (permanently lodged) there. When that happens, blockage of the duct causes a more severe attack of jaundice, and the patient is in constant pain, possibly attended by fever. As prolonged obstruction of the flow of bile is liable to damage the liver cells (Fig. 5-65), surgical treatment may be necessary. Between attacks of pain the patient suffers from a feeling of repletion, flatulence, sometimes also diarrhea and vomiting. Fatty foods, coffee and cold drinks usually aggravate the symptoms. Anger and emotional stress likewise adversely affect the condition.

FIG. 5-65.
Causes and effects of biliary colic.

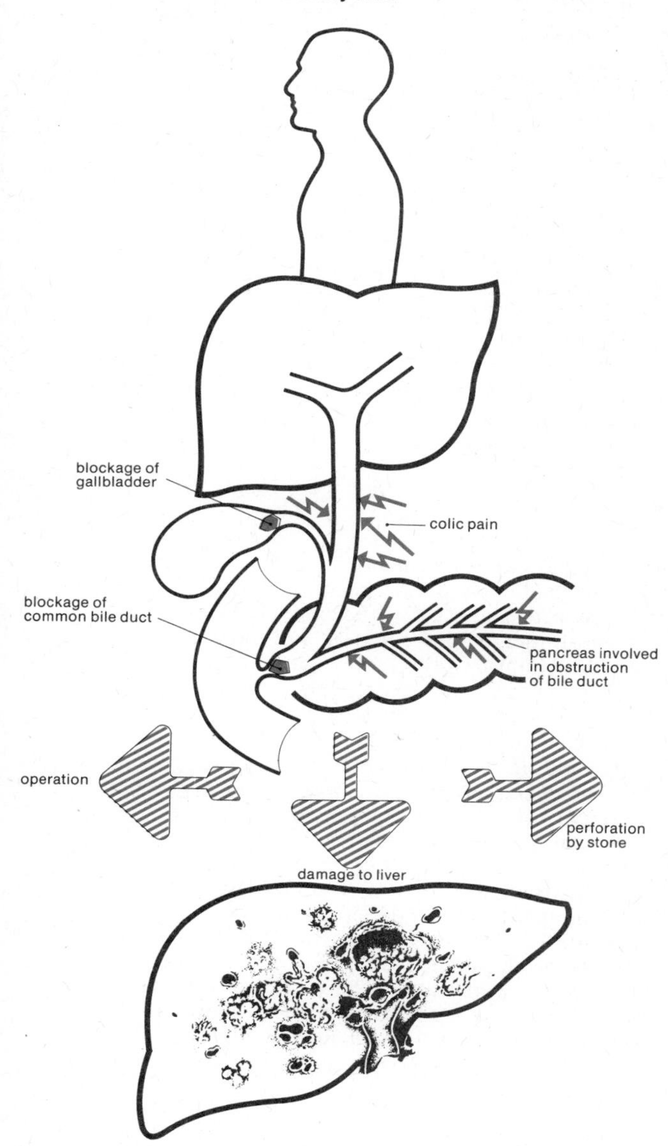

Treatment and prevention of biliary colic are important not only to relieve pain, but also because this disorder may be attended by various complications, such as inflammation of the bile ducts and gallbladder, congestion of bile, liver damage, and perforation of the wall of the duct by a stone. Ordinary biliary colic is treated with antispasmodic and analgesic (pain-relieving) drugs. These include, for example, atropine and certain atropine-type and morphine-type drugs that do not tend to aggravate spasms, as morphine does. Hot wet poultices and purgatives containing salts are traditional remedies which have, on occasion, proved effective. If surgical treatment is judged necessary, the operation will, as a rule, be performed during an interval between attacks.

FIG. 5-66.
Causes of gallstones.

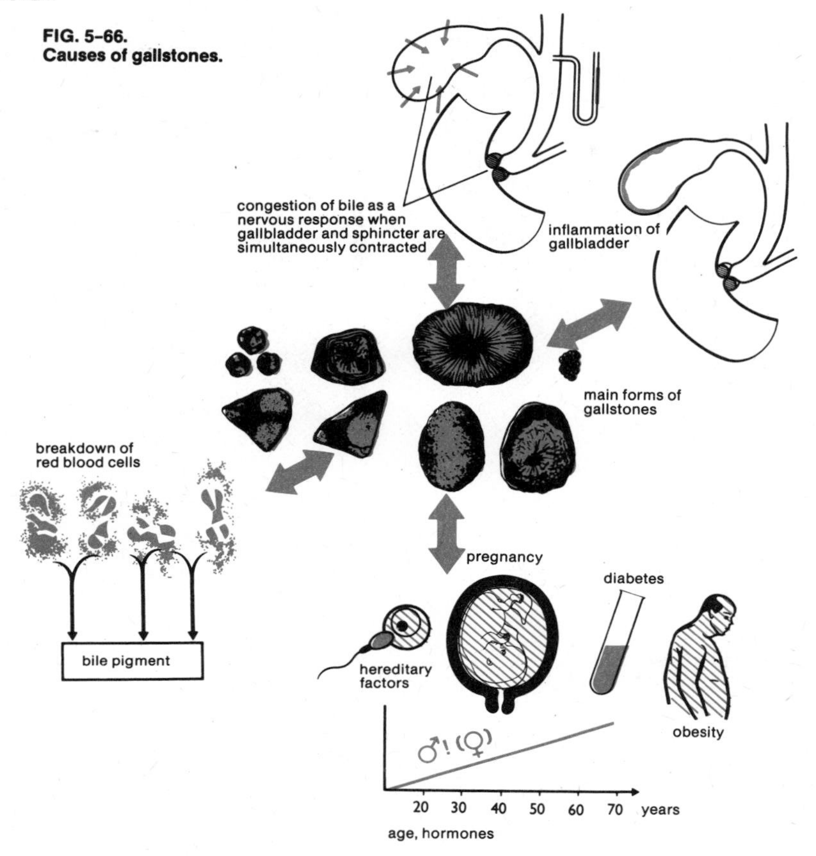

Inflammation of the gallbladder (cholecystitis) may result in perforation, so that the contents of the bladder enter the abdominal cavity. If that happens, antibiotics must be administered and an operation performed without delay.

Gallstones (biliary calculi) are formed as concretions in the gallbladder. They are very common. It is estimated that at least 5–10 percent of the adult population have gallstones; some estimates indeed put the proportion much higher, at between a quarter and third of all people. The frequency is much greater in women than in men. In the great majority of cases there are no symptoms: the stones remain "silent," and the person remains unaware of their presence. Gallbladder attacks usually come on in middle age. Heavy meals with a high content of fat cause the gallbladder to empty itself vigorously, so that the stones are stirred up, as it were. Once they are on the move, stones of a certain size may become wedged in the neck of the gallbladder, thus obstructing its outlet into the cystic duct. Smaller stones, which can readily pass out of the gallbladder, may become wedged farther downstream, e.g., at the sphincter controlling the outlet of the common bile duct into the duodenum. Only very small stones ("gall sand") can easily pass through the ducts into the duodenum and thus be eliminated.

Gallstones usually have a core consisting of a mixture of cholesterol, bilirubin and protein. The exact causes of their formation (Fig. 5-66) have not yet been clearly established; nor is it clear why they are much commoner in women than in men. Hereditary and hormonal factors appear to play a part, as do also the patient's manner of life and emotional stresses. The stones are formed in the gallbladder. Here the bile produced in the liver is concentrated by extraction of water. The tendency for stones to form is greater as the bile is more highly concentrated. Stone formation also depends on the nature of the substances excreted into the bile by the liver. Some gallstones consist predominantly of bilirubin, one of the bile pigments formed by the breakdown of red blood cells; in others there is a predominance of cholesterol, such stones being most likely to develop in circumstances where the cholesterol content of the blood and bile is increased, as in diabetes or during pregnancy. Precipitation of cholesterol as a "stone" will occur more readily if there is less bile acid present. Unless they contain admixtures of calcium, neither type of gallstone shows up on x-ray. There are, however, techniques whereby radiopaque substances (opaque to x-rays) can be introduced into the gallbladder to provide contrast and thus reveal the stones. Inflammation of the gallbladder and congestion of bile flow not only are brought on by stones, but in turn promote stone formation.

FIG. 5-67.

(a) Congestion of bile due to sluggishness of gallbladder.

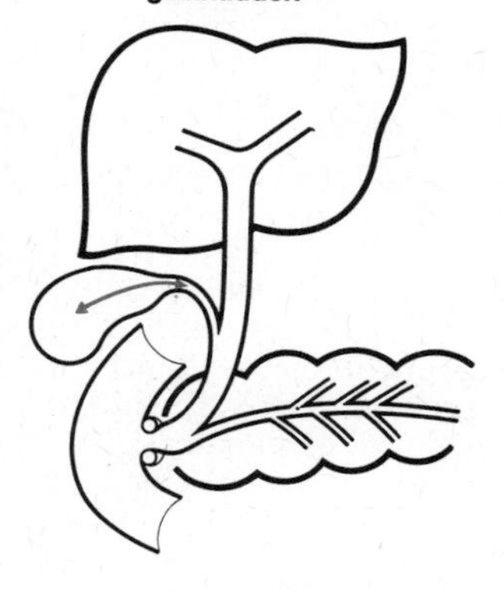

(b) Congestion of bile due to spasm of sphincter.

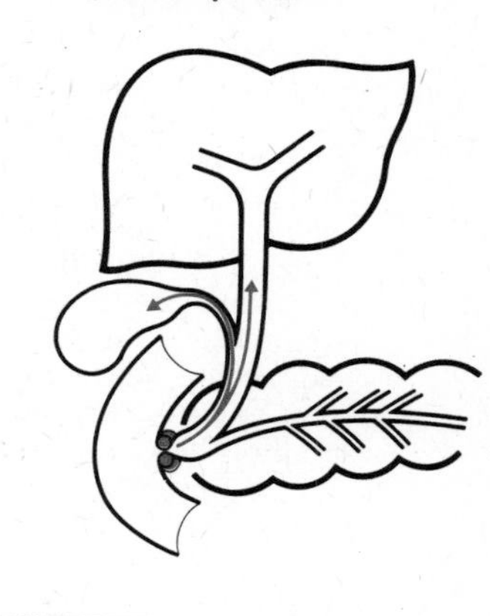

INFLAMMATION OF THE GALLBLADDER AND BILE DUCTS

Although the gallbladder and the ducts that discharge the bile perform their function without conscious control, they are, through their nerve supply, dependent on higher centers in the brain. For this reason gallstone trouble and inflammation of the biliary tract are aggravated by emotional stress and indeed may sometimes even be caused by it. It may occur that the biliary tract responds very sluggishly, thus impairing the proper emptying of the gallbladder and resulting in subsequent inflammation and stone formation (Fig. 5-67a). On the other hand, the tract may be overactive and function in an uncoordinated and illogical manner: the sphincter closes spasmodically and causes congestion of bile, which the muscle of the gallbladder strives to overcome by its own contraction spasms (Fig. 5-67b). Colicky pains occur in consequence (see preceding section).

Inflammation of the bile ducts (*cholangitis*) is often found in conjunction with gallstones. In such cases it is usually difficult to decide whether the stones or the inflammation initiated the disease. Besides stones, another cause of gallbladder inflammation (*cholecystitis*) is deficient emptying of the gallbladder due to faulty nervous control. Decomposition products which form in the stagnant bile irritate the mucous membrane with which the bladder is lined, so that the membrane is inflamed and further restricts the activity of the

FIG. 5-68.

(a) Congestion due to spasm, inflammation.

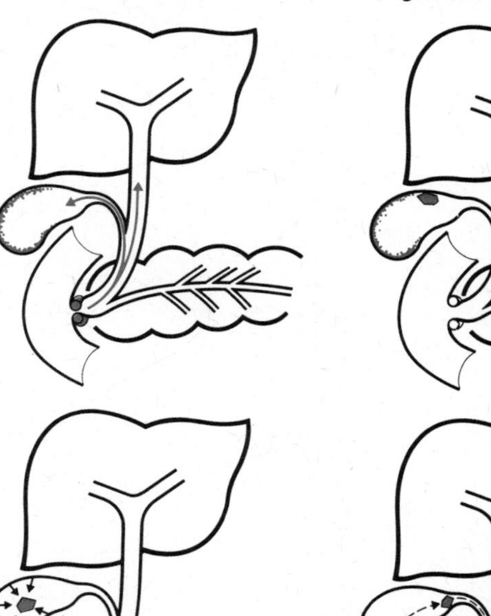

(c) Formation of gallstone.

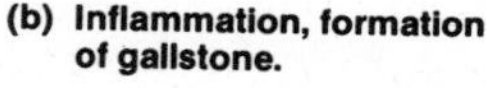

(b) Inflammation, formation of gallstone.

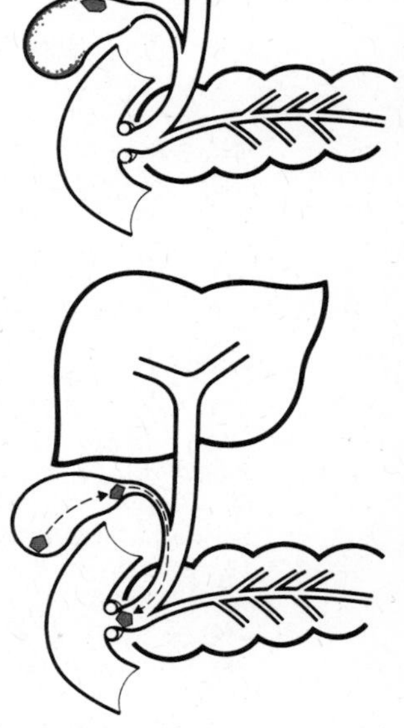

(d) Stone may become wedged at outlet of gallbladder or at sphincter.

bladder. As a result of this vicious circle, gallstones may form, and some of these may become wedged and thus cause colic (Figs. 5-68a to 5-68d; see also preceding section).

Distension and tenderness in the upper part of the abdomen, together with fever and changes in the blood picture, indicate that it is not just a case of wedged gallstone, but an inflammation of the gallbladder. In a case of chronic gallbladder inflammation, fairly long periods relatively free from objectionable symptoms alternate with painful attacks, which are particularly brought on by faulty diet. The inflammation gradually spreads, the wall of the gallbladder thickens, eventually degenerating into a rigid and sometimes calcified bag which can no longer empty itself.

In milder cases of gallbladder inflammation or inflammation of bile ducts, treatment will generally take the form of dieting and drugs. Serious cases will require surgical intervention. An operation will certainly be necessary if bile ducts are completely blocked by impacted stones, a condition that is accompanied by jaundice and fever (see preceding section). Persistent chronic inflammation accompanied by jaundice may also necessitate an operation, as will also a condition in which the gallbladder is overdistended and there is a risk of perforation. In other cases, and also pending an operation, the patient will be placed on a special diet rather similar to a diet prescribed in dealing with liver disease. The main principle of the diet is to avoid all foods that stimulate the gallbladder and ducts (especially roasted products, fat meat or fish, rich sauces, mayonnaise, chocolate, cakes, eggs, cold drinks). In general, alcohol need not be completely prohibited to the patient, except in cases where the liver cells are involved in the diseased condition. Vegetables are tolerated in varying degrees. Strong coffee from ground coffee beans is often less well tolerated than tea without milk or coffee made from soluble powder or extract.

Home procedures that may be helpful for gallstone trouble include hot wet poultices applied to the region of the gallbladder. However, in a case of severe inflammation an ice bag may be better tolerated than a poultice. Antispasmodic and analgesic drugs will help to relieve the pain associated with biliary colic. When the attack has subsided, an attempt can be made to improve the flow of bile by suitable drugs and to achieve more frequent and thorough emptying of the gallbladder.

Part Six

THE GENITOURINARY SYSTEM AND REPRODUCTION

KIDNEYS

The kidneys are a pair of bean-shaped excretory organs, about 11 cm long, which lie at the back of the abdominal cavity, one on each side of the spinal column. Each kidney is embedded in fatty tissue (adipose capsule) and is surrounded by a sheath of fibrous tissue (renal fascia). The principal functions of the kidneys are excretion (i.e., the elimination of unwanted substances) and the maintenance of homeostasis (i.e., constancy of the body fluids, especially the plasma). To this end, the kidneys regulate the water content of the body by varying the amount of urine produced; they also vary the amounts of salts and other substances excreted in the urine, and they maintain the acid-base balance of the plasma and tissue fluid. Also, the kidneys produce two hormones. The adjective *renal* pertains to the kidney.

A kidney (Fig. 6-1) consists of an outer cone (cortex) and an inner zone (medulla) which surround a cavity (sinus). The medulla consists of a number of medullary pyramids; the apex of each pyramid is called a renal papilla and projects into the sinus of the kidney. The sinus contains the renal pelvis, the funnel-shaped upper end of the ureter. The renal artery and vein also pass through the sinus. The kidney is composed of something like a million nephrons, which are the functional units of kidney tissue. A *nephron* (Fig. 6-2b) consists of a *tubule,* about 3 cm long, closed at one end and communicating with a collecting duct at the other. The closed end is enlarged to form a thin-walled bag called a *glomerular capsule* (Bowman's capsule) enveloping the *glomerulus* (Fig. 6-2a), which consists of loops of fine capillary blood vessels connected to an arteriole of the renal artery. The capsule is the collecting chamber for the fluid filtered from the blood circulated through the glomerulus. The capsule and the glomerulus together are called a renal corpuscle (*malpighian corpuscle*). The

FIG. 6-1.

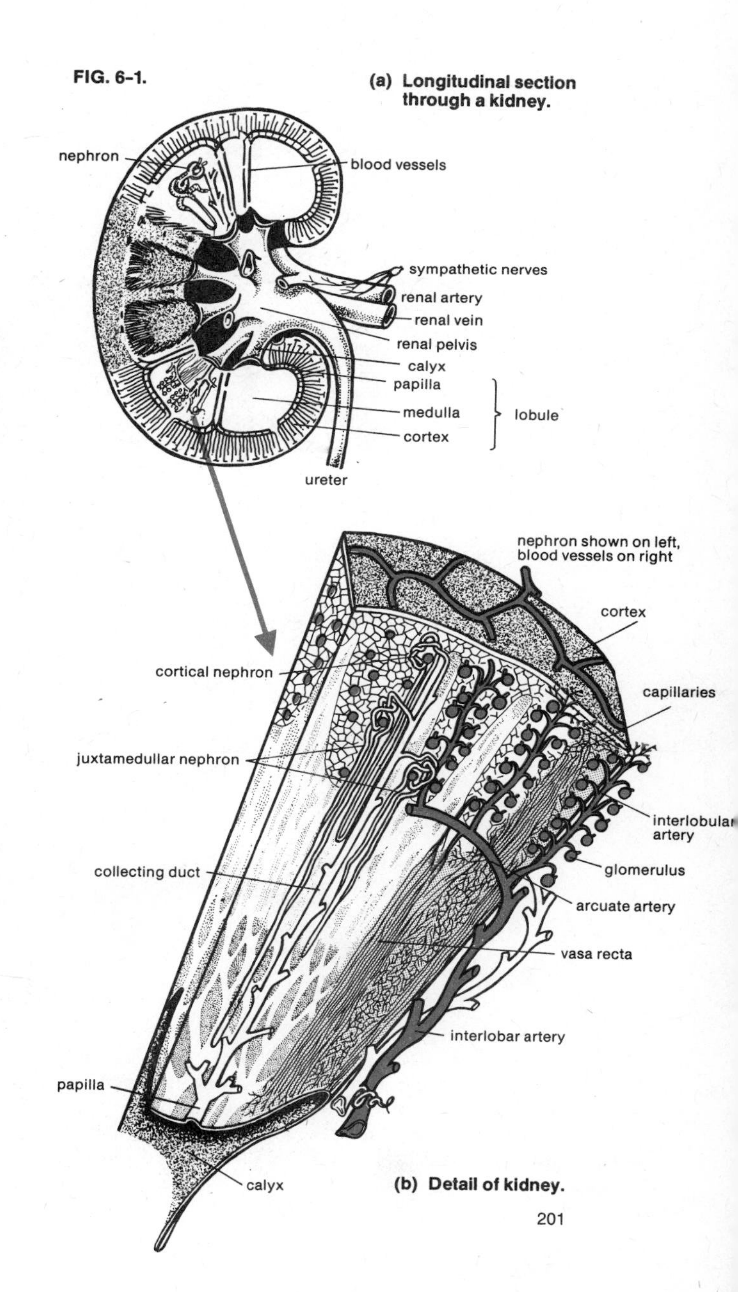

(a) Longitudinal section through a kidney.

(b) Detail of kidney.

FIG. 6-2.

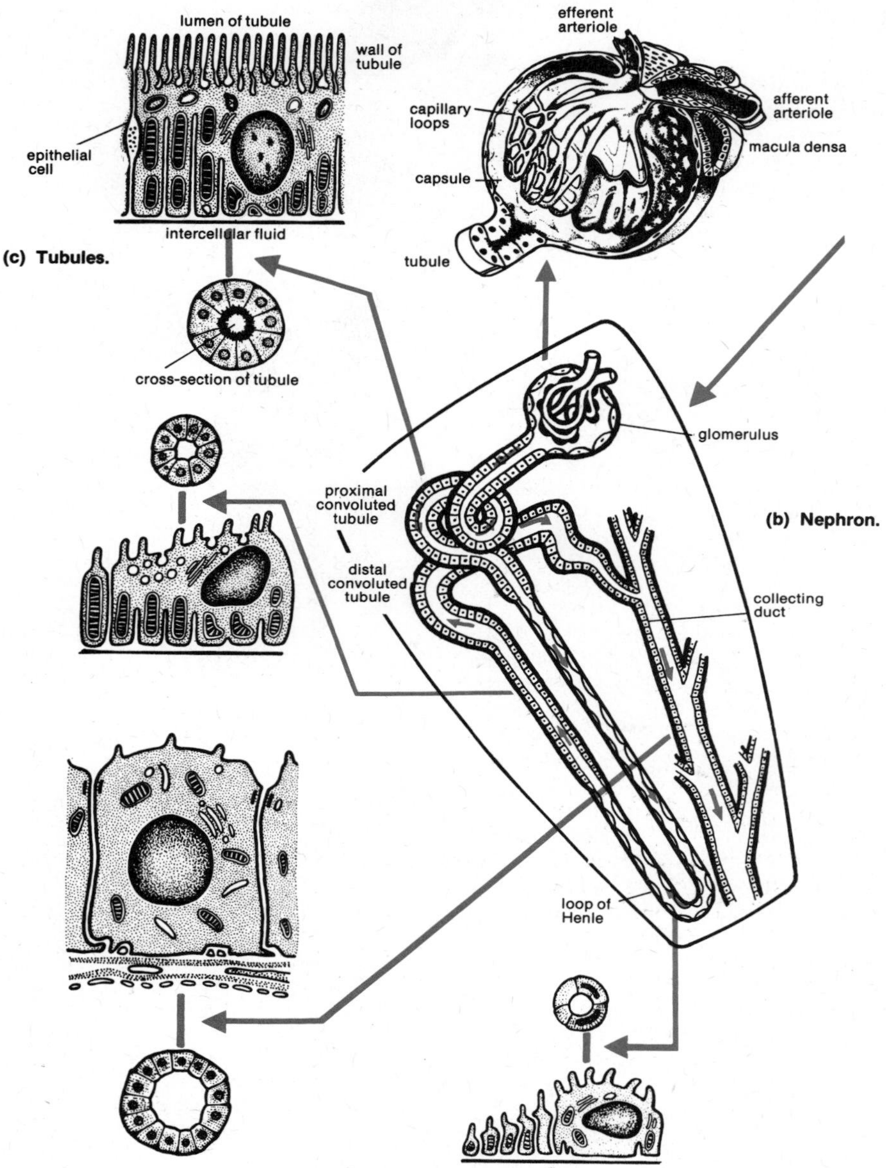

rest of the tubule is coiled and looped. Each collecting duct receives many tubules and passes through a pyramid to drain into a *calyx,* one of the subdivisions of the renal pelvis, and thus into the ureter (Fig. 6-1). The glomeruli and the proximal and distal convoluted tubules (Fig. 6-2b) lie in the cortex of the kidney; the straight portions of the tubules and the collecting ducts are in the medullary pyramids. Each kidney is supplied with blood by a renal artery; in the sinus this artery divides into branches from which an afferent arteriole goes to each glomerulus. After passing through the capillary loops of the glomerulus the blood is carried away by an efferent arteriole and eventually drains into the renal vein. The total volume of blood in the body flows through the kidneys about once every five minutes.

Urine production by the nephron takes place in two stages. First, there is *filtration* in the renal corpuscle: blood pressure in the glomerular capillaries is relatively high, so that plasma is forced through the capillary walls and into the glomerular capsule. The large-molecular plasma proteins are retained in the blood (i.e., are filtered out), as are, of course, also the cellular constituents of the blood. The fluid that makes its way into the capsule, called glomerular filtrate, is produced at a rate of 160–180 liters per 24 hours by the two kidneys. The next stage is that of *selective reabsorption*, which takes place in the tubule. Here substances useful to the body (water, glucose, amino acids, salts, etc.) are reabsorbed by the epithelial cells forming the wall of the tubule and are returned to the bloodstream. Waste products (urea, uric acid, unwanted salts, etc.) pass on and are excreted in the urine. Besides selective reabsorption, a certain amount of active secretion of substances into the urine takes place in the tubule. During its passage through the tubule the glomerular filtrate becomes concentrated and undergoes changes in composition until it is discharged as urine into the collecting duct and thus passes into the ureter.

Although filtration of fluid from the blood takes place at a rate of 160–180 liters per 24 hours, average urine production during that period is only about 1.5 liters, i.e., merely a small fraction of the total volume filtered in the kidneys. Most of the water in the glomerular filtrate is reabsorbed, the actual amount reabsorbed being controlled by the *antidiuretic hormone* secreted by the posterior lobe of the pituitary gland and depending on water need of the body at any particular time. This hormone promotes the absorption of water from the tubules, but if there is an excess of fluid in the body, e.g., after the person has had a long drink of water, secretion of the hormone is temporarily stopped, so that a great volume of water is discharged in the urine. About 95–96 percent of urine is water.

So-called *threshold substances* in the glomerular filtrate are wholly or partly reabsorbed in the tubules. They include water, glucose and salts. Substances that are not reabsorbed, but pass out completely in the urine, are called *nonthreshold substances*. Normally, virtually all glucose is reabsorbed and is therefore not present in the urine. An excess of glucose above the "glucose threshold," however, is eliminated in the urine, as in diabetes (pages 179 ff).

As already mentioned, the kidneys help to maintain the *acid-base balance*, i.e., the mechanism by which the acids and alkalis in the body fluids are kept in equilibrium so that the pH of the arterial blood is maintained at about 7.4, i.e., slightly on the alkaline side of neutral (pH = 7). The kidneys play an important part in the elimination of acids, and fresh urine is slightly acid (pH = 6–6.5). Disturbances in the acid-base balance may result in acidosis or in alkalosis.

About 7.5 kg of common *salt* (sodium chloride) passes through the kidneys per 24 hours. Of this amount about 1.5 kg is filtered into the nephrons and only a few grams are eliminated in the urine. The greater part of the salt is reabsorbed, particularly in the proximal convoluted tubules. A variable amount of salt is further reabsorbed in the distal convoluted tubules under the control of the hormone aldosterone secreted by the adrenal cortex. In this way the rate of salt reabsorption is adjusted to suit the body's needs. The salt concentration in the urine is therefore a variable quantity.

Urea is the final product of protein metabolism and is formed in the liver. It is the chief nitrogenous constituent of urine. A substantial proportion of the urea in the glomerular filtrate is returned from the tubules into the blood by rediffusion. Defective kidney function may cause an excess of urea in the blood (azotemia).

NEPHRITIS

One of the main functions of the kidneys is the excretion of waste products such as urea and uric acid. If the kidneys are defective, these substances will accumulate in the blood (uremia). The kidneys also play an important part in the elimination of water, salts and acids; deficient function in this respect will result in dropsy and acidosis. The latter is a disturbance in the acid-base balance of the body, associated with accumulation of acid substances and excessive loss of bicarbonate, and

FIG. 6-3.
Diagram of kidney with nephron shown much enlarged.

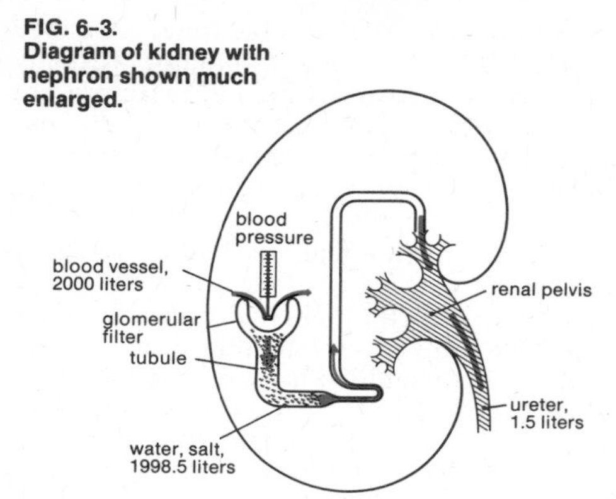

can be a dangerous condition. As already described in the preceding section, the functional unit of the kidney is the nephron, comprising the glomerulus, from which plasma is filtered from the blood into the glomerular capsule, and a convoluted tubule in which water and a large and, in some cases, varying proportion of dissolved substances are selectively reabsorbed into the blood, while the filtrate becomes concentrated and modified to urine. There are many hundreds of thousands of nephrons. Their tubules discharge the urine into collecting ducts, which in turn convey it to the renal pelvis and thus into the ureter, through which the urine flows to the urinary bladder.

The symptoms of kidney disease, even in a fairly advanced stage, are often rather vague and relatively mild. The patient can continue without outwardly serious symptoms for a long time even when a large proportion of the kidney function is impaired. However, if the kidneys cease to function altogether (renal failure, see later), the condition soon becomes dangerous unless effectively treated.

Nephritis is inflammation of the kidney. *Acute nephritis* attacks the nephrons. In the form called *glomerulonephritis* it is more particularly the glomeruli that are affected. As a rule, it is a secondary disease indirectly following a primary disease, usually a bacterial infection with streptococci such as tonsillitis or scarlet fever. The inflammation is apparently due to a kind of allergy. The streptococci or other agents which (as antigens) produce the primary disease initiate the formation of antibodies, so that an antigen-antibody reaction occurs (see pages 41 ff) which has effects that are in some way detrimental to the nephrons of the kidneys. Acute glomerulonephritis occurs mostly in children and young teen-agers. It is frequently accompanied by raised blood pressure (hypertension). Defective excretion causes excess water and salt to be retained in the blood, so that fluid leaks into the tissues, where it causes edema (dropsy). This may manifest itself as visible swelling under the skin, especially in the eyelids and around the eyes (puffiness). There is a leakage of protein and blood cells into the urine, which is then of a dark color and reduced in quantity. The pa-

FIG. 6-4.
Nephritis.

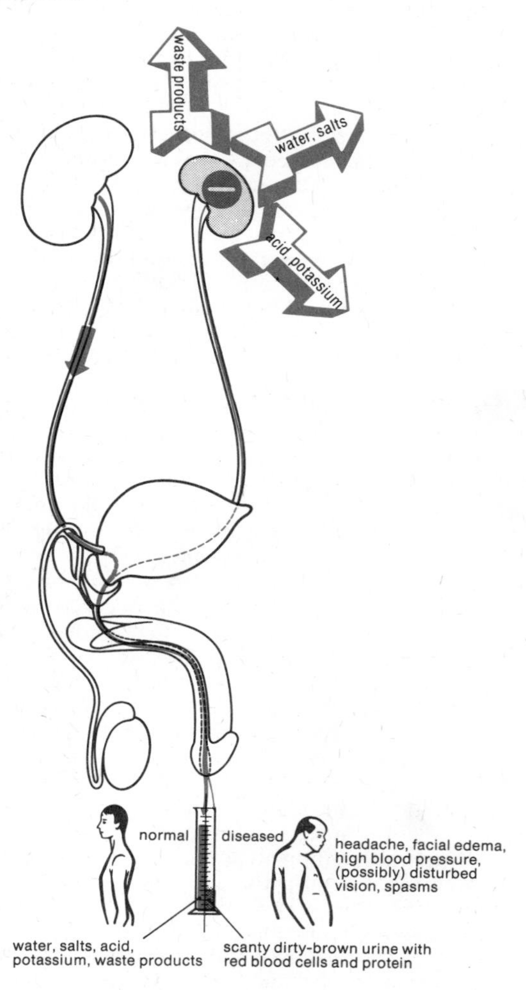

tient feels weak and exhausted, with headache and fever, sometimes also pain in the region of the kidneys. Severe hypertension may be accompanied by vomiting, impaired vision and painful spasms. Treatment will generally begin with two or three days' fasting with minimal intake of fluids, strict confinement to bed, and a low-salt (and sometimes also a low-protein) diet. The patient is given only so much to drink as his kidneys are able to excrete. If they are out of action for a considerable time, artificial dialysis may have to be applied. In most cases of acute nephritis there is full recovery. A minority of cases become chronic. Acute nephritis occasionally passes unrecognized for what it is. Damage to the kidneys may nevertheless occur. For this reason it is advisable to have urine tests carried out after infectious diseases of the upper respiratory tract, especially in children.

Chronic nephritis is a slow but progressive form of the disease, which may cause deterioration of the entire structure of the kidney or be confined just to the glomeruli (chronic glomerulonephritis). It may, but does not necessarily, develop from acute nephritis. In advanced cases the symptoms may include vascular spasms, hypertension, and changes in the blood vessels of the retina. The patient is pale and tired, with headaches and impaired vision. The kidneys tend to become atrophied, and the high blood pressure may cause cardiac trouble. Treatment is centered on a low-salt diet and drugs to lower the blood pressure. The condition of chronic nephritis not directly associated with bacterial infection, and characterized by leakage of protein into the urine, some edema, high blood pressure, and inability to concentrate or dilute the urine, is called *Bright's disease*. The general term *nephrosis* is applied to a condition in which there are degenerative changes in the kidneys without the occurrence of inflammation. It can be caused by the action of bacterial or chemical poisons that attack the tubules or by metabolic disturbances.

PYELITIS, PYELONEPHRITIS

The tubules of the kidney drain into collecting ducts which discharge the urine into the renal pelvis, from where it flows through the ureter to the bladder. If the discharge of urine is obstructed, e.g., by the presence of a stone in the kidney or the bladder (page 208), enlargement of the prostate gland (page 482), narrowing of the ureter or the urethra, paralysis of the bladder, or in pregnancy, there is a risk that pathogenic bacteria make their way into the renal pelvis and cause inflammation there. Sometimes the infection is introduced by a urinary catheter. Inflammation of the renal pelvis is called *pyelitis* (Fig. 6-5). If the inflammation involves not only the pelvis but also the kidney substance, the term employed is *pyelonephritis*. In about two thirds of all cases the infection is due to coli bacilli, which are normally present in the intestine. In general, pyelitis is much commoner in women than in men because the female bladder is more easily accessible to microorganisms from the exterior. More rarely, pyelitis is brought on by bacilli carried to the kidney in the bloodstream. It is estimated that about 3 percent of the adult population have had pyelitis at some time in their lives. In some cases the microbes causing the condition may migrate from the renal pelvis to other parts of the kidney, causing pyelonephritis.

Acute pyelitis starts suddenly, with discomfort, chill, strong fever, fever blisters and headache (Fig. 6-6). There is severe pain in the region of the kidneys which, just as with renal colic that accompanies the passage of stones, radiates from the kidneys around and over the abdomen and into the groin. Usually the kidney region is tender, and urination is frequent and attended by a burning pain. The urine contains numerous bacteria, protein and white blood cells, which give it a cloudy character.

Treatment of pyelitis requires confinement to bed. An adequate intake of fluids is beneficial because of the flushing action. Sulfonamides are usually effective against the bacteria causing pyelitis; to prevent a relapse, these drugs should be continued for a long time. Specific antibiotics are used for the treatment of pyelitis once it has been established by means of bacterial cultures of urine specimens which of the available agents is most effective. Without adequate treatment the condition is liable to turn into pyelonephritis and possibly become chronic, resulting in a condition similar to Bright's disease (see preceding section). Pyelonephritis differs from glomerulonephritis (see preceding section) in being a bacterial primary disease which spreads from the renal pelvis and first attacks the urine collecting ducts, whereas in glomerulonephritis the disease is of a secondary character and initially attacks the

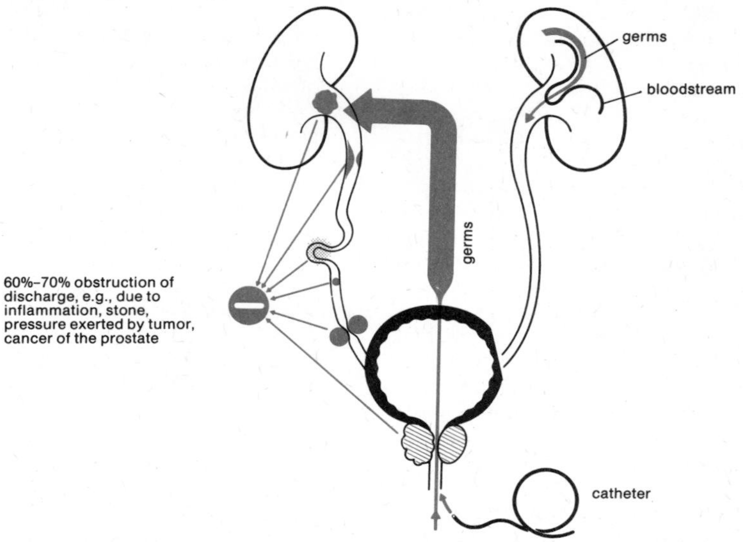

FIG. 6-5.
Causes of pyelitis.

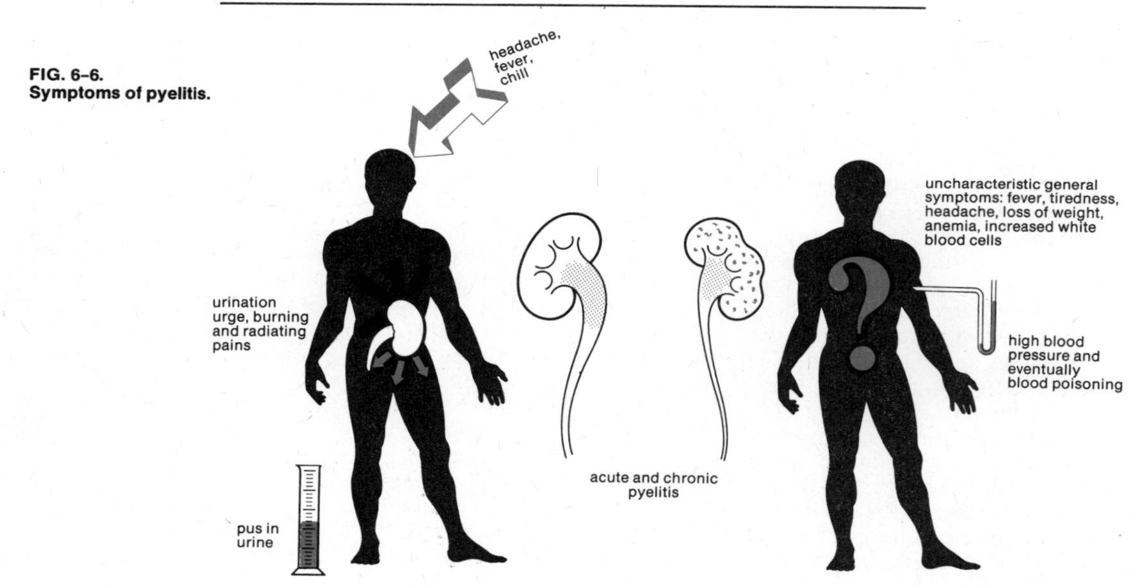

FIG. 6-6.
Symptoms of pyelitis.

filtration system (more particularly the glomeruli) of the kidney.

The symptoms of *chronic pyelonephritis* vary and tend to be rather unspecific (Fig. 6-6). Attacks of fever and an increase in the number of white cells in the blood are indications. In about one third of all cases the disease is insidious, producing somewhat vague general symptoms, but remaining undetected until advanced damage to the kidneys may cause renal failure (see next section). The symptoms may consist merely in tiredness, headache and loss of weight; fever is present in only about a quarter of the cases. The condition is encountered most frequently in middle-aged women. Extensive failure of kidney function results in anemia; restriction of blood circulation through the kidneys causes hypertension (high blood pressure) in a large proportion of cases, which in turn gives rise to cardiovascular disorders; these are the problems that eventually cause the patient to seek medical attention. Dealing with chronic pyelonephritis often requires prolonged treatment with antibiotics. The antibiotic must be regularly checked by bacterial cultures of urine specimens to determine whether it is still effective; if not, a different one will have to be used.

RENAL FAILURE

By far the most important function of the kidneys is the production of urine, which is the medium whereby the body eliminates excess water, waste products, salts, acids, etc. In the event of renal failure, i.e., failure of kidney function, excretion products will accumulate in the body. Although the kidneys can to a great extent be relieved of their load by cutting down the intake of food and liquids to a minimum, the basal metabolism of the body cells must go on, so that even under the most carefully controlled conditions there is still a substantial amount of wastes to be eliminated. These include, besides nontoxic substances, certain substances that will harm the body when present in excess, such as acids and potassium; also, the toxic condition called uremia—the presence of urea and other urinary constituents in the blood—will arise. As a result of these disturbances, death will ensue unless suitable treatment is given.

Depending on circumstances, failure of the kidneys may be acute or chronic, i.e., developing gradually over a fairly long period of time. Such gradual failure *(chronic renal insufficiency)* (Fig. 6-7) may be due to kidney diseases of one kind or another, mostly resulting in damage to kidney tissue, including chronic pyelitis, chronic pyelonephritis, chronic obstruction of the urinary tract by an enlarged prostate gland (page 482), obstruction caused by stones, and kidney damage resulting from high blood pressure, this last-mentioned condition being known as *nephrosclerosis*. Symptoms include anemia, weakness, spasms, shivering, vomiting, diarrhea, deep breathing, hiccup, hypertension, possible failure of the left side of the heart. After initial loss of weight, there is edema (dropsy) due to retention of fluid in the body.

Uremia may develop gradually in conjunc-

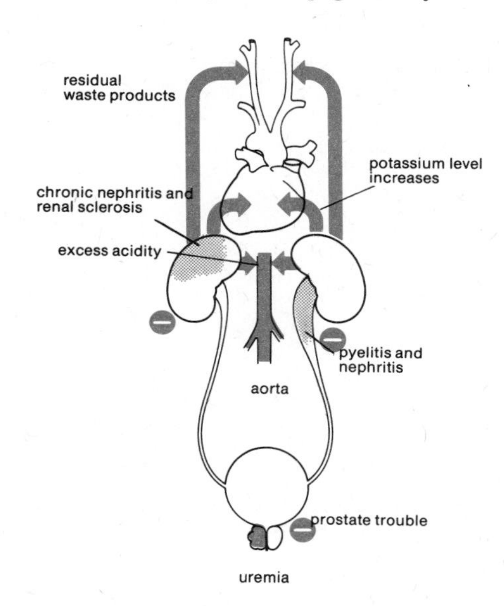

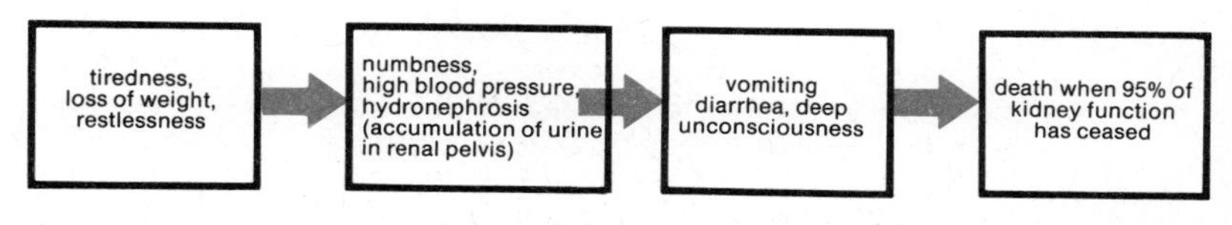

FIG. 6-7. Chronic renal failure.

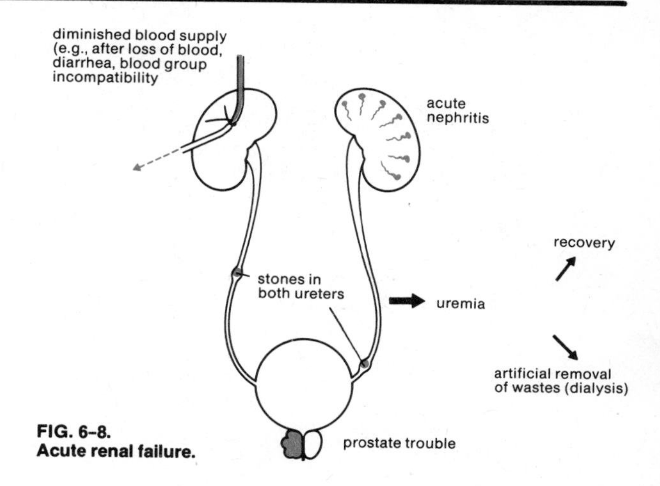

FIG. 6-8. Acute renal failure.

tion with kidney disease over a period of weeks or months, until eventually only a very small portion of kidney tissue is still functioning properly. When that stage is reached, the patient's condition becomes critical. If there is an excess of potassium, he will have to be placed on a low-potassium diet. Accumulated acids can be neutralized by administration of alkalis. Ion-exchange resins may be used to replace positive ions that are present in excess. In an advanced case, dialysis treatment will be needed for the artificial removal of wastes, or a kidney transplant may be indicated. Hypertension (high blood pressure) will have to be reduced in order to prevent further damage to the kidneys.

Acute renal failure (Fig. 6-8) is the sudden inability of both kidneys to perform their functions. It is a serious condition accompanied by reduction or cessation of the flow of urine. There are three main groups of causes. First, the flow of blood to the kidneys may be impaired. This is frequently due to constriction of the renal blood vessels, in response to disturbed nervous impulses, a condition that may result in failure of the filtering system of the kidneys. Such a state of renal shock may occur in consequence of large losses of fluid from the body, hemorrhage (page 116), injury, or blood group incompatibility. Second, the kidney substance may be damaged by toxic agents such as mercury or carbon tetrachloride, bacterial toxins, or inflammation. Third, obstruction of the urinary tract due to such causes as enlargement of the prostate gland (page 482), stones in both kidneys, or tumors may result in renal failure.

For the first few days after acute failure of the kidneys there may be no serious symptoms. But then comes a period in which the patient suffers from weariness, nausea and drowsiness, followed in a few more days by symptoms of uremia, i.e., the toxic condition associated with the retention in the blood of waste products normally excreted by the kidneys. Treatment will be directed at arresting the basic disease. Ion-exchange resins and alkalis may be helpful in neutralizing the condition. Artificial removal of waste products from the blood by dialysis will have to be resorted to in serious cases. In those cases that recover without such assistance, urination is resumed after a few days, and the patient's condition gradually improves; residual disturbances of kidney function may, however, persist for a very long time.

KIDNEY STONES

A *renal calculus* (stone in the kidney) is caused by precipitation of mineral salts (chiefly calcium phosphate and calcium oxalate) which are normally dissolved in the urine. Formation of calculus is liable to occur when certain salts are passed through the kidneys in greater quantity than can be held in solution by the urine at a certain degree of acidity or when the concentration of such salts in the urine increases as a result of too little excretion of water. In highly developed countries about one person in a thousand is afflicted with stones in the kidneys, though the number of "silent" cases, i.e., without detectable symptoms, is certainly much greater.

The actual disturbance that initiates stone formation is probably located within the system of fine tubules of the kidney. The stones, at first of microscopic size, migrate downward through the tubules and collecting ducts into the renal pelvis, where they attach themselves to the wall and grow in size, especially if the renal pelvis is inflamed (pyelitis). In some cases the stone may fill the entire renal pelvis and assume a "staghorn" shape (Fig. 6-9), causing obstruction. However, even quite large stones may produce no noticeable symptoms other than an occasional sensation of pressure in the lumbar region (loins). But when a stone is on the move and becomes wedged in the outlet of the renal pelvis or farther down in the ureter, it is likely to produce agonizing pain (*renal colic*). This may happen quite suddenly, often when the patient is otherwise in the best of health.

The pain is caused by the stone cutting into the mucous membrane of the ureter and by its obstructing the passage of urine. The ureter strives by energetic contractions of the muscle in its wall to expel the stone and thrust it downward into the bladder. The paroxysmal colicky pain, which comes and goes in waves, radiates from the lumbar region into the hypogastric region (lower middle of the abdomen). This occurs particularly when a stone is wedged high up, at the entrance to the ureter. Another bottleneck where stones are liable to be trapped is the point where the ureter is crossed by large pelvic blood vessels, while yet another bottleneck exists at the outlet from the ureter into the bladder. If colic develops at this latter point, pain may radiate into the testicles (Fig. 6-10).

Colic is frequently accompanied by nausea

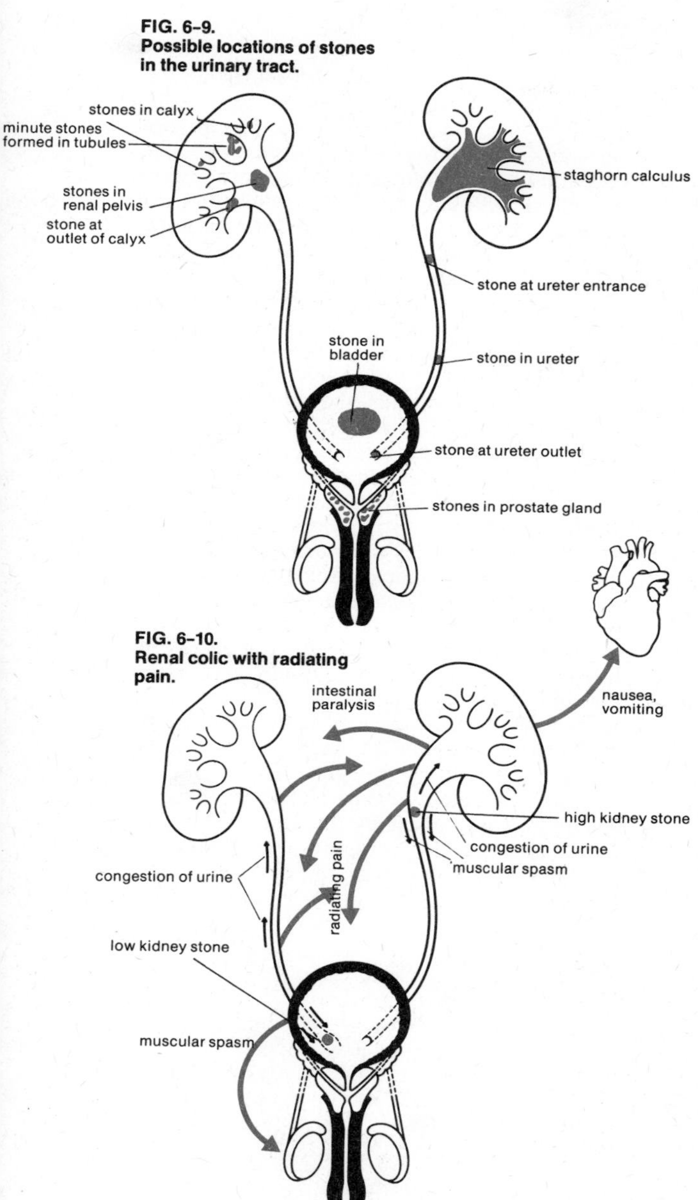

FIG. 6-9. Possible locations of stones in the urinary tract.

FIG. 6-10. Renal colic with radiating pain.

and vomiting. Bleeding may occur from small cuts in the mucous membrane of the ureter, so that red blood cells appear in the urine. On x-ray the dense calcareous stones usually show up quite well; stones containing no calcium can be detected with the aid of radiopaque substances. However if radiography is inconclusive, further laboratory tests will have to distinguish between renal colic and other acute disorders of the abdominal cavity, such as appendicitis, biliary colic, volvulus (twisting of the intestine) or intestinal perforation. The kidney stone patient often strives to make movements which somewhat relieve the pain, whereas in the other acute abdominal disorders mentioned the patient will usually lie still so as not to aggravate his discomfort.

Movements of the body, such as walking or indeed jumping up and down, may in fact dislodge and help to expel the stone. Warmth, hot baths, antispasmodic drugs and increased intake of fluids improve the chance of getting rid of small stones. Sometimes a wedged stone will slip back into the renal pelvis and cause no further trouble for a time. If the prostate gland is enlarged, a kidney stone in the process of being discharged is liable to remain trapped in the bladder, where it continues to grow as a bladder stone (*vesical calculus*). The more general term *urinary calculus* denotes stone in any part of the urinary system, i.e., in the kidney or in the bladder.

If a stone becomes wedged in one of the ureters, thus obstructing the flow of urine from the kidney, congested urine will flood the kidney and impair its function. As this can develop into a serious condition, an attempt will be made to remove a stone wedged near the lower end of the ureter by means of a special catheter guided by visual observation with the aid of an optical device (cystoscope). Sometimes a small incision is made at the inlet of the ureter into the bladder to assist the "birth" of the stone. If such measures fail to get rid of the stone, or if it is wedged higher up in the ureter, it will have to be removed surgically. As the irritation set up by stones in the renal pelvis is liable to cause inflammation (pyelitis) "silent" stones are often also removed by means of an operation, if their presence is detected and considered a potential hazard.

Even after removal of stones, about 45 percent of all cases of urinary calculus suffer a recurrence of the disorder. Measures should be taken to prevent this. Precautions include dietetic measures and, in some cases, treatment for the dissolving of stones. There is no universal means of prevention. Conditions vary according to the material of which the stones consist. Mostly they are composed of calcium oxalate (in 55 percent of cases); stones composed of calcium phosphate are also fairly common (25 percent).

Disturbances in the calcium metabolism promote the formation of oxalate and phosphate stones. The body's calcium reserves are mobilized and calcium excretion is increased, with correspondingly increased tendency to stone formation, as a result of prolonged confinement to bed, after fractures of bones, with overactivity of the parathyroid glands, and with excessive intake of vitamin D. Increased intake of calcium can also raise the calcium concentration in the urine, e.g., as a result of excessive consumption of milk and milk products.

Tendency to form oxalate stones is promoted particularly by foods containing oxalic acid—certain fresh leaf vegetables (e.g., spinach, beets, celery, radishes), certain fruits (e.g., red currants), rhubarb. Oxalate stones relatively seldom give rise to inflammation. They cannot, however, be dissolved, and if they are too large to be discharged, the only way to remove them is by an operation.

Calcium phosphate stones are liable to cause inflammation. When not actually wedged in the ureter, they can sometimes be dissolved by suitable treatment which is centered on reducing the amount of phosphate reaching the kidneys for excretion. Such measures also help to prevent stone formation. Patients with calcium phosphate stones are advised to drink plenty of mineral-free water, but beverages containing phosphoric acid (such as those based on cola nut extract and some artificially acidulated lemonades) should be avoided. Such stones can sometimes also be dissolved by prolonged flushing of the renal pelvis with certain chemicals (such as the compound known as EDTA); this method is confined to special cases, however.

So-called cystine stones are comparatively rare (3 percent of cases). Urate stones are commoner (17 percent); they consist of uric acid or its salts. In some cases of disturbance of cell metabolism there is an increase in the excretion of uric acid (cf. gout, page 148); this is also liable to occur as a result of excessive consumption of anchovies, meat extract and meat (especially liver, kidneys and sweetbreads). These foods can bring on the formation of urate stones in persons so disposed. Stones of this kind can be dissolved at a low degree of urine acidity. Patients afflicted with them are accordingly treated with antacids that reduce this acidity.

In general, plentiful intake of fluids is helpful in the prevention of kidney stones. Rich foods tend to promote stone formation.

URINARY BLADDER

The bladder collects the urine produced by the kidneys, stores it in a quantity of about 500–750 ml, and voids it from time to time through the urethra. *Urination* (*micturition*) is essentially an involuntary action, a reflex that has been brought under voluntary control. Control is absent in babies and also in adults who are in a state of unconsciousness (other than normal sleep). The wall of the bladder is composed of four layers, including a lining of mucous membrane and a muscle coat of smooth muscle, with a sphincter at the outlet to the urethra.

While the bladder fills, the sphincter remains closed under the control of the sympathetic nerves. When the volume of urine reaches about 300 ml, the rise in pressure in the bladder stimulates receptors in the bladder wall. The resulting nervous impulses are interpreted by the brain as a desire to urinate, but inhibitory impulses are transmitted from the cortex to prevent contraction of the muscle in the wall of the bladder, while the sphincter is kept contracted voluntarily, thus preventing urine from flowing into the urethra. At the appropriate time the control is removed, the sphincter relaxes, and the bladder wall contracts by muscular action, causing the bladder to be emptied by reflex action (Fig. 6-11). Urination may occur involuntarily during sleep, especially in young children: bed-wetting (enuresis) (Fig. 6-12). Causes may be emotional or physical, e.g., inflammation of the urinary passages.

FIG. 6-11.
Emptying of bladder under normal voluntary control.

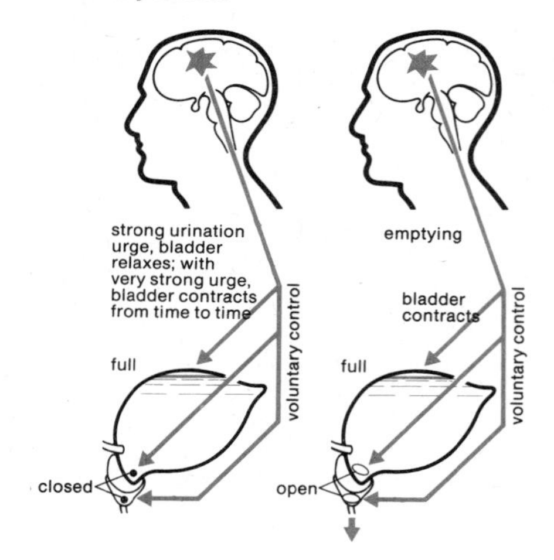

FIG. 6-12.
Emptying of bladder by involuntary action in babies, bed-wetting, or deep unconsciousness.

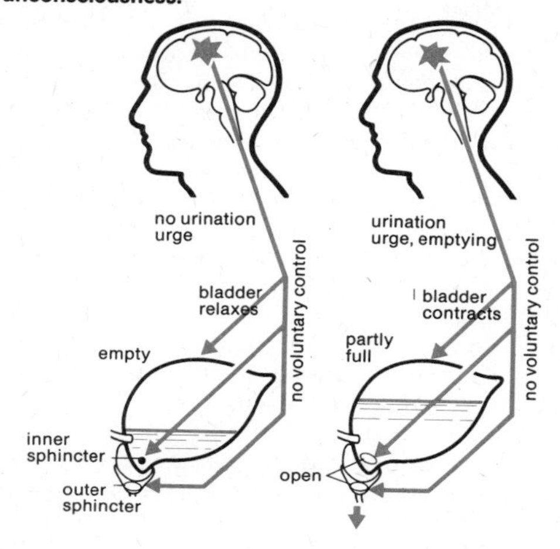

In general, loss of control of urination is called *incontinence of urine*. Damage to the spinal cord may paralyze the bladder. If the nerves below the injury are unaffected, the bladder empties by simple reflex action at intervals, as soon as the pressure of urine begins to increase, as urination is then no longer under voluntary control. Reflex emptying may also be put out of action by spinal injury (Fig. 6-13), so that the bladder fails to contract of its own accord and has to be emptied with the aid of a catheter.

Inability to urinate is called *retention of urine* (Fig. 6-14). This may occur as a result of obstruction of the outlet of the bladder. Acute retention due to spasm of the sphincter may occur after a surgical operation. Obstruction may also be caused by calculus (stone) or by narrowing of the urethra due to disease. In elderly men the urethra may be narrowed as a result of enlargement of the prostate gland (page 482). Retention of urine can be harmful to the kidneys, and the urine is liable to become infected and cause *cystitis* (inflammation of the bladder). The infecting microorganisms usually enter the bladder from outside the body, through the urethra. Cystitis is much commoner in women than in men because the female urethra is much shorter. The disorder may be acute, occurring quite suddenly, accompanied by fever. Urination is frequent and painful. The urine often contains proteins and white blood cells or, less frequently, blood. Treatment is based on sulfonamides and antibiotics, just as in dealing with pyelitis (inflammation of the renal pelvis), which not infrequently develops from cystitis. Tuberculous

FIG. 6-13.
Retention of urine due to spinal injury.

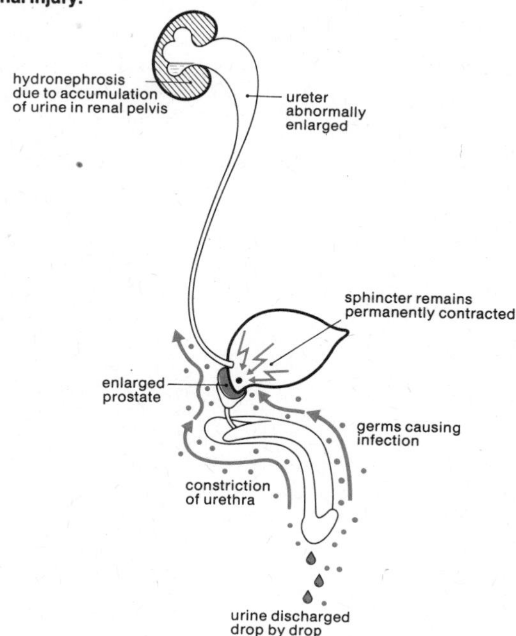

FIG. 6-14.
Causes of retention of urine.

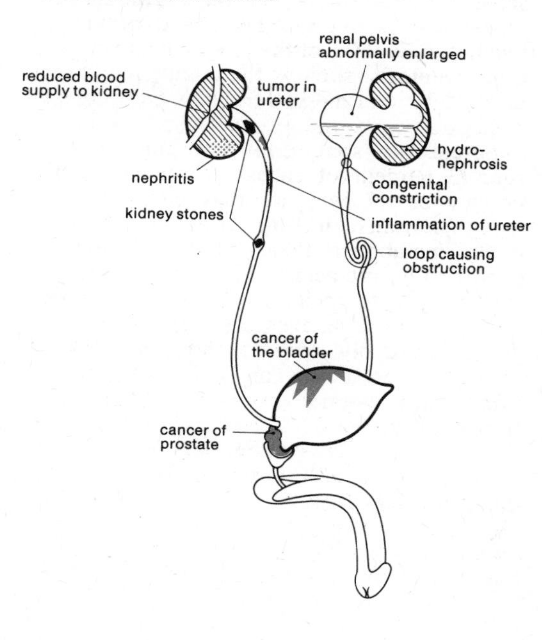

cystitis usually spreads from the kidneys and may cause severe shrinkage of the bladder as a result of cicatrization (scar formation), so that its urine storage capacity is drastically reduced.

Narrowing of the urethra may occur by cicatrization after injury or in consequence of an acute inflammatory condition, e.g., due to untreated gonorrhea. This may cause the bladder to form thickened bands of muscle and finally to become overdistended. To prevent harmful accumulation of urine in the renal pelvis due to obstructed discharge (hydronephrosis), the obstruction may have to be relieved by surgery.

Tumors in the bladder generally cause bleeding and should be given immediate medical attention. Quite often they are benign wartlike tumors (papillomas) which are curable if treated in good time. For cancer of the bladder see page 504.

MALE GENITAL SYSTEM

The male reproductive or germ cell (*sperm* or *spermatozoon*, Fig. 6-15) is of microscopic size, resembling a tadpole in shape and about 0.05 mm (50 microns) in length. It has a flattened oval head consisting largely of the nucleus, a neck, a middle piece and a tail. The front end of the head has a kind of cap (acrosome) which contains special enzymes that enable the sperm to penetrate into the ovum (the female reproductive cell). The neck accommodates two centrioles; the middle piece contains numerous mitochondria, which are believed to supply energy needed by the sperm for its self-propulsion, for which purpose the tail is provided with fiberlike elements called fibrils. The sperm is thus able to swim at a speed of about 3–3.5 mm per minute.

Motility is not the only or indeed the most important criterion of the sperm's fertility, however. Sperms "swim against the current" and thus travel up the female oviduct toward the ovary (see pages 269 and 270). The survival period of the sperms in the female tract is thought to be not more than a day or two. The semen ejaculated by a healthy man normally contains 20–30 percent abnormal sperms (multiheaded, multitailed and other malformed variants); if the number of abnormal sperms exceeds about 60 percent, there is usually infertility. The semen discharged in each ejaculation (about 3–4 ml) contains something like 300 million sperms suspended in the fluid secretion of the reproductive glands. If several ejaculations take place in fairly close succession, the number of sperms per ejaculation rapidly decreases, as the testes are unable to go on supplying such vast numbers at short notice. The number of sperms that a normal man produces in a lifetime runs into millions of millions, whereas a woman produces no more than about 500 reproductive cells during the whole course of her life.

The male reproductive cells are formed in the epithelium of the seminiferous tubules (Fig. 6-16) from *diploid spermatogonia*. The term *diploid* means that the nucleus of a cell contains the normal number of chromosomes characteristically found in the nonreproductive cells of the body (somatic cells); the term *haploid* means that only half this number is present in the nucleus, more particularly in the nucleus of a male or a female reproductive cell, following the reduction divisions by which that cell is produced. In childhood the spermatogonia multiply until the start of puberty; they then develop into cells called *spermatocytes*, which still contain 46 chromosomes (the diploid number). Then the spermatocytes undergo reduction divisions by which cells called *spermatids* are formed, each containing only 23 chromosomes (the haploid number). The spermatids undergo various changes, in the course of which they lose most of their cytoplasm, and develop into spermatozoa (sperms) (Fig. 6-17).

The generic term *gonads* denotes the male or the female sex glands. In the male they are the testes, in the female the ovaries. Each *testis* (*testicle*) (Fig. 6-18) is about 4 cm long, of a laterally flattened ovoid shape, and is enclosed in a thick fibrous sheath (*tunica albuginea*). Internally it is divided into lobules, each of which contains one to three fine coiled tubes called *seminiferous tubules*. The two testes hang in the *scrotum*, a pouchlike receptacle (Fig. 6-19). The seminiferous tubules communicate with a plexus (*rete testis*) from which a number of ducts lead to the *epididymis* (described later). From here the sperms pass through the *vas deferens* to the urethra.

The tissue between the seminiferous tubules is very soft and delicate; its fluid content can vary with the pressure in the testis. The tunica enclosing the testis resists the relatively high internal pressure which apparently is necessary for the production of sperms. The lining of the tubules contains two types of cells: the spermatids and the supporting cells (*Sertoli's*

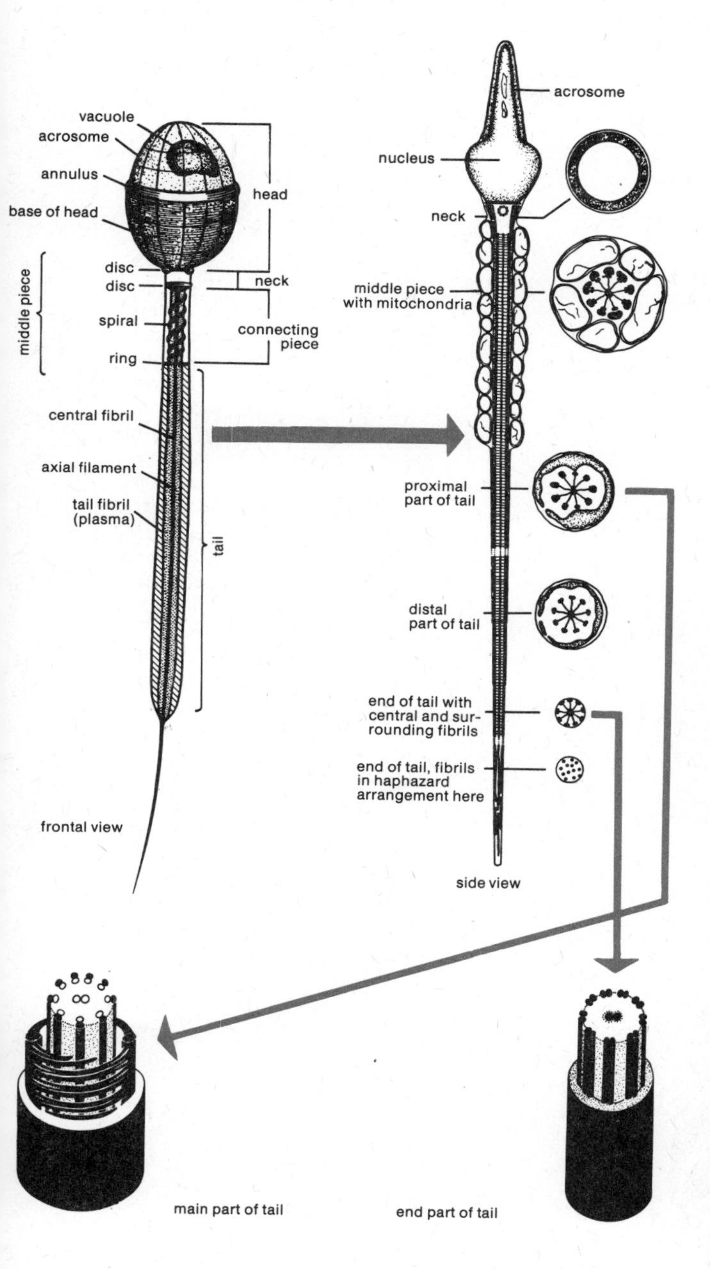

FIG. 6-15.
Frontal and side views of sperm.

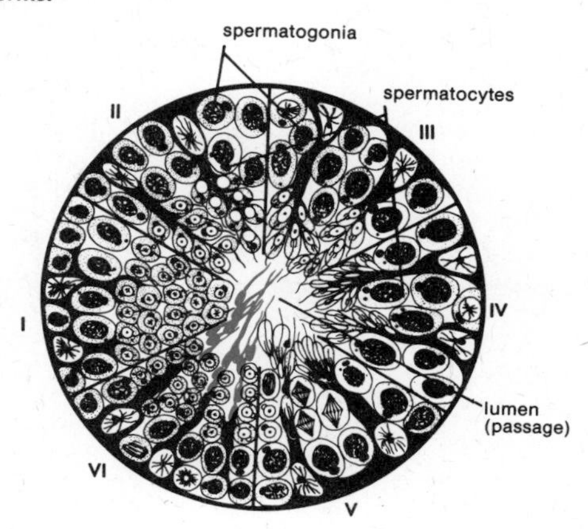

FIG. 6-16.
Section through a seminiferous tubule: the sectors represent different stages of development of sperms.

cells), which nourish the spermatids. The latter, in various stages of development, are disposed layerwise between the supporting cells, by which they are held until they reach maturity (Figs. 6-16 and 6-20).

In between the tubules are epithelium-type interstitial tissue cells (*Leydig's cells*) which produce the male sex hormone called *testosterone*. This hormone is largely responsible for the development of the primary and, especially, the secondary sex characteristics; the latter include growth of hair on the face and chest, deepening of the voice, distribution of body fat, etc. The hormone is secreted into the bloodstream.

The importance of the endocrine function becomes manifest in a case of functional failure or removal of the testes, e.g., as a result of castration. As contrasted with *sterilization*, in which merely the vas deferens on each side is tied off or partly removed, in *castration* the testes themselves are removed. Depending on the man's age at which castration takes place, its consequences will differ. If it is done at an early age, the secondary sex characteristics will wholly or partly fail to develop, while the distribution of the body fat will correspond to the feminine type (eunuchism, i.e., the condition resulting from lack of male hormone). Castration at a later age has less effect on those sex characteristics that have already developed; even the sex urge and the ability to copulate may be retained, though there is, of course, complete sterility.

The scrotum is the pouch containing the two testes, each with its epididymis and the lower end of the spermatic cord containing the vas

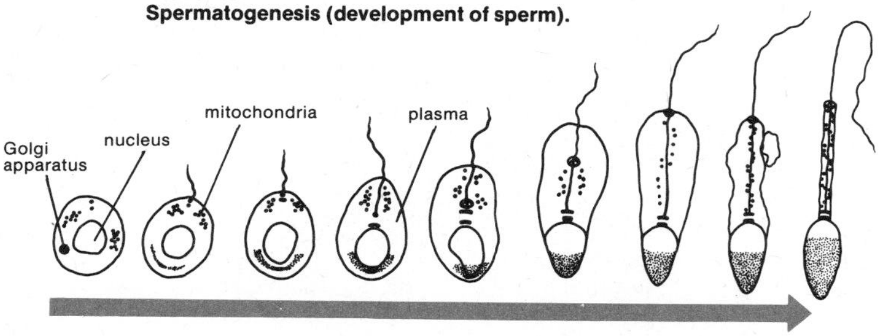

FIG. 6-17.
Spermatogenesis (development of sperm).

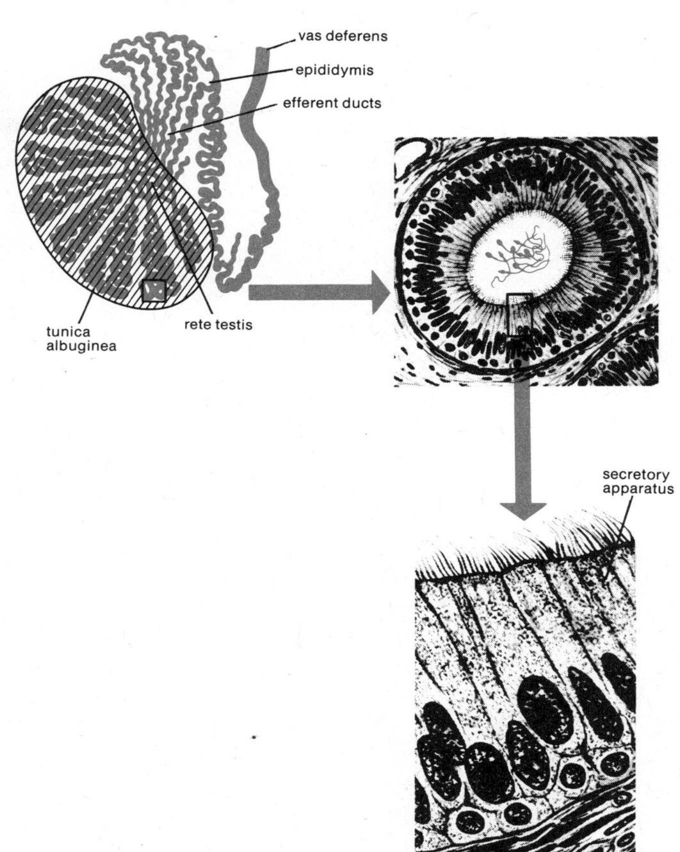

FIG. 6-18.
Testis (testicle).

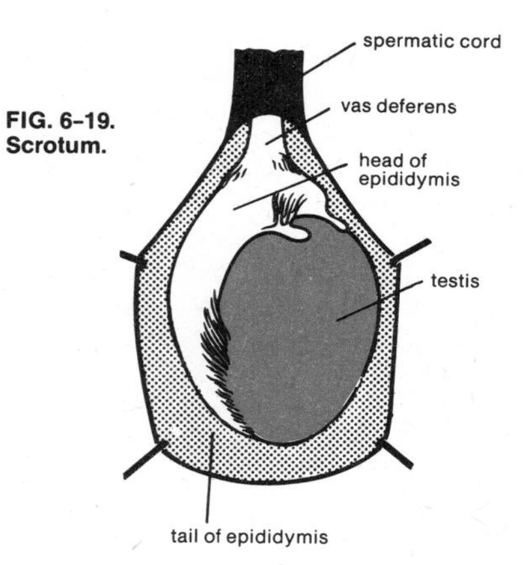

FIG. 6-19.
Scrotum.

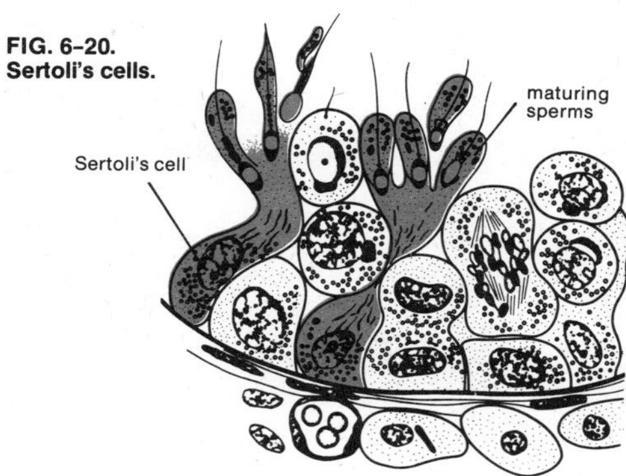

FIG. 6-20.
Sertoli's cells.

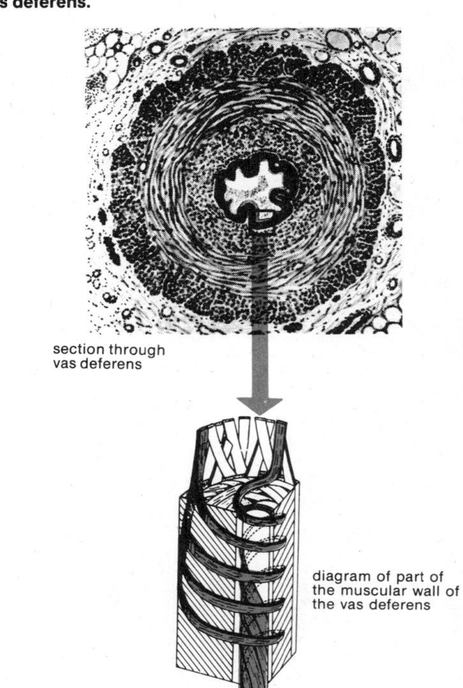

FIG. 6-21.
Vas deferens.

deferens, blood vessels and nerves. The size and shape of the scrotum vary from time to time, mainly because of the presence of subcutaneous smooth muscle fibers. When these relax, the wrinkles in the skin of the scrotum are smoothed out; cold or exercise causes the muscle to contract, so that the skin becomes wrinkled. These movements are thought to assist the thermal regulation of the scrotum, in which the testes are kept at a temperature a few degrees below the temperature in the body generally, this "cooling" being necessary for the production of fertile sperms. In the condition known as *cryptorchidism*, the testis fails to descend into the scrotum (which normally occurs prenatally in the fetus or otherwise in early infancy) and will not produce sperms satisfactorily unless induced to descend by hormone treatment or a surgical operation.

The seminiferous tubules of the testis discharge through small ducts into a long coiled tube, the *epididymis*, ending in the vas deferens, which leads off from the tail of the epididymis and passes upward in the spermatic cord (Fig. 6–19). The epididymis tube, with an overall length of up to 6 m, stores the semen and can be emptied by three or four ejaculations in about 12 hours. It takes approximately two days to refill. It is here that the newly formed sperms undergo a maturing process while they are kept in a state of immobility so as not to exhaust their self-propelling capacity prematurely. To ensure this, the epididymis produces a secretion which raises the acidity of the seminal fluid. Some of the cells lining the sperm ducts are provided with cilia (tiny hairlike projections) which set up a current; this is believed to assist the thorough mixing of the sperms with the secretion.

The *vas deferens* (ductus deferens) (Fig. 6-21) is a continuation of the epididymis; it is a highly muscular thick-walled tube through which the semen, i.e., the seminal fluid containing the sperms, is conveyed from each testis to the prostatic part of the urethra. The muscle wall of the vas deferens performs strong contractions in the course of ejaculation.

Just before it enters the prostate gland the vas deferens has an enlarged portion (ampulla) and is here joined by a duct from the *seminal vesicle* (Fig. 6-22). There are two seminal vesicles, each of which is a coiled tube situated behind the prostate and bladder. They produce an alkaline secretion which stimulates the motility of the sperms, besides increasing the fluidity of the semen and imparting some protection to the sperms to help them survive in the vaginal secretion. The sperms are given their full motility by the secretion of the prostate gland, which surrounds the neck of the bladder and the urethra. The part of the vas deferens passing through the prostate gland is called the *ejaculatory duct* (Fig. 6-22).

The *prostate gland* (often referred to simply as the prostate, Fig. 6-22) consists of glandular tissue embedded in muscle fibers and a fibrous capsule. It secretes a thin, milky and slightly alkaline fluid. With advancing age the prostate may become enlarged and obstruct the prostatic part of the urethra, causing difficulties of urination. Also in older men the prostatic secretion may become thickened and calcified as a result of congestion, so that a kind of stone is formed in the prostate. There are two small subsidiary glands, the bulbourethral glands, one on each side of the prostate, which contribute a secretion to the seminal fluid.

Ejaculation denotes the passage of semen into the urethra by contraction of the muscle of the vas deferens and the expulsion of the semen from the urethra by rhythmic contractions. These are due to the actions of several muscles, assisted by the muscle of the prostate gland. Ejaculation is a reflex, normally preceded by erection of the penis. The *penis* (Figs. 6-22 and 6-23) is traversed by the urethra and is the male organ of copulation. It is composed of cavernous tissue, a spongelike substance which swells and becomes turgid when blood is forced into it; it is therefore also known as erectile tissue. The corpora cavernosa are two lateral columns of such tissue; the corpus spongiosum is erectile tissue surrounding the urethra. The head of the penis is called the glans penis and is covered by the prepuce (foreskin), under which a lubricating substance (smegma) is secreted. The urethra opens at the tip of the glans. *Phimosis* is a term denoting narrowness of the orifice of the prepuce, so that it cannot be pushed back over the glans. This condition can be corrected by *circumcision*, a minor surgical operation for removing the end of the prepuce. This may be done also for hygienic and/or for religious ritual reasons and is considered to reduce the risk of cancer of the penis and of the female cervix.

Erection of the penis is brought about by erectile tissue, particularly the corpora cavernosa, becoming engorged with blood and is assisted by the action of muscles. It is a reflex, under the control of parasympathetic nerves, and is maintained by pressure of blood in the

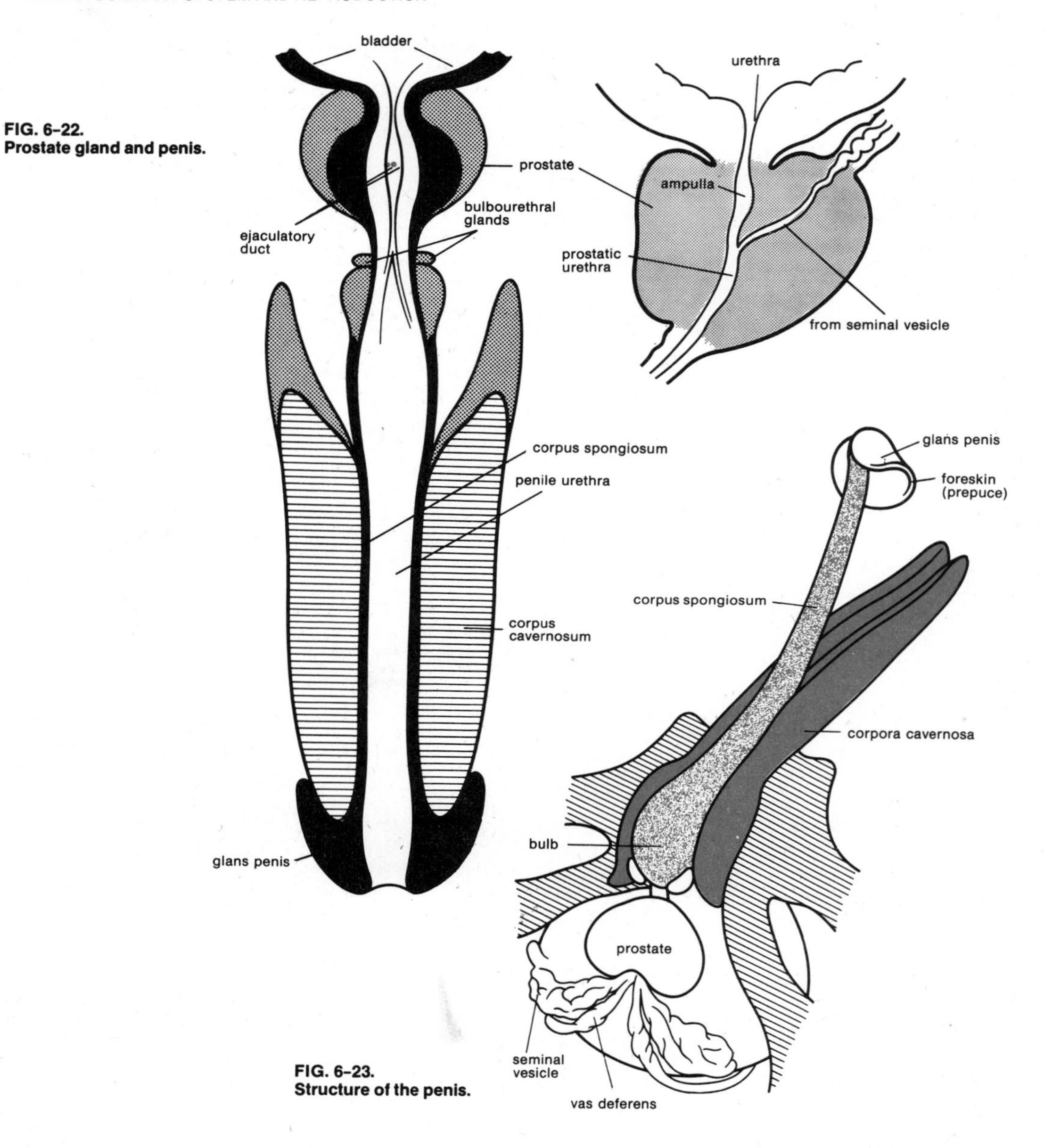

FIG. 6-22.
Prostate gland and penis.

FIG. 6-23.
Structure of the penis.

arteries to the penis, which are distended, while muscles around the veins are contracted to prevent blood from leaving the penis. After ejaculation of semen, the flow in the arteries is reduced and the muscles of the veins relax, allowing blood to flow out. The erection center is situated in the lumbar and sacral region; it is not under direct control of the will and it responds to stimuli (sexual excitement and/or physical stimuli) by transmitting nervous impulses to the corpora cavernosa. The nerve centers associated with erection are kept in readiness to respond to stimuli by sex hormones.

MALE FERTILITY

A distinction must be drawn between (1) infertility (sterility) due to failure of the sperms to fertilize the female reproductive cell and (2) infertility due to inability to copulate (perform coitus), particularly because of failure of erection of the penis or, in some instances, premature ejaculation of semen. Inability to have an erection and thus have sexual intercourse is called *impotence*. Its causes may be physical or psychic (emotional disturbance, neurosis), the latter being the more common and often curable by psychiatry.

Sterility in the male may be due to disturbed production of sperms in the testes. In about a third of such cases the disturbance is attributable to external causes such as injury, inflammatory disease or certain toxic effects. Undernourishment and overexertion may also play a part in making a man sterile, temporarily anyway. In the majority of cases, however, the cause is not known and is probably bound up with hereditary or hormonal factors. Before the days of sulfonamide and penicillin treatment of gonorrhea, the inflammation caused by this venereal disease not infrequently resulted in obstruction of the seminal ducts. Other diseases that may cause such obstruction, as a result of inflammation of the testis or epididymis, are tuberculosis, mumps, malaria, typhus and (rarely) influenza. Ejaculation may be disturbed by malformation of the urethra.

To determine whether a man is fertile an examination of his semen is essential. It should be done as a first step in investigating infertility in a marriage before the wife submits herself to more elaborate examination.

The man should refrain from sexual intercourse for about a week and then collect a specimen of his semen (by masturbation) in a glass receptacle; the glass receptacle is preferable, for this purpose, to a rubber condom (sheath), as the latter may be composed of ingredients that are harmful to sperms. The semen should be examined under a microscope as soon as possible, with particular reference to the number and motility of the sperms. Next, a smear specimen should be prepared and examined for the presence of poorly developed or abnormally shaped sperms.

Sterility is likely to exist if the minimum requirements (according to McLeod and Gold) listed in Figure 6-24 are not fulfilled. The total number of motile sperms in the seminal fluid

FIG. 6-24.
Minimum requirements for male fertility.

	minimum quantity of seminal fluid: 2.5–3.5 ml lower limit of normal quantity: 2 ml
	average concentration of sperms: 60–120 million/ml lower limit or normal concentration: 40 million/ml
	motility of sperms: 2 hours after ejaculation 60%–70% of the sperms should still be motile
	shape of sperms: there should be not more than 20% of misshapen sperms

FIG. 6-25.
Normal sperms compared with abnormally shaped sperms.

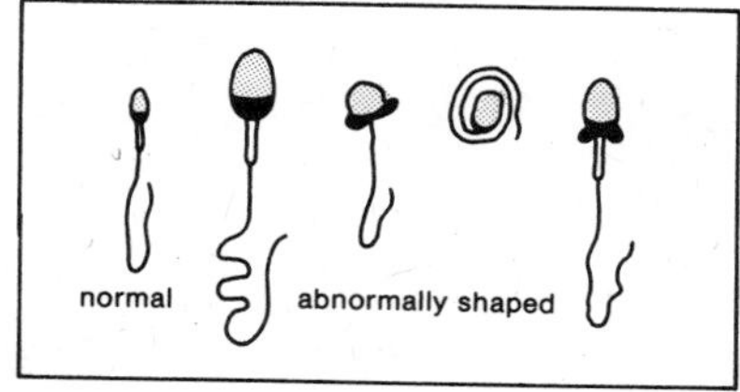

FIG. 6-26.
Removal of a small specimen of tissue from testis for microscopic examination.

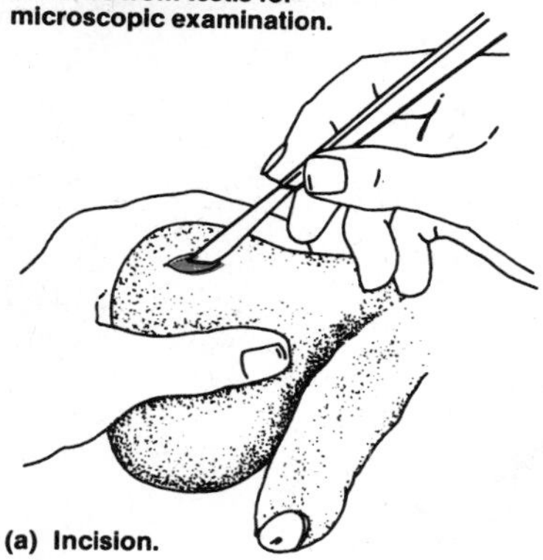

(a) Incision.

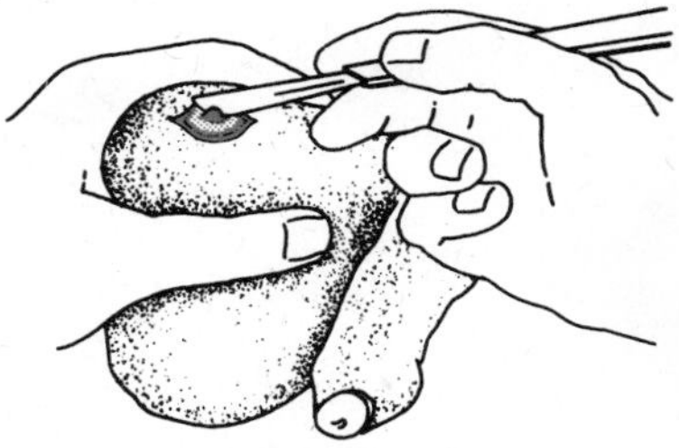

(b) Removal of specimen.

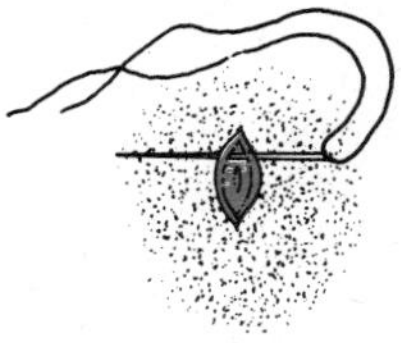

(c) Suturing.

FIG. 6-27.

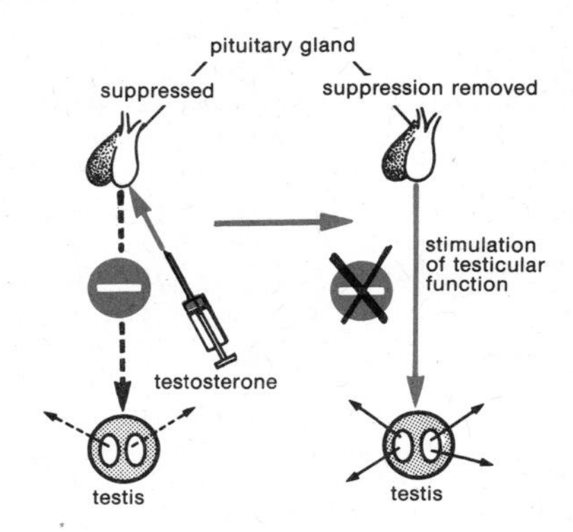

(a) Suppression of pituitary action by injected testosterone. **(b) Outburst of pituitary activity after suppression stimulates function of testes.**

discharged in an ejaculation is an important criterion. For normal fertility there should be at least 50 to 150 million. Generally speaking, if the semen contains below 40 percent of normally developed sperms (Fig. 6-25), the man is very probably sterile. In some cases there is a complete absence of mature sperms or any sperms at all.

Examination of testicular tissue can provide conclusive evidence of male sterility. Obtaining a tissue specimen involves a minor surgical operation under local anesthesia (Fig. 6-26).

In certain cases it may be possible to stimulate sperm production by treatment with a hormone secreted by the pituitary gland or a similar hormone derived from mare serum. Alternatively, a kind of recoil effect of the pituitary gland, with a burst of intensified hormone production, may be induced by first temporarily suppressing the action of that gland by means of testosterone, the hormone secreted by the testis (Fig. 6-27). Some deformations of the reproductive organs can be corrected by surgery.

FEMALE REPRODUCTIVE SYSTEM

The mature *ovum* (egg)—i.e., the female reproductive or germ cell—is one of the largest of all cells in the human body and just visible to the naked eye (0.1–0.14 mm diameter). It consists of the nucleus (germinal vesicle) and the surrounding protoplasm (yolk), which provides initial nourishment for the fertilized ovum (Fig. 6-28). The outer layer of the yolk is the zona pellucida, which protects the ovum from environmental influences in its initial stages of development and is believed to play an important part in the metabolism of the ovum. The zona pellucida is surrounded by a layer of follicular cells, which constitute the corona radiata and are shed in the oviduct.

The developed ovum, before it is released from the ovary, is called a *graafian follicle*. In the ovary something like 400,000 egg cells are formed in all, but only about 400 or 500 of them, at the surface of the ovary, reach maturity during the woman's lifetime. The early or immature ovum, before it has completed its development, is called an *oocyte*; this develops from a primordial cell called an *oogonium*. After a period of growth the oocyte undergoes two successive maturation divisions; in each of these, one of the two daughter cells remains small and does not develop. As a result of these divisions the normal (diploid) number of chromosomes in the nucleus is halved, so that the mature ovum has a haploid number (see page 212). The first maturation division occurs in the graafian follicle, the second occurs after ovulation. Ovulation is the periodic ripening and rupture of the mature graafian follicle, releasing the ovum from the ovary.

There are two *ovaries* (Fig. 6-29), one on each side of the uterus. Their main function is to produce reproductive cells; they also secrete two sex hormones. Each ovary is an almond-shaped organ attached to a fold of peritoneum called the broad ligament and comprises two parts: an outer part (cortex) and an inner part (medulla). The cortex consists mainly of *follicles* in various stages of development—namely, primary, growing, and mature (or graafian) follicles. The surface of the ovary is covered by a single layer of cells, the germinal epithelium, some cells of which form ova (eggs). The cells destined to form ova pass into the interior and embed themselves in the medulla of the ovary, where each cell (oogonium) is surrounded by a layer of follicular cells and is then known as a *primary follicle*. This eventually matures into a graafian follicle containing a nearly mature ovum (oocyte). The graafian follicle comprises a cavity containing the follicular fluid and is surrounded by a layer of cells called the *membrana granulosa*; the oocyte lies in a mass of cells called the *cumulus oophorus,* which protrudes into the fluid-filled cavity. The follicle passes to the surface of the ovary, and the cumulus then faces outward. With further development the surrounding tissue becomes thinner and thinner until the follicle ruptures and liberates the ovum. This process is called *ovulation* and occurs approximately 14 days before the next menstrual period. The cells of the collapsed follicle, after release of the ovum, grow larger, and the cavity becomes filled with cells constituting the so-called *corpus luteum* (Fig. 6-30), which is an endocrine gland secreting sex hormones, including progesterone. If no fertilization of the ovum occurs, the hormone secretion ceases after 14 days.

On liberation from the graafian follicle the ovum enters the funnel-shaped end (infundibulum) of the *uterine* or *fallopian tube* (*oviduct*). Fingerlike projections (fimbria) of the oviduct play a part in assisting the newly released ovum toward the entrance of the tube. The part of the fallopian tube adjoining the infundibulum is called the ampulla; it is here that fertilization, i.e., impregnation of the ovum

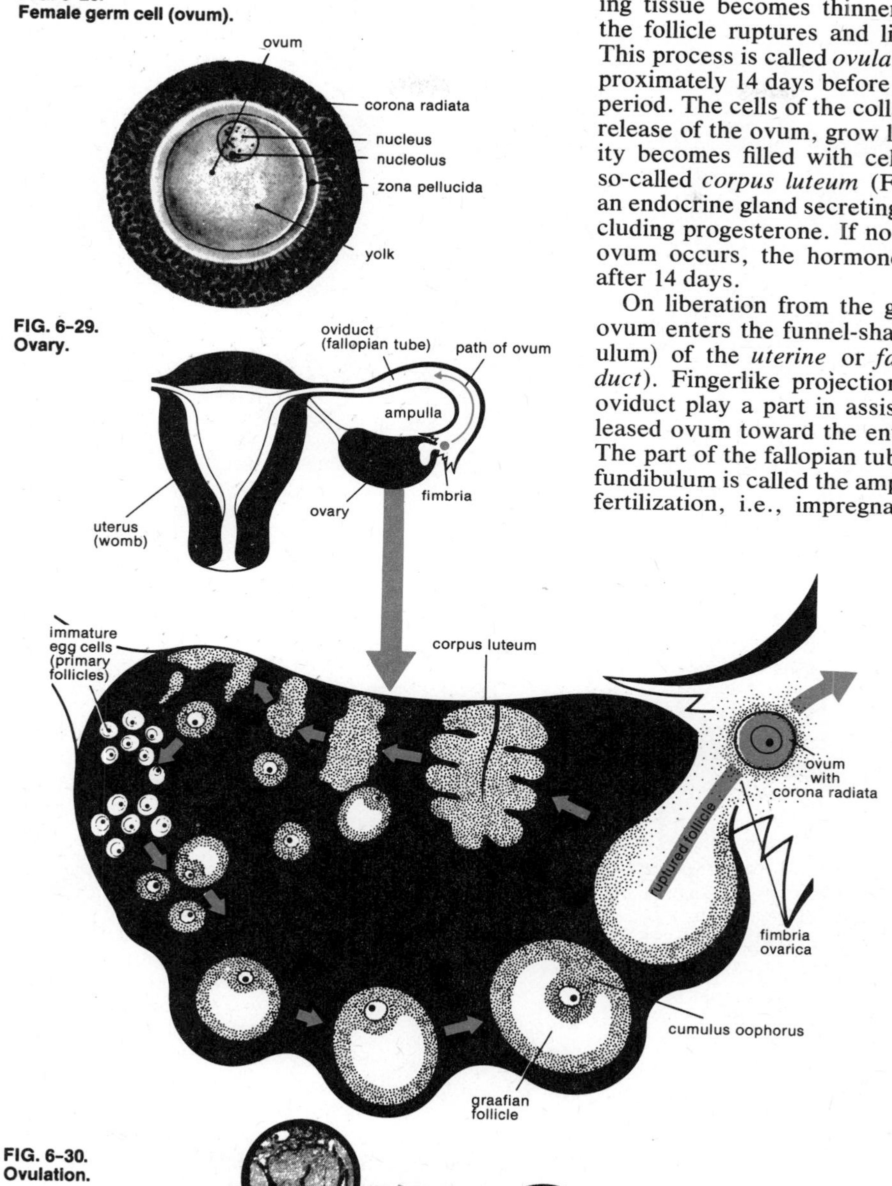

FIG. 6-28.
Female germ cell (ovum).

FIG. 6-29.
Ovary.

FIG. 6-30.
Ovulation.

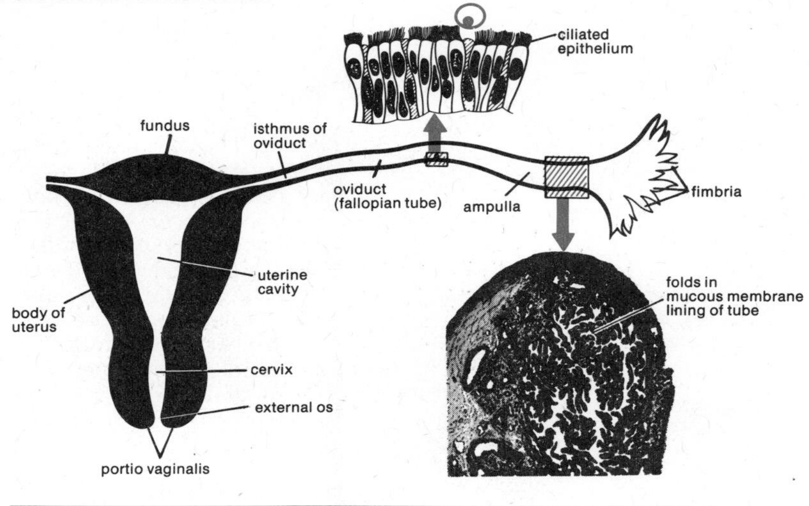

FIG. 6–31.
Uterus.

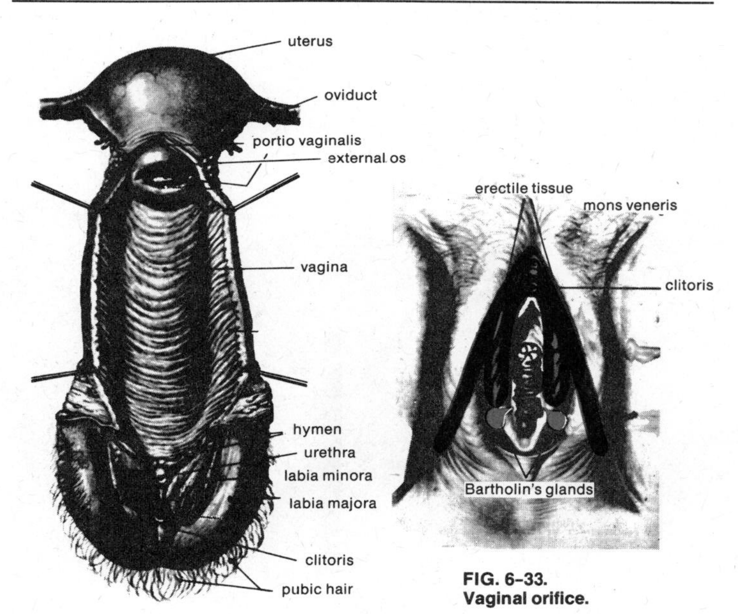

FIG. 6–32.
Female genitalia.

FIG. 6–33.
Vaginal orifice.

with the sperm, normally takes place if sperms are present at the time, these having swum up the tube from the uterus. The inside of the tube is lined with cilia (fine hairlike projections), whose undulatory motion aids the progress of the ovum toward the uterus. The mucous membrane with which the fallopian tube is lined produces a fluid secretion which nourishes the ovum and provides a medium in which the sperms can swim and also eases the movement of the ovum in the tube, while muscular movements of the wall of the tube actively help the ovum along. The two fallopian tubes, one for each ovary, have their outlets in the cavity of the uterus.

The *uterus* (*womb*) (Fig. 6-31) is a pear-shaped, muscular, hollow organ, about 7.5 cm long, 5 cm wide and 3.5 cm thick. In pregnancy it becomes greatly enlarged, to more than four times this size. It comprises two main parts: the *body,* which is the upper part enclosing the uterine cavity (this being of flat triangular

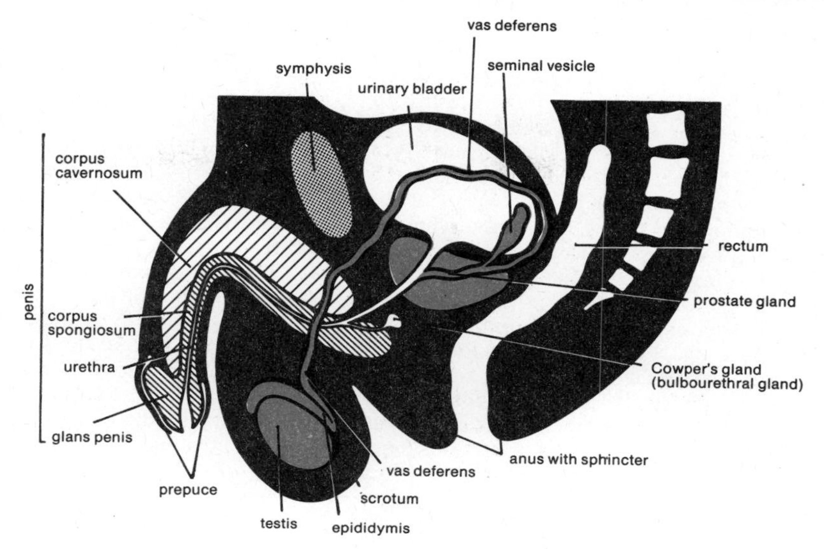

FIG. 6-34.
Male reproductive system.

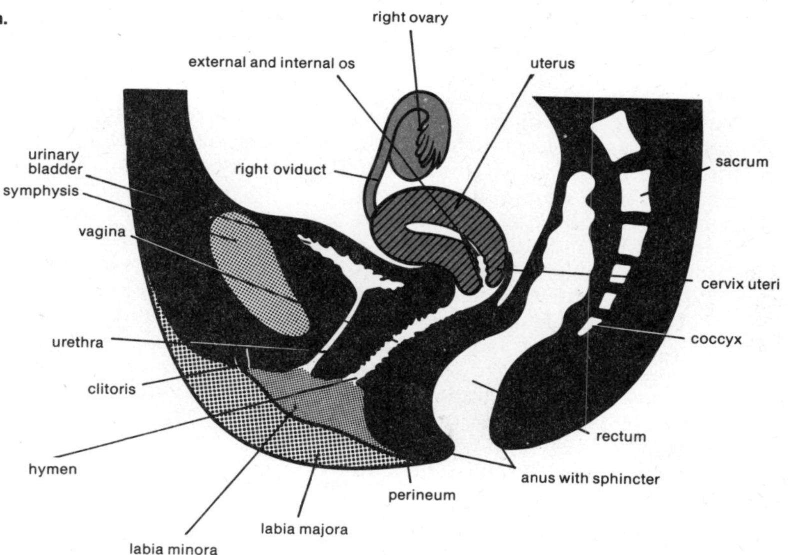

FIG. 6-35.
Female reproductive system.

shape), and the *cervix,* about 2.5 cm long and partly projecting into the vagina. The constricted zone between the two is the isthmus. The fundus of the uterus is the portion of the body above the fallopian tubes. The cervical canal leading into the vagina is constricted at each end and projects backward into the vagina, the part of the cervix within the vagina being called the *portia vaginalis*. The uterus lies in the midpelvis, being supported in this position by various ligaments and muscles, and is normally at right angles to the vagina. The wall of the uterus comprises three coats: perimetrium (formed by the peritoneum), myometrium (thick muscle coat) and endometrium (mucous membrane forming the lining).

The external female genitalia (Fig. 6-32 and 6-33) are collectively termed the *vulva,* bounded at the sides by a pair of folds called the labia majora which enclose two smaller folds, the labia minora, which in turn enclose the clitoris and the openings of the urethra and vagina. The *mons veneris* is the name given to the pad of fatty tissue which, after puberty, is covered with short curly hair (pubic hair). The two *labia majora* consist of fatty material containing connective tissue and smooth muscle fibers. They correspond to the scrotum of the male and close the vaginal orifice (but no longer do so after childbirth); they become enlarged in response to sexual stimulation. The *labia minora* are a narrow pair of folds which enclose the vestibule of the vagina and also, at their front end, partly cover the clitoris. The urethral orifice is in the upper part of the vestibule. Further inside is a fold of mucous membrane, the *hymen* or virginal membrane, which partly covers the entrance to the vagina. Contrary to popular belief, the rupture or absence of the hymen is no reliable evidence of whether or not there has been sexual intercourse, i.e., it cannot constitute a dependable criterion of virginity. Rupture of the hymen on first coitus (defloration) is accompanied by some slight bleeding. In a large proportion of women, however, the hymen is not of such size or condition as to require rupture on entry. Beneath the vestibule, at the base of the labia minora, are a pair of small glands (Bartholin's glands) which secrete mucus (Fig. 6-33). Together with the secretion of sebaceous (oil-secreting) glands they produce a lubricating smegma similar to that secreted under the male prepuce (foreskin of the penis). On each side of the vestibule of the vagina are areas of erectile tissue, which is present also in the *clitoris*. This organ, whose development corresponds to that of the penis in the male, is partly covered by the labia minora, only its tip (glans) being externally visible. It becomes enlarged as a result of sexual stimuli. The clitoris, especially the glans, is highly sensitive, containing sensory cells which respond to tactile and other stimuli. Its physiological function is to initiate and heighten the female sexual response.

The *vagina* is the tube which extends from the vestibule to the mouth of the uterus (Fig. 6-32). It is the female organ of copulation and also of childbirth. Normally its greatly distendable walls, containing muscle fibers and lined with mucous membrane, are in contact with each other, i.e., it is a collapsed tube, flattened transversely over most of its length, at right angles to the vertical slit at the vulva. The functions of the vagina are to serve as a passage for the intromission of the penis and the reception of the semen, for the discharge of the menstrual flow, and for the delivery of the fetus. The cells of the vaginal lining are under the control of estrogenic hormones (female sex hormones) and contain a considerable amount of glycogen, which is converted into lactic acid by bacteria normally present in the vagina. The lactic acid inhibits the growth of harmful microorganisms and thus helps to keep the vagina in a healthy condition. The vaginal secretion thus has an acid reaction, against which the acid-sensitive sperms are to some extent protected by the neutralizing action of the seminal fluid.

INFLAMMATION OF THE FEMALE GENITALIA

Since the female reproductive tract, from the vagina upward to the entrance of the fallopian tubes at the ovaries, is in open communication with the abdominal cavity, any inflammation of the genitalia (genitals) deserves serious attention. Pathogenic organisms that make their way upward can cause inflammation of the uterus, the fallopian tubes and the peritoneum itself (Fig. 6-36). The external entrance to the reproductive tract is protected to some extent by the labia majora and labia minora, with the hymen providing some additional protection, at least when it is present. Furthermore, the acid vaginal secretion, the alkaline solution of the cervical glands, and the secretion of the fallopian tubes together with their tendency to propel bacteria back to the uterus by muscular contraction, and the action of the cilia are all features that help to prevent the ingress of pathogenic organisms (Fig. 6-37).

Vulvitis (inflammation of the vulva, i.e., the external genitalia) is attended by the usual symptoms of redness, swelling and pain, together with itching and sometimes a discharge of pus. Urination causes a burning pain; sexual intercourse is avoided because of the soreness and tenseness of the vaginal orifice (Fig. 6-38). If the defensive capacity of the tissues is lowered by constitutional (general) disease, hormonal disturbance, local injury, or discharge from the internal genitalia, the vulva may be infected and become inflamed by pyogenic cocci (pus-forming bacteria of the coccus

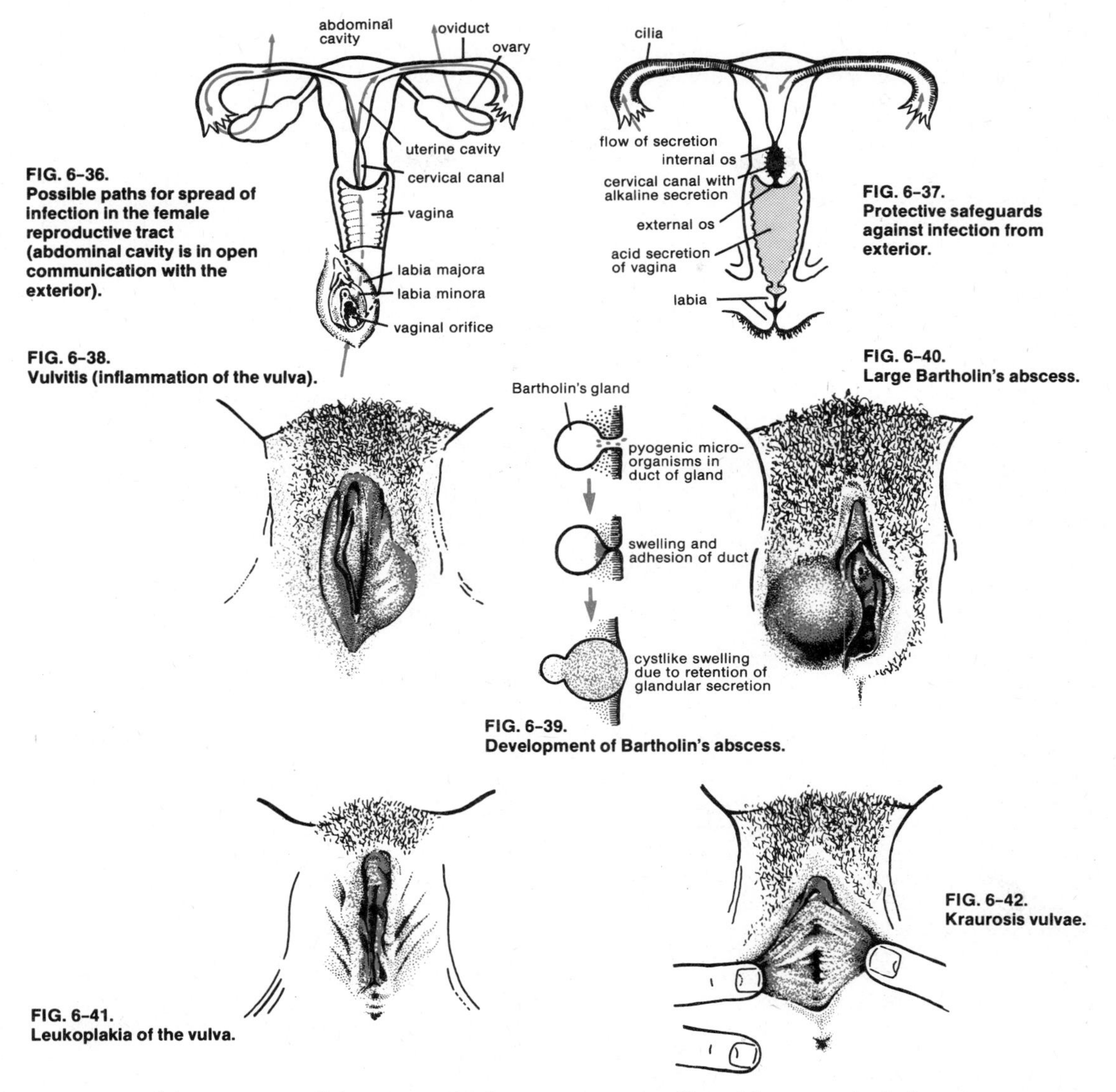

FIG. 6-36.
Possible paths for spread of infection in the female reproductive tract (abdominal cavity is in open communication with the exterior).

FIG. 6-37.
Protective safeguards against infection from exterior.

FIG. 6-38.
Vulvitis (inflammation of the vulva).

FIG. 6-39.
Development of Bartholin's abscess.

FIG. 6-40.
Large Bartholin's abscess.

FIG. 6-41.
Leukoplakia of the vulva.

FIG. 6-42.
Kraurosis vulvae.

type), parasitic protozoa of the genus *Trichomonas*, pathogenic fungi, or coli bacilli from the intestine. Gonorrhea may also cause such inflammation. Some of these invaders can be effectively dealt with by specific chemotherapeutic agents. In general, vaginal irrigation with suitable acid solutions can be helpful. Confinement to bed may be indicated in acute cases.

Inflammation of the vulva may give rise to a condition known as *Bartholin's abscess*, which develops when Bartholin's glands become affected by an acute inflammatory process. Adhesions of the outlet ducts of these glands may cause congestion of their secretion, attended by pus formation (Fig. 6-39). A large accumulation of pus and mucus, sometimes the size of a hen's egg, may develop (Fig. 6-40). The abscess will generally have to be treated surgically.

Itching of the vulva is commonly associated with inflammation and can have a variety of causes, including infection by *Trichomonas* or by pathogenic fungi. Of the constitutional diseases, diabetes especially causes itching. So do retrogressive changes after the menopause.

Leukoplakia of the vulva (Fig. 6-41) is characterized by thickened white spots or patches which may, as a result of degenerative changes, develop into *kraurosis vulvae* (Fig. 6-42), a disease that occurs mostly in elderly women and is characterized by atrophy of the skin and mucous membrane, with severe itching. Without operative intervention this condition may undergo malignant degeneration into a type of cancer. The diseases of this type are often associated with a deficiency of follicular hormone in aging women. Hormone therapy or the application of hormone ointments may be helpful. Severe cases may, however, require surgical treatment, if only to relieve the persistent severe itching.

Vaginitis (inflammation of the vagina) (Fig. 6-43) is mostly caused by infection from outside, as a result either of invasion by a virulent pathogenic organism or of failure of the natural defenses of the vagina. Such lowering of the natural defense system may occur in consequence of inadequate production of protective lactic acid in the vagina, hormone deficiency, or disease of adjacent parts of the reproductive tract, e.g., discharge from the uterus (see next section). Vaginitis can be particularly severe in infants because the vaginal wall in such young children is still very delicate and the secretion is not yet acid. Infections of the vagina are recognizable by swelling and reddening of the skin of the vagina and by the presence of a whitish or purulent coating, which comes away as a discharge (Fig. 6-43a). Another form of vaginal inflammation may be caused by *Trichomonas vaginalis*, a flagellate protozoan (Figs. 6-43b, 6-44). This organism may be present in the vagina without causing disease, but if the protective acid secretion becomes deficient, *Trichomonas* becomes active and then causes vaginitis characterized by burning and itching, with frothy white discharge having a disagreeable sweetish smell. Metromidazole, administered orally, can effectively deal with this disease. In many cases the microorganisms causing vaginitis are bacteria, often pyogenic *cocci* (Fig. 6-43a), which are then found to be present in a smear specimen of the discharge examined under the microscope. The vagina is also susceptible to mycotic infection, i.e., due to *fungus* (Fig. 6-43c) of the type that causes the disease of the mouth known as thrush (*Candida*, see page 150). Like *Trichomonas*, this organism is transmissible by sexual intercourse and may be present in the vagina without causing disease; it is likewise activated by a lowering of the acidity of its environment and also by conditions arising in pregnancy, diabetes, or the destruction by antibiotics of other microorganisms with which it normally would have to live in competition. Vaginitis due to mycotic infection is characterized by a burning sensation, itching and a thickly viscid whitish discharge, generally odorless. Specific antibiotics are available to deal with the disease.

The cervix of the uterus forms the boundary between that part of the female reproductive tract in which potentially pathogenic microorganisms are normally present and the upper or inner part which is normally free from them. In women who have borne no children, the outer orifice (external os) of the cervical canal forms the barrier (Fig. 6-45). After childbirth it often remains open, so that the borderline of microbe penetration will then be situated higher up. This will more particularly be the case if the external os has been torn during delivery of the child. The microbes can then gain easy entry into the cervix and cause inflammation there (*cervicitis*) (Fig. 6-46). It therefore sometimes becomes necessary to repair such damage by suturing it even long after childbirth. Some types of inflammation may exist more or less undetected in the cervical canal, e.g., chronic gonorrhea. Under certain conditions the microorganisms causing them may become activated and ascend farther into the uterine cavity, e.g., after childbirth or during menstruation. Conversely, discharge from the uterus causes a lowering of acidity in the vagina and is thus liable to bring on vaginitis (see next section). For this reason any inflammation of the cervix, however slight, should be treated energetically: with penicillin in a case of gonorrhea or with caustic chemicals or by electrocautery (Fig. 6-47) in dealing with other forms of inflammation.

Inflammation of the mucous membrane of the uterus (*endometritis*) is caused by bacterial invasion and may be acute or chronic. In chronic cases the symptoms are often quite mild; besides some slight fever there are likely to be menstrual irregularities, such as bleeding between periods, sometimes also increased and prolonged menstrual bleeding, or persistent discharge of blood in small amounts. After menstruation the disease may spread quite suddenly to the fallopian tubes; the uterus itself, by shedding its mucous membrane lining,

FIG. 6-43.
Appearance of the external os of the vagina affected by vaginitis (with smear specimen as seen under the microscope).

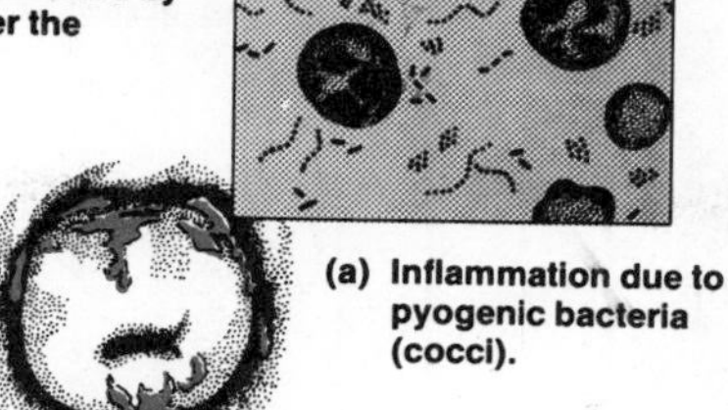

(a) Inflammation due to pyogenic bacteria (cocci).

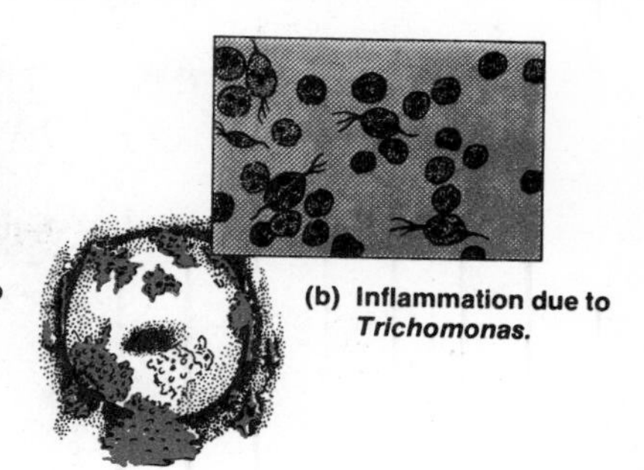

(b) Inflammation due to *Trichomonas*.

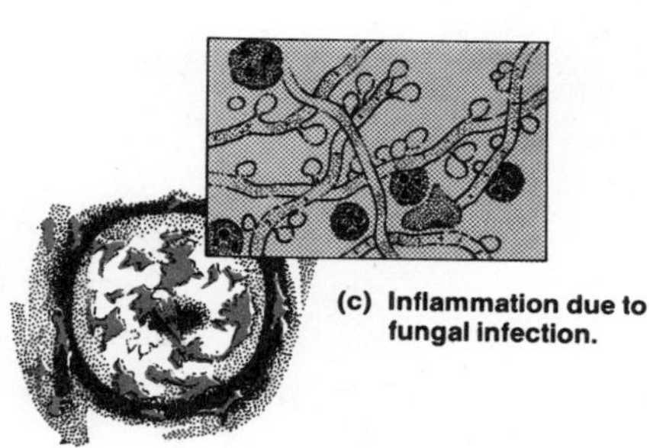

(c) Inflammation due to fungal infection.

FIG. 6-44.
***Trichomonas vaginalis* (greatly magnified).**

FIG. 6-45.
The external os usually forms the boundary between the lower (microbe-infested) and upper (microbe-free) part of the female reproductive tract.

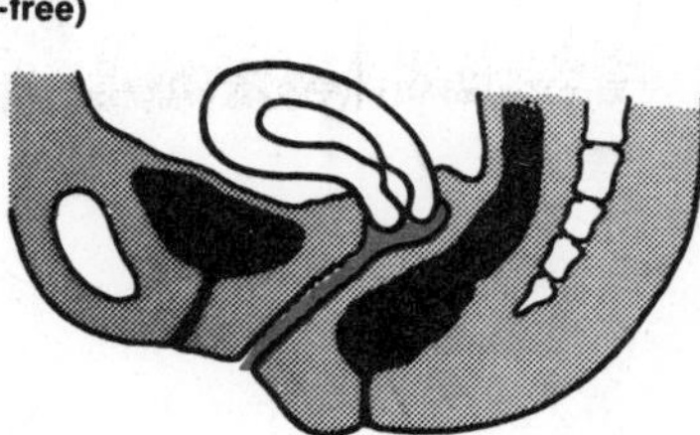

may meanwhile have got rid of the infection. To reduce the risk of such spreading, acute inflammation of the uterus should be energetically treated with antibiotics, the patient meanwhile remaining strictly confined to bed. In some cases of acute inflammation, treatment by hormone therapy may be indicated.

Inflammation of the fallopian tubes (*salpingitis*) is usually caused by bacterial invasion from the uterine cavity or sometimes directly from the vagina. In this respect the woman is particularly vulnerable during menstruation, during the puerperium (the period during and immediately after childbirth, usually 3–6 weeks), and following a miscarriage or abor-

FIG. 6-46.
Purulent inflammation of the cervical canal (the external os is torn and the lacerated mucous membrane provides a path for the microbes).

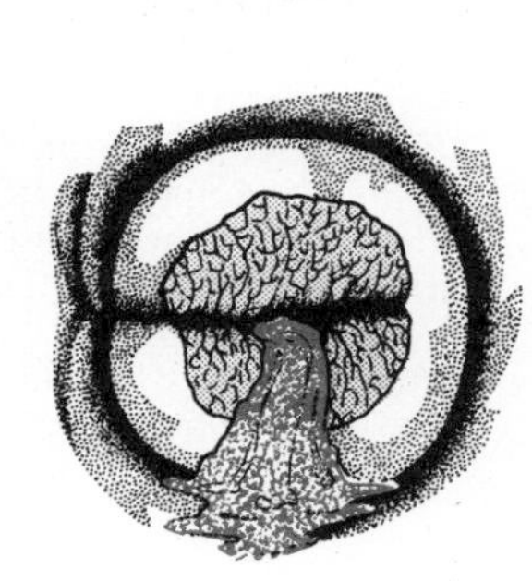

FIG. 6-47.
Electrocautery of chronically inflamed cervical canal.

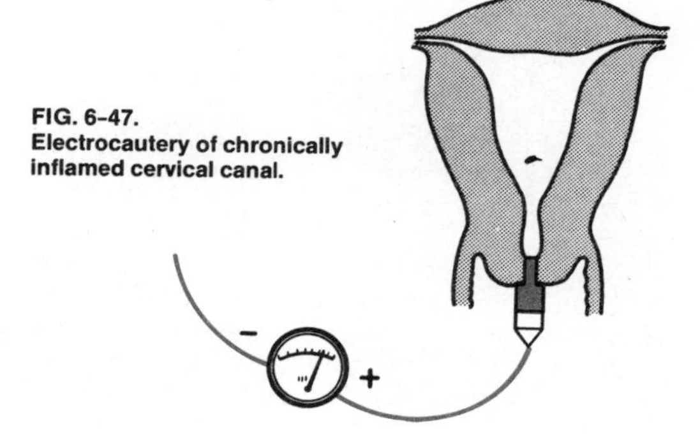

FIG. 6-48.
Adhesions and blind alleys in the lining of the inflamed fallopian tube (oviduct).

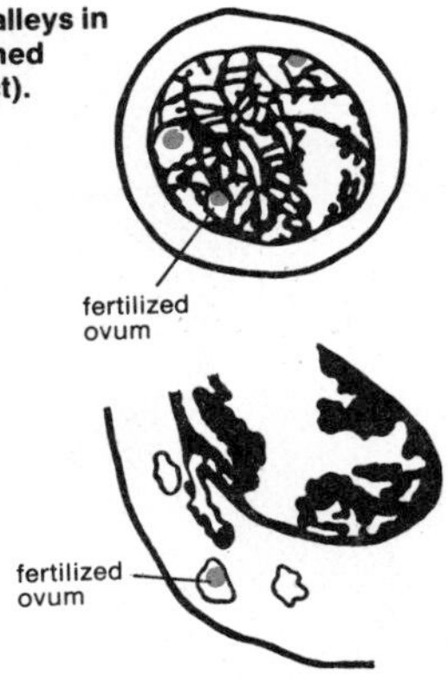

FIG. 6-49.
(a) Healthy oviduct. **(b) Inflamed oviduct.**

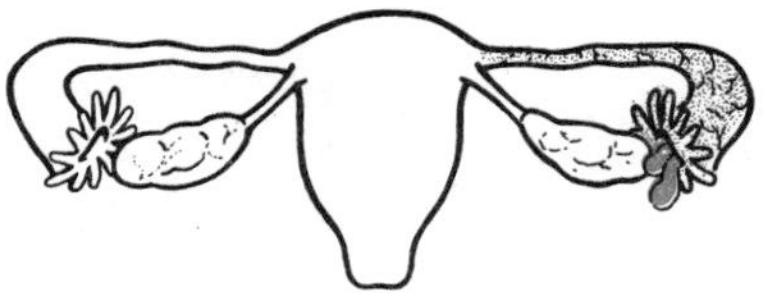

FIG. 6-50.
(a) Inflamed oviduct with entrance closed by adhesion.
(b) Accumulaion of pus in inflamed closed oviduct. 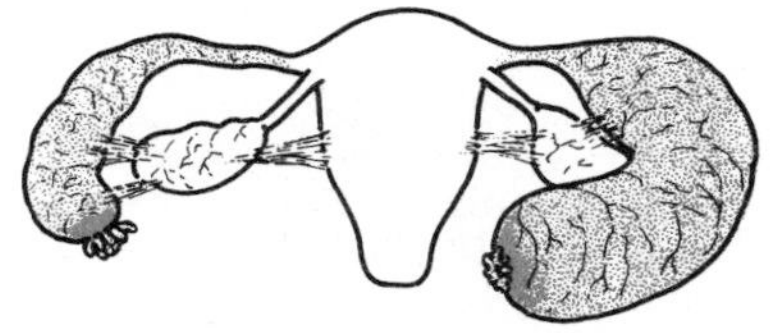

FIG. 6-51.
(a) Inflammation spreads from oviduct to ovary. 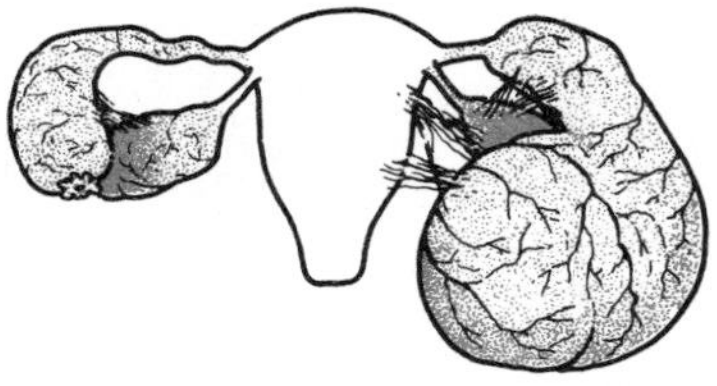
(b) Large pus-filled cavity involving oviduct and ovary.

FIG. 6-52.
Permanent coalescences and adhesions of uterus, oviduct, and ovary.

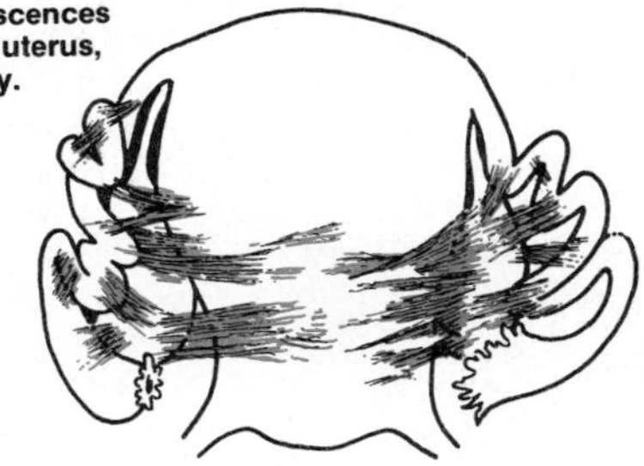

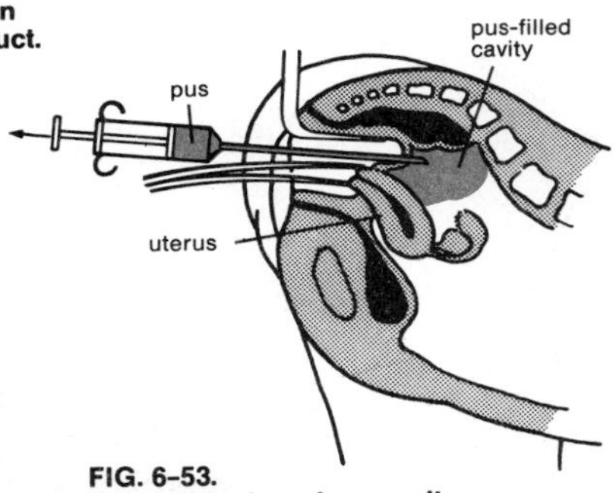

FIG. 6-53.
Extraction of pus from cavity between uterus and rectum.

tion. The fallopian tubes may become infected with bacteria introduced accidentally in the course of an examination or other medical intervention. More rarely, infection may spread from adjacent organs, e.g., from an inflamed appendix. The organisms concerned are usually pyogenic cocci (staphylococcus, streptococcus); in a minority of cases the condition is caused by gonococcus (the organism causing gonorrhea) or, rarely, by the tubercle bacillus, the latter having been carried in the bloodstream. Acute salpingitis may, after an undetected initial stage, start suddenly with fever and severe pains in the lower abdomen, which is tense and tender. Nausea, vomiting, diarrhea or constipation may also occur. This acute stage may gradually turn into a chronic condition, with flare-ups recurring particularly after menstruation. The inflammation of the mucous membrane of the fallopian tube is often followed by adhesions which may prevent conception (see page 247). This inflammation then spreads to the muscle coat of the tube, which thus becomes thickened and rigid (Figs. 6-48 and 6-49). The open entrance to the tube, where it communicates with the abdominal cavity, may become closed by adhesion (Fig. 6-50a) and pus collects in it (Fig. 6-50b). At the same time, inflammatory coalescence (growing together) with adjacent organs may occur, often involving the ovaries (*salpingo-oophoritis,* Fig. 6-51a). In some cases a large pus-filled cavity comprising the ovary and fallopian tube may develop (Fig. 6-51b). Eventually the inflammation spreads to the peritoneum and will, if it is a chronic condition, cause extensive adhesion and coalescence due to scar formation (Fig. 6-52). In about 80 percent of such cases the woman will become sterile, and frequently there is displacement of the uterus as a result of growing together of tissues. In the chronic stage there are tugging pains and tenderness in the abdomen and the lumbar region, discharge, and painful menstrual bleeding. Sometimes the symptoms also include constipation and urination disorders and may continue for a long time.

Treatment of acute salpingitis begins with strict confinement to bed. If there is peritonitis, the patient should be removed to a hospital. Antibiotics are given to prevent the inflammation from spreading. Hormones derived from the adrenal cortex may also be administered to curb excessive inflammatory effect. In the chronic stage, a surgical operation will be necessary in about 10 percent of cases to prevent relapses, detach adhesions or remove cysts which may have developed over a period of time, particularly in the ovaries. A cyst is an abnormal fluid-filled swelling which may be a tumor or may be a gland of which the outlet has become blocked. Accumulations of pus between uterus and rectum can be punctured, opened and drained (Fig. 6-53). Heat should not be applied as treatment.

DISCHARGE

Its normal secretion keeps the vagina internally moist and lubricated. Also, because of its lactic acid content, the secretion kills harmful microorganisms or at any rate keeps them in check. This acid protection is especially important because the abdominal cavity is actually in open communication with the exterior through the fallopian tubes, uterus and vagina. The lactic acid is produced from glycogen, which is present in the cells of the lining, by the action of bacteria normally living in the vagina. The glycogen is released and thus made available to the bacteria by shedding of the epithelium (surface layer of cells) of the lining.

The vaginal fluid may be altered in composition or increased by various causes. If it is increased to the extent that it wets the labia of the vulva and causes stains on clothing, the flow is referred to as a *discharge*. If there is also itching or a burning sensation and if the discharge becomes foul-smelling or so copious as to necessitate the wearing of a bandage, it is obviously time to seek medical advice.

Discharge may be caused by disease not only of the vagina itself but also of parts of the reproductive tract situated higher up, particularly by disease of the uterine cavity. Examination with an optical instrument (speculum) in such cases reveals the emission of fluid from the external os, i.e., the opening of the cervical canal into the vagina. Sometimes the discharge contains pus or blood and may be pink or flesh-colored, which may be an indication of cancer of the uterus (see page 499). In such cases it will be necessary to do a smear test.

Quite often the discharge originates in the cervix of the uterus (Fig. 6-54). If it consists entirely of pus, it is likely to be due to gonorrhea. The gonococci can live in the cervix for a long time and produce a glassy-looking discharge. From this base they can, if conditions are favorable to them (e.g., after childbirth), spread farther upward in the reproductive tract (see page 319). Energetic treatment with penicillin will be necessary to arrest the spread. Discharge may also be caused by a cervical polyp (Fig. 6-55), by injury incurred during childbirth with subsequent protrusion of the mucous membrane of the cervical canal (Fig. 6-56) or by cancer of the cervix (Fig. 6-57a and b, see also pages 499 ff). Timely diagnosis by direct visual inspection, smear tests or examination of tissue specimens is very important in such circumstances (Fig. 6-58a and b).

Thinly liquid discharge from the cervical canal, which is largely harmless in itself, is often due to disturbed nervous or hormonal control—for example, through emotional causes. It is in effect an intensification of normal mucous secretion and may regularly recur

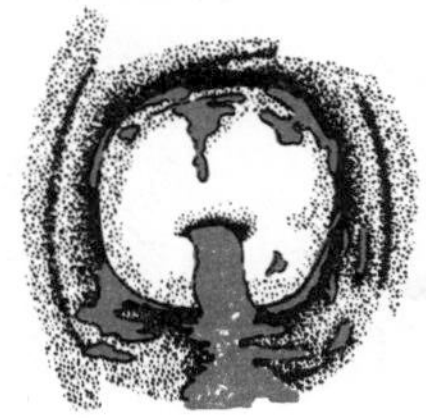

FIG. 6-54.
Purulent discharge from infected cervix.

FIG. 6-55.
Cervical polyp in external os.

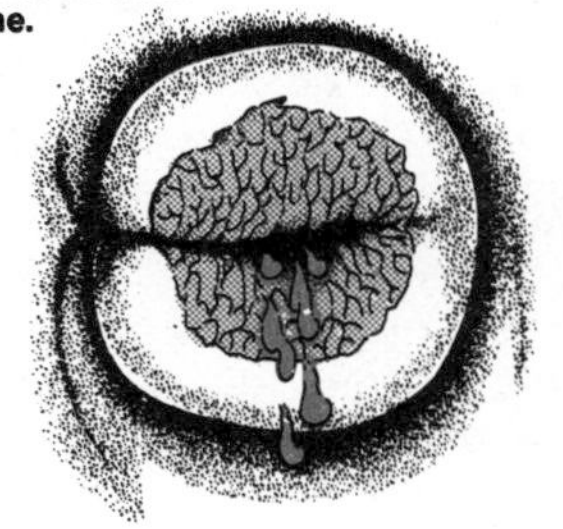

FIG. 6-56.
Discharge from torn external os with protrusion of cervical mucous membrane.

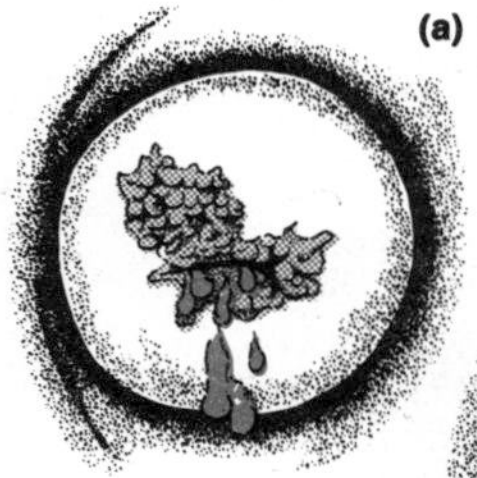

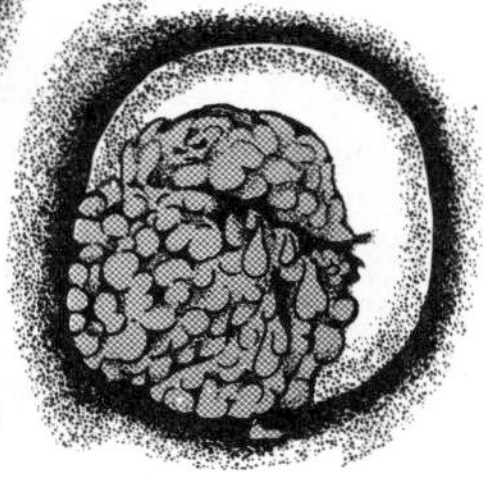

FIG. 6-57.
(a) Discharge in early stage of cervical cancer.
(b) Cancer of the cervix.

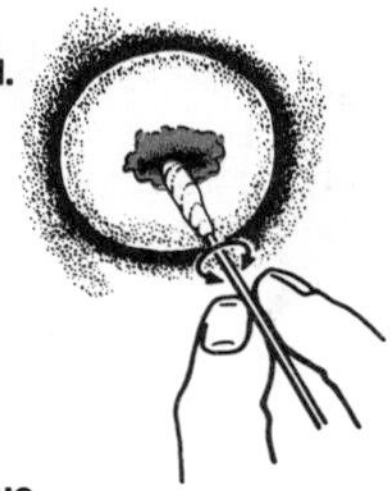

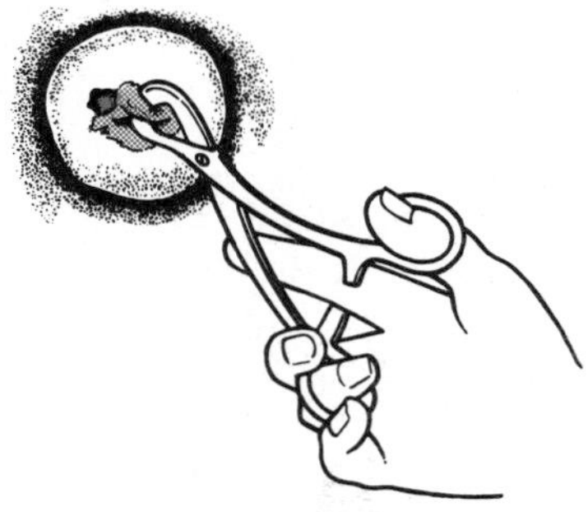

FIG. 6-58.
(a) Obtaining a specimen from the cervical canal.
(b) Biopsy: excision of a small specimen of tissue for microscopic examination.

FIG. 6-59.
Discharge due to presence of a foreign body.

FIG. 6-60.
Discharge due to pressure exerted by contraceptive device.

around the time of ovulation (see page 235). In young girls the presence of threadworms or other foreign bodies may cause a discharge (Fig. 6-59); in adult women it may be caused by the wearing of rubber contraceptive devices for too long a time or by lack of hygiene in the use of such devices (Fig. 6-60).

POLYPS OF THE FEMALE REPRODUCTIVE TRACT

A polyp (polypus) is a usually harmless tumor of mucous membrane. Typically it starts as a small flattish elevation, then grows longer, and finally hangs like an inverted mushroom or berry from a stalk (pedicle). Depending on where they are situated, polyps cause varying symptoms. What they have in common is a tendency to undergo ulcerous degeneration, to irritate the organ to which they are attached, and eventually to be treated by it as foreign bodies. The degenerative tendency is due to the fact that a polyp grows at a faster rate than its stalk is able to support, so that the latter is likely to become twisted and thus cut off the supply of blood and nourishment to the polyp.

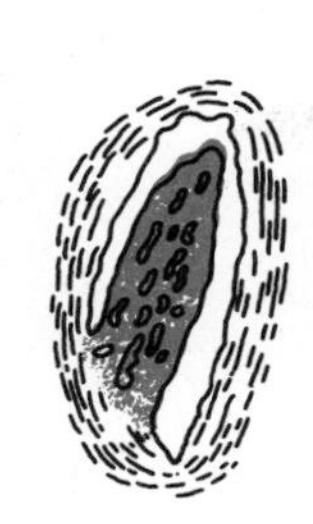

FIG. 6-61.
Polyp in the oviduct arising from displaced uterine mucous membrane (so-called tubal endometriosis).

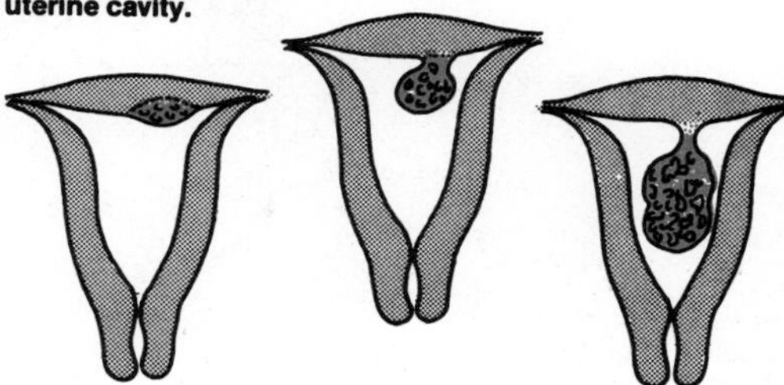

FIG. 6-62.
Development of a polyp in the uterine cavity.

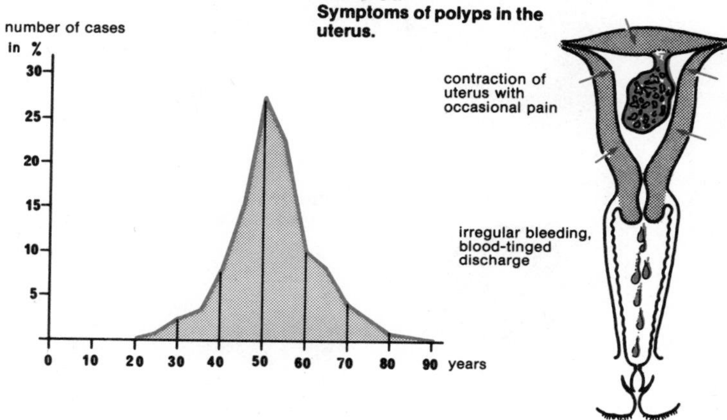

FIG. 6-63.
Incidence of uterine polyps related to age of women.

FIG. 6-64.
Symptoms of polyps in the uterus.

FIG. 6-65.
Polyp in the cervical canal visible at the external os.

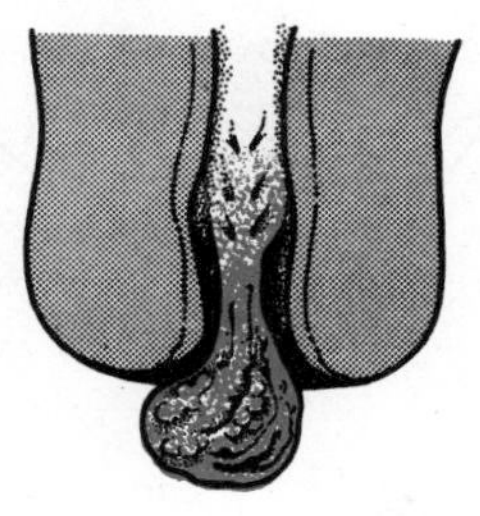

FIG. 6-66.
Polyp in the cervical canal may protrude into the vagina and cause discharge of blood.

Polyps may occur in the urethral orifice, though rarely. In the fallopian tubes (oviducts) they are, as a rule, encountered only in conjunction with intrusions of displaced uterine mucous membrane (Fig. 6-61).

In the uterine cavity itself polyps are of more frequent occurrence (Figs. 6-62 and 6-63). They are mostly benign (nonmalignant) growths of the mucous membrane and are probably caused by an excess of follicular hormone, i.e., hormone secreted by the ovarian follicles. Uterine polyps cause irregular menstrual bleeding, often of long duration, and sometimes also give rise to discharge containing blood at intermediate times between periods. The discharge may contain pus if the degenerating polyp becomes infected. These symptoms are in themselves not dangerous but are likely to cause anxiety because they may be interpreted as a sign of cancer. Such fears

FIG. 6-67.
Floor of the pelvis.

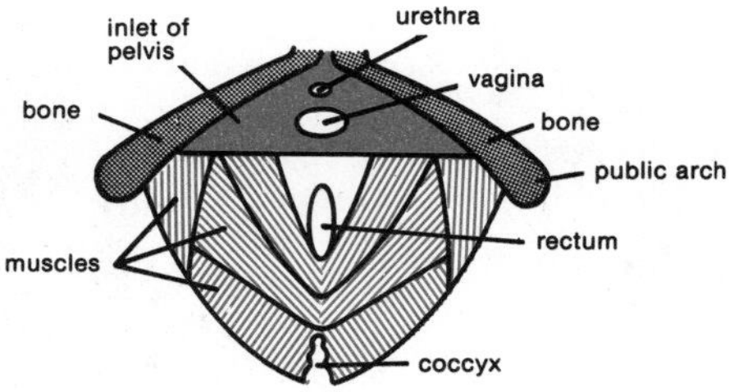

inlet of pelvis
urethra
vagina
bone
bone
public arch
rectum
muscles
coccyx

FIG. 6-68.
Downward displacement of uterus, causing it to protrude abnormally into the vagina.

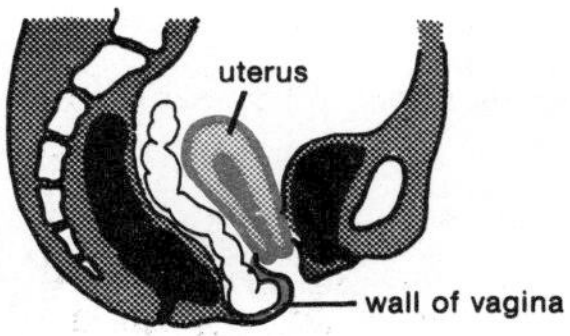

uterus
wall of vagina

FIG. 6-69.
Prolapse of the uterus.

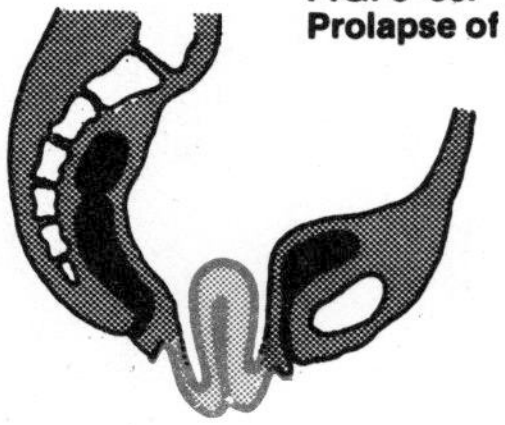

FIG. 6-70.
(a) Cup pessary.

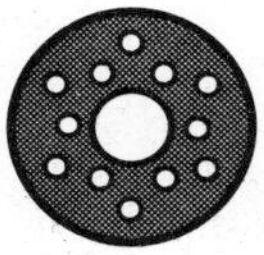

(b) The inserted pessary supports the uterus.

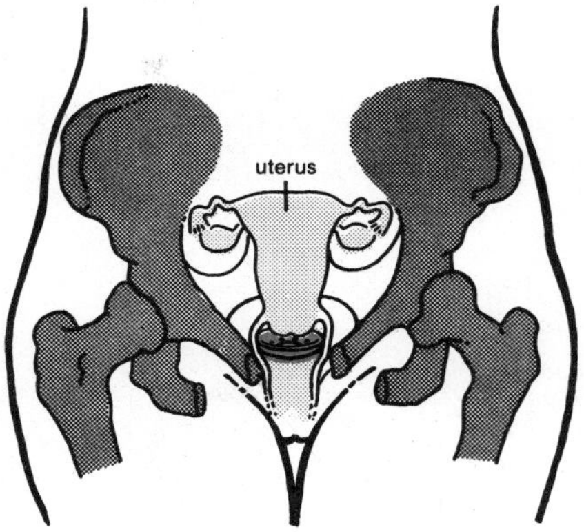

uterus

FIG. 6-71.
Operation for correction of uterine prolapse.

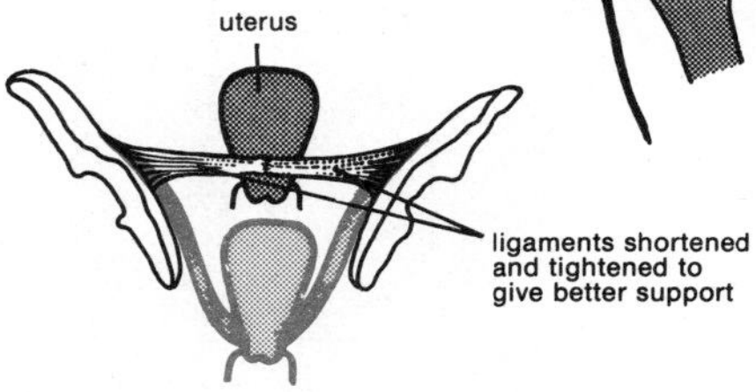

uterus
ligaments shortened and tightened to give better support

can be allayed by histological examination of a tissue specimen obtained from the uterus by curettage, i.e., scraping with an instrument called a *curette*. Occasional pain caused by polyps indicates that the uterus is striving to get rid of these "foreign" growths (Fig. 6-64). Treatment consists in removing the polyps by curettage. The patient should be medically checked for some time after this treatment because polyps, though harmless, may nevertheless be indicative of a tendency for growths, including possibly malignant tumors, to develop in the reproductive tract.

Much commoner than polyps in the uterine cavity are those in the cervical canal, i.e., the cavity within the cervix of the uterus. These growths are usually also benign in character and are generally similar to those already described. A long polyp may be visible at the mouth (external os) of the cervix (Fig. 6-65) or may protrude into the vagina (Fig. 6-66). If the cervical polyp undergoes ulcerous degeneration, it may cause irregular menstrual bleeding and also intermenstrual discharge of blood or blood with mucous. There is moreover a risk that the cervical mucous membrane, already in an irritated condition, will become infected by microorganisms which enter through the external os held permanently open by the polyp growth. Treatment will consist in removing the polyps by curettage. As a precaution, specimens of the polyp tissue and of the other tissue removed should be subjected to histological examination. (Histology is the study of the microscopic structure of tissues.)

DISPLACEMENTS OF ABDOMINAL ORGANS; PROLAPSE

Displacements of the female genital organs are of fairly common occurrence and often require medical treatment. About 10 percent of all women's disorders are due to such displacements, which is not very surprising when it is considered how these organs are accommodated within the relatively restricted pelvic space. They have to be movably attached, so as to enable the uterus to become greatly enlarged during pregnancy and extend from the pelvis to just below the diaphragm (midriff). In addition, because human beings walk upright, the abdominal organs have to be supported by the pelvic floor (Fig. 6-67), which moreover contains apertures for the passage of the urethra, vagina and rectum and must be large enough to allow the child's head to pass at birth.

The uterus, especially if it is a large and easily movable one, is liable to be displaced if the muscles and ligaments that secure it are, as also the lateral supporting tissues, congenitally weak or slack or have become so after childbirth. Thus the uterus may be shifted or tilted in any direction, and its body may be abnormally flexed in relation to the cervix. There are various possibilities, including more particularly *retroflexion*, the condition in which the body of the uterus is bent backward, whereas the cervix retains its normal position; *retroversion*, in which the whole uterus is tipped backward; *anteflexion*, in which the body is excessively bent forward on the cervix; *anteversion*, in which the whole uterus is tipped forward in relation to its normal position (see following section). An important type of displacement of the genital organs is prolapse, especially of the uterus ("falling womb"), as described next.

The genital organs, including particularly the uterus, are liable to undergo downward displacement from their normal position in the body as a result of weakness of the supporting tissues, especially those forming the pelvic floor. If such a displacement is relatively large—so that, for example, the uterus protrudes abnormally far into the vagina (Fig. 6-68) or, as occurs in some cases, the uterine cervix actually protrudes from the vaginal orifice and is thus externally visible (Fig. 6-69)—the condition is called *prolapse* of the uterus (prolapsus uteri). The weakness of the pelvic floor that gives rise to this abnormality often develops after childbirth if the muscles and ligaments have been torn or overstretched. In some respects, prolapse of the female genitals is comparable with hernia. A contributory factor may be an inherent weakness of the supporting system and/or where the abdomen is distended and pendulous, so that the intestines as a whole, instead of being retained within the dome of the diaphragm, have lost their firmness and press heavily on the pelvic floor.

Such displacement of the genital organs usually first manifests itself in a feeling of pressure acting downward. Then follow such symptoms as pain due to tugging at the supporting muscles and ligaments, constipation and discharge. Especially characteristic is the occurrence of bladder trouble, with increased urination urge and, subsequently, inflammation of the bladder. When prolapse develops, hemorrhage and ulceration (due to pressure) of the parts affected are liable to occur, with a risk of cancer developing in the long run. The condition therefore requires proper medical attention.

In mild cases of prolapse, corrective physical exercises and the insertion of a pessary (supporting ring) into the vagina may adequately control it (Fig. 6-70a and b). Pessaries may, however, cause discomfort to the wearer after a time. A surgical operation is therefore often preferred, unless there are compelling reasons against it. Severe cases of prolapse can be treated only by surgery. The operation consists in reconstructing, as far as possible, the supporting tissues so as to restore the genital organs to their normal position and condition (Fig. 6-71). In some cases this may be impracticable, and removal of the uterus may then be necessary.

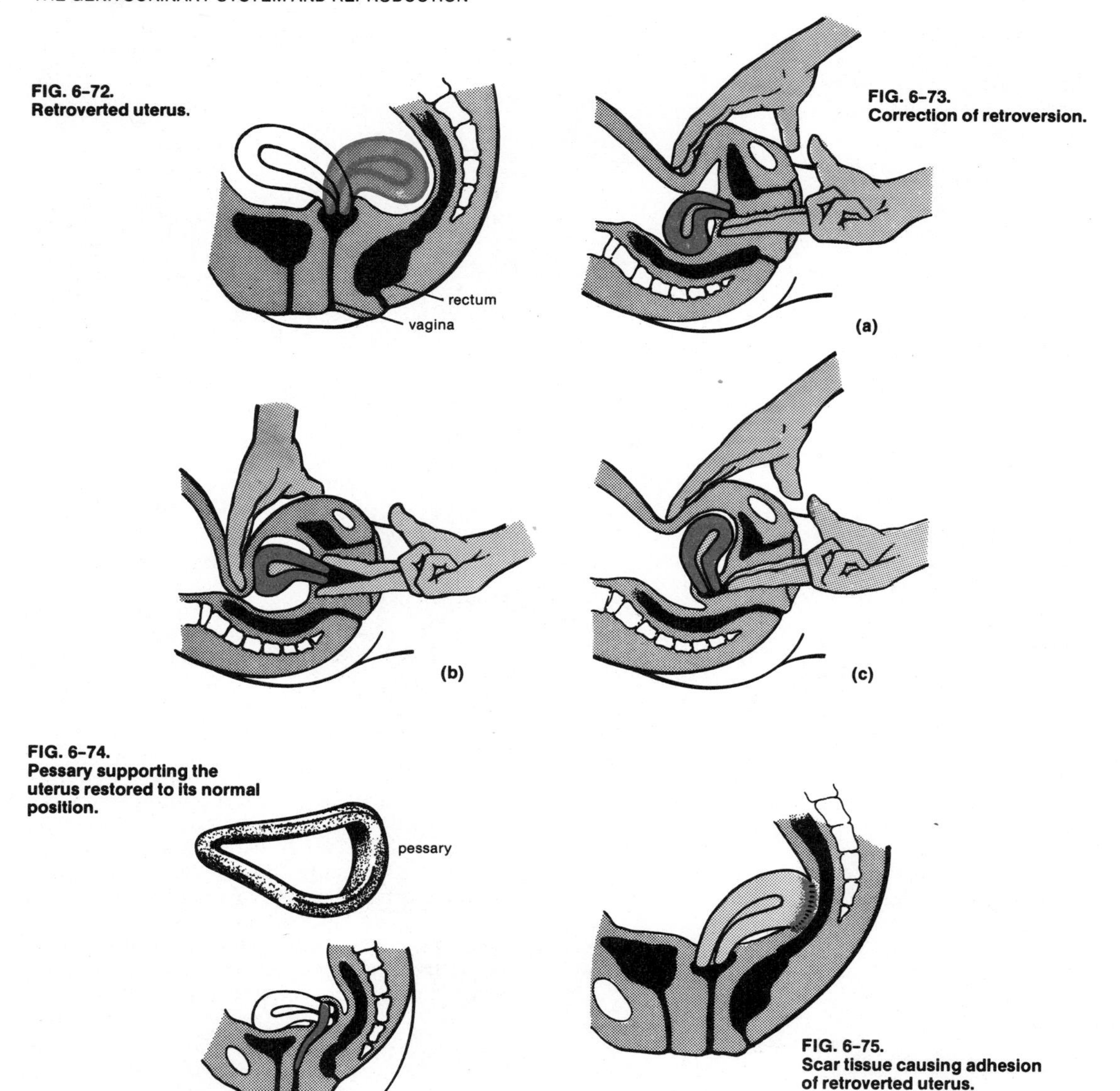

FIG. 6-72.
Retroverted uterus.

FIG. 6-73.
Correction of retroversion.

FIG. 6-74.
Pessary supporting the uterus restored to its normal position.

FIG. 6-75.
Scar tissue causing adhesion of retroverted uterus.

RETROVERSION OF THE UTERUS

The normal position of the uterus is anteverted, i.e., it slopes forward in the sense that the fundus is situated forward from the external os (the mouth of the uterus in the vagina, see Fig. 6-45). In addition, the uterus is normally anteflexed, i.e., the uterine body is bent forward on the cervix. If the uterus is tipped in the opposite direction, so that it slopes backward, it is said to be retroverted (Fig. 6-72). This type of displacement is quite common and usually causes no trouble.

If the patient complains of pains in the lumbar region and the uterus is, on examination, found to be retroverted, the pains are less likely to be caused by the actual uterine displacement than by a general weakening of supporting tissues, including those in the vicinity of the spinal column. If retroversion is present in women with menstrual disturbances, in sterile women or in women with a tendency to have miscarriages, the basic cause of these disorders is frequently traceable to a functional weakness of the ovaries associated with underdevelopment of the genital organs. In such cases the presumed underlying disorder must first be treated. Before this is done, however, it must be established that the retroverted uterus is movable and not coalesced with adjacent organs. This can be ascertained by a trial elevation of the uterus (Fig. 6-73). If lumbar pains and severe menstrual pains are suspected to be indeed due to uterine retroversion, the uterus can be righted by tipping it forward, in which position it can tentatively be maintained with the aid of a supporting device (pessary) inserted into the vagina (Fig. 6-74). If the disorders cease as a result of this intervention, further treatment will often consist in massage and physical exercises. If these methods fail to achieve lasting improvement, an operation to take in (tighten) the supporting ligaments will have to be considered.

Adhesion of the retroverted uterus may occur by scar tissue formation (Fig. 6-75), usually as a result of inflammation in the abdomen. Symptoms include pain in the lumbar region (see page 240), severe menstrual pains (see page 238), constipation, and pain felt during sexual intercourse. When the doctor attempts to right the uterus (as in Fig. 6–73), the adhesions will hurt. They can be surgically severed when the inflammation that has caused them has completely subsided.

In a case of retroverted uterus, pregnancy may present special problems. Growth of the embryo causes the uterus eventually to fill the pelvic cavity, from the fourth month of pregnancy onward. If the uterus fails to right itself from its retroverted position, a dangerous condition may arise, with constriction phenomena and retention of urine. The bladder must then be emptied by artificial intervention and the uterus righted (by surgical means, if necessary). Fortunately, in the great majority of cases, the retroverted uterus will right itself as pregnancy progresses, so that the uterus will move into the position that allows the fetus adequate room for further development. To be on the safe side, however, the position of the uterus should be regularly checked by the doctor from the third month of pregnancy onward. If it fails to right itself by the twelfth week, a pessary will be inserted or surgery may be required.

Excessive anteversion of the uterus (see page 231) is generally not an independent disorder, but is bound up with some degree of underdevelopment of the female genital organs associated with weak functional activity of the ovaries. Some indications of such underdevelopment are that the uterus is small, compact and hard, the vagina is narrow, and menstruation begins at an abnormally late age, or may be absent altogether. Sometimes menstrual pains are severe, and bleeding, when it occurs, may be abnormally heavy. Such women are often sterile. In many cases, however, the condition can be improved and indeed entirely corrected by treatment with ovarian hormones that are deficient in the patient.

MENSTRUATION

Menstruation, also known as the menses or menstrual period, is the discharge of blood from the vagina at approximately monthly intervals. The female sexual functions are centered on the ovaries, which produce not only the reproductive cells (ova) but also sex hormones (estrogens secreted by the follicles, progesterone secreted by the corpus luteum), which are responsible for the development and maintenance of the secondary sexual characteristics, preparation of the uterus for pregnancy, and development of the mammary glands. The functions of the genital system are under the overall control of the sexual center in the midbrain. Associated in a subsidiary capacity are certain cells in the pituitary gland which are stimulated by the sexual center to produce so-called *gonadotropic hormones*, i.e., hormones that stimulate the action of the gonads (more particularly the ovaries in the female, the testes in the male). These hormones control the production of the ovarian hormones which are responsible for the *menstrual cycle* (ovulatory cycle), i.e., a series of periodically recurring changes in the female sex organs associated with menstruation in women of childbearing age.

The lining of the uterus comprises two layers, of which the inner one (the *endometrium*) particularly undergoes cyclic changes, under the control of the ovary, to make it suitable as a nesting place for the implantation of the fertilized ovum. Under this is a layer whose cells initiate the regeneration of those parts of the

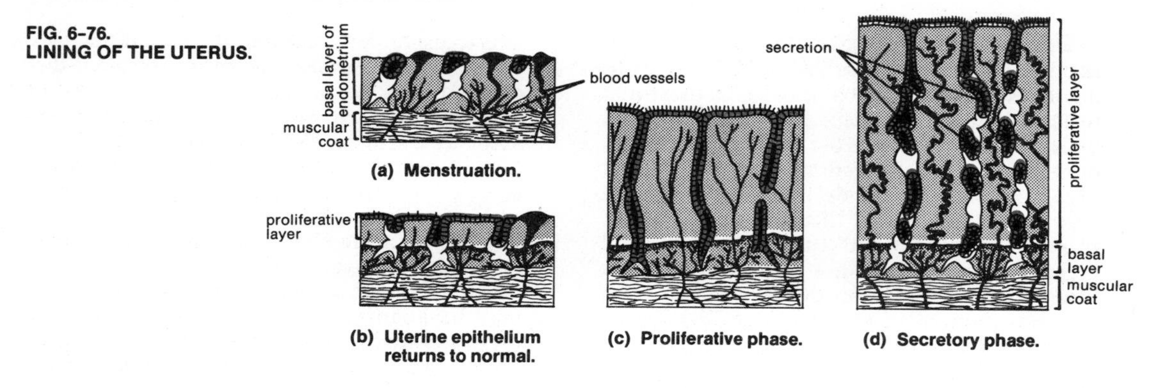

FIG. 6-76.
LINING OF THE UTERUS.

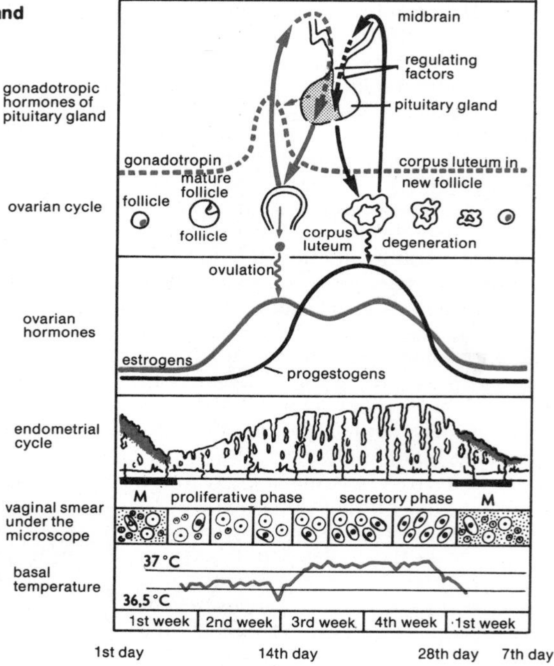

FIG. 6-77.
Menstrual cycle as a function of the central nervous and hormonal systems (M = menstruation).

endometrium that are shed at menstruation if no fertilization of the ovum has taken place (Fig. 6-76).

The uterine mucous membrane contains glands that, among other secretions, produce a special mucus which forms a plug in the cervical canal to bar the ingress of microorganisms from the uterus. Normally this mucus is thick and viscid, but at the time of ovulation it becomes more thinly fluid under the influence of the estrogens and thus more easily penetrable by the sperms seeking entry into the uterus.

The epithelium (surface layer of cells) of the vaginal mucous membrane also undergoes characteristic changes under the influence of follicular hormones. Because of this, examination of a smear specimen from the vagina can provide important clues as to when ovulation takes place (Fig. 6-77).

The *menstrual cycle* comprises a recurrent series of histological changes of the uterine lining and averages 28 days in length, measured from the beginning of menstruation. First, under the influence of the follicle-stimulating hormone (FSH) secreted by the anterior lobe of the pituitary gland, the maturing of the follicle and its production of hormones in the ovary are promoted. During this so-called *proliferative phase*, lasting from the 5th to the 15th day of the menstrual cycle, under the stimulus exercised by the follicular hormones (estrogens) the uterine epithelium is restored and the endometrium as a whole becomes thicker (Fig. 6-76). This phase is terminated by ovulation (rupture of the follicle) under the influence of a sudden increase in the supply of FSH. The ruptured follicle is transformed into a corpus luteum. In the following *luteal or secretory phase* (15th to 28th day) the endometrium further increases in thickness; its glands produce an abundant secretion containing glycogen, while the corpus luteum in the ovary secretes progesterone. If there is no fertilization of the ovum, there follows a *premenstrual phase* of about two days in which the endometrium becomes anemic and shrinks; the corpus luteum begins to diminish, and the secretion of sex hormones is greatly reduced. *Menstruation* now occurs, i.e., a period of uterine bleeding accompanied by shedding of parts of the endometrium, lasting 3 to 5 days on average.

The onset of menstruation, called the *menarche*, occurs between the 10th and 17th year, usually at 12–13 years of age. Normal permanent cessation of menstruation in middle age is called the *menopause;* it may occur between the ages of 35 and 58. Menstrual bleeding in the healthy female usually lasts 3–5 days, but 2–7 days is still within the normal range. Between 20 and 100 ml (average 60 ml) of blood is lost during a menstrual period. Although the average length of the menstrual cycle is 4 weeks, it is in fact quite variable, even in one and the same individual, and may range from 3 to 5 weeks. A few days' irregularity is quite normal. The cycle can be upset by disease or other physical causes or by emotional excitement or stress. Although a missed period may be an indication of pregnancy, it is not necessarily so. Variations in the length of the cycle are usually due to variation in the length of the proliferative phase. The interval between ovulation and menstruation is generally 14 days.

Menstrual Disorders

Disorders or abnormalities of the menstrual cycle may occur as a result of changes affecting the vagina, uterus, fallopian tubes or ovaries. In addition, the pituitary gland and the sexual center in the midbrain are more or less actively involved in controlling and bringing about the menstrual periods.

The most serious of such disorders is *amenorrhea*, i.e., the abnormal absence of menstrual periods in women of childbearing age, as distinct from normal amenorrhea before puberty, after the menopause, and during pregnancy and lactation (breast-feeding). A distinction can be drawn between primary amenorrhea, in which menstruation has never begun, and secondary amenorrhea, in which the menstrual periods, having appeared, subsequently cease. Generally speaking, amenorrhea is not an illness, but a symptom of any one of a number of possible underlying disorders; its direct effects, if any, are more likely to be emotional than physical.

Primary amenorrhea may be due to a congenital abnormality of the reproductive tract. In some cases the hymen is without an opening, so that, when the menstrual bleeding commences in puberty, the blood cannot flow away and instead accumulates and causes congestion, which may extend as far back as the fallopian tubes (Fig. 6-78a,b,c). Treatment will consist in making an incision in the hymen, thereby releasing the pent-up blood and reducing the tumorlike swelling it had caused. Congenital absence of the vagina is very rare. More often there are adhesions of the mucous membrane of the uterus as a result of curettage (see page 281). Such adhesions may be removed by gentle expansion of the uterus, by hormone treatment or by uterine mucous membrane grafts.

One of the commonest causes of amenorrhea is inadequate ovarian function. As a rule, the ovaries themselves are not detectably damaged or deformed, but the stimulus from the sexual center to induce them to function normally appears to be lacking (Fig. 6-79). Such failure of the sexual center may be due to emotional trauma or to severe physical damage. Frequently, women affected in this way are emotionally unstable and fail to establish or maintain close human contacts. Other possible symptoms that may indicate emotional stress include emaciation, constipation and circulatory disorders. Sometimes there are phantom

FIG. 6–78.
(a) Accumulation of blood in the vagina because hymen is closed.

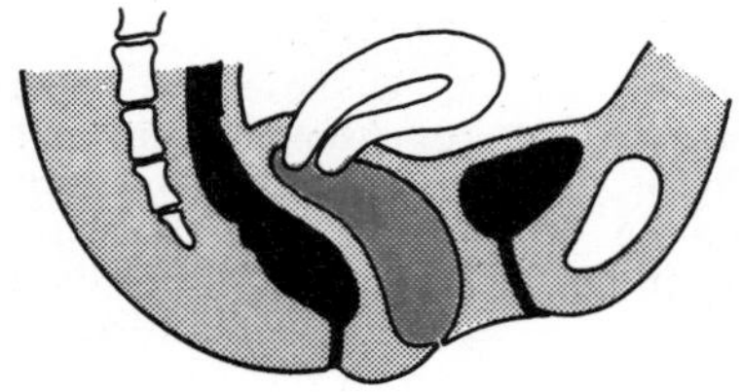

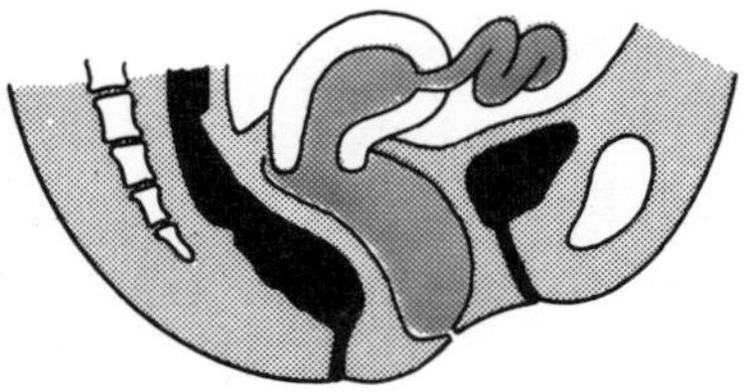

(b) Accumulation of blood in the uterus.

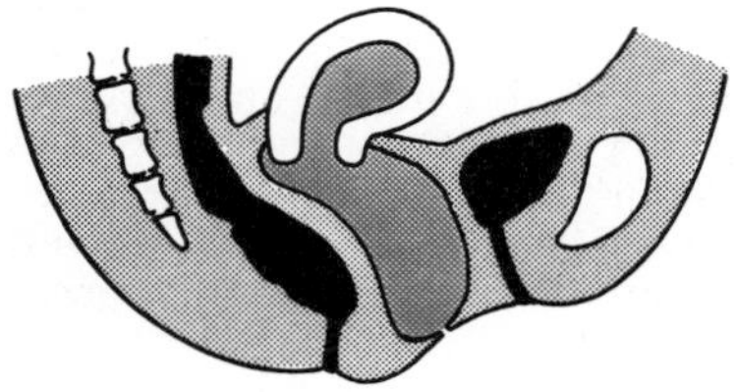

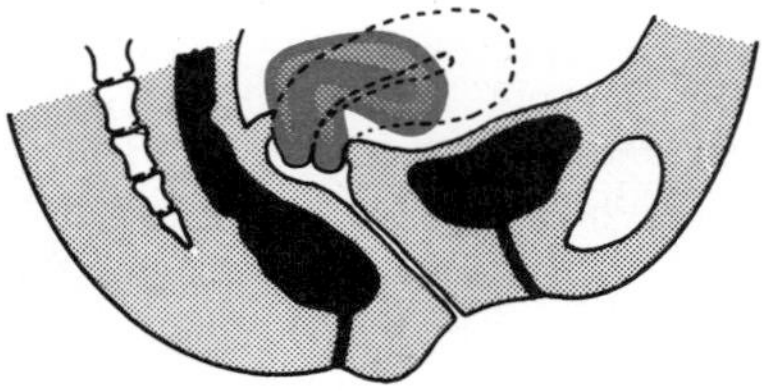

(c) Accumulation of blood extending into the oviduct (fallopian tube).

FIG. 6–79.
Ovarian dysfunction due to sexual center failure.

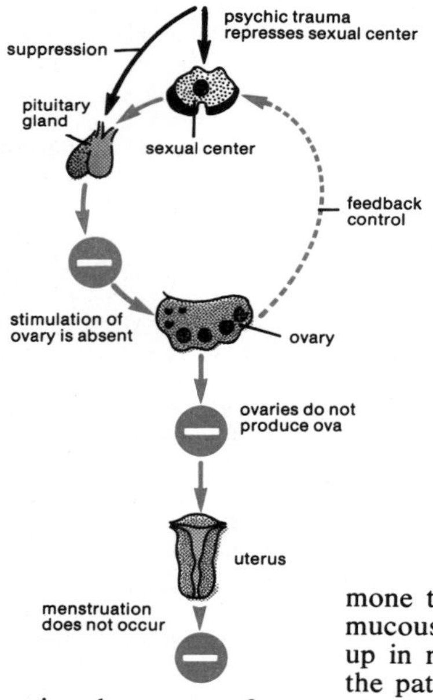

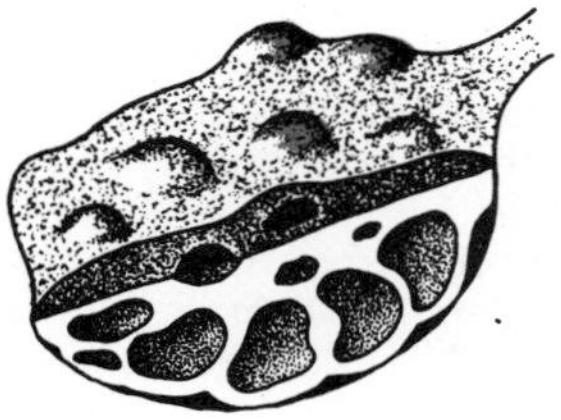

FIG. 6–80.
Small uterus leaning forward at an acute angle.

FIG. 6–81.
Cystlike modified follicles in the ovary.

pregnancies in which the patient has many of the usual signs and symptoms of pregnancy, such as enlargement of the abdomen; they may occur particularly in women who either have an urgent desire to bear children or are extremely anxious to avoid pregnancy. In general, treatment should first aim at removing the external physical or emotional factors causing the menstrual irregularity. Disturbed interaction of the sexual center, pituitary gland, ovary and uterus can sometimes be restored by hormone therapy. In such treatment the uterine mucous membrane is first "artificially" built up in response to hormones administered to the patient; then the hormone supply is temporarily stopped, resulting in shedding of the uterine lining accompanied by menstruation-like bleeding (cf. oral contraceptives, page 268).

Less frequently, amenorrhea is caused by true (physical) damage to the ovaries or the pituitary gland. Thus the ovary may be *underdeveloped* because of defective stimulus from the sexual center. This condition of defective development (*hypoplasia*) results in deficient production of hormone, and this in turn not only will cause menstrual irregularities, but will also give rise to other signs of sexual underdevelopment: small breasts, absent or sparse pubic and armpit hair, narrow vaginal orifice, small uterus leaning forward at an acute angle (Fig. 6-80). Hormone therapy in such cases of amenorrhea is likely to produce results only if adequately developed ovaries exist, which is ascertainable by obtaining a specimen for examination.

Amenorrhea or very infrequent menstruation may also occur in conjunction with obesity and masculinization. It is then due more particularly to malfunction of the ovaries, which are in a condition described as *polycystic*: they are much enlarged, gray, enclosed by a tough membrane of connective tissue and filled with cystlike modified follicles (Fig. 6-81). Treatment may consist in reducing the size of the ovary by surgical means.

The intensity, duration and regularity of menstrual bleeding can be plotted in a menstruation chart. Figure 6-82 (a–g) indicates the possible deviations from regular normal menstruation. Abnormally low frequency (b) with a cycle of more than 31 days is usually due to a slower rate of natural maturing of the follicle and is harmless. High frequency (c) with a cycle of less than 24 days may be due to a shortening of the proliferative phase, i.e., to premature ovulation. In such cases the basal temperature will, for example, rise already on the 10th day and not, as in the 28-day cycle, between the 14th and 16th day (Fig. 6-83). In other cases, ovulation does not take place earlier, but instead the luteal phase is shortened; the basal temperature will accordingly fall more rapidly than normal; there may be insufficient preparation of the endometrium (mucous membrane lining of the uterus) to make it suitable for the implantation of the fertilized ovum. A third form of too-frequent menstrual bleeding may be due to failure of the mature follicle to rupture; it remains in existence for a few days and then undergoes degeneration, which causes bleeding. This bleeding, which arises in the absence of ovulation, may be premature or may take place at the normal regular menstrual intervals. Such so-called *anovulatory cycles* (Fig. 6-84) mostly occur at the time of the menarche (onset of menstruation at puberty) and of the menopause (cessation of menstruation in middle age), also on resumption of menstrual activity after childbirth. Anovulatory cycles are evidently sterile. They are found to occur as a regular feature in about 10 percent of sterile women.

FIG. 6-82. Menstruation and its irregularities in the menstrual cycle.

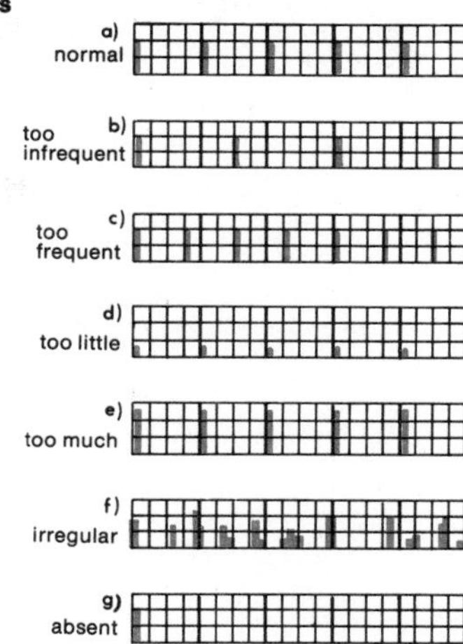

FIG. 6-83. Abnormally short menstrual cycle due to early ovulation (M = menstruation).

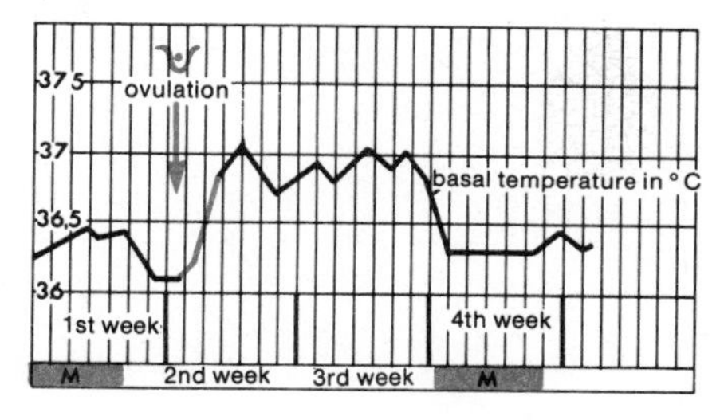

FIG. 6-84. Anovulatory cycle without ovulation, in this case with excessive menstrual frequency (M = menstruation).

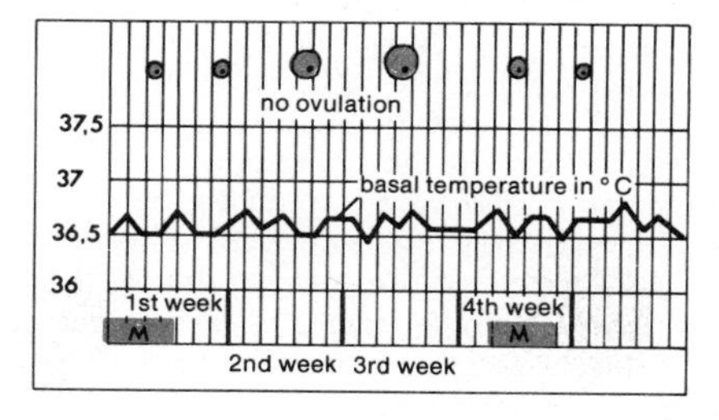

A particular kind of menstrual irregularity occurs when a mature follicle fails to rupture for several weeks. It develops into a cyst, 1–2 cm in size, which produces follicular hormones that continually stimulate the endome-

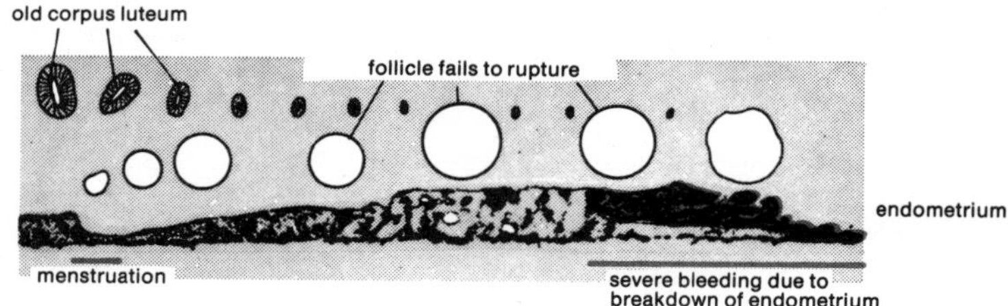

FIG. 6-85.
In the absence of ovulation the follicle stimulates excessive proliferation of the endometrium, resulting in eventual severe bleeding.

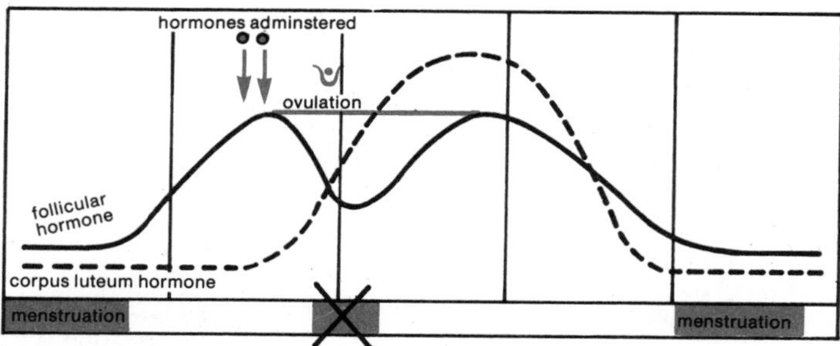

FIG. 6-86.
A fall in follicular hormone level may cause bleeding in between menstrual periods; this can be prevented by hormone treatment.

trium to prolific growth, without menstrual bleeding (Fig. 6-85). Eventually, this membrane becomes riddled with small sacs or cysts of retained uterine secretion, the condition being known as *glandular-cystic hyperplasia*. In the long run the follicular hormone is unable to maintain the membrane in this proliferated condition, and breakdown occurs, accompanied by prolonged and severe bleeding. Like anovulatory cycles, such bleeding is liable to occur particularly at the time of the menarche or at the menopause. It will generally require medical treatment; sometimes it can be stopped by hormone treatment. Otherwise curettage may be necessary.

Scanty menstrual flow (*oligomenorrhea*) is harmless. On the other hand, excessive bleeding (*menorrhagia*), even though it may be of normal duration, is usually a symptom of inflammation of the uterine lining, polyps (page 228) or fibroids (page 247). Treatment will depend on the cause of the bleeding. Actual menstruation is sometimes preceded by persistent slight bleeding, which may be due to hormonal causes or inadequate function of the corpus luteum. Excessively long menstrual bleeding and after-bleeding are often attributable to retarded shedding of the endometrium, this being due to a cessation of corpus luteum activity or sometimes to inflammation of the endometrium. In general, loss of blood between menstrual periods is called *spotting* (*metrorrhagia*). Some women feel pains at the time of ovulation; these may be accompanied by a slight flow of blood from the uterus, lasting a few hours to, in some cases, a couple of days. This abnormality may be due to a sharp temporary drop in the follicular hormone level and is harmless. A short trial course of hormone therapy can show whether this is indeed the cause (Fig. 6-86).

Painful Menstruation (Dysmenorrhea)

Most women feel somewhat unwell during menstruation—a certain, generally slight, temporary diminution of physical and mental energy. If there are symptoms, particularly the occurrence of abdominal pains, so severe as to cause serious discomfort, the condition is called *dysmenorrhea*. Secondary dysmenorrhea is said to exist if the menstrual periods at first were normal but have since become painful, possibly as a result of some pathological condition. Much commoner is primary dysmenorrhea, which starts from a girl's first period at puberty and for which there is no definitely assignable cause. The underlying factors that give rise to such dysmenorrhea are be-

FIG. 6-87.
Fibroids in the uterus.

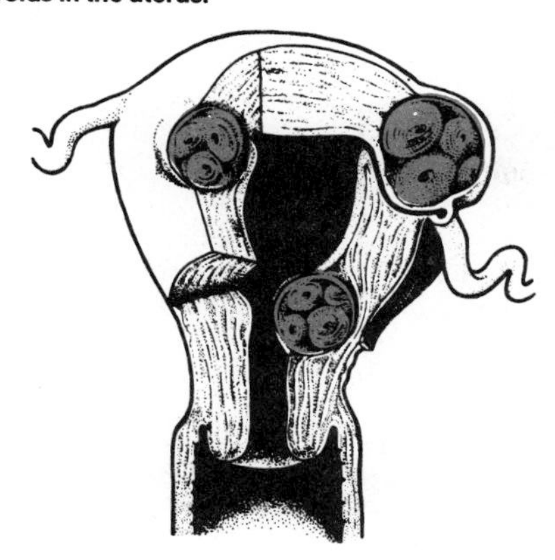

FIG. 6-88.
Polyp in the uterus.

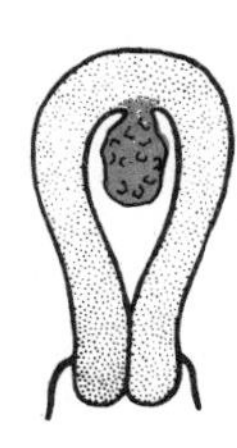

FIG. 6-89.
Scar formation in the cervical canal.

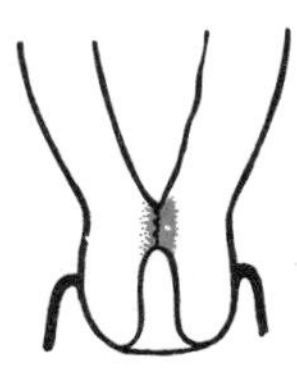

FIG. 6-90.
Deformities of the uterus.

FIG. 6-91.
Underdeveloped uterus.

FIG. 6-92.
Endometriosis.

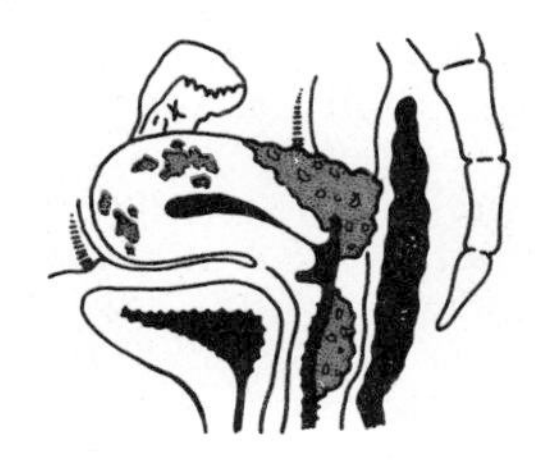

FIG. 6-93.
Painful menstruation due to psychic trauma.

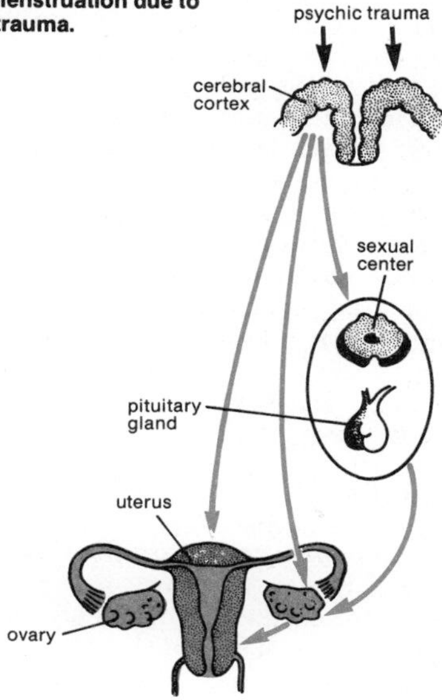

lieved to be hormonal and/or emotional. The general symptoms may include tiredness, headache, loss of appetite, palpitation, irritability, nausea, sometimes diarrhea and vomiting. Furthermore, there are pains in the lower abdomen and in the lumbar region, with a sensation of pressure. Some women have severely painful spasms and feel quite unfit for work at the start of a menstrual period.

Physical causes of dysmenorrhea include fibroids (Fig. 6-87, see also page 247), polyps (Fig. 6-88), inflammation of the uterine lining (endometritis), scar formation at the mouth of the uterus (Fig. 6-89), pelvic inflammatory disease in the vicinity of the uterus, deformities (Fig. 6-90) or underdevelopment of the uterus (Fig. 6-91). Another cause may be *endometriosis* (Fig. 6-92), which is due to endometrium growing in abnormal sites in the pelvis; for example, fragments of endometrial tissue may embed themselves in the muscle coat of the uterus or in the ovaries. These fragments undergo the cyclic changes of the endometrium, i.e., they swell and bleed, but as there is no outlet, severe pains may be caused for several days before a menstrual period. Hormone treatment may be helpful in dealing with this condition. Surgical removal of the offending tissue is sometimes possible.

Dysmenorrhea due to overexcitability of the autonomic (or vegetative) nervous system is not uncommon. Emotional shock, unfulfilled desire to have children, fear of sexual intercourse or of pregnancy, marital difficulties, grief and frustration are all factors that may cause dysmenorrhea, at any rate in women who are so predisposed (Fig. 6-93). Quite probably the symptoms are accompanied by a tendency to exaggerate or overdramatize the natural condition of being slightly indisposed during menstruation.

Preliminary treatment of dysmenorrhea may take the form of drugs to relax spasms and relieve pain. Further treatment will have to deal with the underlying physical or emotional causes. Hormone therapy is often effective. If the uterus is well developed, it may be helpful to suppress menstruation for some months by means of drugs ("the pill"). If the uterus is underdeveloped, simulated pregnancy induced by ovarian hormones over a period of a few weeks may permanently get rid of dysmenorrhea, just as a real pregnancy often does.

LUMBAR PAINS

The term *lumbar* refers to the loins ("small of the back"), i.e., the lower part of the back situated between the top of the pelvis and the lowest pair of ribs (Fig. 6-94). Along with discharge, menstrual irregularities and abdominal pains, pains in the lumbar region are among the commonest female troubles. Something like 20 percent of all abdominal disorders are accompanied by lumbar pains, which are therefore to be regarded as important symptoms. Sometimes, however, the pain may be due to emotional causes, the patient being physically healthy. Furthermore, some cases of lumbar pain are due to disorders of the supporting system that holds the female organs in position, particularly at the lower end of the spinal column.

Lumbar pains may accompany illnesses and disorders of the female genital organs (Fig. 6-94), e.g., dysmenorrhea (page 238) and other menstrual irregularities (page 235), fibroids of the uterus (page 247), inflammations affecting the ovaries and fallopian tubes (pages 225 ff), tumors of the ovaries, adhesion of the retroverted uterus (page 232), and prolapse of the uterus (see also page 231). Generally speaking, these are disorders that cause pain due to local pressure or tension. Irregularities in the menstrual cycle may be painful as a result of temporary overfilling of the uterus with blood. Inflammatory disorders of the intestines or the bladder, or blood clots, may also give rise to painful congestion of blood (Fig. 6-94e, f, and g). Lumbar pains are usually described as "dull." The probable reason that they are felt in this region, although they actually orginate in the pelvic cavity, is that the ligaments supporting the female reproductive organs have their attachments here, where they are amply provided with pain-sensitive nerve endings, so that even quite minor variations in pressure or tension are felt as pain.

In many cases, however, even the most thorough examination fails to reveal any tangible cause of lumbar pains. These are then considered to belong to the large but rather ill-defined category of psychosomatic functional ailments, which are believed often to have their origin in faulty tonicity (state of tension) of intestinal smooth muscle. They are therefore sometimes referred to as neurovegetative ailments or neurovegetatively engendered lumbar pains. They frequently occur in women around the age of 30 who are under nervous or emotional stress due to demands made on them by a young family, a career, etc. The state of stress, in turn, causes painful spasms of smooth muscle in the abdomen, with congestion of blood, sometimes accompanied by increased discharge from the cervical canal (Fig. 6-95), constipation or increased urination urge.

A third set of causes of lumbar pains are associated with various disturbances of the spinal column. Some of these are specifically female disorders, e.g., lack of calcium in bone after the menopause. Others are due to vertebral damage which, although not exclusive to the female sex, more frequently and more seriously affects women because of the greater strains to which their supporting and postural

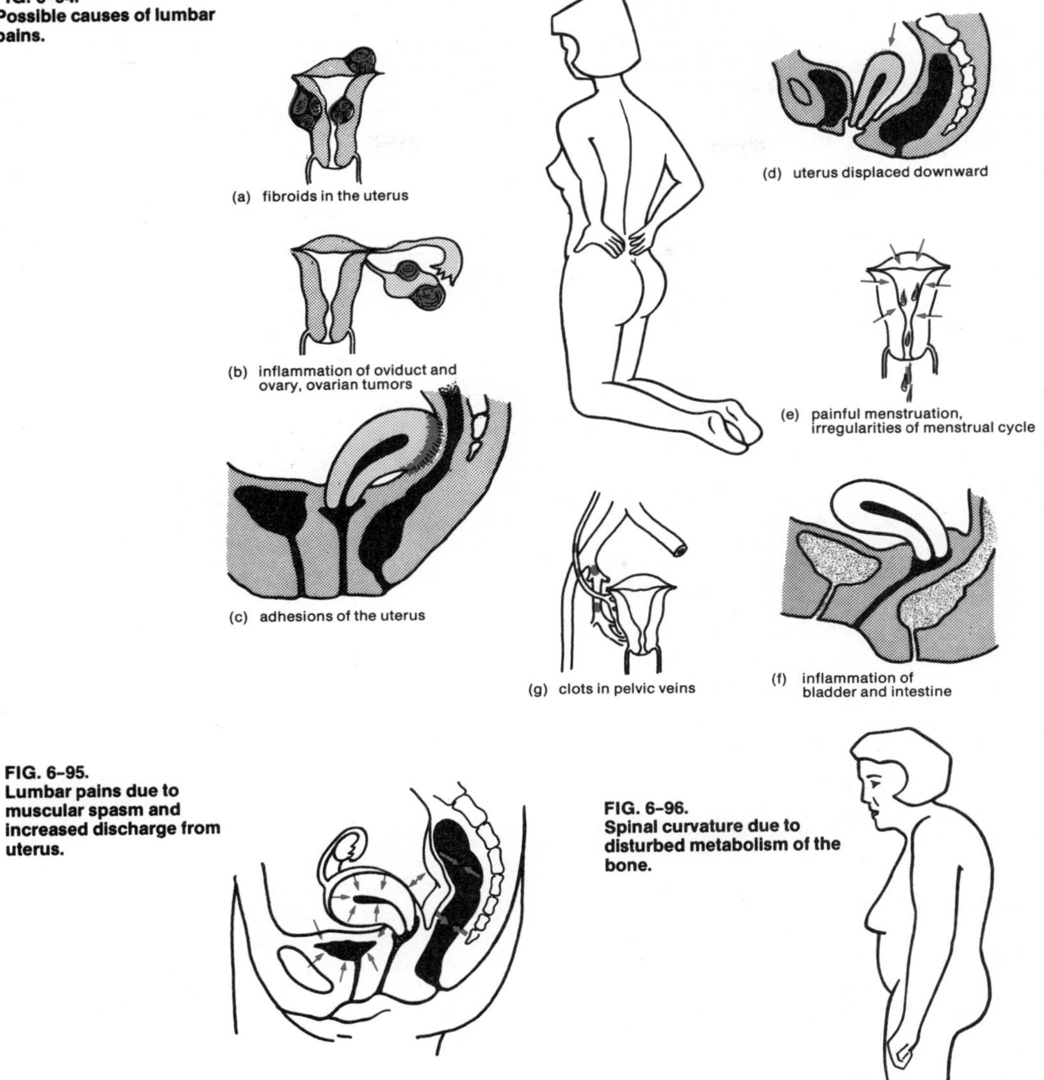

FIG. 6-94.
Possible causes of lumbar pains.

FIG. 6-95.
Lumbar pains due to muscular spasm and increased discharge from uterus.

FIG. 6-96.
Spinal curvature due to disturbed metabolism of the bone.

system (of abdominal muscles, ligaments and spinal column) is subjected. Manifestations may include stooped shoulders, pendulous belly, inflammatory and degenerative disorders of the spinal column, including slipped disc (see page 392). Loss of calcium from the bones is liable to occur about 5 to 10 years after the menopause or even at a later age, around 65. The symptoms are attacks of severe pain in the loins and back; in addition, the attendant softening of the bones may result in spinal curvature and compressive deformation of the spinal column (Fig. 6-96). Treatment with hormones, sometimes in combination with calcium preparations and vitamin D, can be effective in dealing with this condition.

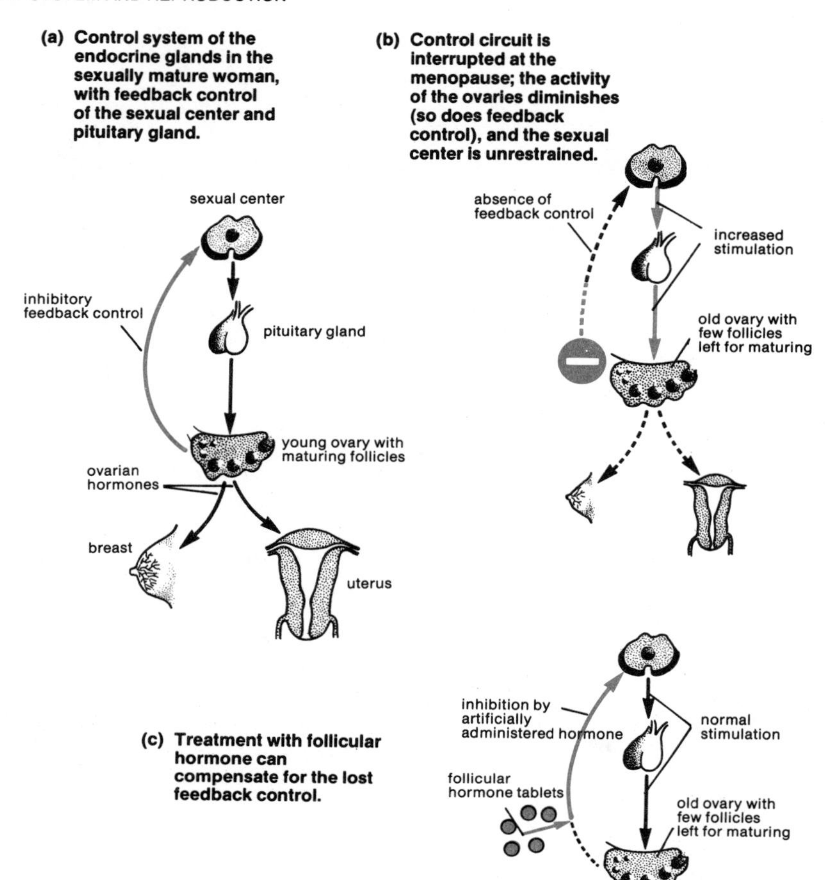

FIG. 6-97. **(a) Control system of the endocrine glands in the sexually mature woman, with feedback control of the sexual center and pituitary gland.** **(b) Control circuit is interrupted at the menopause; the activity of the ovaries diminishes (so does feedback control), and the sexual center is unrestrained.** **(c) Treatment with follicular hormone can compensate for the lost feedback control.**

MENOPAUSE (CLIMACTERIC)

For 35 years or more the life rhythm of the sexually mature woman is under the control of her ovaries, which stimulate the lining of the uterus to undergo monthly menstrual cycles. In middle age, generally around 50 (but with a wide range of individual variation, between 35 and 58 years of age), the activity of the ovaries declines and finally ceases, so that menstruation likewise ceases. This stage in a woman's life, marking the permanent cessation of menstrual activity, is called the *menopause*, *climacteric* or *change of life*. The onset of the menopause may be quite sudden or it may be gradual, involving a transitional stage in which the periods become less and less regular. Such irregularities are generally harmless, and there are relatively few other direct symptoms associated with the menopause, which, after all, is just a natural process of physical aging. Quite often, however, there are disagreeable associated symptoms of an emotional and/or psychosomatic character.

Ovarian activity is monitored and controlled by the sexual center (in the midbrain) and the pituitary gland. Under the influence of this monitoring system the ovaries secrete hormones that act upon the peripheral organs—e.g., the uterus and the mammary glands (breasts)—while, on the other hand, the ovaries will countercontrol and moderate the action of the pituitary gland on the feedback principle (Fig. 6-97a). By the age of 50 more than 90 percent of the approximately 400,000 follicles originally present in the ovaries will have been consumed. When this stage has been reached, the ovaries cease to function, i.e., there are no more ovulations nor menstrual periods, and the woman becomes sterile. Another consequence is that hormone production

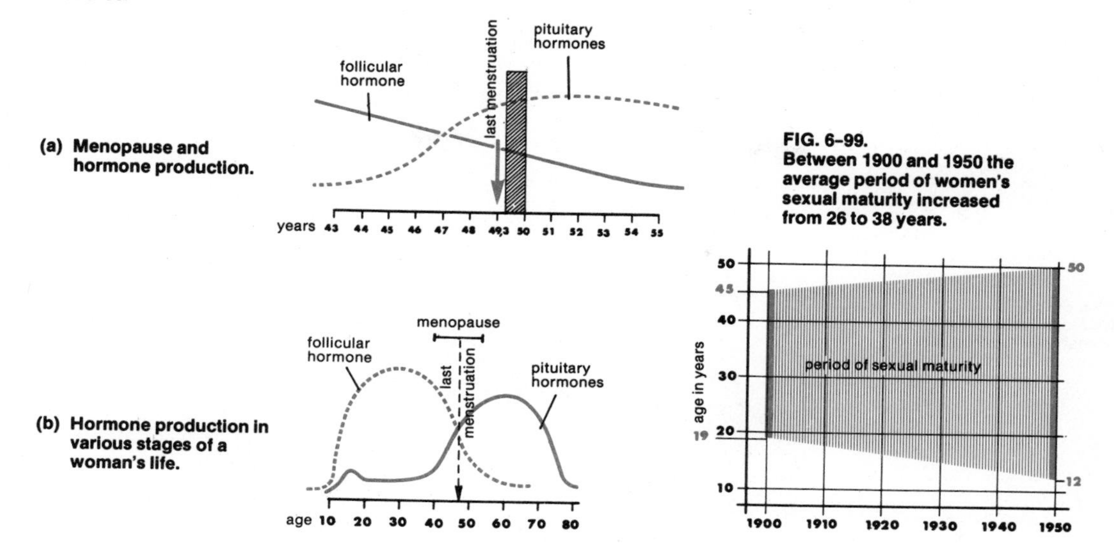

FIG. 6-98.

(a) Menopause and hormone production.

(b) Hormone production in various stages of a woman's life.

FIG. 6-99. Between 1900 and 1950 the average period of women's sexual maturity increased from 26 to 38 years.

by the ovaries ceases, so that the growth-promoting effect of the follicular hormones (estrogens) upon the reproductive organs eventually disappears. Besides, the moderating hormonal countercontrol exercised on the pituitary gland is now reduced (Fig. 6-97b). Figure 6-97c illustrates how medically prescribed hormone tablets can, under certain conditions, provide a substitute for this countercontrolling function. In the absence of the moderating or inhibiting action upon the pituitary gland, the activity of other hormone-secreting glands, such as the thyroid and adrenal glands, may be affected. It is because of this that the effects of the menopause may be widespread and manifest themselves in almost any part of the body. Symptoms may include hot flushes (hot flashes), a feeling of weakness or fatigue, depression, irritability, etc. (see next section). In most women, however, the menopause does not cause any serious amount of discomfort or unusual symptoms other than the cessation of menstruation.

Figure 6-98a illustrates female sex hormone production as related to aging. It shows, for example, that the secretion of follicular hormones gradually diminishes years before menstruation finally ceases and that this curve does not in fact reach its minimum until about seven years after such cessation. In Figure 6-98b essentially the same information is given, but drawn to a different scale and extending over a wider range comprising the whole lifetime. This diagram clearly shows that the menopause is accompanied by, and associated with, a major readjustment of the hormonal balance over a period of some 12 or 13 years.

At the beginning of the present century the menopause, in western Europe, occurred at the age of 44 on average. Now the average age is near 50. As already mentioned, considerable variations from one woman to another may occur, but in about 75 percent of women the menopause occurs between 45 and 55. In comparison with the year 1900 not only has there been a general shift of menopause to an older age, but the onset of menstruation (menarche) has moved backward, i.e., it now occurs in younger girls than it used to. This means that the span of female fertility, over the past 75 years or so, has increased from an average of 26 years to its present average of about 38 years (Fig. 6-99). Of course, not all sexual functions stop so abruptly as menstruation does at the menopause. For example, the capacity for orgasm may continue for many years beyond the menopause; on the other hand, fertility definitely ceases with the last menstrual cycle and is, in most women, greatly reduced or virtually absent for some years before that. The probability of conception is, in broad terms, about 30 percent at the age of 30, about 11 percent at 35, and only 3 percent at 40. These, of course, are only very approximate average figures.

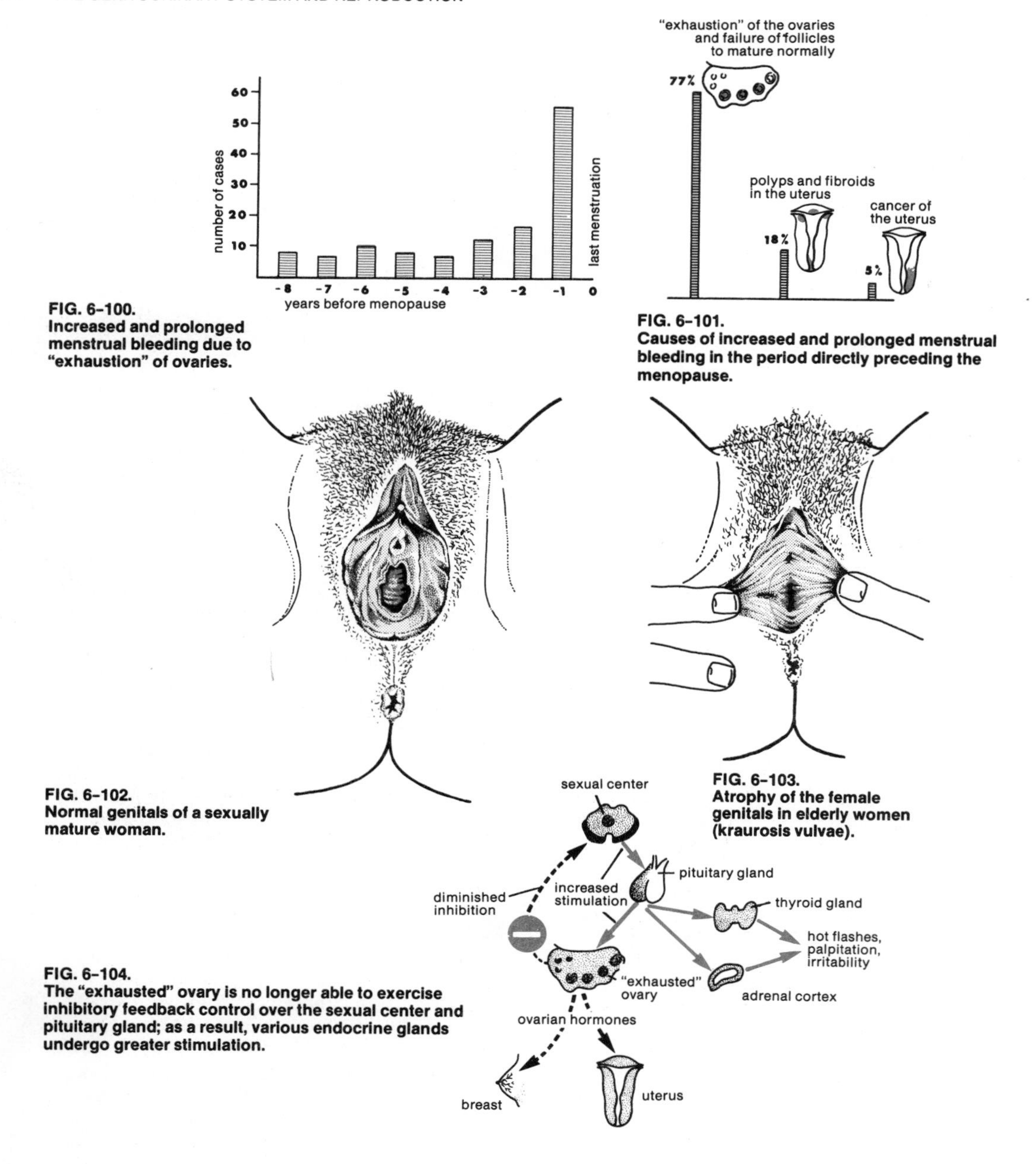

FIG. 6-100.
Increased and prolonged menstrual bleeding due to "exhaustion" of ovaries.

FIG. 6-101.
Causes of increased and prolonged menstrual bleeding in the period directly preceding the menopause.

FIG. 6-102.
Normal genitals of a sexually mature woman.

FIG. 6-103.
Atrophy of the female genitals in elderly women (kraurosis vulvae).

FIG. 6-104.
The "exhausted" ovary is no longer able to exercise inhibitory feedback control over the sexual center and pituitary gland; as a result, various endocrine glands undergo greater stimulation.

Disorders of the Menopause

The menopause or climacteric marks the transition period during which the activity of the ovaries gradually diminishes to a low level and the process of ovulation comes to an end. It is a natural event, not an illness, and the symptoms, if any, associated with it tend usually to be of a subjective and emotional rather than an objective and physical character. In general, the discomforts suffered during the menopause are likely to be fewer and milder as the woman mentally adjusts herself more deliberately, willingly and cheerfully to her "change of life." Mental attitude and a certain amount of self-discipline are therefore important factors in this respect.

Nevertheless, the menopausal years—as indeed any other stage in life—are not without their own particular risks or discomforts. Cessation of the menstrual cycles is sometimes preceded by irregularities of bleeding which, together with diminishing fertility, herald the decline in ovarian activity.

Increased or reduced duration of the cycles is usually harmless. In some instances, however, the follicles in the ovary fail to mature and, instead, develop into enlarged vesicles that continue to be hormonally active and stimulate the lining of the uterus to more prolific growth, attended by prolonged bleeding.

Such increased discharge of blood is particularly frequent in the last year preceding the cessation of menstruation (Fig. 6-100). Increased bleeding may, alternatively, be due to growths in the uterus (fibroids, polyps, cancer) (Fig. 6-101); if these possible causes have been eliminated by curettage (scraping out the uterine cavity) and the bleeding nevertheless continues, hormone therapy may be effective in stopping it.

Bleeding may occur even after the menopause. In this context it is of some importance to decide, in retrospect, when the last true menstrual flow took place. As a rule, this is reckoned to be that menstrual flow which was not followed by any subsequent discharge of blood for one whole year. If bleeding recurs beyond a year after the menopause, medical advice should be sought, as it may be a symptom of cancer of the uterus (see pages 499 ff).

The decrease in follicular hormone production is reflected in a gradual retrogressive development of the uterus, vagina, vulva and breasts. If this process goes too far, it can, in old women, result in inflammatory contraction of the vagina, itching of the vulva or, in extreme cases, chronic atrophy of the vulva (Figs. 6-102 and 6-103).

Apart from irregularities of bleeding, other transitional phenomena of the menopause occur in some women. They are due to the body's adjustment to the dwindling production of hormones by the ovaries. Once this adjustment to a lower hormonal level has been achieved, the symptoms cease. This transitional phase may last about two years, sometimes much longer, even up to ten years. The symptoms may include hot flushes (hot flashes) accompanied by perspiration and followed by chilliness, irritability, fatigue, insomnia, palpitation, formication (a sensation as of ants crawling on the body), disinclination to work, anxiety states, depression. It seems likely that not all these symptoms, if they occur at all, are directly and solely attributable to the decrease in ovarian hormone production. An additional factor is that the moderating or inhibitory action of these hormones on the sexual center is greatly reduced. As a result, the now more or less uninhibited pituitary gland will stimulate the thyroid and the adrenal cortex to excessive activity (Fig. 6-104). The psychic complaints are largely the result of failure to cope adequately (in an emotional way) with the physical consequences of menopause. Medical advice can be helpful in bringing about a readjustment of mental attitude, and treatment with ovarian hormones may be beneficial.

INFERTILITY IN WOMEN

Infertility or sterility can be said to exist if, despite normal sexual intercourse and a desire for children, no pregnancy occurs within two or three years. In 30–40 percent of such cases sterility is due to the man (see page 217); in 40–60 percent of cases the woman is sterile and in the rest no cause can be determined. Female sterility is associated with three possible sets of causes. First, there may be anatomic or functional abnormalities of the reproductive organs; second, constitutional diseases may cause sterility; a third group comprises psychic/emotional factors.

The prerequisite condition for conception leading to pregnancy is, of course, that ovulation duly take place. The ovary not only releases the mature ovum but normally also ensures that ovulation is optimally geared to, and synchronized with, certain cyclic changes in the cervical canal and mucous membrane lining of the uterus, so as to create the right conditions for the fertilized ovum to become embedded in the lining (implantation). Because of this synchronization of the functions within the female reproductive system it is important that the woman's fertile days (Fig. 6-105) are indeed utilized. They can most reliably be determined from the basal temperature chart as used for contraception (see also pages

FIG. 6-105.
The woman's five fertile days.

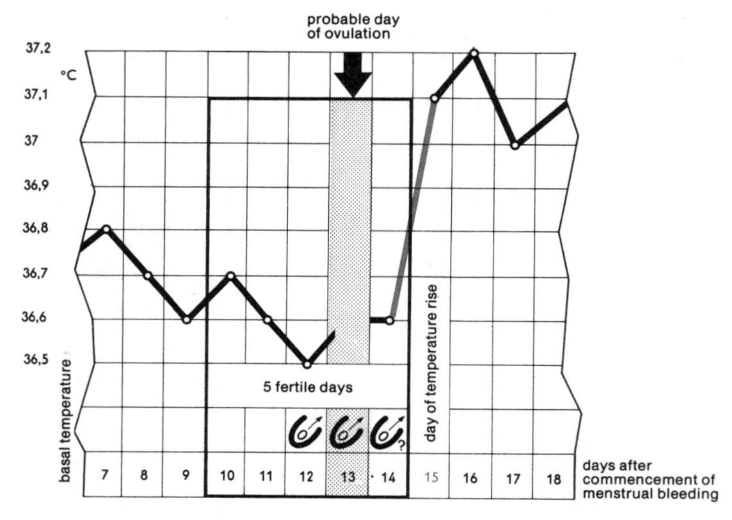

FIG. 6-106.
Deformities of the uterus (mostly a transitional form of double uterus).

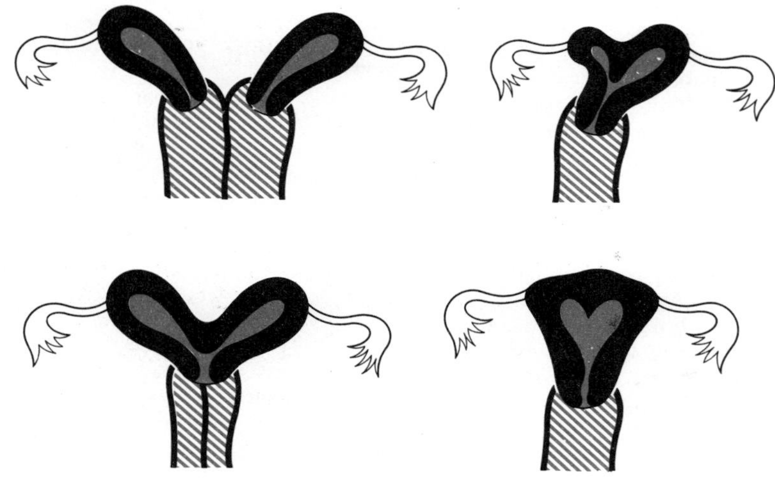

265 ff). Basal temperature records and examination of mucus and vaginal smear specimens can additionally reveal whether ovulation indeed takes place at all and, if so, whether it is properly synchronized with the changes that occur in the uterine lining. Absence of ovulation or lack of coordination with other processes of the reproductive cycle is associated mostly with hormonal malfunction or deficiency of the ovaries.

Disturbed ovarian function is probably the commonest cause of sterility in women. It is sometimes traceable to faulty functioning of the sexual center in the brain; in other cases it may be due to underdevelopment or vesicle-like degeneration of the ovaries (see preceding section). In women in whom the reproductive organs have remained immature—a condition characterized by narrow vagina, small breasts, sparse pubic hair and sparse hair under the armpits—too little follicular hormone is formed. Another indication of deficient ovarian activity is that the uterus is small and bent forward at an acute angle (Fig. 6-80), though admittedly this condition in itself is seldom a cause of sterility. An abnormally short menstrual cycle may be due to weak functioning of the corpus luteum, which deteriorates prematurely and thereby initiates menstrual bleeding before the fertilized ovum has had time to become embedded in the lining of the uterus. The earlier occurrence of ovulation associated with a short cycle may sometimes be accompanied by abnormally prolonged menstrual bleeding. In such cases, coitus can be effective only during a short time at the very end of the bleeding or directly after its cessation. A study of the basal temperature chart can, in such cases too, help to clarify the pattern of cyclic activity. In cases where the ovaries show no discernible anatomic abnormality, but nevertheless fail to perform their functions adequately, hormone therapy is indicated. This can enable pregnancy to be achieved in something like 40–50 percent of women so treated. It is especially promising in dealing with sterility in younger women.

Sometimes, though rarely, sterility may be due to the *uterus*, e.g., if it is deformed (Fig. 6-106) or contains numerous fibroids or perhaps has had its lining damaged by too-drastic curettage. Fibroids (nonmalignant fibrous tumors) can be removed surgically, and in some cases this can be done while leaving the uterus intact and capable of childbearing.

Foremost among the anatomic or functional changes of the reproductive organs likely to cause sterility are abnormal changes affecting the *fallopian tubes* (oviducts), probably in 20–30 percent of all cases. Most frequently there is blockage of the tubes. This may have been caused by a previous miscarriage or by inflammatory processes that have meanwhile healed, e.g., tuberculosis, gonorrhea or unspecific suppurative conditions. It is estimated that between 10 and 20 percent of women remain sterile after a miscarriage. Less frequently encountered causes are underdevelopment or faulty development of the tubes and various noninflammatory changes of the mucous membrane. Examination may reveal a thickening of the tubes, this being an indication of past inflammation. Treatment will depend on whether the abdominal cavity is quite free from inflammation at the time of examination and also on the result of an examination of the man's seminal fluid (see page 217). If both partners are found to be apparently normal in this respect, the next stage will consist in testing and perhaps clearing the fallopian tubes by blowing gas through them and/or examining them by radiography.

A stream of gas is passed through the fallopian tubes to test whether they are open or blocked (tubal insufflation). The gas is introduced through the cervix into the uterus and should pass through the tubes to emerge into the abdominal cavity. The gas pressure required and certain characteristic bubbling noises provide indications of whether the tubes are open or blocked. Gas pressure and gas flow rate can be automatically monitored and recorded. In some cases the insufflation itself will also constitute the corrective treatment by causing internal adhesions in the tubes to separate, thus unblocking them. Coitus following soon after insufflation treatment is likely to result in pregnancy in some 25–30 percent of cases. For this reason, insufflation will generally be carried out shortly before the most favorable time for fertilization, i.e., 10–14 days after the start of the menstrual cycle (commencement of bleeding) or 3–4 days before ovulation.

Like tubal insufflation, radiographic examination of the fallopian tubes, i.e., by means of x-rays, is not suitable in cases where significant traces of inflammatory disorders are still present in the female reproductive system, as they could be reactivated by these techniques and possibly spread into the abdominal cavity. Radiographic examination likewise starts at the cervix of the uterus. Instead of gas, a radiopaque substance (contrast medium) under pressure is blown in, which produces in the x-ray photograph a shadow that indicates how far the tube is unobstructed from its lower end. If there is no obstruction, the contrast medium will make its way to the abdominal cavity (Fig. 6-107a). Figure 6-107b represents a case where the tubes are blocked, while in Figure 6-107c the tubes are blocked at the outlet into the uterus, so that only the latter shows up in the photograph. Radiography involves exposing the ovaries to a certain potentially harmful dose of radiation; besides, entry of the contrast medium into the abdominal cavity is undesirable. For these reasons this method of examination is generally confined to those cases where it has already been decided to have recourse to surgery for the removal of obstructions.

Another technique for examining the female internal reproductive organs is endoscopy. A special optical tube is introduced—through the abdominal wall or the rear wall of the vagina—

FIG. 6–107.

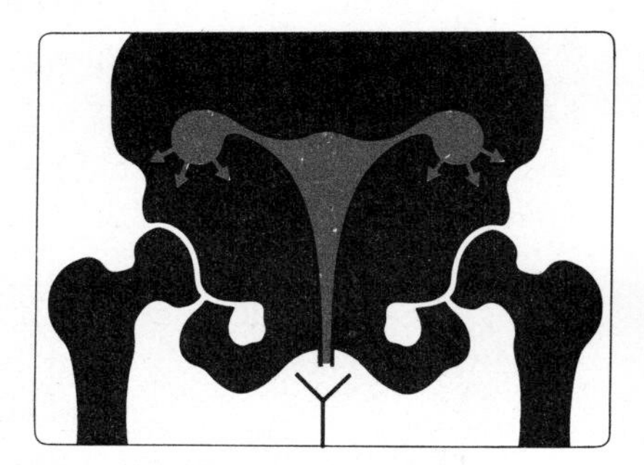

(a) Oviducts are not obstructed.

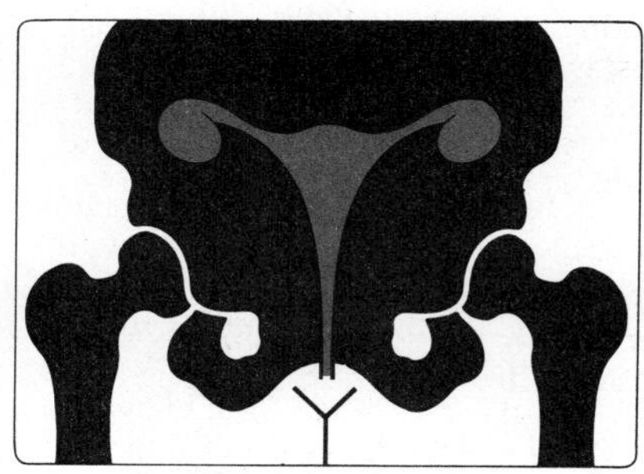

(b) Oviducts are blocked at their far ends.

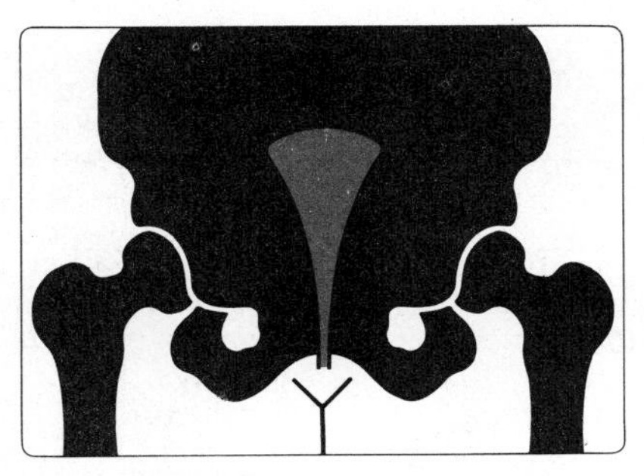

(c) Oviducts are blocked at their outlets.

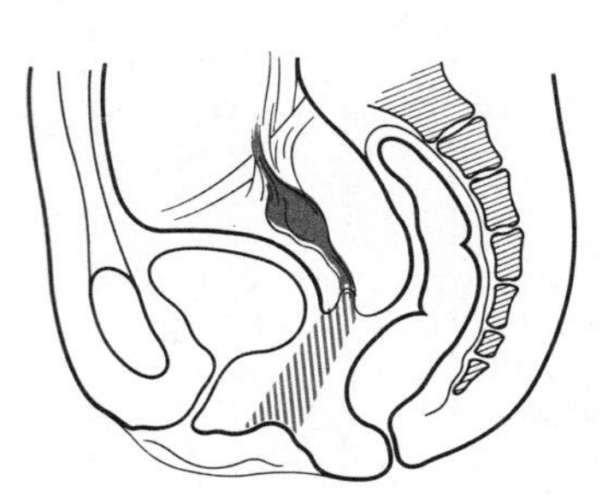

FIG. 6–108.
Congenital absence of vagina.

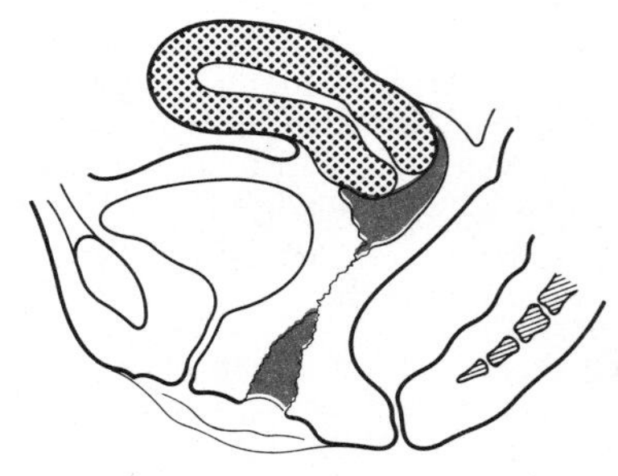

FIG. 6–109.
Coalescence of vaginal walls.

into the air-filled abdominal cavity, enabling the doctor to examine the internal organs by direct vision. A colored marker dye injected into the uterine cavity will be seen to flow into the abdominal cavity from the mouth of the fallopian tube if the latter is free from obstruction. In particular, with this method, residual traces of inflammation such as obstructions and adhesions can be directly detected, enabling the prospects of successful surgical intervention to be more accurately assessed.

Persistent blockage of fallopian tubes due to scar tissue formation can be corrected only by means of an operation. This will consist in removing the obstruction revealed by the examination. If the tubes have external adhesions only, surgery is likely to be successful, in the sense of making the woman fit for childbearing, in a substantial proportion of cases. If the tube is blocked by scarring near its inlet end, i.e., at the abdominal cavity, plastic surgery to repair the damaged inlet aperture may be successful, but only in about 5 percent of cases. The success rate of operations to reconnect a tube that is blocked nearer the uterus is not much better. With advancing surgical techniques, however, the percentage of successful operations is steadily rising.

Sterility may be due to abnormalities or diseases not only of the fallopian tubes but also of other parts of the reproductive system. Thus the *vagina* may be absent altogether (Fig. 6-108), blocked by coalescence of its walls (Fig. 6-109) or severely narrowed. Much more frequent causes of sterility are, however, inflammatory conditions of the vagina; these greatly increase the paralyzing action that the acid vaginal secretion exercises upon the sperms even under normal conditions. The presence of microorganisms such as *Trichomonas* (flagellate parasitic protozoa) and *Candida* (a fungus, see page 150) in the vagina is also harmful to sperms. If there is also a purulent discharge, the sperms will moreover have to run the gauntlet of phagocytic white blood cells.

Another cause of female sterility may lie in a defect of the cervical canal, e.g., as a result of local inflammations or lacerations, so that the canal cannot properly fulfill its function as a receptacle for sperms. It may also occur that the necessary cyclic changes of the plug of mucus (see page 234), which is normally impermeable to sperms, fail to take place. Microscopic examination of mucus from the cervical canal between the 10th and the 13th day of the menstrual cycle, 2–10 hours after coitus, will normally reveal plentiful and energetically motile sperms. If this is found not to be so, sterility due to a defective condition—harmful to sperms—of the cervical mucus can be suspected. Failure of the sperms to penetrate into the mucus will indicate that the latter is "wrong" for them. Additional confirmation can be obtained by cross-testing, i.e., with a drop of undoubtedly fertile seminal fluid or with mucus from the cervix of a woman of undoubted fertility. From a routine study of vaginal smear specimens, cervical mucus and the basal temperature, together with regular examination of the external os (the inlet to the cervical canal from the vagina), it can finally be ascertained whether there is a disturbance of the menstrual cycle due to causes arising from the ovaries (see pages 237 ff). In such cases treatment with follicular hormone can be helpful.

Constitutional diseases which may result in sterility are particularly those affecting the thyroid gland, the adrenal glands, and diabetes.

Psychic/emotional causes such as feelings of inferiority of the woman who believes herself to be sterile, matrimonial difficulties, anxiety and worry can aggravate an existing tendency to sterility or may indeed cause sterility. In the majority of such cases disturbed ovarian function due to faulty control by the central nervous system is the underlying cause; more rarely there may be spasms of, for example, the fallopian tubes, as are believed to occur especially in women in whom an obsessively ardent desire for children has the self-defeating consequence of producing a kind of sexual and emotional blockage. Whether, and if so, to what extent, female frigidity can cause sterility is not entirely certain. The absence of libido (instinctive sexual drive) and failure to reach orgasm can undoubtedly have an indirect effect, e.g., in lower frequency of intercourse and possibly in matrimonial difficulties which may, through the sexual center, in some way affect the function of the ovaries. It is supposed by some that the female orgasm assists and accelerates the progress of the sperms by uterine contractions. The importance of orgasm in this respect is not an established fact, however, and should not be overrated. Vaginal cramps or spasms may originally develop in response to physical pain, e.g., as felt by the woman when she has sexual intercourse for the first time, and may continue recurrently as a psychic/emotional condition that may form an obstacle to conception. A confidential talk with the doctor can be helpful in such circumstances. Many women feel relieved and relaxed on being informed that absence of orgasm probably has little direct effect on fertility. Hormone treatment may be beneficial in some cases of frigidity.

THE SEXUAL RESPONSE CYCLE

The physiological and emotional processes associated with sexual activity and coitus cannot be conveniently schematized, but a subdivision into a number of more or less well-defined stages or phases can nevertheless be

FIG. 6-110.
Sexual response cycle of the male and the female.

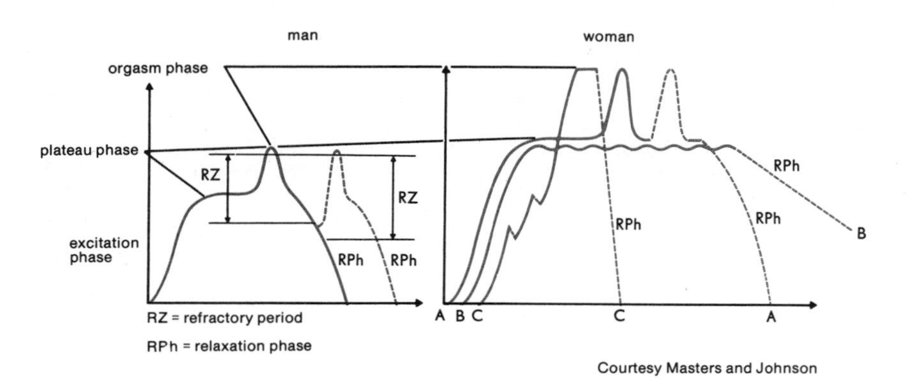

helpful in understanding these processes. First comes what may be termed the *excitation phase*, which is brought about by physical or mental stimuli. The intensiveness of this phase will usually depend at least to some extent on the responsiveness of the sexual partner in any sexual activity where two people are involved. Then comes the *plateau phase*, in which the sexual tension builds up to the now following *orgasm phase*. The latter proceeds involuntarily, i.e., independently of the conscious will, its duration usually being no more than a few seconds. Although the subjective sensations associated with orgasm are confined to the pelvic region, in some respects the entire organism is involved. Apart from artificial insemination, the male orgasm is indispensable in achieving fertilization; not so the female orgasm. After orgasm comes the *relaxation phase*, in which sexual stimulation dies away.

Of course, many responses are determined by sexual stimuli arising from the physical differences between the sexes. Recent research has revealed, however, that the responses to sexual stimuli proceed in the two sexes with a degree of similarity not previously suspected. Fundamental differences between men and women do nevertheless exist with regard to the intensity and duration of the responses. Thus it is known that the stimulation or excitation phase in the male develops more rapidly than in the female and also passes away more rapidly. Also, in the male the whole response pattern is of a more stereotyped kind, with only minor variations between one individual and another. After orgasm a man is unresponsive to any further sexual stimuli for a certain length of time (*refractory period*), whereas a woman may be capable of repeated stimulation and sexual response cycles in close succession. Besides, in the woman the relaxation phase lasts longer than in the man (see Fig. 6-110, the curves A, B and C represent various response patterns).

Male Sexual Response Cycle

The man's physical response to sexual stimulation involves all parts of his body (Fig. 6-111), in some extreme cases manifesting itself in involuntary convulsions of the hands and feet and quivering of the whole body.

Excitation phase: Erection of the penis occurs, so that it increases in length and volume by engorgement of the corpora cavernosa with blood (see page 215). At the same time the testes are raised as a result of tightening and thickening of the wall of the scrotum (Fig. 6-112a). Erection is initiated by the erection reflex center in the sacral region of the spinal cord through parasympathetic nerves and is not under the direct control of the will (Fig. 6-111). Increase in size of the penis is accompanied by a rise in blood pressure and temperature. The erection reflex may be initiated by stimulation of the nerve endings situated in the glans (head) of the penis; these act upon the erection center through the nervus pudendus. The latter is one of the spinal nerves and also produces the feeling of sexual lust, while parasympathetic nerves from the erection center control the process of erection. Stimulation of nerve endings in the so-called erogenous zones can also bring about erection of the penis. Although in the male these zones are confined mainly to the sexual organs themselves, a man's lips, hands, hair, arms and legs may

FIG. 6-111.
Nerve pathways of male sexual response.

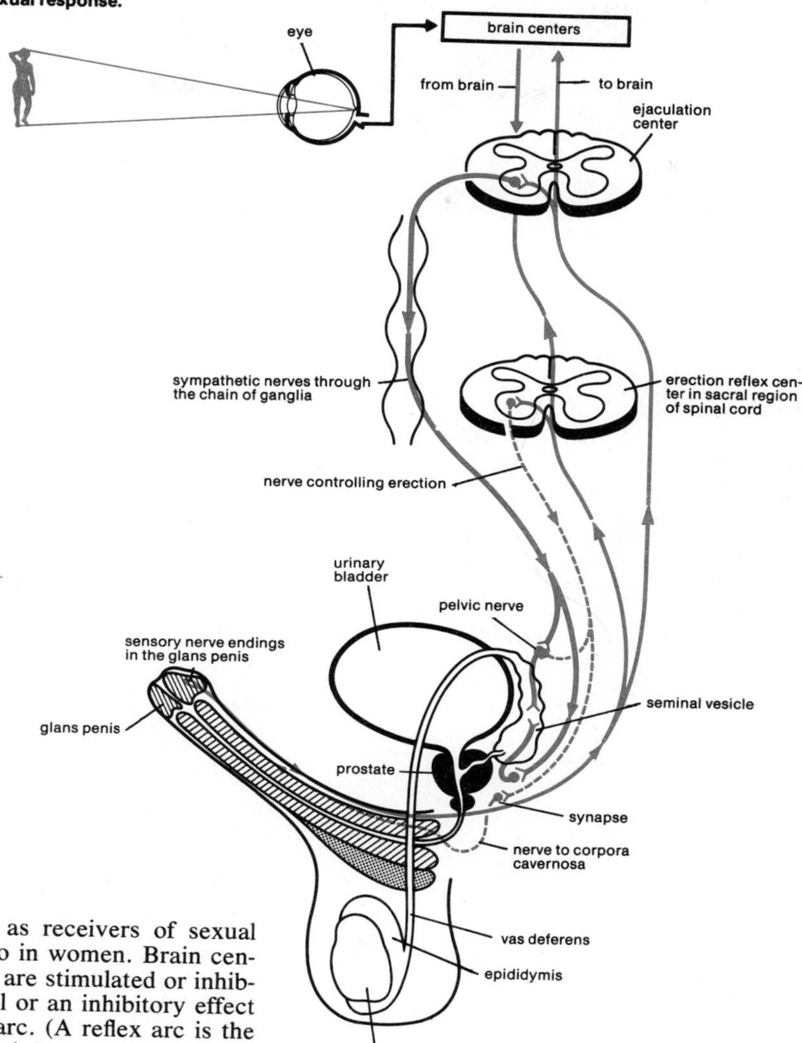

nevertheless also act as receivers of sexual stimuli, just as they do in women. Brain centers which themselves are stimulated or inhibited can have a helpful or an inhibitory effect on the erectile reflex arc. (A reflex arc is the neural pathway or circuit between the point of stimulation and the responding organ in a reflex action.) The brain centers may be stimulated, e.g., by the sight of a woman or by erotic photographs or mental images, or they may be inhibited, e.g., by sexual taboos or fear of sexual inadequacy.

Plateau phase: The glans of the penis undergoes a further increase in volume. The testes also swell and are further raised. Interruption of this sequence may cause pain in the testes. At the same time, Cowper's glands discharge their mucous secretion into the end part of the urethra, thus preparing it for the ejaculation of semen and neutralizing any traces of urine harmful to sperms (Fig. 6-112b). In this phase there is usually also sudden reddening of the skin (sex flushes).

Orgasm phase: This directly follows the plateau phase and is entirely outside the control of the will. Ejaculation takes place: the semen is first delivered in rhythmic spurts into the urethra by muscular action occurring particularly in the vas deferens. As a result of the mechanical stimulus exercised by the arrival

FIG. 6-112.
Phases of male genital response.

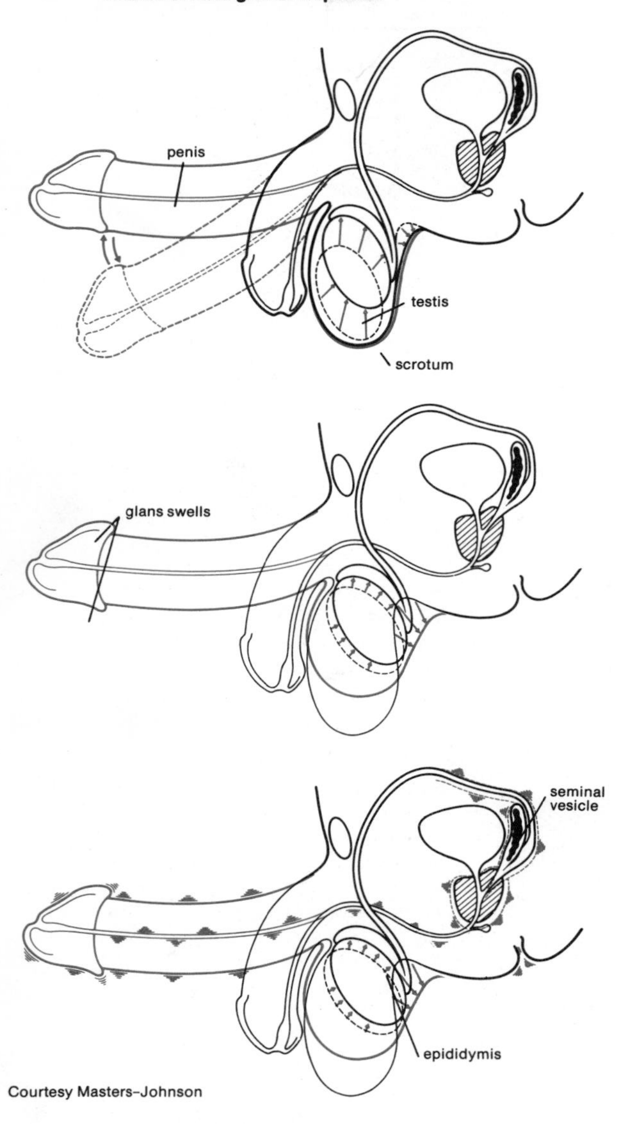

Courtesy Masters-Johnson

of the semen in the urethra, another muscle, the musculus bulbocavernosus, is stimulated to rhythmic contractions. Acting in conjunction with other striped muscle of the penis, it discharges the semen forcefully through the now fully straightened and distended urethra into the vagina as far as the entrance of the uterus. At the same time, the inner sphincter of the bladder is closed, so that flowback of semen into the bladder is prevented. During ejaculation, first the secretion of the prostate gland, then the sperm-containing fluid from the epididymis and finally the secretion of the seminal vesicles are discharged (Fig. 6-112c). Ejaculation is assisted by the action of the cremaster muscle, which is one of the muscles suspending and enveloping the testes and spermatic cord. This muscle further raises the testes and thereby brings them into a more favorable position for discharging the seminal fluid.

Relaxation phase: In this phase the penis loses its turgidity, the testes go back to their normal (lower) position and become smaller, as does the scrotum itself. The refractory period generally falls within this phase.

Female Sexual Response Cycle

The woman's response to sexual stimulation and excitation involves her whole body to a greater extent than does the man's response. Orgasm in the female is likewise a psychophysical experience in which the emotional aspect and the extensive involvement of the body are of major importance. Corresponding to the wide spectrum of female sexual response, the woman's erogenous zones are not confined essentially to the pelvic region, as in the man, but are distributed over all parts of her body (Fig. 6-113). Those parts situated farthest from the sexual organs contribute least to arousing her sexually, but first have to be stimulated in order to enable those erogenous zones situated nearer the sexual organs in their turn to become responsive to stimuli. The palm of the hand not only imparts but also receives stimuli; more responsive are the bends of the elbows (walking arm in arm) and the outsides of the thighs (walking close together). The inside of the upper arm and especially also the borderline of the hair low down on the neck can, when caressingly stimulated, arouse the woman's whole body to sexual excitement. The lobe of the ear itself is usually less sensitive than the little groove behind the ear. Most receptive to stimuli are the breasts and, of course, the sexual organs themselves. Premature erotic stimulation of the latter, however, is often experienced as disagreeable by the woman. The waist, hips, loins and buttocks may require stronger stimuli. In a woman already sexually aroused, the insides of the thighs are especially receptive to touching and stroking.

Excitation phase: The woman's first re-

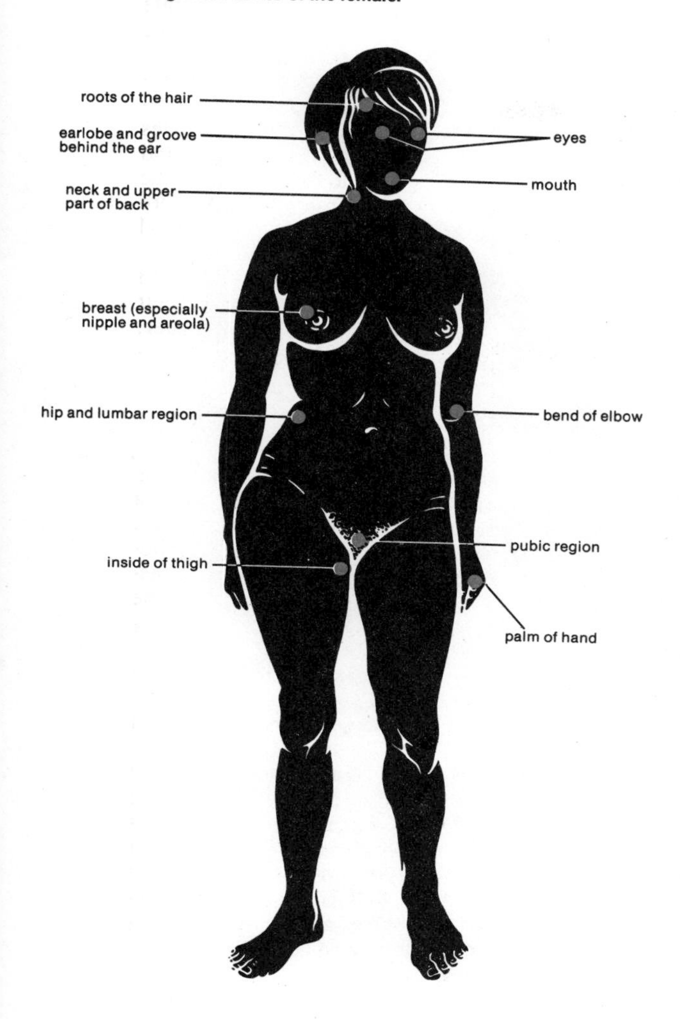

FIG. 6-113.
Erogenous zones of the female.

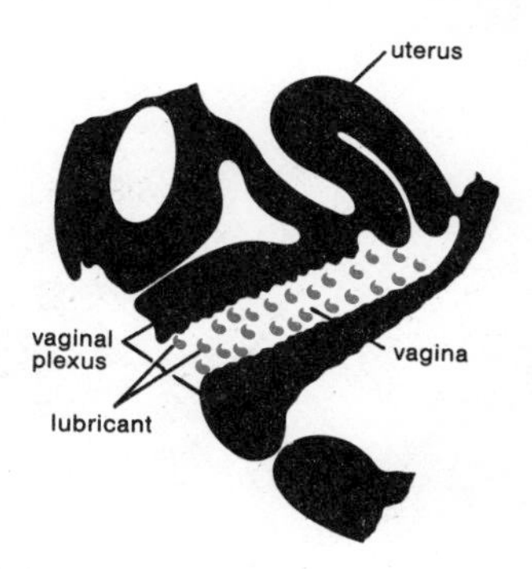

FIG. 6-114.
Excitation phase.

sponse to erotic stimulation is the discharge of a mucous secretion that lubricates the interior of the vagina (Fig. 6-114). The secretion serves also to modify the acidity of the vagina and thus make it more suitable for the survival of the sperms, at least during their progress to the mouth of the cervix (see page 252). At the same time, the rear part of the vagina becomes enlarged as a result of upward and rearward withdrawal of the body and cervix of the uterus (Fig. 6-115a). Stronger stimulation causes erectile enlargement of the clitoris and swelling of the labia. These phenomena correspond in principle to erection of the penis in the male. The nerve center for the clitoral erection and for increased secretion of mucus is located in the sacral region of the spinal cord. It receives its excitation from sensory stimuli arriving from nerve endings in the clitoris and labia through the nervus pudendus, and then, acting through a parasympathetic nerve (nervus erigens), it brings about the responses described. In the female this reflex arc, however, can more readily be affected and diverted by subsidiary psychic or physical stimuli than in the male.

Plateau phase: In this phase erection of the clitoris fully develops, while the enlargement of the rear of the vagina increases (Fig. 6-115b). The labia minora and the front third of the vaginal wall become further distended by an inflow of venous blood, in preparation for the orgasm.

Orgasm phase: The vaginal enlargement is most pronounced during this phase and forms a receptacle for the seminal fluid. In a woman who has borne no children (nullipara) the semen-retaining function is assisted by the generally high, firm "dam" that is formed by the perineum (Fig. 6-117). The perineum is the bridge of muscle and fibrous tissue between the anus and the genitals; it is liable to be injured or weakened during childbirth. In women who have borne children the seminal receptacle is generally shallower. This function, whereby a pool of semen is collected and retained near the inlet to the uterus, is believed

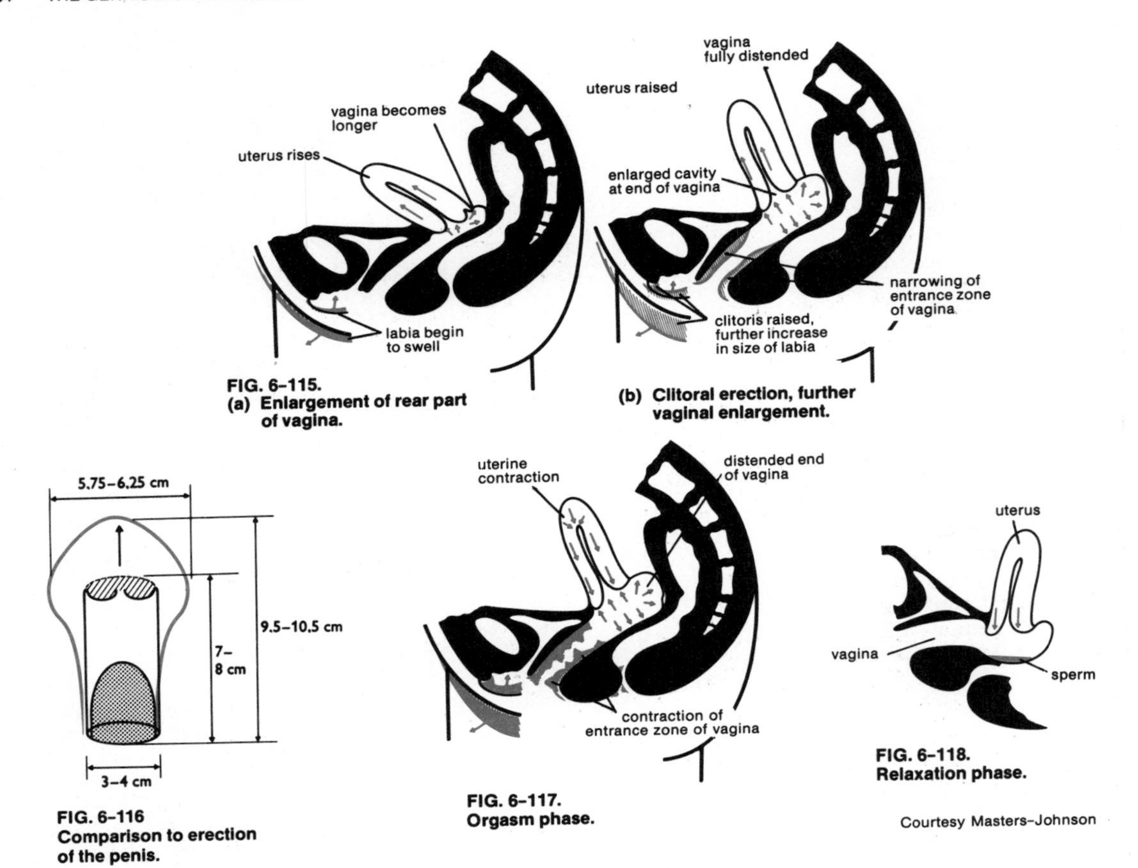

FIG. 6–115.
(a) Enlargement of rear part of vagina.
(b) Clitoral erection, further vaginal enlargement.

FIG. 6–116
Comparison to erection of the penis.

FIG. 6–117.
Orgasm phase.

FIG. 6–118.
Relaxation phase.

to be an important one. Excessive pelvic movements or getting up soon after intercourse will reduce the likelihood of pregnancy. While these processes in the vagina are taking place, the uterus becomes considerably enlarged and undergoes contractions. It used to be supposed that these contractions exercised a suction effect, but this is now known not to be so. Research has come up with the surprising discovery that something very like orgastic uterine contractions occur even in women who have had the uterus removed. Not only the uterus but also the blood-distended wall of the front part of the vagina performs contractile movements during the orgasm and assists the ejaculation of semen from the penis. At the same time, the woman's "sex flush" reaches its maximum intensity. Her whole body begins to move rhythmically, in time with the man's movements, by the coordinated action of all the various parts controlled by the sympathetic nervous system and involving also the heart, blood circulation and respiration. The muscles of the neck and limbs respond with spasms so that the woman strives to cling with hands and feet to her partner. Just as in the man, so also in the woman is the onset of orgasm experienced as something irresistible, beyond any voluntary control. The female orgasm is indeed the counterpart of the orgastic ejaculation of the male, except that a woman can reach the maximum sensation of sexual satisfaction more frequently and at much closer intervals of time than a man can. On the other hand, the female orgasm is not essential to fertilization and pregnancy. The associated nervous phenomena are likewise basically similar to those that accompany and control the male orgasm. Summation of stimuli which are conducted by the nervus pudendus causes stimulation of a nerve center in the lumbar region of the spinal cord, which in turn, acting through the nervus hypogastricus, brings about the contraction of the genital musculature.

Relaxation phase: In this phase the foregoing changes take place in reverse order of their initial occurrence. At the same time, the uterus descends, and its inlet (the external os of the cervical canal) dips into the pool of semen at the base of the vagina (Fig. 6-118). This reaction does not occur if the uterus is retroflexed or retroverted (see page 232), which may account for the somewhat diminished fertility of women in which these uterine conditions are present.

IMPOTENCE AND FRIGIDITY

Impotence, i.e., the inability of a man to perform coitus, may be a congenital or an acquired defect; it may be permanent or temporary; and it may be due to physical causes or be the result entirely of some psychic disturbance. Physical causes may include actual defects of the genitals themselves (anatomic or organic impotence); psychic causes may be of various kinds (functional or psychogenic impotence), to be discussed.

Impotence is an obvious consequence in certain cases of severe injury to the penis, possibly necessitating its amputation. More often, however, impotence is due to disturbed function of the testes, with inadequate or absent testosterone, the principal hormone secreted by the normal testes. In this condition no sexual intercourse is possible if only for the reason that erection of the penis fails to occur. The testes may be directly affected by defective development, inflammation, or degenerative changes, or they may be insufficiently stimulated by the hormones of the pituitary gland on account of degenerative changes in the latter (see next section). If testosterone production by the testes fails before puberty, the male external genital organs will remain underdeveloped and infantile. If the production of this hormone fails after puberty, these organs will have developed to maturity, but the sexual responsiveness of the reproductive system will be impaired. Such cases of impotence due to hormone deficiency can sometimes be successfully treated by surgery and/or by hormone therapy. The impotence that comes on with advancing age in most men is also due to the (gradual) cessation of testosterone production. The degree of success achieved with hormone therapy in the treatment of such cases of normal aging is somewhat controversial (see next section). There is always some degree of overlapping with psychic/emotional factors. Impotence in the aging male is often due to the development of sexual and emotional apathy in the too-familiar pattern of married life.

Constitutional diseases may also adversely affect the ability or desire to have sexual intercourse, e.g., weakness, undernourishment, metabolic diseases, some types of chronic poisoning, and alcoholism. Disorders of the spinal cord sometimes interrupt the nerve pathways that control the phenomena of normal sexual response in the male (Fig. 6-119, see also Fig. 6-111).

Psychic causes of male impotence may include hatred or dislike of the sexual partner or of the sexual act itself, sexual apathy or indifference in an emotionally drab marriage (marital impotence), lack of confidence or shyness

FIG. 6–119.
Disorders affecting nerve pathways of male sexual response.

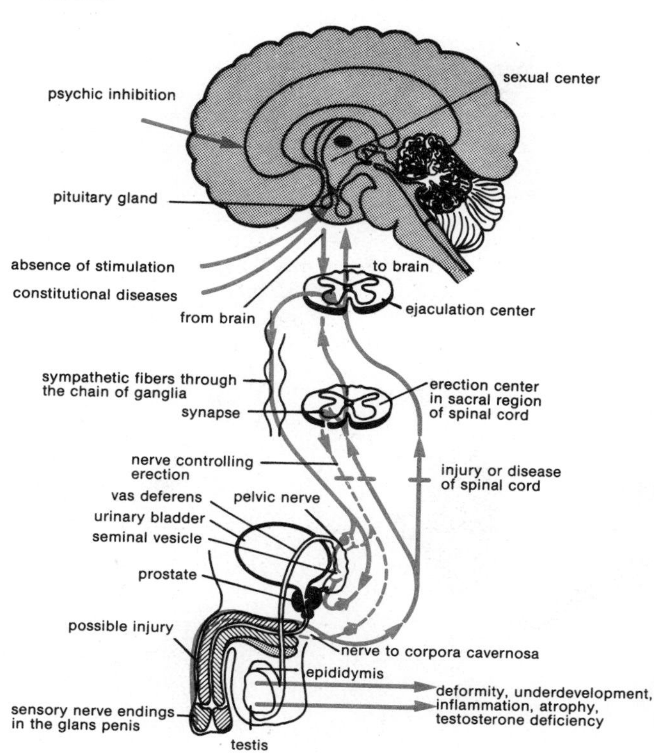

in the sexually inexperienced, etc. Unrecognized homosexuality which is latently present may also constitute an obstacle to sexual response. In many cases the man's individual experiences in life cumulatively promote impotence: for instance, disappointed expectations, deficient or misleading education in matters of sex, fear of the imagined consequences of masturbation or "sexual excesses," fear of venereal disease, fear of personal involvement or commitment and/or of fathering an unwanted child. Subconscious inhibitions and conflicts may come to the surface when confronted with the actual prospect of sexual intercourse and may thus give rise to disturbed general behavior. In addition, there may be overlapping with depression and states of mental exhaustion from other causes or with basically physical causes of impotence. Some cases of psychogenic impotence respond well to psychotherapy. Aphrodisiacs are drugs that claim to stimulate sexual desire and thus counteract impotence. In reality they have little, if any, real effect and may even be harmful. Prevention and treatment of marital impotence must consist in bringing about a change in mental attitude in one or both of the partners, so that they take a fresh interest in each other and explore new possibilities in their life together.

A particular type of impotence is due to *ejaculatio praecox*, the man's inability to prevent ejaculation of semen before or at the beginning of copulation, so that the penis prematurely loses its stiffness. It is caused by a too-rapid reflex action in the sexual response cycle and occurs not infrequently in men of a nervous temperament. Psychological anxiety conditions are often involved in this disorder, just as they are in true impotence; sometimes it is associated with—usually subconsciously—defensive attitudes toward the other sex.

Frigidity, which can in some respects be regarded as the female counterpart of impotence, can have various causes. In some cases it may be due to the man's lack of understanding of the woman's slower sexual response. Her sexual desire may have to be aroused gradually and subtly, especially at the beginning of a relationship. If the man is too impatient or unwilling to adjust to the woman's different rate of response, she may become and remain frigid. Such women usually derive no pleasure from sexual intercourse and experience no orgasms. Although reliable statistical evidence of such matters is scarce, results obtained by some investigators suggest that something like 50 percent of all women fail to achieve full sexual satisfaction. If the man cannot or will not "make love" to her in a way that is in tune with her temperament and rate of response, she will not be sexually aroused but will instead feel unsatisfied, cheated and perhaps resentful. This may develop into a state of frigidity characterized by indifference if not actual aversion to sexual activity. Of course, female frigidity differs radically from male impotence in that it does not render a woman physically incapable of intercourse.

Not all cases of frigidity can, however, justly be blamed on lack of psychological insight, lack of patience or tenderness, or lack of an acceptable lovemaking technique on the man's part. A woman may be frigid because her ovaries are underdeveloped and their secretion of sex hormones is deficient (female sexual infantilism), or she may have mental inhibitions due to faulty sex education or to moral taboos inculcated at an early age. Frigidity may also be due to anxiety (about becoming pregnant, etc.) or to unrecognized homosexuality manifesting itself in an aversion to sexual dealings with a man. The subconscious rejection of intercourse may bring about, as its physical counterpart, tension in the abdomen and sometimes spasms of the vagina, so that coitus is experienced as a painful activity (dyspareunia). This condition will, of course, tend to confirm and perpetuate the woman's frigidity.

SEX HORMONES AND THEIR DEFICIENCY

The sex hormones (Fig. 6-120) are specific internal secretions of the gonads (ovary and testis), the adrenal cortex and, during pregnancy, also the placenta. Although the sexes are characterized by their own respective dominant sets of hormones, there is no sharp dividing line between them in this respect: each sex produces both male and female hormones, but in different proportions. They are all closely related compounds called *steroids*.

The *estrogens* (follicular hormones) are the primary female hormones responsible for typical changes in the uterus and vagina (see page 218). They also bring about the development of the female secondary characteristics. There are several of these hormones, including

FIG. 6–120.
Hormonal control of gonads in childhood and in sexual maturity.

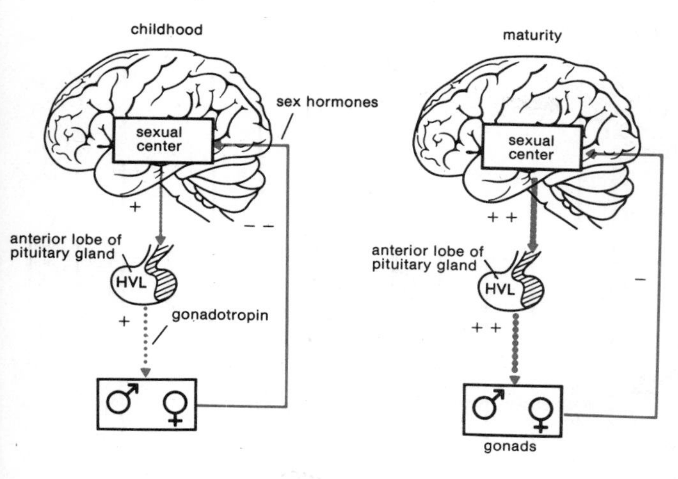

FIG. 6–121.
Interactions between estrogens and progestogens.

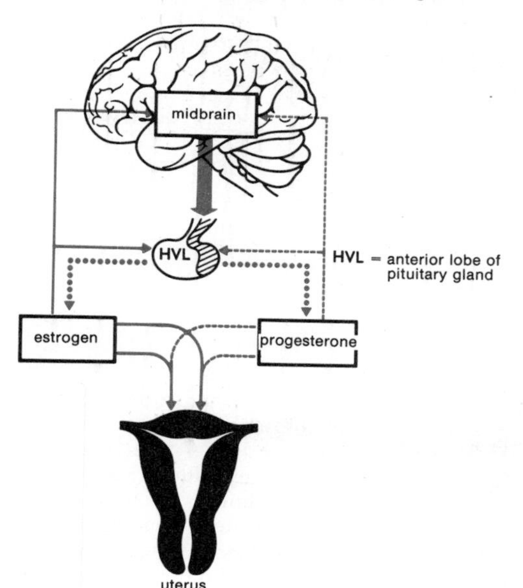

mainly *estradiol*. The second group of female hormones are the *progestogens* (gestagens, corpus luteum hormones), represented especially by *progesterone*. They are concerned with the preparation and maintenance of pregnancy. In particular, they cause changes in the mucous membrane lining of the uterus in readiness for the implantation of the fertilized ovum. The estrogens and progestogens exist in an interactive relationship with one another (Fig. 6-121).

The male sex hormones are collectively called *androgens*, the most important of which is *testosterone*, and are responsible for the male primary and secondary sexual characteristics. In both sexes androgens are produced in the adrenal cortex, and in the male they are formed also and more particularly in interstitial tissue cells (Leydig's cells) in the testes (testicles). In the female reproductive system it is mainly the clitoris, the labia and the pubic hair growth that are affected by the androgens.

Puberty, the onset of sexual maturity or fertility, begins at about the same age in boys and girls—between 12 and 14. In girls, puberty is considered to begin with the first menstrual period (menarche); in boys it cannot be pinpointed so precisely. Sexual development is controlled by certain centers in the brain which, acting through the midbrain and pituitary gland, exercise their influence on the maturing process. In boys the increased production of gonadotropins (gonad-stimulating hormones) by the pituitary gland stimulates the testes to secrete more testosterone, thus bringing about the typically male secondary characteristics (broken but deepening voice, facial hair, muscular physique, etc.), while in girls it stimulates estrogen secretion by the ovaries, resulting in the typically female secondary characteristics (growth of breasts, etc.) (Fig. 6-122). Broadly speaking, female characteristics may be regarded as representing the basic human form, with the male as a superimposed modification. Thus the estrogens influence bodily characteristics less than the androgens do. Yet the female requires the estrogens in order to achieve sexual maturity; they determine the sexual cycle on which fertility depends.

The testes are of major importance to the attainment of male sexual maturity. At the beginning of puberty they are only 0.5–2 cm^3 in size and practically functionless; from an age of between 15 and 18 they grow to their full size of 5–25 cm^3.

The gonadotropic hormones secreted by the pituitary gland include, first, the so-called *follicle-stimulating hormone* (FSH), which, in the male, stimulates spermatogenesis (formation of sperms) in the testes. In the female, FSH stimulates development of the ovarian follicles. A second pituitary hormone is the *inter-*

FIG. 6-122.
Development of sexual characteristics during puberty.

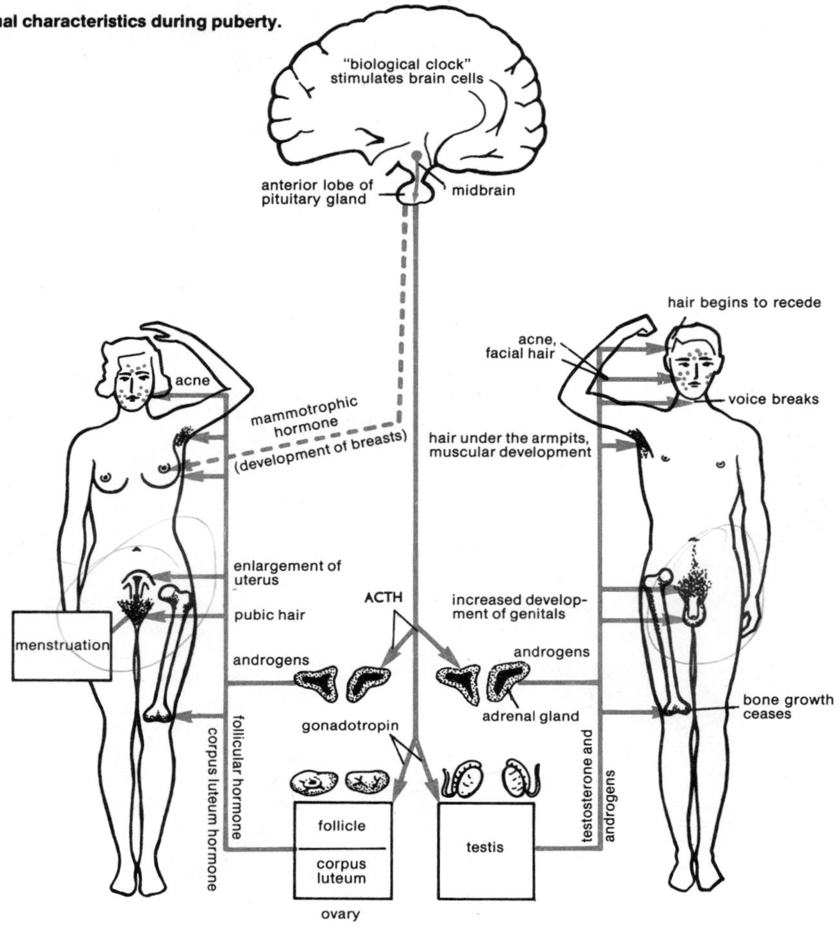

stitial-cell-stimulating hormone (ICSH), which stimulates the formation of testosterone by the interstitial cells of Leydig in the testes. In the female this hormone is called the *luteinizing hormone* (LH) and stimulates the development of the corpus luteum.

In contrast with the sexual cycle in the woman, which is controlled by two groups of sex hormones (estrogens and progestogens) secreted by the follicle and the corpus luteum, in the man the sexual functions are, in the main, controlled by only one group of hormones, the androgens. Most important among the latter is testosterone, which ensures that during the course of puberty not only the testes but also the other reproductive organs, including particularly the penis and prostate gland, reach maturity and that, moreover, the secondary sexual characteristics of the male are fully developed (see Fig. 6-122). After the attainment of maturity, testosterone serves to maintain the sexual functions.

A deficiency in sex hormone production will have different consequences for sexual development and activity according to whether the deficiency arose before or after sexual maturity was attained. If the hormonal function of the gonads fails to develop adequately by the onset of puberty (Fig. 6-123), the genitals remain immature, the secondary sexual characteristics fail to develop properly or do not develop at all, the voice does not deepen, the muscular physique and skin retain their infantile delicacy and softness, and the growth activity distinctively associated with puberty is absent. However, after normal puberty the increase of sex hormones tends to inhibit bone growth, whereas eunuchs, in whom these hormones are lacking, continue to grow for a longer time and therefore tend to become taller than normal males. In the eunuch the arms and legs are too long in relation to the body; there are pads of fat on the shoulders, chest, belly, thighs and hips; facial hair is absent or underdeveloped.

Early deficiency in the hormonal function of the testes, resulting in the suppression of puberty, may be due to a direct disturbance in the function of the testes, e.g., by inflammatory disease, injury in prepuberty or failure of the testes to descend into the scrotum (cryptorchidism). Deficiencies in pituitary function have entirely similar effects insofar as the gonadotropic hormones are concerned, for it is these hormones that stimulate and control normal maturing development of the gonads (more particularly the testes in the male). If the pituitary gland as a whole fails to function properly, the result is likely to be stunted growth due to absence of the growth hormone. Sometimes only the gonadotropic hormones may be deficient.

Treatment of such cases of eunuchoidism (the condition characterized by deficient production of male hormone by the testes) may advantageously consist in hormone therapy with testosterone. Although this will promote the development of normal male sexual characteristics, it cannot compensate for deficient growth of the testes and for their failure to produce sperms. Premature treatment with testosterone can be harmful because it causes premature inhibition of bone growth and may, if the pituitary function is disturbed or retarded, prevent any possibility of subsequent spontaneous puberty. For this reason, after commencement of testosterone treatment, an interval is interposed during which the treatment is suspended in order to observe whether, at 15–17 years of age, the testes will not develop spontaneously after all. If the disorder is located in the testes themselves rather than in the sexual central system, such reservations do not apply. In the former case it will often be necessary to continue the testosterone treatment indefinitely in order to sustain the gains achieved.

Klinefelter's syndrome is a particular form of disturbed sexual development. It is a congenital condition of primary testicular failure which usually does not manifest itself before puberty. Typical symptoms may include abnormally long legs, and subnormal intelligence. There are many variant forms. Treatment with testosterone may improve the condition, but such individuals nevertheless remain sterile.

Cryptorchidism is characterized by failure of the testes to descend into the scrotum shortly before birth. This abnormality is found in about 4 percent of newborn male children, in 0.7 percent of one-year-olds and in 0.5 percent of adult males. In order to rule out the risk of permanent damage to the testes by the higher temperature prevailing in the abdominal cavity, hormone treatment may be attempted when the child is 8 or 9 years old in order to induce the testes to descend of their own accord. If this is unsuccessful, as it is in about 50 percent of such cases, an operation will be performed to transfer the testes into the scrotum. Unfortunately, in a substantial proportion of cases it subsequently turns out that the testes have remained underdeveloped from the outset and have become sterile, so that the operation produces no positive result.

The effects of functional deficiency of the gonads in the adult male as a result of injury or removal of the testes are indicated in Figure 6-123b. It is evident that in such castrates many of the normal sexual characteristics acquired in puberty will be preserved even after the function of the testes ceases—such as the characteristic male physique, the fully developed penis, the deep voice, and facial hair. On the other hand, the instinctive sexual drive (libido) and male potency will be preserved only if the supply of testosterone can be maintained. As a result of castration, the prostate gland and seminal vesicles will undergo atrophy, and the production of sperms will cease. Castration may be necessitated by medical considerations, e.g., as treatment for malignant tumors of the testes. Directly afterward, phenomena associated with "male climacteric" may manifest themselves, such as hot flushes, anxiety states, depression and emotional instability. Quite often, too, the psychic trauma of castration has consequences that are difficult to separate from those of absent sex hormone secretion.

Apart from the direct impairment or cessation of the testicular function due to injury, removal, inflammation or degeneration (Fig. 6-124a), similar deficiency phenomena will occur also as a result of failure of the pituitary

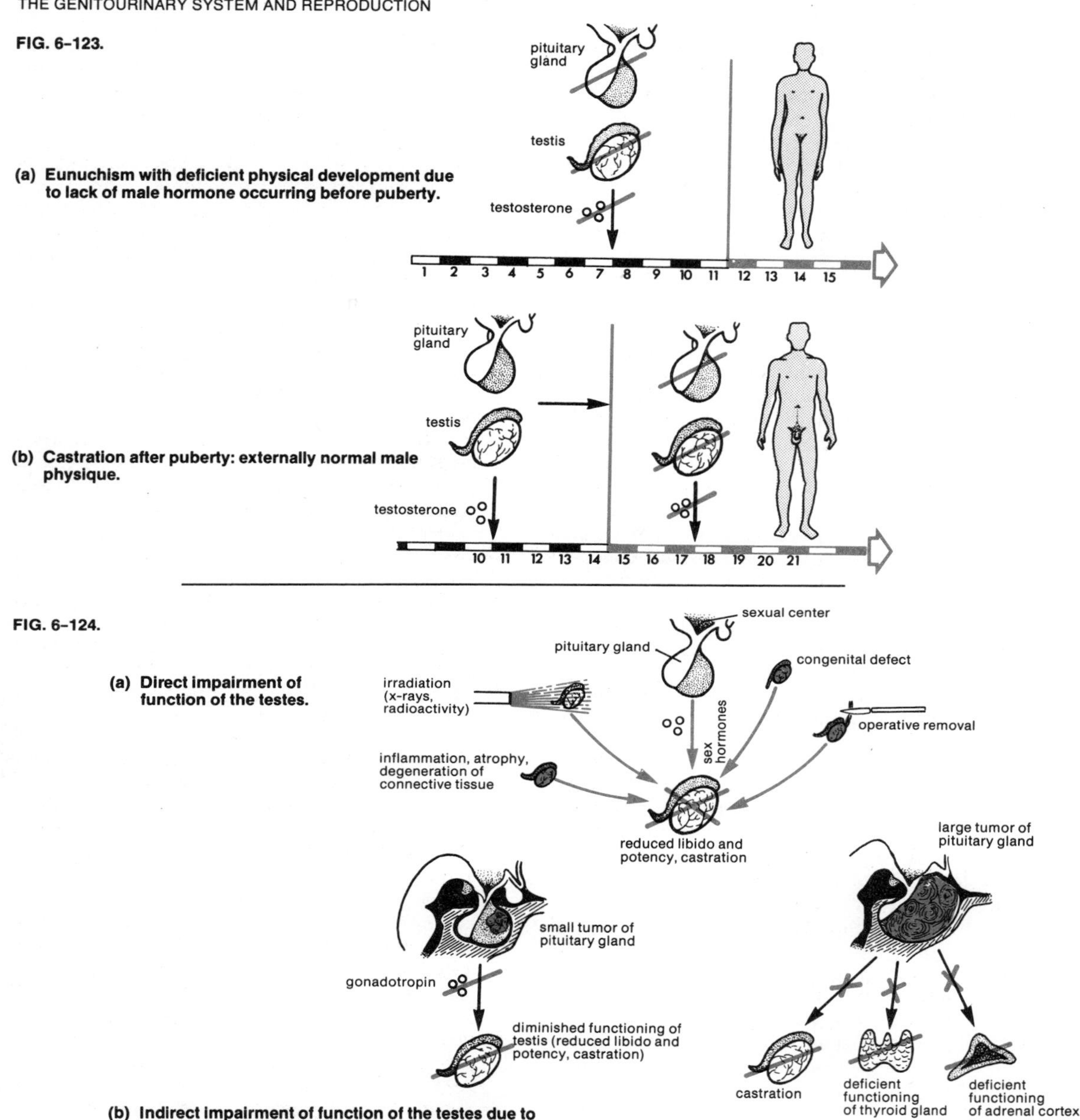

FIG. 6-123.

(a) Eunuchism with deficient physical development due to lack of male hormone occurring before puberty.

(b) Castration after puberty: externally normal male physique.

FIG. 6-124.

(a) Direct impairment of function of the testes.

(b) Indirect impairment of function of the testes due to deficient stimulation with gonadotropic hormones.

gland to perform its controlling function with regard to the testes (Fig. 6-124b).

It would appear that disturbed testicular function occurs in about 5 percent of all men. Minor disturbances in male fertility or potency can sometimes be corrected by hormone treatment. In a case of varicose veins of the testes or occlusion of seminal passages, an operation may be necessary. Generally speaking, the loss of potency due to testicular deficiency can more easily be restored by means of hormones than can loss of fertility.

With advancing age the production of testosterone by the testes diminishes, accompanied by a decline in libido and potency and a withering of muscular strength and physique. In comparison with the woman, the man is capable of reproduction for a substantially longer time. His sex hormone production also diminishes at a more gradual rate than hers and usually does not cease within a more or less well-defined period of time, as occurs in the female climacteric (menopause). Nevertheless there can, broadly speaking, be said to exist a male climacteric between the ages of 45 and 60, though there are considerable individual variations both in the timing and in the intensity of the symptoms. Typical symptoms are restlessness, irritability, disturbed sleep, discontent, headache, formication (a sensation as of ants crawling on the arms and legs), lapses of memory, poor concentration, hot flushes, circulatory instability, moodiness, anxiety states, weakness, and waning sexual powers. Such cases may respond quickly to testosterone treatment, just as the disorders associated with the female climacteric can sometimes be removed by treatment with estrogens. The advisability of giving testosterone to the aging male has its controversial aspects and may, in the opinion of some experts, have harmful consequences. For example, it is asserted that testosterone may have a growth-promoting effect on cancer of the prostate.

INTERSEXUALITY, HERMAPHRODITISM

A *hermaphrodite* is an individual who, in theory anyway, is both male and female, with the distinctive characteristics of both sexes more or less equally represented in his/her physical and mental construction. Male or female sex in human beings is essentially determined at four levels. Already at the instant of fertilization of the ovum by a sperm it is genetically determined whether the new individual's cells will each contain two X chromosomes (female pattern) or an X and a Y chromosome (male pattern). These sex chromosomes carry the genes for sex-linked hereditary characters and particularly determine the development of the gonads toward the male (testes) or the female (ovaries) type in the embryo. In the male embryo the visible rudiments of testes appear in approximately the seventh week of pregnancy; in the female embryo the ovaries make their appearance in approximately the tenth week. The gonads constitute, as it were, the second level of sex determination, and this in turn controls, at a third level, the development of the external genital organs and what may thus be termed "organic sex." The second and third levels are also responsible for the so-called secondary sexual characteristics—physique, hair development (facial hair, pubic hair), development of the breasts, etc.—constituting the fourth level at which an individual's sex is manifested. In normal individuals there is definite concurrence among chromosomal or genetic sex (XX or XY), hormonal sex (ovary or testis), organic sex (vagina or penis), and secondary sexual characteristics.

The term *hermaphroditism* is somewhat loosely applied to persons in whom hormonal and organic sex is ambiguously developed or to those in whom the chromosomal sex pattern, hormonal sex and organic sex are not in agreement with one another. Whereas true hermaphroditism in human beings, characterized by the possession of ovarian and testicular gonads in one and the same individual, is extremely rare, more than 0.2 percent of all people, i.e., roughly 1 in every 500, can be said to have intersexual characteristics. Since intersexuality is often accompanied by physical and/or emotional disturbance, the diagnosis and treatment of such conditions constitutes a medical problem.

Determination of sex in individuals who appear to have both male and female characteristics can be difficult, requiring a study of the chromosomes as well as microscopic and macroscopic examination of anatomic features. Deciding the sex of the external genitals and the secondary sexual characteristics generally presents no difficulty. On the other hand, reliable examination of the internal reproductive organs may be possible only with the aid of an abdominal incision, while determination of hormonal sex will generally require microscopic examination of tissue specimens. Direct determination of the genetic sex of an individual requires examination of the sex chromosomes. A simple procedure is to ascertain the so-called nuclear sex, by which is meant the diagnosis of that person's sex from simple properties of the nuclei of the cells, readily observable under the microscope, particularly the detection of the presence or absence of sex chromatin. In a smear specimen obtained from the mucous membrane of the mouth, for example, female cells are seen to contain sex chromatin, a dark-staining mass at the periphery of the nuclei, which is thought to be due to

FIG. 6-125.
Determination of nuclear sex from smear specimens of mucous membrane of the mouth and blood cells.

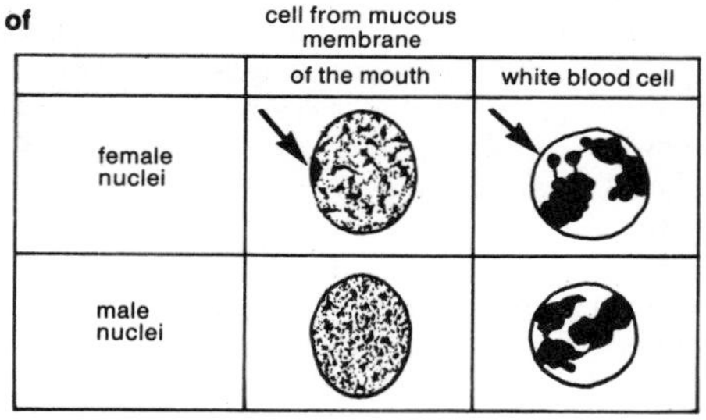

	cell from mucous membrane of the mouth	white blood cell
female nuclei		
male nuclei		

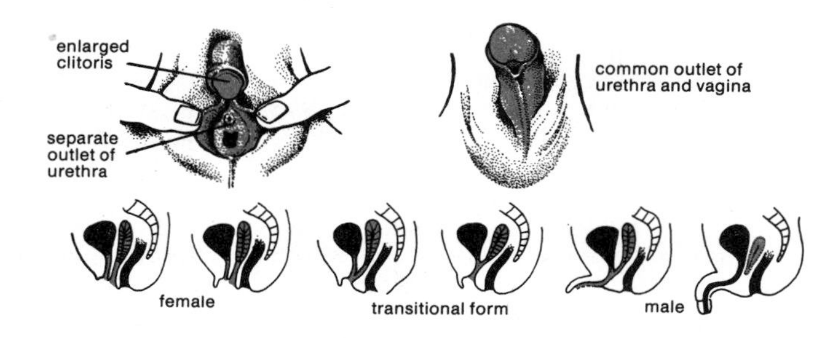

FIG. 6-126.
Transitional forms of the genital organs in true and in apparent hermaphrodites.

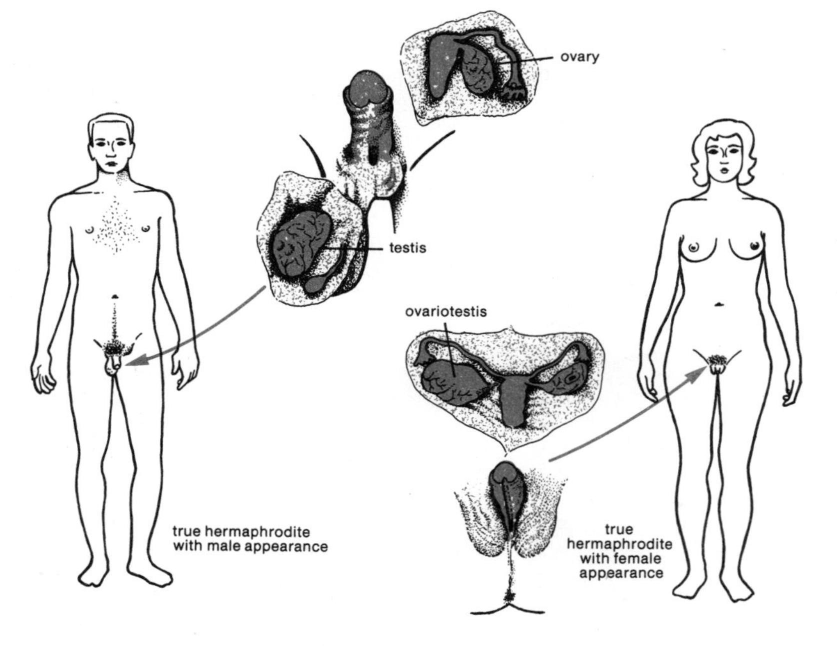

FIG. 6-127.
True hermaphrodites.

the X chromosome pair present in females and not present in males. The nuclei of female white blood cells contain drumstick-shaped appendages (Fig. 6-125); these are absent in male white blood cells.

True hermaphroditism, if indeed it exists at all, is a rarity, as already mentioned. It is characterized by the presence, in one and the same individual, of true ovaries and testes or an organ in which ovaries and testes are merged.

Such individuals are invariably sterile, a fact that in itself is a negation of the concept of "true" hermaphroditism. The external and internal genital organs are usually bisexual, though sometimes with the male or with the female characteristics predominating (Fig. 6-126). The individual's physique and secondary sexual characteristics tend to conform to whichever sex develops the dominant hormonal activity (Fig. 6-127). It would appear best to bring up and educate a hermaphrodite child as a member of the sex that it most closely resembles in its external genital organs. In many cases corrective surgery may be helpful. Sometimes, though exceptionally, hormone therapy may assist the development of secondary sexual characteristics. Complete removal of the gonads a short time before the onset of puberty may improve the condition of some hermaphrodites. Treatment with hormones directed at bringing out the characteristics of one particular sex can then help to reduce the patient's physical and sexual ambiguity.

The great majority of so-called hermaphrodites are cases of *pseudohermaphroditism* (false hermaphroditism), in which the individual possesses the gonads of one sex (ovaries or testes), but has the secondary sexual characteristics and external genitals of the opposite sex. Thus there is a discrepancy between organic sex and hormonal sex. It is estimated that about 1 in 5000 physically male children, i.e., with testes, are in fact pseudohermaphrodites. The opposite type of pseudohermaphroditism, i.e., in individuals with ovaries and therefore essentially female, is much rarer. In general, the development during puberty will depend on how active the testes are in producing male sex hormones and thus governing the individual's sex pattern (Fig. 6-127).

So-called sex-change operations are plastic surgery procedures to establish harmony between the pseudohermaphrodite's physical appearance and emotional make-up. In some cases the breaking of the voice and the growth of facial hair at puberty are the first indications that a person previously regarded as a girl is in fact (hormonally) a boy. However, male hormone production in such cases usually remains at a low level, and pseudohermaphrodites mostly retain a more or less feminine appearance, particularly in their secondary sexual characteristics. Experience has shown that it is generally better to classify young pseudohermaphrodites as girls and to educate and treat them socially as such, to remove their male gonads before puberty, and to give female sex hormones rather than perform a sex-change operation. Depending on the condition of the external genitals, additional plastic surgery may be advisable, however. Such operations cannot, of course, establish the reproductive function, and such women are sterile.

In the *Ullrich-Turner syndrome* (Fig. 6-128) the Y sex chromosome is absent from the nuclei of the individual's cells. Thus the combination XO exists instead of XY (XY is the genetic pattern of the normal male, XX of the normal female). In such individuals, who are classed as women, the external physical characteristics are weakly developed female. The breasts and uterus are small; there is no menstruation; the physical stature is short and is sometimes characterized by a thick neck; in some cases underdeveloped male genitals may also be present. Treatment with female sex hormones may be helpful.

Klinefelter's syndrome (Fig. 6-129) was mentioned in the preceding section. It is a congenital condition of primary testicular failure in which the sex chromosome combination XXY occurs instead of XY. This abnormality is found in something like 0.1–0.2 percent of the male population as a whole and is five times more frequent among the mentally subnormal. In these individuals the testes, though small, do indeed bring about puberty phenomena, but subsequently become atrophied. Breasts tend to be abnormally large (gynecomastia), though such individuals are classed as males.

A particular form of hermaphroditism is caused by overactivity of the cortex of the adrenal glands. The cortex, an endocrine gland, produces not only hormones for glucose, fat and mineral metabolism, but also sex hormones. In the rare condition called *adrenogenital syndrome,* which occurs in girls, the sex hormones are secreted in excess, with a tendency to develop male characteristics. It may be a consequence of disease or it may be a congenital abnormality; depending on the degree of overactivity, the genitals may display various intermediate or transitional variants (see Fig. 6-126). Such girls mature at a relatively early age and eventually become more or less completely masculinized, with facial hair and a penislike enlarged clitoris (Fig. 6-130). The congenital condition can be treated

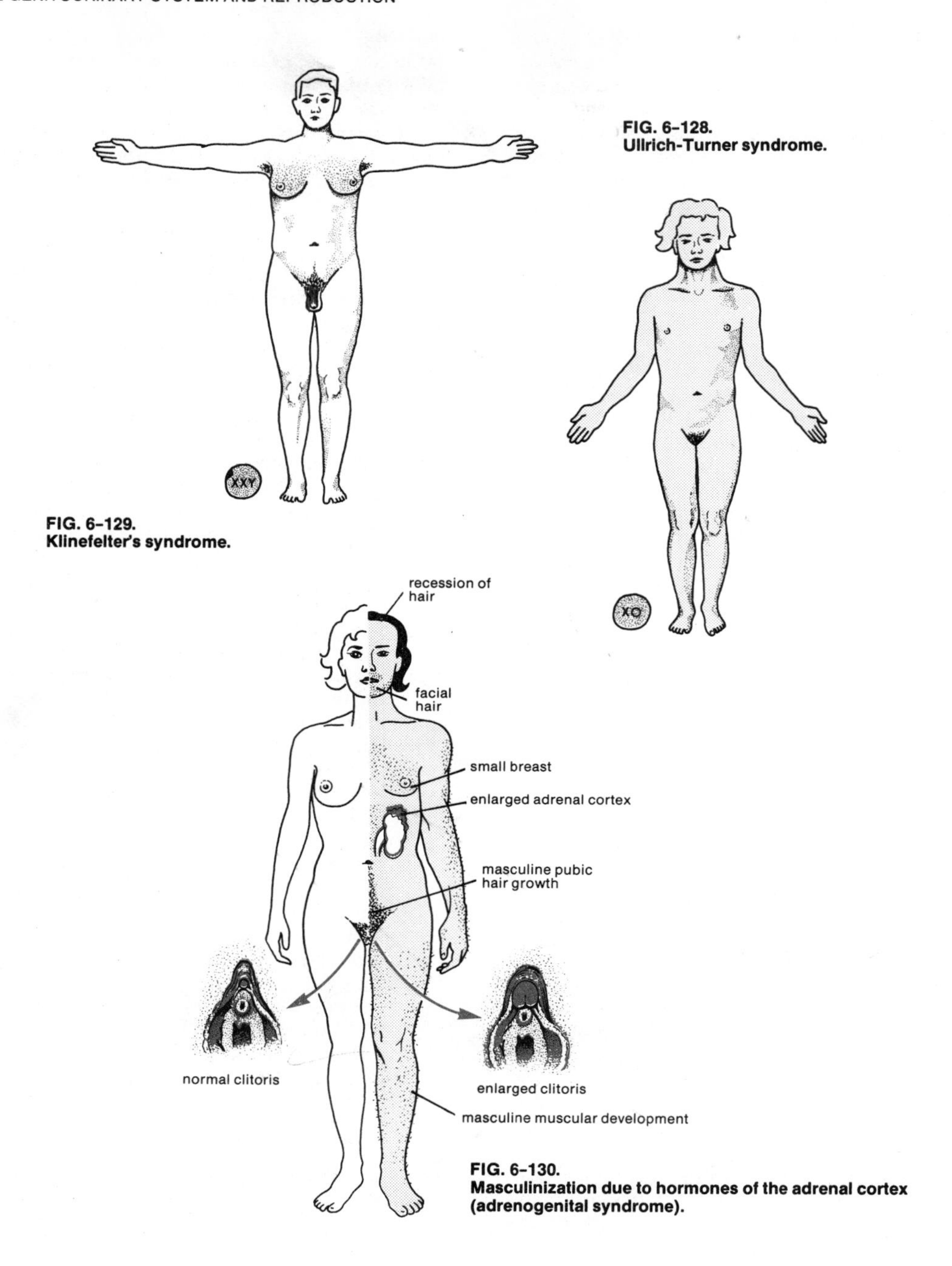

FIG. 6-128.
Ullrich-Turner syndrome.

FIG. 6-129.
Klinefelter's syndrome.

FIG. 6-130.
Masculinization due to hormones of the adrenal cortex (adrenogenital syndrome).

with corticosteroids, i.e., steroid hormones formed in the adrenal cortex. A similar condition may develop in an adult woman as a result of the cortex becoming abnormally overactive, e.g., due to the effects of a tumor. Treatment will consist in surgical removal of the tumor or, in extreme cases, of the adrenal glands themselves.

CONTRACEPTION (BIRTH CONTROL)

Contraception comprises all deliberate measures for the prevention of conception and must therefore be clearly distinguished from termination of pregnancy, i.e., abortion. Medically speaking, pregnancy can be said to exist when the fertilized ovum has implanted itself in the lining of the uterus. Implantation occurs approximately one week after fertilization (see next section).

Throughout human history numerous methods and techniques of contraception, ranging from the crude to the sophisticated and from the unreliable to the highly reliable, have been devised and used.

The *rhythm method* or *Knaus method* (also known as the Ogino-Knaus method) is based on the theory of "safe periods"—i.e., certain days during the menstrual cycle on which conception is very unlikely—and presupposes menstruation occurring with a high degree of regularity. The most favorable conditions for conception arise when an ovum on its way through the fallopian tube encounters sperms. By reference to a menstruation chart or rhythm chart, this point of time can be retrospectively determined by counting backward from the start of menstruation to find the time of ovulation. This provides the basis for the prediction of future high-fertility ("unsafe") and low-fertility ("safe") days in the menstrual cycle. As a rule, menstrual bleeding begins 14 days after ovulation. Assuming a cycle of 26–30 days and counting 14 days backward from the start of the cycle, i.e., from the first day of menstruation, ovulation can be reckoned to occur between the 12th and the 16th day of the cycle. Adding a few days to each end of this time interval as an extra precaution, we thus arrive at a high-fertility ("unsafe") period extending from the 8th to about the 19th or 20th day of the cycle. For contraception, sexual intercourse should therefore be avoided on those days (Fig. 6-131). This method can be reduced to the following rule of thumb: take the woman's shortest known menstrual cycle, say 26 days, and subtract 18 to find the first fertile day = 8th day of the cycle; then take her longest known cycle, say 30 days, and subtract 10 to find the last fertile day = 20th day of the cycle. The method is based on probabilities and therefore cannot be regarded as entirely reliable. Though apparently scientific, it is based on some unverifiable assumptions as to the effective life-span of the ovum after ovulation and of the sperms. Also, various factors (illness, emotional stress, etc.) may cause menstrual irregularity, thus significantly altering the time of ovulation, so that the calculation of the safe period becomes inaccurate. Premature ovulation, in particular, cannot be

FIG. 6-131.
Chart for determining the "fertile days" by the rhythm method (Knaus method): the small squares denote days on which conception risk is high.

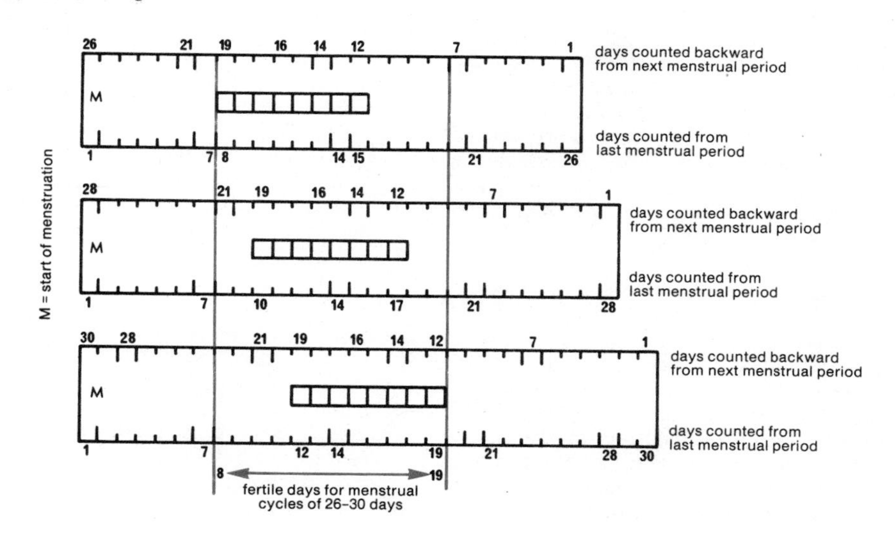

FIG. 6-132.
Chart for determining the "fertile days" by the basal temperature method.

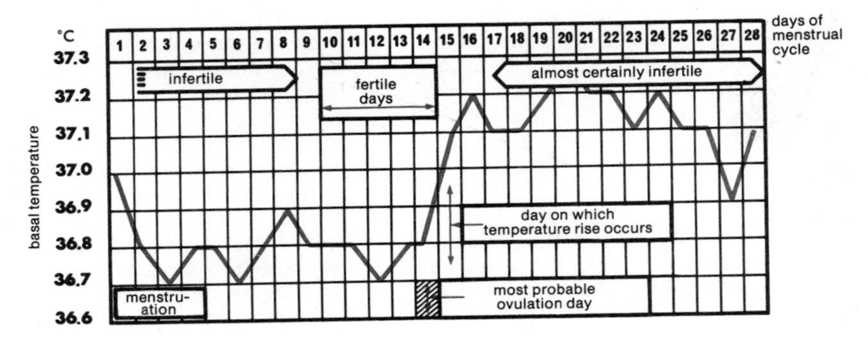

ruled out. Generally speaking, the only really "safe" period in the cycle is the last week prior to menstruation, though even during this period and indeed during menstruation itself the possibility of conception, though remote, cannot be completely ruled out. Women who menstruate very irregularly will find the Knaus method of contraception highly unreliable. In any case, for the method to be of any use at all, a woman will have to observe her menstrual periods carefully for at least six months, preferably longer, and record them on a menstruation chart. Despite its shortcomings, the method can be usefully applied as an additional safeguard with some other method of contraception, such as condom or pessary.

Greater refinement and reliability are claimed for the *basal temperature method*, which is based on the relationship between the menstrual cycle and body temperature. The woman wishing to use this method will take her temperature every morning on awakening and record it on a basal temperature chart. After several months some women are able to predict the time of ovulation by analyzing the rhythm and character of the temperature variations during the course of the menstrual cycle. Figure 6-132 represents a typical basal temperature chart. During the first half of the cycle the temperature fluctuates around 36.5°–36.7°C (97.6°–98.0°F). Then, in a space of 1–2 days, the temperature undergoes a rather steep rise of about 0.5°C (0.9°F) to around 37.0°–37.2°C (98.6°–99°F). It remains at this higher level until the next menstrual bleeding. What is important is that, on average, ovulation occurs 1–2 days before the steep rise in temperature. Since the ovum released on ovulation remains capable of fertilization only for a few hours, no risk of conception will exist between the second day after the temperature rise and the start of the next menstruation (Fig. 6-132). The advantage claimed for this method is that the time of ovulation can be determined even if the menstrual cycle is not of constant length, i.e., if it becomes longer or shorter. It does, however, require regular and accurate measurement of the basal temperature every morning on awakening, and preferably always at the same time, after a night's rest of at least 6 hours. Rectal temperature readings of 5 minutes' duration are considered to be most reliable for the purpose. Whereas the reliability of the Knaus method is estimated at 14 unwanted pregnancies per hundred years of regular and conscientious use of the method, for the basal temperature method the corresponding statistical risk of pregnancy is put at only 1 per hundred years.

Another common method of contraception that does not rely on chemical or mechanical aids is *coitus interruptus*, i.e., the man withdraws his penis and thus terminates copulation just before ejaculation. It is an unsatisfactory method for various reasons. For one thing, it is unreliable because some sperm cells may get into the vagina before ejaculation occurs (though some experts doubt whether their number can be sufficient to cause conception). Also, it is emotionally frustrating, possibly even harmful, because it involves physical separation of the partners at the very instant of intense sexual emotion, so that the woman usually fails to achieve orgasm. The reliability of this method is difficult to estimate, but can tentatively be put at 15–38 unwanted pregnancies per hundred years.

One of the most widely used and generally

reliable contraceptives is the *condom*, a thin rubber sheath worn over the penis. Its reliability has been estimated at 7 unwanted pregnancies per hundred years. It can, moreover, be used in combination with chemical contraceptives or as a protection against venereal disease. The main disadvantage of the condom is that it requires preparatory manipulations shortly before intercourse, and it largely prevents pleasurable intimate sexual contact.

Contraceptives worn by the woman include rubber or plastic caps and pessaries inserted into the vagina sometime before intercourse and used preferably in combination with a chemical spermicide or with the Knaus method of contraception. The *cap* (Fig. 6-133) fitted over the external os of the cervix prevents sperms from entering. It should be removed during periods of menstrual bleeding. If correctly fitted, preferably each time by a doctor, and with proper use, this method is claimed to be as reliable as the condom. The vaginal *pessary* is more widely used and is easier to fit by the woman herself (Fig. 6-135). It is a bowl-shaped soft rubber diaphragm which is placed high up inside the vagina. It has an elastic solid rim which is in tight contact with the wall of the vagina, thus presenting a barrier to the sperms. Used in combination with a spermicide (Fig. 6-136), it is considered to be highly reliable: about 4 unwanted pregnancies per hundred years. Depending on the internal dimensions of the vagina, the pessary ranges in diameter from 5 to 9 cm. It is inserted sometime before sexual intercourse and removed several hours afterward. Its effectiveness will depend primarily on its tight fit behind the pubic bone, and the woman should initially be instructed by a doctor in choosing the correct size of pessary and in correctly fitting it.

So-called *intrauterine devices*, sometimes

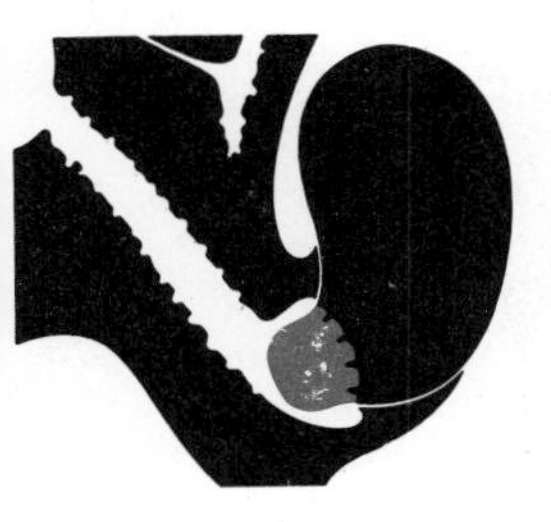
FIG. 6-133. Contraceptive cap.

FIG. 6-134. Intrauterine spiral.

FIG. 6-135. Vaginal pessary or diaphragm.

FIG. 6-136. Chemical foam contraceptive.

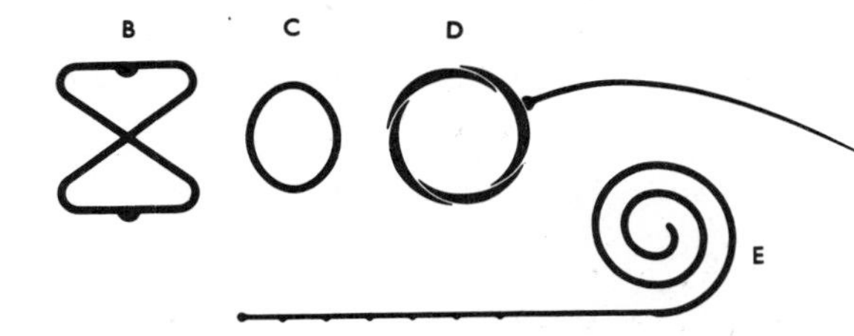

FIG. 6-137. Intrauterine contraceptive devices.

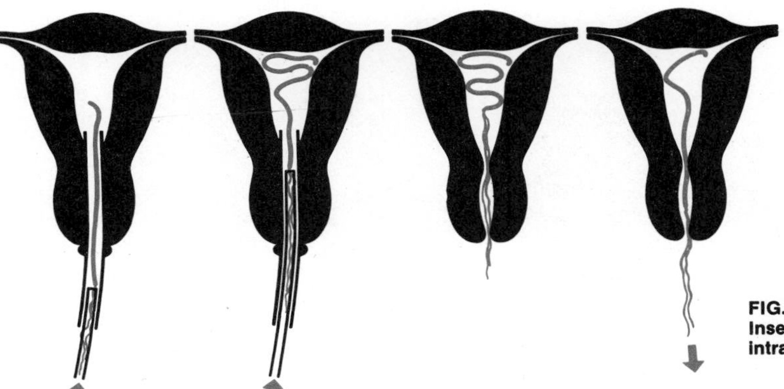
FIG. 6-138. Insertion and removal of intrauterine device.

also referred to as intrauterine pessaries, are made of plastics and are of varying shapes and types (Fig. 6-137). The device is inserted into the uterus by a doctor, e.g., with the aid of a thin tube, and has synthetic fiber threadlike appendages for checking its proper positioning and for its eventual removal (Fig. 6-138). Such a device can be left in position for a long time. Its function is not to prevent conception but to prevent implantation of the fertilized ovum in the uterus. The reliability of these devices is claimed to be about the same as for the vaginal pessary. They have some disadvantages. In about 15 percent of women, they are liable to cause irritation, pain, inflammation and bleeding. Also, in about an equal percentage the protective action fails because the device is spontaneously rejected from the uterus.

The conventional *chemical contraceptives* (spermicides) (Fig. 6-136) include creams, jellies, ointments, etc., for use particularly in combination with rubber devices. There are also various foam-producing sprays or tablets which are introduced into the vagina prior to intercourse. The foam presents a physical barrier to the sperms and moreover contains chemical agents to kill or immobilize them. Depending on the type of contraceptive employed, the effectiveness is rated at between 7 and 42 unwanted pregnancies per hundred years.

A more reliable and more convenient modern method is *oral contraception* (Fig. 6-139) by means of "*the pill*," of which various types are available. These pills or tablets contain hormone preparations (estrogens and progestogens) which act through the pituitary gland and inhibit the secretion of gonadotropins by that gland, so that ovulation is suppressed and no ovum is released (see page 233). Suppression of ovulation is achieved by strengthening the normal inhibitory processes: supply of extra estrogens and progestogens, resulting indirectly in a decrease in production of luteinizing hormone, i.e., the pituitary hormone that induces ovulation and the formation of the corpus luteum. The contraceptive action of estrogens and progestogens is due, in addition, to certain changes they bring about in the mucous membrane lining of the uterus, so that implantation of the fertilized ovum (if ovulation should nevertheless occur) is inhibited. Furthermore, under the influence of progestogens, the mucus at the mouth of the cervix becomes more impervious to sperms.

Progestogens and estrogens are given either in combination with each other (combined method) or separately in succession (two-stage method). In the combined method the woman takes a pill every day, starting on the 5th day after the commencement of menstruation, and continues taking a daily pill until the 24th or 25th day. She then stops and waits for the menstrual bleeding to begin. She again starts counting from the first day of this bleeding and resumes her daily pills from the 5th day on-

FIG. 6-139.
How oral contraceptives work.

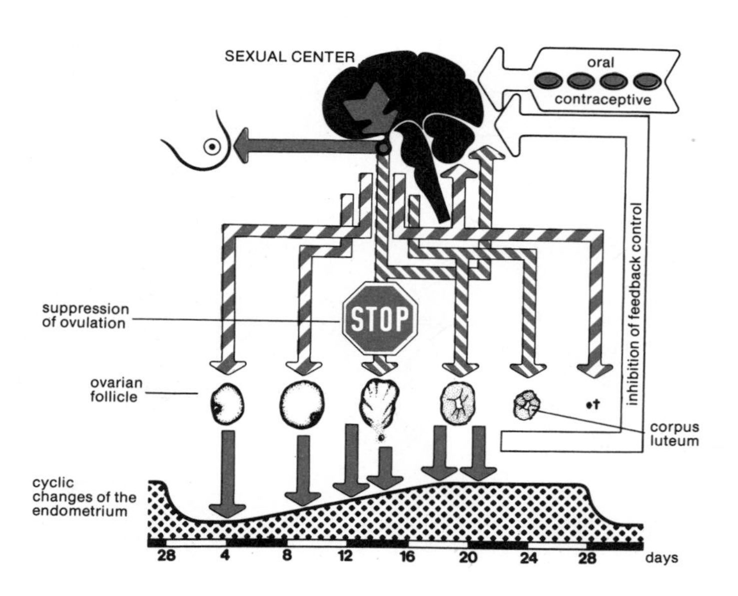

ward. It is important to take the pills regularly, not missing a day. In the two-stage method the woman takes only estrogen pills from the 5th to the 19th day of the menstrual cycle; then for 5 days she takes combined estrogen-progestogen pills. This method claims to simulate more closely the natural cyclic hormone production sequence, but is considered to be not quite so reliable as the combined method.

Oral contraceptives are considered to become fully effective only from the second cycle of pill-taking onward. During the first cycle the possibility of premature ovulation exists, especially in women who tend usually to have a short menstrual cycle. If the woman accidentally misses taking a pill on any particular day, she should take two pills on the following day, one in the morning, the other in the evening. If the interval between two successive pills exceeds 36 hours, effective contraception cannot be guaranteed. In that case the woman should stop taking the pills for 7 days. If no menstrual bleeding has occurred by the end of that interval, she is likely to have become pregnant. In view of possible harm to the developing embryo, particulary if it is a female child, no more contraceptive pills should be taken.

The menstrual bleeding that occurs (if there is no pregnancy) in the week during which no pills were taken is not a normal menstruation, but is due to the fact that the mucous membrane of the uterus has become modified by artificial supply of hormones and is now shed as a result of the withdrawal of this supply. Sometimes this withdrawal does not result in bleeding ("silent menstruation"), however, even though there is no pregnancy. Correctly used, oral contraception is superior to all other methods. With the combined method the failure rate is estimated at between 0.5 and 1 unwanted pregnancies per hundred years, and with the two-stage method it is estimated at 1.3.

Although the amounts of active constituents in oral contraceptives are kept as small as possible, these preparations do occasionally produce side effects. Some of these resemble the typical symptoms associated with the early stages of pregnancy, such as putting on weight, nausea, headache. These symptoms generally disappear in the course of time. As the sex hormones are broken down in the liver, the use of oral contraceptives should be viewed with caution in a case where the user suffers from a disease liable to harm the liver. The same warning applies to women who have had phlebitis (inflammation of a vein) or occlusion of a vein, though the risk of thrombosis due to the use of these contraceptives even in such cases is generally held to be slight. It has also been asserted that oral contraceptives entail an increased risk of cancer of the uterus and breasts, but this is by no means certain, and the prevailing medical view is that these hormone preparations are quite reasonably safe to use from the standpoint of the woman's general health. They have in fact been regularly used by large numbers of women for well over a decade now, without having produced ascertainably harmful side effects in any significant number of cases. All the same, it is desirable to have regular medical checks when these contraceptives are used for long periods at a stretch. This is especially advisable in women who have had phlebitis or occlusion of a vein, severe varicose veins, epilepsy, otosclerosis, high blood pressure and other cardiac or circulatory disorders, diabetes, and tetany (a nervous affliction characterized by intermittent paroxysmal spasms involving the extremities).

FERTILIZATION

Fertilization is the union of the male and the female germ cell, i.e., the sperm and the ovum. Following coitus (sexual intercourse), fertilization normally occurs in an enlarged portion of the fallopian tube, the so-called ampulla, where the ovum making its way down the tube from the ovary encounters the sperms on their way up the tube. Since the ovum as well as the sperms have only a limited life-span, it is, for fertilization to take place, essential that their encounter be correctly timed. The sperms in the ampulla remain capable of fertilization for a day or two; the survival period of the ovum is less, perhaps only a few hours.

The *ovum* is about 0.1 mm in size—just visible to the naked eye. It comprises a nucleus and surrounding protoplasm (yolk), which provides initial nourishment for the fertilized ovum. In all, a woman's ovaries form about 400,000 ova, but only a few hundred of these actually reach maturity in her lifetime—at a rate of about one per month for 30–40 years. Exceptionally, an ovum that fails to make its way into the entrance of the fallopian tube is nevertheless fertilized (abdominal pregnancy, see page 278). If the ovum is prevented from passing down the fallopian tube by some obstruction, e.g., an inflammatory adhesion, pregnancy may occur in the tube itself (see page 278). In such cases of so-called extrauterine pregnancy normal development of the fertilized ovum into an embryo is generally not possible.

The male germ cells or reproductive cells, commonly called *sperms,* are formed in the testis (testicle). They are discharged through the epididymis and vas deferens into the urethra, which traverses the penis, the male organ of copulation (pages 215 ff). In the course of this journey, various secretions are added to the semen (seminal fluid), which is finally ejaculated into the vagina during sexual intercourse. The sperms then have to make their own way from the vagina through the uterus and into the fallopian tubes. They are well equipped to do this, being provided with a motile tail for self-propulsion, so that they can "swim against the current," i.e., against the flow of fluid secretion that carries the ovum along the tube.

When sperms arrive in the vicinity of the ovum, they are lured toward it by certain chemical substances formed and excreted by the ovum (chemotaxis). On meeting the ovum, the head of the sperm with the middle piece penetrates it, and the tail of the sperm drops off. Subsequently the middle piece becomes detached from the head, which swells by absorption of fluid. The head, consisting mainly of the nucleus of the sperm, travels to the center of the ovum, where it fuses with the nucleus of the ovum. As a result of this fusion of the male and the female nucleus a new nucleus with the full (diploid) set of 46 chromosomes is obtained (see page 212). The *fertilized ovum,* i.e., the ovum into which a sperm has penetrated and in which nuclear fusion has occured, is called the *zygote*. Within a few hours of fertilization the zygote starts to divide (cleavage); repeated divisons result in a cluster of cells called *blastomeres*. The zygote subsequently acquires a cavity filled with fluid and is then known as the *blastocyst*. While these changes are taking place, it is carried along through the fallopian tube and into the uterus. The blastocyst's journey through the tube takes about four or five days. After a further two or three days the blastocyst becomes embedded in the lining of the uterus, this process being call *implantation* (Fig. 6-140). As a rule implantation occurs high up in the uterus; if it occurs low down, complications are liable to occur because the placenta may obstruct the cervical canal and prevent normal birth (see page 287).

Whether an embryo will develop into a boy or a girl is predetermined by chance (with a roughly 50–50 probability) at the time of fertilization. It depends on the pattern of chromosomes, more particularly the sex chromosomes, present in the nucleus of the fertilizing sperm. In the ordinary nonreproductive (somatic) cells of the human body, the nucleus contains 46 chromosomes (23 pairs), comprising 44 somatic and 2 sex chromosomes (Fig. 6-141). The sex chromosomes are of two types, designated X and Y respectively, and carry the genes for the sex-linked characters. The woman possesses a pair of identical sex chromosomes (2 X), whereas in the man the two sex chromosomes are nonidentical (X + Y). In ordinary cell division (*mitosis*) the chromosomes themselves each divide longitudinally, so that each daughter cell of such a division contains the complete set of 46 chromosomes with the full range of genes (the basic units of heredity). Formation of germ cells (sperms in the male, ova in the female) occurs by a different process of cell division (*meiosis*), which produces daughter cells in which the nucleus contains only 23 chromosomes, i.e., only half the normal number. This is because in meiosis the chromosomes do not themselves divide, but merely separate into two equal groups, one for each daughter cell. For this reason the process is also known as *reduction division*. Each female germ cell (ovum) will thus contain 22 somatic chromosomes and 1 X chromosome. The male germ cells (sperms), on the other hand, will be of two kinds—namely, those with 22 somatic chromosomes and 1 X chromosome and those with 22 somatic chromosomes and 1 Y chromosome. (See Fig. 6-141).

Now if the ovum is fertilized by a sperm containing an X chromosome, the resulting zygote will contain 44 somatic chromosomes (22 from each parent) and 2 X chromosomes (1 from each parent); the child resulting from this fertilization will be a girl. But if fertilization is accomplished by a sperm containing a Y chromosome, the zygote will have 44 somatic chromosomes, an X chromosome (from the mother) and a Y chromosome (from the father); the child will be a boy. There are equal numbers of sperms with X chromosomes and

FIG. 6–140.
Reproductive tract: fertilization normally occurs in the ampulla of the oviduct.

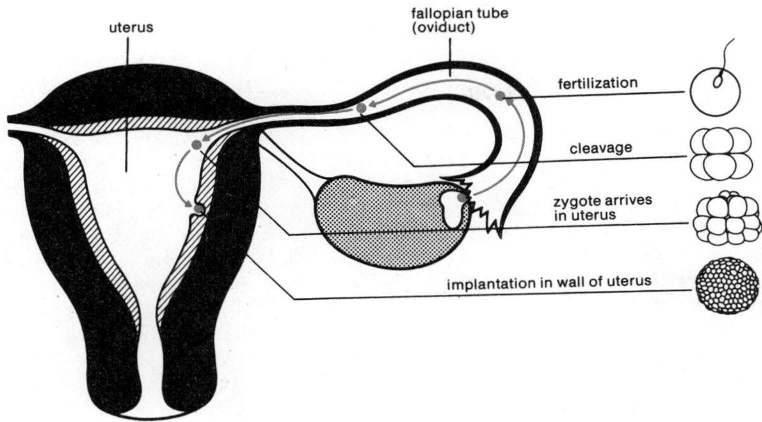

FIG. 6–141.
Possible chromosome combinations in fertilization.

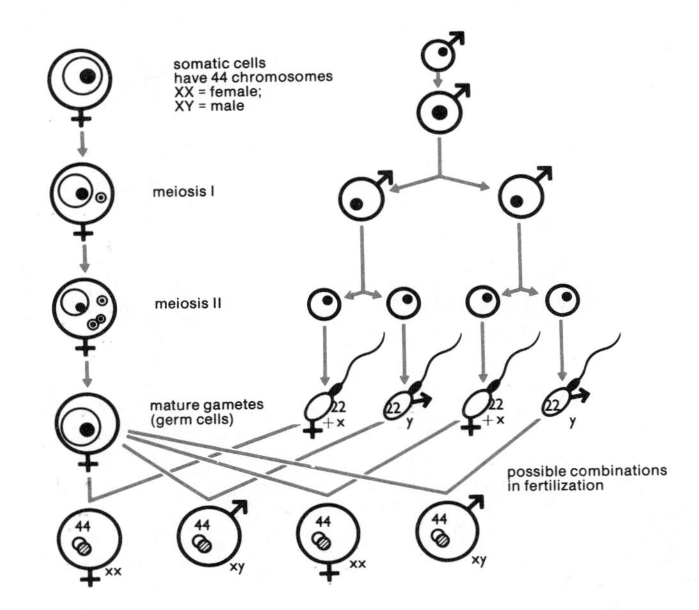

with Y chromosomes, respectively, and since only one sperm fertilizes the ovum, there are approximately equal probabilities of male or female sex in the child. A person's sex is thus determined from the instant of fertilization. Subsequent influences, such as the action of sex hormones, can merely strengthen or weaken the sex-linked characters, but cannot fundamentally alter them, so that although hermaphroditism may occur in extreme cases, the hermaphrodite will nevertheless almost invariably be of basic male or basic female type.

PREGNANCY TESTS

Reliable conventional signs of pregnancy do not manifest themselves until pregnancy is fairly well advanced. If it is considered undesirable to wait so long before having certainty, a pregnancy test can provide the answer. The object of such tests is to detect the presence, in the woman's urine, of *chorionic gonadotropin,* a special hormone produced by the developing placenta and excreted in the urine (Fig. 6-142).

So-called *biological tests* for pregnancy have long been used. In the Aschheim-Zondek test a quantity of the woman's urine is injected into an immature white mouse; certain developmental changes take place in the reproductive organs of the mouse if the urine is that of a pregnant woman; in the absence of such changes the test is negative. In the simpler and quicker Hogben test, female toads are used; if they ovulate within 12 hours after injection with the urine, it indicates a positive result.

Nowadays *immunological tests* are widely used. These, too, are based on the detection of chorionic gonadotropin in the woman's urine. In particular, they rely on the occurrence of an antigen-antibody reaction (see page 44) and have the advantage of being more convenient than biological tests. In one procedure, for example, a rabbit is injected with chorionic gonadotropin, which acts as a "foreign" antigen in the rabbit's blood and gives rise to the formation of an antiserum containing antibodies. If suspected urine is added to a quantity of this antiserum taken from the rabbit, an antigen-antibody reaction will occur if the urine contains chorionic gonadotropin. As a result, the antiserum becomes inactive, which means that the pregnancy test is positive. If the antiserum remains active, it means that there was no chorionic gonadotropin in the urine and that there is no pregnancy. The residual activity of the rabbit antiserum can be verified in various ways. In one method, specially treated sheep's red blood cells are used, which react with active antiserum, in which case no agglutinative deposit will form in the test tube. On the other hand, if the antiserum is inactive, agglutination of the red blood cells will occur, meaning that the woman is pregnant (Fig. 6-143). Another method is based on the fact that particles of plastic with chorionic gonadotropin adhering to them will undergo agglutination only in the presence of active antiserum (Fig. 6-144).

Lately the possibility of establishing a simple temperature test for the early diagnosis of pregnancy has received a good deal of attention. It is based on the measurement of the basal temperature every morning before getting up. Normally the temperature shows a rise from about 36.5°C (97.7°F) to about 37°C (98.6°F) on the 15th or 16th day of the 28-day menstrual cycle; it then goes down again with the degeneration of the corpus luteum shortly before the start of menstruation. If pregnancy occurs, however, the corpus luteum remains in existence and the temperature continues at the higher level (Fig. 6-145).

For this test it is merely necessary—when the woman has missed a menstrual period and suspects that she is pregnant—to measure the basal temperature rectally for five minutes on three successive mornings and to repeat these measurements after one week. If the temperature in each set of measurements is found to be above 37°C (98.6°F), and provided that the woman is not suffering from some feverish infection or from thyroidal hyperfunction, it is very probable that she is pregnant.

PREGNANCY HORMONES, MORNING SICKNESS

The production of hormones is monitored through feedback circuits between the pituitary gland and the endocrine (hormone-secreting) glands under its overall control. The so-called gonadotropic hormones, i.e., the pituitary hormones that specifically act upon the ovaries, and their interaction with the follicular and corpus luteum hormones (estrogens and progestogens) produced by the ovaries, come within this hormonal control system. In this way, during the course of each menstrual cycle, the uterine lining (endometrium) is prepared for pregnancy (Fig. 6-146a) in a carefully regulated and coordinated sequence. This occurs in the proliferative phase of the cycle (see page 235) with the aid of follicular hormones (estrogens) and in the then following luteal phase with the aid of the corpus luteum hormones (progestogens), including more particularly progesterone. If fertilization of the ovum has occurred, the cycle is interrupted, and the zygote (fertilized ovum) embeds itself in the endometrium, this process being called

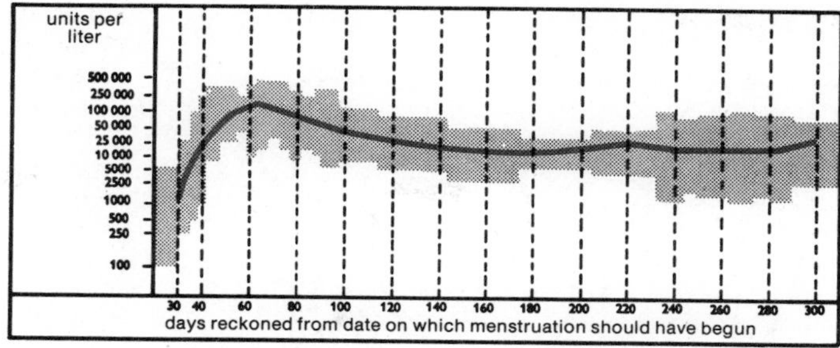

FIG. 6-142.
Content of chorionic gonadotropin in the urine of pregnant women.

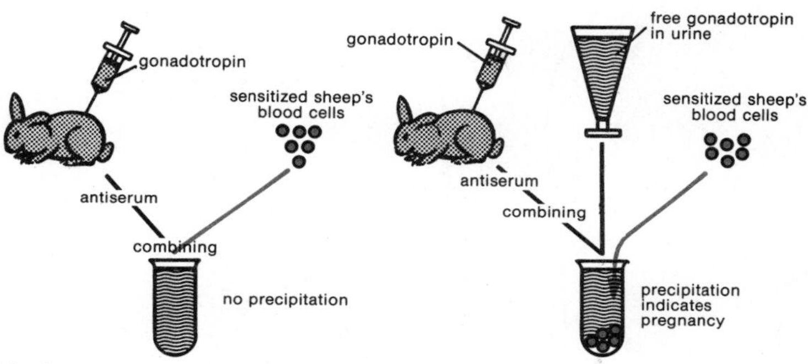

FIG. 6-143.
Immunological pregnancy test with sheep's blood cells.

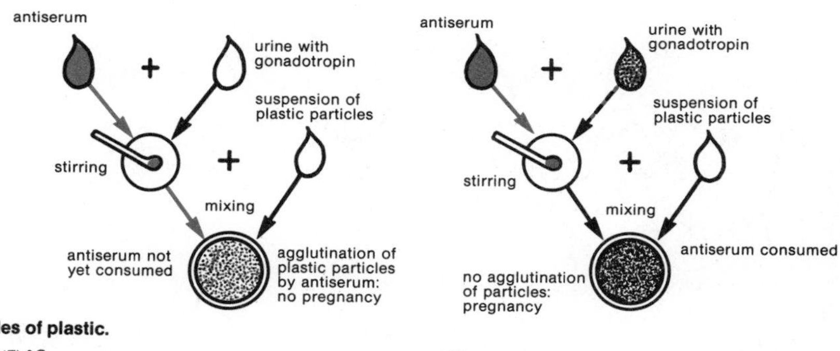

FIG. 6-144.
Immunological pregnancy test with particles of plastic.

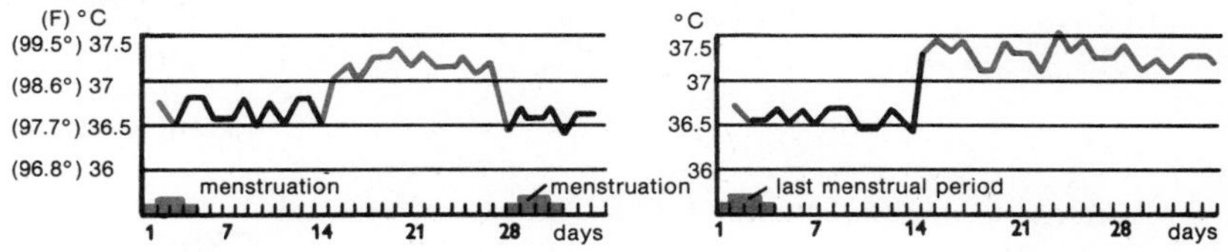

FIG. 6-145.
If the basal temperature remains high for more than 16 days, pregnancy is likely.

implantation or *nidation*. The developing fetus and particularly the placenta (see next section) take over the control function. The placenta produces hormones of its own, including chorionic gonadotropin (see preceding section), which are closely allied to the luteinizing hormone, i.e., the pituitary hormone that induces ovulation and formation of the corpus luteum.

FIG. 6-146.

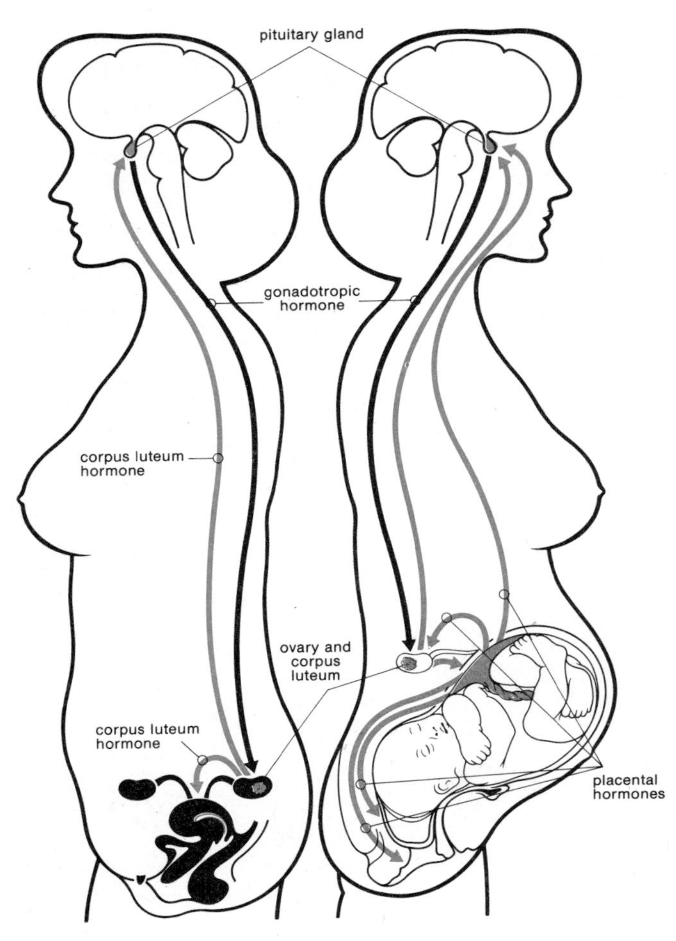

(a) Endometrium ready for pregnancy (the secretory phase is under the control of the pituitary gland).

(b) Pregnant uterus (the uterus is conditioned for pregnancy first by the corpus luteum, subsequently by the placenta).

Because of this hormonal activity the corpus luteum in the ovary does not degenerate (as it would do in the absence of pregnancy), but instead grows larger and increases its secretion of progestogens. In the fourth or fifth month of pregnancy it diminishes and eventually becomes inactive. By that time, however, the placenta is producing substantial amounts of estrogens and progestogens (Fig. 6-146b), the former being responsible for the development and growth of the uterus and breasts, the latter for maintaining the pregnancy.

The hormonal changes occurring in the first three months of pregnancy, which can in general terms be described as "chemical changes in the maternal blood," may cause symptoms such as loss of appetite, aversion to particular foods, and nausea, especially in the form of *morning sickness,* which usually starts between the 5th and 12th week of pregnancy and disappears after about the 16th week. A light snack—for instance, tea and a few biscuits—before rising may relieve the symptoms. Morning sickness is most likely to occur in women who are pregnant for the first time. Although believed to be associated with hormonal changes, the cause of morning sickness is not known with certainty. Psychic/emotional factors—anxiety over an unwanted pregnancy, the prospect of childbirth, etc.—may play a part, and in such cases reassurance and diversion of the woman's attention to other matters can be helpful. It is interesting to note that an imagined pregnancy may sometimes also be attended by morning sickness.

Exceptionally, more serious and persistent vomiting occurs in pregnancy (hyperemesis gravidarum). In such cases the patient may have as many as ten attacks of vomiting in the course of a day, after every intake of food or drink and also at other times. This can be a serious condition, involving losses of salt and water, rapid loss of weight, metabolic disorders, acidosis, and pathological changes in the liver causing jaundice. The condition is mostly encountered in highly sensitive, neurotic women, but the underlying cause is not known. Treatment will consist in keeping the patient resting quietly in bed, on a carefully controlled diet. Drugs to stop the vomiting may be given. Serious cases will require treatment in a hospital, with an intravenous drip to compensate for loss of fluid. In severe cases the pregnancy may have to be artificially ended by abortion.

PREGNANCY

Pregnancy begins with implantation, i.e., the embedment of the fertilized ovum in the uterine lining (endometrium) (see page 221). From this instant until birth, a period of 260 days elapses on average. As the fertilized ovum spends about 5 days on its journey through the fallopian tube, the entire length of time between fertilization and birth can be put at 265 days, i.e., about 38 weeks. However, this is a rather theoretical figure, since the precise date of conception is often uncertain. As a rule, there is a fairly constant interval of 15 days between the first day of menstruation and the occurrence of fertilization. Hence, for practical purposes, it is convenient to date pregnancy from the first day of the woman's last menstrual period, making a total of 280 days, i.e., 40 weeks. Naegele's rule gives the same result, without having to consult a calendar: to calculate the probable date of birth, add 1 year to the date on which the last menstruation commenced, subtract 3 months and add 7 days. For example: last menstruation started on 6/10/1977; probable date of birth is 3/10/1978 + 7 days = 3/17/1978. Calculations of this sort are, however, seldom "spot-on" accurate. A range of variation of about 3 weeks is quite normal, i.e., the birth may occur at any time between 10 days before and 10 days after the predicted date, and statistics show that a substantial proportion of births occur even outside this 3-week period. The calculations obviously become even more unreliable in cases where the woman has not kept a proper count of her menstrual dates or where conception occurred after a prolonged interval in menstrual activity, e.g., after childbirth. In such cirumstances the doctor has to rely on his own judgment and on such information as the mother can give on the first movements of the fetus that she can feel in the uterus; known as *quickening*, these movements usually manifest themselves at some time from the 18th to the 20th week of pregnancy. Inaccuracies and mistakes, e.g., due to confusion with bowel movements, are liable to affect predictions based on these phenomena, however. Also, wide variations are possible. Movements have been reported as early as the 10th week, whereas in other cases, though rarely, no movement is felt at any time throughout the pregnancy. Another indication of the probable time of childbirth can be obtained from the position of the uterus (see later).

After being fertilized in the fallopian tube, the ovum, now known as the zygote (see page 270), continues its journey through the tube toward the uterus. Within about 30 hours after fertilization the zygote begins to divide, a process called cleavage, as a result of which a progressively larger number of smaller cells (*blastomeres*) are formed. When there are about 16 blastomeres, the zygote is known as a *morula*. A cavity filled with fluid forms in the morula, which thus becomes a *blastocyst* by the time it reaches the uterine cavity.

In the uterus the blastocyst sinks into the endometrium and becomes embedded in it (*implantation* or *nidation*). After the blastocyst is implanted, the endometrium is known as the *decidua* (Fig. 6-147). In its early stages the blastocyst consists of a hollow ball of cells which is thickened in one place by the presence of an inner cell mass. Cells from this mass spread round the inside of the blastocyst, which thus becomes a two-layered structure, the enclosed cavity being called the primary *yolk sac*. A split then forms in the inner cell mass and develops into another cavity, the *amniotic cavity* (Fig. 6-148). Subsequently another split develops and extends almost the whole way round the yolk sac and amniotic cavity. The latter subsequently expands and becomes filled with a clear fluid (amniotic fluid or liquor amnii), which surrounds the embryo or fetus and serves to protect and support it. This fluid, which is continually being absorbed and renewed, increases in quantity as pregnancy progresses and amounts to about 1 liter at the time of birth. The tail end of the embryo remains connected to the chorion by the so-called body stalk. The *chorion* is the outer membrane which, in early development,

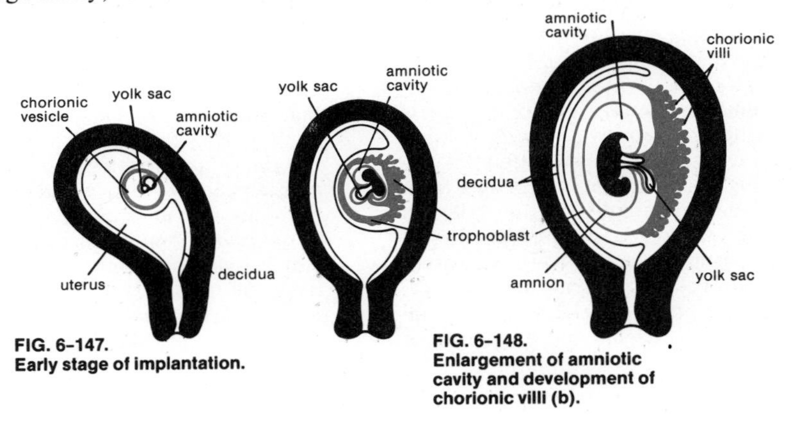

FIG. 6-147. Early stage of implantation.

FIG. 6-148. Enlargement of amniotic cavity and development of chorionic villi (b).

is the outer wall of the blastocyst; it completely surrounds the embryo or fetus and separates it from the maternal tissues. (In general, the term *fetus* is used when the embryo has developed to a stage where it is recognizably human, i.e., from about the 8th week of pregnancy.)

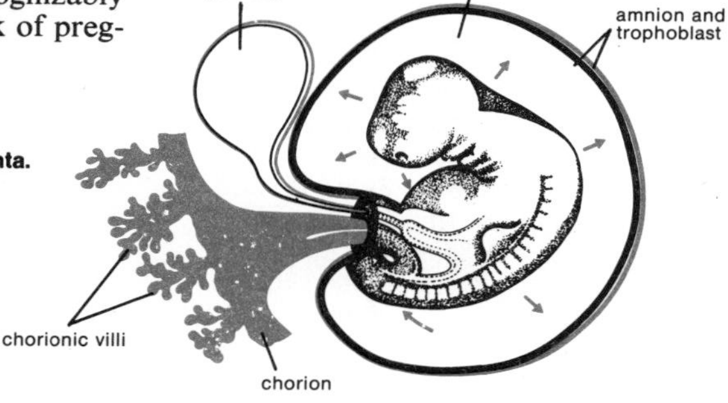

FIG. 6-149. Enlargement of embryo and formation of placenta.

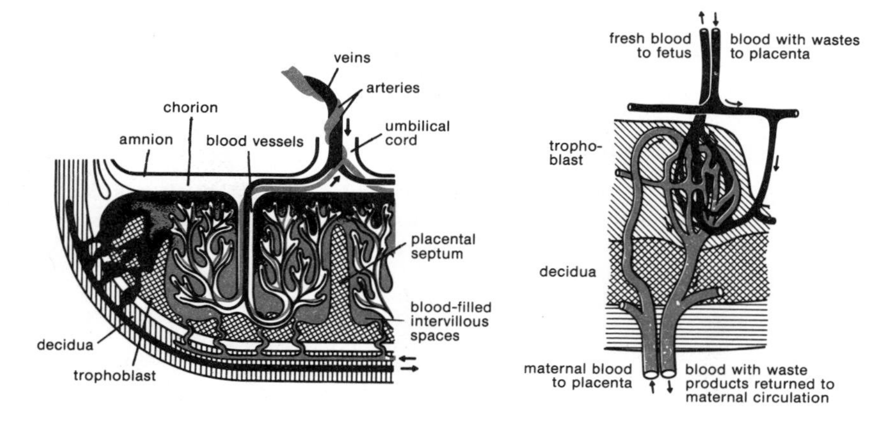

FIGS. 6-150 and 6-151. Fetal blood supply.

The outer layer of cells of the chorion develop tiny branching projections, called *chorionic villi,* which establish an intimate connection with the decidua (Figs. 6-148b and 6-149) and help to form the *placenta,* which is the organ by which the unborn infant obtains its nourishment. The placenta consists of a fetal portion comprising chorionic villi, which interlock with the decidua (the lining of the uterus) constituting the maternal portion. A dense network of blood vessels in the placenta communicates through the umbilical blood vessels (two arteries and one vein) with the circulation of the fetus; these vessels pass through the umbilical cord (Fig. 6-150), which develops from the body stalk and eventually reaches a length of about 50 cm. Although the fetal portion of the placenta is firmly attached to the maternal portion, no blood passes between the two, i.e., there is no intermixture of fetal and maternal blood (Fig. 6-151). Nutrients and oxygen from the maternal blood pass by diffusion into the fetal bloodstream in the placenta, while carbon dioxide and other waste products from the fetus diffuse in the opposite direction. The diffusion takes place through the chorionic villi.

The placenta also functions as an endocrine organ, producing hormones (chorionic gonadotropin, estrogens, progesterone) serving to promote and sustain pregnancy.

The fetus has its own blood and own blood circulation. Because of its separation from the maternal blood, most pathogenic microbes cannot cross the placenta from the mother's to the infant's bloodstream. There are some exceptions, however, including notably the spirochetes of the genus *Treponema,* especially *T. pallidum,* which causes syphilis. This organism can be transmitted to the unborn infant from the mother, causing congenital (not "hereditary") syphilis.

The mature placenta is a disc-shaped object,

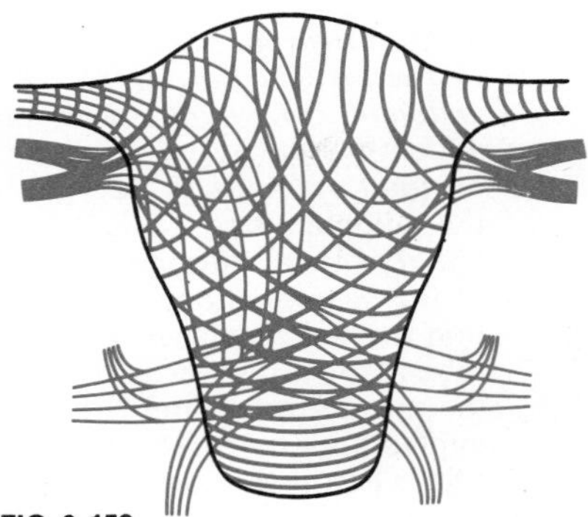

FIG. 6-152.
Configuration of muscle fibers in the uterus.

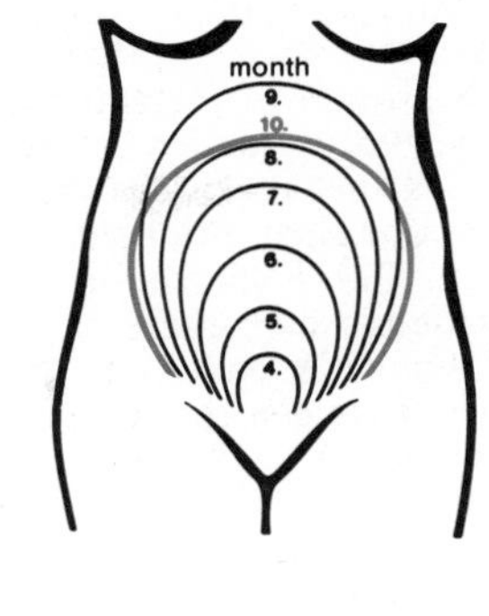

FIG. 6-153.
Progressive enlargement of the uterus during pregnancy.

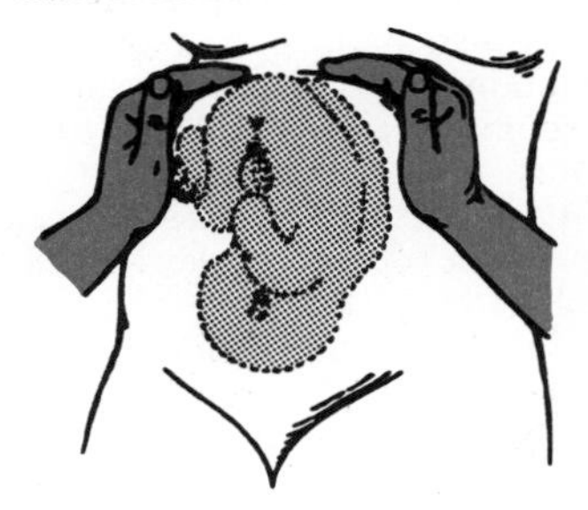

(a) Determining the position of the pregnant uterus.

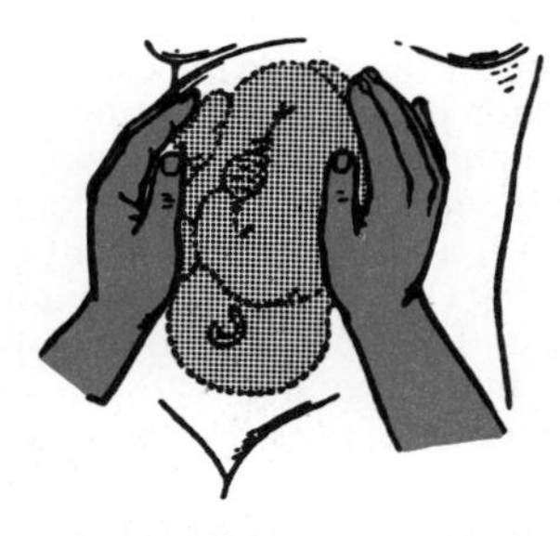

(b) Determining on which side the baby's back is situated.

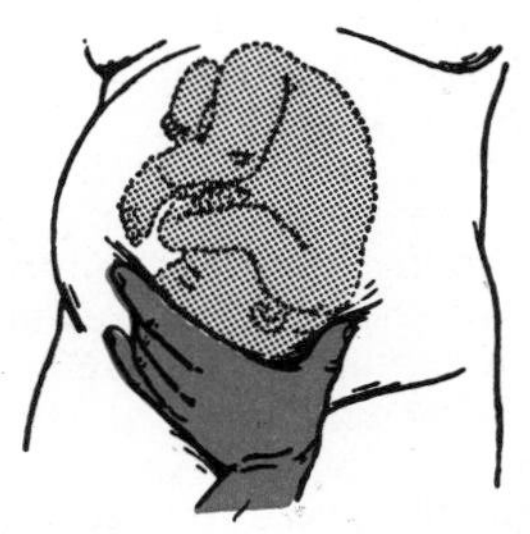

(c) Determining whether the baby's head is foremost and how deep it has descended.

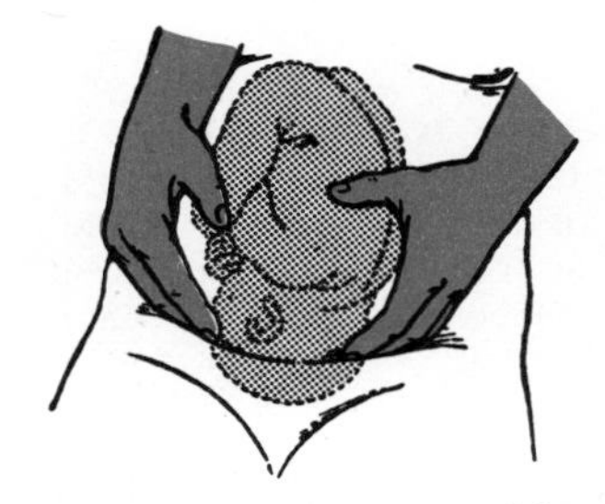

(d) Determining the progress of descent.

FIG. 6-154.

15–20 cm in diameter and 1.5–3 cm thick, weighing 450–500 g. There are about a hundred villi per square centimeter of placenta. Their total surface area is about 7 m^3; this is the contact area through which the transfer of nutrients, waste products, etc., to and from the fetus takes place by diffusion.

The *uterus* (womb) normally has a length of 6–8 cm and weighs about 50 g. During the course of pregnancy its weight increases 20- to 30-fold. Whereas the virgin uterus has an internal volumetric capacity of 2–3 cm^3, the capacity of the pregnant uterus at full term (just before childbirth) is 5000–7000 cm^3. This tremendous increase in size occurs not merely as a result of muscular growth; it is also due to the special configuration of the muscle fibers in the wall of the uterus (Fig. 6-152). They are arranged in a spiral pattern extending from the exterior to the interior; this enables the uterine cavity to become greatly enlarged without at first necessitating an increase in length of the fibers themselves. Subsequently, under the influence of hormones and the stimulus emanating from the growing fetus, the uterine muscles do undergo growth.

The combined growth of the fetus and the uterus causes the uterus to rise in the abdomen, i.e., extend farther upward, thus providing an approximate indication of the progress of the pregnancy. For instance, in the 24th week the top of the uterus is more or less on a level with the woman's navel; in the 32nd week it is midway between the navel and the lower part of the sternum (Fig. 6-153). About four weeks before birth the infant's head descends some distance, so that the uterus correspondingly goes down. This often brings a feeling of physical relief, enabling the woman to breathe more easily. The doctor can, by placing his hands on the woman's abdomen, feel the size and position of the uterus (Fig. 6-154a). In advanced pregnancy he can also perform certain other external manipulations to determine on which side the baby's back is (Fig. 6-154b), which part of its body will come out first (Fig. 6-154c), and how the descent of this part is progressing (Fig. 6-154d). Normally the baby lies with its head downward and its body longitudinally in the uterus.

A woman will, of course, put on weight during pregnancy, but the total increase should in a normal case not exceed about 12 kg (26½ lb). The following table shows how the weight is distributed (in a typical example):

	kg	*lb*
mature fetus	3.5	7¾
amniotic fluid, placenta	2	4½
increase in size of uterus	1	2¼
increase in size of breasts	1.5	3¼
other increases, especially fluid	4	8¾
	12	26½

EXTRAUTERINE PREGNANCY

Fertilization of the ovum normally takes place in the upper part of the fallopian tube. The fertilized ovum (zygote) then continues its progress along the tube and eventually, after about 5 days, embeds itself in the endometrium (lining of the uterus), this process being called implantation. If viable sperms already happen to be waiting near the ovary in readiness for the ovum to be released on ovulation, fertilization may occur (though rarely) before the ovum enters the fallopian tube. This may result in an extrauterine pregnancy (i.e., a pregnancy outside the uterine cavity), in the form of *ovarian pregnancy,* with implantation of the fertilized ovum in the ovary itself, or *abdominal pregnancy,* with implantation in the abdominal cavity in the vicinity of the ovary, or *tubal pregnancy,* i.e., pregnancy in one of the fallopian tubes, this being the commonest variety of extrauterine pregnancy (Fig. 6-155). In this last-mentioned case, although fertilization has taken place in the upper part of the tube in the normal way, the further progress of the zygote to the uterus is obstructed, so that implantation occurs in the wall of the tube itself. The obstruction may be due to an adhesion of the mucous membrane of the tube (Fig. 6-156) as a result of inflammation, e.g., due to gonorrhea, abortion or puerperal sepsis. It may also happen that the lining of the tube contains gaps or cavities in which the zygote becomes trapped (Fig. 6-157). Another possibility is that the fallopian tube is abnormally long and its muscles are too weak to help effectively in propelling the zygote along (Figs. 6-158 and 6-159).

In a case of tubal pregnancy the embryo nearly always dies prematurely, involving a serious risk to the pregnant woman. If the embryo is situated in the upper, relatively wide part of the fallopian tube, which is in fact by far the commonest form of tubal pregnancy, the inadequately nourished embryo will die in about 6–8 weeks and then be expelled from the tube into the abdominal cavity (so-called tubal abortion, Fig. 6-160a). On the other hand, if the zygote has become implanted farther down, in the narrower part of the tube, the developing embryo will burrow deeper into the wall of the tube and eventually perforate it (tubal rupture, Fig. 6-160b). This is a serious condition which may arise quite unexpectedly, without the woman even suspecting that she is pregnant. There is sudden abdominal pain, with severe internal bleeding into the abdominal cavity, which can quickly be fatal. The patient should be removed to a hospital at once for surgical treatment to stop the bleeding.

The symptoms associated with tubal abortion are, fortunately, not so acute as with tubal rupture: there is only a relatively small amount of bleeding into the abdominal cavity, occurring in spasms accompanied by pains somewhat resembling labor pains felt on one side of the abdomen and perhaps continuing intermittently for weeks. Accompanying symptoms are weakness, rapid pulse, pallor and sweating. A blood sample extracted through an exploratory puncture will show that surgical in-

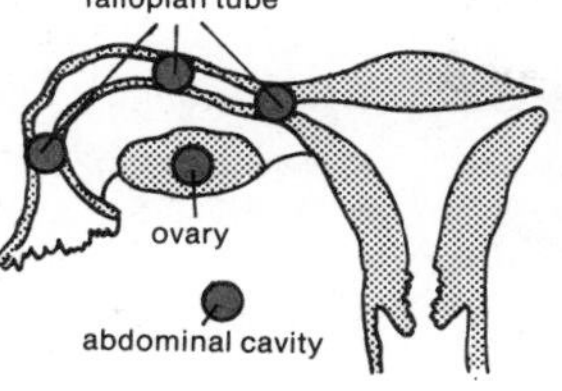

FIG. 6-155.
Possible places where implantation may occur.

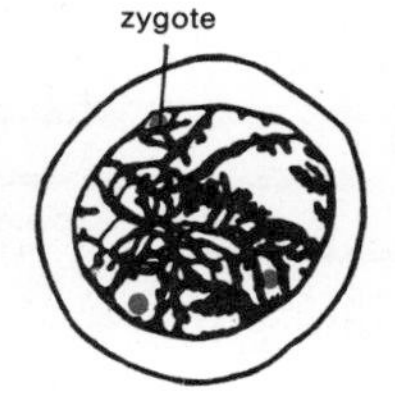

FIG. 6-156.
Adhesions in the oviduct (fallopian tube).

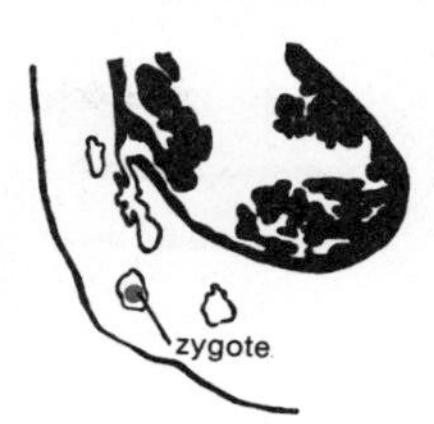

FIG. 6-157.
"Blind alleys" in which the fertilized ovum may become trapped in the wall of the fallopian tube.

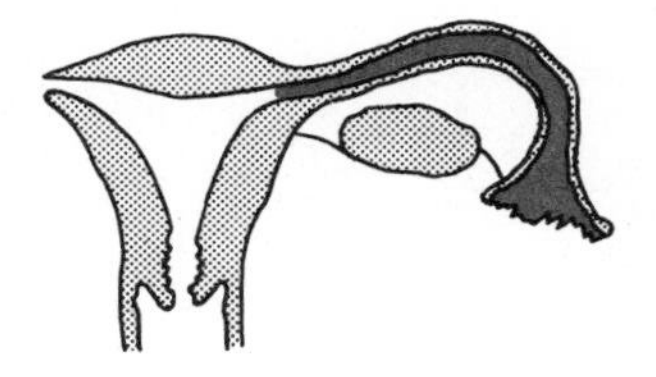

FIG. 6-158.
Normal fallopian tube.

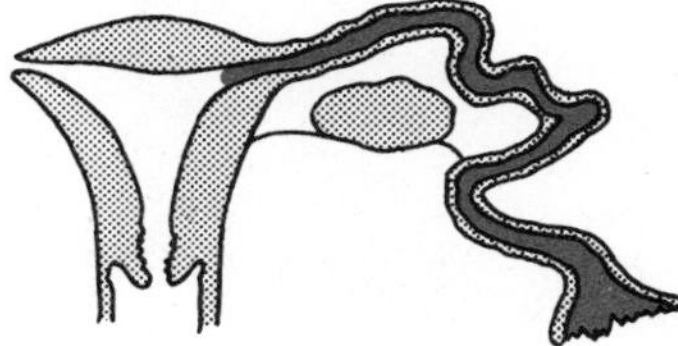

FIG. 6-159.
Elongated and weakened muscles of fallopian tube.

FIG. 6-160.
CONSEQUENCES OF TUBAL PREGNANCY.

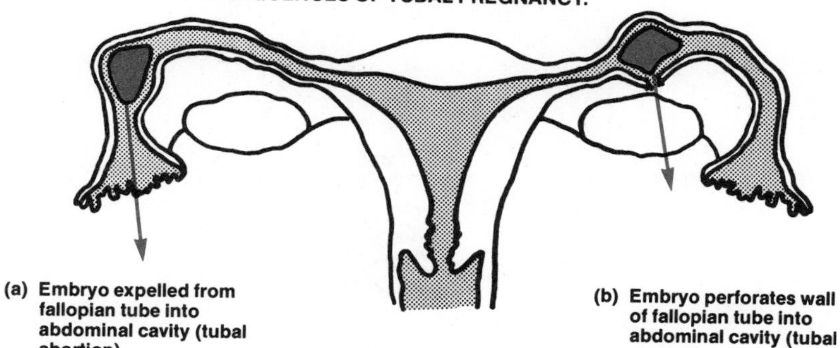

(a) Embryo expelled from fallopian tube into abdominal cavity (tubal abortion).

(b) Embryo perforates wall of fallopian tube into abdominal cavity (tubal rupture).

tervention is necessary. As in a case of tubal rupture, the operation will consist in removal of the affected fallopian tube, without damaging the ovary, if possible. Blood transfusions may help to save the patient's life.

Close attention to warning signs can be very important. During the first few weeks of an extrauterine pregnancy there are no symptoms other than the general ones of cessation of menstruation and raised basal temperature (see page 278). Extrauterine pregnancy may be suspected if blood is discharged from the vagina 6–8 weeks after the last menstruation (i.e., 2–4 weeks after the "missed" menstruation). In most of these cases only small amounts of blood are involved, usually not enough to be mistaken for delayed menstrual bleeding. Such bleeding, together with labor-like pains felt on one side (particularly in a case of tubal abortion), may indicate extrauterine implantation and death of the embryo.

MISCARRIAGE (ABORTION)

Medically the terms *miscarriage* and *abortion* are synonyms denoting the termination of pregnancy before the fetus is viable. In nonmedical usage, however, *miscarriage* is generally used to denote accidental or sponta-

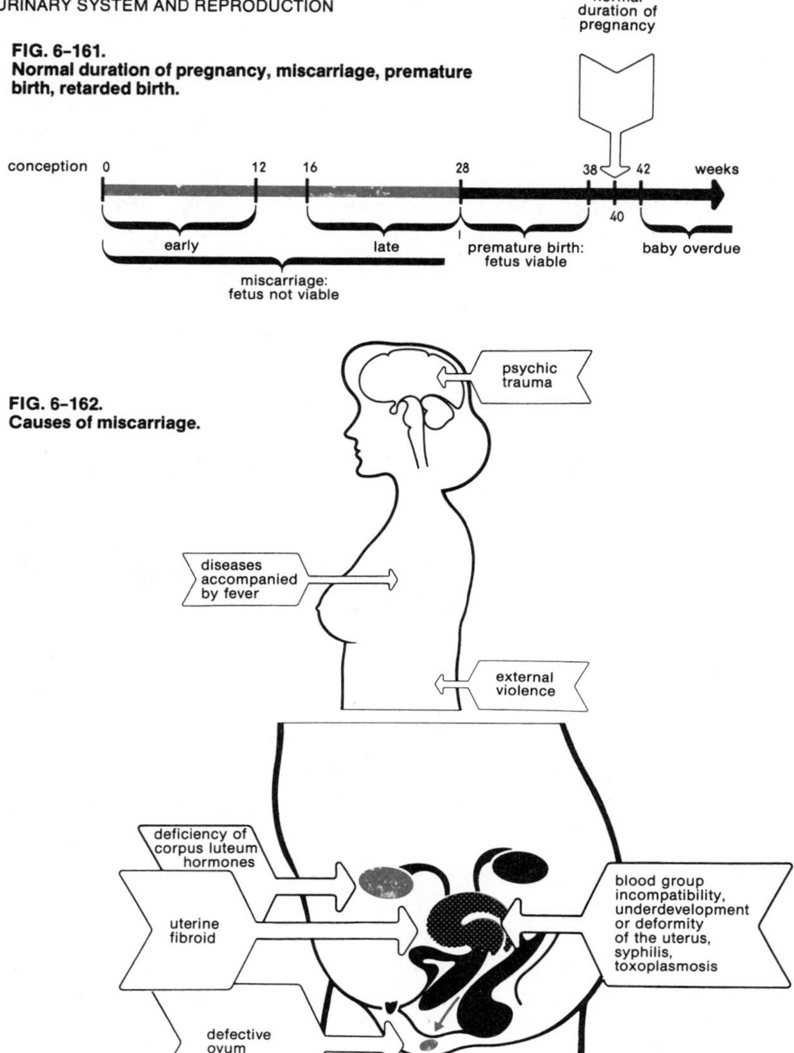

FIG. 6-161.
Normal duration of pregnancy, miscarriage, premature birth, retarded birth.

FIG. 6-162.
Causes of miscarriage.

neous abortion, while *abortion* is reserved for deliberate termination of pregnancy. The average normal duration of pregnancy is 280 days. If it continues for more than two weeks beyond this period, there is said to be retarded (overdue) childbirth. If pregnancy ends after 20–38 weeks, the child is said to be born prematurely. A premature baby is one that is at least potentially viable, whereas if pregnancy ceases within the first 28 weeks, the fetus is generally not viable, in which case the terms *miscarriage* and/or *abortion* are applicable (Fig. 6-161).

Miscarriage is a fairly common complication of pregnancy. There are many possible causes (Fig. 6-162), arising from some defect either in the mother or in the developing embryo or fetus. Defects of the mother include underdevelopment of reproductive organs, deformity of the uterus, narrowness of the pelvis, tumors in the birth passage, etc. Defective physical condition due to such causes as hunger, vitamin deficiency, hormone deficiency, diseases accompanied by high fever, toxoplasmosis and syphilis may also cause miscarriage. Other causal factors are blood group incompatibility (see page 282) and psychic trauma, such as severe fright or overexcitement.

More often the cause of miscarriage is in the embryo itself. Upward of 50 percent of miscarriages are believed to be due to some defect that prevents the embryo from developing normally. This may be due to hereditary factors, but more often there is some extraneous cause such as oxygen deficiency, radiation damage or poisoning.

In the first three months of pregnancy the embryo or fetus is still small and not highly developed. Its attachment to the wall of the uterus is formed by fairly short and as yet not very firmly anchored chorionic villi. In this stage it can relatively easily be expelled as a single whole, this being accompanied by bleeding from that part of the uterine lining where the embryo was embedded. After the fourth month the placenta, fetal membranes and amniotic fluid have formed. From this stage onward, miscarriage resembles full-term childbirth, i.e., with labor pains, rupture of the membranes, and stages of labor.

The first symptoms of an impending miscarriage are discharges of blood, usually in small amounts, sometimes attended with lumbar or abdominal pains. In a more advanced stage of pregnancy these may resemble labor pains. Such symptoms are a warning and do not necessarily mean that miscarriage is inevitable; the situation is what is known as a *threatened miscarriage*. With rest, sedation and abstinence from sexual intercourse, the pregnancy may continue for its full normal term. Hormone treatment is sometimes given, but its value in such circumstances has been disputed. Purgatives and the local application of heat or cold tend to promote miscarriage and should therefore be avoided. The pregnant woman should remain in bed for at least a week after any bleeding has ceased. Death of the fetus can be diagnosed by pregnancy tests, which in that case give a negative result. If bleeding continues, miscarriage usually becomes inevitable. In most cases the bleeding is not severe, but if the fetus is not entirely detached, surgery may be needed to complete the abortion and stop the bleeding. Sometimes scraping-out treatment (curettage) of the uterus is indicated after a miscarriage in order to remove all traces of placenta and prevent recurrence of bleeding. Fever is a symptom of infection similar to puerperal sepsis. Such complications are usually treated with antibiotics.

Habitual Miscarriage

Habitual miscarriage can be said to occur when a woman has had three or more consecutive miscarriages. At least 50 percent of early miscarriages, in the first three months of pregnancy, are believed to be due to defects in the embryo, either from hereditary or from extraneous causes. At present there is no known method of treating such defects. In other cases of habitual early miscarriage it is quite likely that hormone deficiency (especially of progesterone) is a causal factor. Hormone therapy may therefore be helpful as a preventive.

Habitual late miscarriage (fourth to seventh month of pregnancy) can be due to any one of a number of possible causes, generally the same as those already mentioned (see also Fig. 6-163). Toxoplasmosis, a disease due to infection with protozoa (*Toxoplasma gondii*), is nowadays thought to be a more frequent cause

FIG. 6-163.
Causes of habitual miscarriage.

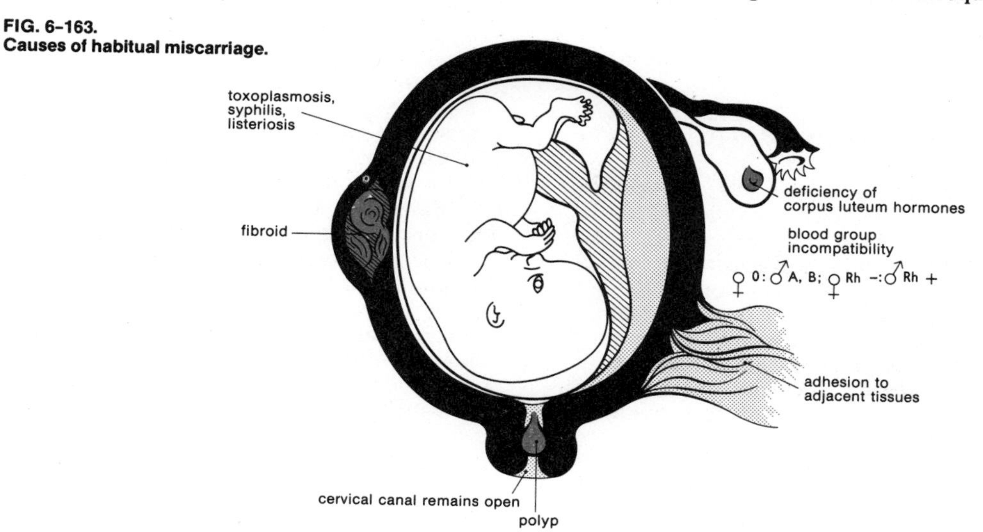

of miscarriage than used to be supposed. It can be treated with drugs. Much less frequently, syphilis and listeriosis may cause miscarriage; these, too, are treatable diseases. Listeriosis is caused by a microorganism that lives in soil and may become pathogenic for animals or man, causing meningitis. The importance of blood group incompatibility with regard to miscarriage is still a matter of some controversy. Lead poisoning and prolonged vitamin deficiency are other possible causes, but are rare. In many cases of habitual miscarriage, no diagnosable cause can be found.

Uterine tumors (fibroids, polyps) or a retroverted uterus which has become coalesced with adjacent tissues and is unable to right itself during the pregnancy, may likewise cause miscarriage, but these defects can be corrected by surgery. Surgical intervention is less likely to be successful in a case where the uterus is actually deformed. An underdeveloped uterus can sometimes be induced to develop to fuller maturity by hormone treatment. A uterine defect that is intrinsically suspect as a cause of miscarriage is failure of the cervix to remain closed during pregnancy. Some time after a miscarriage this defect can be surgically corrected by the application of a constricting suture around the cervix to reduce the aperture of the cervical canal.

During a pregnancy a woman who has had habitual miscarriages will have to take precautions as mentioned earlier (for threatened miscarriage) and continue with them for a greater length of time. The risk of miscarriage arises particularly at times when, in the absence of pregnancy, she would have menstruated. If premature labor pains occur, the patient may be given a general sedative together with progesterone.

BLOOD GROUP INCOMPATIBILITY BETWEEN MOTHER AND CHILD

In modern blood transfusion practice the risk of incompatibility between the recipient's and the donor's blood has been virtually eliminated by tests for the determination of blood groups (see pages 113 ff). The combination of mutually incompatible blood groups of mother and unborn infant can, however, cause serious complications. More particularly the so-called *rhesus factor* or *Rh factor* is important in this connection (see page 113). This factor is present in the red blood cells of about 85 percent of the population (who are Rh-positive) and is absent in the other 15 percent (Rh-negative). When an Rh-negative mother becomes pregnant by an Rh-positive father, their child may be hereditarily Rh-positive and its blood therefore different, in this respect, from that of its Rh-negative mother. Although the mother's circulation is separate from the fetus's, red blood cells may nevertheless sometimes pass from the fetal into the maternal bloodstream. Antibodies (anti-Rh agglutinins) are then formed in the maternal blood to counteract and destroy the "foreign" Rh-positive intruders. This can present a complication in connection with subsequent blood transfusions that may be given to the mother. What is of more immediate importance is that these antibodies may cross the placenta in the opposite direction and thus enter the fetal blood circulation (Figs. 6-164 and 6-165), where they may destroy so high a proportion of the red blood cells that the fetus suffers severe anemia. Experience has shown that in a first pregnancy the protective effect of the placenta in keeping the fetal blood uncontaminated is usually adequate to prevent damage to the fetus. With subsequent pregnancies of an Rh-negative woman by an Rh-positive man, the risk of having an Rh-positive baby that will suffer serious damage or die prematurely as a result of the action of the antibodies becomes progressively greater. The degree of hazard to the fetus can be assessed on the basis of tests on the mother's blood.

When an Rh-negative woman has children by an Rh-negative man, all their children will be Rh-negative, so that there will be nothing to fear from the Rh factor. If the woman is Rh-negative and the man Rh-positive, the question arises whether the man is, in respect to the Rh factor, of a "pure" strain or is a "hybrid" (Fig. 6-166). In the former case, i.e., if the man is "pure," the Rh-positive character will be dominant in all the children; in the case of a "hybrid" man, however, the children will have a 50–50 probability of being Rh-positive or Rh-negative. Thus, in such a marriage, if the woman has already given birth to Rh-damaged or stillborn babies, there will be only a 50–50 chance of having subsequent healthy children.

In general terms the risk of complications due to the Rh factor theoretically exists in about 15 percent of marriages. Fortunately, the anti-Rh agglutinins are formed in only

FIG. 6-164.
Antigen-antibody reaction in blood group incompatibility in the Rh system.

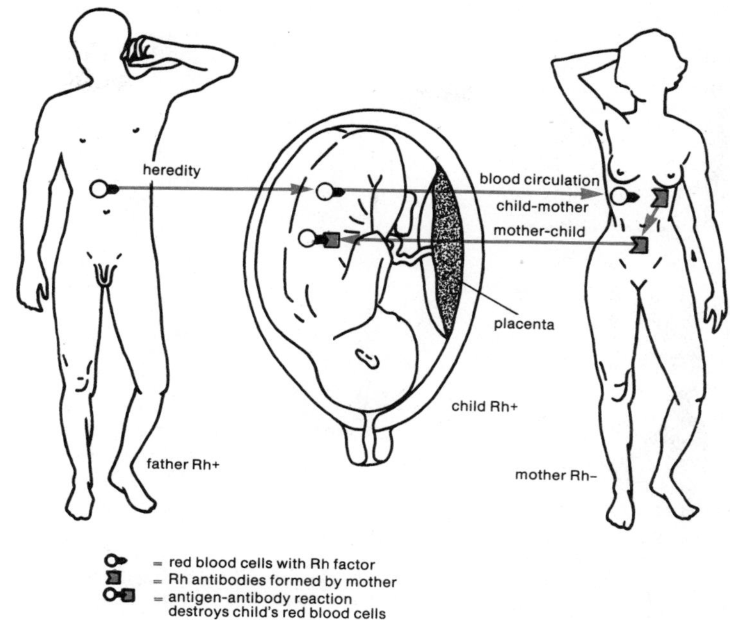

FIG. 6-165.
Red blood cells (Rh antigen) and Rh antibodies may cross the placenta.

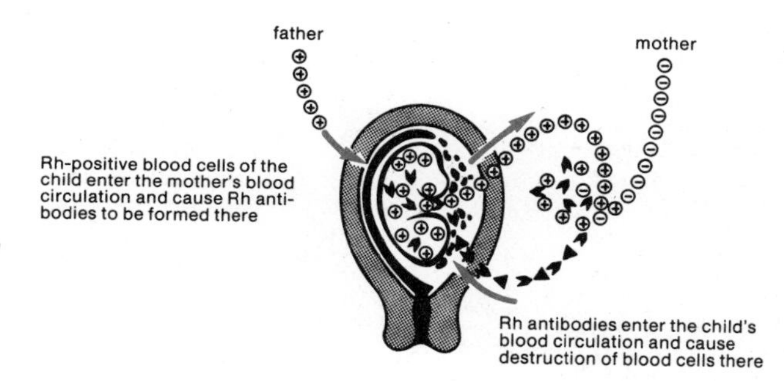

about 5 percent of all women, and even in these the amounts formed are not necessarily always seriously dangerous to the fetus. Also, even a damaged fetus can, with modern medical resources, often be saved. Precautionary measures include early admission to the hospital in suspect cases, examination of a specimen of the amniotic fluid to assess the degree of damage to the fetus, transfusion of Rh-negative blood into the peritoneal cavity of the fetus, premature termination of pregnancy (not abortion in the nonmedical sense) to protect the baby, and replacement of the newborn baby's blood with Rh-negative blood by exchange transfusion.

The best results are obtained when precautions are taken in good time. Foremost among these is the determination of the woman's blood group. If she is found to be Rh-negative, the next step will be to determine the father's blood group. If he is Rh-positive, it means that a risk of blood group incompatibility exists. It is then advisable to determine the presence and the concentration of antibodies in the mother's blood and to continue checking at four-weekly intervals. For assessing the haz-

FIG. 6-166.
Hereditary pattern of blood groups in the Rh system.

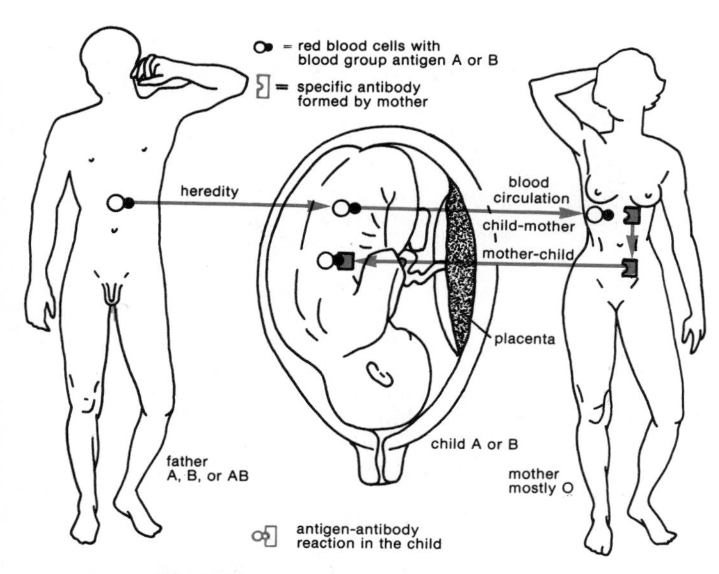

FIG. 6-167.
Antigen-antibody reaction in blood group incompatibility in the ABO system.

ard of fetal damage in subsequent pregnancies, i.e., after the birth of the first child, it is desirable to determine whether the Rh-positive father is genetically pure or hybrid in respect to the Rh factor, as already mentioned.

There are three degrees of severity with which newborn infants may be affected by Rh incompatibility: In the first degree, destruction of the baby's red blood cells causes anemia, to which the bone marrow responds with increased production of red blood cells, some of which appear as immature cells in the blood. This is known as hemolytic disease of the newborn infant (hemolysis is breakdown of red blood cells). Exceptionally, this anemic condition may develop even several weeks after birth. The second degree of severity is characterized by accumulation in the baby's body of bile pigment, as the breakdown product of blood pigment, causing jaundice (hemolytic jaundice of the newborn). The bile pigment is poisonous, especially to the nervous system, and in serious cases it may cause permanent damage to the baby's brain or indeed kill the baby. The third degree of severity occurs when the antigen-antibody reaction also damages the fetal capillary blood vessels, so that fluid from the blood seeps into the tissues and cavities. Such babies, and those that are born with heart failure due to severe anemia, are stillborn or die soon after birth. The standard method of treatment for babies born with hemolytic disease is exchange transfusion: the baby's blood is removed by draining it from an artery, while at the same time compatible Rh-negative blood is given intravenously. Transfusion of blood into the fetus before birth is now possible, and methods of suppressing the mother's antibodies before they can harm the fetus are also available.

Besides the Rh system, the conventional *ABO system* of blood groups (A, B, AB and O), as described on page 113, can in certain circumstances also cause incompatibility problems in connection with pregnancy. In this system the natural antibodies against "foreign" blood cells are already hereditarily present in the individual. However, there may also be a process of sensitization, just as in the Rh system, when "foreign" blood cells enter the bloodstream; additional antibodies may then be formed in the ABO system too. Such sensitization may occur as a result of a transfusion or during pregnancy. In the latter case the mother's antibodies may enter the fetal circulation and cause hemolysis, as in the Rh system already described (Fig. 6-167). Serious incompatibility symptoms in the ABO system are, however, much rarer than in the Rh system and usually also less serious. On the other hand, even a firstborn infant is liable to be affected by ABO incompatibility. The risk is greatest in babies with blood group A born to mothers with group O. If high concentrations of blood pigment breakdown products are found in such babies, exchange transfusion may be necessary.

ECLAMPSIA

Eclampsia may occur during the last three months of pregnancy, during labor or, though rarely, after delivery. It is characterized by fits which may be fatal to the baby and sometimes also to the mother. Preliminary symptoms may comprise a rise in blood pressure, excretion of protein in the urine (albuminuria), and swelling of the ankles and other parts of the body due to accumulation of water in the tissues (edema). Eclampsia is generally attributed to "toxemia of pregnancy," suggesting the presence of toxic substances in the blood, though in fact no such substances have definitely been identified. It is thought that defective nervous-hormonal control and failure of the organism to adapt itself properly to pregnancy are underlying causes. The rise in blood pressure is believed to be produced by vascular spasms which reduce the flow of blood to certain organs, particularly the kidneys, liver, uterus and placenta, which respond by releasing chemical agents that enter the bloodstream and cut down the supply of blood to other organs by constricting the blood vessels in order to restore the supply to the organs originally affected. This pattern of events is liable to produce a vicious circle. The placenta, in particular, is thought to secrete such vasoconstricting agents which may cause many of the symptoms associated with eclampsia. Loss of protein in the urine occurs in connection with tubal nephritis in the kidneys; the edema is due to the kidneys' failure to eliminate salt, causing retention of water; the spasms are due to deficient blood supply to the brain. A hazard to the fetus arises from deterioration of the placenta. In some cases the mother's and occasionally the baby's life can be saved only by premature termination of the pregnancy.

Early symptoms of eclampsia are swelling of the legs and feet which does not disappear upon rest in bed. Subsequently there is swelling of the hands or puffiness of the face. Retention of water causes the patient to gain more weight than would normally be expected. During the final months of pregnancy the increase in weight should not exceed about 0.5 kg (1 lb) per week, and the total increase in the entire pregnancy should not be more than about 10–12 kg (22–26½ lb). Besides albuminuria, a rise in blood pressure is an early warning sign of eclampsia. The imminent onset of eclampsia is heralded by severe headache, dizziness, vomiting, epigastric pain, dimness of vision, spots and flashes of light before the eyes. Unless preventive action is taken in time, convulsions occur which resemble epilepsy, involving the whole body, and are followed by coma, which may last up to several hours.

Treatment of eclampsia is prophylactic, with careful watching of the patient's blood pressure, urine and weight. She should be kept quiet and put on a suitable diet (low-salt, with easily digestible proteins, carbohydrates, no fat). There is then every chance that her condition will settle down and that her pregnancy will run its normal course. At a later stage, however, drugs to lower the blood pressure and promote the excretion of salt may be given, as well as antispasmodics and sedatives. If eclamptic fits occur, the patient will have to be removed to a hospital and put under sedation, as recurrence of fits can be dangerous. All stimulation by sudden noise, light or touch should be avoided.

Infant mortality formerly occurred in 30–50 percent of cases of eclampsia, but with modern methods of treatment of the mother's condition and delivery of the baby by Cesarean section, this has been reduced to 3–4 percent. Eclampsia itself is uncommon, occurring in only about 1 in 500 pregnancies.

RETARDED BIRTH

The average duration of pregnancy, reckoned from the beginning of the last menstrual period, is 280 days. Birth is said to be retarded if it is delayed more than about 14 days beyond the calculated date of birth and there is still no spontaneous labor. If mere errors of calculation can be ruled out and if, moreover, the estimate of the approximate time of birth, based on the occurrence of the first movements of the fetus (quickening), is in agreement with the doctor's diagnosis of the due date, so that the baby really is overdue, steps are usually taken to induce labor by artificial means, in order to prevent possible damage to the baby.

In order to avoid wrong decisions and not to induce labor unnecessarily, attempts have long been made to find reliable criteria for determining whether or not the unborn infant is at hazard due to overlong pregnancy. The modern technique of amnioscopy provides such a criterion. It is based on visual inspection of the amniotic cavity, through the enveloping fetal membranes, with the aid of a special instrument called an amnioscope which is introduced into the uterus through the cervical canal. An early sign of danger to the fetus is that a discharge of fecal matter occurs from its intestine, observable as a greenish discoloration of the amniotic fluid. A decrease in the amount of fluid is also an indication of impending danger to the fetus.

When such signs are observed, there is every justification to induce labor. This can sometimes be achieved by a warm bath followed by an enema. If such simple methods fail, releasing the amniotic fluid from its sac may bring about the desired result, i.e., set off the contractions of the uterus which expel the baby. This can be done with a cannula under visual inspection through the amnioscope.

If the uterine contractions still fail to develop, an injection of oxytocin, a pituitary hormone that stimulates the uterus to contract, may be given after a few hours. Meanwhile the baby's condition can be monitored by listening to its heart sounds. In doubtful cases it is possible to obtain samples of the baby's blood by means of a special instrument. If oxytocin fails to induce labor, it usually becomes necessary to deliver the baby by Cesarean section.

At birth the baby is normally covered with a deposit of fatty material called the *vernix caseosa*. In cases of retarded birth this deposit may be absent, and the skin of such babies is sometimes yellowish and wrinkled. These are

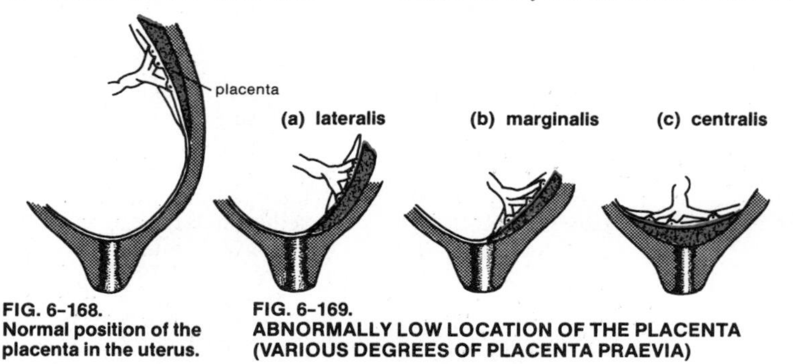

FIG. 6-168.
Normal position of the placenta in the uterus.

FIG. 6-169.
ABNORMALLY LOW LOCATION OF THE PLACENTA (VARIOUS DEGREES OF PLACENTA PRAEVIA)

FIG. 6-170.
If the placenta covers the internal os, normal birth is not possible without detachment of the placenta.

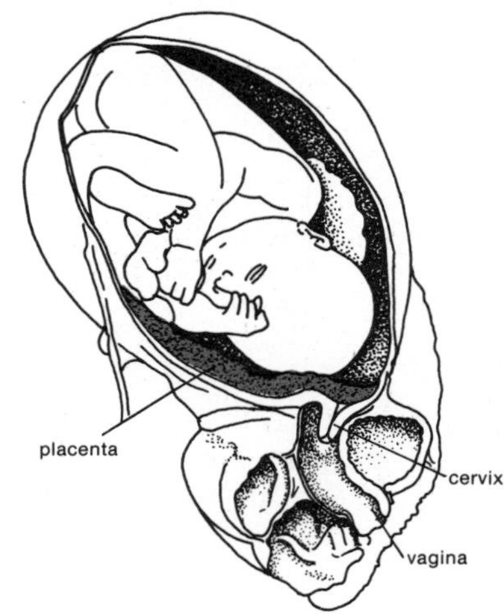

FIG. 6-171.
Hemorrhage due to detachment of the placenta in a case of placenta praevia.

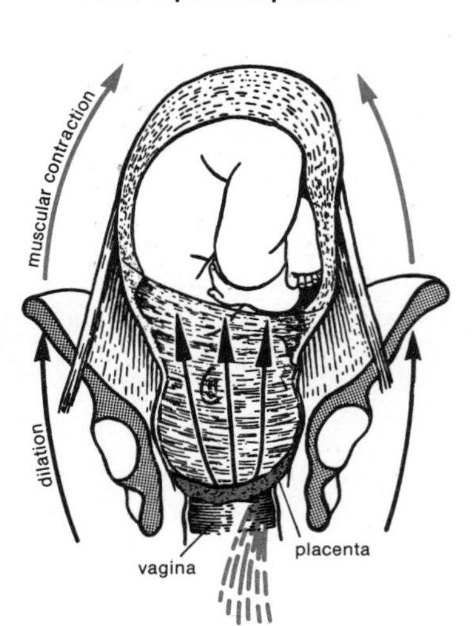

merely temporary blemishes.

The underlying causes of overlong pregnancy are not fully known. It does appear certain that infant mortality is higher in such cases, probably because the function of the placenta declines, so that in the end the baby suffers damage from a deficiency of oxygen and nutrients.

PLACENTA PRAEVIA

Normally the placenta is attached to the wall of the upper part of the uterus (Fig. 6-168), this being the area where implantation of the fertilized ovum, on arrival in the uterus after its journey through the fallopian tube, is most likely to occur. In a small proportion (0.2–0.5 percent) of pregnancies, however, implantation is located lower down in the uterus, near or indeed at the opening (internal os) of the cervical canal. The reason for such abnormally low location of the placenta, a condition known as *placenta praevia*, is probably not that the fertilized ovum travels exceptionally fast, but that it fails to find adequate nutritional conditions in the upper part of the uterus. Placenta praevia is indeed more likely to occur in women who have borne several children, especially in quick succession, as a result of which the mucous membrane lining (endometrium) higher up in the uterus has deteriorated and is thus relatively poorly supplied with blood. Effects of inflammatory disease of the uterus, repeated miscarriages, and curettage adversely affect the normal implantation possibilities.

Three types, or degrees of severity, of placenta praevia are to be distinguished: placenta praevia lateralis (Fig. 6-169a), marginalis (Fig. 6-169b: the placenta partly covers the internal os) and centralis (Fig. 6-169c: the placenta completely covers the internal os). In the last-mentioned case, normal birth, whereby first the baby and then the placenta emerge, is not possible (Fig. 6-170). Either the baby must be delivered through the injured and severely bleeding placenta, or the life-sustaining placenta must detach itself before the baby can descend through the pelvis. It is evident that such complications are dangerous and that a Cesarean section may be the only way to save the lives of both mother and child.

In all cases of placenta praevia there is a serious risk of bleeding, since the placenta is in a part of the uterus that has to undergo considerable expansion at childbirth. When uterine contractions commence, the placenta is unable to accommodate itself to this expansion and becomes detached from its site, which is, as it were, torn away from under it (Fig. 6-171). Slight hemorrhage, recurring with increasing severity, usually manifests itself in the seventh or eighth month of pregnancy. Sometimes, however, these warning signs are absent, and severe bleeding occurs fairly suddenly with the onset of labor. In general, dangerous uterine bleeding in the last three months of pregnancy may arise not only with placenta praevia (though this is the most frequent cause) but in any situation where the placenta becomes prematurely detached from the wall of the uterus. In any such circumstances the patient should be removed to a hospital without delay. Treatment will consist in conserving the blood supply before and during delivery, measures to control bleeding after childbirth and to combat anemia, and prevention of sepsis. In about a third of these cases it is possible to deliver the baby without recourse to a Cesarean section. If the bleeding in the final stages of pregnancy is severe, if it is not a case of placenta praevia centralis and if the mouth of the cervical canal has already opened a little, labor may be induced by intentionally puncturing the membranes to release the fluid. This intervention also causes the baby's head to descend deeper and helps to stop the flow of blood. With placenta praevia centralis, a Cesarean section is the only way to deliver the baby.

PREMATURE BIRTH

The normal duration of pregnancy is 40 weeks on average. If pregnancy is terminated between the 29th and 38th week, the baby is said to be premature (Fig. 6-172a). In practice it is difficult to obtain reliable information as to the precise length of the gestation period in any particular case. For this reason a premature infant is defined by international standards as one having a weight at birth of 2500 g (5½ lb) or less, irrespective of the actual period of gestation. This is a rather arbitrary definition which could, and does, include a proportion of full-term babies that happen to be of low birth weight but are otherwise mature. The average length of a baby born prematurely is

FIG. 6–172.

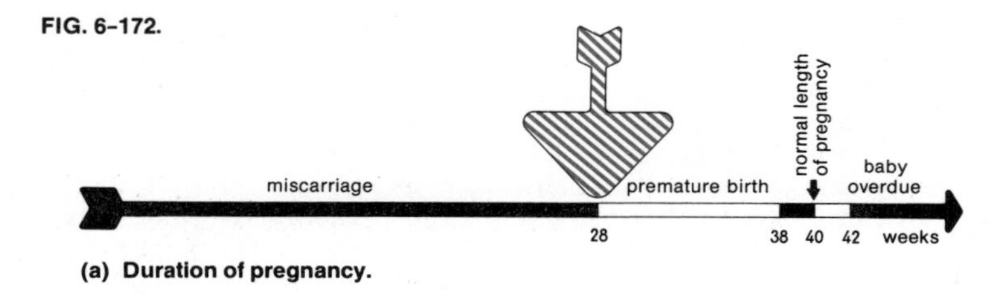

(a) Duration of pregnancy.

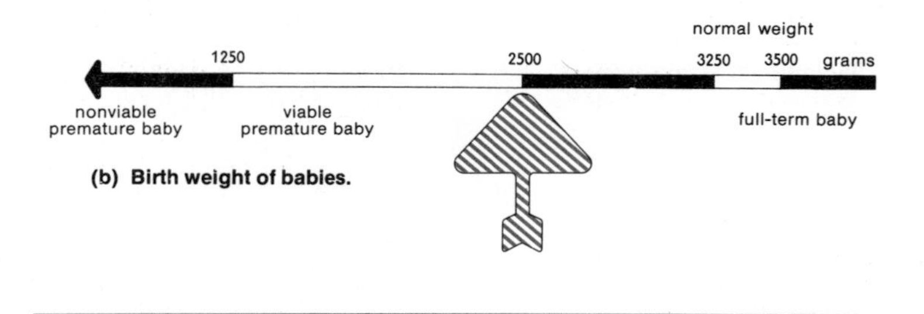

(b) Birth weight of babies.

FIG. 6–173.
Danger factors in premature birth.

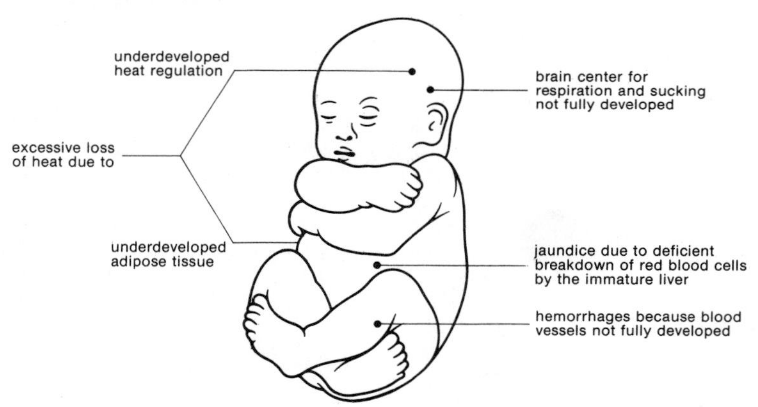

about 35 cm at 28 weeks, 47 cm at 37 weeks. Prematurity occurs in 5–10 percent of all births and is therefore quite common.

In something like 50 percent of all such cases, the precise cause of premature birth is not known. It is, however, to some extent associated with poor environmental conditions, being especially frequent in countries where nutritional standards and living standards in general are low. It would also appear that the incidence of prematurity is statistically higher in cases of unwanted pregnancy. A hereditary tendency is also present in some cases. Physical causes include diseases affecting the mother, such as acute infections (including syphilis, diabetes and hyperthyroidism). Physical or emotional trauma may also bring on premature birth (overexertion, accidents, fright, etc.). Prenatal disease affecting the fetus (syphilis, toxoplasmosis), blood group incompatibility, and the presence of twins in the uterus are other possible causes.

In purely mechanical terms a premature delivery is generally easier than a full-term one, because the baby is smaller and its skull softer. This easier delivery is a boon to the baby's chances of survival. Premature babies are frequently handicapped by anatomic and physiological limitations, which are generally more severe in proportion to the degree of immaturity. They include impaired kidney and liver function, immature development of the lungs, respiratory weakness, and inadequate regulation of body temperature due to sluggish circulation. For this last-mentioned reason the baby must be kept warm. During labor the mother must not be given pain-killing drugs or anesthetics that tend to suppress respiration, because they are liable to paralyze the baby's immature respiratory center.

Premature babies have a bright red color and look thin because their adipose tissue is still underdeveloped. The body is covered with down hair (lanugo), and the finger- and toenails are short. In boys the testes have not yet descended into the scrotum; in girls the labia minora protrude from between the labia majora.

Generally speaking, the premature baby's chances of survival are lower than those of a full-term baby. Prematurity is the main cause of infant mortality in the neonatal period (the first four weeks after birth). On average, mortality among babies weighing less than 2500 g is about 20 times as high as among those weighing more than 2500 g at birth. Causes of mortality include, particularly, defective breathing, infection (the premature baby cannot produce antibodies and is therefore very vulnerable), intracranial hemorrhage, abnormal blood conditions, and congenital abnormalities. The survival chances range from about 95 percent in infants with a birth weight of 2000–2500 g to no more than 10 percent in those weighing under 1000 g at birth.

The premature baby may have to be reared in an incubator, an apparatus in which the temperature and humidity can be regulated and stabilized. Oxygen may have to be given to assist the baby's breathing; sometimes artificial respiration is necessary. The intake of nourishment by the premature baby is rendered difficult in various ways: weakness of sucking and swallowing reflexes, small stomach capacity, lowered tolerance of the alimentary tract, etc. A feeding tube is generally used.

CHILDBIRTH (PARTURITION, DELIVERY)

The uterus is situated in the midpelvis; it is supported in this position by various ligaments and especially by the muscular floor of the pelvis. The pelvis is the bony structure comprising mainly the lower part of the backbone (sacrum and coccyx) and the two hipbones. In front the hipbones meet at the symphysis pubis, where they are connected by fibrous tissue. The lower part of the pelvis is called the true pelvis, whose bottom aperture is called the pelvic outlet and whose top aperture is called the inlet (Fig. 6-174). All diameters normally are larger in the female than in the male; also, in the female the joints between the pelvic bones allow some slight movement during childbirth, whereas in the male they are rigid. The floor of the pelvis (Fig. 6-175) comprises three layers of muscles and ligaments and constitutes the bottom closure of the abdominal cavity. At childbirth the floor has to open wide enough to permit delivery of the baby. When the baby's head penetrates it, the muscles and ligaments of the pelvic floor stretch to form an extension of the birth passage (or birth canal), through which the baby emerges. The pelvic outlet is not simply a downward opening, but faces partly in the forward direction, and the birth passage itself is curved (Fig. 6-176).

As a rule, especially in a first pregnancy, several weeks before the onset of labor the head of the fetus descends into the pelvis (this

FIG. 6-174.
Inlet of the pelvis as seen from above.

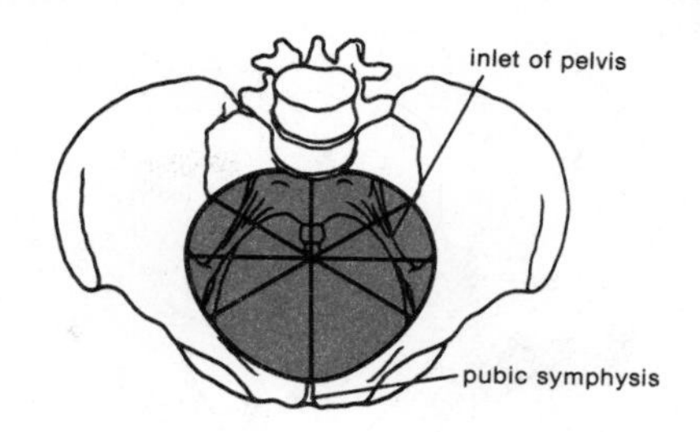

FIG. 6-175.
Floor of the pelvis.

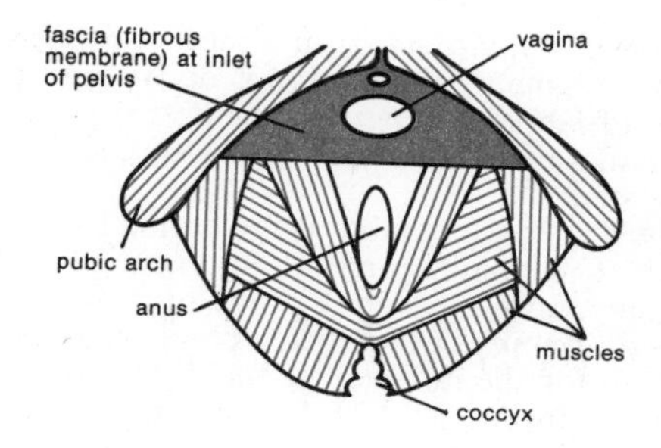

FIG. 6–176.
Birth passage fully dilated at start of second stage of labor.

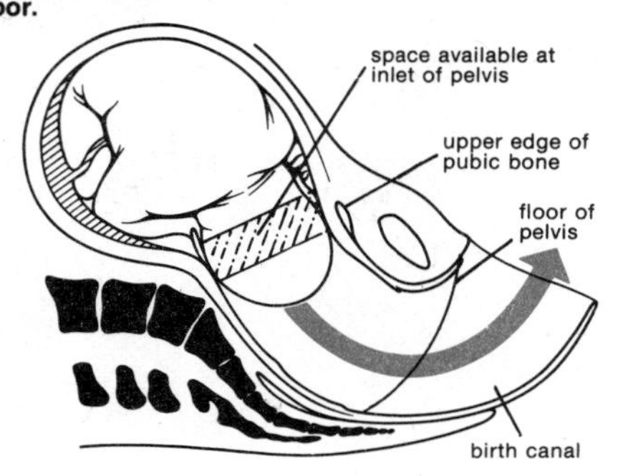

is known as *lightening*). In a second or subsequent pregnancy this may not occur until the onset of labor. *Labor,* i.e., the actual process of giving birth, is traditionally divided into three stages.

The *first stage* of labor is the period from the onset of rhythmic contractions, accompanied by pains, until the cervix of the uterus is fully dilated, as a result of which a continuous clear passage for the delivery of the baby from the uterus is formed (Fig. 6-177). In a primigravida, i.e., a woman in her first pregnancy, this dilation of the cervix may proceed somewhat reluctantly and with difficulty, whereas in a multipara, i.e., a woman who has previously borne children, the cervical canal is already somewhat dilated and therefore opens more quickly and easily (Fig. 6-178). Normally the baby's head is downward and therefore emerges first. With a first baby this stage of labor lasts, on average, about 12 hours, but is shorter with subsequent pregnancies (8 hours). Identification of this stage is important, especially when the woman is having her first baby. Diagnosis may be complicated by false labor pains, which may begin several weeks before the onset of true labor. A reliable sign of impending labor is the discharge of a slight amount of blood-tinged mucus from the vagina. It usually indicates that labor will begin within the next 24 hours.

The *second stage* of labor lasts from the complete dilation of the cervix to the complete expulsion of the baby. It lasts up to about one hour in a first pregnancy, usually less than half that time in subsequent pregnancies. Labor pains are severe, occurring at intervals of two or three minutes and lasting about one minute. Rupture of the amniotic sac (bag of waters) occurs mostly in the early part of this stage, with discharge of amniotic fluid from the vagina. Premature rupture of this membrane results in so-called dry labor. The contractions become more methodical, and the woman instinctively assists by a "pushing" effort (called "bearing down"). In a normal case the head of the baby appears through the vulvar opening, advancing a little more with each recurrence of pain. When the head appears, it may be necessary to make an incision in the perineum to facilitate delivery.

When the head is free from the vagina, it rotates as the shoulders turn to come down through the pelvic outlet. There is a gush of amniotic fluid while this occurs. On complete removal from the birth passage the baby is still attached to the placenta by the umbilical cord. The cord is tied off and cut soon after birth (when the umbilical blood vessels have ceased pulsating); the remaining stump withers and falls off. The navel is where the cord was attached to the fetus.

After the birth of the baby, the *third stage* of labor comprises some further contractions to expel the afterbirth, i.e., the placenta and membranes, from the vagina. The remainder of the amniotic fluid also escapes. This stage

FIG. 6–177.

(a) Uterus at time of conception.

(b) Uterus just before onset of labor.

(c) Birth passage fully open.

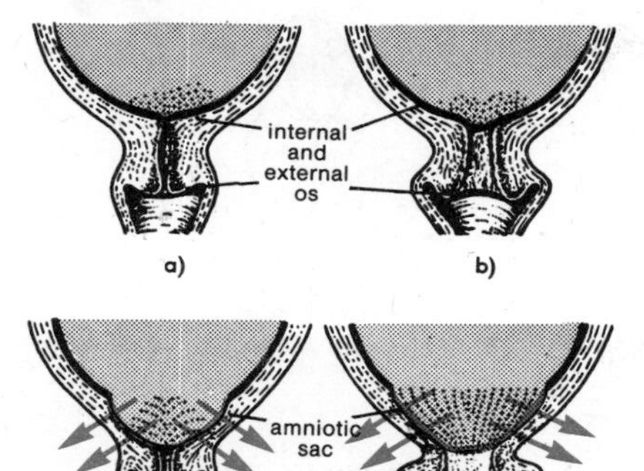

FIG. 6-178.
Dilation of cervical canal in a primigravida (a, c) and a multipara (b, d).

usually lasts about ten minutes. There is some bleeding from the uterus, usually between 100 and 500 ml. Postpartum hemorrhage, i.e., severe bleeding from the uterus after parturition, is a serious complication requiring urgent medical attention.

The factors that bring about the onset of labor at the end of pregnancy are not fully known. Contraction of the uterus is stimulated by oxytocin, a hormone secreted by the pituitary gland. It acts specifically on the smooth musculature of the uterus. During the earlier part of pregnancy the corpus luteum hormone helps to maintain pregnancy and suppress uterine contractions. When the level of this hormone in the blood goes down, the uterus becomes responsive to oxytocin. The secretion of that hormone is at first counteracted by oxytocinase, an enzyme from the uterus. Toward the end of pregnancy the supply of the enzyme diminishes; in response to that, the amount of oxytocin in the blood increases and thus stimulates the uterine muscles to contract.

The pelvic inlet is not circular, but somewhat elongated in the transverse direction. To ease its way through the inlet, the baby's head is in the transverse position on entry (Fig. 6-179). The outlet of the pelvis is also somewhat elongated, but in the front-to-rear (anteroposterior) direction. As the head descends farther into the pelvis, it gradually rotates through 90 degrees to emerge from the pelvic outlet "back-to-front," i.e., with the back of the head and neck facing forward in relation to the mother's body (Figs. 6-180 and 6-181). The baby's shoulders likewise rotate during their descent through the pelvis and, because of the flexibility of the neck, can do this independently of the movements of the head. When the head is finally free from the vagina, the shoulders rotate to the anteroposterior position to emerge from the pelvic outlet (Figs. 6-182 and 6-183), usually assisted by the obstetrician.

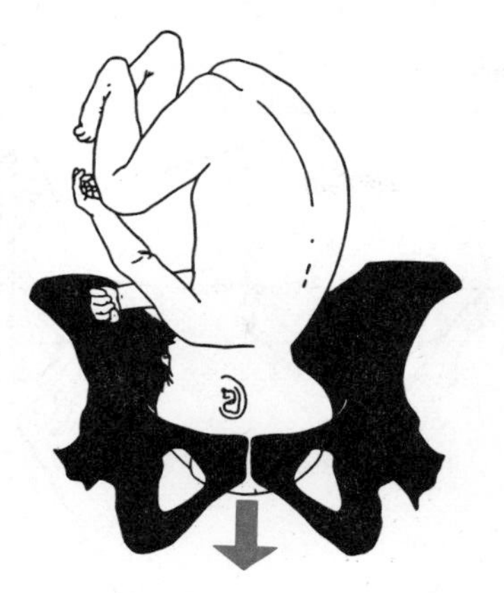

FIG. 6-179.
First phase of the expulsion of a baby in normal childbirth: baby's head in transverse position.

FIG. 6-180.
As the head descends, it gradually rotates so that the back of the head is facing forward in the pelvis.

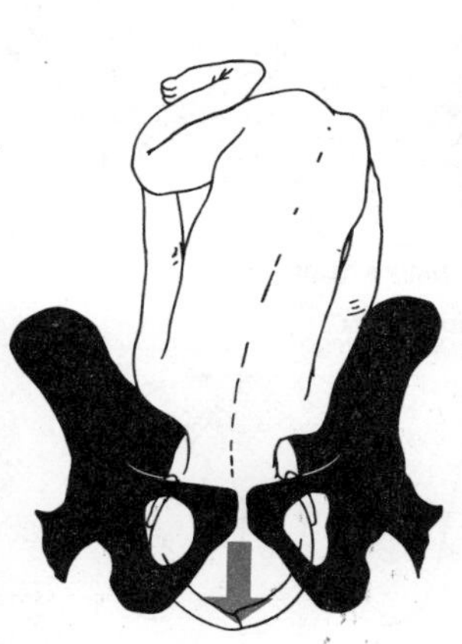

FIG. 6-181.
Rotation completed and head bent backward for emergence through birth passage.

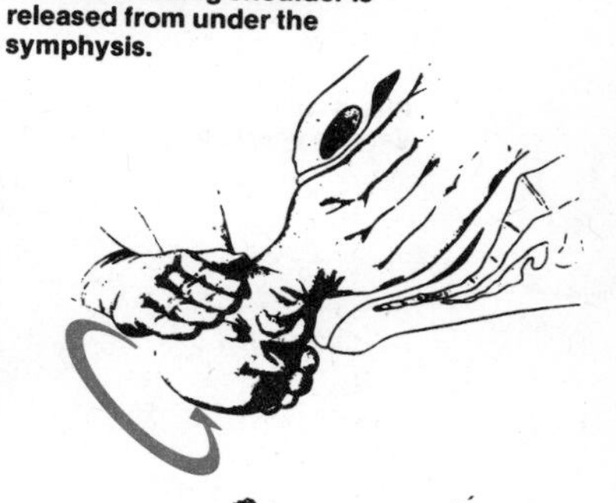

FIG. 6-182.
Obstetrician may assist the emergence of the baby's shoulders by manipulations. Here the leading shoulder is released from under the symphysis.

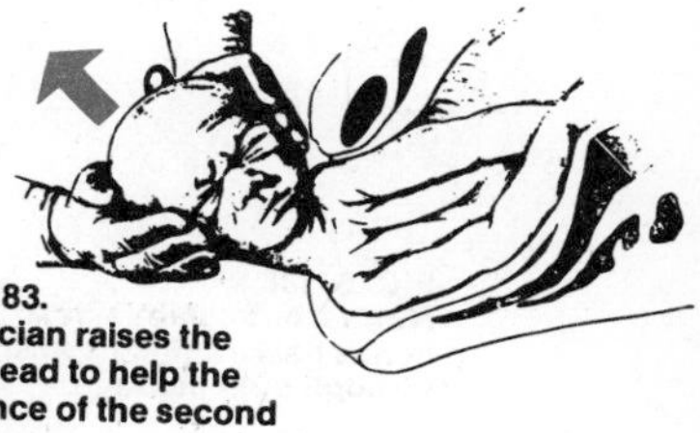

FIG. 6-183.
Obstetrician raises the baby's head to help the emergence of the second shoulder.

FIG. 6–184.
Various positions of the baby before birth.

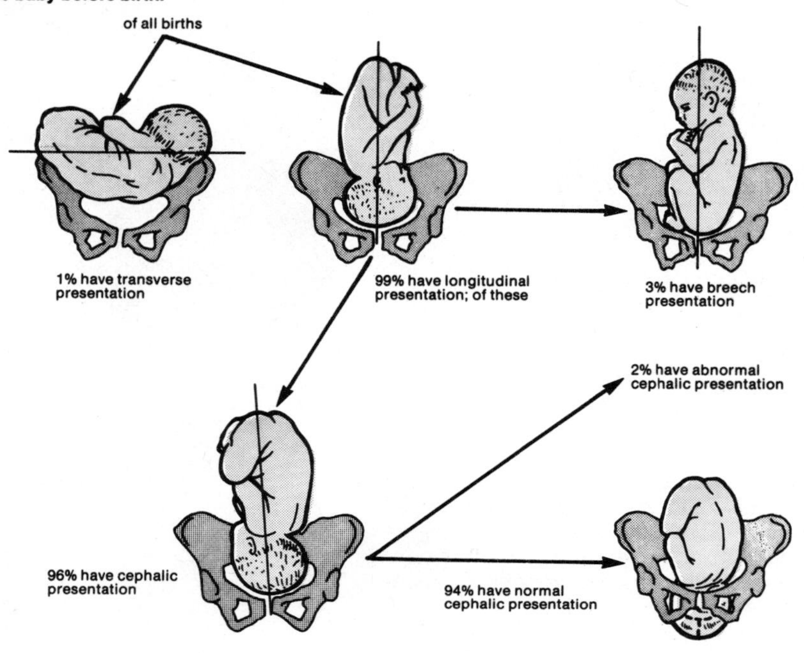

FIG. 6–185.
Various possible positions of the baby's head.

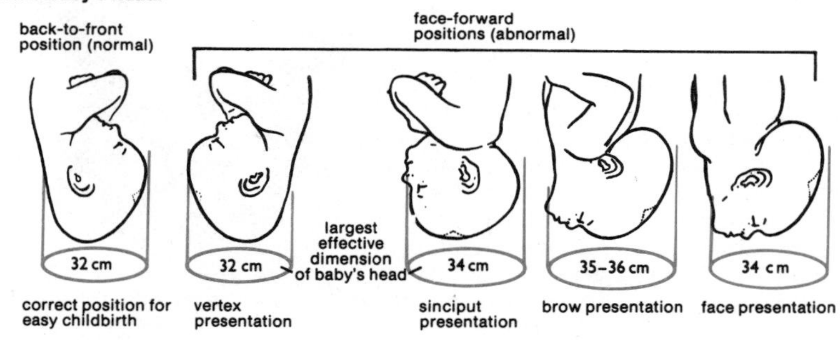

FIG. 6–186.
Normal birth: baby's head bends back during expulsion through birth passage.

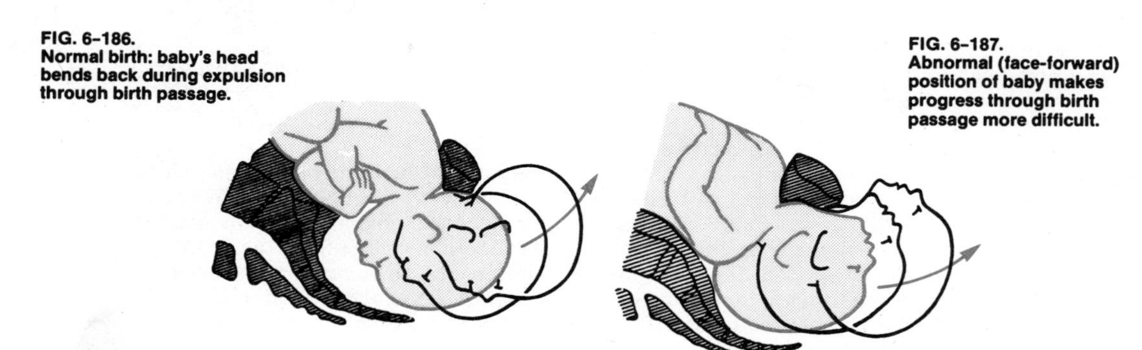

FIG. 6–187.
Abnormal (face-forward) position of baby makes progress through birth passage more difficult.

VARIOUS POSITIONS OF THE BABY

About 99 percent of babies are in the longitudinal position at birth, enabling them to pass through the mother's pelvis (longitudinal presentation), whereas in the remaining 1 percent the baby lies crosswise (transverse presentation) (see Fig. 6-184). By far the commonest presentation is head foremost (cephalic presentation, 96 percent) and more particularly with the back of the baby's head and neck facing forward, i.e., toward the symphysis pubis of the pelvis (94 percent), while in about 2 percent of births the head, though emerging foremost from the birth passage, does not conform to the normal position. Thus in about 1 percent of cases the baby's head is not in the transverse position, as it should be (see page 291), on descending into the inlet of the pelvis, so that without corrective intervention the head usually cannot descend at all. A frequent cause of this abnormality is thought to be that the pelvis is too narrow. Delivery by Cesarean section is often necessary in such circumstances. Another complication, though less common, is that the baby's head, after duly descending transversely into the pelvic inlet, fails to rotate through 90 degrees to enable it to pass through the outlet, i.e., it remains positioned transversely. This abnormality is easier to correct, so that in the majority of such cases spontaneous delivery is (with help from the obstetrician) still possible. In other cases the baby's head, during its progress through the pelvis, does not rotate to the normal "back-to-front" position, but instead to the "face-forward" position, with several possible variants (Fig. 6-185). With these various face-forward positions the baby's head cannot, as in the normal case, bend back on the neck and thus make it easier for the head to be delivered from the birth passage (Figs. 6-186 and 6-187).

In general, in a case of abnormal presentation it will be attempted (if this can safely be done) to manipulate the baby into the natural position before birth. The main hazard associated with these wrong positions is that labor is liable to be difficult and unduly protracted, with attendant risk to the baby and sometimes also to the mother. In the event of danger to mother and/or child, obstetric forceps may be used to ease the head past the pelvic outlet (see page 308). It is sometimes necessary for the obstetrician to make an incision in the woman's perineum at the end of the second stage of labor to facilitate delivery (episiotomy, see page 304). The perineum is the region between the anus and the genital organs, which is liable to laceration during childbirth.

The cause of the *face-forward position* with vertex presentation (see Fig. 6-185) is thought to be that the baby is too small (e.g., because of prematurity) or that, in a woman who has already borne children, parts of the birth passage are too flabby and thus fail to give mechanical guidance to the baby's head. On the other hand, the three face-forward positions illustrated in the right-hand part of Figure 6-185 (sinciput, brow and face presentation, respectively) are attributed to narrowness of the woman's pelvis. Low position of the placenta in the uterus (placenta praevia), or a fibroid tumor in the cervical canal, may likewise obstruct normal delivery. Sometimes, too, the shape of the baby's head will significantly impede its delivery.

In a case of face presentation the head must bend forward on the neck during delivery (Fig. 6-187). This position is not particularly unfavorable, and although it will make labor more difficult, it will generally not prevent spontaneous childbirth. With sinciput presentation or with brow presentation, however, the scope for head movement to adapt itself to the shape of the birth passage is more limited, and in such cases spontaneous delivery is usually not possible. For this reason it used to be frequently necessary, in the circumstances, to use obstetric forceps for delivering the baby. This method, particularly in cases of brow presentation, entailed a considerable risk of infant mortality. Nowadays such babies are nearly always delivered by Cesarean section. Even with the "easier" sinciput presentation, the duration of labor, if it can be achieved spontaneously at all, is greatly increased in comparison with a normal labor. Besides, with these three face-forward positions the baby's head presents a larger effective dimension and is thus less likely to pass comfortably through the pelvic outlet, e.g., 34–36 cm front-to-rear measurement as against 32 cm in the vertex position (Fig. 6-185). These unfavorable face-forward positions are, however, quite rare; for example, brow presentation occurs in only something like 0.1–0.15 percent of all pregnancies.

Breech presentation, i.e., the position in which the baby's buttocks show first, occurs in about 3 percent of pregnancies. Three types are to be distinguished (Fig. 6-188): frank breech, complete breech, footling. Footling presentation may be single, double, or (if the leg remains flexed) knee presentation. With breech presentation the woman often feels painful fetal movements low down in the abdomen caused by the fetus kicking against the lower part of the uterus. Sometimes, upon stooping, she can feel the baby's head as a large and hard object in the upper abdomen. In most cases a doctor can diagnose breech presentation (Fig. 6-189) even by external examination. Rectal or vaginal examination can confirm this. It is, with breech presentation, usual to make a vaginal examination before the onset of labor so as to enable proper precautions to be taken in good time.

Breech presentation does indeed differ in some important ways from normal cephalic presentation. The baby's buttocks are softer, i.e., more yielding, than its head and will pass more easily through the birth passage, but will expand the mouth of the uterus less effectively than the head would (Fig. 6-190). After the buttocks, the trunk and shoulders are delivered (Fig. 6-191). Finally the head, with the back of the neck facing forward in relation to the mother's body, reaches the pelvic outlet. This is the critical phase, not only because the head, being large and hard in comparison with the rest of the baby's body, is the most awkward part to deliver, but also because it presses against the umbilical cord, which extends from the placenta—past the head—to the baby's navel. This pressure, constricting the cord, together with the uterine contractions which in themselves tend to reduce the blood supply, may deprive the baby of adequate blood and therefore of oxygen. Thus only a few minutes are available in which to release the head from the birth passage before the baby is asphyxiated. Since it is doubtful if the head can indeed be spontaneously delivered quickly enough, the obstetrician will assist the process by manipulation. For example, he may bend the baby's back forward over the mother's symphysis pubis (the front junction of the pubic bones) to release the head, while an assistant presses down on the abdomen (Fig. 6-192). In a first pregnancy it is, in such cases, usual to make an incision in the perineum (episiotomy) to facilitate delivery; this is done in advance, under anesthetic, and is stitched after the birth. Sometimes it is necessary to perform more elaborate manipulations or to deliver the baby by Cesarean section, e.g., in a case where the pelvis is too narrow, or if the umbilical cord is prolapsed (premature expulsion of the cord) while the mouth of the uterus is insufficiently dilated, or in delivering an unusually large baby born to a woman of comparatively advanced age in her first pregnancy. Despite these precautions, stillbirth occurs in something like 10 percent of cases of breech presentation.

In *transverse presentation*, which occurs in about 1 percent of pregnancies, the baby lies crosswise in the uterus (Fig. 6-193). This abnormality is most likely to occur in a multipara (a woman who has previously borne children) in whom the uterus is in a stretched condition and the abdominal wall is flabby. It also occurs in cases where a woman has twins for the second time or has a small premature baby. Generally speaking, transverse presentation is attributable to an excess of space in the uterus, so that the head of the fetus is not pushed into the pelvis and held there by the firmness and elasticity of the uterine wall. Alternatively, correct head-down positioning of the fetus (cephalic presentation) may be rendered impossible by narrowness of the pelvis or by the placenta being abnormally low down in the uterus (placenta praevia).

In the great majority of cases the transversely positioned fetus is incapable of being born in the normal way. With the onset of labor it becomes wedged in the pelvic outlet and is, as it were, smothered by the contractions of the uterus. In the end, the only possible way to save the mother is to remove the dead baby by an operation to dismember it. In other cases, tearing of the overdilated uterus may occur. It is obviously a dangerous abnormality, but modern obstetric practice has greatly improved the chances of the baby's survival. Timely intervention to correct or improve the baby's position in the uterus by appropriate obstetric maneuvers ("turning the baby," technically known as *version*) can be important. Even so, about 50 percent of such babies used to die in the past. Nowadays, thanks to Cesarean section, infant mortality in cases of transverse presentation has been reduced to about 5 percent.

Obviously, early diagnosis of transverse presentation and removal of the woman to a hospital in good time are important. The abnormality can often be diagnosed by external examination. The pregnant woman's body presents a somewhat transversely instead of longitudinally oval appearance. In comparison with a normal pregnancy the uterus is still in a very low position, scarcely above the level of

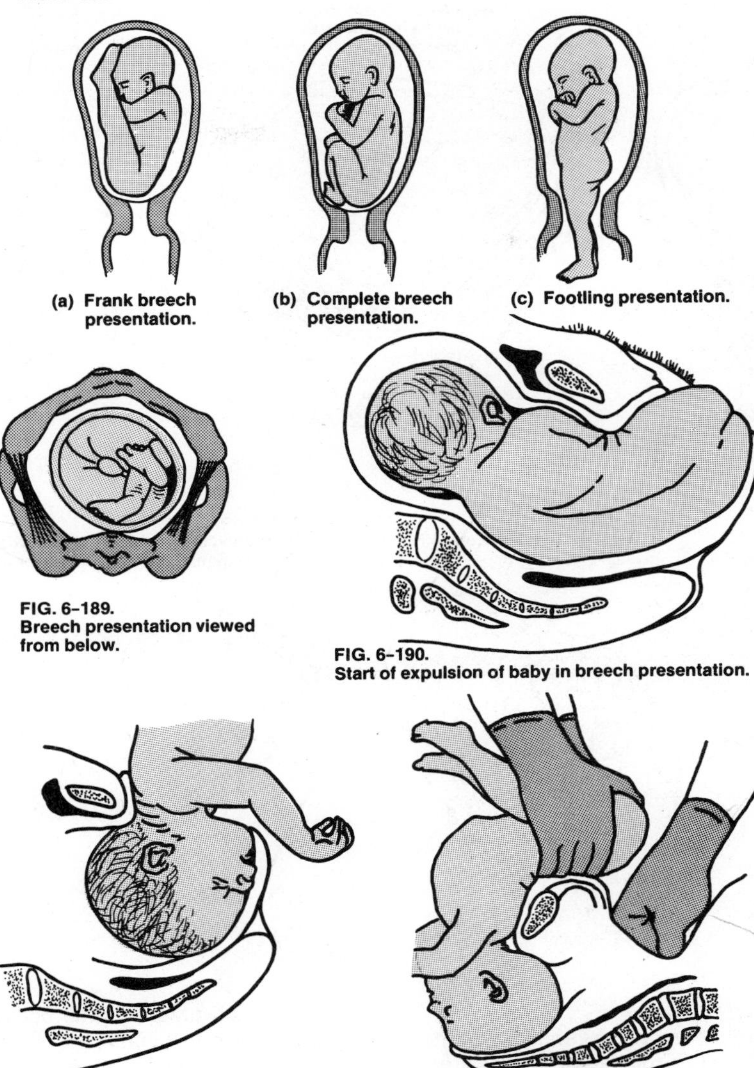

FIG. 6-188.

(a) Frank breech presentation.

(b) Complete breech presentation.

(c) Footling presentation.

FIG. 6-189.
Breech presentation viewed from below.

FIG. 6-190.
Start of expulsion of baby in breech presentation.

FIG. 6-191.
Emergence of baby's body in breech presentation.

FIG. 6-192.
Emergence of baby's head in breech presentation.

the navel. Initially it is preferable not to carry out any vaginal or rectal internal examination, as this can cause rupturing of the membranes, so that the baby's shoulders may enter the lower part of the pelvis and become wedged there (Fig. 6-194).

At the onset of labor pains an attempt may be made to turn the baby by external manipulation through the abdominal wall (external version) and thus bring it into a position that will permit normal delivery (Fig. 6-195). But after rupture of the membranes the baby often

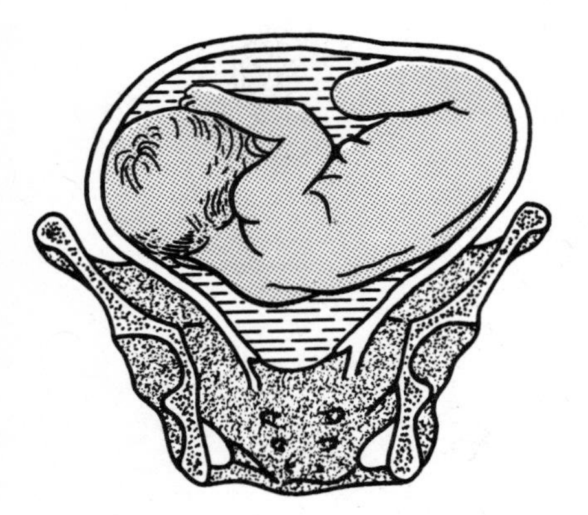

FIG. 6-193.
Transverse presentation.

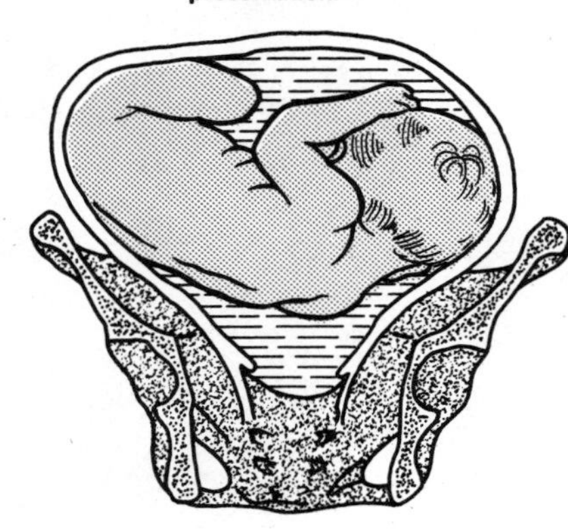

FIG. 6-194.
(a) Start of expansion of cervix in transverse presentation.

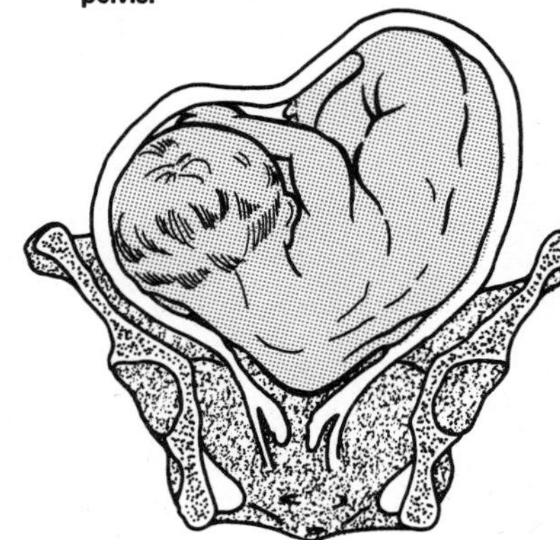

(b) After rupture of the amniotic sac the baby's shoulder enters the pelvis.

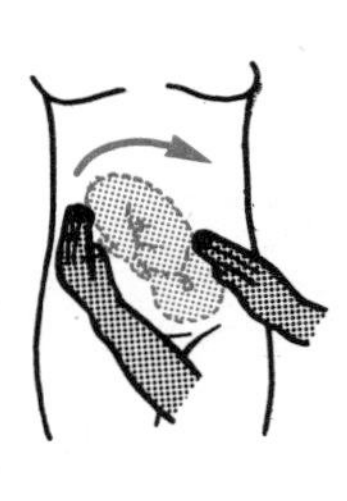

FIG. 6-195.
Attempted external manipulation of baby.

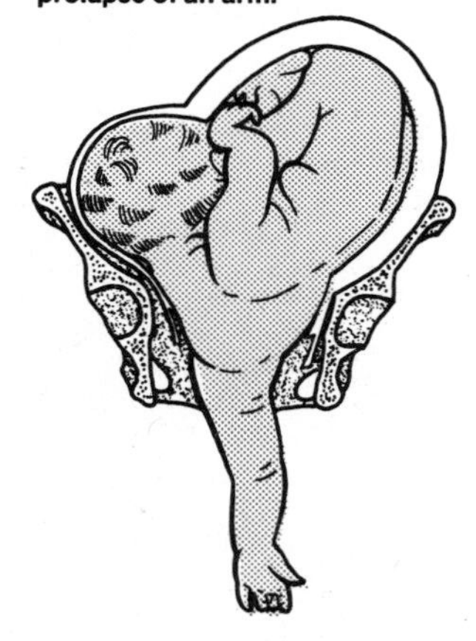

FIG. 6-196.
Transverse presentation with prolapse of an arm.

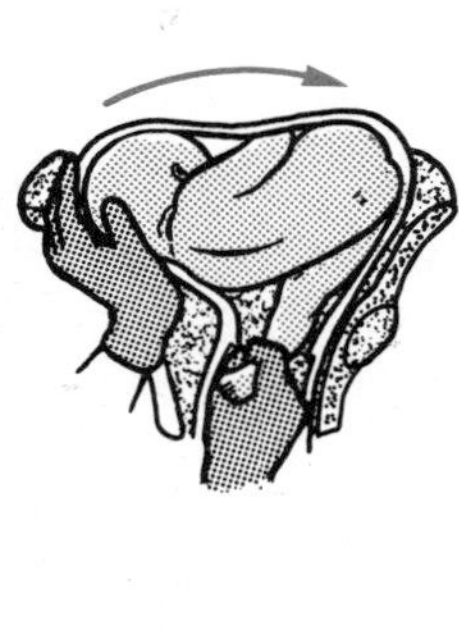

FIG. 6-197.
Combined external and internal manipulation.

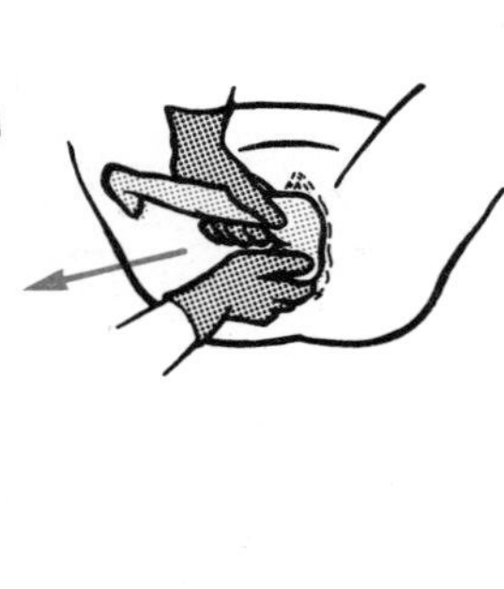

FIG. 6-198.
Extraction of baby after turning it.

becomes immovably wedged by the shoulders, especially in a case of prolapse of an arm (Fig. 6-196); or prolapse of the umbilical cord may occur, or an infection may reach the uterus from the exterior if labor is prolonged. A major hazard arises from damage to the placenta, on which the fetus depends for its blood and oxygen supply.

After rupture of the membranes has occurred, the obstetrician will investigate by vaginal examination to what extent the cervix of the uterus has dilated. If there is no appreciable dilation, delivery by Cesarean section will be necessary. This operation may be performed at an earlier stage in a case where there are signs of danger to the baby. On the other hand, natural vaginal delivery is preferred if there is no immediate danger to the baby and the membranes remain unruptured until complete dilation of the cervix; it is also preferred for the birth of twins or for that of a premature or a nonviable baby. In such cases it is usual to apply a combination of external and internal version followed by extraction of the baby (Figs. 6-197 and 6-198). These manipulations, especially the extraction of the baby, require full dilation of the cervix and are performed under anesthesia. Since version is not without danger to the mother and extraction can be dangerous to the baby, delivery by Cesarean section will be preferred in all other cases.

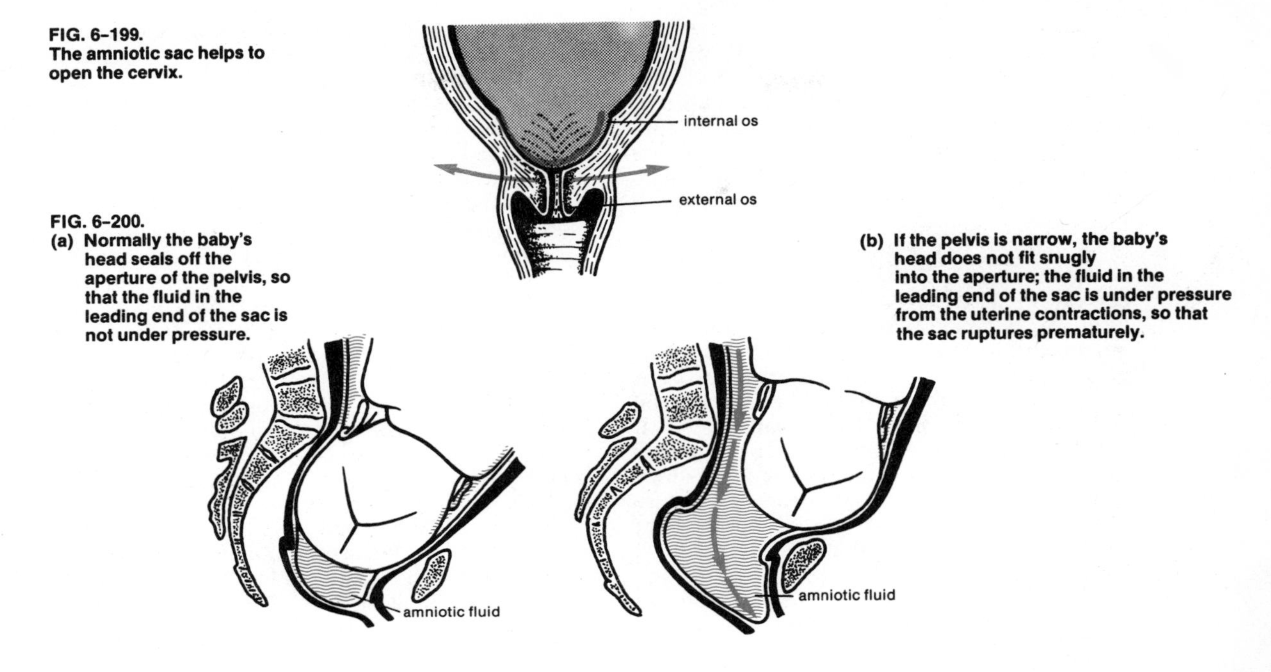

FIG. 6-199.
The amniotic sac helps to open the cervix.

FIG. 6-200.
(a) Normally the baby's head seals off the aperture of the pelvis, so that the fluid in the leading end of the sac is not under pressure.

(b) If the pelvis is narrow, the baby's head does not fit snugly into the aperture; the fluid in the leading end of the sac is under pressure from the uterine contractions, so that the sac ruptures prematurely.

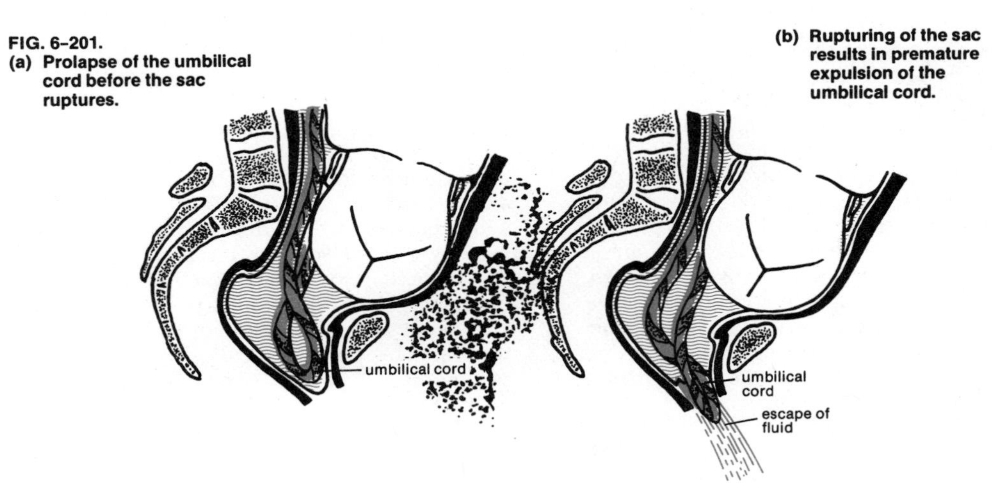

FIG. 6-201.
(a) Prolapse of the umbilical cord before the sac ruptures.

(b) Rupturing of the sac results in premature expulsion of the umbilical cord.

PREMATURE RUPTURE OF THE MEMBRANES

Until birth the developing fetus is surrounded by protective membranes. At an early stage some of these form the placenta, on which the fetus depends for its nourishment; the others, particularly the inner membrane (amnion), form a bag containing a watery fluid (amniotic sac or bag of waters). During the first stage of labor (see page 300) the bottom part of the sac performs an important function: under the action of the uterine contractions it forms a fluid-filled pocket in advance of the baby's head and exerts gentle pressure at the internal end of the cervical canal, thus assisting the latter to dilate (Fig. 6-199). Normally the sac then ruptures, usually during the early part of the second stage of labor, leaving the passage clear for the expulsion of the baby (Fig. 6-200a). Delayed rupturing of the membranes will hold up delivery, and for this reason, if the sac fails to discharge its fluid spontaneously when full dilation of the cervix has been achieved, the obstetrician will puncture it painlessly. In rare cases a baby may be born enveloped in unruptured fetal membranes or with membranes covering its head ("born with a caul"). These membranes must quickly be torn open to prevent asphyxiation.

Sometimes the amniotic sac ruptures between the onset of labor and full dilation of the cervix, so that the fluid flows away prematurely ("dry labor"). Such a birth is somewhat more painful and also delayed because now the pressure to assist the dilation of the cervix has to be exerted by the baby's head itself instead of by the pocket of fluid in advance of it. If premature rupturing of the sac occurs even before the onset of labor, the consequences may be more serious. The cause of such premature loss of fluid may be that the membranes are particularly delicate or that the baby's head does not fit snugly into the aperture of the pelvis and thus does not enable a separate pocket of fluid to form in advance of the head (Fig. 6-200b). This difficulty may arise if the pelvis is narrow. In such a case even quite minor preliminary contractions of the uterus can so increase the pressure of the fluid in the leading end of the sac that it ruptures. When this happens, the woman should rest and the doctor be sent for. If labor does not now soon begin, she should be removed to a hospital, because there is a risk of prolapse (premature expulsion) of the umbilical cord due to the outflow of fluid (Fig. 6-201). A further hazard is that the placenta, which remains the baby's vital source of blood and oxygen supply until it has been safely delivered, may be too severely squeezed by the powerful contractions of the "dry" uterus. Also, with the premature loss of fluid, the protective action of the intact membranes is largely absent, while there is moreover a risk of infection spreading upward into the birth passage along the discharge path of the fluid. Such infections are more liable to ensue when rupturing of the membranes occurs earlier and the interval of time between their rupturing and delivery of the baby is therefore longer. If premature loss of fluid occurs, the pregnant women should remain in bed and not take baths. Evacuation of the bowels should be done with the least possible amount of pressure. It may be necessary to induce labor, i.e., stimulate uterine contractions prior to the time they normally would occur.

UMBILICAL CORD: COMPLICATIONS, PROLAPSE

The umbilical cord connects the fetus to the placenta and contains two arteries and one vein embedded in a jellylike connective tissue. The fetus derives its nourishment from the blood supplied through these arteries. In the full-term fetus the umbilical cord is on average about 50 cm long, but may be substantially longer or shorter. It leaves a depression, the navel or umbilicus, on the abdomen of the child. As it is literally the lifeline of the fetus, any defect or injury to the cord is liable to have serious consequences to the fetus.

The umbilical cord is normally attached to the placenta at or near the middle thereof (Fig. 6-202), but is sometimes attached in an off-center position (Fig. 6-203) or at the edge of the placenta (Fig. 6-204) or even to the chorion (see page 000) beyond the edge of the placenta (Fig. 6-205). This last-mentioned abnormal attachment is especially undesirable, because at birth the blood vessels of the umbilical cord are liable to be squeezed shut or the branching blood vessels at the attachment of the cord are liable to be damaged when the amniotic sac ruptures. In this latter case, severe hemorrhage of the baby occurs, and its life can be saved only by rapid delivery by Cesarean section.

Exceptionally, the umbilical cord may be very short or its free length may be reduced because it encircles the baby. In such circumstances the labor pains may be particularly acute because of tugging of the cord, while moreover the descent of the baby's head may be prevented by the restraining pull of the cord, and the supply of oxygen to the baby may be at risk. Encirclement of some part of the baby by the umbilical cord occurs quite commonly, especially encirclement of the neck (Fig. 6-206). The hazard to the oxygen supply arises chiefly as a result of compression of the cord between the baby's head and the edge of the mother's pelvis, so that the arteries in the cord are squeezed shut. Another hazard, fortunately rare, is that a knot forms in the cord when the fetus at some stage slips through a loop that happens to have formed (Fig. 6-207). Rupturing of the umbilical cord occurs in about 25 percent of precipitate deliveries, i.e., when childbirth is relatively sudden, usually as the result of rapid labor in the absence of expert obstetric assistance. If the cord is prematurely severed in such circumstances, the baby will bleed to death unless the

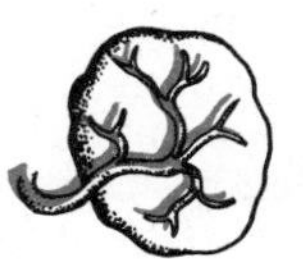

FIG. 6-202.
Normal attachment of umbilical cord to placenta.

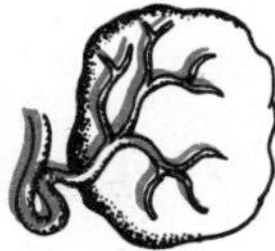

FIG. 6-203.
Off-center attachment of umbilical cord.

FIG. 6-204.
Edge attachment of umbilical cord.

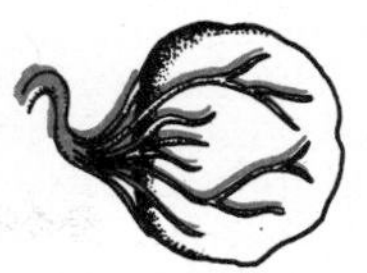

FIG. 6-205.
Attachment to chorion.

FIG. 6-206.
UMBILICAL CORD ENCIRCLES THE BABY'S NECK.

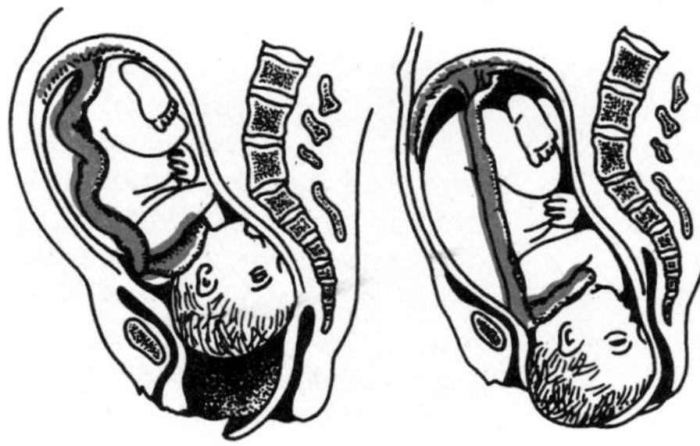

(a) Between uterine contractions.

(b) During uterine contraction.

FIG. 6-207.
Knot in umbilical cord.

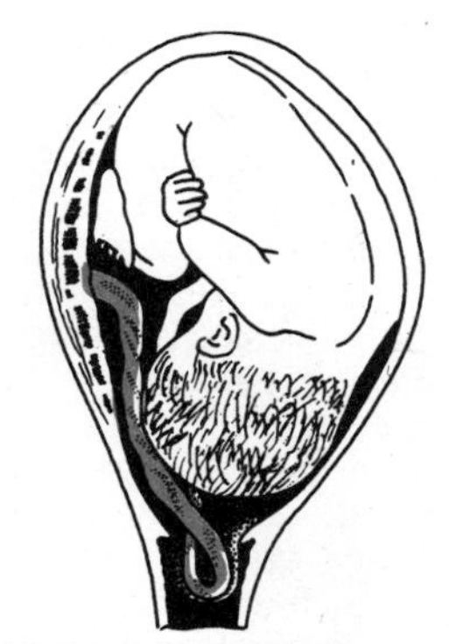

(a) Prolapse of umbilical cord in the amniotic sac.

FIG. 6-208.

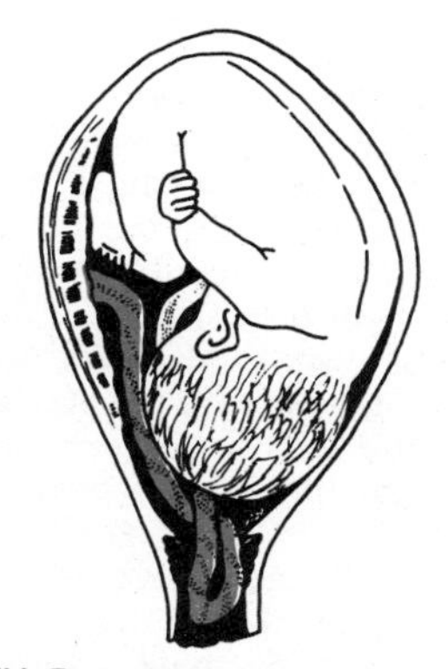

(b) Premature expulsion of umbilical cord when the sac ruptures.

remaining length of cord attached to it is immediately tied to stop the escape of blood.

Prolapse of the umbilical cord is liable to occur especially in a case where the baby's head does not descend snugly into the pelvis. The cord may then hang down beside the head (Fig. 6-208a); when the membranes rupture, the escape of amniotic fluid can cause prolapse (Fig. 6-208b), i.e., the cord comes down ahead of the baby into the cervix and vagina and may even be externally visible. It is sometimes possible to make the cord withdraw back into the uterus by placing the woman in a certain position. Prolapse is particularly likely to arise if the baby is in a transverse or other abnormal position in the uterus, if the mother's pelvis is too narrow or too wide, if the baby is premature, if the membranes rupture prematurely (see page 297), or if the woman gives birth to twins.

Since prolapse of the umbilical cord is most likely to occur as the direct result of the outflow of amniotic fluid, the baby's condition must be carefully watched from that moment onward. If there is serious weakening of its heartbeat with each uterine contraction, an internal examination through the vagina will be carried out, and if this reveals prolapse of the cord, steps will be taken to deliver the baby as speedily as possible. If the pulsations of the blood vessels in the cord cease completely, the obstetrician has only about 5–10 minutes in which to save the child's life. In such cases, delivery by Cesarean section will be the usual course of action, especially if the cervix is not fully dilated. Prolapse of the umbilical cord nearly always causes pressure to be exerted on the cord, with reduced oxygen supply and therefore danger to the baby. There is indeed a high risk of infant mortality due to asphyxiation in such cases. The risk of constriction of the prolapsed cord is greatest in cases of cephalic (i.e., normal) presentation because of the hardness of the head pressing on the cord.

LABOR

Labor is the physiological process by which childbirth (parturition), i.e., expulsion of the fetus from the uterus, takes place. More particularly, it refers to the rhythmic contractions of the muscular wall of the uterus, accompanied by sensations of pain.

Slight and often unperceived uterine contractions may occur long before the actual time of childbirth. In the last few weeks before childbirth these gradually intensify into preliminary labor pains and then, under the influence of certain hormonal changes, develop into the true labor pains associated with dilation of the cervix and expulsion of the baby. Stimulated by oxytocin, a hormone from the pituitary gland, the retractions of the powerful musculature in the upper part of the body of the uterus pull back the dilatable lower parts of the uterus (particularly the cervix) over the baby. These retractions alternate with the uterine contractions that actually force the baby out of the birth passage.

The expulsive effort developed by the uterus is assisted by the conscious effort that the woman can develop with her abdominal muscles, this supporting action being known as "bearing down." So long as the leading part of the baby's body, normally the head, is still high up in the birth passage, this bearing-down action can be started at will. Later on, the process becomes automatic: as a result of pressure exerted by the baby on certain nerve endings, it proceeds as a reflex, so that eventually the musculature of the woman's whole body is involved in the rhythmic expulsive movements. Even so, the woman in labor can still make a deliberate contribution to speedy delivery by energetic and correctly timed "pushing" efforts. In particular, she should develop these efforts to coincide with the climax of each uterine contraction. During this stage of labor she will involuntarily draw back her legs. The overall effort developed by the contractions aided by pushing can be further intensified by the woman's holding onto some support, even her own thighs, while in labor. Correct control of breathing is important: breathe deeply at the onset of each contraction and then, at the right time, with the chin down on the chest, push energetically as when evacuating the bowels with difficulty.

The progress of labor shows whether the expulsive effort developed by the uterus and by the woman's own conscious active cooperation is sufficient. In particular, the intensity of the contractions can be felt by placing the hand on the abdomen. In the first stage of labor the contractions initially occur at intervals of 10–15 minutes, subsequently at 5-minute intervals (Fig. 6-209). In the second stage the frequency further increases, the contractions now coming about every 2 minutes. If they are substantially weaker, shorter or less frequent than normal, delivery is likely to be delayed or indeed not achieved spontaneously at all. Such terms as *missed labor* (labor begins normally, but the contractions stop after a time) and *arrested labor* (failure of labor to proceed through the normal stages) are used to describe abnormalities of this type. If delivery is too long delayed, there are risks to both mother and child. Sometimes arrested labor is characterized from the outset by failure of the cervix to dilate adequately or in good time. If the uterus is very slack during the intervals between contractions, an injection of oxytocin may stimulate it to more energetic action (Fig. 6-210). On the other hand, the uterus may be very tight between contractions, and the cervix may be in a state of spasm and remain closed. This condition may be relieved by an antispasmodic drug. Arrested labor may also occur later on, as a result of fatigue, so that the uterine contractions and the bearing-down

FIG. 6-209.
Normal contractions, and contractions of insufficient duration in labor.

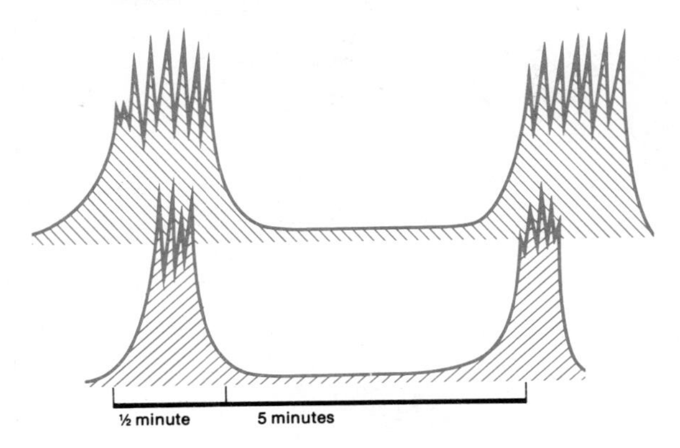

FIG. 6-210.
Stimulation of labor by oxytocin.

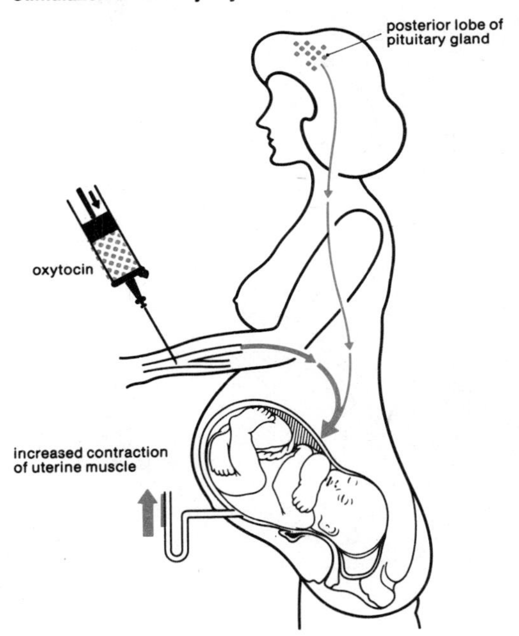

effort are weakened. In such cases an oxytocin injection may be helpful in speeding up delivery. Frequently, however, the slackening of the expulsive action is due to some major obstruction of the birth passage, e.g., if the pelvic outlet is too narrow. In such circumstances a hormone injection could be harmful, with possible rupturing of the uterus by excessive exertion.

Induction of Labor

The fetus may be in danger at any time during pregnancy. If it is not yet viable, it will generally be attempted—provided that this does not involve any serious risk to the mother—to maintain the pregnancy. On the other hand, once the baby has reached full term and the onset of labor is unduly retarded, steps will generally be taken to induce labor artificially and thus bring on the delivery of the baby. Danger to the baby exists when pregnancy lasts for more than about 42 weeks (see page 285) or with a second or subsequent Rh-positive child of an Rh-negative mother (see page 282). Premature rupturing of the membranes, resulting in loss of amniotic fluid, may also necessitate induction of labor if the spontaneous contractions fail to develop vigorously enough within a few hours (see page 286). Disease in the mother, e.g., diabetes, or an attack of eclampsia (Fig. 6-211), may likewise present a threat to the baby's life.

To induce labor, the membrane of the amniotic sac is seized with special forceps (which are introduced through the vagina) and punctured (Fig. 6-212). Only a small hole is made in the membrane, so that the fluid can drain away slowly; a sudden rush could cause prolapse of the umbilical cord or of an arm of the baby. At the same time the baby's head should move correctly downward into the pelvis and thus close the pelvic outlet, which can be verified by external examination. Greenish discoloration of the amniotic fluid indicates discharge of fecal matter by the baby and is a danger sign (see page 286). In such circumstances artificial puncturing of the sac can be done by means of a cannula under visual inspection through an amnioscope (Fig. 6-213). Puncturing the sac will, with a full-term baby, generally initiate labor because, when the fluid starts to drain away, the baby will exert stronger and more direct pressure on the nerve endings that set off the reflex action of uterine contractions. If sufficiently energetic contractions still fail to develop, an injection of oxytocin, the hormone that stimulates uterine activity at childbirth, is given. Failure of the uterus to respond to the hormone will generally necessitate delivery by Cesarean section.

FIG. 6-211.
Symptoms that make induction of labor advisable.

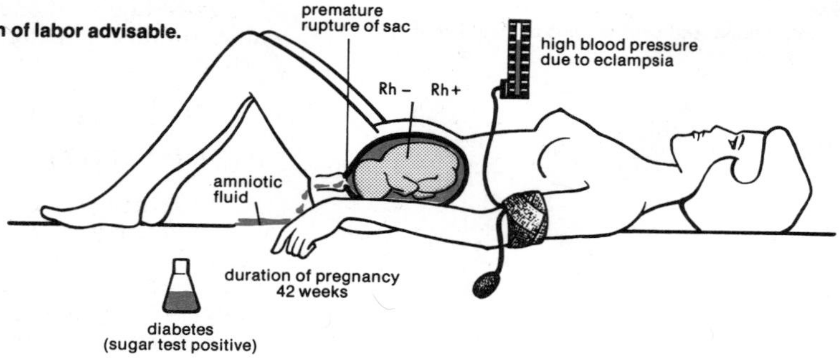

premature
rupture of sac
high blood pressure
due to eclampsia
Rh – Rh +
amniotic
fluid
duration of pregnancy
42 weeks
diabetes
(sugar test positive)

FIG. 6-212.
Puncturing the amniotic sac with forceps.

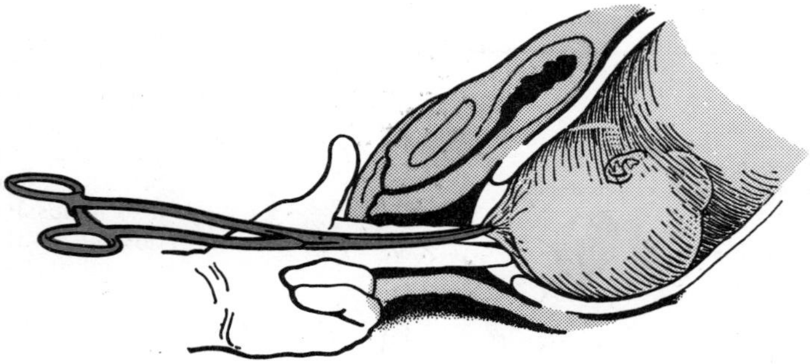

FIG. 6-213.
Puncturing the amniotic sac with a cannula.

amniotic sac
amnioscope
cannula
amniotic fluid

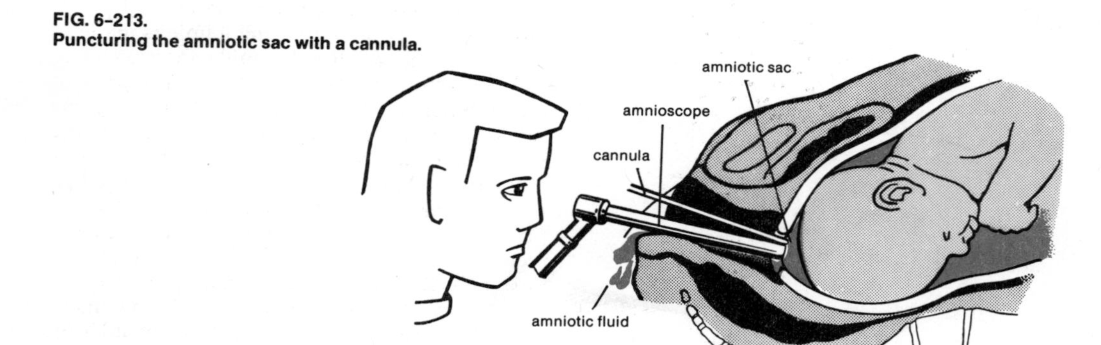

PAINLESS CHILDBIRTH

There is divergence of opinion on "painless" childbirth. The suppression of labor pains by anesthetics is by no means a new idea. It dates back to the early days of chloroform in the middle of the last century. The older criticism on moral grounds—that such pains were "natural," even "divinely ordained," and must therefore be endured—is hardly taken seriously in modern civilized societies, but there are medical grounds for caution.

Broadly speaking, the problem with drugs is that, while they suppress pain, they must not suppress labor itself nor injure the baby. So although it is now generally recognized that every effort should be made to ease the pains of childbirth, different experts hold different views on how this can best be achieved. Ad-

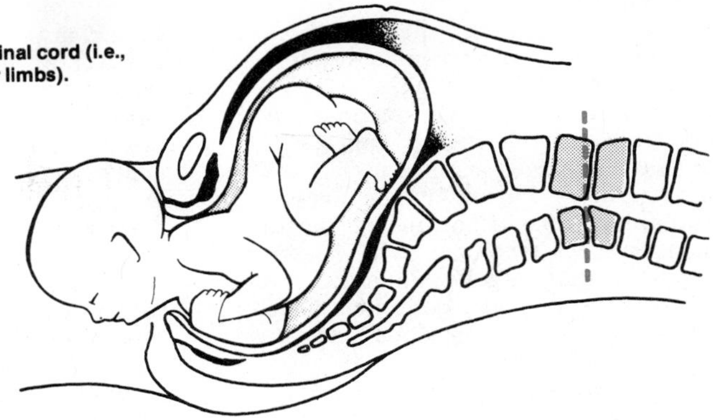

FIG. 6-214.
Normal childbirth despite severance of spinal cord (i.e., complete paralysis of abdomen and lower limbs).

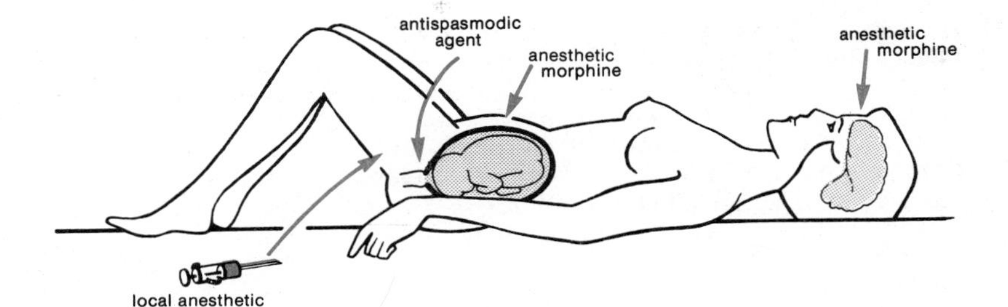

FIG. 6-215.
Drugs to suppress pain.

vocates of completely painless childbirth point out that pain is not at all an essential accompaniment of labor and that, for example, in many cases where women with spinal paralysis have given birth without any sensation of pain, the delivery of the baby has proceeded quite normally. From such examples, particularly those relating to cases of complete severance of the spinal cord (Fig. 6-214), it is evident that the processes associated with childbirth can take place independently in the abdomen even when the latter is separated from the upper part of the body insofar as the nerve functions are concerned. The natural purpose of labor pains would, however, appear to be that they prompt the woman to help delivery by "pushing" and also to give a warning of possible tearing or rupturing of overdilated parts of the birth passage. The advocates of painless childbirth nevertheless consider that fear and painful suffering should have no place in modern obstetrics and that therefore every effort should be made to prevent pain.

Drugs such as morphine can indeed prevent pain, but when used in large doses they tend to suppress the uterine contractions and also to inhibit the baby's breathing immediately after delivery. These drugs must therefore be used with caution.

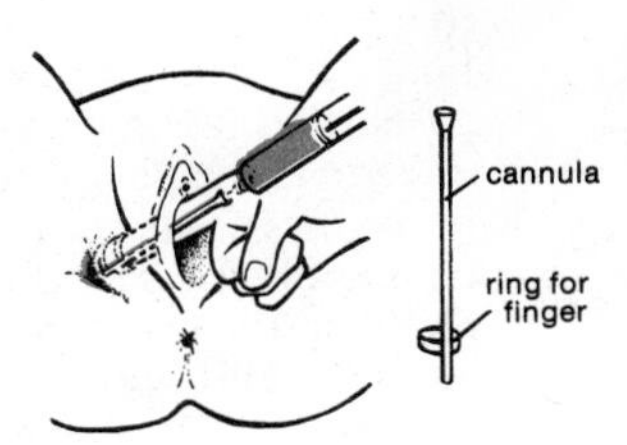

FIG. 6-216.
Local anesthetic administered to pudendal nerve (the obstetrician determines the correct point of injection by probing with his finger inserted through the ring).

Pain in the first stage of labor is often due to spasm of the cervix. Modern obstetric sedatives and analgesics incorporating an antispasmodic agent, but without some of the drawbacks of morphine, may be used in such circumstances (Fig. 6-215). In a case of arrested labor due to fatigue, a sedative can ensure a period of rest to enable the woman to regain her strength.

The second stage of labor, in which the actual expulsion occurs, is especially painful when the baby's head traverses the perineum. To relieve the pain the woman may be lightly anesthetized. The most famous early instance of the use of obstetric anesthesia was for Queen Victoria, who was given intermittent doses of chloroform, just enough to ease the labor pains, but not to make her unconscious. In modern practice, anesthetics other than chloroform would be used, as the latter is now known to be harmful to the liver. The aim is to remain in the first stage of anesthesia, the so-called analgesic stage (see page 32). The sensation of pain is thus largely eliminated, but the woman is still sufficiently conscious to cooperate in "pushing" (bearing down). On the other hand, if surgical intervention becomes necessary for delivering the baby, a more advanced stage of general anesthesia will have to be applied. In some cases, in order not to harm the baby, an injection of a local anesthetic may be given in the pelvic region, e.g., the cervix, the tail end of the spinal canal, or the pudendal nerves (Fig. 6-216, see also page 35). So-called twilight sleep, a state of partial anesthesia, can be induced by the injection of morphine and scopolamine or by infusing the rectum with an oil-ether or other anesthetic solution.

A different approach to painless childbirth is based on the view that to many women the fully conscious process of giving birth is a particularly rewarding and joyful experience. Efforts are therefore directed at allaying the patient's anxiety and fears and at educating her, by instruction and exercises, in preparation for the happy event. With the removal of fear her body will be more relaxed, spasms will be eased and labor pains, as a result, greatly reduced. Special gymnastic exercises during pregnancy and training in correct breathing technique are important aids. In the first stage of labor the woman should relax completely and concentrate on breathing, but in the second stage she should cooperate actively by pushing energetically. Other methods that claim to relieve labor pains are hypnosis and suggestion. On the whole, the best approach would appear to consist in a combination of a relaxed nonanxious state of mind and the judicious use of pain-relieving drugs.

PERINEUM, EPISIOTOMY

The perineum is the region between the vulva and anus, consisting of a bridge of muscle and fibrous tissue. When the baby's head emerges from the birth passage during childbirth, the perineum is liable to suffer injury by laceration due to overstretching. Malposition of the baby increases the risk of this occurring. Precautions to prevent it are therefore of some importance during labor. The obstetrician will perform manipulations to slow down the delivery of the baby's head and let it take place as gently as possible, while also guiding the head in such a way that it will present its least diameter and thus encounter the minimum resistance, which is generally the case with normal cephalic presentation (see page 293). To ensure a slow and smooth exit of the baby's head from the birth passage, the obstetrician places one hand from above on the head and restrains it, while his other hand seizes the woman's perineum from below (Fig. 6-217). Both hands then guide the occiput (the back part of the head) gently forward, so that the head is extended from its bent-forward position only when the back of the neck passes through the pelvic outlet (Fig. 6-218).

If the vagina and perineum are comparatively inflexible and difficult to stretch (as is more particularly likely in primiparas over 30 years of age), overstretching of the floor of the pelvis or tearing of the perineum is liable to occur despite the aforementioned precautions. To prevent this, and also in cases where difficult delivery is to be expected, e.g., with malposition of the baby's head, the outlet of the birth passage can be artificially enlarged by *episiotomy,* i.e., surgical incision of the perineum. The deliberate cut is much easier to control than the random tearing of the tissues. The incision may extend straight back from the vagina to about 2 cm in front of the anus (Figs. 6-219 and 6-220a); it may extend farther by tearing when the baby's head emerges. Alternatively, an oblique incision may be made through the muscle around the vaginal entrance. After the baby has been delivered, the incision is stitched (Fig. 6-221). This is done under anesthesia.

If episiotomy cannot be performed in time, tearing of the perineum will usually occur in a case of rapid delivery accompanied by overstretching of the birth passage. Various degrees of severity can be distinguished, ranging from the short tear in the vaginal mucous membrane and partial tearing of the perineum (Fig. 6-222) to complete perineal tearing in which even the sphincter of the anus is severed (Fig. 6-223), in which case the floor of the pelvis may be so seriously damaged as to result in a prolapse of the uterus in later years. The torn perineum is repaired by stitching under anesthesia.

FIG. 6-217.
Protecting the perineum.

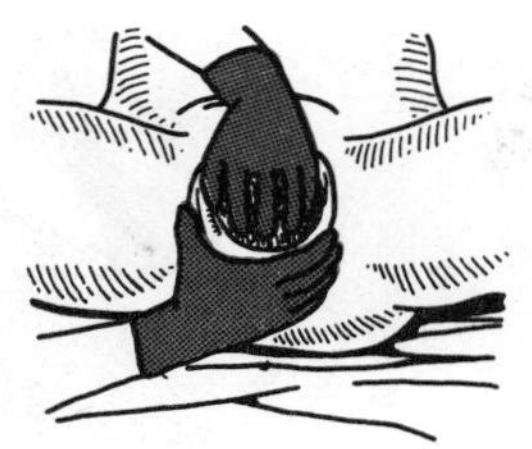

FIG. 6-218.

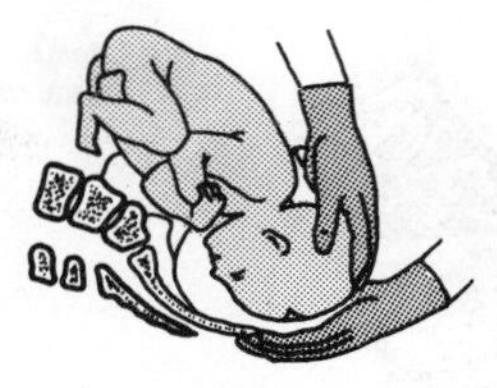

(a) The baby's head should come low down into the birth passage . . .

(b) and only then be swung back.

FIG. 6-219.
Episiotomy with incision straight back from vagina.

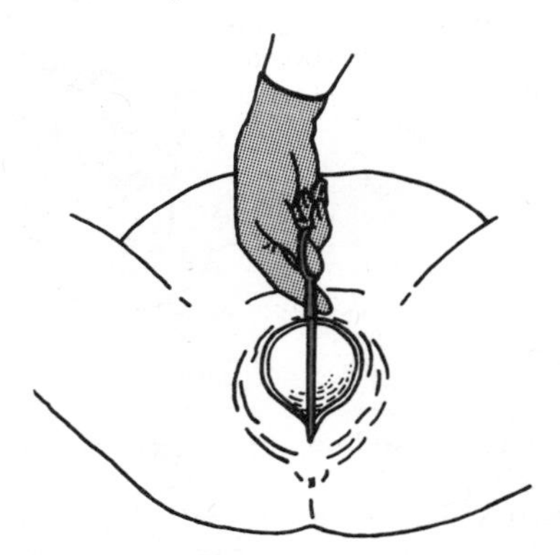

FIG. 6-220.
(a) Episiotomy with straight incision.
(b) Oblique incision.

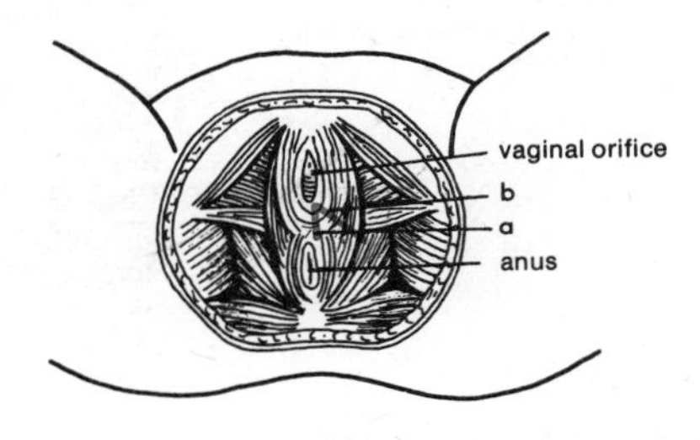

FIG. 6-221.
Incision is sutured (stitched) after childbirth.

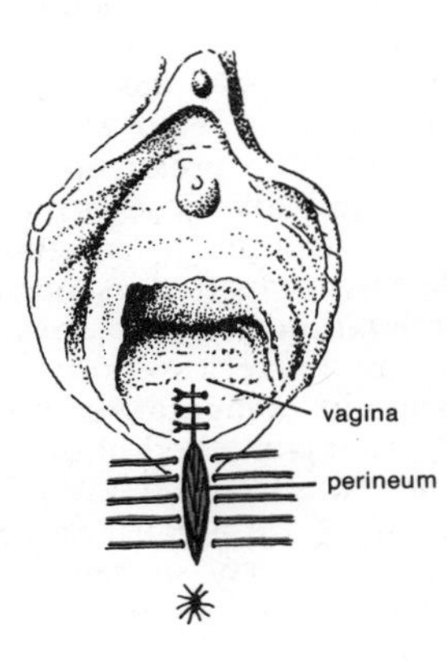

FIG. 6-222.
Perineum slightly torn.

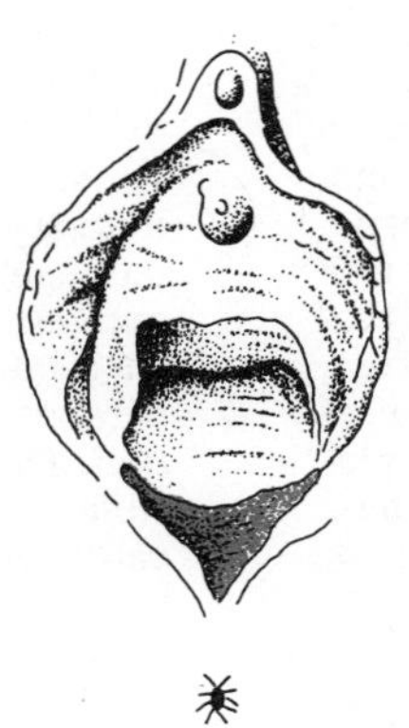

FIG. 6-223.
Complete tearing of perineum involving anal sphincter.

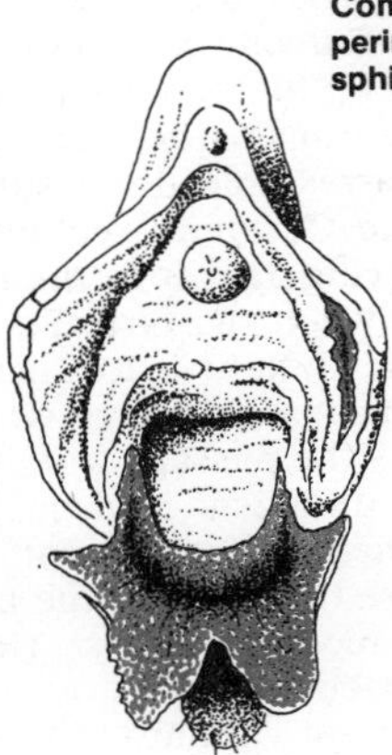

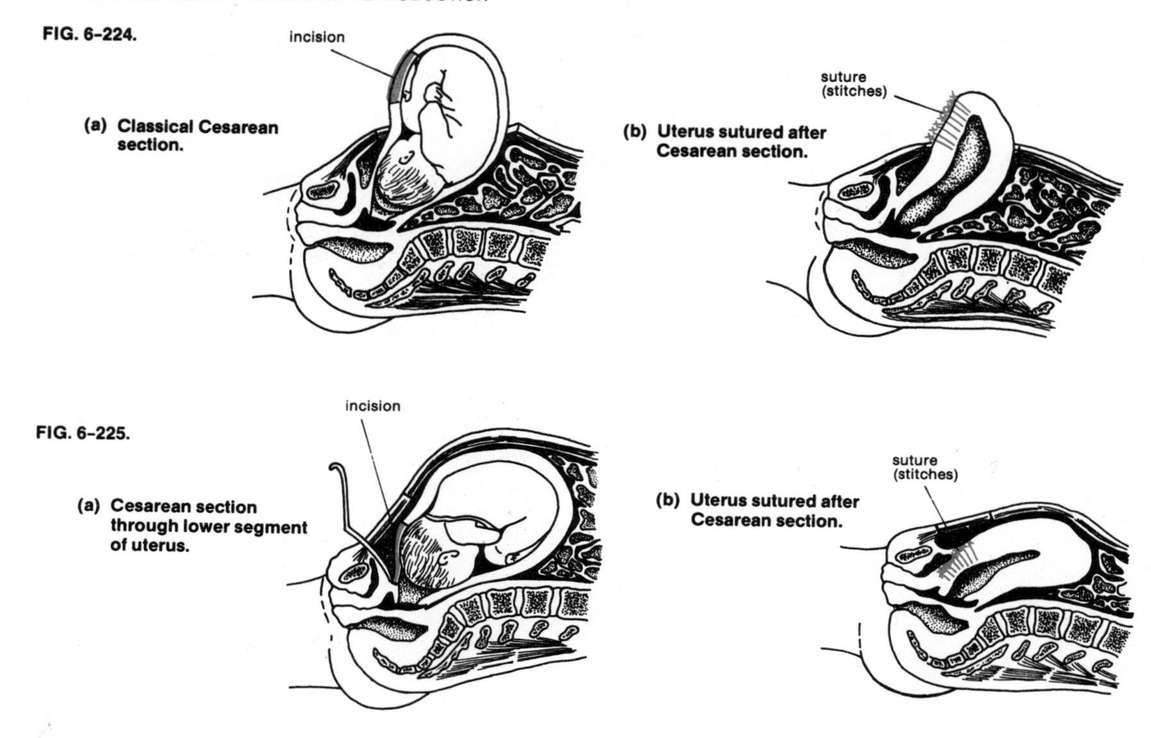

FIG. 6-224.

(a) Classical Cesarean section.

(b) Uterus sutured after Cesarean section.

FIG. 6-225.

(a) Cesarean section through lower segment of uterus.

(b) Uterus sutured after Cesarean section.

CESAREAN SECTION

The Cesarean section is a surgical operation for the delivery of the baby through an incision into the uterus. More particularly, it consists in cutting through the abdominal wall and the wall of the uterus, taking out the baby and removing the placenta and membranes, then suturing the incision. Though technically straightforward, it used to be a hazardous operation until the advent of present-day aseptic surgery backed up by modern anesthetics, antibiotics, blood transfusion resources, coagulant drugs to arrest bleeding, suturing techniques, etc. The Cesarean section is used in situations involving danger to the mother and/or baby—e.g., when the birth passage, especially the pelvic outlet, is too narrow, or in some cases of placenta praevia (where the placenta lies across the passage from the uterus), prolapse of the umbilical cord, malposition of the baby, premature detachment of the placenta, premature rupturing of the membranes, blood group incompatibility (Rh factor), eclampsia, arrested labor.

There are several variants of the operation, differing in the surgical technique employed. In the so-called classical Cesarean section the woman's abdominal cavity is opened, the uterus is eased out, and an incision is made across the fundus (upper part) of the uterus. After the baby has been delivered, the placenta and membranes are removed by hand, and the uterus and abdominal wall are sutured (Fig. 6-224). A more modern variant, which has the advantage of being "kinder" to the uterus, is the Cesarean section with incision through the noncontractile lower segment of the uterus after stripping back the bladder flap (Fig. 6-225).

Present-day surgery has turned the Cesarean section into a relatively safe operation which is now often preferred in cases where delivery of the baby through the natural birth passage would, though possible, involve a greater hazard than that associated with the surgical operation. Even so, less than about 1 percent of all deliveries are by Cesarean section. Despite its comparative safety, the decision to apply this method of delivery requires careful and expert consideration. If a woman has once had a Cesarean section, any other children that she may subsequently have will most probably have to be delivered by the same method.

FIG. 6-226.
Vacuum extraction.

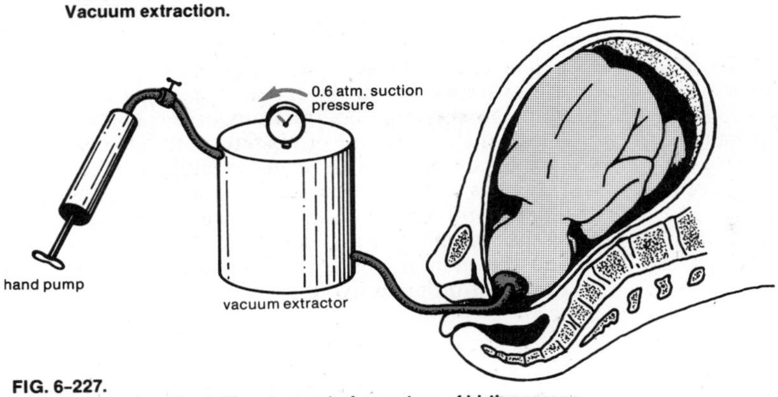

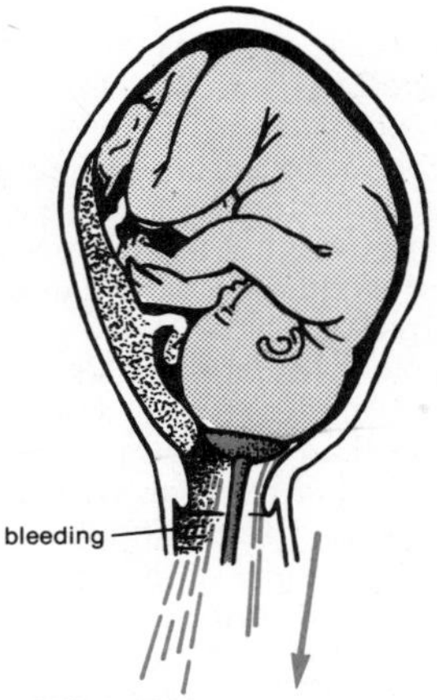

FIG. 6-227.
Direction of extraction taking account of curvature of birth passage.

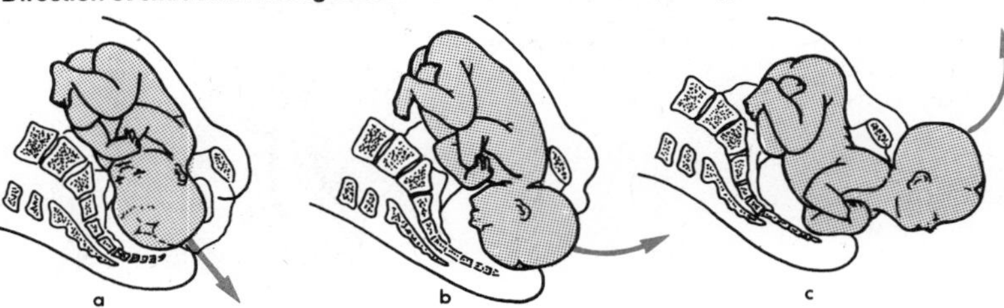

FIG. 6-228.
Vacuum extraction in a case of placenta praevia.

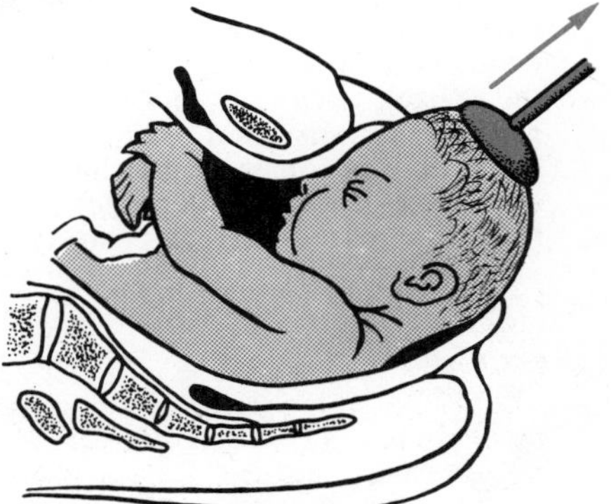

FIG. 6-229.
Vacuum extraction in a case of sinciput presentation.

VACUUM EXTRACTION

The object of vacuum extraction in childbirth is to apply a pull from below to assist the pushing pressure that the uterine contractions exert on the baby from above. The device called a vacuum extractor is equipped with a suction cup which is introduced into the birth passage and applied to the baby's head. An advantage of the vacuum method is that the woman does not have to be anesthetized. Furthermore, a small suction cup can be applied even if the cervix is not fully dilated, so that, in contrast with other obstetric aids to delivery, this technique can be employed in the first stage of labor. When the cup has been placed in contact with the baby's head, air is evacuated from the tube by means of a hand-operated pump, so that the cup attaches itself by suction to the head (Fig. 6-226). The obstetrician operating the device can then assist delivery by pulling the tube at correctly timed intervals "in step" with the uterine contractions and guiding the baby along the birth passage by suitably varying the direction of pulling (Fig. 6-227). This method can be a useful aid in a case of placenta praevia (Fig. 6-228) and of malposition of the baby's head, e.g., with sinciput presentation (Fig. 6-229, see also page 293). Though simple in principle as a convenient means of getting a hold on the baby and assisting its delivery, the method is safe only in the hands of an expert. A harmless "bump," which disappears in a day or so, is raised on the baby's head by the suction cup. In general, the latter should not be applied for more than about half an hour.

DELIVERY WITH FORCEPS

Obstetric forceps are surgical instruments designed to assist the delivery of a baby in an advanced stage of labor if spontaneous delivery cannot be accomplished. There are various types of forceps, but the essential principle is that they enable the baby's head to be gently seized so that the baby can be eased out of the birth passage.

Typical forceps consist of a pair of spoon-shaped blades which are inserted separately and are then connected together by means of a lockable and releasable pivot or similar device (Fig. 6-230). The "spoons" are of a size and shape appropriate to the baby's head, which they should grip transversely (Fig. 6-231). First one and then the other half of the forceps is inserted (Fig. 6-232). When they are in position, gripping the baby, the obstetrician exerts a gentle downward pull with both hands, until the baby's head appears at the mouth of the vagina (Fig. 6-233). He then extracts the baby carefully, making its head and body follow the curvature of the birth passage as in spontaneous childbirth. If the baby's head is not in the "back-to-front" (normal) position (see page 293) on insertion of the forceps, the manipulation becomes somewhat trickier. For one thing, it is then more difficult to get the forceps into the correct position around the baby's head, and furthermore it will be necessary also to rotate the head into the desired back-to-front position to extricate it from the pelvic outlet. In general, the pull exerted by the obstetrician will be timed to keep in step with the rhythmic contractions of the uterus.

For the proper use of obstetric forceps certain conditions will have to be fulfilled. It always has to be done under anesthesia. To enable the forceps to be inserted without danger to the woman, the cervix must be fully dilated and the membranes of the amniotic sac ruptured. In order to avoid danger to the baby's

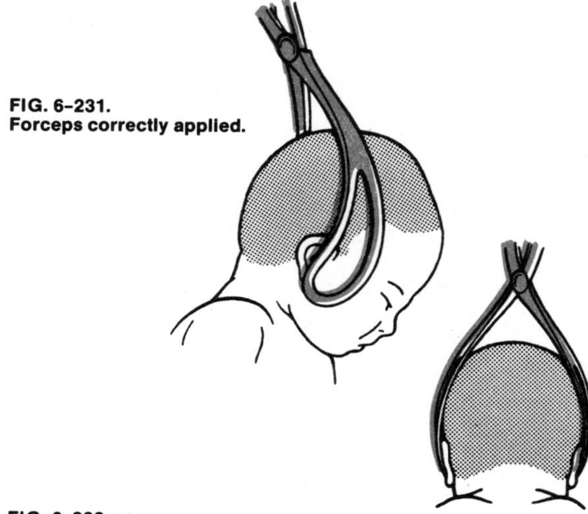

FIG. 6-231.
Forceps correctly applied.

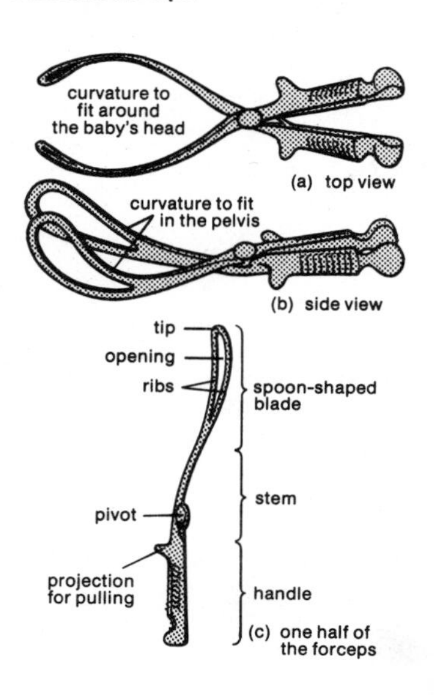

FIG. 6-230.
Obstetric forceps.

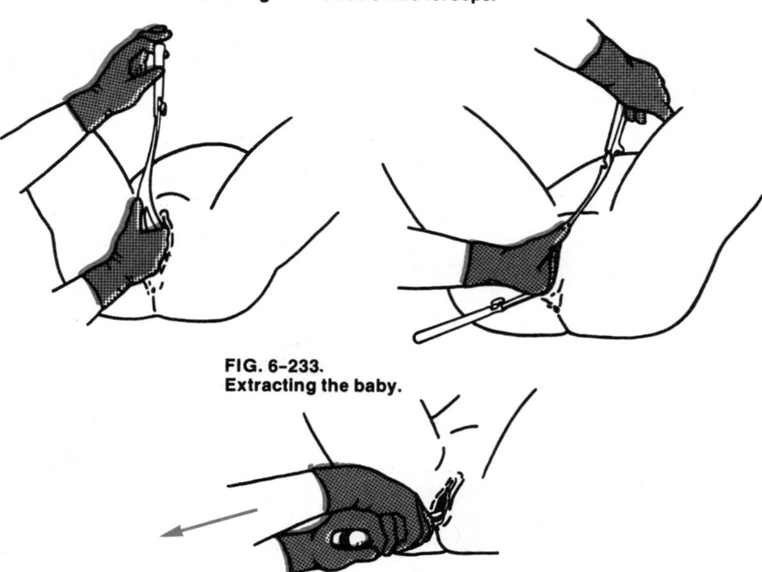

FIG. 6-232.
Inserting the blades of the forceps.

FIG. 6-233.
Extracting the baby.

head by pressure and tension exerted by the forceps, the latter should be of the right size to fit snugly round it, and the head should have descended far enough into the mother's pelvis to enable it to be properly seized by the forceps. And for forceps delivery to be possible at all, the pelvic outlet must, of course, be sufficiently wide to permit the head to pass. The forceps can be applied low, medium or high, according to how far into the birth passage they have to be inserted in order to grip the baby's head. The use of forceps always involves a risk of injury to the mother (tearing of the cervix, vagina and perineum) and especially also to the baby (head injuries, brain damage, paralysis, etc.). In modern obstetric practice the use of forceps is discouraged unless absolutely essential.

DELIVERY OF TWINS

Twins are born in about 1 in every 80–90 pregnancies, while the probability of triplets is only about 1 in 10,000.

Fraternal twins (dizygotic or biovular twins) are those from two separate ova fertilized at the same time. As a rule, a woman's ovaries release only one ovum each month, but it may happen that two (or more) are released simultaneously and are fertilized together. Such twins may be of the same or different sex and do not necessarily resemble each other more than brothers or sisters of different ages. Roughly 80 percent of all sets of twins are of this type. It is a well-known fact that a tendency to have twins "runs in families," this being due to an inherited tendency to shed more than one ovum. A case is on record where a woman, herself one of quadruplets, gave birth to no fewer than 32 children in 11 pregnancies.

Identical twins (monozygotic or uniovular twins) develop from a single fertilized ovum which splits into two separate halves in an early stage of its development. Such twins have the same genetic makeup, are of the same sex, and show striking similarity in physical, physiological and mental traits. They have a common placenta, but each embryo usually develops its own amnion and umbilical cord. Fraternal twins usually have separate placentas (Fig. 6-234a); but even with only a single placenta the twins may nevertheless be of the

FIG. 6-234.

fraternal twins

identical twins

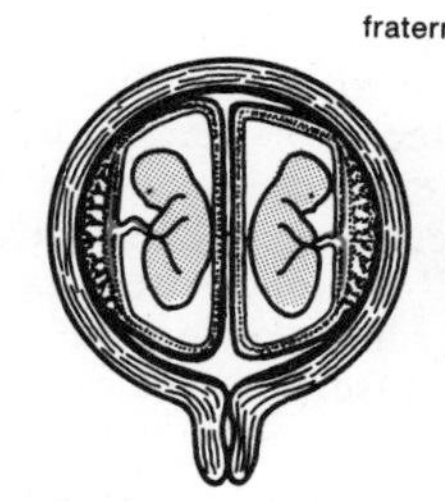

(a) Separate placentas.

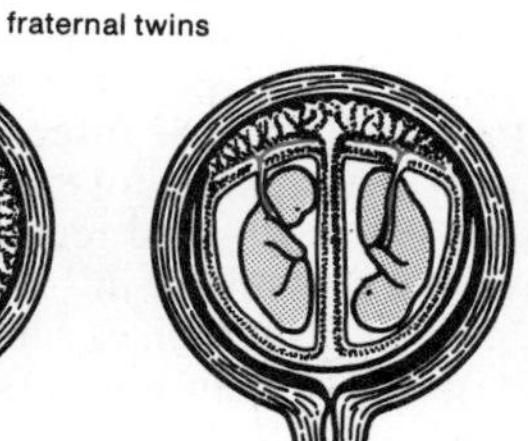

(b) Dividing wall has more than two membranes.

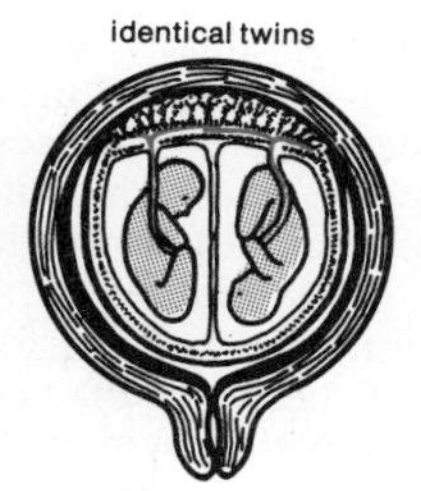

(c) Dividing wall has only two membranes.

FIG. 6-235. POSITION OF TWINS.

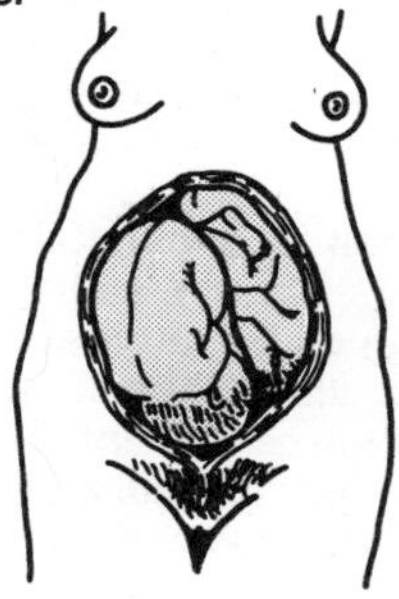

(a) In 45% of cases.

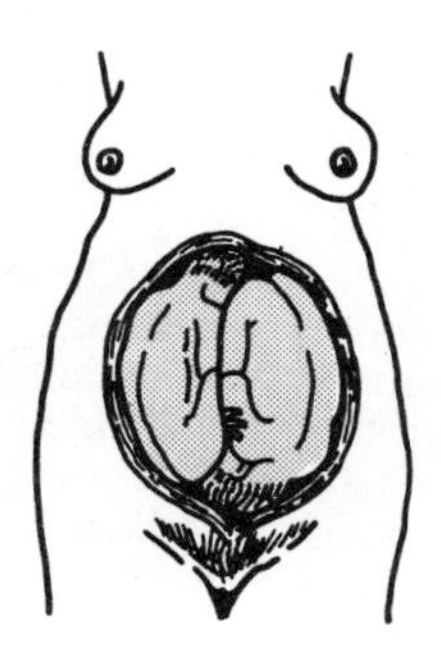

(b) In 35% of cases.

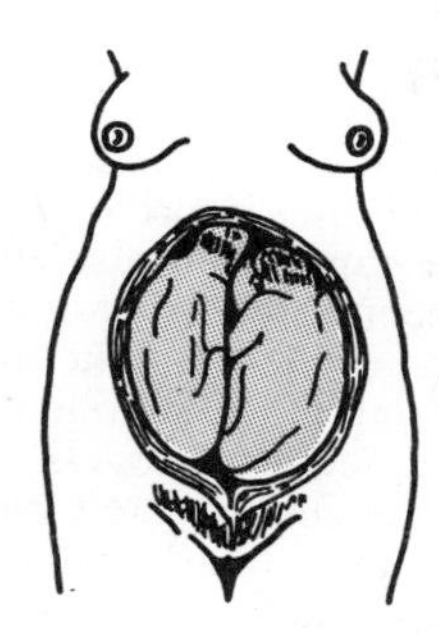

(c) In 10% of cases.

fraternal type, in which case the embryos are, however, separated by a dividing wall comprising more than two membranes (Fig. 6-234b), whereas with identical twins there are only two such membranes (Fig. 6-234c). In cases of doubt it is nowadays possible to establish by means of blood group tests and other criteria whether twins are fraternal or identical.

Despite their identical genetic characters, identical twins show some individual differences in their development from infancy to adulthood. Since such differences must be due to environmental factors, it is evident that such twins are of great importance to research into the question of whether and to what extent personal characteristics are genetically determined or are modifiable by factors in the individual's environment.

A twin pregnancy is to be suspected if the woman's abdomen is conspicuously enlarged and the uterus extends very high up. More reliable clues are obtained when the doctor can feel the major parts of two babies through the abdominal wall. If two separate heartbeats can be heard or two distinct electrocardiograms are recorded, the question can be definitely settled, and, of course, an x-ray photograph can provide complete certainty.

The most frequent position of twin fetuses in the uterus is with their heads down, one behind the other, and the separating membranes slanting (45 percent of all cases, Fig. 6-235a). Alternatively, the fetuses may lie head up and head down respectively, with the separating membranes approximately vertical in the middle (35 percent, Fig. 6-235b), or with both heads upward (10 percent, Fig. 6-235c).

In about 50 percent of all twin births the delivery of the babies proceeds spontaneously and without incident. In many cases, however, the overexpanded uterus is unable to develop very energetic contractions. Nevertheless, once the cervix is dilated, expulsion of the first baby often proceeds more quickly than in normal circumstances. Its umbilical cord has to be efficiently tied so that no blood can escape from the placenta and thus deprive the second baby of an adequate supply in a case where the two fetuses have been attached to a common placenta. After a short interval in the labor contractions, rupturing of the second amniotic sac occurs and the second baby is expelled fairly easily from the fully dilated birth passage. Complications may arise because the babies sometimes get in each other's way at the pelvic outlet. In such cases premature loss of amniotic fluid, premature birth, or prolapse of the umbilical cord may occur.

PUERPERIUM

The puerperium is the period following childbirth, more specifically the period following the third stage of labor and lasting until involution of the pelvic organs has occurred (especially the return of the uterus to its normal nonpregnant size), usually in 3–6 weeks. In the period directly after giving birth to her baby, the mother should have a good night's sleep. Before the baby is brought to her on the following day, she should begin to take normal meals. It is important that she should regularly empty her bladder, because soon after childbirth a considerable amount of fluid is discharged as urine. Also, involution of the uterus is helped by an empty bladder. Urination may at first be difficult because of slackness of the bladder and be painful because the urine comes into contact with lacerations or sore areas. It may be necessary to give a drug to assist urination or, more rarely, to empty the bladder with the aid of a catheter.

Urination and emptying the bowels are assisted by getting the woman on her feet as soon as possible after childbirth, generally within two or three hours, letting her perform certain postnatal exercises, thus rather belying the traditional concept of the puerperium as the "lying-in" period. After about a week these exercises can take a more energetic form, to train the muscles of the pelvic floor, abdomen and back (Figs. 6-236 and 6-237). Besides toning up the muscles and helping the woman to regain her figure, the exercises stimulate the circulation and help to prevent thrombosis. Constipation is very common in the puerperium because the bowels are still slack and sluggish, so that it may be necessary to give an enema a day or two after childbirth. Any surgical sutures are removed on the fifth or sixth day.

Quite soon after childbirth it is permissible to take a shower standing up, but bathing should be avoided until at least 10 days after childbirth, while vaginal douches are on no account allowed except on medical advice. Cleanliness is especially important in the earlier part of the puerperium.

A vaginal discharge (*lochia*) occurs during

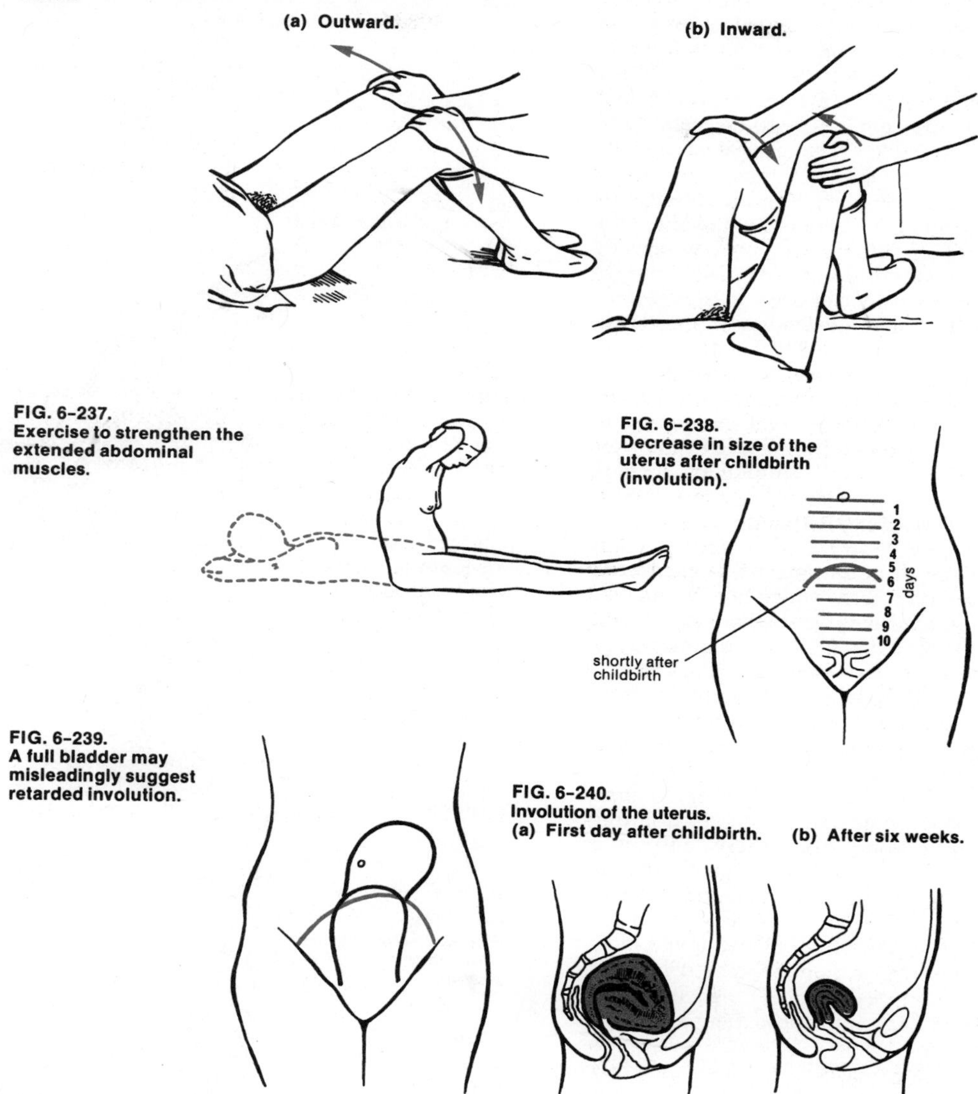

FIG. 6-236.
Exercising the pelvic floor muscles by moving the knees.
(a) Outward. (b) Inward.

FIG. 6-237.
Exercise to strengthen the extended abdominal muscles.

FIG. 6-238.
Decrease in size of the uterus after childbirth (involution).

FIG. 6-239.
A full bladder may misleadingly suggest retarded involution.

FIG. 6-240.
Involution of the uterus.
(a) First day after childbirth. (b) After six weeks.

the puerperium. For the first 6 days or so it is blood-tinged, then becoming brownish, yellowish and finally white. The lochia normally ceases after 4–6 weeks. As it picks up microbes in the vagina, it is important not to allow this discharge to come into contact with the breasts or the baby. Pads or towels used to absorb it should be changed several times a day and not be touched by hand.

If there are no complications, 7–10 days' convalescence after childbirth is generally sufficient. Menstruation does not occur until the sixth week, by which time the ovaries have resumed their cyclic function. In women who breast-feed their babies, menstruation usually remains absent until the end of lactation. To avoid risk of infection, sexual intercourse should not be resumed until the seventh week. The woman's temperature should be regularly checked, as a rise in temperature in the first three weeks after childbirth could signify the onset of puerperal sepsis. For this reason she should continue to take her temperature at regular intervals for some time after leaving the maternity hospital. The doctor can tell how the involution of the uterus is progressing by checking its position (Fig. 6-238), while taking account of the presence of urine in the bladder, as a full bladder may misleadingly suggest retarded involution (Fig. 6-239). It takes about six weeks for the woman's body to return to the shape it had before pregnancy and for the uterus to go back to its normal (nonpregnant) size, involving a reduction in its weight from about 1000 g to a mere 50 g or so (Fig. 6-240).

POSTPARTUM HEMORRHAGE

Hemorrhage from the internal reproductive organs directly after childbirth—as distinct from the normal discharge (lochia, see preceding section)—can be a dangerous complication. It may arise from the site of the placenta in the uterus or from lacerations to the birth passage. Besides tearing of the perineum (page 304), tearing of the cervix may occur in childbirth that requires surgical or manipulative intervention (e.g., forceps, version of a malpositioned baby) while the cervix is not fully dilated. Lacerations from these and other causes tend to bleed profusely and persistently. Damage to the cervix of the uterus is treated by suturing (Fig. 6-241) as soon as possible after expulsion of the placenta and membranes. Hemorrhage can be temporarily arrested by compressing the big abdominal artery (Fig. 6-242) until the patient's removal to hospital.

Sometimes the uterus is too slack and enfeebled to contract sufficiently to seal the wound left by the placenta after its detachment and expulsion. In such cases bleeding from the uterus will occur immediately after childbirth and will continue in intermittent gushes. Normally about 200–300 ml, seldom more than 500 ml, of blood is lost in the postpartum discharge (lochia). Larger losses of blood (1000 ml and more) must be regarded as serious, potentially dangerous. A blood transfusion will generally have to be given and immediate steps taken to stop the bleeding.

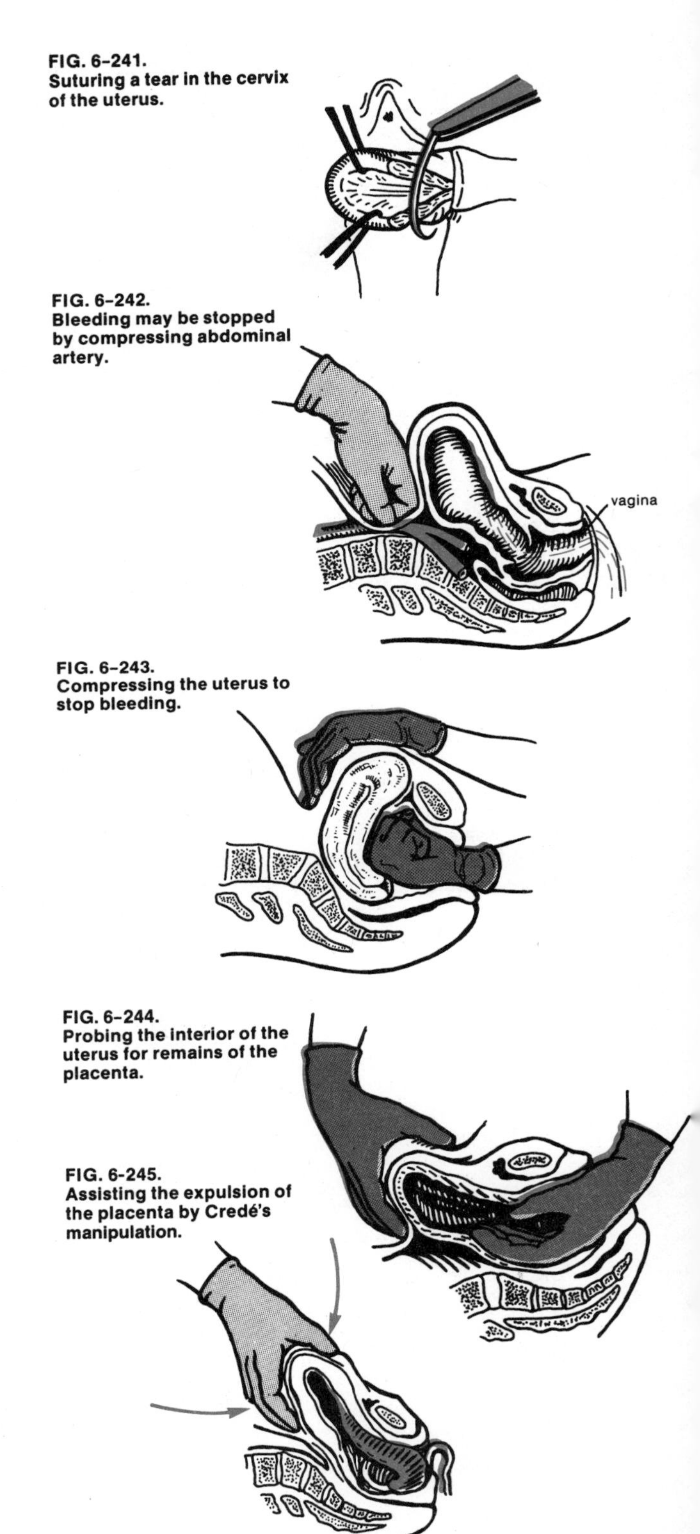

FIG. 6-241. Suturing a tear in the cervix of the uterus.

FIG. 6-242. Bleeding may be stopped by compressing abdominal artery.

FIG. 6-243. Compressing the uterus to stop bleeding.

FIG. 6-244. Probing the interior of the uterus for remains of the placenta.

FIG. 6-245. Assisting the expulsion of the placenta by Credé's manipulation.

Fortunately, the obstetrician or midwife is often able to tell in advance that postpartum hemorrhage is likely to occur, e.g., in a case where the capacity for contraction of the uterus during labor has been weak from the very beginning, the uterus has been overstrained, or delivery of the baby has been abnormally rapid (twin birth, Cesarean section). It is sometimes possible to prevent the hemorrhage by giving injections to promote contraction of the uterus. If it is certain that the bleeding is not due to a laceration and if the placenta has been completely expelled, such injections may usefully be given also after childbirth.

Massaging the uterus with ice applied externally to the abdomen and/or packing the lower part of the uterus with gauze or other suitable material are methods of controlling postpartum hemorrhage. In some cases the uterine bleeding can be stopped by compressing the uterus (Fig. 6-243) or the abdominal artery (Fig. 6-242). Another method is for the doctor, wearing long gloves, to insert one hand into the fundus of the uterus and massage the uterus with the other hand on the abdomen.

In a certain percentage of cases bleeding will continue even after the uterus has contracted tightly after childbirth. This bleeding may be due to a deficiency of a specific clotting agent in the blood or to failure of the uterus to expel all of the placenta, so that part of it has remained behind in the uterine cavity. As even quite small residual pieces can be dangerous in this way, it is important always to inspect the placenta after childbirth, to make sure that all of it has been duly expelled. Any pieces that remain in the uterus must be located by manual probing (Fig. 6-244) and carefully removed under aseptic conditions. In certain other cases the placenta, though intact, fails to detach itself completely from the wall of the uterus and thus, by its presence, prevents the subsequent hemorrhage-arresting contraction of the uterus from taking place. Such failure to dislodge and expel the placenta is often due to slackness and enfeeblement of the uterus. If injections or other means to induce spontaneous expulsion of the placenta do not produce the desired result, Credé's manipulation can be applied: the placenta is expelled by downward pressure exerted on the uterus through the abdominal wall (Fig. 6-245). If this also fails, the placenta will have to be removed by direct internal manipulation. Any remains of the placenta left behind in the uterus may, besides presenting a hemorrhage hazard, result in puerperal sepsis. It is essential to maintain strictly aseptic conditions in all measures aimed at arresting postpartum hemorrhage.

PUERPERAL SEPSIS

After childbirth the birth passage is vulnerable to infection until healing is complete. Puerperal sepsis (puerperal fever, childbed fever) is a toxemia, accompanied by a rise in temperature, which may affect a woman during the first three weeks of the puerperium. In general, toxemia denotes the distribution throughout the body of poisonous products of localized bacterial infection, thus producing generalized symptoms. It is due to invasion by microorganisms of the wounds caused by childbirth; these wounds may be those at the placental site (Fig. 6-246) or be the result of injury or laceration of parts of the birth passage during labor. Most frequently, the microorganisms are streptococci, but other kinds of bacteria may also cause serious infection.

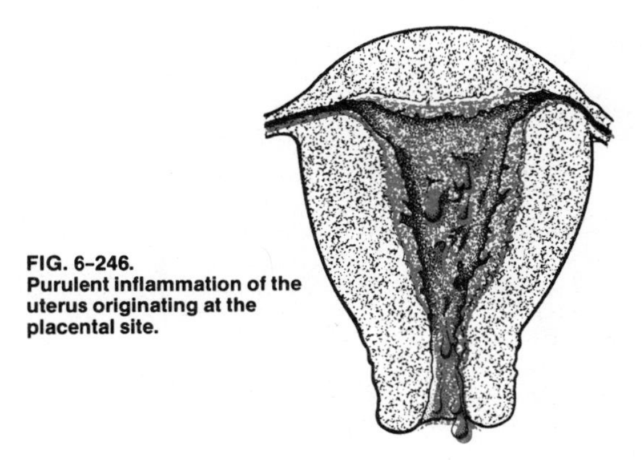

FIG. 6–246.
Purulent inflammation of the uterus originating at the placental site.

Until comparatively modern times, puerperal sepsis caused by unhygiènic midwifery practices was always a major hazard. The risk of infection has now been largely eliminated by strict asepsis, including such precautions as the wearing of surgical masks by all those who come into contact with the patient, while modern drugs (antibiotics, sulfonamides) are moreover available to combat any infection that may nevertheless occur. So-called broad-spectrum antibiotics give good results. In severe cases of infection of the abdominal cavity, surgical intervention may be necessary. In general, lowered resistance is a danger. Correct

FIG. 6-247.
Spread of uterine inflammation.

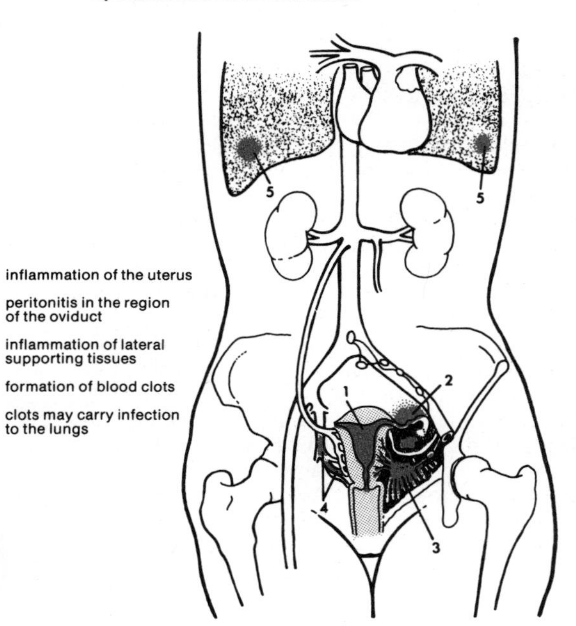

FIG. 6-248.
Swelling of a leg due to congestion caused by clots in the veins (thrombophlebitis).

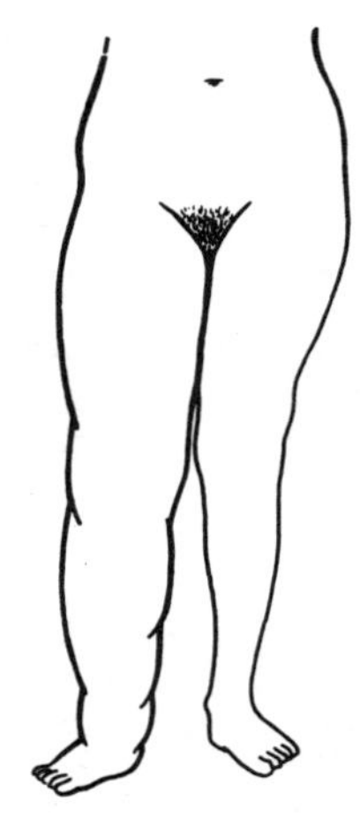

care of the patient during long labor, use of the least traumatizing method of delivery, avoidance of blood loss, blood transfusions, etc., are some of the present-day precautions that help to eliminate puerperal sepsis.

The infection may remain localized at the site of the wound in the uterus, but there is a risk that it will spread upward through the fallopian tubes into the abdominal cavity and cause peritonitis. It may spread through the lymphatic vessels to the lateral supporting tissues of the uterus and/or may develop into blood poisoning with further spreading of the infection by blood clots to other parts of the body such as the lungs or the heart valves (Fig. 6-247). Clots may also form in the veins of the legs, which become inflamed and swollen (thrombophlebitis) (Fig. 6-248). Such spreading of the infection is attended by a high rise in the patient's temperature, rapid pulse, and pain in the fallopian tubes and in the abdomen generally.

BREASTS, LACTATION

The female breasts (mammary glands) consist of fatty tissue and 15–20 glandular lobes, each of which is drained by a separate lactiferous duct which opens on the tip of the nipple (Fig. 6-249). The nipple contains erectile tissue and is surrounded by a pigmented area called the areola, which becomes more heavily pigmented in pregnancy. The erectile tissue causes the nipple to protrude and thus be grasped by the baby's mouth in sucking. The breasts are under the control of sex hormones. Up to puberty they are merely a rudimentary duct system. During puberty, estrogens from the ovary stimulate growth and development of the duct system (Fig. 6-250), but the growth in size is mainly due to accumulation of fat. The size of the breasts is therefore not an indication of their efficiency in feeding a baby; small breasts can be just as effective as large ones.

FIG. 6-249.
Structure of the mature breast.

FIG. 6-250.
Factors influencing development of breasts.

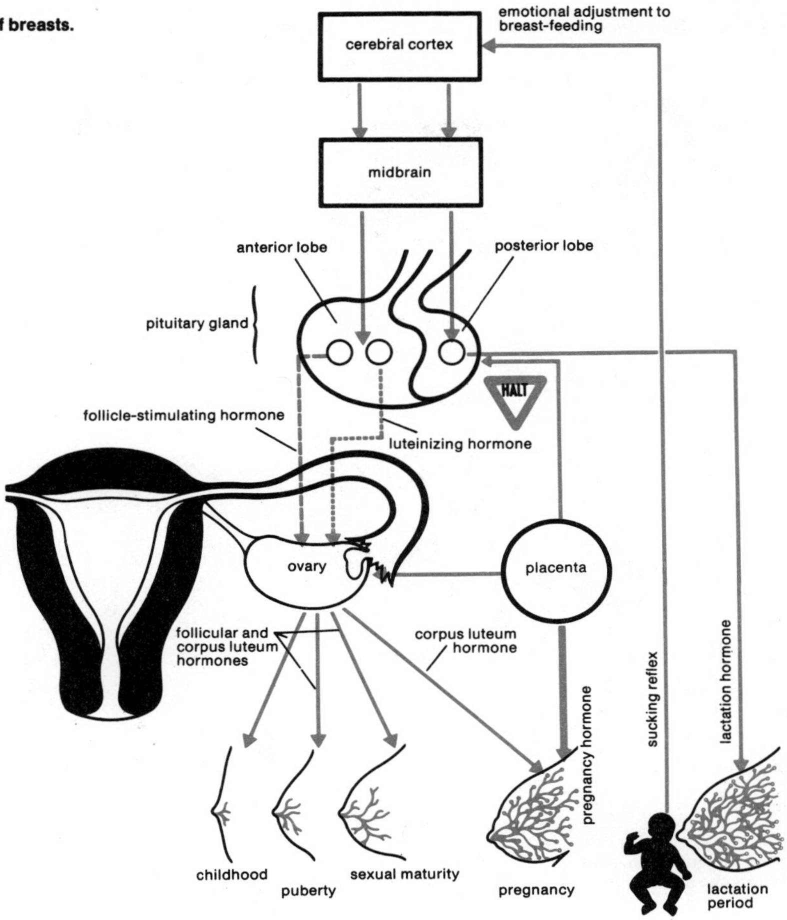

It is only when a woman becomes pregnant that the breasts become capable of secreting milk. This happens under the influence of hormones secreted by the corpus luteum and the placenta. The ducts proliferate and the secretory (milk-forming) tissue develops from their ends. These are surrounded by contractile cells which help to discharge the milk. Secretion of milk after childbirth is initiated by the action of a hormone (prolactin) from the pituitary gland in conjunction with adrenal corticoids; another pituitary hormone (oxytocin) induces ejection of milk. Suckling stimulates both milk secretion and discharge of milk by a reflex action involving the release of oxytocin from the pituitary gland. When the baby is weaned, the suckling stimulus is withdrawn, with the result that secretion of milk soon ceases. If necessary, hormone injections to stop secretion more quickly can be given.

Lactation refers to the period of suckling; it also denotes the actual function of secreting milk. More than 80 percent of mothers are capable of lactation, i.e., breast-feeding their babies. Mother's milk has the advantage of providing the baby with ideally balanced nourishment in terms of protein, fat, carbohydrate and salts. Human milk differs from cow's milk in containing more carbohydrate but less inorganic substances and protein; in addition, the proportions of proteins are different. Statistics indicate that, on the whole, breast-fed babies are more resistant to infection and other health hazards and are more likely to survive than babies that are bottle-fed from birth. Besides, suckling is good for the mother, too, and

FIG. 6-251.
Sucking action.

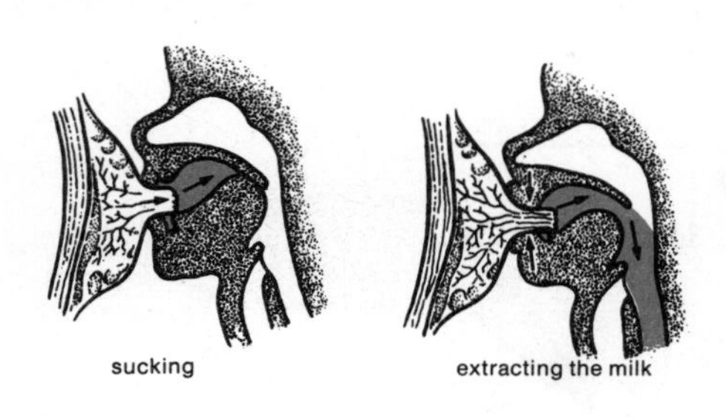

FIG. 6-252.
Emptying the breast after feeding the baby.

FIG. 6-253.
Pumping out the milk to empty the breast.

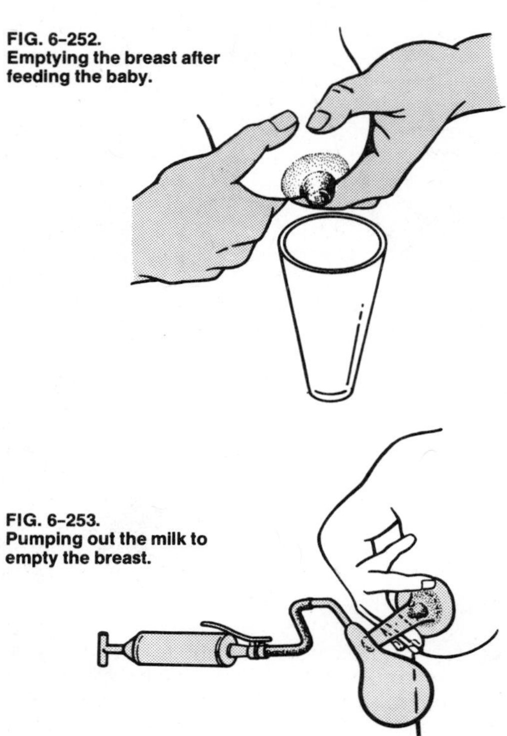

stimulates involution of the uterus, i.e., its return to its normal nonpregnant size. The lactation period extends from birth until the baby ceases to be breast-fed. As a rule, a baby should be breast-fed for not less than 3 and not more than 9 months. Weaning, i.e., the gradual cessation of breast-feeding, is usually begun when the baby is 5–6 months old.

The baby should be given its first trial feed 12–20 hours after birth. To begin with, it is necessary to be very patient until it has learned to grasp the nipple with its lips and perform the correct rhythmic movements (Fig. 6-251). While doing this the baby must be able to breathe through the nose. Later on, the baby should be suckled for periods generally not exceeding 12–15 minutes, certainly not more than 20 minutes. On each occasion it should be fed from one breast only. If the baby is suckled for too long periods, it may get into the bad habit of going to sleep with the nipple in its mouth. As a result, cracks or soreness of the nipples may occur, with risk of inflammation of the breast (mastitis) due to bacterial infection (see next section); this can usually be cleared up with antibiotics, but sometimes an abscess forms which requires surgery.

It is important always to give the baby the same number of regular feedings at fixed times. For most babies five feedings per 24 hours are sufficient, e.g., at 6 and 10 A.M. and at 2, 6 and 10 P.M. If the baby does not appear to be thriving, the number of feedings may be increased to six or seven. After each feeding, the baby should be held upright for a short time to enable it to belch and so expel any air that it has swallowed. If a considerable amount of milk is regurgitated with belching after each feeding, medical advice should be obtained. To prevent congestion of milk, the breast should be properly emptied at each feeding. If this is not achieved by suckling the baby for up to 20 minutes, the breast should be emptied by manual massage or with the aid of a breast pump (Fig. 6-252 and 6-253). This prevents the ducts from becoming plugged and also stimulates milk secretion.

After the second day the quantity of milk that a baby drinks will increase by about 50–70 g per day. Despite this, the newborn baby undergoes a loss of weight of about 10 percent during the first 3–5 days of its postnatal life because it loses moisture. Then its weight gradually increases, regaining its birth weight at 10–14 days after birth. From then onward the breast-fed baby should gain weight at a rate of about 170 g per week in the first three months, 150 g per week in the second three-monthly period, 110 g per week in the third, and 90 g per week in the fourth. Like most of such figures, these are, of course, averages.

To prevent infections, cleanliness of the breasts is very important. Aseptic precautions should be used in breast care to avoid infection of the breasts and of the baby's mouth, especially in the puerperal period. The breasts should be cleaned with sterile (boiled) water before and after each feeding. Early treatment of any soreness, cracks or fissures is essential.

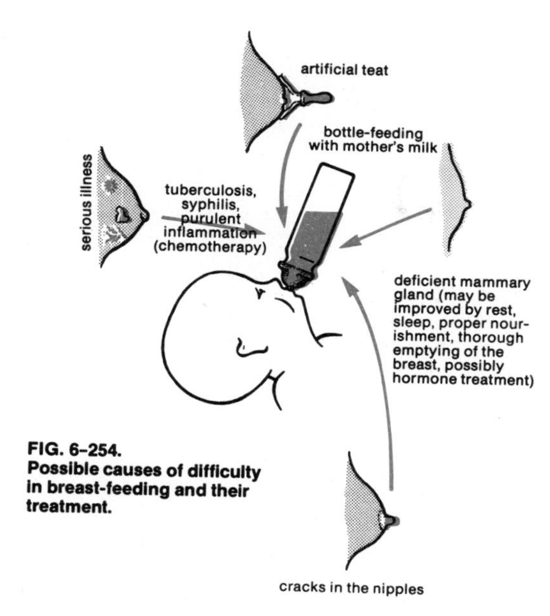

FIG. 6-254. Possible causes of difficulty in breast-feeding and their treatment.

Also, the woman should thoroughly wash her hands before each feeding and should then take care not to touch any part of her genitals or pads applied to them, for there is a risk of infection from the lochia (see page 311) to the breasts.

In certain circumstances breast-feeding may be difficult or impossible or may be undesirable for medical reasons (Fig. 6-254). Thus, a woman with tuberculosis should not suckle her baby, and should avoid contact with the baby until it has been vaccinated against the disease. On the other hand, if the mother has syphilis, the baby will also have the disease and there will generally be no objection to breast-feeding, with this exception: if she was infected with the disease in the last few months of pregnancy, the baby may be healthy; if so, it should not be fed direct from the breast, but be given its mother's milk pumped from the breast and boiled. Inflammation of the breast, especially if it is accompanied by a discharge of pus, is a reason to stop breast-feeding, as it involves an infection hazard to the baby. While suckling a baby should not impair the health or vigor of a healthy woman, it imposes too great a strain on a woman who is seriously ill, e.g., suffering from puerperal sepsis or from serious disease of the heart, liver or kidneys.

Sometimes the baby has difficulty in grasping the nipple because it is too flat, i.e., not protruding sufficiently for the baby to get its lips firmly round it. An artificial teat placed over the nipple may provide the remedy (Fig. 6-254). Or the milk may be extracted by pumping (Fig. 6-253) and fed from a bottle.

Cracks in the nipples may be caused by suckling the baby for too long periods or by lack of proper hygiene and care for the breasts during lactation. Such cracks or fissures make feeding more difficult and consitute an infection hazard. It may then be necessary to interrupt breast-feeding for a time in order to give the cracks a chance to heal. In the meantime the baby will be fed with milk pumped from the breasts. Moist dressings or wrappings for the nipples are unsuitable, as they tend to make the cracks worse. Sterilized clean cloths should be placed over the nipples after each feeding; antiseptic ointments, powder, etc., may also be used on medical advice. Regular washing with cold water and gentle massage of the nipples during the final months of pregnancy are very helpful antenatal precautions against subsequent cracking and fissuring.

It is only in a relatively small minority of cases that a mother's breasts fail to give enough milk from the outset. Sometimes the milk is there, but fails to flow easily from the nipples. Usually the baby will manage to improve the flow by its sucking action. If not, a short course of hormone treatment may be indicated. True deficiency of the mammary glands is often recognizable by an abnormally small amount of glandular tissue in the breasts. In some cases a deficiency of milk may be due to lack of sleep, undernourishment of the mother, or indeed sheer lack of patience or of will to apply the proper breast-feeding technique. The best stimulus to milk production is to empty the breast thoroughly at each feeding, perhaps with the aid of a pump. Quite often, breasts that at first appear unable to give an adequate supply of milk can, with a little patience and perseverance, be made to increase the yield.

Whether a baby is getting enough nourishment can be judged by its weight curve and its behavior. Babies that get too little milk tend to be rather dull and drowsy, and there are changes in the condition and consistency of their stools. In such cases breast-feeding should not be stopped entirely, but the baby should be put on mixed breast- and bottle-feeding. At each meal it is first given the breast, then the bottle. The teat hole of the bottle should be small enough to ensure that the baby will not prefer the "easier" bottle to the breast.

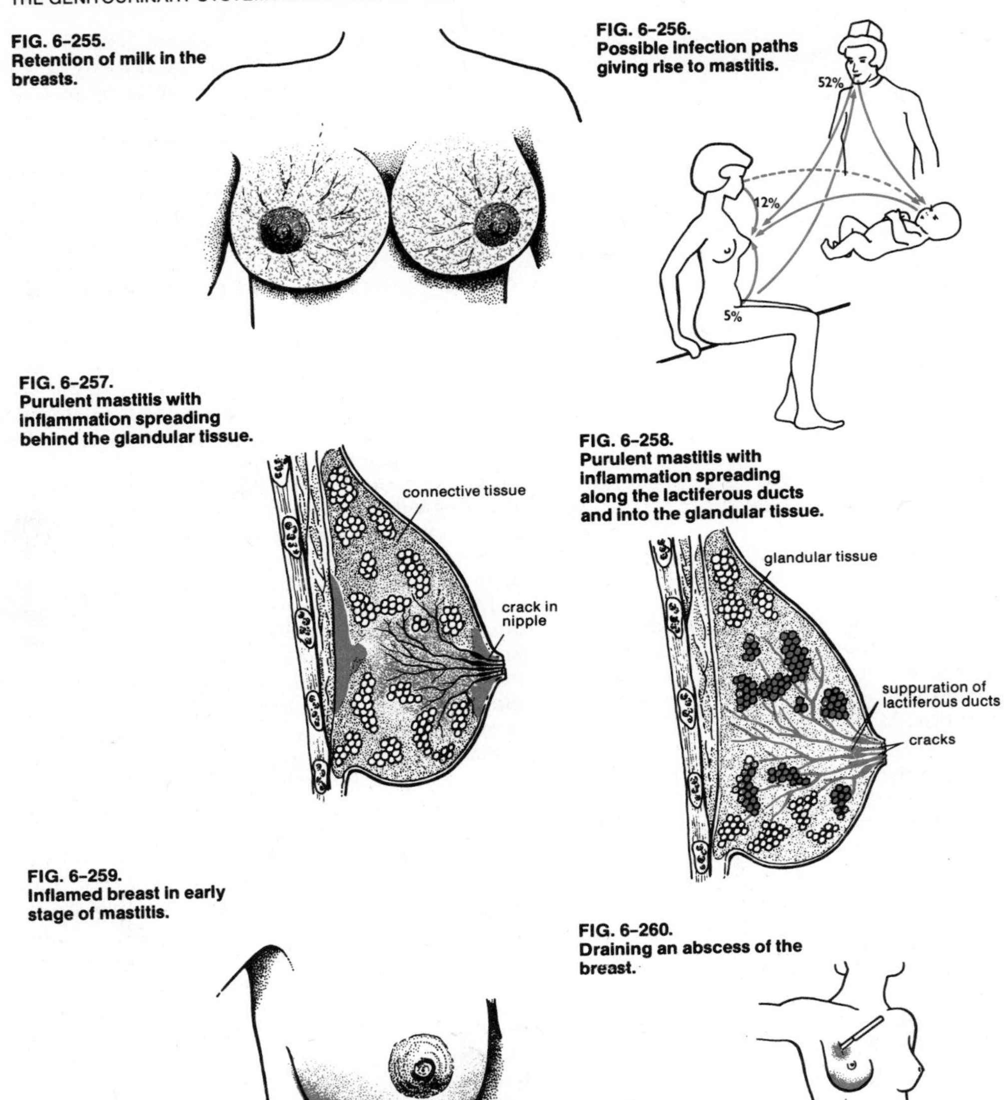

FIG. 6-255. **Retention of milk in the breasts.**

FIG. 6-256. **Possible infection paths giving rise to mastitis.**

FIG. 6-257. **Purulent mastitis with inflammation spreading behind the glandular tissue.**

FIG. 6-258. **Purulent mastitis with inflammation spreading along the lactiferous ducts and into the glandular tissue.**

FIG. 6-259. **Inflamed breast in early stage of mastitis.**

FIG. 6-260. **Draining an abscess of the breast.**

INFLAMMATION OF THE BREAST

Inflammation of the breast (*mastitis*) manifests itself especially in soreness and swelling of the breast affected. It may be confused with retention of milk, which can be painful and sometimes precedes mastitis. Retention may occur when the breast is not properly emptied after feedings. Severe and painful swelling may develop on the fourth or fifth day after childbirth; it may also develop later on, if there is a prolonged interruption of regular feeding. The breasts may become engorged with milk (Fig. 6-255), enlarged, heavy, warmer than normal, hard and very painful. In contrast with mastitis, retention of milk usually occurs in both breasts simultaneously. In such circumstances it is necessary to empty the breasts with the aid of a breast pump. If weaning is desirable, the breasts should be given proper—but not too tight—support with a binder which pulls upward and inward. Hormone injections to stop the secretion of milk may be given.

Mastitis may develop from retention of milk as a result of the retained milk becoming infected with microbes. Most cases of mastitis are not caused by retention, however, but by the entry of microbes through a crack or abrasion of the nipple. It occurs usually between the second week and the fifth month after childbirth. The risk of mastitis is highest in the second and third weeks. In the great majority of cases it is caused by pyogenic (pus-producing) germs, mostly staphylococci. These come mainly from the nose and throat of the mother or of other persons, enter the baby's nose and throat, and are thus transmitted to the nipple (Fig. 6-256). In a minority of cases infection may be due to contamination by the lochia.

Infection usually penetrates into the connective tissue between the milk-secreting glands. More rarely the inflammation is localized under the skin of the areola. The inflammation may spread deep behind the glandular tissue (Fig. 6-257). Alternatively, it may spread inward along the lactiferous ducts, especially if the breast is engorged with milk, which provides an excellent nutrient medium for the germs, resulting in inflammation of the lobes of glandular tissue (Fig. 6-258). All these forms of inflammation may merge into one another, suppuration may follow, there may be a high temperature and pulse rate, the breast becomes deeper red, more painful and edematous. The suppuration will eventually become localized into an abscess. With mastitis, as a rule only one breast is affected, though it can, of course, spread to the other, unless hygienic precautions are taken. The earliest sign is a triangular flush, generally underneath the breast (Fig. 6-259). This area becomes red, hard and painful.

In severe cases it may be necessary to wean the baby and stop milk secretion by means of hormone injections. The breast should be supported with a binder. Alcohol poultices or ice bags should be applied and antibiotics given to check the spread of the infection and prevent abscess formation. If abscesses nevertheless develop, they will usually have to be drained by an incision (Fig. 6-260).

GONORRHEA

Gonorrhea, the commonest of the venereal diseases, is an infectious inflammation of the genital mucous membrane of either sex and is caused by a bacterium (*Neisseria gonorrhoeae*), commonly referred to as *gonococcus*. In particular it infects the mucous membrane of the urethra, vagina, cervical canal, uterus and fallopian tubes. It may, however, also affect the mucous membrane of other parts, such as the rectum, the mouth or the conjunctiva of the eyes, if these happen to be penetrated by gonococci. The disease is transmitted almost invariably by sexual intercourse, which is, of course, the reason it is a disease mainly of the sexual organs. Babies may, however, become infected at birth, involving a particular hazard to the eyes. For this reason the eyes of newborn babies are treated with an antiseptic solution (silver nitrate or penicillin) as a routine precaution.

The incubation period of the disease is 2–8 days. In the male the first signs of infection are itchings and a slight burning sensation when urinating. A few hours later, a mucous discharge appears from the urethra; in a day or two this discharge becomes yellow, containing pus, due to inflammation. Urination is slow, difficult and painful, and the symptoms grow worse as the infection spreads upward through the urethra (Fig. 6-261). When it reaches the sphincter of the bladder, acute pain is felt also after urination. If the infection spreads outside the urethra, it may affect the membrane under the foreskin. A slight fever develops, and inflammation of affected lymph nodes causes painful swelling in the region of the groin. Infection of the rear part of the urethra may cause a constant urge to urinate and sometimes also painful induration (stiffness) of the penis. It may also spread to the prostate gland, seminal vesicles and epididymis. These and other complications are comparatively rare, however. In either sex the disease may eventually clear up without serious consequences, or it may become chronic, this condition being accompanied by a urethral discharge called *gleet*. A chronic infection may be reactivated by transmission of the infection to the sex partner, in whom the disease becomes virulent, so that the original carrier of the disease is reinfected with this virulent form which overcomes the immunity that this person's urethral mucous membrane had previously acquired.

FIG. 6-261.
Infection paths of gonorrhea in the male.

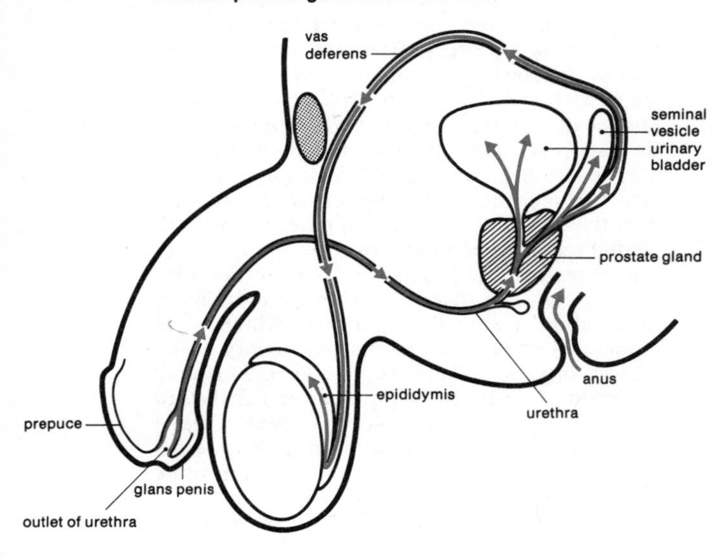

FIG. 6-262.
Infection paths of gonorrhea in the female.

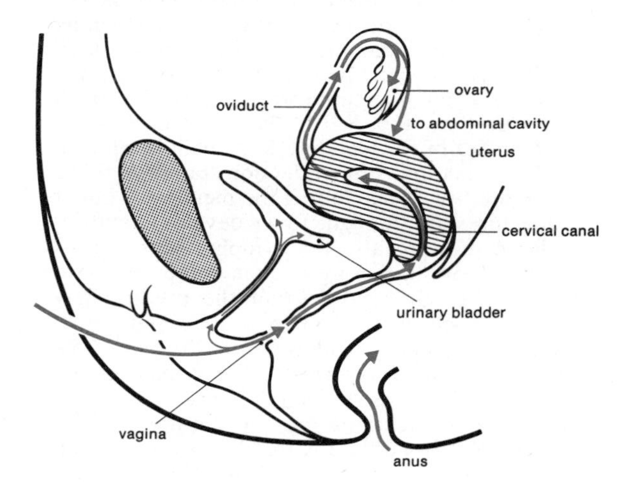

Early symptoms of gonorrhea in the female are inflammation of the vulva and vagina, painful or frequent urination, urethral or vaginal discharge, while Bartholin's glands are usually infected and painful. In some cases there are hardly any symptoms, however, or the symptoms are not objectionable enough for the patient to seek medical attention. On the other hand, serious complications may occur, with the infection eventually spreading up from the cervix through the uterus and fallopian tubes to the ovaries (Fig. 6-262). This may be accompanied by acute abdominal pains, and there is a risk of peritonitis. Gonorrhea affecting the uterus and tubes causes adhesions of tissue and thus often results in sterility. Complications are more likely to occur in women than in men, especially if the disease is left untreated, is inadequately treated or is discovered when it has already reached a very advanced stage. Gonorrhea can be caught again and again; there is little natural immunity.

Effective treatment of gonorrhea became possible with the introduction of sulfonamide drugs, but eventually the gonococcus developed resistant strains. Present-day treatment is centered on antibiotics, notably penicillin. Some strains of gonococcus have become resistant to this, too, in which case certain other antibiotics can be used instead, e.g., erythromycin.

SYPHILIS

Syphilis is an infectious venereal disease caused by a spirochete (a spiral-shaped bacterium) called *Spirochaeta pallida* or *Treponema pallidum,* which is transmitted by direct contact between human beings—in adults almost invariably by sexual intercourse—and enters the body through mucous membrane (vagina, male urethra). In certain circumstances it may, however, enter through a scratch or abrasion on the skin. The spirochete moves along by a kind of corkscrew motion and owes its name "pallidum" to the fact that it is difficult to stain for microscopic examination (Fig. 6-263). Outside living human tissue the spirochete can survive only for a very limited time. It is for this reason that transmission of the disease by other than sexual contact (in the more intimate sense), e.g., through a small wound or scratch that comes into contact with an infectious patient or through kissing, is rare though by no means impossible.

Syphilis runs a well-ordered course that can conveniently be subdivided into three stages (Fig. 6-265). In the *first stage* the spirochetes remain at the site of infection and multiply there for about three weeks. This is the period of initial incubation. They then invade the adjacent blood vessels and lymph nodes, and a so-called *primary lesion* develops at the site of infection, usually on the external genitals. This lesion develops into a *hard chancre*, a small roundish and painless sore or ulcer. It contains large numbers of spirochetes, whose presence is the basis of the dark-field test for syphilis. The patient is highly contagious in this stage of the disease. The primary lesion is usually the first sign of syphilis, though in a few cases it does not occur and there is no hard chancre. Also, it may happen that a person is infected simultaneously with gonorrhea and syphilis. The symptoms of gonorrhea appear first and cause the patient to seek treatment, usually in the form of penicillin therapy. This therapy may at the same time suppress the development of the syphilitic chancre, without curing the syphilis itself, so that this disease may thus remain undetected for a long time.

As already stated, the hard chancre develops as a primary sore at the site where the spirochetes entered the body. If the disease has been transmitted by sexual intercourse, as it nearly always is, the chancre will, in the male, occur on any part of the penis; in the female, it will usually be found on the vulva (especially the labia) or at the external os of the cervix. Infection of the anus or the mouth may occur as a result of sexual practices involving those parts. Nonsexual contact with syphilis infection may result in the primary lesion appearing on any other part of the body, wherever the spirochetes happened to effect their entry, e.g., on the fingers or lips. Any painless, dark red, hard-edged sore that undergoes little noticeable change for two or three weeks must be viewed with suspicion. The chancre gradually disappears even without treatment, but this does not mean that the disease itself is in retreat. (The hard chancre of syphilis is quite distinct from the so-called *soft chancre*, or *chancroid*, which is a sore due to a different type of venereal bacillus infection, effectively treatable with sulfonamide drugs.)

The symptoms of the *second stage* of syphilis appear, on average, about 4–6 weeks after the appearance of the primary lesion. This stage begins with enlargement of regional lymph nodes, especially in the region of the groin if the disease has been transmitted by sexual intercourse. After a few more weeks the enlargement of the lymph nodes may become widespread throughout the body and can be felt as hard, painless, pea-sized or larger swellings under the skin, especially on the arms and legs. Signs of constitutional disturbance—headache, fever, loss of appetite, malaise—commonly occur, but may be absent. About 9–10 weeks after infection, the first eruptions of the skin and mucous membranes appear, the so-called *secondary lesions*. These may be localized or widespread and may take a variety of forms. Thus they may arise suddenly as red patches which fully develop in a few days, or they may appear as a rash of reddish spots or papules which become yellowish when pressure is exerted on them. Papules form more particularly on moist surfaces, e.g., on the external genitals, around the anus, on the mucous membranes of the mouth, nose, where they soon turn into weeping papules, i.e., exuding fluid. Papules on the tonsils cause sore throat and may be mistaken for tonsillitis, while lesions that appear on the palms of the hands are sometimes wrongly diagnosed as eczema. A characteristic feature of syphilitic secondary lesions is that they do not itch. Large numbers of spirochetes are present in the lesions. More particularly the genitals and anus may be affected by broad flat papules, called *condylomata*, which are larger than ordinary papules formed earlier on. Eventually, after several weeks, the secondary lesions also subside, even without treatment.

There follows a period during which the disease, still lurking, becomes latent for several years, during which time the patient is not manifestly ill. At the end of this period the disease, now called *tertiary* syphilis, enters its *third stage*. It now does permanent damage, with possible cardiovascular symptoms, damage to the nervous system, etc. The spirochetes establish themselves in colonies in various parts of the body, where they set up persistent inflammation, with obstruction of small blood vessels and destruction of tissue. Any organ in the body may be affected, and chronic abscesses (*gummata*) may arise anywhere, causing symptoms by producing masses of fibrous tissue and destroying adjacent healthy tissue. Mucous membranes, skin and bones are most likely to be attacked by tertiary syphilis. Gummata produce large hard tumors in the long bones. Necrosis, perforation and deformity of the bones of the palate and nose (saddle nose) may occur. The disease may attack the small arteries that nourish the aorta, so that the latter loses its resilience and becomes locally overdilated (aneurysm); this in turn results in faulty operation of the aortic valves, and thus the heart no longer pumps efficiently. In this stage of the disease the nervous system is also liable to be attacked, especially in men (neurosyphilis). This may

FIG. 6-263.
Spirochetes revealed by dark field microscopic examination.

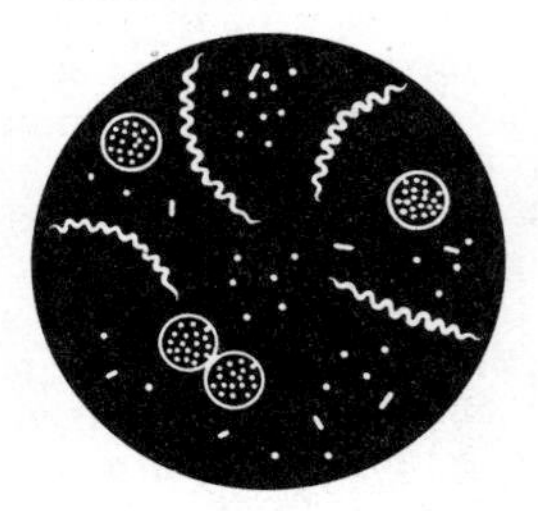

FIG. 6-264.
Gonococci in white blood cells.

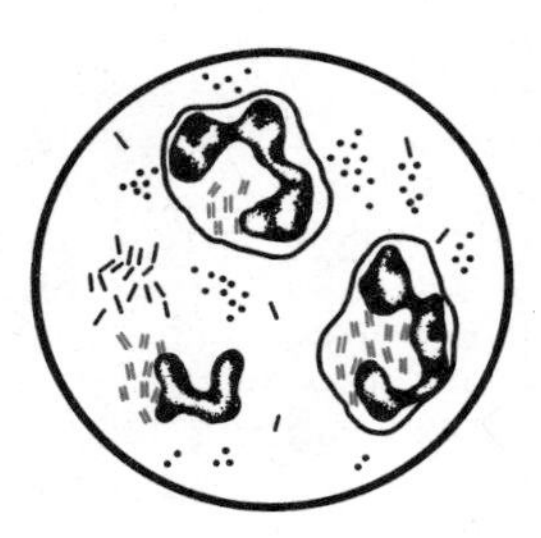

FIG. 6-265.
Successive stages of syphilis.

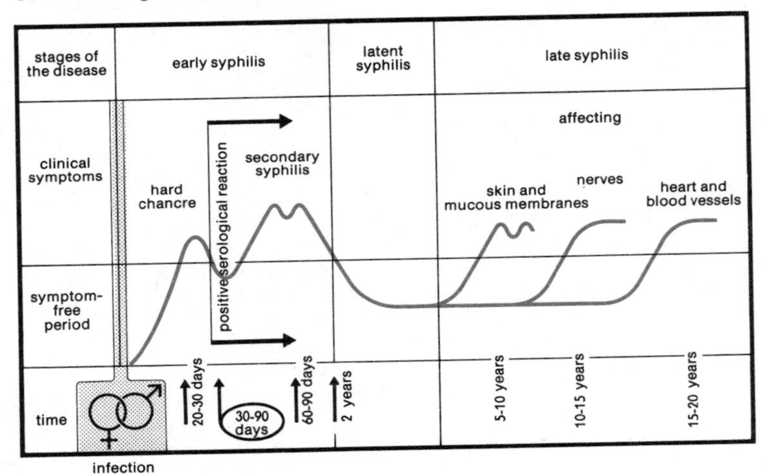

cause the condition known as *tabes dorsalis* (locomotor ataxia), with degeneration of nerve fibers in the spinal cord, so that the reflexes which coordinate voluntary movement and posture are lost. Paralysis of the intestine and bladder may occur, as well as blindness and eventually general physical deterioration. Another manifestation of neurosyphilis is *paresis* (dementia paralytica, general paralysis of the insane), characterized by deterioration of personality, progressive dementia with delusions of self-importance or grandeur alternating with depression, generalized paralysis, total insanity and death. Without treatment, paresis may run its course in 2–5 years, though in some cases death may occur in a matter of months rather than years. Spontaneous recovery, though rare, is not unknown, though generally at the expense of some permanent deterioration or change of personality. With modern treatment, tabes and paresis can be arrested, though actual physical damage that has occurred in this stage of the disease cannot be undone.

Syphilis cannot be inherited in the true sense of the term, but there is such a thing as *congenital syphilis*, i.e., present in a baby at birth. In such cases the fetus has been infected with spirochetes from the diseased mother, which have crossed the placenta into the fetal bloodstream. The spirochetes are exceptional among pathogenic agents in being able to do this. Such prenatal syphilis often causes miscarriage, especially in the latter part of pregnancy. If the infection is not severe, the baby may survive, however, and be delivered at the normal time, though infected with the disease. Syphilitic lesions and other symptoms may be present at birth, or they may appear after several weeks, or not until the child is several years old. A syphilitic baby apparently healthy at birth will fail to gain weight, will develop a rash, and will have difficulty with its nasal breathing (''snuffles''). If the disease manifests itself when the child is several (usually at least 8) years old, the symptoms will be generally similar to those of tertiary syphilis: gummata, impaired vision, deafness, retarded mental development, dementia; notched teeth and saddle nose are typical outward signs.

Laboratory tests for the *diagnosis of syphilis* comprise three main procedures: microscopic dark-field detection of spirochetes (Fig. 6-263) in specimens obtained from a chancre or other lesion; examination of cerebrospinal fluid; serological tests performed on blood serum or spinal fluid and based on the detection of antibodies formed as defensive reaction to the spirochetes. A number of serological tests have been developed over the years. These include what are technically known as serum *complement fixation tests,* more particularly the *Wassermann reaction* (the earliest and best-known blood test for syphilis) and its modifications, all of which are based on the detection of an antibody called *reagin*. There are also what are known as flocculation tests, such as the Kahn test and several others. Another group of tests is based on the so-called *TPI* (*Treponema pallidum* immobilization) test.

For centuries there was no effective cure for syphilis until, in 1910, Ehrlich and Hata discovered salvarsan, an arsenical compound. This drug and others, including those incorporating bismuth instead of arsenic, had the drawback of producing side effects, sometimes of a serious character. In modern practice these drugs have been superseded by penicillin, which is now used for the treatment of all types and stages of syphilis. Should allergic reactions occur, other antibiotics are available instead, such as oxytetracyline or erythromycin. The main object of the treatment is to maintain an effectively high level of penicillin in the patient's blood for a sufficient length of time. The progress of the cure is monitored by means of regular serological tests.

Part Seven

THE NERVOUS SYSTEM

The nervous system has two main divisions: (1) the *central nervous system,* which comprises the brain and the spinal cord; (2) the *peripheral nervous system,* which comprises the remainder of the nervous tissue and consists of nerves emerging from the central nervous system (cranial nerves from the skull, spinal nerves from the spine) and which also includes the organs of special sense (e.g., the retina of the eye). Certain groups of nerve cells (ganglia) outside the central nervous system are likewise part of the peripheral system. Those parts of the nervous system that control the internal organs, glands and all the smooth muscle in the body are together referred to as the *autonomic nervous system,* subdivided into the *sympathetic* and the *parasympathetic nervous systems*.

The purpose of the nervous system is to coordinate, in cooperation with the hormones, the functions of all parts of the human body. To enable it to perform this task it is equipped with what are generically termed *receptors,* which are specialized cells or groups of cells for the reception of external and internal stimuli and are associated with so-called *afferent* nerve endings. In the receptors the signals received are converted and encoded into the "language" of the body's communication system. Afferent nerves transmit the stimuli to various nerve centers located in the brain and spinal cord, where they are processed. The commands or instructions issued by these centers in response to the stimuli are transmitted through *efferent* nerves to the peripheral organs of the body, where they produce the required reactions, such as muscular movements (Fig. 7-1). The efferent nerves are, for this reason, also known as *motor nerves,* while afferent ones, which transmit the sense stimuli to the central nervous system, are called *sensory nerves*.

The nervous system is made up of nerve cells (*neurons*) and specialized connective tissue cells (neuroglia). A neuron consists of the actual cell body and all its threadlike projections (technically called processes); there are generally several short branched processes (*dendrites*) and a single long one (*axon*). The dendrites convey impulses toward the cell body; the axon conveys impulses away from it. The axons of the nerve cells are the actual nerve fibers, each of which is surrounded by a sheath of fatty substance called *myelin*. Outside the central nervous system the nerve fibers have an additional covering (*neurolemmal sheath*) containing special cells (*Schwann cells*) which play an important part in nerve fiber regeneration, which is thus possible only outside the central nervous system. A junction between two neurons in a neural pathway, where the end of the axon of one neuron links with a dendrite or with the cell body of another, is called a *synapse*.

Nerve endings are basically of two kinds: (1) *afferent* or *sensory* or *receptor* nerve endings; (2) *efferent* or *motor* or *effector* nerve endings. The afferent nerve endings are concerned with the appreciation of temperature, touch, pressure and pain (pain receptors). Receptors concerned with smell, taste, sight, hearing and appreciation of movement constitute the special sense organs. Afferent nerve fibers transmit the impulses from the afferent nerve endings to nerve centers in the central nervous system. Efferent nerve fibers transmit impulses from the central nervous system; when these cause contractions of muscles, they are more particularly referred to as motor nerve fibers; their nerve endings at striated muscle fibers are called *motor end-plates*. Not all efferent nerves are motor nerves within the meaning of transmitting impulses to muscles; some efferent nerves also, or alternatively, have a secretory action (i.e., causing glands to secrete) or an inhibitory action (i.e., curbing the activity of certain organs).

The *central nervous system*—the brain and the spinal cord—contains the very great majority of all the nerve cells present in the body, together with their dendrites and part or the whole of their axons. It is composed of what is conventionally called *gray matter,* consisting of the nerve cell bodies, and *white matter,* consisting of nerve fibers (axons). In the spinal cord the gray matter is located centrally and is

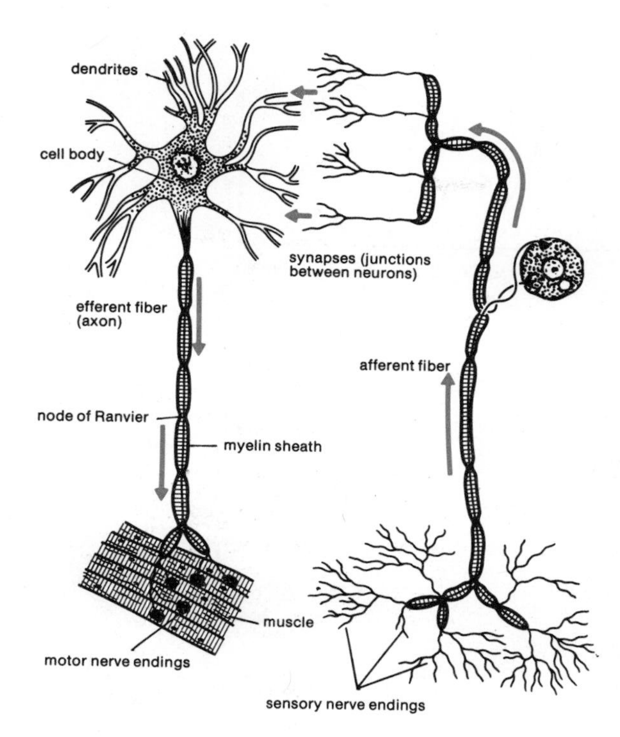

FIG. 7-1.
Afferent and efferent nerves.

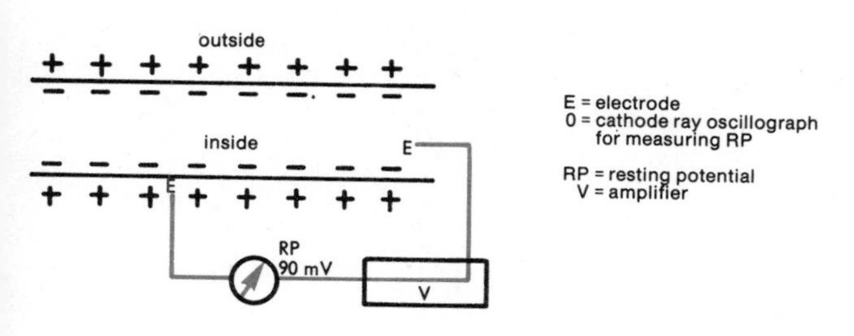

FIG. 7-2.
Measuring the membrane potential.

surrounded by white matter; in the brain some of the gray matter covers the surface of certain areas and is known as *cortex*. Some of the nerve fibers run part of their course in white matter and then leave the central nervous system to continue within peripheral nerves. Nerve fibers are grouped in bundles. A nerve commonly denotes a bundle or a group of bundles of nerve fibers. A bundle of fibers located within the central nervous system and performing a particular set of functions is called a *tract*.

The *peripheral nervous system* comprises the parts of the nervous system outside the brain and spinal cord, namely, the peripheral nerves and certain groups of nerve cells known as *ganglia*. Some of the fibers of the peripheral nerves have their cell bodies in the central nervous system. Other fibers (sensory or afferent fibers) have their cell bodies in a sensory ganglion. The peripheral nerves are usually mixed nerves, containing both efferent and afferent (i.e., motor and sensory) fibers. There are 12 pairs of cranial nerves, which spring from the underside of the brain, and 31 pairs of spinal nerves from the spinal cord. There are two kinds of ganglia: sensory and autonomic. All sensory nerve fibers are the axons of cells situated in a sensory ganglion. Autonomic ganglia are associated with the autonomic nervous system; they consist of motor nerve cells and function as relay stations.

The *conduction of impulses* along nerve fibers is based on a momentary change (the *action potential*) in the electrical state of the cell membrane which, when at rest, is characterized by a potential difference of about 80 millivolts between the outside and the inside of the membrane (the *resting potential*). The resting potential is due to the presence of potassium ions in a much higher concentration inside the cell membrane than outside it, whereas sodium ions are more highly concentrated outside than inside it. Besides, when at rest, the membrane is highly permeable to potassium ions and almost impermeable to sodium ions (selective permeability). In making their way out of the cell through the membrane, positively charged potassium ions impart a positive electrical charge to the outside of the membrane in relation to the inside thereof. This is the normal at-rest state, characterized by the resting potential difference between the inside and outside of the membrane, which can be measured by special intracellular techniques using tiny electrodes (Fig. 7-2).

In response to a stimulus, this potential difference undergoes a brief temporary change. First, there is a decrease in potential (depolarization), followed by a very short reversal in polarity, so that the inside of the membrane momentarily becomes electrically positive in relation to the outside. This reverse potential (the action potential) is about 20–50 millivolts. It is caused by a sudden very great increase in

FIG. 7-3.

permeability to
potassium
sodium
action potential
0
1
2
3
4
m/s

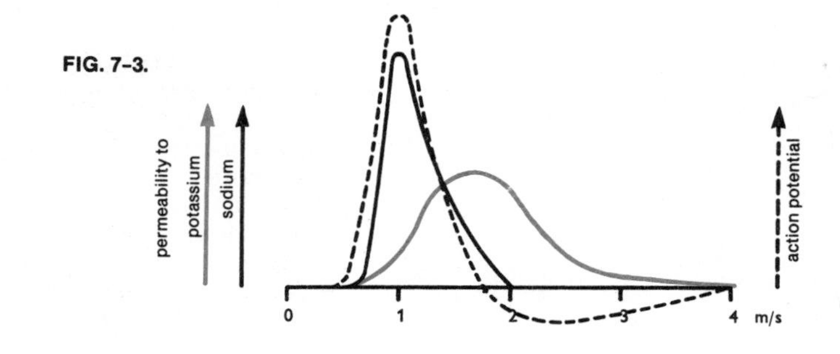

FIG. 7-4.
Conduction of impulses in an unmyelinated nerve fiber.

stimulated
not yet stimulated
small local currents
fractions of a millimeter

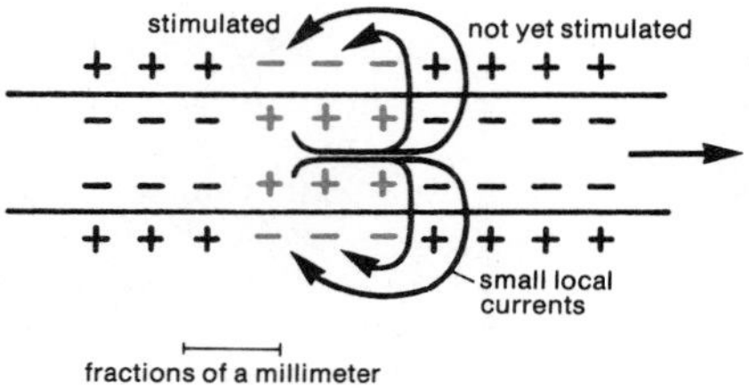

FIG. 7-5.
Conduction of impulses in a myelinated nerve fiber.

nodes of Ranvier
myelin sheath
stimulated
not yet stimulated
Axon
small currents
current skips from node to node
1mm

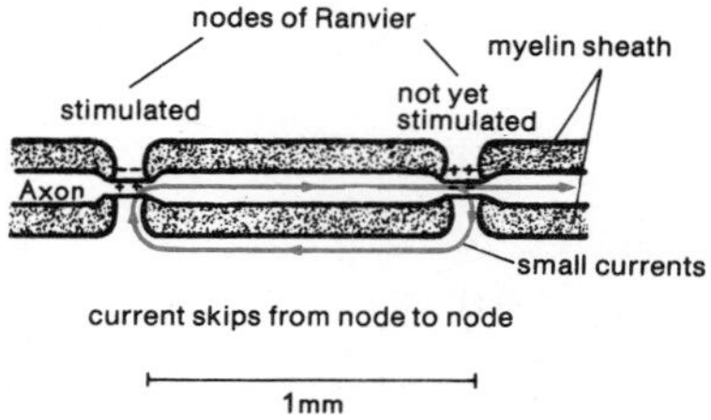

FIG. 7-7.
Stretch reflexes.

FIG. 7-6.
Movements in response to a pain stimulus.

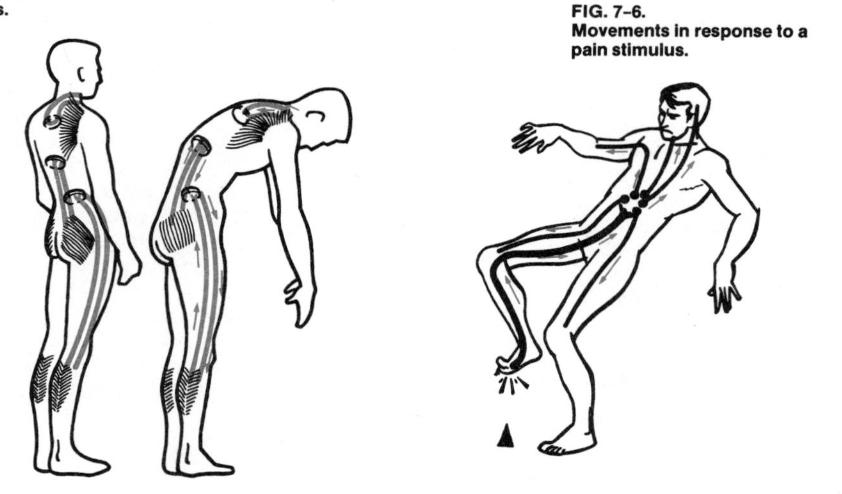

the sodium permeability of the nerve cell membrane, so that sodium ions can rush into the cell at a greater rate than potassium ions can get out, with the result that the inside of the membrane becomes positively charged. This reversal of potential, i.e., the action potential, lasts only a small fraction of a second; the permeability of the membrane quickly reverts to its original condition, which is the normal at-rest condition characterized by the resting potential, with the outside of the membrane positive in relation to the inside (Fig. 7-3).

An impulse is conducted along a nerve fiber as a wave of depolarization. So-called unmyelinated (nonmedullated) nerve fibers do not possess a myelin sheath; in such fibers the speed of conduction of impulses is relatively low (0.5–2 m/sec), the action potential being propagated by small local currents which continuously and progressively depolarize small adjacent areas of the membrane (Fig. 7-4). In the myelinated (medullated) nerve fibers, i.e., those possessing a myelin sheath, the membrane is responsive to stimuli only at the so-called *nodes of Ranvier* which form notches in the sheath at regular intervals along the fiber (Fig. 7-1). In between the nodes the fiber is insulated by the sheath. Conduction of impulses takes place by the depolarization current skipping from node to node, so that the action potential proceeds in a succession of rapid leaps; this is known as *saltatory conduction* (Fig. 7-5). It is a very rapid form of conduction, attaining upward of 100 m/sec in some of the thickest fibers communicating with skeletal muscles, while even higher speeds have been found in certain spinal nerves.

The motor function of the central nervous system is closely linked to the sensory function. Interaction between the two functions can be achieved in two ways. Environmental influences acting through the receptors and the sensory system may produce conscious sensations and these in turn may bring about, as a response, *voluntary movements* initiated by impulses transmitted to the muscles through the motor system. However, only a relatively small proportion of the body's movements are under the direct and continuous control of the conscious will. The nervous system functions for the most part by *reflex action*, this being the second type of interaction between sensory and motor function. A reflex is a specific and predictable involuntary response to a stimulus. Many reflexes are inborn; others can be acquired as completely new automatic responses (so-called conditioned reflexes). A simple *reflex arc* comprises a receptor (sensory nerve ending), an afferent neuron, a synapse with an efferent neuron, and an effector organ, i.e., a muscle or a gland which, when stimulated, produces an effect. Such a reflex involving only two neurons, an afferent and an efferent, is called a *monosynaptic reflex*. Many reflexes are more complicated and include intermediate neurons interposed between the afferent and efferent neurons (*multisynaptic reflexes*). Most bodily functions are controlled by inborn reflexes, e.g., constriction of the pupils in response to bright light, salivation in response to eating, sweating in response to heat, shivering in response to cold. Some reflexes such as breathing can be modified under voluntary control.

A pain stimulus is transmitted from pain-sensitive sensory nerve endings (pain receptors) through afferent neurons to the spinal cord, where the sensation is transmitted through intermediate neurons to motor neurons. The latter transmit the impulse to the effector organ, in this case a muscle or a number of muscles (Fig. 7-6), which respond by contracting and thus causing the limb to be withdrawn from the object causing the pain. All this takes place in a fraction of a second. At the same time, nerve fibers ascending in the spinal cord report the pain stimulus (and also the muscular movement in response to it) to the brain, where it becomes a conscious sensation. Multisynaptic reflexes of this kind are so-called sensorimotor reflexes, as they are both sensory and motor in character.

On the other hand, there are a number of purely motor reflexes associated with, for example, maintaining the correct muscular tone—enabling a muscle to resist a force for a long period of time, particularly in the extensor muscles of the limbs and body which enable us to maintain an upright posture in opposition to the pull of gravity—and with channeling and modulating the voluntary movements. Such movements include a reflex element: they are voluntary to the extent that the will determines what movement is to be made, but the choice of which muscles are to contract and which are to be stretched is reflex. Monosynaptic reflexes play an important part here; the synapse is located in the spinal cord, forming the simplest possible link between the afferent and the efferent neurons of the reflex arc. In many of these reflexes the receptor consists of, for example, nerve endings in the skin, and the effector organ is a muscle. In others the receptor and the effector are one and the same muscle. A well-known example of a monosynaptic reflex of this latter kind is the *knee-jerk reflex*, which is a spinal stretch reflex in which extension of the leg occurs in response to percussion of the patellar tendon. To demonstrate this reflex, the knee is supported and the leg is allowed to hang loosely in a partially flexed position. The patellar tendon is tapped with a hammer; in response to this stimulus the quadriceps muscle contracts, causing the foot to kick forward as this muscle extends the leg at the knee joint. The receptor is situated in the quadriceps muscle, which is also the effector organ that produces the response. Balance and upright posture are maintained by the tendency of muscles automatically, by reflex, to resist any action that tends to stretch them. Thus, the muscles resist any disturbance of position so long as a movement is not actually needed. So-called stretch reflexes (Fig. 7-7), i.e., contraction of a muscle in response to stretching the same muscle, play an important part in automatically maintaining the muscular tone and correcting the body's position and balance.

As already mentioned, a *synapse* is the point of junction where part of one neuron links with an adjoining one (Fig. 7-8). It is the junction where the end of the axon of one neuron comes into close proximity with the cell body or dendrites of another neuron. The impulse traveling in the first neuron initiates an impulse in the second neuron. A synapse is polarized, i.e., the impulses pass in one direction only. There is no protoplasmic continuity between neurons at a synapse; there is a "gap" across which the nerve impulse is conveyed by a special substance, called a *chemical transmitter,* which is released by the first neuron and excites a fresh impulse in the next neuron. Chemical transmitters occur not only at every synapse of the central nervous system but in the ganglia of the peripheral nervous system. A chemical transmitter is liberated by a sensory receptor to stimulate an impulse in an afferent neuron; similarly, an impulse in an efferent neuron releases a chemical transmitter at its motor end-plate to stimulate the muscle fiber to contract. Various chemical transmitters are known, including acetylcholine (to which muscle cells and most nerve cells respond) and adrenaline (epinephrine).

The *autonomic nervous system,* also known as the *vegetative* nervous system, is concerned with the control of involuntarily occurring functions, more particularly of the glands, internal organs and smooth muscle. It thus regulates the organs of blood circulation, respiration, digestion, excretion and reproduction. The autonomic motor nerve cells are grouped in ganglia outside the central nervous system. The activity of the autonomic nervous system is wholly reflexive, apparently independent of the brain and of voluntary control. The autonomic nerves, however, also convey impulses to the brain, where they stimulate reflexes but do not reach consciousness. For example, a rise in blood pressure is reported to the brain and this, in turn, causes the vagus nerve to decrease heart activity. Also, the activity of the whole autonomic system is modified and coordinated by the brain, particularly the part called the hypothalamus, which is also concerned with emotions that have their physical expression in reflexes of the autonomic system (dilation of the pupils, sweating, increased pulse rate, etc.). Functionally the autonomic system comprises two main divisions: the sympathetic system and the parasympathetic system.

FIG. 7-8. Synapses.

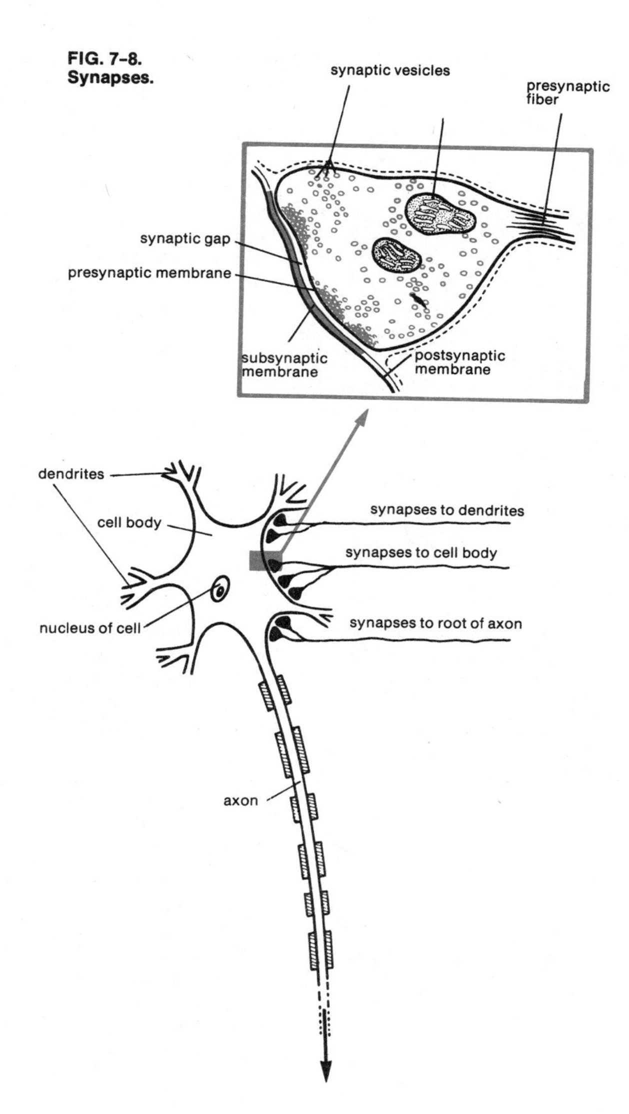

The *sympathetic nervous system* (Fig. 7-9) comprises mainly the sympathetic trunk (a chain of autonomic ganglia connected by nerve fibers), its connections within the thoracic and upper lumbar parts of the spinal nerves, and certain ganglia in the abdomen. The *parasym-*

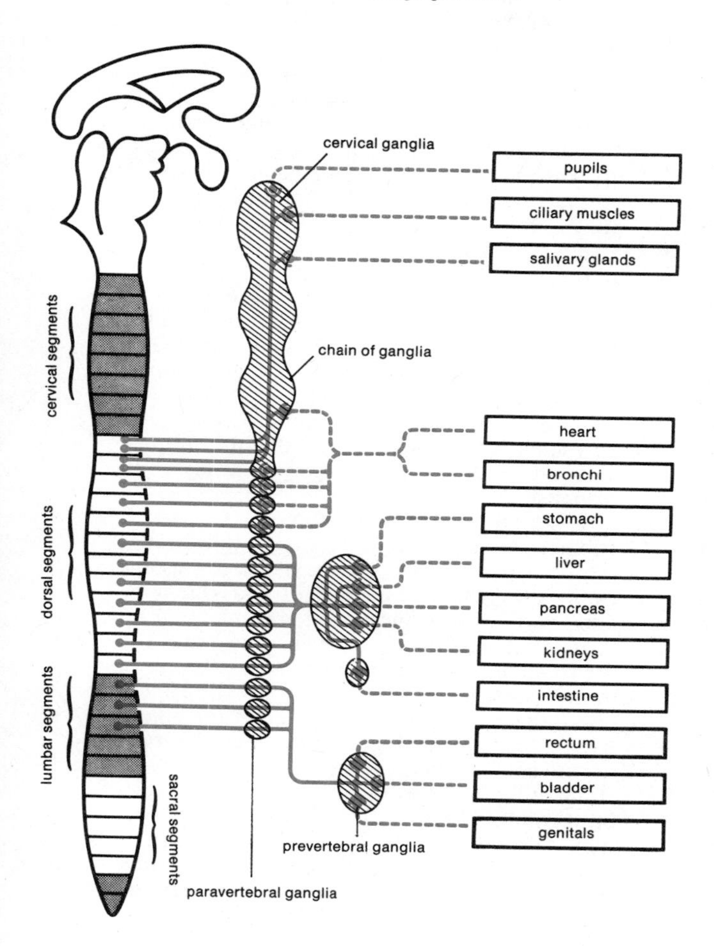

FIG. 7-9.
Sympathetic nervous system.

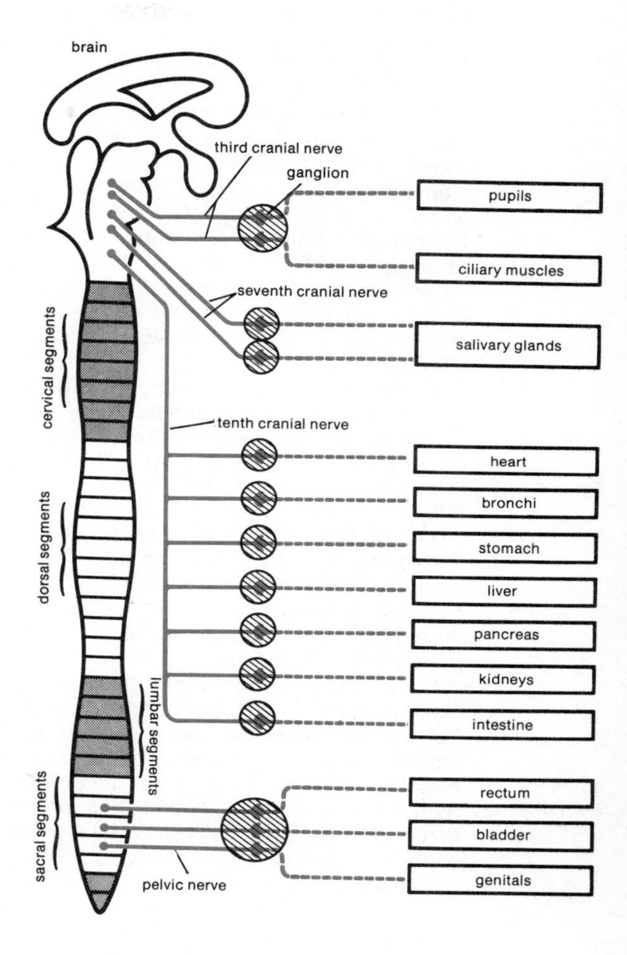

FIG. 7-10.
Parasympathetic nervous system.

pathetic system (Fig. 7-10) comprises fibers of certain cranial nerves, of which the most important is the vagus, and other fibers connected with the sacral part of the spinal cord. The *vagus nerve* arises from the brainstem, passes through the base of the skull, and has branches to the main digestive organs, heart and lungs. It comprises sensory fibers, which convey impulses from these organs and initiate reflexes to control and modify their action, and motor fibers, which convey the controlling and modifying impulses to those organs, e.g., to slow the heart, constrict the bronchi and bronchioles of the lungs. The vagus also comprises nerve fibers that control speech and swallowing and are not connected with the parasympathetic system.

Broadly speaking, the *sympathetic nervous system* comes into action in preparing the body for "fright, flight or fight." Thus sympathetic stimulation causes dilation of the pupil of the eye, as in fright, and increases the rate and force of the heartbeat. In emergencies blood vessels to the skin and certain internal organs are constricted, but vessels supplying blood to skeletal muscles are dilated, so that these muscles are better able to cope with emergencies (flight or fight). Similarly, the smooth muscle in the walls of the bronchi and bronchioles is relaxed, allowing more air to enter the lungs and ensuring better oxygenation of the blood. The smooth muscle in the walls of the gastrointestinal tract is inhibited, so that peristalsis stops; the sphincters are closed; digestion is halted in an emergency. The sphincter of the bladder also contracts, and the bladder wall relaxes. Muscles in the skin make the hair "stand on end," and sweating is increased. The chemical transmitter by which the sympathetic nerves act is adrenaline, which transmits the impulses to the effector organs. Secretion of this hormone by the adrenal medulla has the same effects as stimulation of the sympathetic nervous system. The nerves of the sweat glands are an exception in that they release acetylcholine as the chemical transmitted.

Stimulation of the *parasympathetic nervous system* constricts the pupil of the eye and accommodates it for near vision, slows the heart rate and reduces the force of its contraction; it also constricts the bronchi and bronchioles of the lung. Parasympathetic stimulation increases the secretion of digestive juices (saliva, gastric juice, etc.); sphincters are relaxed to promote digestion; defecation and urination are facilitated; and arterioles supplying blood to erectile tissue in the genital organs (penis, clitoris) are dilated, thus promoting erection. The chemical transmitter in all the actions of the parasympathetic system is acetylcholine.

The sympathetic and parasympathetic nervous systems can, in a way, be described as mutually "antagonistic," but there are many exceptions, and their interaction can more correctly be described as a pattern of meaningful cooperation in establishing and maintaining a suitable balance in any particular set of circumstances. Disturbances in the balanced cooperation of the sympathetic and parasympathetic systems occur in some people. Depending on which of these two systems becomes overdominant, the condition is referred to as *sympathicotonia* and *vagotonia*, respectively. *Sympathicotonia* denotes increased tonus of the sympathetic system, with tendency to vascular spasm and heightened blood pressure. *Vagotonia* (vasomotor instability) denotes overreactivity of the parasympathetic system. These conditions seldom occur in their extreme forms, but can be said to correspond broadly to two distinct types of physiological and psychic makeup. Sympathicotonics are lively and oversensitive to sensory stimuli, their blood pressure is unstable, they tire easily and are restless, and are prone to palpitations, nausea, fainting fits and travel sickness. Vagotonics show symptoms of a more or less opposite type: they tend to have low blood pressure, sometimes with slow and irregular pulse, often suffer from gastric ulcers and migraine-type headaches, and are likely be quiet and reserved.

A mixed pattern is much commoner. It is characterized by alternate imbalance in favor of the parasympathetic and the sympathetic system, so that symptoms of vagotonia alternate with those of sympathicotonia. Because of its instability and continually varying character, the condition is difficult to treat with drugs that suppress particular nervous responses.

So-called sympathomimetic drugs produce effects similar to those of stimulation of the sympathetic nervous sytem. These drugs include adrenaline (epinephrine) and amphetamine. Sympathetic nerve action is inhibited by sympatholytic drugs, which include drugs for the treatment of high blood pressure. Some of these drugs are counteractive to all the effects of adrenaline, whereas others act more selectively, so that they can, for instance, be used to prevent the sympathetic system from overstimulating the heart (as in dealing with a case of angina) and not disturb the rest of the system. There are also drugs that produce effects similar to those of acetylcholine in promoting the action of the parasympathetic nerves, while other drugs, such as atropine, inhibit that type of action.

NEURALGIA

Neuralgia means pain in a nerve, usually a severe pain along the course of a specific peripheral nerve. It is a rather loosely used term, but more particularly denotes paroxysmal pains originating in the nerve affected. Neuralgia may be an accompaniment of *neuritis* and is often, though not correctly, used almost as a synonym for this. Neuritis is inflammation of a nerve, usually characterized by a more or less constant pain, which may be due to infection or other causes, often obscure. The term *neuritis*, in turn, is often somewhat misapplied

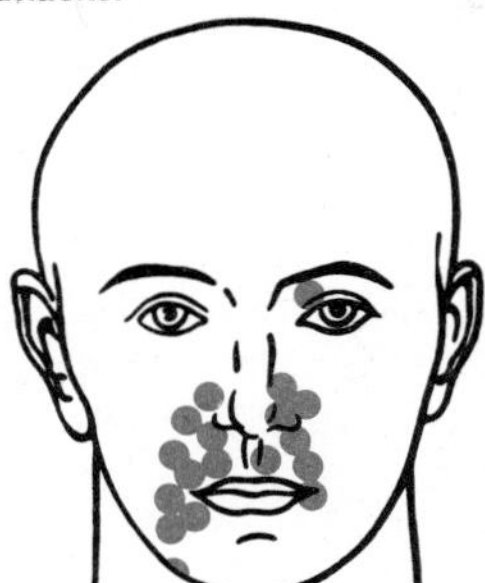

FIG. 7-11.
Points of stimulation that set off trigeminal neuralgia attacks.

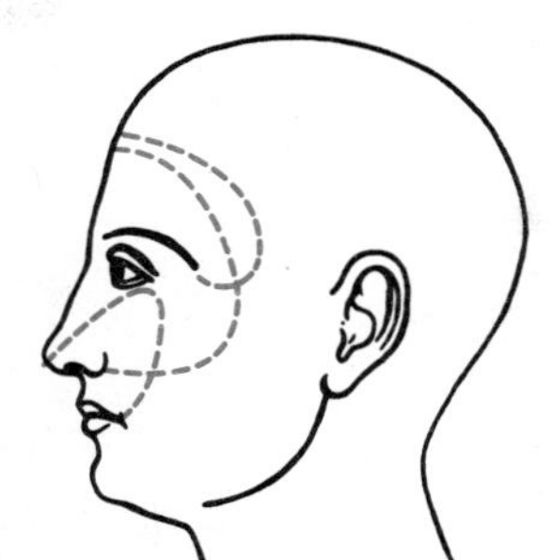

FIG. 7-12.
Zones from which an attack may be initiated.

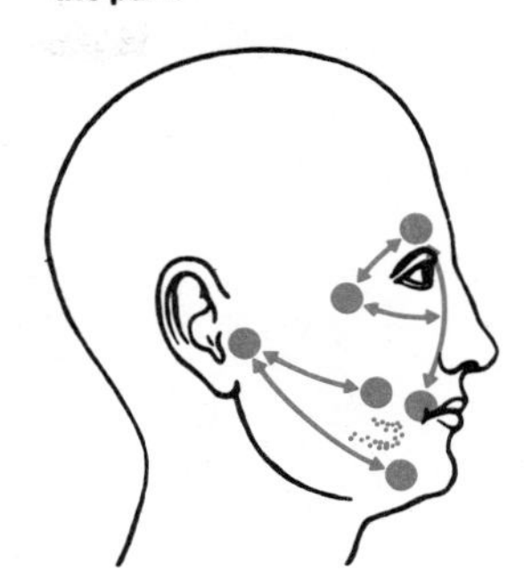

FIG. 7-13.
Localization and spread of the pain.

to a noninflammatory condition that should more properly, though less specifically, be referred to as *neuropathy,* which in a general way can denote any disorder of the peripheral nerves, but more especially one without inflammation, usually caused by pressure which impairs conduction, such as "slipped disc," or by constitutional disease (e.g., diabetes), circulatory disorders, vitamin deficiency, etc.

Neuralgia properly so called is characterized by acute *paroxysmal* pain, i.e., coming in spurts. The onset of pain sometimes occurs with surprising suddenness, whereas in other cases it may be heralded by sensations of heat, tenseness or formication (a sensation as of "crawling ants"), the pain itself being variously described as cutting, stabbing, burning or piercing. Between the paroxysms the patient is usually free of pain or feels only a dull and not very intense pain. Two familiar examples of neuralgia are trigeminal neuralgia and sciatica.

Trigeminal neuralgia (tic douloureux) is severe pain originating in the trigeminal nerve, which is the cranial nerve that carries sensations to the face. The pain usually first occurs in the region of the upper and lower jaw, for which reason the patient sometimes wrongly supposes it to be due to toothache. The pain then moves to one side of the face, where it always affects one particular area. Onset of pain is very sudden; the attack lasts only a few seconds, but is intensely painful. Attacks recur at intervals which may be very long, perhaps several weeks, but which are likely to become more frequent in advanced cases and recur every few minutes. Any sort of stimulus, however slight, may bring on an attack, so that ordinary activities such as washing, shaving, combing one's hair, eating, drinking or indeed speaking may trigger off acute suffering (Fig. 7-11, 7-12, 7-13). Periods of frequently recurring attacks may alternate with months and even years free from pain. Trigeminal neuralgia occurs mainly in the middle-aged and elderly and is more frequently encountered in women than in men. In most cases it has no recognizable cause and is apparently of spontaneous origin. Sometimes it disappears without treatment, but in other cases adequate treatment will be necessary to relieve pain. Pain-killing drugs are usually prescribed, but often have little effect. Injections of alcohol into the nerve can provide longer periods of relief. New drugs which are more effective than the conventional analgesics have also been developed. A permanent though rather drastic cure can be effected by cutting the nerve near its origin. Another technique producing a similar result is to destroy the nerve at its root by heat generated by a high-frequency electric current (electrocoagulation).

Intercostal neuralgia affects the nerves that run from the spinal cord along the ribs to the front of the chest. The pain is sometimes mistakenly attributed to a lung disease such as pleurisy. This form of neuralgia often precedes or follows an attack of shingles (herpes zoster).

Another well-known form of neuralgia is *sciatica,* characterized by persistent severe pain along the course of the sciatic nerve, which runs down the back of the thigh and into the leg (Fig. 7-14), though the pain does not necessarily occur throughout the whole of this area. Sciatica cannot be regarded as a disease; it is a symptom of an underlying cause, the commonest being pressure exerted on the sciatic nerve or its roots as a result of a "slipped" or ruptured disc in the spine. The

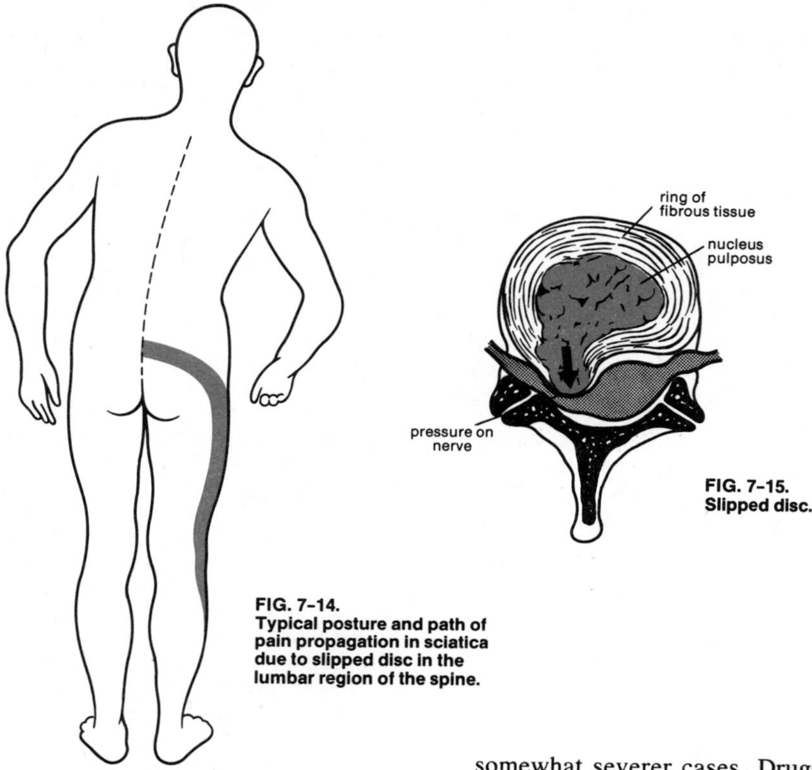

FIG. 7-14.
Typical posture and path of pain propagation in sciatica due to slipped disc in the lumbar region of the spine.

FIG. 7-15.
Slipped disc.

intervertebral discs each consist of a dense fibrous tissue enclosing a soft pulpy core (nucleus pulposus). Severe strain may cause the core to be squeezed out and thus exert pressure on the nerve (Fig. 7-15).

Less frequently, sciatica may be due to inflammation of the sciatic nerve resulting from infectious or metabolic diseases (e.g., diabetes), toxic disorders (e.g., alcoholism), etc., which cause irritation of nerves. Sciatica may begin abruptly or gradually; it is characterized by a sharp pain shooting down the back of the thigh. The symptoms usually grow worse at night and may be aggravated by cold damp weather. In most cases of sciatica the attacks of severe pain are brought on by some sudden movement or strain, as in lifting a heavy load. Between the attacks the patient usually feels no more than a slight backache. In general, movements of the legs and lower part of the body will intensify the sciatic pain. Mild cases clear up with aspirin, rest and application of heat. Bed rest, preferably on a hard mattress, and avoidance of exercise may be necessary in somewhat severer cases. Drugs such as morphine may have to be given to control the pain. If the cause is a ruptured intervertebral disc, surgical intervention to correct this may be necessary.

HEADACHE

Headache (cephalalgia) denotes any diffuse pain in different parts of the head. It may be acute or chronic and is not confined to any particular nerve distribution area. The pain itself may be a dull ache or acute, almost unbearable; it may be continuous, intermittent or throbbing. Generally speaking, headache is a symptom that may accompany almost any illness. In itself it is therefore a very unspecific indication of some disorder or disturbance, which may range from the transient and harmless to the chronic and dangerous.

Headache may be caused by toxic factors, such as poisonous fumes or gases, drugs, alcohol (hangover headache), tobacco, wines with a high histamine content, bacterial toxins produced in all sorts of acute or chronic infection. Systemic diseases such as nephritis, diseases of the liver, and diabetes are also often associated with headache. Gastrointestinal disturbances (dyspepsia, hyperacidity, constipation, etc.) and cardiovascular disturbance (high or low blood pressure, myocardial and valvular insufficiency, etc.) can cause headache, as also can endocrine disorders (tumors of various endocrine glands) and gynecological factors (puberty, pregnancy, menopause, etc.). Other causes of headache include organic diseases of the brain, especially those causing pressure (tumor, cyst, hemorrhage, intercranial vascular disease, arteriosclerosis, encephalitis, meningitis, etc.). Besides these there are many other possible causes, e.g., injury to the head, sunstroke, motion sickness, insomnia, eyestrain, hay fever. Even in a simple classification a list of the possible underlying causes of headache would run to several hundreds.

Numerous headache remedies are available. Generally they are mild analgesic drugs, mostly derivatives of salicylic acid, of which the most widely used is aspirin (acetylsalicylic acid). Headaches that do not respond to these drugs and persistent headaches should be investigated, as they may be a symptom of a serious illness which should not be treated symptomatically without an attempt to find its cause. It should be remembered that analgesics do not cure diseases, but merely suppress pain symptoms.

Migraine is a form of headache which occurs in paroxysmal attacks, usually accompanied by nausea and sometimes by vomiting (hence the name "sick headache"). The headache is generally confined to one side of the head. An attack is often preceded by warning symptoms such as flickering before the eyes, sometimes other visual disturbances. It occurs more frequently in women than in men. The underlying causes of migraine are not known. Heredity may be an important factor, and the condition appears to be associated with certain types of personality. The immediate cause is thought to be dilation of arteries in the neck and skull, which in turn is possibly attributable to overactivity of substances such as histamine and serotonin that control the dilation and constriction of arteries. Psychic strain and emotional conflict also play a part. In many cases there are general disturbances of the autonomic nervous sytem, with sweating, abdominal pain (sometimes severe), palpitation, diarrhea and other symptoms. An attack may last several hours, and the attacks may range in frequency from very few and far between to almost daily recurrence. During the attack the patient cannot stand bright light; rest in a quiet darkened room is desirable. Various drugs are available for the treatment of migraine, and different cases respond differently to them. Ergotamine tartrate is often effective; caffeine (sometimes even a few cups of strong coffee), amphetamine, barbiturates or aspirin may also bring relief in particular cases.

TETANUS

Tetanus (lockjaw) is an acute infectious disease due to the toxin of the tetanus bacillus (*Clostridium tetani*), which lives harmlessly in the intestines of horses, cows, sheep and occasionally man. The bacilli can multiply only under anaerobic conditions, i.e., in the absence of air. When conditions are unfavorable, they form spores, particularly when they are discharged with the feces and thus come into contact with the air. A spore is a dormant form of the bacillus, enabling it to survive for long periods under adverse conditions. Bacterial spores are often highly resistant to antiseptics and those of some species are difficult to destroy even by boiling. With the return of favorable environmental conditions, the spore becomes active again. Tetanus spores are commonly found in soil, especially in soil on farms or in gardens treated with horse manure. They are also found in decaying wood. Tetanus bacilli or their spores cannot penetrate the skin or the mucous membrane lining of the intestine. To do harm, they must enter through a wound. But even if a wound is contaminated with tetanus spores, these will not necessarily become active and produce the disease. For this to occur the conditions must be exactly right for anaerobic growth. If a wound is in healthy tissue, well supplied with blood (and therefore with oxygen), it will not provide a suitable environment for the bacillus to thrive in. Bacterial activity can develop only if the spores become embedded in tissue that is poorly supplied with blood, especially tissue that has been killed by other bacteria, by injury (e.g., due to burns) or by chemical poisons. If the conditions favor it, the bacillus grows and multiplies at the site of the injury, where it does not harm the tissues, so that the wound will not necessarily show any unusual signs. As the incubation period is 1–3 weeks, the wound may even have superficially healed by the time the disease manifests itself.

What makes the tetanus bacillus so dangerous is the extremely poisonous toxin that it produces—similar in its action to strychnine, but tens of thousands of times more powerful (though, of course, the quantities released by the bacilli are extremely small). The toxin makes its way along nerves to the spinal cord and medulla oblongata, where it suppresses the automatic impulses that restrain the motor nerve cells. In the absence of this restraint the muscles receive numerous random impulses and respond with convulsive twitching or with sustained contraction.

Tetanus usually begins gradually, though sometimes the onset is sudden. Early symptoms may be unspecific—restlessness, tiredness, sleeplessness, outbreaks of sweating. The first more definite sign is stiffness of the jaw, esophageal muscles and some of the muscles of the neck. Soon the jaws become rigidly fixed by a spasm of the jaw muscles (trismus). Contraction of the facial muscles produces a peculiar grin (risus sardonicus). The muscles of the back and limbs also become tetanic, i.e., go into a state of sustained contraction, so that the patient may be bent back in a bow (opisthotonos) or sideways or forward. The convulsions are reflex actions and are triggered off by noise, touch, bright light, a current of air or indeed any other sensation. They usually last several seconds and may recur at intervals of a few minutes. There are usually a high temperature and severe pain. As the respiratory muscles are also affected, the patient may die from asphyxiation, pulmonary congestion, physical exhaustion or heart failure.

With modern treatment the majority of patients recover and suffer no permanent effects. The tetanus patient should be nursed under sedation in quiet surroundings and subdued light. In severe cases a drug such as curare is given to suppress muscular spasms, and mechanical respiration will have to be applied. Treatment will consist in giving the patient injections of tetanus antitoxin, though this can only neutralize toxin that has not yet invaded nerve cells, and in opening up the wound to expose the bacilli to air and thus inhibit their activity. To assist breathing, tracheotomy may be used, i.e., an incision into the trachea (windpipe) for the insertion of a breathing tube. The patient will be fed through a stomach tube, and losses of water and salt must be replaced. An attack of tetanus does not confer immunity against the disease. Preventive vaccination is effective, however, for a period of about five years, after which it will have to be repeated.

RABIES

Rabies (hydrophobia) is an acute infectious disease of the brain and spinal cord, caused by a virus that is transmitted with the saliva from an infected animal, usually as the result of a bite, sometimes also by licking of a scratch or other small wound in the patient's skin. Dogs, cats and cattle are the domestic animals particularly susceptible; they may have been infected by wild animals (foxes, martens, squirrels, etc.) with which they have been in contact or whose dead bodies they have fed on. Rabid wild animals lose their natural fear of man and tend to be aggressive and quick to bite. Suspect animals should, if possible, not be immediately destroyed, but be captured and caged for observation.

From the site of a bite the virus travels along nerves to the spinal cord and brain, where it destroys nerve cells, resulting in paralysis and finally death. The incubation period between infection (i.e., in general, being bitten) and the first symptoms is usually 1–3 months, though shorter or much longer periods are possible. The disease then begins with low fever, headache, feelings of depression and anxiety. There may also be pain at the site of the wound. This condition passes into one of excitement, irritability and aggressiveness. Even quite minor external stimuli, such as sound or touch, can trigger off spasms and outbursts of excitement. There is difficulty in breathing and swallowing, the voice becomes husky, there is a choking sensation, and the patient is deeply agitated. Painful spasm of the throat prevents him from drinking, and even the sight of water sets off these spasms (hence the name *hydrophobia*). If he does not die from asphyxiation, the patient becomes delirious, loses consciousness and dies—usually within 1–4 days from the onset of the symptoms. Postmortem examination of the brain following death from rabies reveals the presence of so-called Negri bodies, characteristic small particles, inside some of the brain cells. In the brains of suspect animals these particles provide a means of identifying the disease.

Without treatment, rabies is invariably fatal. As a normal precaution, bites or scratches made by an animal should be thoroughly cleaned with strong medicinal soap solution; deep puncture wounds should be opened to permit access of the solution. Although an antirabies vaccine was developed by Pasteur as long ago as 1885, no completely effective

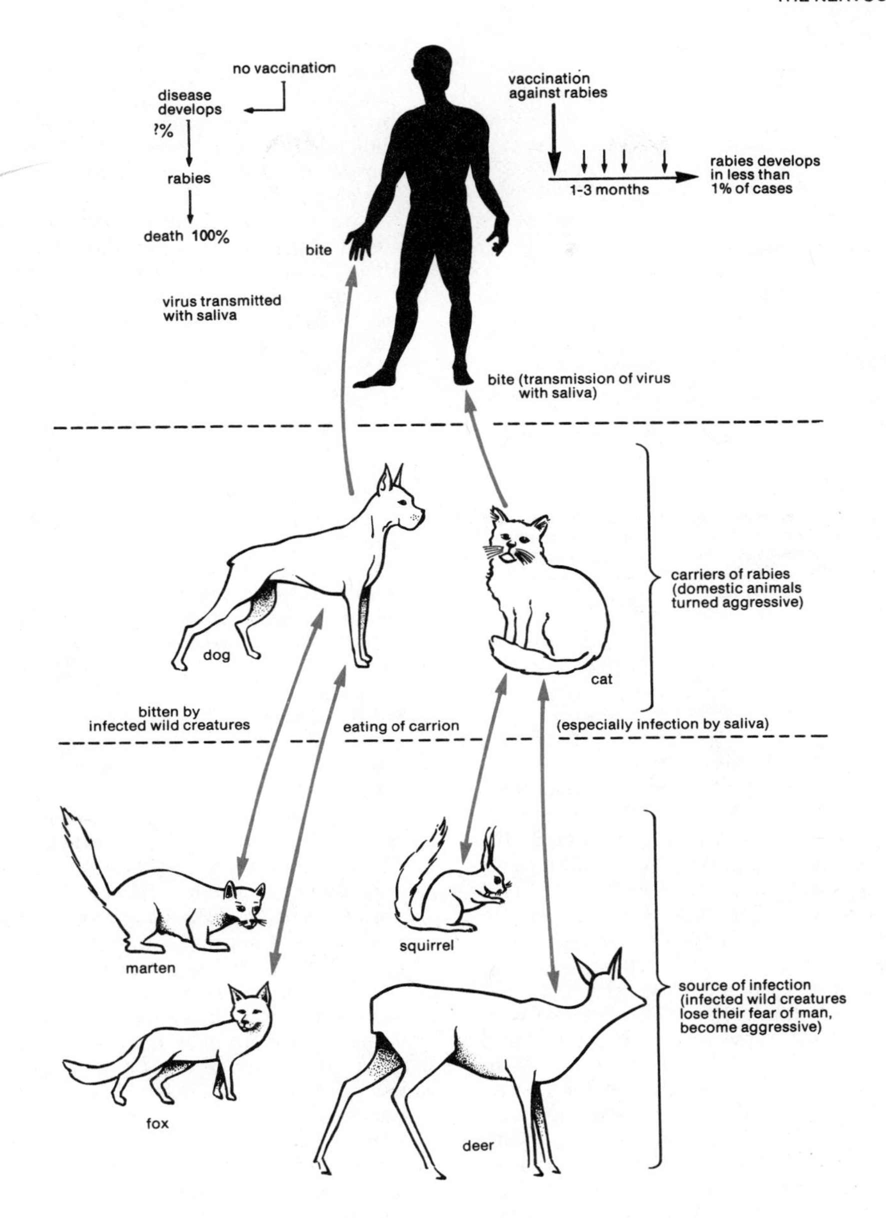

FIG. 7-16.
Transmission of rabies and its consequences.

treatment for an infected person as yet exists. The vaccine treatment is not without risk and not always successful, but its success rate is high. In the British Isles rabies has been eliminated by strict quarantine regulations for imported dogs, cats and other animals. In other countries, where the disease is endemic, dogs are usually vaccinated against rabies.

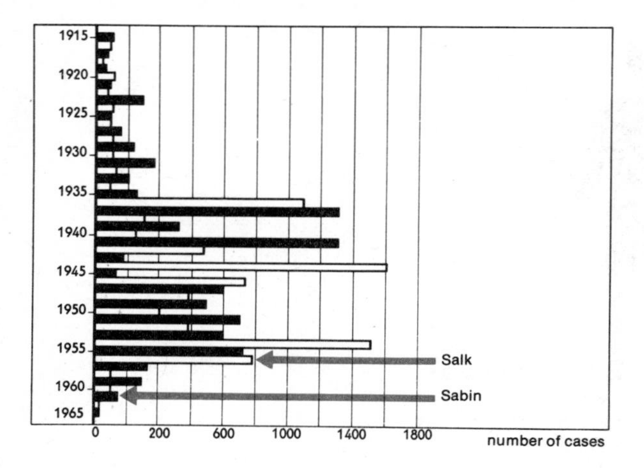

FIG. 7-17.
Incidence of poliomyelitis in Switzerland during the period from 1915 to 1964. After the Salk vaccine was introduced there was a sharp decrease in the number of cases, and with the introduction of the Sabin vaccine in 1965 the disease was eliminated entirely.

POLIOMYELITIS

Poliomyelitis (infantile paralysis)—*acute anterior poliomyelitis*, to give it its full name—is an acute infectious inflammation of the anterior (front) horns of the gray matter in the spinal cord. It is caused by a virus that attacks particularly those nerve cells in the cord that transmit the motor impulses from the brain to the peripheral nerves which stimulate the skeletal muscles. Three different types of the disease are to be distinguished: type I is the most malignant and manifests itself especially in major epidemics; type II occurs in isolated cases; type III generally causes minor epidemics. The virus can be transmitted by human excrement. Infants living in surroundings with poor sanitation usually become infected with poliomyelitis at an age when they still have maternal antibodies; in such children the disease generally causes no symptoms, and they acquire lifelong immunity. In highly developed countries with their sophisticated hygienic conditions, on the other hand, there are many children and adults who have never been in contact with the disease and are therefore not immune. In such societies, first contact with poliomyelitis, if it occurs at all, generally comes at a more advanced age in life, and in these older patients the disease tends to have more serious effects. For instance, Figure 7-17 shows the increase in poliomyelitis in Switzerland since 1914 and its subsequent suppression by mass vaccination of the population with Salk and Sabin vaccines in 1957 and 1960. Similar results have been achieved in most other developed countries.

Vaccination provides active immunization, i.e., the protective antibodies are produced in the vaccinated person's own body. The Salk vaccine, which is given by injection, is prepared by killing the virus with formalin. The dead virus cannot cause the disease, but it stimulates the production of antibodies that confer immunity (Fig. 7-18a). The Sabin vaccine, given by mouth, contains attenuated virus which can no longer cause disease symptoms, but stimulates the production of antibodies (Fig. 7-18b).

The polio virus enters the body through the mouth, and in the majority of cases the infection is confined to the throat and intestines, with only very minor symptoms. Even in the major form of the disease, involvement of the spinal cord occurs only in a small proportion of cases. An attack of the disease is often preceded by some other disorder which, as it were, "opens the door" to the polio virus: tonsillitis, measles, scarlet fever, etc. Surgical removal of the tonsils, pregnancy or muscular fatigue due to overexertion may have a similar effect of making a person more susceptible to polio infection. The incubation period of the disease is usually 2–3 weeks, but may be somewhat longer or shorter. Onset may be abrupt, but is usually gradual, with fever, tiredness, sometimes coughing or sore throat, the symptoms being generally similar to those of influenza or a severe cold. Or there may be gastrointestinal disturbance with diarrhea, vomiting and stomach ache. Usually the initial fever goes down after 3–4 days, and the early symptoms of the disease disappear (latency period). The patient's temperature rises again after 1–3 days; this is the stage when the virus invades the central nervous system. The patient complains of headache, pain in the arms and legs, followed by stiffness in the neck. After a further 2–3 days this stiffness disappears, but he now has difficulty in raising the upper part of his body. The temperature has meanwhile risen to 39°–40°C (102.2°–104°F), the limbs are very painful, and the patient will try to avoid all unnecessary movement.

Within a period ranging from a few days to a week after the second rise in temperature, paralysis may develop, though only in a small

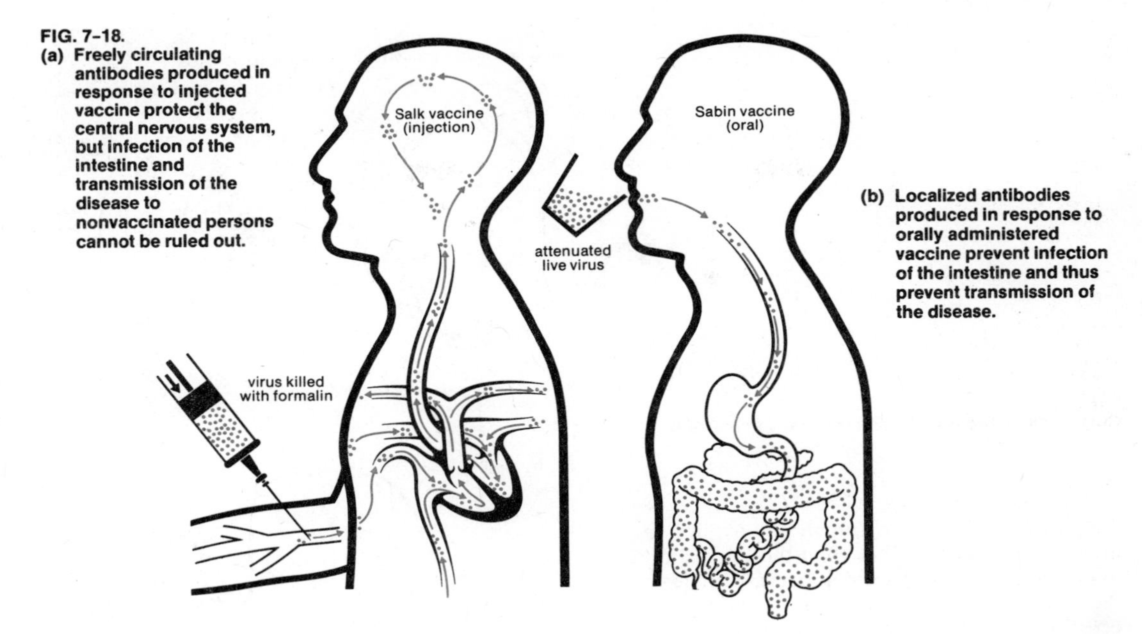

FIG. 7-18.
(a) Freely circulating antibodies produced in response to injected vaccine protect the central nervous system, but infection of the intestine and transmission of the disease to nonvaccinated persons cannot be ruled out.
(b) Localized antibodies produced in response to orally administered vaccine prevent infection of the intestine and thus prevent transmission of the disease.

minority of cases. Even then, there is little danger to life unless the respiratory muscles, especially the diaphragm, are paralyzed; artificial respiration then becomes necessary to keep the patient alive. The extent of the paralysis depends on the degree of nerve involvement; it may be confined just to one small group of muscles. Sensation is not affected. Paralysis, if it occurs, affects the muscles on one or both sides of the body, usually in the legs, but also in the arms and at the shoulder girdle.

The progress of the disease may stop at any stage. As already mentioned, in the great majority of (nonvaccinated) persons infected with the virus the disease passes without giving rise to any significant symptoms, but conferring immunity for life. In a certain relatively small proportion of cases the disease is halted after the influenzalike initial stage. In only about 1 percent of persons infected does the disease go into the "stiff neck" stage, and only in a small proportion of these is the spinal cord involved, with weakening or paralysis of groups of muscles. Indeed, despite the name *infantile paralysis*, the occurrence of paralysis is to be regarded, not as a normal feature of the disease, but as a comparatively infrequent complication. Also, although children are more susceptible than adults, persons of any age can contract poliomyelitis.

There is no specific drug treatment for the disease, once it has gained a hold. Treatment must aim at relieving the symptoms and preventing deformities. In minor cases it is generally sufficient to keep the patient in bed for a week or so. Severe cases require careful nursing, especially if there is paralysis. Proper positioning of the patient can relieve muscle tenderness and pain. Gentle passive movement, hot moist packs or hot baths are also helpful. In many cases of paralysis, the condition soon improves, usually within a week or so, when the spinal inflammation subsides. Convalescence of the paralyzed patient will often require physical and occupational therapy and orthopedic treatment.

SPINAL CORD

The spinal cord (Fig. 7-19) is a column of nervous tissue which lies protected within the vertebral canal of the spinal column. It extends from the medulla oblongata to the second lumbar vertebra. Like the spinal column itself, the spinal cord is of segmental construction, this being manifested in the paired emergence of the spinal nerves. There are 31 pairs of spinal nerves: 8 cervical, 12 thoracic, 5 lumbar, 5 sacral and 1 coccygeal, corresponding to the vertebrae. Since the spinal cord is shorter than

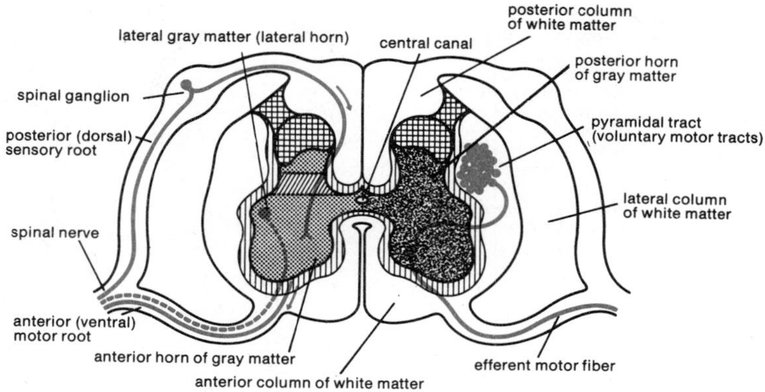

FIG. 7-19.
Cross section through the spinal cord and nerve roots.

the vertebral canal in which it lies, its segments are located higher than the corresponding vertebrae. Each spinal nerve still leaves the vertebral canal through its appropriate opening between the vertebrae (intervertebral foramen), so that the roots of the lower nerves go down a considerable distance within the canal before emerging. All nerves to the trunk and limbs originate in the spinal cord, which is also the center of reflex action and contains the conducting paths to and from the brain. The spinal cord is narrower than the vertebral canal; the space around the cord contains three membranes (meninges), blood vessels, and the cerebrospinal fluid. This fluid is a water cushion protecting the spinal cord and brain from shock.

The spinal cord is composed mainly of the so-called gray matter, which contains large numbers of cell bodies of neurons. In cross section the gray matter is approximately H-shaped, with a posterior (or dorsal) and an anterior (or ventral) horn in each half. The anterior horn is composed of motor (efferent) neurons from which the fibers making up the motor portions of the peripheral nerves arise. The posterior horn contains the cell bodies of connector neurons and the termination of sensory (afferent) neurons. In the thoracic region there is moreover a lateral horn which accommodates the visceromotor neurons for the outgoing sympathetic impulses. The gray matter is surrounded by the white matter, which is divided into three columns (posterior, lateral and anterior) and which contains ascending and descending nerve tracts which serve to connect the brain and the spinal cord in both

FIG. 7-20.
Pyramidal tract.

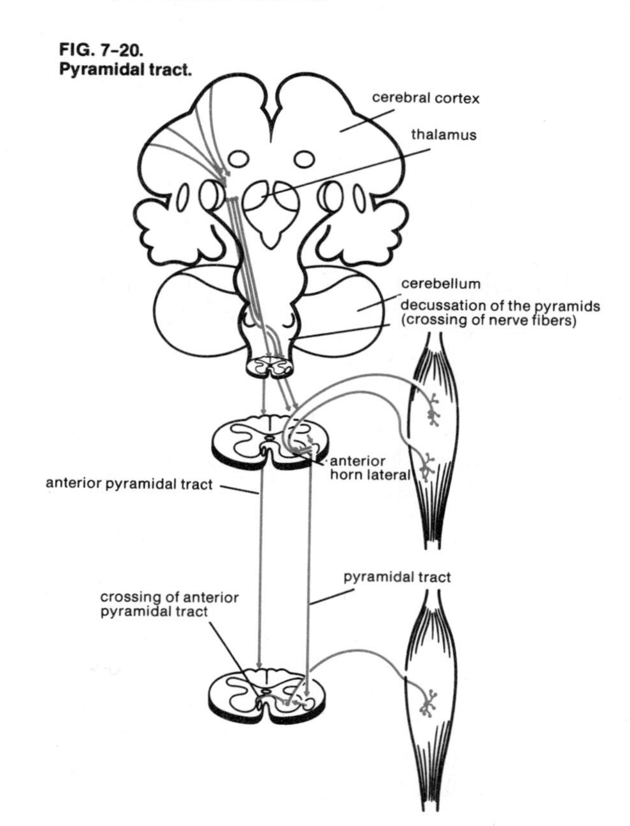

directions (i.e., comprising afferent and efferent nerves) as well as various portions of the cord itself. The efferent (descending) tracts conduct motor impulses from the brain to the periphery, while the afferent (ascending) tracts in the white matter conduct sensory impulses from the periphery to the brain. The main descending tract is the pyramidal tract (Fig. 7-20) and is so called because it arises from the giant pyramidal cells of Betz in the motor area of the cerebral cortex.

Each spinal nerve is connected to the spinal cord by means of a posterior (or dorsal) and an anterior (or ventral) root; the former consists of afferent fibers conveying impulses to the cord, the latter of efferent fibers conveying impulses from the cord. The anterior root consists of the axons of motor neurons situated in the anterior horn of the gray matter; the posterior root possesses a swelling (spinal ganglion) which contains the cell bodies of somatic and visceral afferent neurons.

SPINAL INJURIES

Although the spinal cord is well protected in the vertebral canal, it can nevertheless suffer damage as a result of accident or disease. In particular, the fracture of a vertebra or the breakdown of vertebrae by bone tuberculosis or cancer (Fig. 7-21) presents a major hazard to the cord. The consequences of injury will depend on how high up in the spinal cord the injury is located and what its transverse extent is. If only one half of the cord in cross section suffers concussion, or is crushed or severed, the result will be paralysis of movement and loss of the sense of position on the side affected, while sensations of temperature and pain are abolished on the other side (the "intact" side). This is because the afferent nerve tracts conducting these last-mentioned sensations to the brain "cross over to the other side" of the spinal cord from the side at which they enter it (Figs. 7-22 and 7-23).

In a case of total spinal injury, all the nerve tracts at the site of injury are severed. As a result, communication between the brain and the parts of the body below the injury is interrupted, with loss of sensation and loss of voluntary movement in those parts. The patient feels them as "cut off," though the spinal cord can still maintain reflex actions below the injury. Also, there are disturbances in blood pressure control, in the regulation of sweating and in emptying the bladder (see page 211) in consequence of damage to the autonomic nerve tracts in the lateral columns of white matter of the spinal cord. At first the paralyzed muscles are slack because shock of the spinal cord puts the entire nervous motor system out of action. Subsequently, spastic paralysis develops, i.e., characterized by muscular rigidity, with increased reflex activity. Spastic paralysis occurs because peripheral nerve impulses normally maintain a certain "resting tension" in the muscles and when the spinal cord is injured, the inhibitory action of the brain in controlling this tension is absent, so that now the tension and the reflex responsiveness increase unchecked. This may be so acute that even the slightest touch of the limbs, e.g., by bedclothes, will automatically trigger off violent movements.

The aspects of the disease will vary with the height at which the injury is located. This is not surprising when it is considered that the

FIG. 7-21.
Causes of total spinal injury.

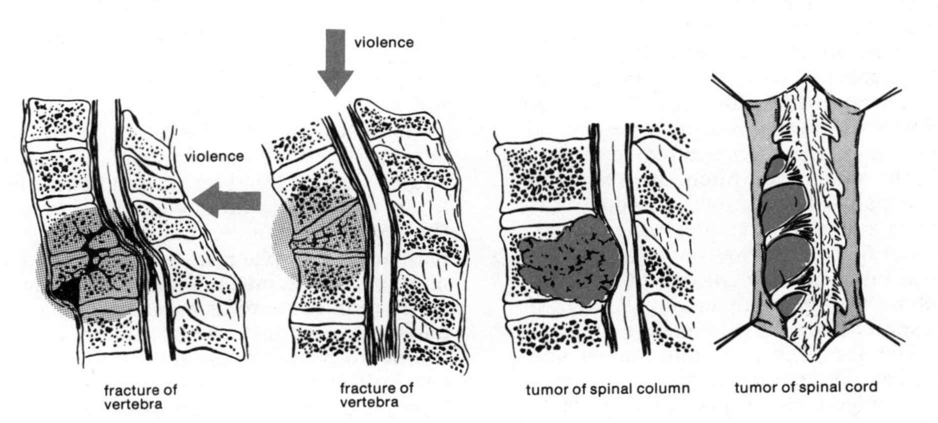

FIG. 7-22.
Nerve tracts in the spinal cord.

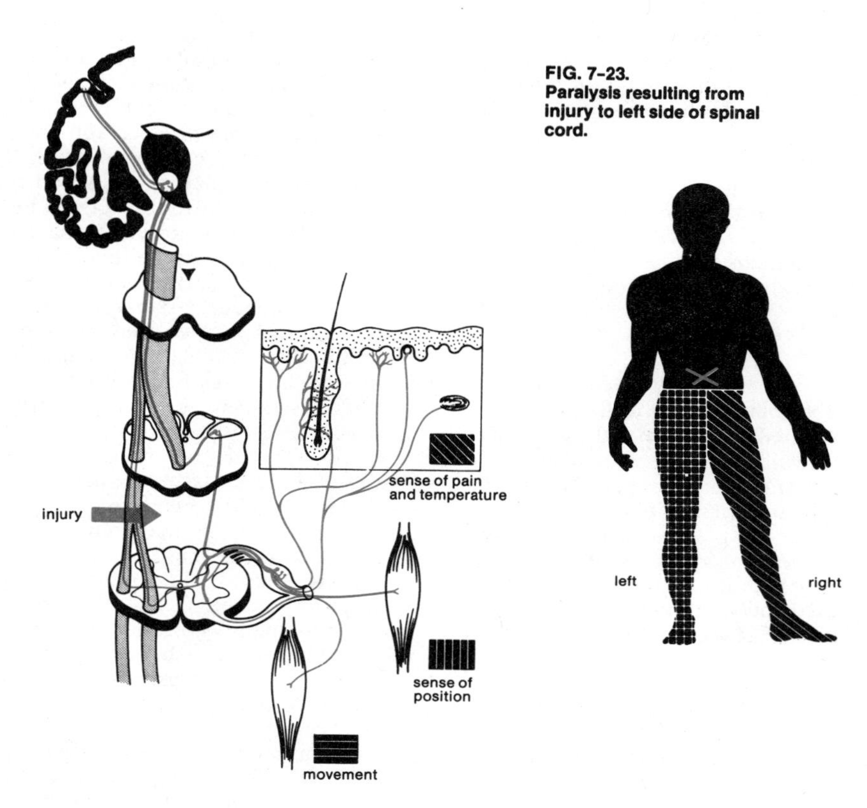

FIG. 7-23.
Paralysis resulting from injury to left side of spinal cord.

spinal nerves communicating with the peripheral parts of the body emerge from the spinal cord at different levels. Injuries in the vicinity of the fourth cervical vertebra are particularly dangerous because they affect the diaphragm, so that not only thoracic but also abdominal (diaphragmatic) respiration is affected. Injury high up in the spinal cord often also results in muscular atrophy of the shoulder girdle, arms and fingers. Paralysis of the intestine and bladder occurs at first in all cases of injury to the spinal cord, but is permanent only if the cord is injured in the lumbar region. Paralyzed limbs are soon affected by deficient blood circulation. The legs become cold, bluish and puffily swollen. There is a risk of bedsores (decubitus), i.e., ulceration and gangrene of localized areas, especially over a bony prominence (buttocks, heels, insides of the knees), due to pressure from prolonged confinement to bed.

Injury of the spinal cord is always a serious condition. In the event of an accident involving spinal injury, the patient should be laid flat on his back and preferably not be moved except under the supervision of a doctor. He should then be lifted by several people in such a way as to cause no movement or bending of the spinal column. Unskillful handling of the patient can displace a fracture and thus do more serious damage. Treatment of spinal injuries requires careful nursing in special hospitals. Severed or torn nerve fibers in the central nervous system do not regenerate as they can in peripheral nerves. The patient will be trained to make the best possible use of his remaining functions.

MULTIPLE SCLEROSIS

Multiple sclerosis (disseminated sclerosis) is a slowly progressive chronic disease of the central nervous system, characterized by degeneration of small scattered areas. In particular, the nerve fibers lose their insulating myelin sheaths, which are replaced by scar tissue occuring in grayish hard (sclerotic) areas. There is also degenerative growth of the neuroglia, the interstitial supporting tissue of the nervous system. Similarities exist between the sclerosis in this disease and the scarring that remains after a viral infection such as poliomyelitis; but the cause of multiple sclerosis is not known. It is not contagious. Some scientists believe it to be caused by a virus or to have some other infectious origin; others attribute it to disturbances of metabolism (excessive intake of fats, disturbed function of the liver). There are indications that it may be some form of "misguided" defensive reaction of the nerve tissue against certain of its own constituents. It is thought not to be hereditary in the ordinary sense, though there are indications that the disease is more frequent in some families than in others.

Multiple sclerosis is one of the most widespread of neurological diseases, about 0.05 percent of the population being affected. It is about twice as common in women as in men. In about two thirds of cases it affects young adults, between 20 and 40 years of age. In persons suffering from the disease, nerve messages from the brain to muscles and organs are "short-circuited," so that control over various parts of the body is lost. Particularly the muscles of locomotion are affected in this way.

The symptoms of the disease depend on where the areas of sclerosis are located in the central nervous system. An early symptom in some cases is a loss of sharpness of vision in one eye, occurring rather rapidly, within a few days; yet examination of the fundus of the eye with an ophthalmoscope does not reveal any change. Further symptoms include blurred or double vision, tremors (Fig. 7-24), numbness, and slurred speech. Frequently there is trigeminal neuralgia on both sides of the face (Fig. 7-25). When the patient is asked to bring his finger to his nose while his eyes are closed, he is unable to do this directly, but only in a zigzag fashion (Fig. 7-26). During the further course of the disease the limbs often become convulsively paralyzed, while reflex actions are intensified. Bowel and bladder disturbances occur, with increased urination urge or incontinence. The patient develops a staggering gait (Fig. 7-27). There may also be marked psychic changes.

Multiple sclerosis is often characterized by remissions, i.e., periods during which there is a marked improvement in the patient's physical condition, sometimes for weeks, months or perhaps years. In general, however, the disease nevertheless runs a progressive course, and the patient will eventually have to use a wheelchair and may become bedridden. In a small number of cases the disease may run a rapid course, resulting in death within weeks. As a rule, it develops gradually. About 10 years after the onset of multiple sclerosis about 80 percent of patients are still alive; 25 years after the onset only 6 percent survive, and of these the great majority are severely handicapped. The survival chances in any individual case will depend to a great extent on the quality of nursing care that the patient receives, including precautions to prevent cystitis (inflammation of the bladder) and bedsores in bedridden patients. In the earlier stages, much can often be achieved by physical ther-

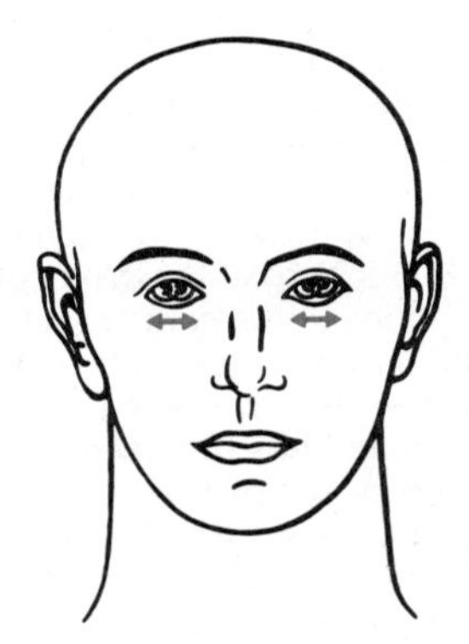

FIG. 7-24.
Eye tremors.

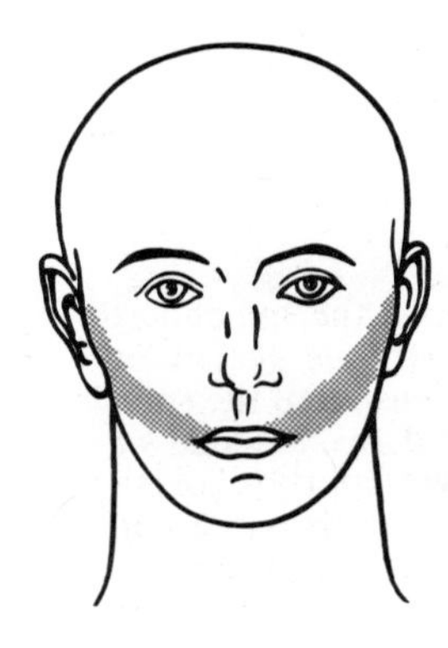

FIG. 7-25.
Trigeminal neuralgia on both sides of the face.

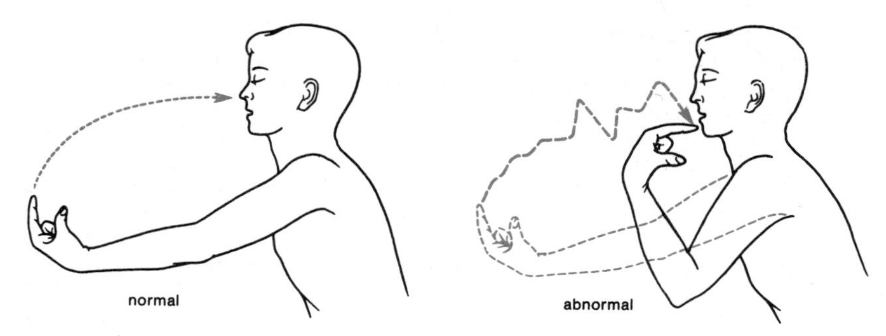

FIG. 7-26.
Tremor as demonstrated by patient bringing his finger to his nose while his eyes are closed.

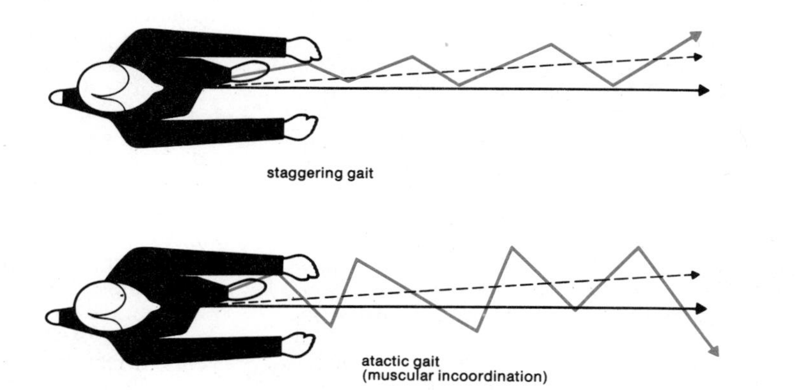

FIG. 7-27.
Patient walking with eyes closed.

apy, exercise and rehabilitation procedures to enable the patient to lead a reasonably normal life within his possibilities. There is no specific therapy for multiple sclerosis. Adrenocorticotropic hormone, which is secreted by the pituitary gland and regulates the functional activity of the adrenal cortex, has been tried as a means of halting the course of the disease.

BRAIN

The brain, a continuation of the spinal cord, is the part of the central nervous system enclosed by the skull. The extension of the spinal cord into the skull is called the medulla oblongata, which is the lower portion of the brainstem; the latter also comprises the pons and the midbrain. The brain has three main divisions: hindbrain, midbrain and forebrain.

The *hindbrain* (rhombencephalon) is made up of the *pons* (pons varolii), the *cerebellum* and the *medulla oblongata*. In the spinal cord the nerve cells form the gray matter, which is surrounded by white matter composed of nerve fibers. In the medulla the gray matter is concentrated in groups of cells (nuclei) which include the cells of cranial nerves and also contain the vital centers that govern the reflexes for breathing, heart rate and blood pressure. The pons appears as a broad band of transverse fibers across the front of the brainstem and contains fibers connecting the medulla and cerebellum with upper parts of the brain. The cerebellum, which is the largest portion of the hindbrain and fills the lower back region of the skull, comprises two lateral hemispheres and a narrow medial portion (vermis). It is involved in the synergic (cooperational) control of skeletal muscles and helps to coordinate voluntary muscular movements. It receives and discharges impulses; though it is not a reflex center, it reinforces some reflexes and inhibits others. The cerebellum consists of an outer layer of gray matter, the cerebellar cortex, and a core of white matter containing central nuclei.

The *midbrain* (mesencephalon) contains some cranial nerve cells, centers for the coordination of movements, and reflex centers for the senses of sight and hearing. It includes the reticular formation, which consists of groups of cells and fibers arranged in a diffuse network which extends throughout the brainstem and is important in controlling or influencing alertness, waking, sleeping and various reflexes. More especially, it plays an important role in facilitating and inhibiting impulses coming down from or traveling up to the cerebral cortex.

The *forebrain* (prosencephalon) contains the highest centers for the control and integration of the activities of the nervous system; it is here that sensory impulses "reach consciousness" and motor impulses are initiated by the will. It comprises the *cerebrum* consisting of the two cerebral *hemispheres*, one on each side, which are attached to a midline part called the *diencephalon*. Each hemisphere has a thin outer layer of gray matter, the cerebral *cortex*, enclosing white matter, within which is a central mass of gray matter (corpus striatum or basal ganglia). The hemispheres are joined together by a bridge of transverse fibers (corpus callosum) which pass across the midline. The cortex forms a convoluted covering for the cerebral hemispheres, which are characterized by numerous ridges and grooves. Main grooves divide the hemispheres into lobes. The diencephalon consists mainly of the two thalami, each thalamus being a mass of gray matter, one on each side of the midline. Below the thalami lies the hypothalamus, to which the pituitary gland is attached. The *thalamus* is an important relay station on the ascending sensory tracts and has connections with the hypothalamus and with the basal ganglia, through which it plays a part in the control of muscular activity. The *hypothalamus* controls the function of the pituitary gland and also contains centers for controlling the autonomic nervous system. It is moreover an important relay center for tracts concerned with the expression of emotion (changes in facial expression, changes in the heart rate and blood pressure, sweating, etc.) and is also concerned with the mechanism of sleep.

The *cerebral cortex* contains rounded or granular cells at which the sensory tracts terminate and pyramidal cells which initiate

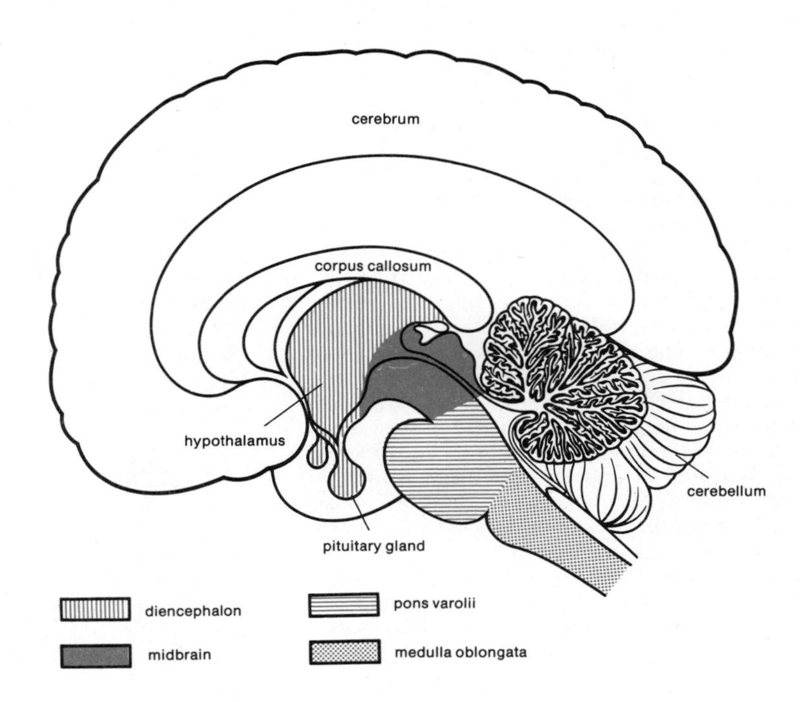

FIG. 7-28.
Parts of the brain.

motor impulses. The conscious appreciation of sensory impulses is localized in various parts of the forebrain: hearing (in the temporal lobe), smell (frontal lobe), sight (occipital lobe). There is a motor speech area on each side, only one of which is used: the left one in right-handed individuals, and vice versa. Furthermore there are association areas which integrate the various functions, e.g., areas that integrate tactile sensation with vision, and vision with hearing. The frontal lobe contains association areas that affect motor activity but are also important in determining the personality of the individual. No particular areas can be allotted to such functions as reason or memory. Activity of the cortical cells is associated with electrical activity which varies as a result of physical action or mental processes. The recording of the electrical activity of the brain is called *electroencephalography*; the record obtained in this way is an *electroencephalogram*. Electrodes are placed on the scalp in various positions, and the difference in electrical potential between the two sites is recorded. Normally, under resting conditions, the so-called alpha rhythm is obtained; a characteristic change in the wave occurs during sleep, on opening the eyes, and during mental attention.

The *brainstem* (comprising the medulla, pons and midbrain) contains important ascending and descending tracts and also the nuclei of the cranial nerves, of which there are twelve pairs: olfactory, optic, oculomotor, trochlear, trigeminal, abducens, facial, auditory, glossopharyngeal, vagus, accessory, hypoglossal. Within the brain the central canal of the spinal cord is expanded into interlinked chambers called *ventricles*. Inside each cerebral hemisphere is a lateral ventricle, which communicates with the so-called third ventricle, a narrow midline chamber between the two thalami.

The brain, spinal cord and nerve roots attached to them are covered by three membranes (*meninges*): dura mater, arachnoid, and pia mater. The *dura mater* is the tough and fibrous outer covering; inside the skull it comprises an outer and an inner layer and provides support for the brain substance. Folds of dura mater partly separate the cerebral hemispheres from each other and the cerebrum from the cerebellum. The *pia mater*, the innermost membrane of the meninges, follows every contour of the brain and spinal cord. The *arachnoid* is a cobweblike membrane which joins the pia to the dura; its interstices form the so-called subarachnoid space, which contains cerebrospinal fluid. The *pia mater* is provided with blood vessels which run within the subarachnoid space.

Nervous tissue, the brain substance in particular, requires a constant and plentiful supply of oxygen and glucose, conveyed to it by the blood. It is very sensitive to any failure in this supply; if the supply of blood to the brain ceases for more than a few minutes, irreversible damage will result. Blood circulation in other organs of the body is variable, but the brain receives a steady supply of about 750 ml per minute. The cerebral arteries are of two types: superficial arteries on the surface of the brain, and central arteries which enter the hemispheres through the base of the brain. These are end arteries, i.e., their branches do not link with one another. Blockage of one of these arteries causes necrosis of the part of the brain affected (cerebral thrombosis); alternatively, one of the arteries supplying the interior of a hemisphere of the brain may burst (cerebral hemorrhage).

Cerebrospinal fluid is a clear watery fluid which fills the subarachnoid space in the skull and spinal column, the ventricles of the brain and the central canal of the spinal cord. It is secreted in the ventricles, more particularly in the so-called choroid plexuses, which are networks of capillaries in folds of the pia mater. Cerebrospinal fluid acts as a tissue fluid for the central nervous system and also serves to support and protect the jellylike brain substance. The spinal cord reaches down only to the upper two lumbar vertebrae, but the subarachnoid space at the base of the spinal column reaches farther down. Lumbar puncture consists in inserting a hollow needle into this lower part of the subarachnoid space by passing it between the third and fourth lumbar vertebrae. This may be done in order to obtain a specimen of fluid for examination or to measure its pressure. Fluid from inside the skull can be obtained by the same method, the needle being inserted into the cisterna magna, an enlarged part of the subarachnoid space (cisternal puncture).

APOPLEXY

Apoplexy (stroke or cerebrovascular accident) is a more or less sudden loss of consciousness due to an interruption in the blood supply to the brain. If the supply to the brain as a whole fails, death will ensue in about 10

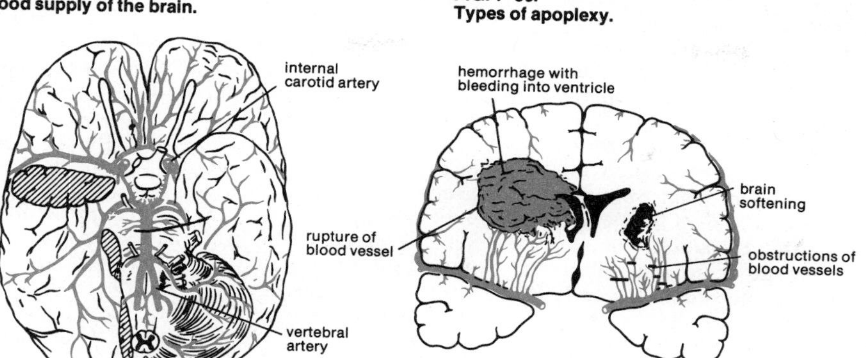

FIG. 7-29.
Blood supply of the brain.

FIG. 7-30.
Types of apoplexy.

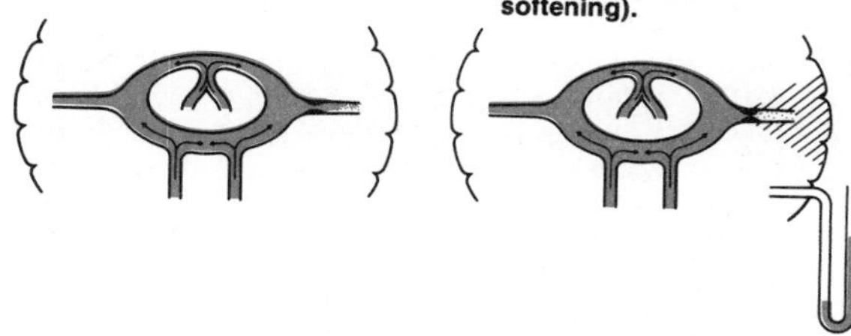

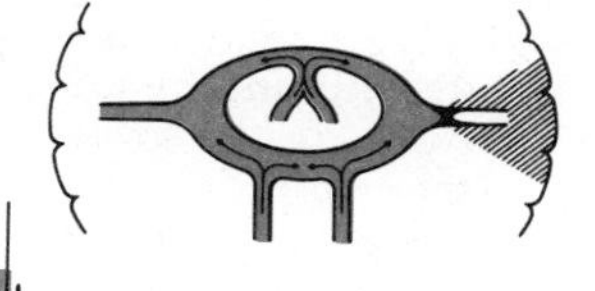

FIG. 7-31.
(a) Narrowing of blood vessels without symptoms of illness.
(b) Lowering of blood pressure occurring when blood vessels are narrowed by arteriosclerosis may cause localized damage to brain tissue (brain softening).
(c) Complete blockage of blood vessels resulting in brain softening.

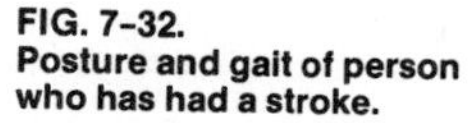

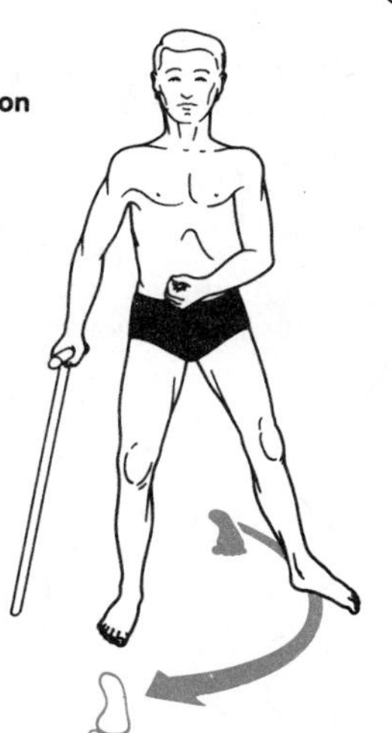

FIG. 7-32.
Posture and gait of person who has had a stroke.

minutes. Failure affecting only certain of the brain's blood vessels will not necessarily cause loss of consciousness. Particularly at the base of the brain there are major interconnections between the arteries that supply it with blood (Fig. 7-29). Apoplexy is caused by cerebral hemorrhage (i.e., bleeding from an artery in the brain) or, in about 75 percent of cases, by disturbed circulation due to sudden blockage of an artery (embolism) by a fragment of material detached from some other part of the circulatory system or more gradual blockage resulting from the formation of a clot (thrombus) in a diseased artery of the brain itself. Sometimes hemorrhage is due to a weak spot in the wall of an artery. Apoplexy due to circulatory disturbance occurs mainly in older people, especially when the blood vessels of the brain are affected by arteriosclerosis, which is a predisposing condition. The stroke may occur at night, while the patient is asleep, or it may occur after a heavy meal or severe exertion, for example. Under these conditions there is a lowering of blood pressure, and it is this that initiates the actual stroke (Fig. 7-31).

The blood vessel damage often results in localized areas of softening in the brain. Apoplexy due to cerebral hemorrhage is less common than that due to blockages in blood supply and occurs mostly in the middle-aged; about 80 percent of such patients have high blood pressure. In such cases the escape of blood from a weakened artery into one of the ventricles of the brain will cause a sudden onset of symptoms, usually in the daytime, when the patient is engaged in his ordinary occupation.

A severe stroke is typically characterized by sudden paralysis affecting one side of the body (hemiplegia); face, arm or leg, or all three, may be paralyzed. Paralysis may occur while the patient is asleep or indeed at any time. The onset is often quite sudden. In other cases, however, an attack may be preceded by headache and malaise. Sometimes the paralysis develops gradually, over a period of hours. Only in about 50 percent of patients does loss of consciousness occur. At first the muscles in the paralyzed parts are relaxed; most of the reflexes are normal or only slightly diminished. The pupils of the eyes may be unequal, the larger one being on the affected side of the brain; the eyeballs at first are turned away from the paralyzed side of the body. The skin may be covered with clammy sweat. Breathing is laborious, and there may be speech disturbances. After a time the paralysis becomes characterized by muscular spasm; the head and eyeballs are now turned toward the paralyzed side.

In general, about 20 percent of patients die as a result of a first stroke; another 40 percent are killed by subsequent strokes. In the survivor of a stroke, paralysis of a leg usually clears more quickly than paralysis of an arm. At a later stage of convalescence the patient drags the paralyzed leg with a characteristic semicircular motion, while the paralyzed arm is bent a little and does not swing during walking (Fig. 7-32).

EPILEPSY

Epilepsy is a disorder, or group of disorders, characterized by fits (seizures) associated with recurrent attacks of disturbed brain function. There are various combinations of motor, sensory or psychic malfunction, with or without convulsions, and attended by disturbances or loss of consciousness. Next to apoplexy it is the commonest neurological disease.

Most cases of epilepsy are *idiopathic*, i.e., apparently spontaneous, without assignable cause, though various possibilities have been suggested, such as susceptibility of the brain cells to minor irritation, or some unknown agent that triggers off the fits, or a minute undetected injury that left a scar in a particularly sensitive area of the brain. Heredity plays an important part. There is a direct relationship between changes in electrical brain potentials and the occurrence of epileptic fits; this is revealed by electroencephalograms (Figs. 7-33 and 7-34). A fit is initiated by disorganized and excessive activity in some part of the brain, causing simultaneous discharge of impulses from the brain along a number of pathways and thus resulting in uncoordinated convulsive movements in various parts of the body.

Some forms of epilepsy, however, are *symptomatic*, i.e., there is a detectable physical agent. Such causes may be of many kinds, e.g., degenerative changes in tissues (due to scars, tumors, etc.) or metabolic disturbances with deficiency of blood sugar or oxygen supply to the brain (due to a blood clot, for example). All sorts of causal factors may produce symptomatic epilepsy, such as injuries followed by scar formation or inflammation of the brain or the meninges. Sometimes the fits occur as an accompaniment of specific diseases, e.g., the convulsions that children sometimes have at the beginning of a fever, as in measles. In other cases the cause persists and produces attacks which recur at indefinite intervals.

More than half of all cases of epilepsy begin in childhood (Fig. 7-35). Possible causes of such early epilepsy are believed to be prenatal brain damage (due to oxygen deficiency, syphilis, German measles, toxic agents) or brain damage at birth (arrested respiration, hemorrhage inside the skull, injury from obstetric forceps). The other factors already mentioned may cause symptomatic epilepsy in middle age. Similar symptoms may be produced by cerebral hemorrhage, tertiary syphilis, brain tumors, alcoholism, etc. It is therefore difficult to speak of epilepsy as a disease in its own right.

The distinction between "symptomatic" and "idiopathic" is merely a distinction between those cases where the cause is known (or believed to be known) and those where it

FIG. 7-33.
Normal propagation of nerve impulses.

(a) normal insulated brain pathway.

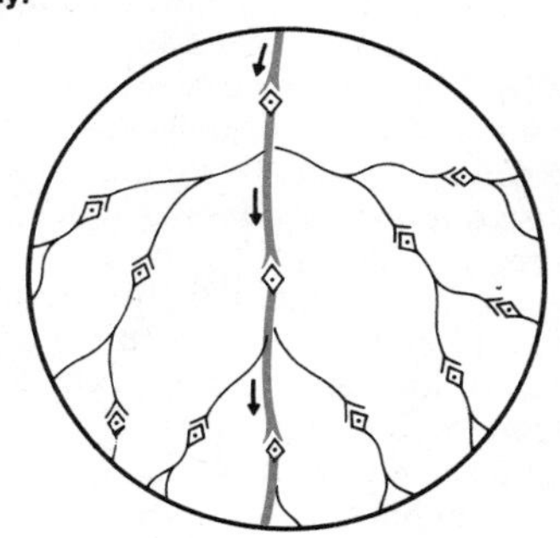

(b) With corresponding EEG printout.

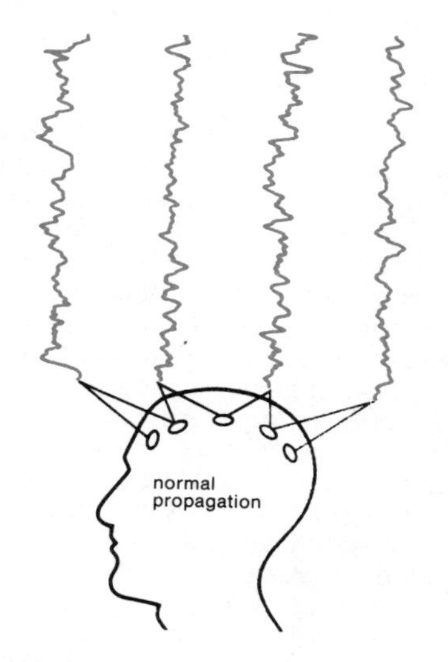

FIG. 7-34.
Diffuse excessive propagation of nerve impulses in epileptic seizure.

(a) additional stimulated nerve pathways.

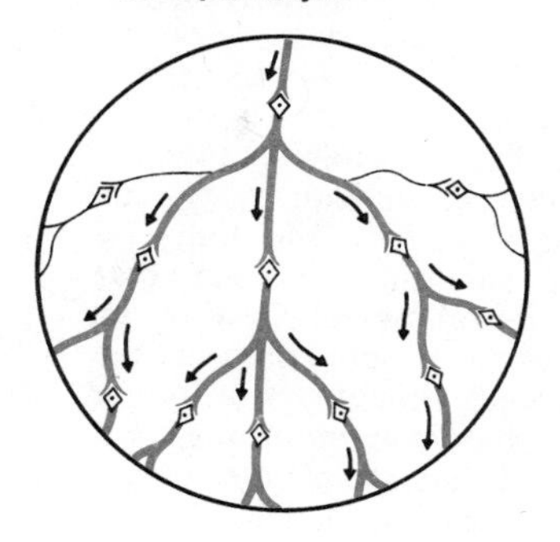

(b) With corresponding EEG printout.

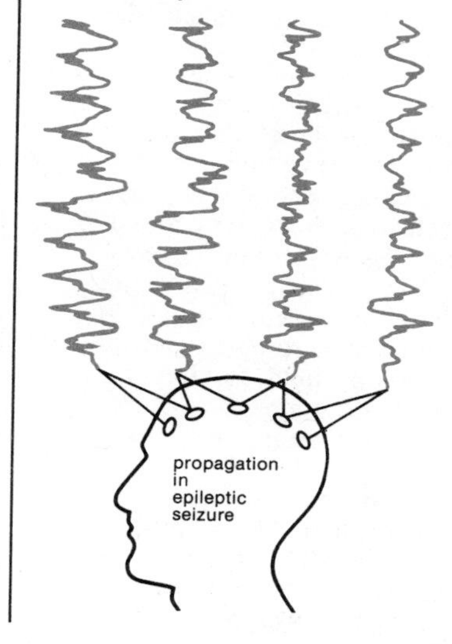

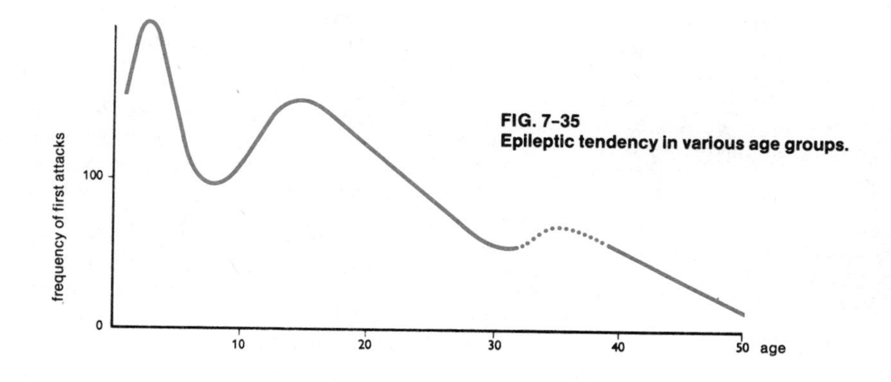

FIG. 7-35
Epileptic tendency in various age groups.

is unknown. Idiopathic (or "genuine") epilepsy typically manifests itself in early childhood between the ages of 5 and 8 or in early puberty or between 20 and 25. Epilepsy that appears after the age of 30 is usually of the symptomatic type. In general, there is no essential difference between the seizures associated with these two types of epilepsy.

The well-known epileptic fit in which the patient falls unconscious, often with a cry (due to sudden expulsion of air from the lungs), is known as *grand mal*. Sometimes there are preliminary symptoms (aura), which may be an uncontrolled movement in some part of the body, a peculiar sensation in a finger or toe, or a sensory hallucination (visual, auditory, etc.). The aura may precede an attack by several hours or by only a few seconds; it gives an indication of the area of the brain where the disturbance begins. The patient falls during the attack and may injure himself. His muscles stiffen and twitch violently (tonic spasm); he may bite his tongue and pass urine. The muscles of the chest are involved, and breathing is hampered, so that the patient's complexion turns blue and he appears to be choking. This first stage of the attack lasts 1–2 minutes. The patient's body jerks, and the second stage then begins. His legs thrash with short jerky movements, the breath is expelled in short gasps, the head is often thrown to one side, and the eyeballs (with dilated pupils) are rotated in their sockets. The tongue is pushed out in jerks, and the jaws are brought forcibly together, so that biting of the tongue and injury to the lips may occur. The patient froths at the mouth, and the saliva may be tinged with blood from these self-inflicted wounds. He may involuntarily pass urine and perhaps also evacuate the bowels. After 2–3 minutes the convulsions grow less frequent and the patient's bluish color gradually disappears as he draws deep stertorous (snorting) breaths. He briefly goes into a deep sleep with complete relaxation of his limbs. He soon regains consciousness, but may feel exhausted and go to sleep again, this time for several hours; in other cases the patient may feel reasonably fit within minutes after recovering from the attack. In general, he has no recollection of the fit itself and may become aware of what has happened only when he finds he has hurt or wet himself.

In the minor epileptic condition known as *petit mal*, the seizure is of very short duration, involving merely a momentary loss of consciousness, which may even pass unnoticed by the patient (usually a child). Young children may have *salaam convulsions*, characterized by muscular spasms resulting in a slow bowing movement of the head and upper part of the body, while the arms are raised forward or sideways. In other cases there may be jerking movements, and in the variant called *nodding spasms* only the head moves up and down. This condition occurs particularly in babies between the ages of 3 and 18 months. So-called *absences* occur chiefly in children of school age and consist merely in fleeting intervals of unconsciousness lasting from 5 to at most 30 seconds; the only outward sign is a rigid facial expression and absent gaze. *Jacksonian epilepsy* is characterized by convulsions confined to certain groups of muscles or to one side of the body. It is due to a disorder in motor areas of the cerebral cortex. As a rule, it begins with twitching of one half of the face or body, followed by loss of consciousness, sometimes culminating in a grand mal epileptic fit, of which the Jacksonian attack is essentially the aura.

Epileptic attacks may occur at any time. In women, however, they may tend to come particularly during menstruation (menstrual epilepsy). In general, the frequency of attacks varies greatly in different cases. Attack may occur several times in the course of a day, then remain absent for weeks. In mild forms of epilepsy the attacks may be years apart. If attacks occur with great frequency, the condition can be serious. In severe cases the seizures may come in rapid succession over a period of hours, during which the patient does not regain consciousness (status epilepticus).

Symptomatic epilepsy can sometimes be cured by removal of the underlying cause, e.g., surgical removal of a brain tumor or scar tissue. Various modern anticonvulsant drugs are effective in upward of 80 percent of cases of grand mal, but their effectiveness varies from one patient to another, so that often the most suitable drug (or combination of drugs) has to be determined by trial and error in any given case. Treatment should continue until the patient has been free from attacks for at least three years. Petit mal, which is essentially a disorder of childhood, is also treated with drugs, though of a different kind; most patients make a complete recovery from this type of epilepsy. In general, an epileptic should not do work involving a risk from falls or the use of dangerous machinery.

FIG. 7-36.
Cross section through vertebra and spinal cord.

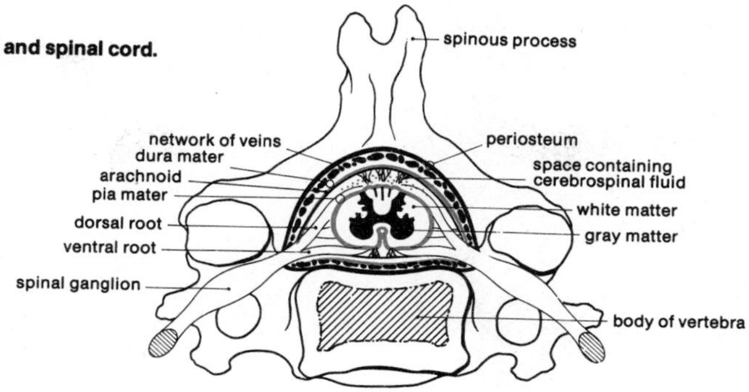

MENINGITIS

Meningitis is inflammation of the meninges, which are the three membranes enclosing the brain and spinal cord: pia mater (internal), arachnoid, and dura mater (external). They provide resilient support for the central nervous system and protect it from the entry of harmful agents (Fig. 7-36). These membranes may, however, themselves become infected with bacteria or viruses which set up inflammation. Meningitis is not confined to any age, but it tends to be particularly serious in patients whose resistance is low because of immaturity (in very young children), old age or an attendant debilitating disease. It displays the following typical symptoms: severe headache, vomiting, spasm of the muscles in the neck and back, fever, intolerance to light and sound, delirium, convulsions and coma.

Inflammation of the meninges can be subdivided into four main categories, which differ to some extent in their symptoms and especially in the type of treatment they require:

Acute purulent meningitis is an infection caused by bacteria, mainly meningococci or pneumococci. It may develop as a complication of an inflammatory condition affecting adjacent parts, be caused by infection due to a head injury, or be transmitted through the bloodstream from a remote focus of inflammation (in the heart or lung, for example). This form of meningitis is characterized by very high fever. The presence of pus in the cerebrospinal fluid and other changes in the fluid's condition confirm the diagnosis of the disease; it is usually possible to identify the infecting microbe in the fluid, which may contain large numbers of white blood cells. Acute purulent meningitis can be effectively treated if diagnosed in time. Sulfonamides or specific antibiotics may be employed; if the microbes have not been identified, a so-called broad-spectrum antibiotic, effective against a wide variety of organisms, may be given.

Serous meningitis comprises a number of milder inflammatory disorders of the meninges due to various causes. Viral infection in some cases may, in addition to affecting the meninges, be accompanied by signs of inflammation of the brain itself (e.g., paralysis). There is a serous exudation into the cerebral ventricles. Examination of the spinal fluid reveals a marked increase in protein content but only a slightly increased cell count. There is no specific therapy for serous meningitis, and treatment will consist in confinement to bed, careful nursing, and the use of antipyretics.

Chronic meningitis may develop from acute meningitis, especially if it is inadequately treated. In other cases, however, the chronic form of the disease may be a manifestation of involvement of the meninges in some inflammatory constitutional disease such as syphilis or toxoplasmosis. In general, this form of meningitis runs a benign course. Sometimes there are scarcely any symptoms at all, and the disease may remain undetected for a long time.

Tuberculous meningitis is a complication of tuberculosis, an acute inflammation of the cerebral meninges caused by the tubercle bacillus. It starts insidiously. Early symptoms include loss of weight, gradual waning of strength, evening rise of temperature, restlessness, sleeplessness and irritability. Simultaneous involvement of other organs, particularly the lungs, may provide an indication, and the presence of tubercle bacilli in the cerebrospinal fluid will confirm the diagnosis. The disease used to be invariably fatal until the introduction of modern tuberculostatics, i.e., drugs that arrest the growth of the bacillus, such as streptomycin.

Part Eight

THE ENDOCRINE GLANDS

HORMONES

A hormone is a substance that originates in a gland and is conveyed through the blood to other parts of the body, stimulating them by chemical action to increased functional activity. Together with the nervous system, the hormones control and regulate the functions of metabolism, growth and reproduction. Hormones vary in their chemical structure, and they also differ greatly from one another in the manner in which they act. Most of them are secreted in so-called *endocrine glands*, known also as ductless glands, which release their products directly into the bloodstream (internal secretion). A characteristic feature of hormones is that they are active in very small amounts and do not supply energy. The most important endocrine glands are the pituitary gland, the adrenal glands, the thyroid gland, the parathyroid glands, the pancreas, and the gonads (ovary, testis). The body's functions that are under the control of hormones can be performed properly only if the glands produce the correct amounts of their respective hormones—neither too little nor too much.

The secretions of *exocrine glands* are not released into the blood and are in this respect classed as "external," even though many of them, such as gastric juice, are released in the interior of the body. Some glands are both exocrine and endocrine. The pancreas, for example, produces pancreatic juice as its external secretion, while certain cells in this organ produce the hormones insulin and glucagon as internal secretions. The ovary and testis are other examples of combined exocrine and endocrine glands. The pituitary, thyroid, parathyroid and adrenal glands are wholly endocrine.

Large sections of the hormonal system are under the overall control of the cerebrum, i.e., the largest part of the brain, consisting of the cerebral hemispheres. Next in importance is the diencephalon, especially the part called the hypothalamus, which produces neurosecretions (hormonelike substances secreted by neurons) and emits nervous impulses by means of which it controls the secretions of subordinate endocrine glands. The third stage in this hierarchy is formed by the pituitary gland (hypophysis cerebri), which is sometimes called the master gland of the body and which secretes a number of hormones that in turn stimulate peripheral endocrine glands. Fourth in this system come these last-mentioned glands with their hormones, which not only act upon "target" organs but also exercise feedback control upon the pituitary and other controlling centers. For instance, a pituitary hormone stimulates the thyroid gland to secrete thyroxine; but if too much thyroxine is secreted, it suppresses this stimulating action of the pituitary gland, so that thyroxine secretion is adjusted to the correct required level. This feedback mechanism, also known as "push-pull" mechanism, is the general regulating principle for keeping the various hormones at their correct level in the blood, which, in turn, is determined by the nervous system in any given circumstances. Most hormones are selective in their action and affect only certain organs or tissues. Overfunctioning (hyperfunction) or underfunctioning (hypofunction) of endocrine glands can cause a number of diseases and disorders. Most complex in this respect are the consequences of malfunction of the pituitary gland (Fig. 8-1).

THYROID GLAND

The thyroid is an endocrine gland consisting of two lobes, on each side of the trachea, in the base of the neck. The lobes are connected by a narrow band (isthmus) across the front of the trachea. The secreting cells form hollow vesicles, called follicles, in which the hormone *thyroxine* is produced. Between the follicles

FIG. 8-1.
Hormones of the pituitary gland.

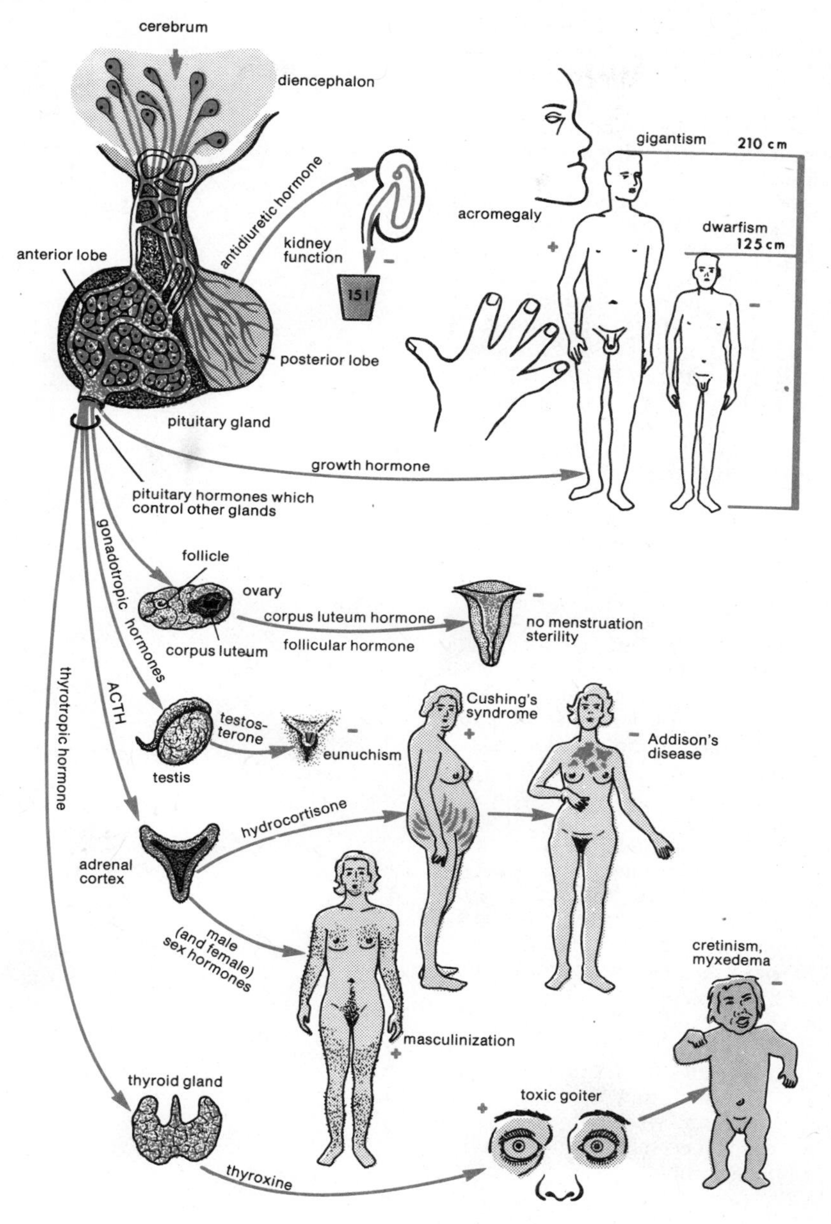

are small groups of cells that secrete a second hormone, called *calcitonin*. Thyroxine, the principal thyroid hormone, increases the metabolic rate of the organs and tissues of the whole body, including the promotion of growth and development in the young. It accelerates the combustion of glucose in the tissues and thus promotes the release of energy.

At the same time such functions as breathing and blood circulation to meet the higher demand for oxygen are increased, while physical and mental activity are stimulated. Thyroxine is an amino acid with a high proportion of iodine in its composition. The other hormone, calcitonin, lowers the blood calcium level and, with parathyroid hormone, helps to control the body's calcium metabolism. The release of thyroxine into the bloodstream is controlled by the thyroid-stimulating hormone (thyrotropin) secreted by the pituitary gland, which itself is controlled by autonomic cells in the diencephalon, this in turn being controlled by the cerebral cortex. When the concentration of thyroxine in the blood increases, the release of thyroid-stimulating hormone by the pituitary gland is inhibited (feedback). Through this system not only the body's physical needs but also psychic conditions (fear, excitement, etc.) can stimulate the thyroid gland to greater activity.

Hyperfunction of the thyroid gland (hyperthyroidism) is often associated with the condition called *toxic goiter*, also known by such names as thyrotoxicosis, exophthalmic goiter, Graves' disease or Basedow's disease. It is characterized by moderate enlargement of the thyroid gland and a general speeding-up of the body's metabolism. The patient loses weight, dislikes heat, has an increased pulse rate, and is nervous and excitable, with palpitation, fine tremor of the hands, and exophthalmia (protrusion of the eyeballs). It is much commoner in women than in men, occurring predominantly in individuals approaching, or in early, middle age. The underlying cause is not known. The disease may be initiated by emotional stress or by hormonal changes (puberty, pregnancy, menopause). The fault is mostly not in the thyroid itself, but in overstimulation of the thyroid by an excess of the thyroid-stimulating hormone released by the pituitary or by an abnormal hormone (long-acting thyroid stimulator). Laboratory tests reveal an enhanced level of basal metabolism and increased iodine uptake by the thyroid gland, which always needs a certain amount of iodine to synthesize its hormone.

Mild cases of toxic goiter can be treated by rest and sedatives. Hyperthyroidism that develops during pregnancy or menopause may clear up without special treatment. In more serious cases the use of antithyroid drugs which reduce the production of thyroxine can be helpful, but they are liable to have objectionable side effects. For older patients, past the age of fertility, treatment with radioactive iodine is available; this substance is absorbed into the thyroid gland, where the radiation destroys some of the overactive tissue. The most effective treatment is surgical removal of most of the thyroid gland. Before this can safely be

FIG. 8-2.
Treatment of hyperthyroidism.

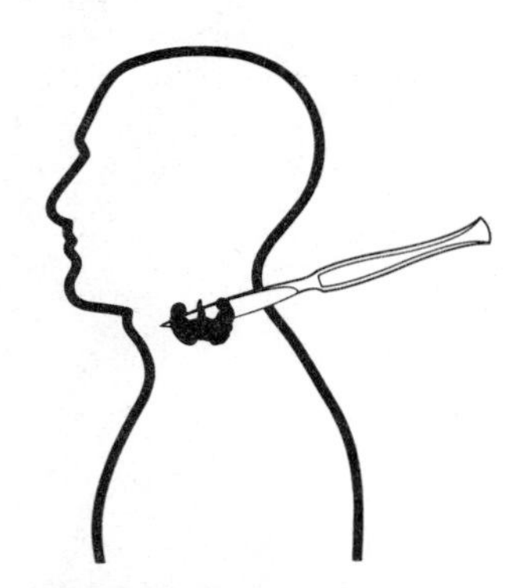

(a) Removal of thyroid gland (thyroidectomy).

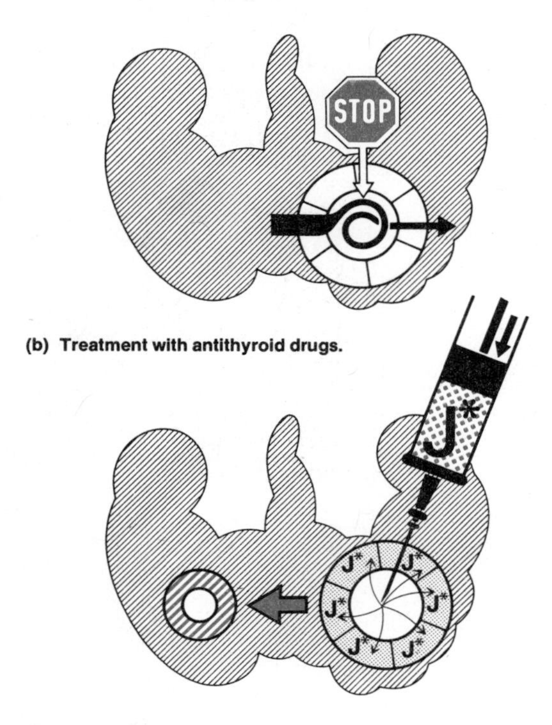

(b) Treatment with antithyroid drugs.

(c) Treatment with radioactive iodine.

done, however, the activity of the gland must be damped down by suitable preparatory treatment. This operation is indicated also in dealing with the condition called *toxic adenoma*, which is essentially a disease of old people. It is characterized by enlargement of part of the thyroid gland, with considerable overactivity that can result in serious cardiac damage, even heart failure; there is no exophthalmia.

Hypofunction of the thyroid gland (hypothyroidism) is characterized by deficiency of thyroid secretion, lowered basal metabolism and general sluggishness of the body's functions: low blood pressure, slow pulse, depressed muscular activity. Treatment will generally consist in replacement therapy with natural or synthetic thyroid hormone preparations.

Cretinism is the most important congenital disorder of the thyroid; it affects children and is due to deficiency in the secretion of the thyroid hormones. At birth the cretin appears normal, having obtained thyroxine from the maternal bloodstream. The symptoms of cretinism manifest themselves after 6–12 months, and the child's physical and mental development is retarded. Growth is stunted, hearing is impaired, the facial features are repellent (flat wide saddle nose), the tongue protrudes, the belly is enlarged, the skin is dry and coarse, the hands and fingers are thick. The child is of low intelligence. Without treatment the cretin remains a dwarf and mentally incapable of education. If the child is given thyroxine therapy from an early age, he will develop normally. As the disease is difficult to recognize in the first few months, there is the risk that treatment will be delayed, so that normal mental and physical development can then no longer be achieved.

Cretinism and simple goiter (see next section) tend to have the same geographic distribution, occurring in regions where there is (or was) a deficiency of iodine in the diet. Since the introduction of iodized table salt and the iodizing of drinking water, there has been a marked reduction of these diseases, and it can be concluded that iodine deficiency is an important underlying causal factor. Usually there is congenital absence or defect of the thyroid gland, or its ability to absorb and utilize iodine is disturbed.

Acquired (noncongenital) hypothyroidism, like hyperthyroidism, is very much more frequent in women than in men. In extreme cases, thyroid deficiency starting in adult life causes *myxedema*. Broadly speaking, the symptoms of this disease are the reverse of those associated with hyperthyroidism. The patient, usually a woman, becomes sluggish, with loss of energy and appetite, and a slowing down of the basal metabolism; laboratory tests show an abnormally low rate of iodine uptake by the thyroid gland. The hands and face become puffy, coarse and thickened, with dry skin; the tongue is enlarged, speech is slow, the mind is dulled (lack of interest, lowered concentration) and there is loss of hair; the heartbeat is slowed down, and the patient is sensitive to cold. The disease is caused by iodine deficiency in the diet or by atrophy of the thyroid gland itself as a result of circulatory deficiency or an inflammation. In some cases it is brought about by excessive use of antithyroid drugs or occurs as the undesired result of surgical removal of the thyroid to cure hyperthyroidism. Defective stimulation of the thyroid by the pituitary may also be a cause. Myxedema is treated by giving the patient thyroxine in suitably controlled amounts; the treatment is completely effective in most cases.

GOITER

As a general term, *goiter* (struma) denotes abnormal enlargement of the thyroid gland. Toxic goiter has just been described. The type referred to as *simple goiter* is more particularly due to deficiency of iodine in the diet. As this disease tends to occur most commonly in regions where the soil and drinking water lack iodine—especially in inland mountainous regions in Europe (Fig. 8-3) and other parts of the world—it is known also as endemic goiter (Fig. 8-4b). In such "goiter belts" upward of 10 percent of the population of some areas used to be affected. In modern times, however, the prevalence of goiter in these regions has been greatly reduced by the use of iodized table salt (i.e., with added traces of iodine in the form of sodium or potassium iodide); iodine may also be added to the drinking water. In most parts of the world the thyroid gland is adequately supplied with its iodine requirements from the amounts of iodine normally present in the food and water consumed. In simple goiter the thyroid gland, though abnormally enlarged, usually produces the normal amount of hormone, so that this disease is not accompanied by symptoms of hyperthyroidism or hypothyroidism.

FIG. 8-3.
Dissemination of goiter in central Europe.

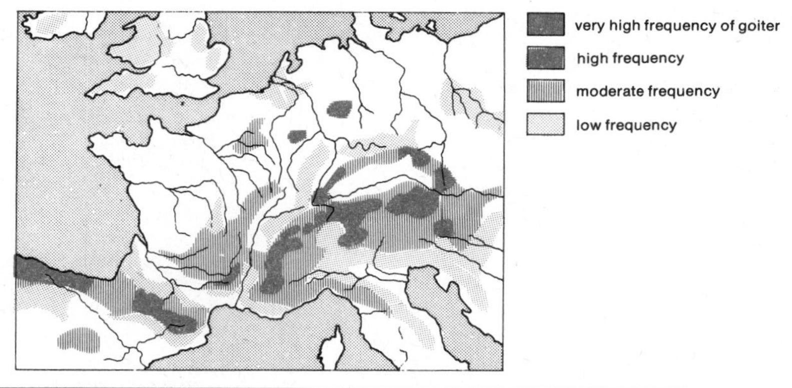

FIG. 8-4.
Development of goiter to compensate for inadequate hormone production.

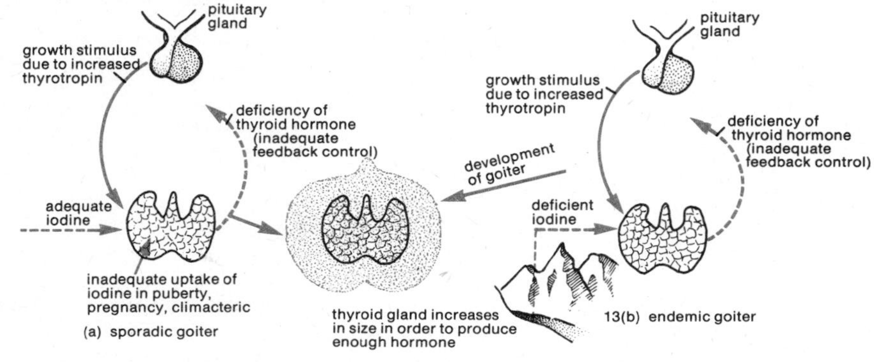

FIG. 8-5.

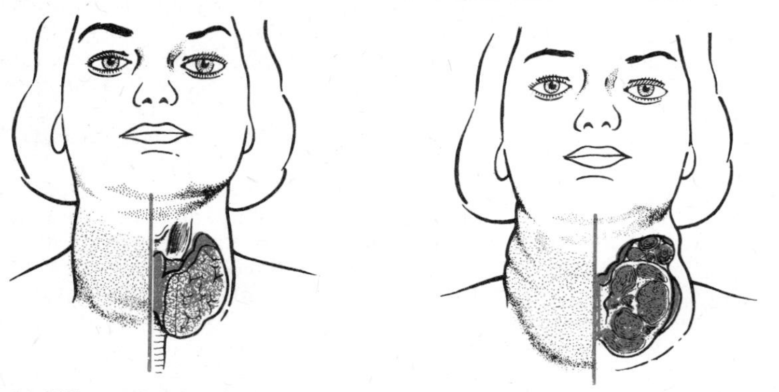

(a) Diffuse early stage of goiter. **(b) Goiter which has become nodular.**

A sporadic (nonendemic) form (Fig. 8-4a) of simple goiter is not confined to particular geographic regions. It occurs mainly in women and may be due to hormonal changes as, for example, in puberty, pregnancy or the menopause. Such goiters may disappear spontaneously after a time. Some congenital disorder of the thyroid gland preventing adequate iodine uptake may cause sporadic goiter, while antithyroid substances present in certain foods, such as cabbage, may be a contributory factor in persons with a predisposition to this disorder.

The thyroid gland needs iodine for the production of thyroxine, its principal hormone. If the daily iodine intake falls below the critical minimum of about 50–75 micrograms, the thyroid is unable to obtain enough of this element from the blood to meet its requirements. As a result, the production of thyroxine decreases, so that the normal feedback control that this hormone exercises on the pituitary gland is diminished (see page 358). The pituitary, thus relieved of this inhibiting effect, releases larger amounts of thyroid-stimulating hormone, so that the thyroid gland is stimulated to produce more thyroxine. Lack of iodine prevents it from doing this, however; instead, it responds by overgrowing. In some cases the goiter may grow to a very large size indeed and may then compress the trachea or blood vessels, so that breathing becomes difficult or the return flow of blood to the heart is obstructed. This is particularly likely to occur when a goiter spreads from the neck downward behind the breastbone into the chest. In exceptional cases a simple goiter may develop into hyperthyroidism.

To prevent such complications, simple goiter is treated with thyroxine, which relieves the thyroid of the need to produce its own and also inhibits thyroid-stimulating pituitary activity. If the goiter is very large and will not yield to hormone therapy, its surgical removal is indicated in order to relieve congestion and compression of adjacent structures.

In the early stages of goiter the enlarged gland is still soft (Fig. 8-5a); later it becomes nodular (Fig. 8-5b) due to encapsulated follicles.

ADRENAL GLANDS

The adrenal glands (suprarenal glands) are a pair of small triangular bodies adjacent to the upper ends of the two kidneys. Each gland is essentially a double organ comprising two different tissue systems—an outer cortex and an inner medulla—each of which secretes hormones.

The *adrenal medulla* secretes the hormones *adrenaline* (epinephrine) and *noradrenaline* (norepinephrine), which are chemically related. Adrenaline stimulates certain areas of the central nervous system. Its effects are generally similar to those brought about by stimulation of the sympathetic division of the autonomic nervous system. The release of adrenaline (and noradrenaline) is under the control of the sympathetic nerves, and the adrenal medulla is therefore to be regarded as an auxiliary and intensifier of the sympathetic nervous system. Adrenaline promotes glycogenolysis, i.e., the conversion of glycogen into glucose in body tissues (especially in the liver) and thereby raises the blood sugar level (in counteraction to insulin, which has the effect of lowering that level). Both the medullar hormones have a stimulating or alternatively an inhibiting effect on organs under the control of the autonomic nervous system. Noradrenaline also functions as a transmitter substance in the transmission of stimuli from sympathetic nerves to their effector organs. Adrenaline mobilizes the body's resources in an emergency—the "fight or flight" response to fear: the heart rate is increased, the blood supply to organs vital for coping with the emergency is increased by vasodilation (widening of the blood vessels), while the supply to other organs is reduced by vasoconstriction. In this way the adrenal medulla prepares the organs for physical effort and thus helps in the struggle for survival. More particularly, the medullar hormones act upon the relevant stimulating or inhibiting receptors of the visceral (smooth) muscle so as to produce the appropriate contraction or relaxation response to meet the situation or to perform certain functions. Thus, so-called *alpha receptors* have a stimulating function (vasoconstriction, contraction of the uterine muscles), whereas *beta receptors* have an inhibitory or relaxing effect (vasodilation, relaxation of the uterine muscles and bronchial tree, but stimulation of the heart muscle). Noradrenaline acts preferentially on alpha receptors and is thus chiefly a vasoconstrictor, acting in some respects in opposition to adrenaline.

The *adrenal cortex* is not directly connected with the nervous system. It secretes three different groups of hormones called *corticosteroids*. The first group comprises the *mineralocorticoids*; these regulate the sodium and potassium balance; the principal hormone in the group is *aldosterone*. The *glucocorticoids* form the second group, with *hydrocortisone* (cortisol) as the principal hormone, which promotes the storage of glycogen in the liver and the conversion of protein to carbohydrate and

FIG. 8-6.
Feedback control of adrenal cortex.

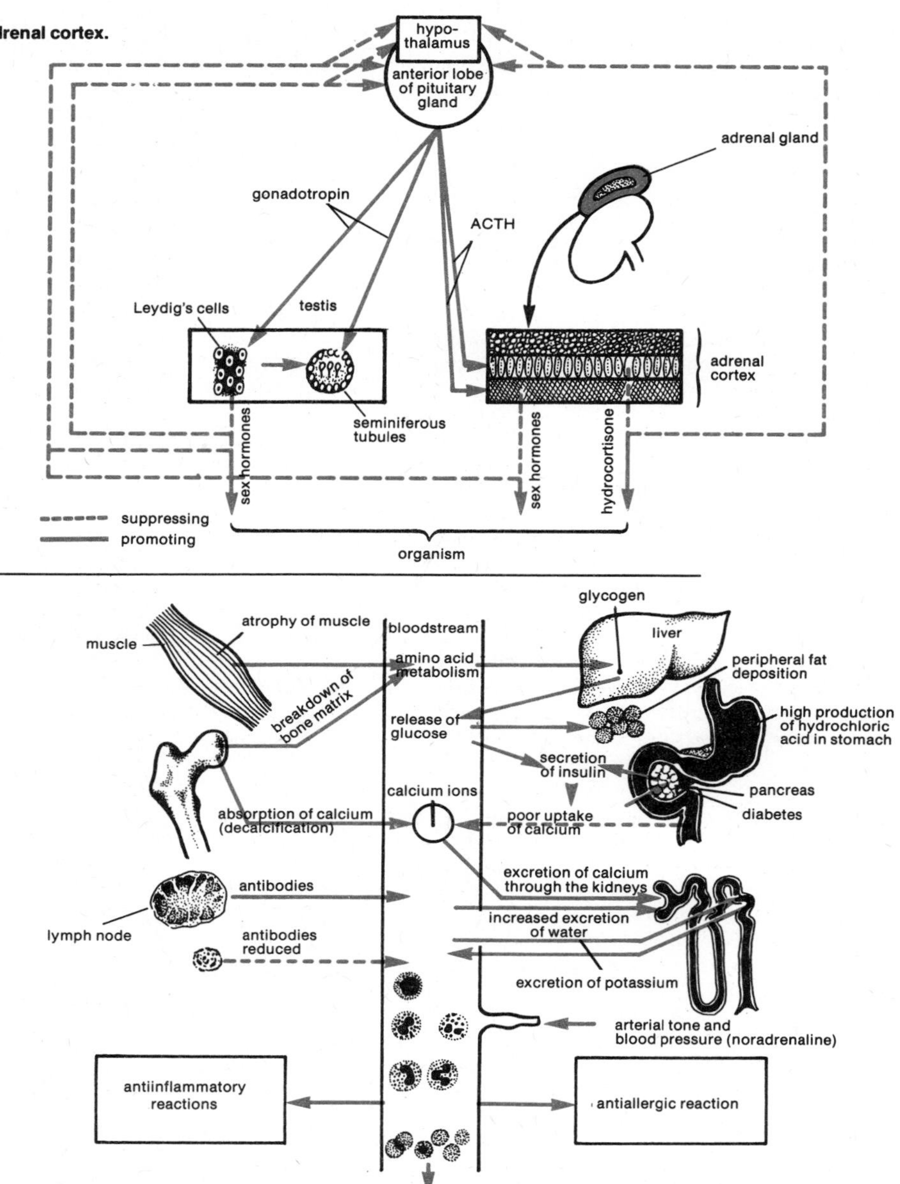

FIG. 8-7.
Action of adrenal cortical hormones and sex hormones.

which has an inhibitory effect on allergic phenomena and inflammation. The glucocorticoids are also of importance to the body's specific (immunological) defensive reactions. The third group are *sex hormones*, male and female, identical with those produced by the testes and ovaries, which they supplement. The hormones secreted by the adrenal cortex, especially hydrocortisone and aldosterone, are essential to life; their overall effect is to maintain the chemical stability of the body despite changes in environment, i.e., they enable the tissues to withstand various "stresses" due to such causes as injury, disease, severe exertion and mental strain.

The functional activity of the adrenal cortex is regulated by adrenocorticotropic hormone (ACTH) secreted by the anterior lobe of the pituitary gland. On the feedback principle already described in connection with the thyroid gland, secretion of ACTH is stimulated when the level of corticosteroids in the blood is low and is suppressed when this level becomes high. Stress causes an increase in hydrocortisone by stimulating the pituitary to release ACTH. The gonads (ovaries, testes) and the hypothalamus are associated with this feedback control (Fig. 8-6), in which gonadotropins (gonad-stimulating hormones) from the pituitary stimulate the gonads to produce sex hormones, which in turn have a feedback action on the pituitary. Thus there exists a complex control system comprising the adrenal cortex and the gonads, under the control of the pituitary gland. The latter is governed by the hypothalamus, linking the endocrine system to the autonomic nervous system. The secretion of aldosterone is not under the control of ACTH, but is regulated by the sodium, potassium and chloride content of the body fluids.

Chronic deficiency in the secretion of hormones by the adrenal cortex causes the rare disorder called *Addison's disease*. This may be due to progressive destruction of the cortex by a chronic infectious disease such as tuberculosis or to malfunction of the pituitary gland. In most cases, however, there is no ascertainable cause to explain why the cortex stops functioning. In general, Addison's disease is characterized by increased pigmentation of the skin (bronzing, very dark freckles, whitish patches) and a lowered blood sugar level, with tiredness, loss of appetite and loss of weight. Without treatment the disease runs a chronic course with progressive deterioration of the patient's condition, eventually resulting in death. It can, however, be effectively treated with hormone therapy. The symptoms are mainly due to deficiency of sugar in the blood and to disturbed balance of sodium and potassium through loss of salt by excretion through the kidneys. This imbalance is caused mainly by the absence of aldosterone. Loss of salt causes increased loss of water through the kidneys, so that the volume of blood is reduced, the circulation impaired and the heart weakened. With a lack of glucocorticoids the glucose formed in the blood from the breakdown of carbohydrates cannot be stored in the liver and is instead immediately burned. Also, the

FIG. 8-8.
Action of hydrocortisone in Cushing's syndrome.

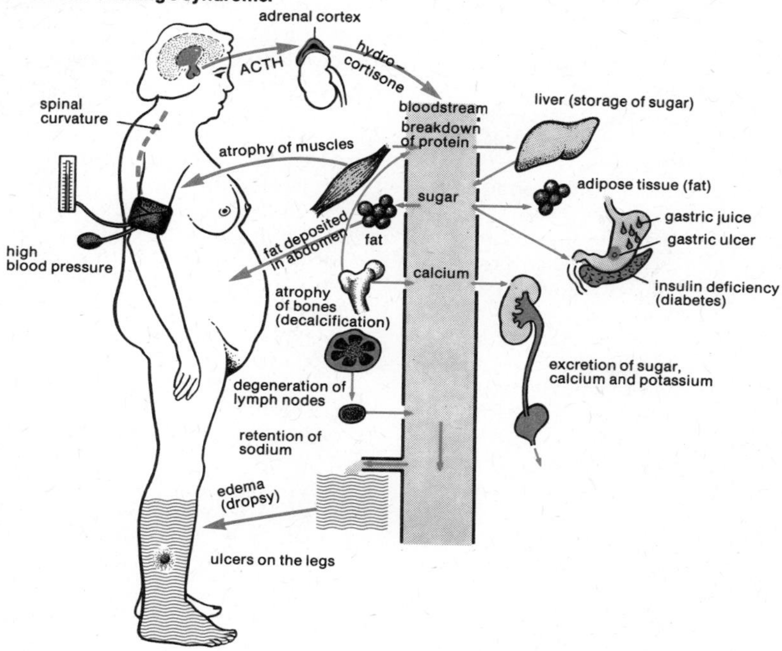

body's reserves of fat, which provide a source of energy, cannot be converted back into carbohydrates, and the body is unable to produce glucose from protein. As a result the blood sugar level, which is important for maintaining the vital functions, is lowered. This, in turn, manifests itself in muscular weakness and a tendency to tire easily. Also, deficiency of the adrenal cortex lowers the body's resistance to infectious diseases in general. In addition, lack of glucocorticoids removes inhibitions normally imposed on the pituitary gland, which now secretes larger amounts of ACTH. It is this that causes the skin pigmentation abnormalities associated with Addison's disease, especially the vary dark freckles on the face, neck and shoulders, with blackish discoloration of the mucous membranes of the mouth and the lips.

Hyperfunction of the adrenal cortex may be due to a tumor or to excess stimulation of that gland as a result of hyperfunction of the pituitary. Increased secretion of hydrocortisone is attended by *Cushing's syndrome* (Fig. 8-8), which is commoner in women than in men. Symptoms are adiposity, fatigue, weakness, loss of protein, edema, excess hair growth, diabetes mellitus (see page 179), skin discoloration, decalcification of the bones (especially in the spinal column) so that they fracture easily, high blood pressure, amenorrhea in women, impotence in men. Female patients tend to develop male secondary characteristics (virilism). Treatment may consist in radiation therapy of the pituitary gland; if there is a tumor of the adrenal gland, surgical removal of this gland may be indicated.

Overproduction of male hormones by the adrenal cortex causes the *andrenogenital syndrome,* which may be due to a tumor of the adrenal cortex or be congenital. In young children the congenital disorder results in stunted growth brought about by accelerated maturing of the bones without proportionate growth, while there is increased muscular development ("infant Hercules"). In girls it causes pseudohermaphroditism; in boys it causes precocious puberty (pseudopubertas praecox) associated with atrophy of the testes and absence of sperms in the seminal fluid. In adult women the androgenital syndrome manifests itself as adrenal virilism: the woman develops a masculine build and appearance, with increased growth of hair (hirsutism) (Fig. 6-130). If the condition is due to a tumor, treatment will consist in an operation to remove the adrenal gland.

PITUITARY GLAND

The pituitary gland (hypophysis cerebri), the "master gland" of the body, is a small rounded object, about the size of a large pea, attached to the hypothalamus (at the base of the brain) by the pituitary stalk (Fig. 7-28, page 343). It is the most complex of the endocrine organs and secretes a number of hormones which regulate many processes of the body, including growth, reproduction and various metabolic activities. Actually it comprises two distinct organs: the anterior lobe (adenohypophysis), which is a typical endocrine gland, and the posterior lobe (neurohypophysis), which is really a modified part of the nervous system.

The *anterior lobe* of the pituitary gland produces a whole range of hormones: (1) *growth hormone,* which stimulates skeletal growth and promotes synthesis of protein and other components of tissues; (2) *lactogenic hormone* (prolactin), which stimulates and maintains milk secretion after pregnancy; (3) *tropic hormones,* by means of which the pituitary controls other endocrine glands. This control is associated with a feedback mechanism whereby an increase in the concentration in the blood of the hormone of the controlled gland causes reduced production of the controlling tropic hormone by the pituitary. There are at least three types of tropic hormone: adrenocorticotropic hormone (ACTH), which stimulates the adrenal cortex (see page 355); thyrotropic hormone (TH, thyrotropin), which stimulates the thyroid gland (see page 350); gonadotropic hormones, comprising the follicle-stimulating hormone (FSH) and the luteinizing hormone (LH), both of which are important in regulating the sexual cycle in women; in men the luteinizing hormone (or an equivalent of this) is called the interstitial-cell-stimulating hormone (ICSH). The gonadotropic hormones stimulate the production of sex hormones and reproductive cells by the gonads (testes, ovaries).

Oversecretion of growth hormone in childhood causes *gigantism.* If it occurs in an adult, only the extremities gradually become enlarged, especially the hands, feet and face, and the resulting condition is known as *acromegaly,* which is further characterized by enlargement of the nose and lips, thickening of facial tissues, general muscular weakness and visual impairment. Treatment is by x-ray therapy or surgery of the pituitary gland. Deficient secretion of growth hormones causes *dwarfism;*

FIG. 8-9.
Pituitary hormonal activity.

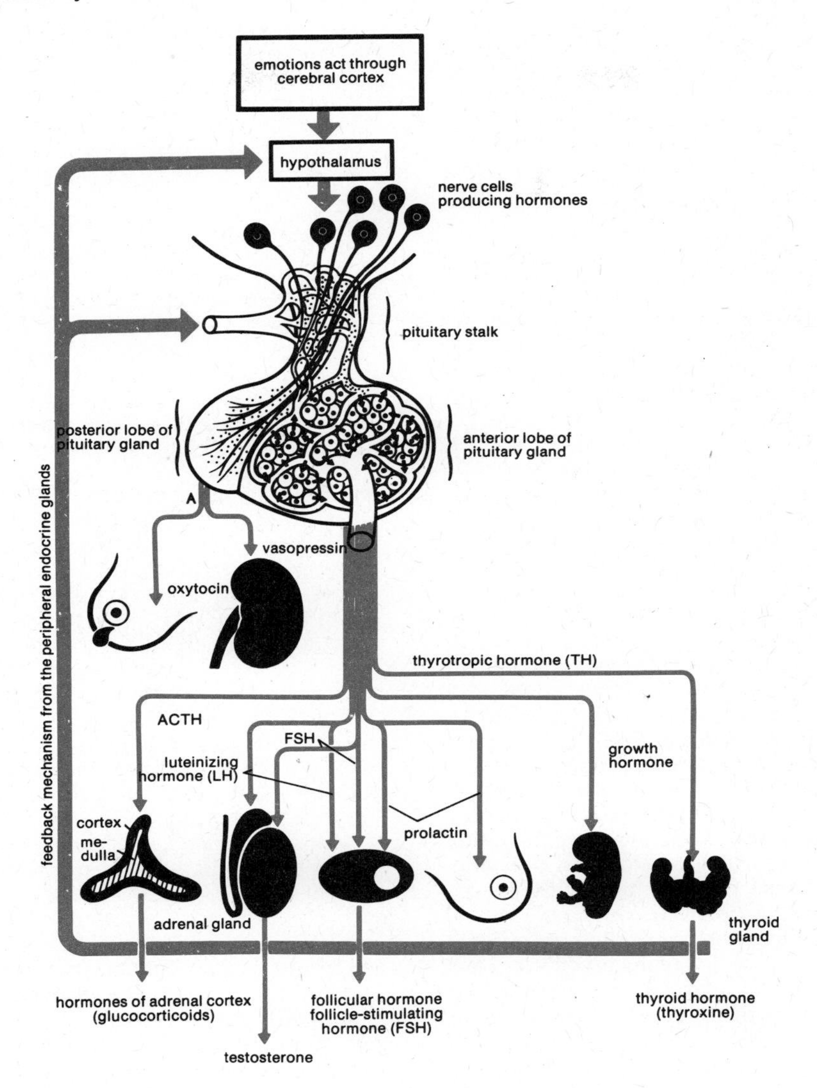

such persons are, however, well proportioned and of normal intelligence. *Simmonds' disease* is caused by complete failure of the pituitary gland and is characterized by premature senility.

The *posterior lobe* of the pituitary gland produces two hormones: (1) *antidiuretic hormone* (ADH, vasopressin), which controls loss of water from the body through the kidneys; (2) *oxytocic hormone* (oxytocin), which stimulates the muscles of the uterus to contract in labor and subsequently promotes uterine involution, stops bleeding from the site of the placenta, and stimulates ejection of milk from the breasts. Deficiency of ADH causes diabetes insipidus, a disease in which very large quantities of dilute urine containing no abnormal constituents are discharged.

Part Nine

MUSCLES AND BONES

MUSCLES

The muscles of the human body perform three main groups of functions: (1) locomotion and maintaining the erect posture; (2) contraction of thin-walled hollow organs (e.g., intestine, bladder, blood vessels); (3) circulation of the blood (pumping action of the heart). The innervation (nerve supply) and fine structure of the muscles correspondingly differ according to these respective functions. Locomotion is controlled by voluntary nerves (see page 327); the actual movements of the limbs are performed by striped skeletal muscles. Contraction of the hollow organs within the body is effected through the autonomic (involuntary) nervous system (see page 328). The blood is pumped through the circulatory system by cardiac muscle (see page 61), which is functionally and anatomically intermediate between skeletal and visceral muscle (Fig. 9-1). Despite many differences, these three types of muscle have one property in common: contractility, i.e., the ability to contract, which is achieved by conversion of chemical energy into the mechanical work done in shortening the muscle fibers. All muscle derives its energy from the nutrients supplied to it by the blood.

Striped (striated, skeletal, voluntary) muscle consists of large cylindrical cells marked with cross striations and containing myofibrils (long contractile chains of protein molecules) running the length of the cell. It is designed for strong contractions over relatively short periods of time, with considerable expenditure of energy. *Smooth* (plain, visceral, involuntary) muscle consists of small spindle-shaped cells containing myofibrils. This type of muscle performs comparatively slow contractions over long periods, with little expenditure of energy. *Cardiac* (heart) muscle fibers have striations, but contract rhythmically without any nervous impulses supplied to them from the nervous system. These fibers are interconnected in such a way that an impulse can spread from one fiber to another as well as along the length of a fiber.

FIG. 9–1. Types of muscle.

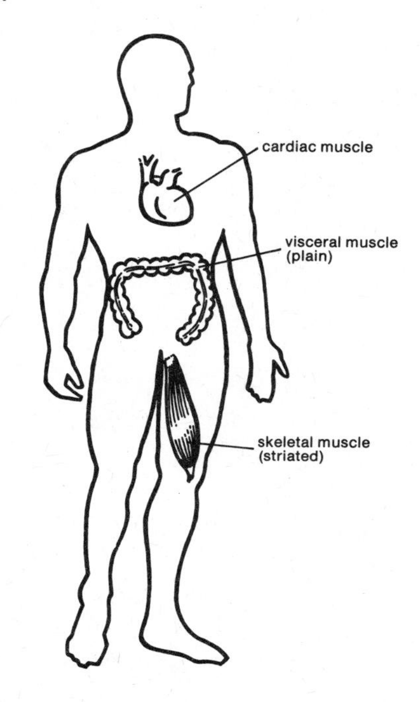

A muscle fiber contains two proteins—actin and myosin—which are present in the form of threadlike structures. When these proteins combine to form actimyosin, the fiber is shortened. The energy required for this chemical reaction is obtained from the conversion of adenosine triphosphate (ATP) to adenosine diphosphate (ADP). A nerve impulse arriving at a motor end-plate causes release of acetylcholine, which initiates this conversion—a rapid breakdown of high-energy ATP into ADP and phosphorus (P). ADP subsequently has to be converted back to ATP. (See Fig. 9-2.) The necessary energy is ultimately derived from the combustion of glucose with oxygen to form carbon dioxide. When oxygen is not immediately available in sufficient quantity, as

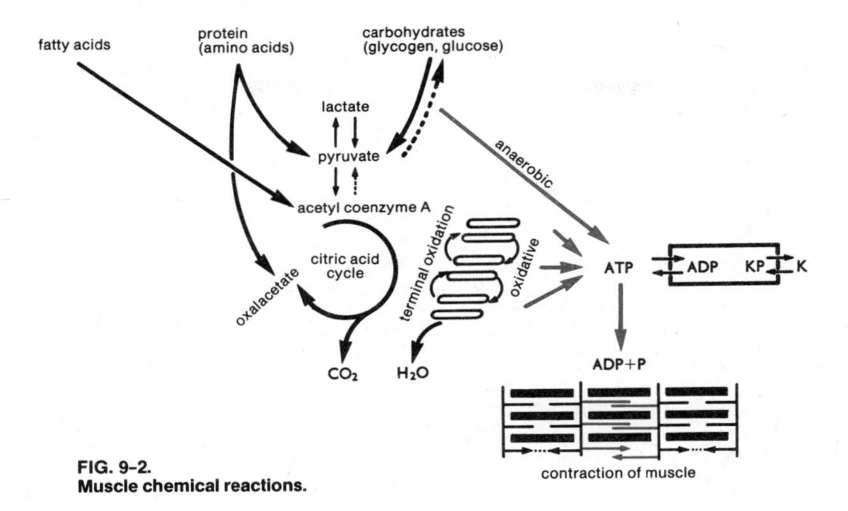

FIG. 9-2.
Muscle chemical reactions.

when a person holds his breath for a short time during muscular exertion, a certain amount of energy is temporarily obtained from the conversion of glucose to lactic acid, a reaction that does not require oxygen but cannot continue for any great length of time because accumulation of lactic acid in the muscle causes it to tire quickly. Oxygen is always required to complete the conversion to carbon dioxide, with further release of energy. The source of energy for the heart muscle is fat instead of glucose, so that variations in the glucose level in the blood cannot affect the action of that muscle.

The internal structure of *skeletal muscle,* i.e., striped muscle, is illustrated in Figure 9-3. Each diagram shows an enlarged detail of the preceding one. The muscle consists of elongated cells, usually referred to as fibers, which are 0.01 to 0.1 mm thick and 1 to 40 mm (exceptionally up to about 120 mm) long. Each muscle fiber is enclosed in a close-fitting membrane (sarcolemma). The cytoplasm of the muscle cell (sarcoplasm) is a semifluid substance through which 0.001 mm thick myofibrils extend longitudinally; these consist of long chains of protein molecules which are responsible for the contraction of the fiber. In cross section the skeletal muscle fibers are seen to contain several nuclei under the sarcolemma (Fig. 9-4a). Under the microscope each myofibril in such muscle is seen to consist of numerous short segments (sarcomeres), each of which is marked by light and dark transverse bands. These bands are exactly in line with one another in the adjacent myofibrils, so that the whole cell, i.e., the whole muscle fiber, appears transversely striped (Fig. 9-4b). The nature of these stripes or striations is shown in greater detail in Figure 9-3(e,f). The electron microscope reveals that the muscle contains two kinds of protein filaments. The thick ones, which are found only in the dark bands, consist of myosin; the thin ones consist of actin and are fixed to the transverse membrane or diaphragm that extends through the middle of each light-colored band. Depending on the functional state of the muscle at any given time, the thin filaments penetrate to a varying depth between the thick ones.

Each muscle fiber is reached by a branch of the motor nerve fiber serving a group of muscle fibers. There is a so-called motor end-plate at the point of contact between each nerve fiber branch and its associated muscle fiber. With its branches, a nerve fiber thus supplies anywhere from 5 to 200 muscle fibers, so that the latter receive their stimulating impulses simultaneously and thus all act together as a "motor unit." Muscles that have to perform differentiated and finely controlled movements consist of smaller motor units, i.e., comprising fewer muscle fibers per unit (as, for example, in the muscles of the eyes or the fingers), while those that only have to perform relatively coarse movements consist of large motor units, i.e., comprising large numbers of muscle

FIG. 9-3.
Skeletal muscle.

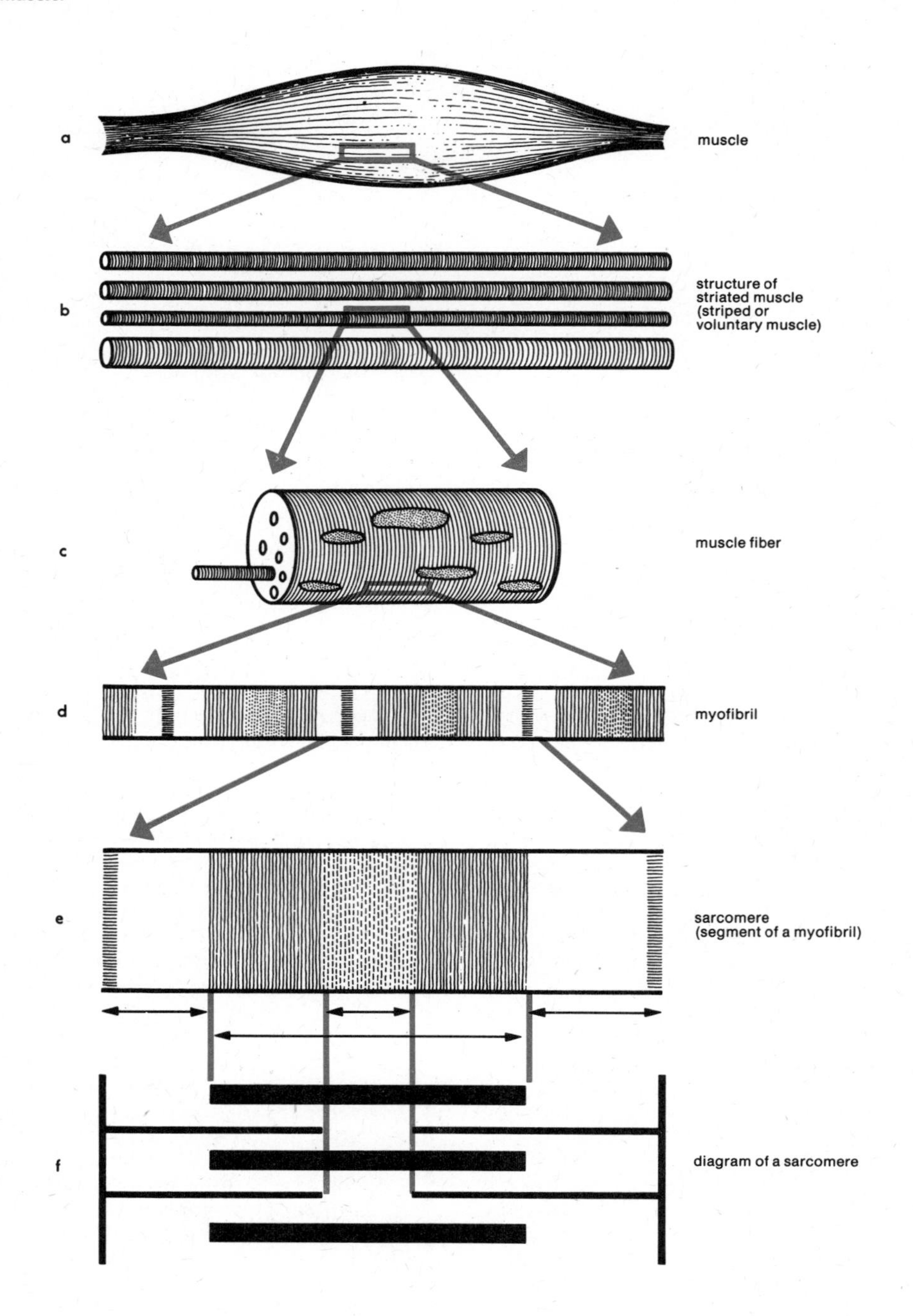

FIG. 9-4.

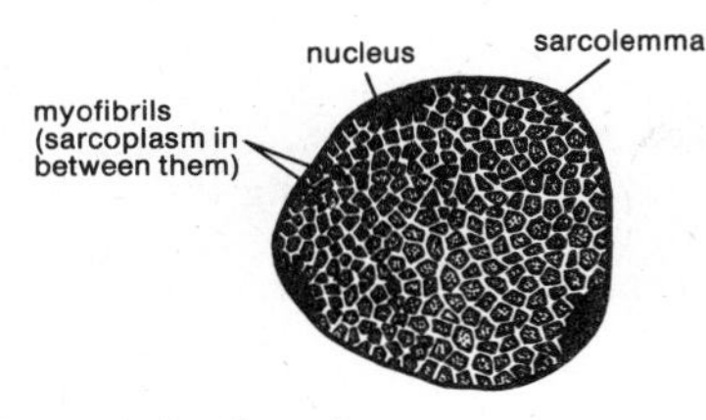

(a) Cross section through a muscle fiber.

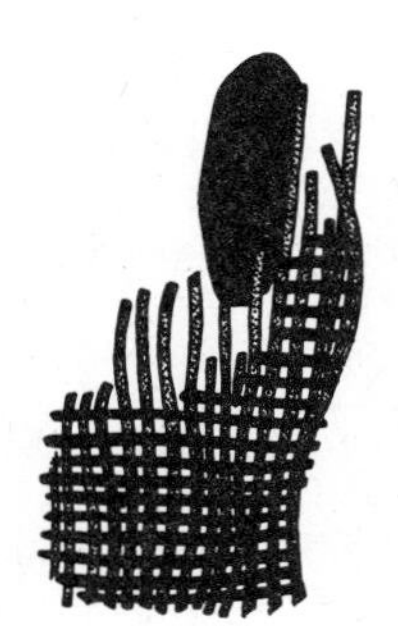

(b) Muscle fiber with nucleus (the myofibrils are shown pulled apart).

FIG. 9-5.
Synaptic transmission of a stimulus at the motor end-plate.

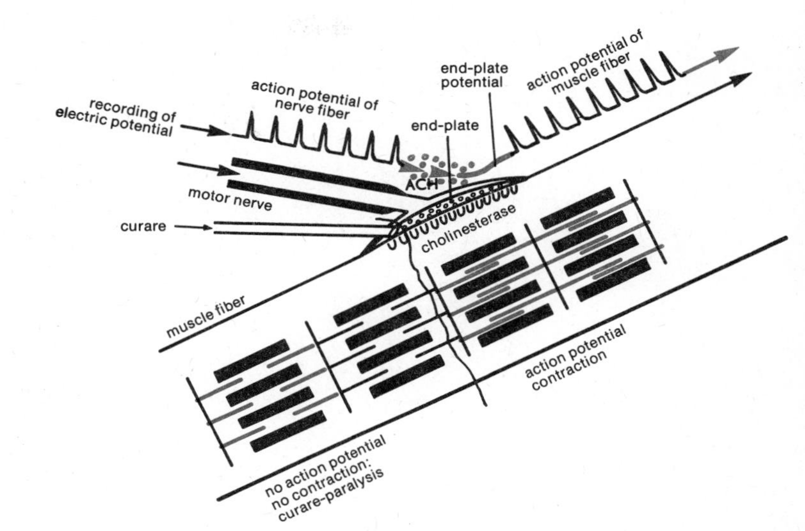

fibers stimulated by the branches of one motor nerve fiber (as exemplified by the extensor muscles of the back).

Contraction of a skeletal muscle is triggered by electrical stimuli (action potentials) transmitted to it by efferent (motor) nerve fibers. The motor end-plate constitutes the myoneural junction, i.e., the special synapse between nerve fiber and muscle fiber (Fig. 9-5). Transmission of the stimulus at the motor end-plate is accomplished by acetylcholine as the chemical transmitter, which is released from storage vesicles on arrival of a motor nerve impulse. At the subsynaptic membrane of the muscle fiber, this substance produces a stationary "local potential" (end-plate potential) which causes an action potential to be generated on the muscle fiber membrane. This new action potential, which is peculiar to the muscle fiber (as distinct from the action potential in the motor nerve fiber), travels as a signal along the muscle fiber and finally, by an indirect process, stimulates it to contract. The drug curare renders the subsynaptic membrane insensitive to acetylcholine and thus prevents the end-plate potential from developing: the muscle is thus paralyzed by a *neuromuscular block,* a term that in general denotes a disturbance in the transmission of impulses from motor end-plate to muscle. A muscle is never completely relaxed, in the sense of being "flabby," even when at rest. A slight tension (muscular tone) is maintained at all times as a result of continuous activity of the nerve cells. If some external influence tends to stretch a muscle, it automatically increases its tone to resist the tendency. When a muscle undergoes an increase of tone without shortening, this is known as *isometric* contraction; on the other hand, when a muscle shortens against a steady resistance, this is known as *isotonic* contraction. After each stimulation the acetylcholine is broken down to acetic acid and choline by an enzyme called cholinesterase.

The action potential generated in the muscle fiber at the motor end-plate brings about the contraction of the fiber by causing the contractile protein filaments to "telescope" together more deeply so that there is an overall shortening of the muscle. This *electromechanical coupling* (Fig. 9-6) comprises four successive stages: (1) the action potential travels along transverse outgrowths of the network in the fiber (sarcoplasmatic reticulum) and thus en-

FIG. 9-6.

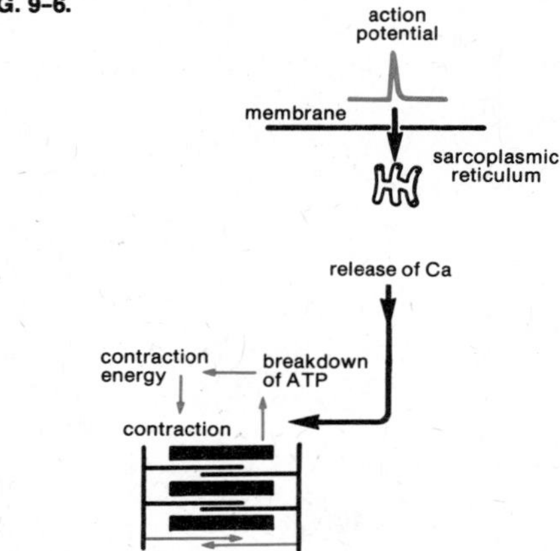

(a) Electromechanical coupling and contraction.

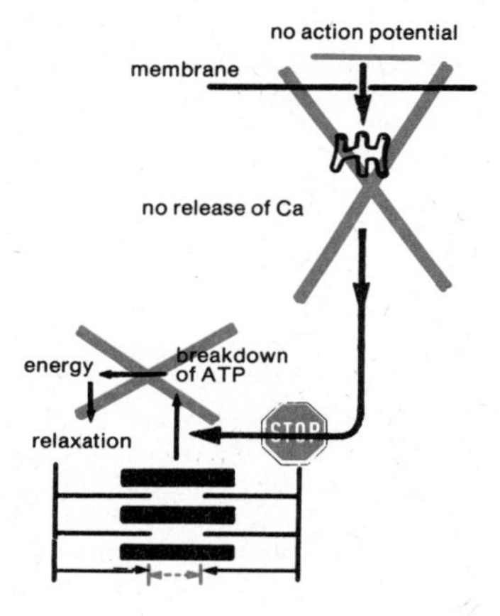

(b) Relaxation of muscle, resting state.

ters the interior of each muscle fiber; (2) the variation in electric potential causes calcium ions to be released from the longitudinal outgrowths of the reticulum; (3) the calcium ions activate an enzyme which breaks down ATP; (4) the energy liberated by this chemical breakdown of ATP induces the actin and myosin filaments to slide telescopically together (Fig. 9-7).

When the action potential has passed, calcium is put back into the reticulum, and the breakdown of ATP is reversed, so that resynthesis of ATP now occurs (electrochemical decoupling, see Figure 9-6b). The bonds that draw the contractile filaments together are released, so that the latter can now "detelescope," allowing the muscle to relax to its resting position. An interesting feature is that ATP breakdown is catalyzed by the contractile proteins themselves (which possess also an enzymelike character). In this way very favorable conditions are created for the transmission and conversion of the chemical and mechanical energy released.

Single stimuli of the muscle cause individual contractions (Fig. 9-9a), which can be recorded by means of an apparatus called a *kymograph* (Fig. 9-8). Each peak appearing on the chart corresponds to a contraction wave which travels along the muscle fiber—and along the muscle as a whole when an entire motor nerve is stimulated. If a second stimulus is applied before the first contraction wave has had time to die away and before the muscle has fully relaxed, the contractions become cumulatively superimposed, this effect being known as summation (Fig. 9-9b–d). If the individual contractions are still separately distinguishable, summation is said to be incomplete. If a large number of stimuli occur in rapid succession, a corresponding number of contraction waves travel along the muscle; in such circumstances complete summation may

FIG. 9-7.
When the muscle contracts, the actin and myosin filaments slide telescopically together without themselves undergoing any appreciable change in length.

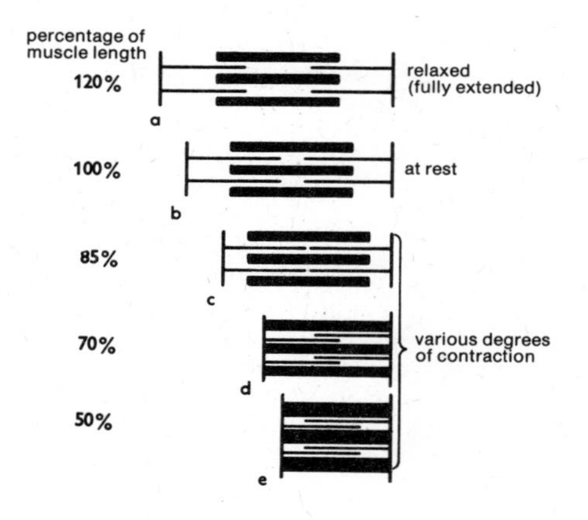

FIG. 9-8.
Kymograph.

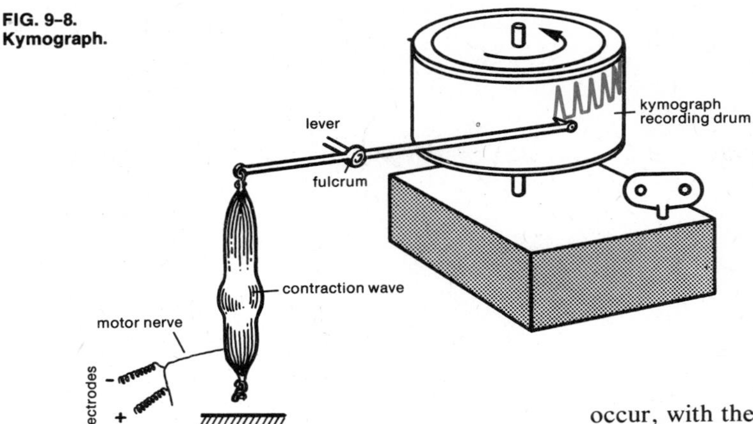

FIG. 9-9.
Summation of contractions.

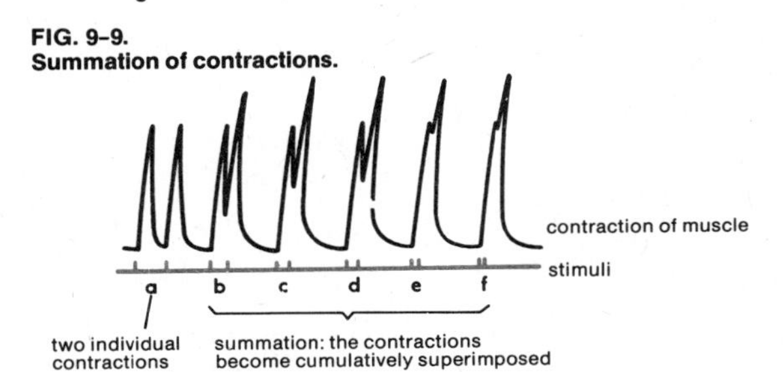

FIG. 9-10.
State of tetanus.

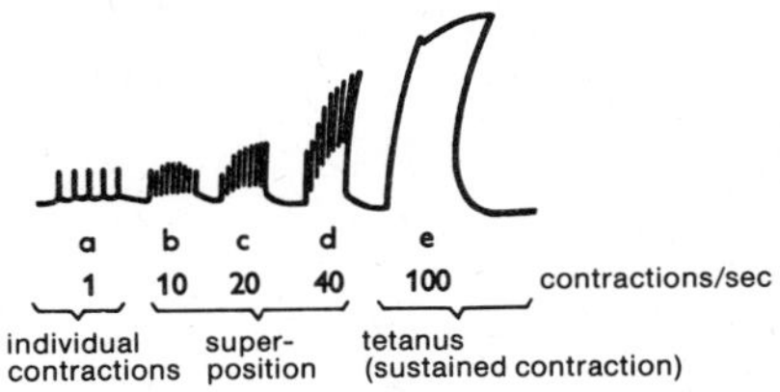

a b c d e
1 10 20 40 100 contractions/sec
individual contractions
super-position
tetanus (sustained contraction)

FIG. 9-11.
Muscle attachments.

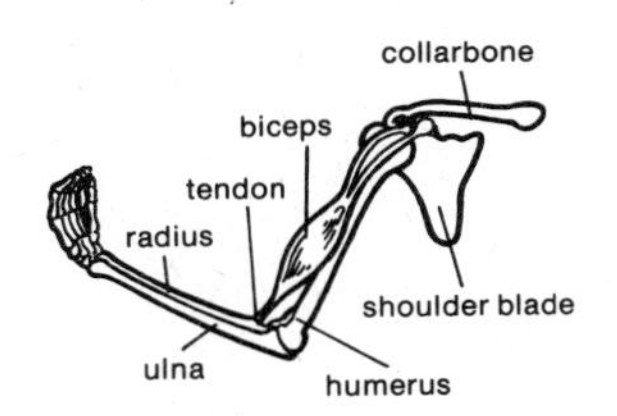

occur, with the result that a state of sustained contraction (tetanus) develops (Fig. 9-9e,f). Figure 9-10 shows in more detail how, with increasing frequency of the stimuli, the individual contractions merge with mounting intensity into a state of tetanus. In human muscle this develops when the stimuli occur at a frequency of between 50 and 350 per second. Any further increase in frequency of stimulation has no effect, because during and for a short time after each stimulus the muscle is *refractory,* i.e., resistant to stimulation. The intensity of contraction of a muscle fiber can be controlled through the frequency of stimulation (the number of contraction waves that are sent along the muscle fiber per second); besides, in the muscle as a whole the number of fibers simultaneously stimulated can progressively increase. This phenomenon is known as *recruitment:* the condition in which the response increases to a maximum when a stimulus is prolonged, even though its strength remains unchanged; it is due to the activation of increasingly large numbers of motor neurons.

Skeletal muscles are attached to the bones by means of *tendons,* consisting of fibrous connective tissue (Fig. 9-11). Each muscle fiber is surrounded by a sheath; fibers are bound together in bundles, and a number of bundles form the muscle, which is also surrounded by a sheath (epimysium).

Muscles that contract to produce a given movement are called *prime movers;* the muscles that produce the opposite movement are called *antagonists.* A *flexor* is a muscle that bends a part; an *extensor* is one that extends a part. The flexors in the upper arm, for example, are the prime movers in bending at the elbow joint; the antagonists in this case are the extensors in the upper arm. While some mus-

FIG. 9-12.
Action potential of a skeletal (striped) muscle compared with the action potential of cardiac muscle.

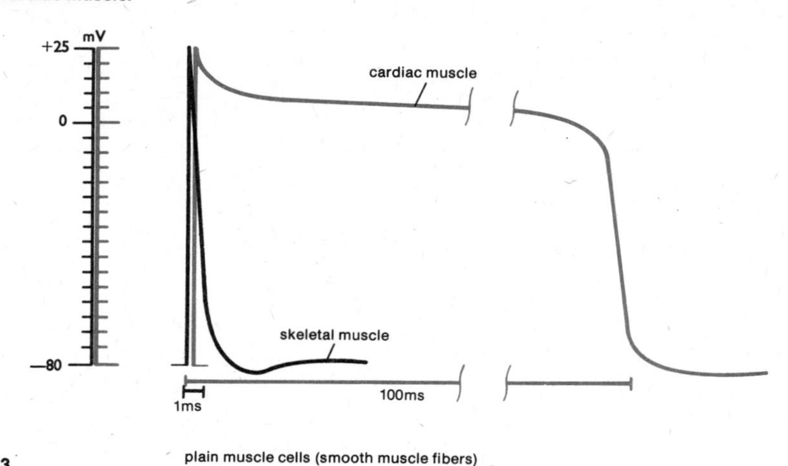

FIG. 9-13.
Smooth muscle.

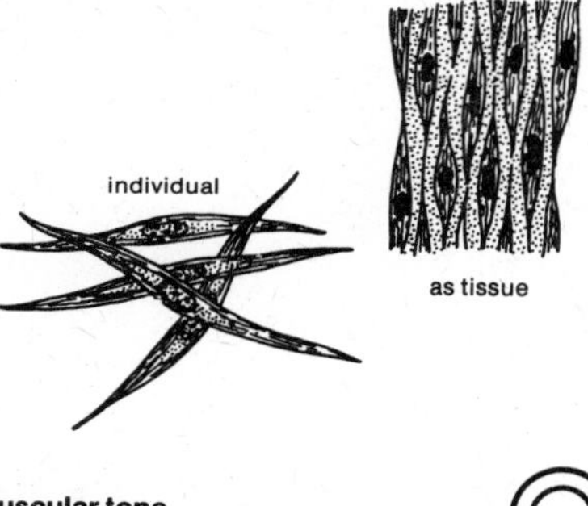

FIG. 9-14.
With varying degree of filling of the heart its muscular tone is varied so as to maintain the internal pressure constant.

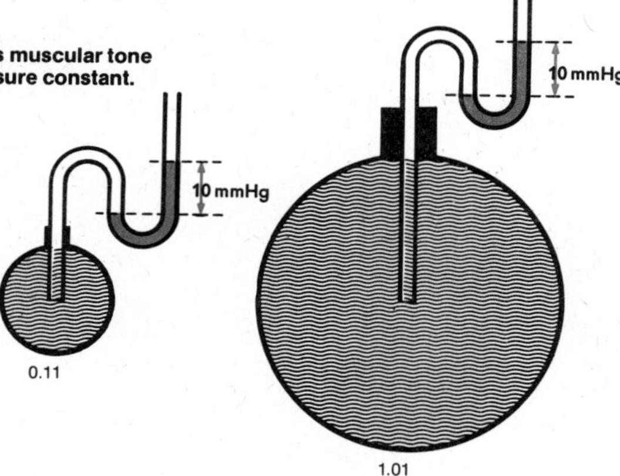

cles act as prime movers, others have the function of holding the skeleton steady (fixation muscles).

The *cardiac muscle* is striped like skeletal muscle; its fibers contain fibrils consisting of actin and myosin. The contraction mechanism of this muscle is no different in principle from that of skeletal muscles. There are nevertheless some important distinctions:

1. The cardiac muscle fibers are branched and interconnected, so that action potentials in this muscle are propagated through more and more fibers, until the stimulus extends

to whole regions of the muscle and eventually involves the heart as a whole.

2. The heart has the ability to generate its own action potentials, without external stimuli; this ability is localized particularly in the sinoatrial node, which functions as the "pacemaker."
3. The heartbeat is thus not stimulated by motor nerve fibers from outside—neither by voluntary nor by autonomic nerves. There are autonomic fibers (sympathetic and parasympathetic) extending to the heart but they only modify the action of the cardiac muscle (promoting or inhibiting).
4. The action potentials in the cardiac muscle fibers are of much longer duration than those in ordinary striped muscle—100 milliseconds as against 1 millisecond (Fig. 9-12). The refractory period of the cardiac muscle is correspondingly longer. This is important for the synchronization and coordination of the many thousands of cardiac muscle fibers into one powerful heartbeat. Because of the long refractory period, the risk of extrasystole (premature contraction of the heart due to some extraneous stimulus) is largely obviated.

The *smooth muscle* of the internal organs (viscera) is less highly specialized than skeletal muscle. It consists of small spindle-shaped cells with centrally placed elongated nuclei (Fig. 9-13), not the large cylindrical cells (fibers) found in striped muscle. The cells of smooth muscle are often interconnected to form a so-called syncytium, an assembly of cells in which there is continuity of protoplasm between adjoining cells. This makes it unnecessary for each and every cell to be supplied with stimuli from its own individual branch of a nerve fiber. Instead, the motor nerve fibers to the viscera are connected only to some of the smooth muscle fibers. Another characteristic of smooth-muscled organs is that they (like the heart) often contain "pacemaker" cells which initiate rhythmic contractions or contraction waves, while the autonomic nervous system merely performs an overall modifying function. Smooth muscle possesses a particular mechanical property, namely, *plasticity*. This enables hollow organs with smooth-muscled walls, such as the bladder, to vary their muscular tone so that they can adapt themselves to changes in the volume of their contents while maintaining the internal pressure at a constant value (Fig. 9-14). Another characteristic of smooth muscle is its relatively low energy requirement, correspondingly low fatigability and long contraction period.

Visceral (smooth) muscles are arranged in bundles that encircle the hollow organs and others that run longitudinally: in the blood vessels, the respiratory tract, the alimentary tract, and the urogenital system. In addition, there is smooth muscle in the interior of the eye, in the skin and in the outlet ducts of glands. The activity of smooth muscle, particularly in the walls of hollow internal organs, is called *peristalsis:* the encircling muscle fibers squeeze while the longitudinal ones shorten a segment of the organ, pushing its contents forward to the next section, which is relaxed to receive them. Waves of contraction travel along the organ. Smooth muscle usually has double innervation: one set of nerves causes contraction and the other promotes relaxation. Striped muscle has only one set of motor nerves, and relaxation is merely the absence of contraction. Smooth muscle is able to perform "tonic" actions through the agency of two fundamentally different mechanisms: contractile tone, involving consumption of energy, and plastic tone, which is an energy-saving and more or less passive condition.

Certain orifices of the body, such as the anus, the outlet of the bladder, the opening of the esophagus into the stomach, the pyloric opening of the stomach into the duodenum, and the opening of the ileum into the rectum, are closed by sphincters. A sphincter is a circular muscle constricting an orifice; in normal contraction it keeps the orifice closed. The sphincters of the alimentary tract, for example, are under the control of the autonomic nervous system: sympathetic impulses cause them to contract, parasympathetic to relax.

Electrophysiologically the smooth muscle cell is characterized by a low and unstable membrane potential of, on average, only about 50 millivolts, as compared with the 70–80 millivolts in striped muscle. There are, however, considerable rhythmic and spontaneous variations in this potential, so that quite often merely a slight additional depolarization will suffice to stimulate rhythmic contraction of the muscle. The additional depolarization may be initiated by transmitter substances and be facilitated by hormones such as adrenaline, noradrenaline, acetylcholine and the sex hormones. Similar "humoral" factors may, alternatively, have the effect of making the response of the muscle more difficult to initiate. Characteristic and especially important is the depolarization and contraction of smooth muscle in response to extension, as exemplified in the intestine: stretching of its wall as a result of increased filling initiates contraction of the muscle, causing the intestine to be emptied (autonomic contraction brought about by pressure).

SKELETAL SYSTEM

The bones interconnected by joints and attached to muscles form the apparatus of locomotion, enabling the individual to move from one place to another. In addition, certain parts of the skeleton serve to protect and support delicate organs. The skeletal system is highly adaptable and possesses considerable powers of regeneration (healing of bone fractures). It comprises the bones of the head and trunk and the bones of the limbs. In youth the buildup of bone predominates over breakdown, so that bone growth occurs. In adult life the two processes are in equilibrium. Finally, in old age the process of breakdown gradually outstrips growth. Bone growth is under the control of many influencing factors, the most important of which is the growth hormone secreted by the anterior lobe of the pituitary gland. The secretion of this hormone is, in turn, regulated by the sex hormones and thyroid hormones. The adrenal glands, the parathyroid glands and vitamin D also play a part in bone growth control (Fig. 9-15).

FIG. 9-15.
Bone growth control factors.

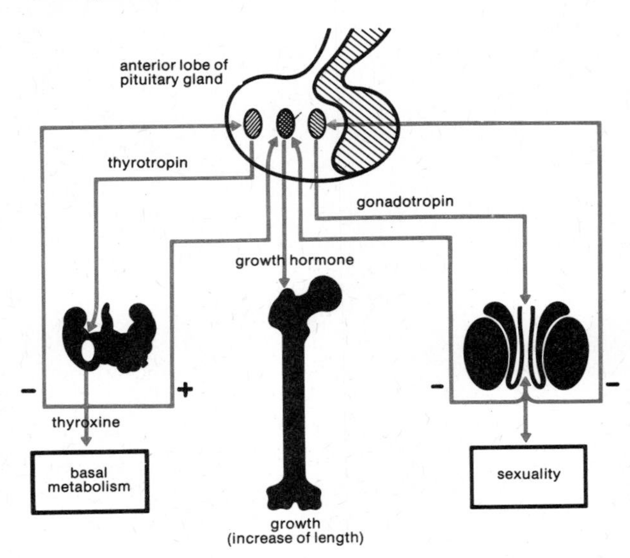

Cartilage (gristle) consists of cells (chondrocytes) separated by an amorphous intercellular substance in which fibers are embedded. In general, cartilage is a soft and resilient material that serves to support the softer tissues. Hyaline cartilage, which has a translucent appearance, covers the articular surfaces of bones. Fibrous cartilage (fibrocartilage), which is found in the intraarticular discs and at the attachments of tendons to bone, contains larger numbers of collagen fibers and possesses greater mechanical strength. Collagen is a fibrous insoluble protein found in connective tissue, including skin, bone, ligaments and cartilage.

Bone is similar to cartilage in that it consists of cells (osteocytes) surrounded by a large amount of intercellular substance. An important difference is that bone contains numerous blood vessels, whereas cartilage does not. The osteocytes lie in spaces called lacunae which are interconnected by tiny channels (canaliculi) in the intercellular substance; tissue fluid seeps through them to the osteocytes. Bone consists of about 50 percent solid matter and 50 percent water. The solids are chiefly cartilage hardened by impregnation with inorganic salts, especially calcium carbonate and calcium phosphate. The proportion of calcium salts in bone gradually increases with age, so that in old people the bones become brittle and break easily. The intercellular substance consists mainly of collagen in which the inorganic salts are deposited. Bone is surrounded by a fibrous membrane (periosteum) except at articular surfaces; this membrane consists of a dense external layer containing numerous blood vessels and an inner layer of connective tissue cells (osteoblasts). There are two types of bones: compact bone (cortical bone), which is traversed only by very fine channels (haversian canals), and spongy bone (cancellous bone), which consists of slender beams and plates (trabeculae) partly enclosing numerous intercommunicating spaces filled with bone marrow.

The skeleton consists of 206 bones. They are of four main types: long, short (e.g., bones of the wrist joint and ankle), flat and irregular (e.g., vertebrae, certain bones of the skull) bones. A *long bone* (Fig. 9-16) has a cylindrical shaft, enlarged at the ends for articulation and for attachment of muscles. The shaft is hollow, consisting of compact bone enclosing a cavity

FIG. 9-16.
Long bone.

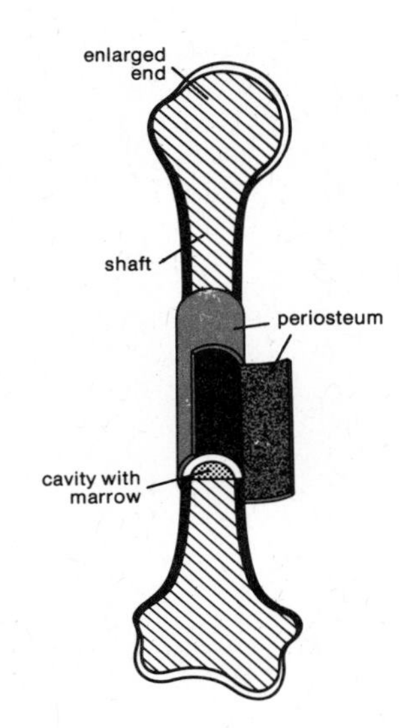

FIG. 9-17.

(a) Stress trajectories of bone.

(b) Altered trajectories in fracture healing.

(c) Continuity of trajectories in adjacent bones.

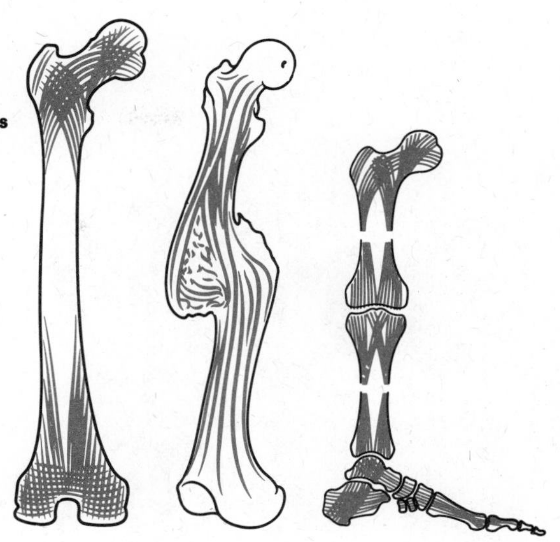

filled with yellow marrow. The enlarged ends have a covering of compact bone enclosing spongy bone containing red marrow. The femur (thighbone) is an example of a long bone. The criterion is the shape of the bone, however, not its actual length. Thus even quite small bones, as in the hand, rate as "long" bones. A *flat bone* is composed of two plates of compact bone with a layer of spongy bone, containing red marrow, sandwiched between. Examples are the bones of the top of the skull, also the shoulder blade (scapula) and the ribs. Red marrow produces red blood cells, platelets and granular leukocytes. The spongy bone in the interior of particularly the long bones forms a system of interpenetrating trabeculae following the principal lines of compression and tension within the bone structure and corresponding to so-called compressive and tensile stress trajectories, a concept familiar to structural engineers (Fig. 9-17a). In this way the bone is efficiently equipped to transmit forces and carry loads with maximum economy of material. If the bones were all completely solid, the skeleton would be impossibly heavy and yet not possess very much more structural strength than it actually does. When a bone fracture heals, the trajectories are restored and rearranged to make the best of the altered situation (Fig. 9-17b). The trajectories are not confined to individual bones, but continue in adjacent ones (Fig. 9-17c). Figure 9-18 shows a section through the shaft of a long bone.

Bone development takes place in two ways: by *intramembranous ossification* (Fig. 9-19), i.e., the formation of bone in or underneath a fibrous membrane, as in the formation of the cranial bones, or by *enchondral ossification* (Fig. 9-20), i.e., the formation of bone in cartilage, as in the formation of long bones. In both of these processes the bone develops and grows by the activity of osteoblasts, which are special cells in the inner layer of the periosteum and at the ends of long bones. These cells produce an enzyme that causes insoluble calcium salts to be precipitated from the soluble salts in the blood; this starts at the so-called center of ossification. Once the osteoblasts have surrounded themselves with their special intercellular substance (collagen impregnated with calcium salts), they turn into osteocytes within bony lacunae. In enchondral ossification a model, made of hyaline cartilage, of the future long bone is formed. The periosteum enclosing it then produces (by means of the osteoblasts in its inner layer) a sleeve of compact bone round the middle of the shaft, and capillaries and osteoblasts grow inward from the periosteum into the interior of the cartilage model to form bone there. From the primary center of ossification, where this bone-forming process begins, it extends toward the two ends of the future long bone, as yet consisting of

FIG. 9-18.
Section through shaft of long bone.

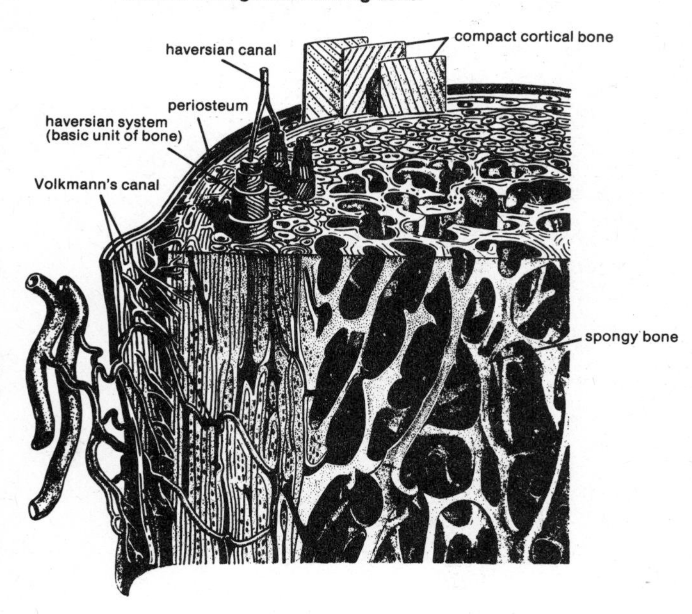

cartilage. At a later stage, blood vessels and osteoblasts also penetrate into the cartilaginous ends of the developing bone, where they set up a secondary center of ossifaction (epiphysis). On the end surfaces of the bone the cartilage persists as articular cartilage. The cartilage remaining between the epiphysis and the shaft of the bone forms a disc (epiphyseal cartilage) which continues to grow and thus makes bone growth possible. Eventually this cartilage also ossifies, and bone growth stops. Normally this occurs on reaching adulthood. Although a long bone has an epiphysis at each end, one appears earlier than the other, and the bone grows more at this end (the "growing end"). During the course of its growth a bone has to be remodeled.

In addition to the formation of new bone, a certain amount of old bone has to be removed by a process called *resorption*. This is done by cells called osteoclasts. Bone is a living tissue in which resorption of intercellular substance takes place continuously and is equaled by the activity of osteoblasts producing new bone (Figs. 9-21 and 9-22).

FIG. 9-19.
Intramembranous ossification.

FIG. 9-20.
Enchondral ossification.

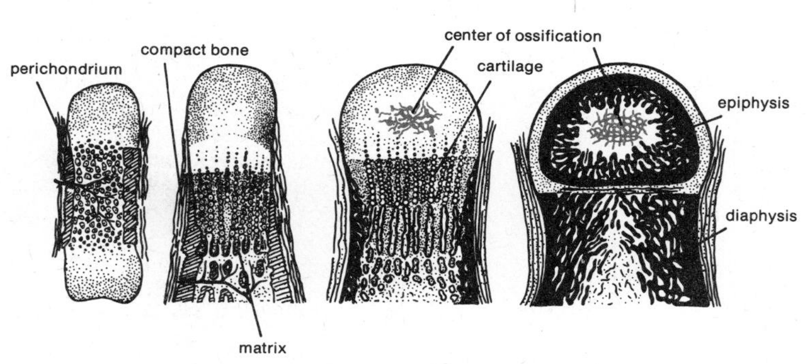

FIG. 9-21.
Bone formation.

FIG. 9-22.
Bone resorption.

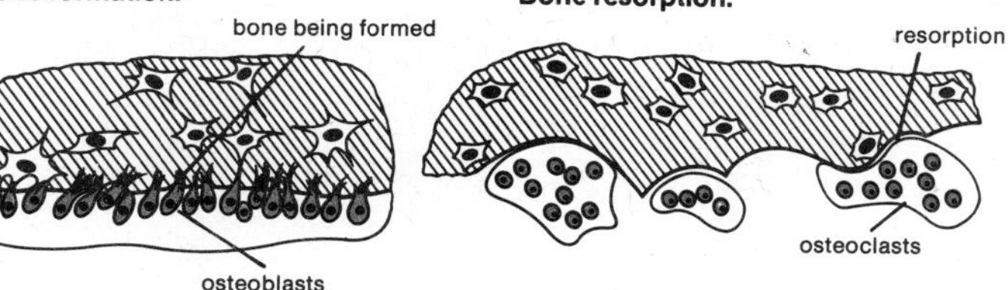

FRACTURES

Bones that are weakened or damaged by disease (e.g., osteomalacia, syphilis, osteomyelitis) fracture relatively easily under the action even of normal loads and forces (spontaneous or pathological fracture). Otherwise a substantial degree of violence by application of an external force is generally needed to cause fracture. In certain circumstances a sudden violent contraction of muscles may also cause fracture.

If a bone is broken but there is no external wound, it is a case of *simple* (or closed) fracture. In a *compound* (or open) fracture the bone is broken and there is an external wound extending to the site of the fracture. A *complicated* fracture is one where the broken ends of the bone have injured some internal organ, e.g., a nerve, an artery or a lung (by a broken rib). In a *comminuted* fracture the bone is splintered or broken into pieces. As a rule, a bone breaks completely, but sometimes there is incomplete fracture in which the fracture does not extend through the whole bone. So-called greenstick fracture may occur in young bones (in children): the bone is then partly bent and partly broken. Fractures may be caused by various mechnical actions: flexure (bending), twisting, compression or shearing (Fig. 9-23). External violence and/or the pull exerted by muscles may cause displacement of the broken ends of the bone in relation to each other (Figs. 9-24, 9-25a and 9-25b).

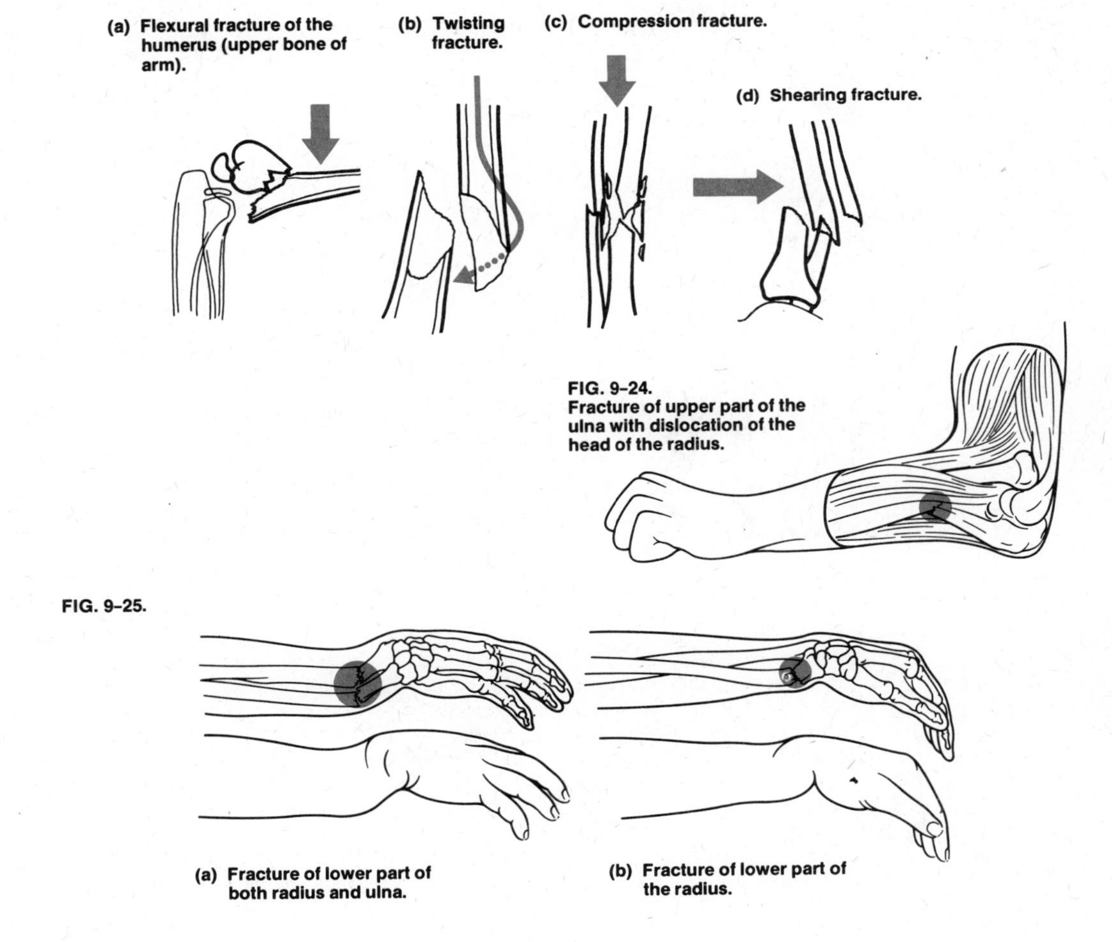

FIG. 9–23.
Various types of fracture of long bones:

(a) Flexural fracture of the humerus (upper bone of arm).
(b) Twisting fracture.
(c) Compression fracture.
(d) Shearing fracture.

FIG. 9–24.
Fracture of upper part of the ulna with dislocation of the head of the radius.

FIG. 9–25.

(a) Fracture of lower part of both radius and ulna.
(b) Fracture of lower part of the radius.

FIG. 9-26.
Continuous traction applied to fractured femur (thighbone).

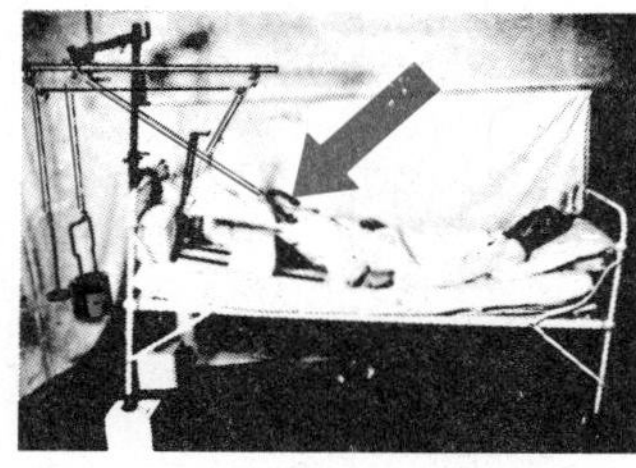

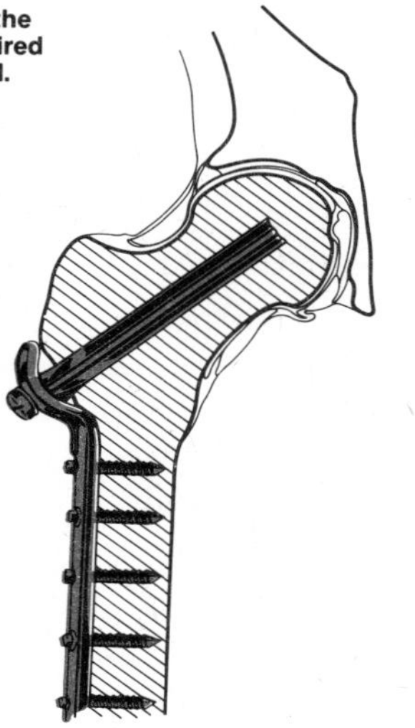

FIG. 9-27.
Fracture of the neck of the femur (thighbone) repaired by means of a metal nail.

Signs of fracture are loss of power of movement, with pain, acute tenderness over the site of fracture, swelling and bruising. There may be deformity and shortening of a limb. Other outward signs may include unnatural mobility of parts of bones, and a grating sound may be heard when the broken ends rub together. The type of fracture and its precise position can be determined from x-rays.

To treat a fracture, it first has to be *reduced,* i.e., the broken ends have to be restored to their proper position (setting the bone). As a rule, this is done under anesthetic. The bone must then be held immovably in position by means of splints until union has taken place and the fracture has healed. Under these conditions normal bone will heal like any other tissue. If the fragments of a fractured bone grow in a faulty position, forming an imperfect union, so-called malunion will result.

When they have been properly reduced and are then kept immovable, e.g., in a plaster cast, most fractures will remain in position and heal satisfactorily. In some bones, however, notably the thighbone, an external cast cannot hold the reduced fracture sufficiently immovable; in such cases continuous traction (Fig. 9-26) has to be applied. For this purpose a system of weights and pulleys is employed, and the traction is applied to the limb or directly to the fractured bone through a wire passed through its enlarged lower end. By this method it is nearly always possible to achieve satisfactory union and prevent any shortening or deformity of the limb.

Surgical repair of fractures may be necessary in cases where natural healing is not possible. There are various techniques involving the use of metal nails, wires, screws or plates in the bones (Fig. 9-27). In one such method a steel pin is inserted into the marrow cavity of each part of the bone so as to form a dowel across the fracture and hold the two parts firmly together.

In general, first aid for fractures should consist in keeping the patient warm and quiet, lying down, awaiting the arrival of a doctor. The patient should not be transported, unless this is absolutely necessary, until splints have been applied. An emergency splint can be made from any long rigid object, which should be padded and tied fairly loosely so as not to obstruct blood circulation.

AMPUTATION

In modern surgical practice, amputation is generally a last resort, to be used only if a diseased or injured limb or other part cannot be saved. In some cases amputation is the only way to save the patient's life, e.g., where there are extensive open wounds with crushing of bones, blood vessels and nerves, as frequently occurs in road accidents and industrial accidents. If the wound is severely infected and inflamed, so that general blood poisoning is likely to develop, amputation may be necessary as a lifesaving intervention. It will also be necessary when parts of limbs develop gangrene as a result of deficient blood supply. If limbs are affected by malignant tumors of the bone, periosteum or muscle tissue, amputation of the limb will be meaningful only if the tumor has not already spread to other parts of the body.

In an emergency after an accident, or in

FIG. 9-28.
Drawings showing those parts of the bones of the arm that are essential, nonessential, or (in certain cases) troublesome when part of the arm or leg has to be amputated.

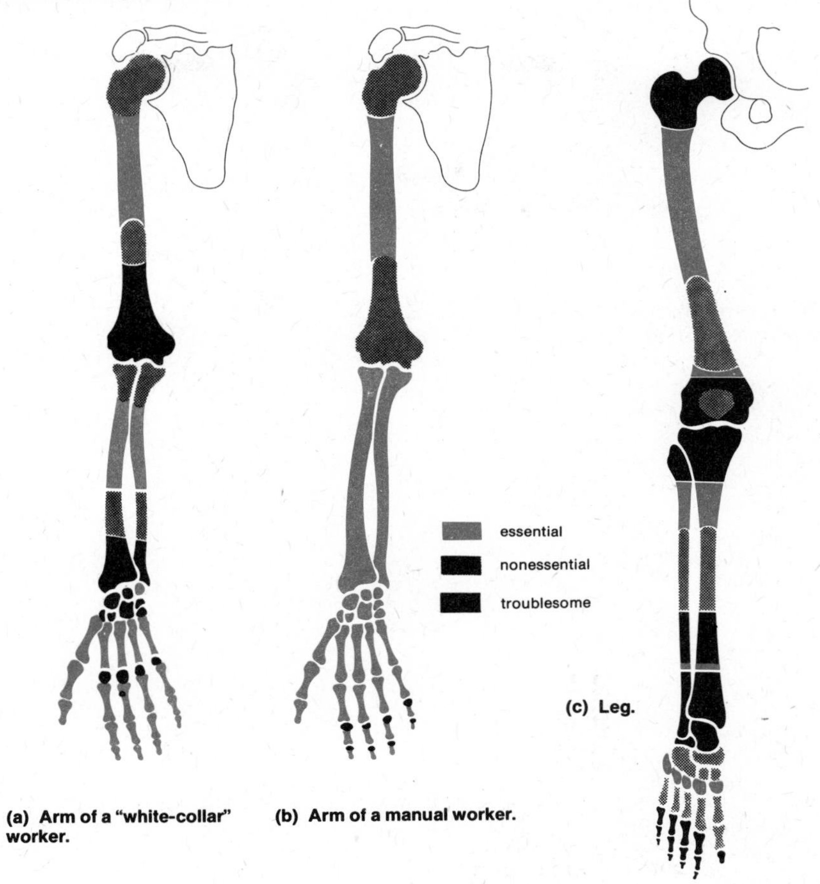

(a) Arm of a "white-collar" worker. **(b) Arm of a manual worker.** **(c) Leg.**

order to deal with a dangerously inflamed wound, a provisional amputation will first be performed. This must always be done in healthy tissue, otherwise the stump will not heal. When the patient has fully recovered and the wound has healed, the final amputation will be done in such a way that the remaining part of the limb remains as functionally serviceable as possible to enable the patient to derive maximum benefit from an artificial limb. The most "conservative" amputation is not necessarily the best solution, for a stump that is too long may be a disadvantage to the proper functioning of the articulations of the artificial limb. By way of example, in Figure 9-28 those parts of the bones of the arm are indicated which—depending on the patient's occupation or profession—are essential, nonessential or troublesome, respectively. If two thirds of the forearm is preserved, the two bones (radius and ulna) can be modified into a tonglike gripping device.

The general procedure for the amputation of a limb is as follows: First, the patient is anesthetized. Then the flow of blood above the amputation is arrested by means of a rubber bandage, so that there is no blood at the site of the operation. Next, the soft tissues—skin, muscles, sinews—are severed in such a way that they can later be pulled over the bone stump to cover its end and be stitched in position; so always more skin and tissues must be preserved than bone (Fig. 9-29). For this purpose an amputation flap has to be formed by cutting

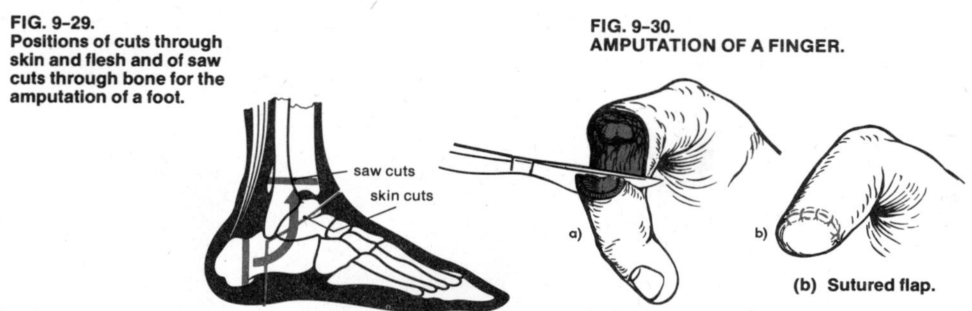

FIG. 9-29.
Positions of cuts through skin and flesh and of saw cuts through bone for the amputation of a foot.

FIG. 9-30.
AMPUTATION OF A FINGER.

(a) Forming a flap to cover the end of the stump.

(b) Sutured flap.

off some extra tissue (Fig. 9-30). The nerve trunks are pulled out of the wound and cut back; the larger blood vessels are ligatured. The severed nerve ends must be well embedded in the soft tissues, otherwise they will cause pain after the wound has healed. Then the bone is sawed at a predetermined spot (chosen on the basis of an x-ray), and the flap is stitched over the end of the bone. The actual suture must not cross the area on which pressure will subsequently be exerted, since the scar remains sensitive to rubbing and pressure. The final amputated stump should be painless, movable, muscular and suitable for fitting an artificial limb. After the wound has healed, the stump should be exercised and "toughened up." When it is ready to take loads, a suitable artificial limb can be fitted.

In some cases the amputated stump becomes very painful about 6–10 weeks after the operation, even if it has been performed correctly. This pain is due to pressure of the severed nerve trunks against the amputation flap. In such cases it will usually be necessary to perform a second operation to further shorten the nerves. In other cases the patient may feel pain which seems to be in the cutoff part of the amputated limb (phantom pain), especially if the nerve was damaged or crushed in the injury that necessitated the amputation.

JOINTS

A joint or articulation is a junction between bones of the skeleton. It may be immovable (*synarthrosis*), slightly movable (*amphiarthrosis*) or freely movable (*diarthrosis*). In a synarthrosis the bones are united by a continuous intervening substance (cartilage, fibrous tissue, or bone)—e.g., the joints (sutures) between cranial bones. An amphiarthrosis is a joint in which there is a fibrocartilaginous disc (e.g., the joints of the vertebrae, the joint at the angle of the sternum) or a ligament between the bony elements. In a diarthrosis, also known as a synovial joint (Fig. 9-31), the contact surfaces of the bones are coated with smooth cartilage and the whole joint is enclosed in a fibrous capsule lined with synovial membrane, which secretes a viscous fluid (synovia or synovial fluid) for lubricating the joint. The capsule, which is a continuation of the periosteum (the membrane surrounding the bone), contains strong ligaments (bands of fibrous connective tissue) which serve to limit movement and to bind the ends of the articulating bones together. If the cartilage coating of the contact surfaces is destroyed by disease or wear, bone-to-bone contact occurs, producing a grating noise when the joint moves; this condition may cause arthrosis.

If the ligaments of the joint are overstrained, injured or torn, the condition is called a *sprain*, characterized by swelling or an effusion of blood. A *dislocation* is the disruption of a joint, with displacement of a bone, so that the joint surfaces become separated from each other. A dislocated bone can usually be replaced fairly easily, even without anesthetic, within a few hours of the injury; with delay, adhesion of tissues and swelling will occur at the joint, so that the dislocation becomes increasingly difficult to correct. Dislocation is uncommon in most joints, except as a complication of fracture. The jaw and the shoulder joint dislocate fairly easily, however. Dislocation of the hip requires considerable violence and is always a serious injury.

Various types of synovial (freely movable) joints are to be distinguished: the *ball-and-socket joint* (Fig. 9-32), in which the spherical

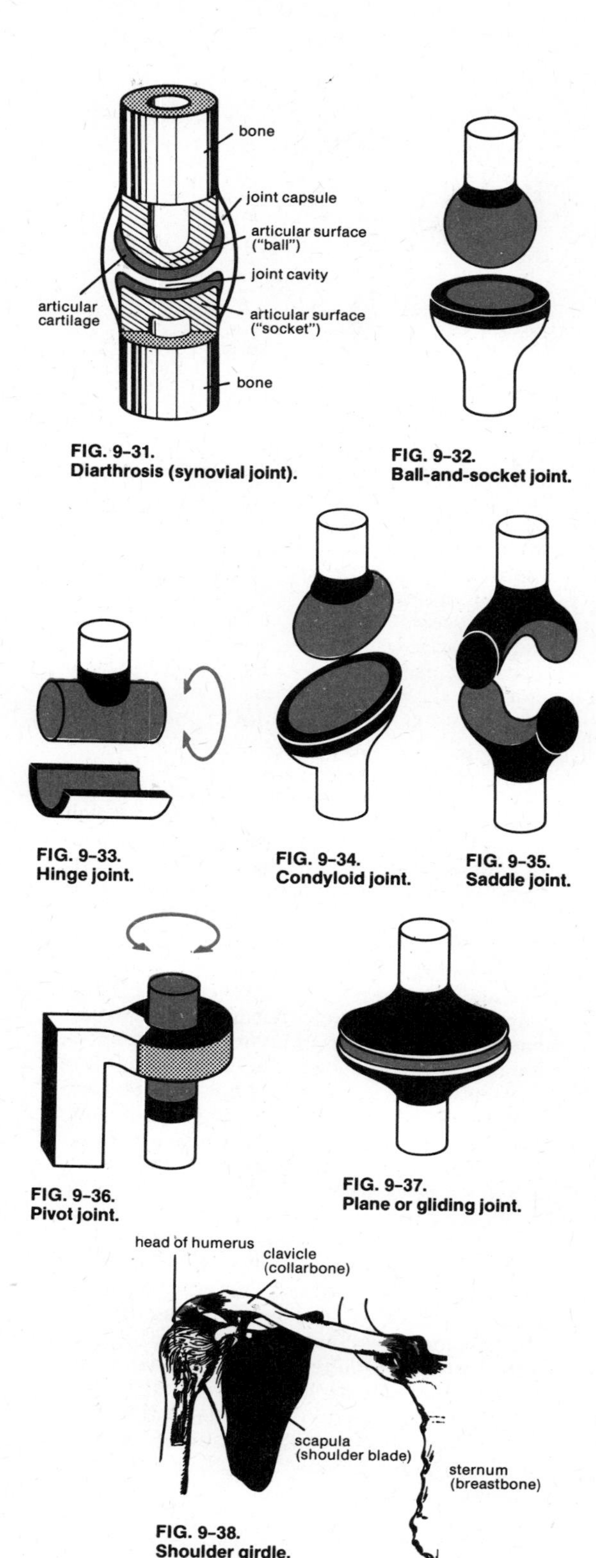

FIG. 9-31.
Diarthrosis (synovial joint).

FIG. 9-32.
Ball-and-socket joint.

FIG. 9-33.
Hinge joint.

FIG. 9-34.
Condyloid joint.

FIG. 9-35.
Saddle joint.

FIG. 9-36.
Pivot joint.

FIG. 9-37.
Plane or gliding joint.

FIG. 9-38.
Shoulder girdle.

end of one bone fits into a cup-shaped socket on the other, allows movement in any direction (example: shoulder joint); the *hinge joint* (Fig. 9-33) allows angular movement (flexion and extension) only in one plane (examples: joints of the fingers, elbow, knee); the *condyloid joint* (Fig. 9-34) resembles a ball-and-socket joint but has an ovoid instead of spherical ball, so that only angular movement but not rotation of one bone in relation to the other is possible (example: wrist joint); the *saddle joint* (Fig. 9-35) allows movement in two perpendicular planes (example: thumb joint); the *pivot joint* (Fig. 9-36) is formed by a pivotlike projection which turns within a ring (example: joint between first and second cervical vertebrae; a simpler form of pivot joint, where one bone pivots on its own longitudinal axis, is the superior radioulnar joint of the forearm); the *plane* or *gliding joint* (Fig. 9-37) has two flat surfaces in articulation and allows sliding movement of the two opposing surfaces in any direction in that one plane (example: joints between the bones of the carpus). The hip joint is a ball-and-socket joint in which the range of movement is restricted by the fact that the socket largely encloses the ball, i.e., the rounded head of the upper end of the femur (thighbone). The ankle joint (Fig. 9-45) is a hinge joint whose range of rotational movement is resiliently restricted by tendons and ligaments.

The *shoulder girdle* comprises the *scapula* (shoulder blade), which is a flat triangular bone, and the *clavicle* (collarbone), a slender long bone which articulates with the scapula and the sternum (breastbone), respectively. The humerus (upper arm bone) has a rounded head which articulates with a socket in the scapula; this is the *shoulder joint*. Because of the large ball and relatively small socket it is very mobile, but also unstable. The wide-ranging movements that this joint can perform are enhanced by movements of the shoulder girdle itself. The joint where the clavicle articulates with the sternum is a modified ball-and-socket joint. The shoulder girdle (Fig. 9-38) derives great mobility from the fact that it is almost independent of the skeleton, as the scapula is attached to the skeleton of the trunk by muscles only.

The *knee joint* (Fig. 9-39) is a hinge joint in which two rounded protuberances (condyles) at the lower end of the femur (thighbone) articulate with two oval surfaces on the upper end of the tibia (shinbone). Interposed between the

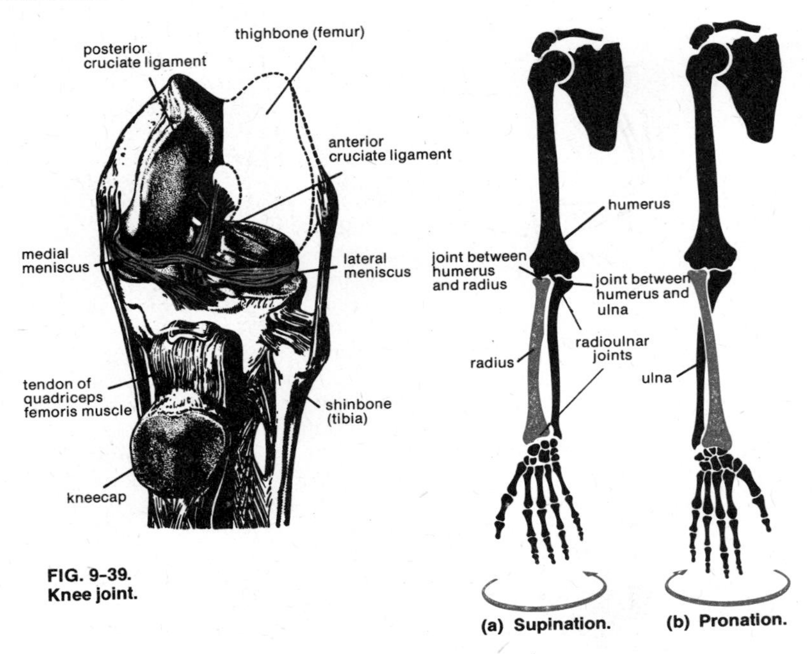

FIG. 9-39.
Knee joint.

FIG. 9-40.
ELBOW AND RADIOULNAR JOINTS.

FIGS. 9-41.
Skeleton of the hand.

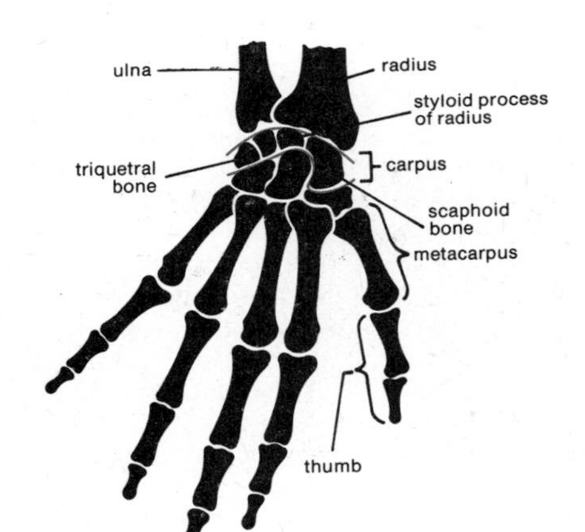

condyles and these opposing surfaces of the tibia are the medial meniscus and the lateral meniscus (also known as the medial and lateral semilunar cartilages), which are wedge-shaped crescents of fibrocartilage. They act as shock absorbers to resist impact in running, jumping, etc. Violent rotational movements of the thigh or leg while the knee is flexed (football, skiing) may damage the menisci. Especially the medial meniscus is liable to be nipped between the bones and thus suffer damage. This can be cured by resting the knee; but if the trouble recurs, it may become necessary to remove the damaged meniscus by surgery. The capsule of the joint encloses the upper end of the tibia and the lower end of the femur. These two bones are held together mainly by two ligaments (the anterior and posterior cruciate ligaments), which are essential to the stability of the joint. They also prevent backward and forward movements between the bones. There are also strong ligaments at the sides. The patella (kneecap) is situated in the tendon of insertion of the quadriceps femoris, which is the muscle that produces extension at the knee joint; flexion is produced by the hamstrings, which are three muscles at the back of the thigh. In the final stage of extension, the femur rotates in relation to the tibia, causing the knee joint to "lock," so that it can be held rigidly in extension with a minimum of muscular effort.

The *elbow joint* (Fig. 9-40) is another hinge joint, strengthened by ligaments at the sides. The muscles that produce flexion at this joint are biceps and brachialis; the triceps produces extension. The *wrist joint* is of the condyloid

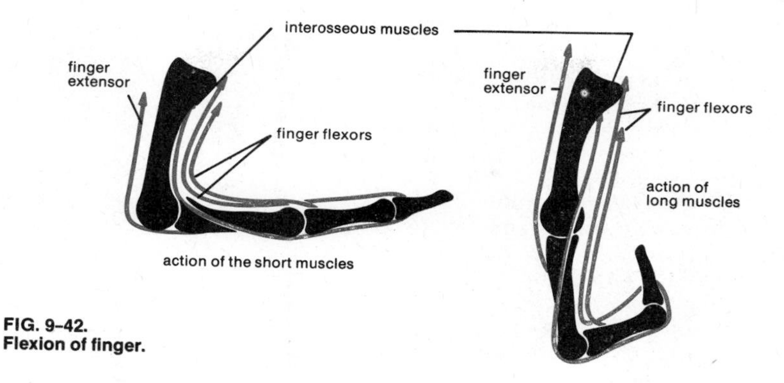

FIG. 9-42.
Flexion of finger.

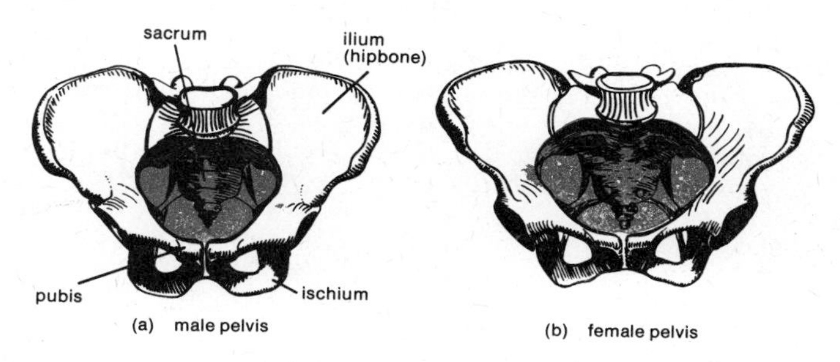

FIG. 9-43.
Pelvic girdle of male and female.

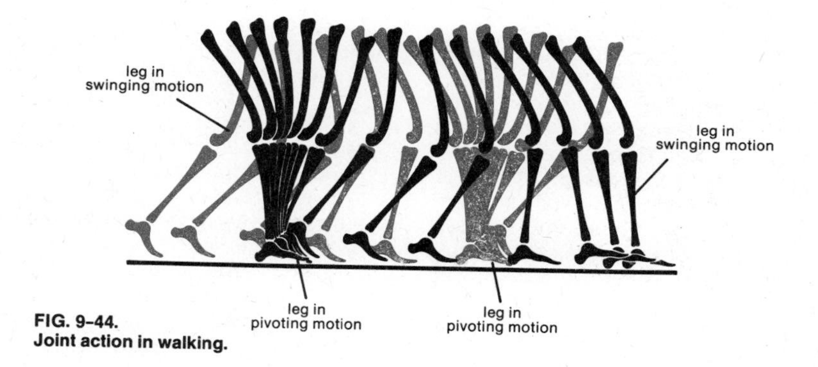

FIG. 9-44.
Joint action in walking.

type. The bones of the forearm (radius and ulna) are interconnected at both ends by the radioulnar joints. These joints always move simultaneously; the movements are known as pronation (Fig. 9-40b), in which the hand is rotated until the palm faces backward, and supination (Fig. 9-40a), which is the opposite movement that rotates the hand to the position with the palm facing forward and the shafts of the forearm bones laying parallel. In these movements the radius moves and the ulna remains stationary.

The bones of the *hand* (Fig. 9-41) comprise the carpus (wrist) with 8 bones, the metacarpus (body of the hand) with 5 bones, and the phalanges with 14 finger bones. The joints between the carpal and metacarpal bones allow small gliding movements. The metacarpals articulate by means of condyloid joints with the phalanges; the interphalangeal joints are hinge joints. The joint between the first metacarpal bone and the first carpal bone has more mobility, giving the thumb the important facility of apposition, i.e., enabling the thumb to be brought into contact with all the other fingers, so that grasping movements are possible. Flexion and extension of a finger (Fig. 9-42) are in part governed by the participation of the other fingers in these movements and by the position of the hand at the wrist joint.

The *pelvic girdle* (Fig. 9-43) connects the lower limbs with the trunk; it comprises the two hipbones and the sacrum. The hipbones articulate in front at the symphysis pubis, and each articulates with the sacrum at the sacroiliac joint. The upper end of the femur (thighbone) articulates with the hipbone at the *hip joint,* which is a synovial joint of the ball-and-socket type and possesses great stability. The acetabulum, a rounded cavity of the hipbone, forms the socket with which the rounded head (the "ball") of the femur engages. The socket is further deepened by a rim of fibrocartilage which fits closely round the head. In man the line of the vertebral column and that of the lower limb bones lie close to the body's center of gravity, so that the erect posture can be maintained with relatively little muscular effort. Movements of the lower limbs, as in walking, alter the state of equilibrium, which is constantly restored by compensatory movements of the vertebral column and pelvis. The body maintains its balance by control of the position of joints through reflexes of the nervous system which act on the muscles. Thus the positions of the different joints of the vertebral column and lower limb are interrelated. In walking, the leg is moved in relation to the hip, alternating with movement of the hip in relation to the leg (Fig. 9-44).

The skeleton of the *foot* is similar in principle to that of the hand, but is modified to form a strong and resilient arch structure to support vertical loads. The *ankle joint* is a hinge joint, but its range of movement is limited. The arched form of the foot gives it strength and resilience, enabling it to develop leverage for propelling the body forward in locomotion. The most important movements of the foot are inversion (turning the sole inward) and eversion (turning the sole outward), and the powerful flexion of the big toe which gives the final thrust at each stride. The arched form of the foot is maintained by muscles and ligaments (Fig. 9-45). The most important factor in maintaining the arch is the tension in the muscles. If the arches sag as a result of deficient muscular tone, the condition is called *fallen arches*. It may not be painful and will not necessarily cause significant inconvenience. If the arches have sagged so that the entire sole touches the ground, the condition is called *flatfoot*. It may be painful and require treatment, though in some cases there is not much discomfort.

FIG. 9-45. Ankle joint.

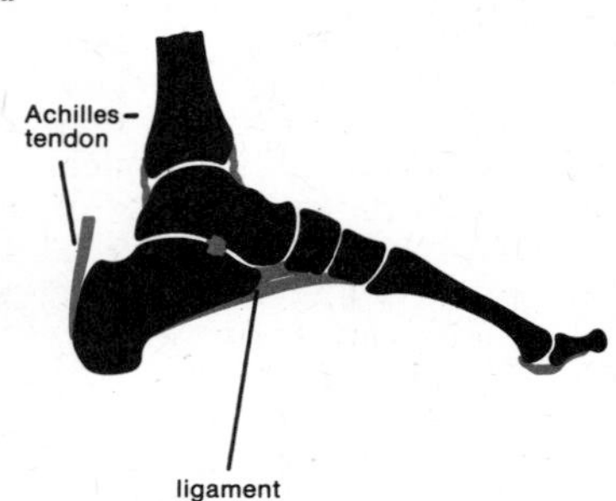

TENOSYNOVITIS; BURSITIS

A tendon forms the attachment of a muscle to a bone or other part. It consists of fibrous connective tissue—bundles of strong and tough collagen fibers. Sheaths which enclose tendons and give them lateral guidance and support are found particularly on the flexor side of the limbs. The inner layer of the tendon sheath secretes a lubricating fluid that assists the smooth action of the tendon. *Tenosynovitis* is inflammation of a tendon sheath, often associated with inflammation of the tendon itself. In an acute form it may be due to inflammation spreading from surrounding tissues, usually a bacterial infection attended by pus formation. Very often, too, it arises as a result of overexertion or strain of a tendon; in such cases, where there is no infection, no pus formation occurs, merely exudation of serum. There is then usually not much swelling, but pressure or movement will cause pain. Chronic inflammation of a tendon sheath may occur as a consequence of rheumatism, tuberculosis, syphilis or some unknown cause. In the long run, thickening of the sheath may impair sliding of the tendon within it. Purulent or tuberculous inflammation may destroy the tissues of these parts to such an extent that the affected limb becomes stiff. Treatment will vary according to circumstances. In the purulent form the limb is immobilized with the aid of bandaging and splints; if necessary, the tendon sheath can be opened by a number of small incisions for the insertion of drainage tubes to enable the pus to escape. In addition, antibiotics will be given. In the simple form of tenosynovitis it will usually suffice to immobilize the limb by splints or a plaster cast. In obstinate cases, drugs to inhibit inflammation (e.g., corticosteroids) may be given. Sometimes, in advanced cases, severe constriction of the sheath develops; complete stiffening of the limb can then be prevented only by surgery.

A bursa is a padlike sac or cavity found in connective tissue in the vicinity of joints and lined with synovial membrane. The function of the bursa, which contains synovial fluid, is to minimize friction between tendon and bone or between tendon and ligament. *Bursitis* is inflammation of a bursa. This may be an acute condition caused, for example, by a direct open injury or by inflammation spreading from surrounding tissues or by pathogenic microorganisms carried by the boodstream from a remote site of infection. Typical symptoms are reddening, swelling, restriction of movement in the affected part, local pain. In severe cases there may be fever and a general feeling of malaise. Chronic bursitis may develop from the acute form, but is more often caused by repeated pressure, friction, excessive movement or injury, and is usually associated with a particular type of occupational activity, e.g., housemaid's knee (due to long periods of kneeling), miner's or tennis elbow, dustman's shoulder. A *bunion* is bursitis at the base of the big toe, associated with deformity of the joint and caused by unsuitable footwear. In general, chronic bursitis is characterized by thickening of the bursa, with increased fluid manifesting itself as swelling. Treatment of bursitis will consist in resting and immobilization of the affected part in the acute stage. Acute bursitis characterized by pus formation can be treated as an abscess, which may have to be opened and drained. Antibiotics will then be given. Chronic inflammation of a bursa may be treated by puncture, extraction of fluid, and injection of corticoid preparations. In persistent cases, surgical removal of the bursa may be indicated (bursectomy).

SPRAIN, DISLOCATION

Sprain and dislocation are injuries affecting a joint, generally as a result of violence. The simplest, though often painful, injury arising from this cause is *contusion* (bruising) of a joint, accompanied by bleeding under the skin from damaged blood vessels of the articular capsule. The joint is swollen and painful. The effusion of blood causes the skin locally to undergo bluish or greenish discoloration, this being due to decomposition of the spilled blood in the process of its absorption, which is usually accomplished in a few days.

When the supporting tissues of a joint are injured by violence, e.g., torn or overstretched by twisting or wrenching, the condition is described as *sprain*. The term *strain* is used more particularly to denote the overstretching of muscles, ligaments and tendons that bind and operate a joint. In general, in a sprain the degree of disability and the intensity of the pain will depend on the degree of injury to the ligaments. These may be completely torn in a severe case. Signs of sprain are rapid swelling, pain, disability, often discoloration due to effusion of blood. Treatment will consist in the application of cold compresses, bandaging, elevating the limb, immobilization of the joint. The ankle is more often sprained than any other joint. A severe sprain may give trouble for weeks and months if it is not properly treated. If the effusion of blood is not completely absorbed, prolonged and painful inflammation may develop. If there is a large accumulation of blood, it may have to be drawn off with a needle.

Disruption of a joint, with displacement of a bone, so that the articular surfaces become separated, is called *dislocation*. Dislocations are produced by the same type of violence that causes sprains and fractures. As a result, the joint becomes unstable or immobilized. Any attempt to move the joint is usually very painful. The affected limb may be in an odd position in consequence of the dislocation. Most joints do not dislocate at all easily, except as a complication of fracture, but dislocations of the shoulder, elbow, fingers and toes are fairly common. Dislocation of the jaw may occur as a result of opening the mouth too wide, as in yawning; it can sometimes be snapped back into place without difficulty.

In general, the reduction of a dislocation, i.e., restoring the bones to their proper relative position, requires skilled medical attention. An expert can usually replace a dislocated bone fairly easily, even without anesthetic, if the case is brought to him for treatment within a few hours of the injury. Delay is attended by swelling at the joint and adhesion of tissues, making effective treatment more difficult. When the dislocated bones are reset in their correct position, the joint will have to be immobilized—by means of a cast, splints or bandaging—to give the torn ligaments and the capsule of the joint time to heal. After a few days, massage and passive exercise may be advised, and gentle active movements may be allowed after a week or two, depending on the severity of the case. If a dislocation is left untreated for a long time, the capsule may meanwhile have contracted and partly healed around the dislocated bone, which can then no longer be eased back into its correct position by external manipulation. Treatment of such cases may require surgery. Even so, stiffening of the joint cannot always be prevented.

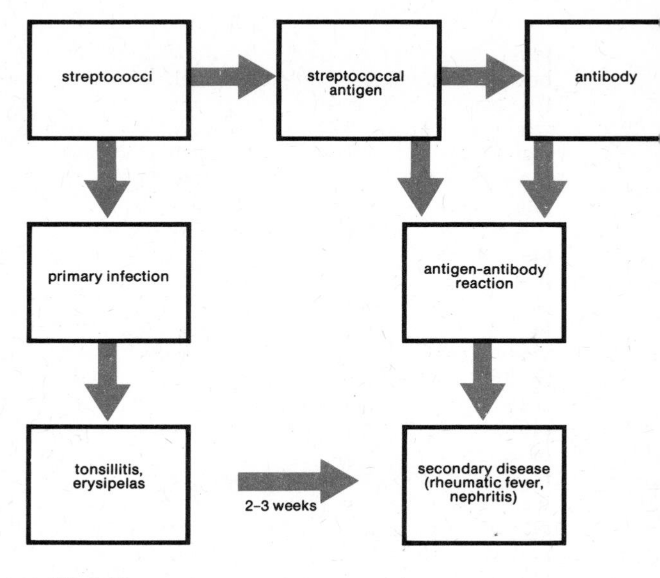

FIG. 9-46.
Development of rheumatic fever.

RHEUMATISM, RHEUMATIC FEVER

Rheumatism is a general term for acute or chronic conditions characterized by painful joints and soreness and/or stiffness of muscles not directly due to injury or infection. The term includes rheumatic fever (acute articular rheumatism), rheumatoid arthritis, osteoarthritis, gout and fibrositis (this last condition is a nonsuppurative inflammation of fibrous connective tissue anywhere in the body, also known as muscular rheumatism).

A feature common probably to all forms of rheumatism is the so-called antigen-antibody reaction. In general, an antigen is usually a "foreign" protein, e.g., a bacterial toxin. When it enters the human body, there is a defensive response characterized by production of proteinlike substances called antibodies. When antigens and antibodies encounter one another, a reaction takes place, accompanied by inflammation. Normally such processes serve to improve the body's immunity or resistance to infection. In rheumatism, however, they overshoot the mark, as it were, and develop into a diseased condition.

Some acute rheumatic disorders are initiated by certain pyogenic (pus-forming) microorganisms, particularly those that produce tonsillitis, scarlet fever and erysipelas: i.e., beta hemolytic streptococci. Rheumatic fever, for example, is not an ordinary infectious disease,

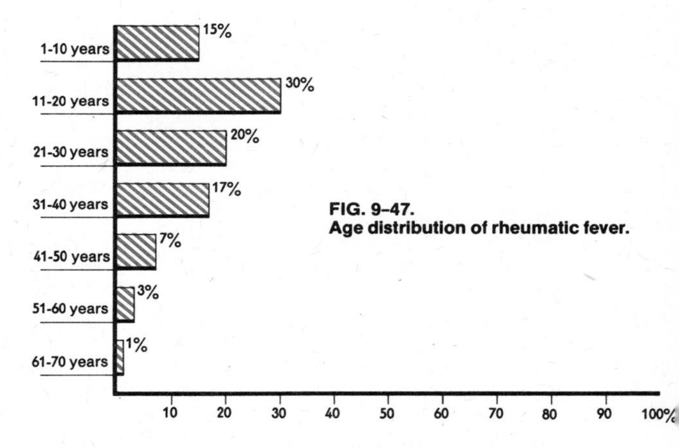

FIG. 9-47.
Age distribution of rheumatic fever.

but the manifestation of an antigen-antibody reaction provoked by streptococcal infection and arising, as a secondary disease, some time after the original infection, when sufficiently large amounts of antibodies have been formed (Fig. 9-46). The symptoms associated with the antigen-antibody reaction then manifest themselves particularly in acute inflammation of certain joints. In another group of rheumatic disorders the initiating antigens are unknown. They are, however, believed to be proteins that are produced within the human body itself. They are not necessarily abnormal proteins, although they appear incompatible to the body's immune mechanisms, which are defective and thus unable to distinguish between what is a normal substance and what is foreign, so that antibodies are formed against those proteins (autoimmune diseases: see pages 44 ff).

Rheumatic fever commonly affects children and adolescents. It is usually preceded by repeated tonsillitis, sometimes erysipelas or scarlet fever. Predisposing factors are an inherited tendency to the disease, age (11–20 years, see Fig. 9-47), undernourishment, overtiredness, wet and cold weather, catching cold, living in close proximity with other people in camps, barracks, etc. In about 3 percent of cases the allergic secondary disease, i.e., rheumatic fever, manifests itself two or three weeks after the primary infection with pyogenic streptococci. The major joints become painfully swollen and red; the pain moves about from joint to joint, sometimes even affecting the small joints such as those in the fingers and toes. The patient has fever and headache, feels ill, and lies motionless so as to avoid all painful movements.

In something like a quarter of adult cases and three quarters of cases where children are affected, the rheumatic inflammation involves the heart valves, heart muscle and pericardium. In particular, the mitral valve is likely to suffer damage, which may take the form of mitral stenosis (narrowing of the valve) or mitral insufficiency (leakage, see pages 57 ff). Symptoms may include rapid or irregular pulse, heart murmurs, dilation of the heart, and breathlessness. As a rule, the joints recover completely, but the heart may be permanently affected, with rheumatic defects of the valves persisting in about a third of all cases and possibly contributing to heart failure many years later.

Without treatment, the acute rheumatic symptoms affecting the joints usually disappear in 3–6 months; with proper treatment, they clear up in a few weeks. In some cases the disease runs a mild course, with only fleeting pains (mistakenly called "growing pains") in the joints and hardly any fever. About 50 percent of children affected by rheumatic fever suffer relapses, which can again endanger the heart, but seldom develop into chronic rheumatism. In some cases rheumatic fever affects the skin, the pleura, the peritoneum and the kidneys. In a small proportion of children, especially girls, affected by the disease the condition known as *Sydenham's chorea* (chorea minor) may develop; it is characterized by curious involuntary movements due to purposeless contractions of the muscles of the trunk and limbs; for this reason it is sometimes also called *St. Vitus's dance*.

Although the precise nature of rheumatic fever has not yet been established, prompt and effective treatment with sulfonamides or penicillin to destroy the streptococci of the primary infection will generally also prevent the secondary disease from developing. The symptoms of rheumatic fever are relieved by salicylates such as aspirin. Basically, treatment aimed at minimizing heart damage will consist in prolonged bed rest.

RHEUMATOID ARTHRITIS

As explained in the preceding section, rheumatic fever is believed to be caused by an antigen-antibody reaction associated with a streptococcal infection. Such reactions are also involved in a group of chronic rheumatoid diseases which are more particularly of an autoimmune character (Fig. 9-48). One of the most widespread of these is *rheumatoid arthritis,* which is a chronic disease of connective tissue manifesting itself especially in inflammation of the fibrous connective tissue around the joints, though any organ may be affected. Small nodules of inflamed tissue are characteristic of the disease, which attacks something like 1–3 percent of the population and occurs most frequently between the ages of 30 and 40; it is three to four times commoner in women than in men. It tends to run in families, so that a hereditary factor may be involved.

Rheumatoid arthritis may vary greatly in the manner in which it manifests itself and develops. In most cases it comes on gradually, over a period of years. Early general symptoms are tiredness, loss of weight, circulatory disorders (tingling sensations, cyanosis of fingers and toes, etc.), oversensitivity to cold water, morning stiffness of fingers and toes. Subsequently the small joints of the fingers and toes become swollen and painful. In many cases the inflammation subsides without causing much permanent harm; but sometimes a chronic inflammation persists which causes degenerative changes in the cartilage of the articular surfaces, so that the joints stiffen. The muscles around the joints may also be affected, becoming tightened and thus contributing to defor-

FIG. 9-48.
Autoimmune disease due to autosensitization. (Theoretical bacterial induction.)

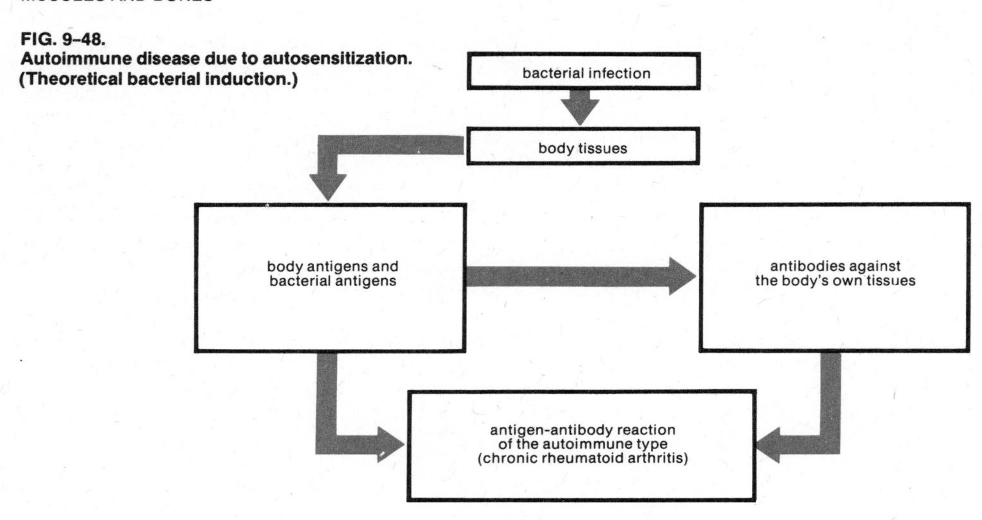

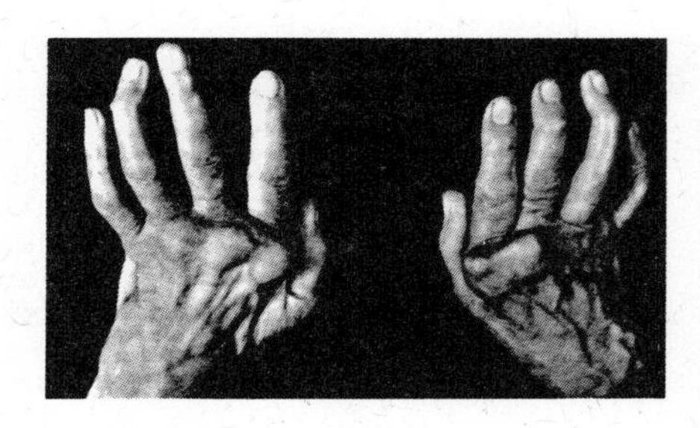

FIG. 9-49.
Advanced rheumatoid arthritis of the hands.

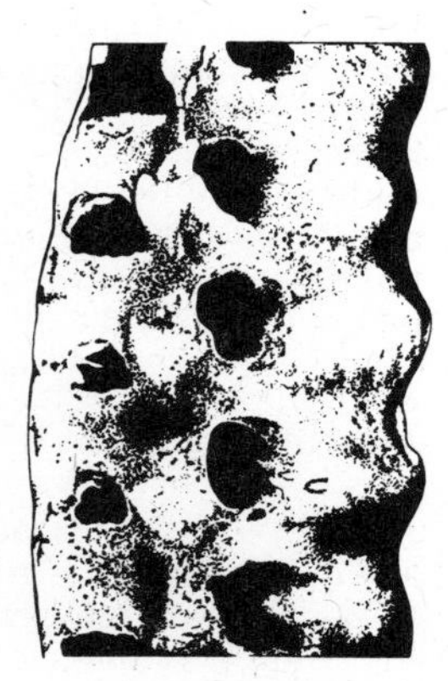

FIG. 9-50.

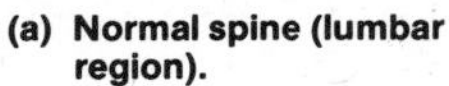

(a) Normal spine (lumbar region). **(b) "Bamboo spine" due to rheumatoid spondylitis.**

mity of the joint. Crippling deformities and invalidity are the end result of the diseases in some 10–20 percent of cases. Sometimes, though rarely, rheumatoid arthritis manifests itself as a generalized feverish illness resembling rheumatic fever, but hardly ever with involvement of the heart. In children, however, chronic rheumatoid arthritis is often associated with severe general symptoms, fever and swelling of lymph nodes. In adolescents the disease often affects one particular joint, usually a knee joint. Rheumatic inflammation may in certain circumstances also affect the eye.

There is no specific cure for rheumatoid arthritis, but pain and inflammation can be controlled with analgesics, the safest of which for

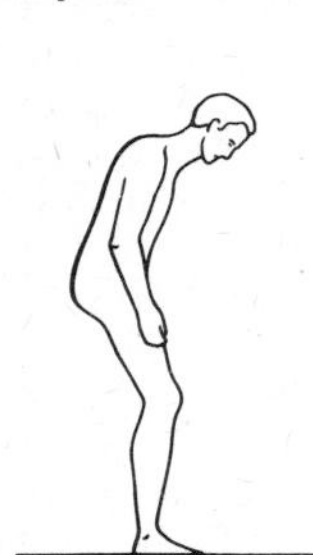

FIG. 9-51.
Posture in rheumatoid spondylitis.

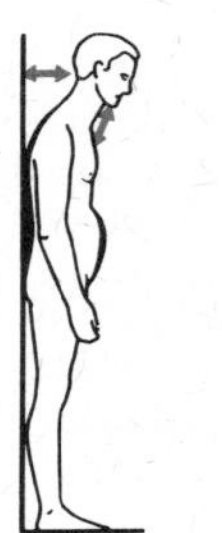

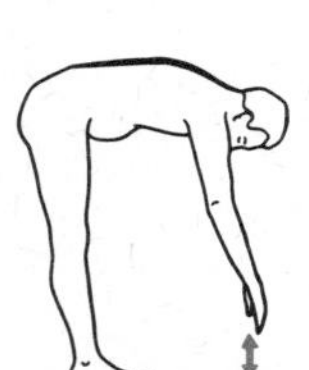

FIG. 9-52.
(a) Altered posture. **(b) Inhibition of bending in spondylitis.**

long-term use is probably aspirin; other frequently used analgesics are phenylbutazone and indomethacin. In some cases good results are obtained with gold salts. Steroid hormones (corticosteroids) often relieve acute symptoms, but may produce undesirable side effects. Hot baths, poultices and other forms of heat application can bring relief. Massage and suitable exercises are important in preserving and restoring mobility of affected joints.

Rheumatoid spondylitis (ankylosing spondylitis) is a chronic progressive disease related to rheumatoid arthritis and characterized by inflammation of the joints between the vertebrae, which gradually lose their mobility. The ligaments that support these joints lose their flexibility and harden to an almost bonelike consistency, so that the spine becomes rigid (bamboo spine, poker spine, Fig. 9-50). In an advanced case the patient has a typical forward-leaning posture (Fig. 9-51). He cannot lower his chin and is unable to stoop in the way that a healthy person can (Fig. 9-52). Pressure may act on the spinal nerves emerging between the vertebrae and may cause pain resembling that of sciatica. The disease, which occurs mostly in young adults, is about ten times commoner in men than in women. Treatment is similar to that of rheumatoid arthritis.

ARTHROSIS

Arthrosis is a degenerative affliction of joints which is due to an imbalance between the load on a joint and its resistance capacity. It is a very common disorder. To some extent it occurs in nearly all people who live to a great age. It is initiated by damage and wear affecting the cartilage surfaces of a movable joint. When ankylosis develops, i.e., the joint becomes stiff and immobilized, it cannot become arthrotic. A number of factors can cause arthrosis. These include congenital defectiveness of the articular cartilage, faulty load conditions due to unnatural positions of bones (bowlegs, knock-kneed legs), or defectively healed fractures. Overloading of joints due to overweight or to sports exertions is a common cause. Hormonal influences may also play a part (particularly, defective functioning of the pituitary gland or the gonads), especially in women past the menopause.

The degenerative arthrotic changes begin in the matrix of the articular cartilage, which loses its elasticity, becomes brittle, splinters and is eventually destroyed. The bone itself then becomes involved: breakdown of the bone tissue occurs in those parts which are subjected to pressure, while the adjacent unloaded areas of the joint are affected by irregular proliferation of bone, accompanied by the formation of bulges and jagged projections. This condition is called *arthrosis deformans* (Fig. 9-53). It affects particularly the large joints of the lower limbs, as these are subjected to the heaviest mechanical loads by the weight of the body.

The first symptoms of arthrosis are stiffness and a feeling of tension in the joint, soon fol-

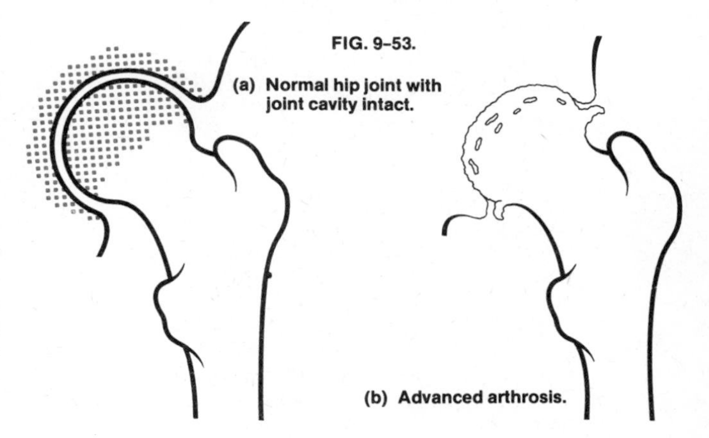

FIG. 9-53.
(a) Normal hip joint with joint cavity intact.
(b) Advanced arthrosis.

FIG. 9-54.
Plastic surgery of hip joint.

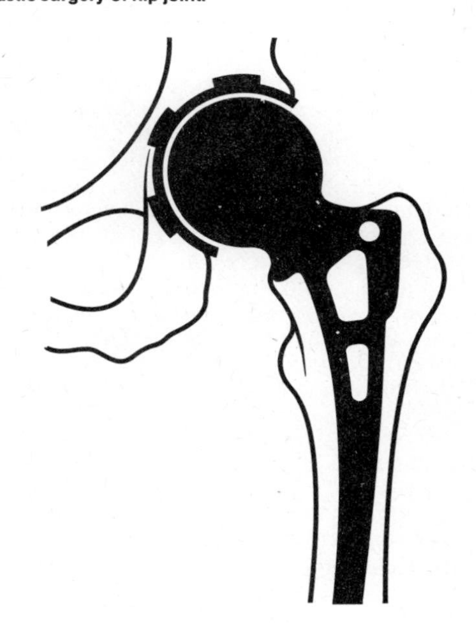

lowed by pain, which is most acute directly after physical activity. Sometimes the affected joint emits a grinding or grating sound. As a result of bulging bone formation at the edges, the joint will swell and eventually undergo a loss of mobility. Projections of bone may break off and come adrift in the joint, so that they may cause sudden jamming, with restriction of movement and severe pain.

Treatment of arthrosis will consist in the application of heat by means of warm baths or by shortwave or microwave diathermy, which can be effective in relieving pain. Therapy with antirheumatic preparations, hormones, drugs to promote the supply of blood to affected parts, etc., can also be helpful. If the pain is very acute, the only way to alleviate it is by relieving the joints of load—by bed rest or the use of special supporting devices—and injecting the patient with a local anesthetic. In serious cases surgical intervention will be necessary. Depending on the circumstances of the case, an operation can produce an immobilized (stiff) joint, which will at least be free from pain, or can graft an artificial joint onto the bone. This latter operation has, in recent years, been extensively and often successfully used for repairing the hip joint (Fig. 9-54). Like other joints, those between the vertebrae may also be affected by arthrosis, the condition being known as spondylarthrosis.

MALFORMATIONS AND DEFORMITIES OF THE FOOT

The commonest deformity is *flatfoot,* which occurs when the foot has to support a load beyond its capacity. This is likely to happen when a person has to stand for long periods, so that the muscles of the foot tire and a large proportion of the load has to be resisted by the ligaments. These may become overstretched as a result, thus further increasing the strain on the muscles. Factors liable to cause flatfoot are congenital weakness of the muscles and ligaments, overweight in persons whose daily pattern of life involves much standing, degenerative disease affecting the bone (e.g., rickets), chronic inflammation of bones and joints, injuries due to accidents, etc. In severe cases, deterioration of the articular surfaces and contraction of the ligaments may result in stiffening of the foot. Quite often, however, people with flatfoot do not suffer any serious disability from the defect. Besides longitudinal weakness of the arch of the foot, there may be

FIG. 9-55.
Abnormalities of the foot.

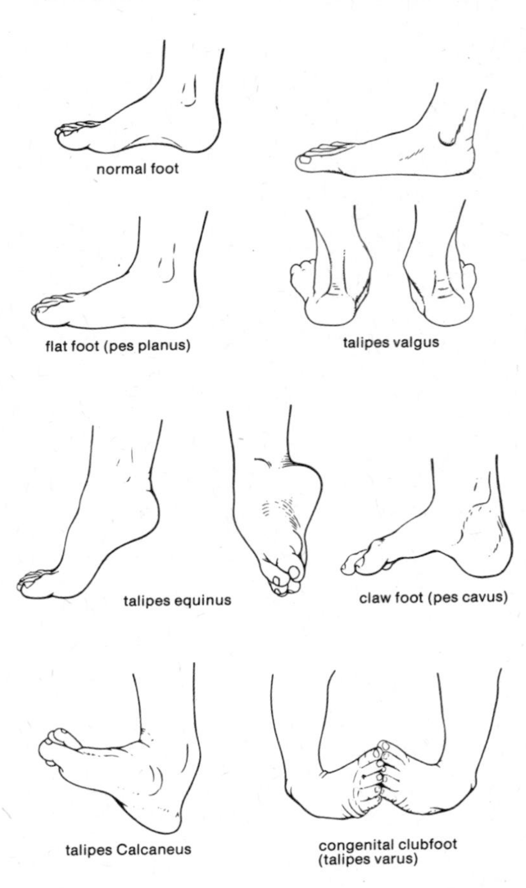

Courtesy Exner and Kaiser

weakness in the transverse direction, resulting in the condition known as *splayfoot,* which usually takes the form of flatfoot accompanied by extreme eversion of the foot, i.e., with the sole turned to face outward.

Clubfoot (talipes) is usually a congenital malformation, but sometimes develops in later life as a result of poliomyelitis. There are many varieties of clubfoot. For instance, in *talipes calcaneus* the foot is twisted at the ankle so that the heel alone touches the ground; in *talipes equinus* the foot is twisted forward so that only the toes touch the ground; in *talipes varus* the sole is turned inward so that the person walks on the outer edge of the foot; *talipes valgus* is the opposite condition, i.e., the person walks on the inner edge of the foot because the sole is twisted outward. Clubfoot in infancy can sometimes be effectively treated by means of splints and bandages, massage, manipulation, etc. Talipes equinus, for example, may be corrected by stretching the calf muscles by means of special traction devices or by an operation to lengthen the Achilles tendon. Surgical reconstruction of the joints of the foot may be necessary in other cases.

Clawfoot (hollow foot, pes cavus) is a malformation characterized by exaggeration of the inner longitudinal arch of the foot, often accompanied by dorsal contraction of the toes. The condition may be congenital or it may arise in consequence of an accident or paralysis of the muscles of the front of the leg. In children it can often be corrected without an operation, but in adults some form of surgical intervention to improve the condition will generally be necessary.

SKULL: FRACTURE, CONCUSSION

The skull is composed of a number of bones connected to one another by immovable joints called sutures (Figs. 9-56 and 9-57). There are two main regions of the skull: the *cranium* or braincase (also referred to more specifically as the neurocranium) and the skeleton of the *face,* which is suspended in front of the cranium. The vault of the skull (*calvaria*) consists of the *frontal bone* in front, the *parietal bones* at the top and sides, the *temporal bones* around the ears, and the *occipital bone* at the back. These are curved flat bones. The frontal bone is joined to the two parietal bones by the *coronal suture;* the *lambdoid suture* is between the parietal bones and the occipital bone, and the *sagittal suture* unites the parietal bones. The temporal bone consists of the *squamous* part, which is on the side of the cranium, and the *petrous* part, which helps to form the base of the skull and contains the middle ear and internal ear. The *external auditory meatus,* an oval hole below the squamous temporal bone, comprises the canal extending from the eardrum to the external ear. In front of the meatus the temporal bone has a forward extension which is joined with an extension of the malar bone to form the *zygoma* (or *zygomatic arch*).

The floor of the cranium is called the *base of the skull* and is formed by part of the occipital bone, by the petrous part of the temporal bones, by the sphenoid bone, and by backward projections from the frontal bone. The base of the skull contains holes for the passage of nerves and blood vessels. Through the largest of these holes, the *foramen magnum* in the occipital bone, the spinal cord enters the skull. The two occipital condyles are situated on each side of the foramen magnum; they are rounded protuberances with articular surfaces which articulate with the atlas (the first cervical vertebra). The floor of the interior of the cranium is divided tierwise into three cranial fossae. The foremost and highest of these is the anterior fossa, which supports the frontal lobes of the cerebral hemispheres and forms the roof over the orbits (the cavities that hold the eyeballs) and nose; the middle fossa behind the eyes supports the temporal lobes of the cerebral hemispheres; the posterior fossa, bounded by the petrous temporal bones and the occipital bone, contains the cerebellum, medulla oblongata and pons.

The bones forming the skeleton of the *face* are arranged around the openings for the mouth, nose and eyes. The main bone of the face is the *maxilla,* which forms the floor of the orbits, the side of the internal part of the nose, and the upper jaw; its lowest part carries the upper teeth and forms the hard palate. The lower jaw (*mandible*) comprises a curved horizontal part, which carries the lower teeth, and at each end a vertical part, which articulates with the skull.

The cranium is molded to the contours of the brain, and the size of the brain is usually expressed in terms of the *cranial capacity,* i.e., the volume of the cranial cavity. In men it averages 1450 cm^3, in women rather less. At birth the human head is strikingly large in relation to the body, and the cranium is relatively large compared with the face. This is because the brain, eyes and ears develop rapidly before birth, whereas the mouth and jaws are still quite small and do not develop until later. At birth the average circumference of the head is about 33 cm; in an adult it is about 58 cm.

There are two unossified portions in the cranial vault at birth, called *fontanelles:* the anterior fontanelle, which is about 2.5 cm wide at birth, closes during the second year of life; the posterior fontanelle is smaller and is closed by the age of 3 months, i.e., the original cartilage has turned into bone.

The hard bone of the mature skull still possesses some elasticity. If the vault of the skull is subjected to violence, such as sharp blow or knock, the bone will deflect up to its limit of elasticity and will fracture only when this limit is exceeded. The kind of fracture that occurs will depend on the area affected by the impact. If the contact area is small, e.g., when the skull is struck by a stone or a blow from an axe, a *flexural fracture* will occur in which the inner surface of the bone is fractured over a larger area than the outer contact area and is moreover often splintered, whereas the outside of the bone may show no more than a small depression. On the other hand, if the blow is delivered by a relatively large object, a bursting action will develop whereby the cranium as a whole undergoes deformation, with internal pressure transmitted in all directions, so that irregular fracture of the enclosing bone is liable to occur in several places (Fig. 9-58). The lines of fracture may be confined to the vault of the skull, but often also involve the base (see next section).

The seriousness of a skull fracture will depend on the degree of damage to the brain when parts of the bone are depressed and driven inward. The severity of such damage will decide whether the patient will live or die and whether, if he survives, he will make a complete recovery or suffer permanent brain injury. The least severe form of damage to the brain is *concussion,* which may or may not be accompanied by fracture of the vault of the skull. The violence that causes concussion generally also causes the patient to lose consciousness for a length of time ranging from a few minutes to several hours. Return of consciousness often is accompanied by vomiting, with the attendant risk that the patient suffocates by inhalation of vomit. To prevent this, he should be laid on his side or recline face downward. As a rule, there is retrograde amnesia, i.e., the patient does not remember events immediately before concussion. In mild cases, however, there may be no loss of consciousness; there may be no more than dizziness and headache that pass off in a few hours.

If the skull has fractured, there will be a risk of infection caused by dirt in the wound, hairs or bone splinters. Inflammation of the meninges (meningitis) may develop.

Not infrequently the temporal artery located close to the inner surface of the temporal bone is ruptured. Even quite minor fractures which are not externally visible may cause this, as the blood vessel is so closely associated with the bone. Often such patients regain consciousness after concussion and at first feel reasonably well. As not all minor concussions are treated in a hospital, the patient may go home or indeed carry on with his normal activities. After a few hours he will feel unwell, complain of increasingly severe headache, begin vomiting again, speak incoherently, and eventually lapse into unconsciousness; the heart rate goes down to about 40–50 per minute. Such signs of a buildup of pressure in the

FIG. 9-56.
A child's (left) and an adult's (right) skull, viewed from above.

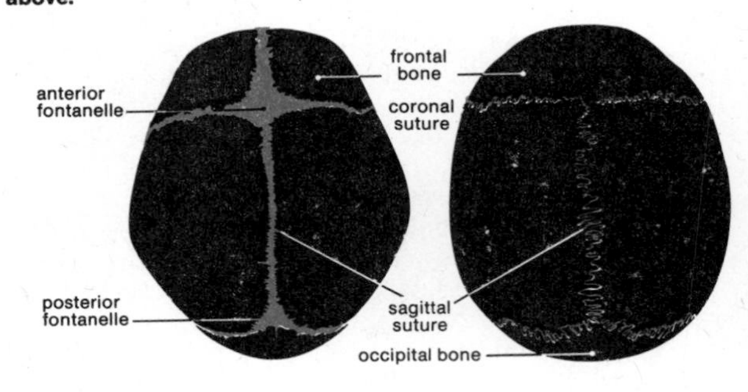

FIG. 9-57.
Skull viewed from the left side.

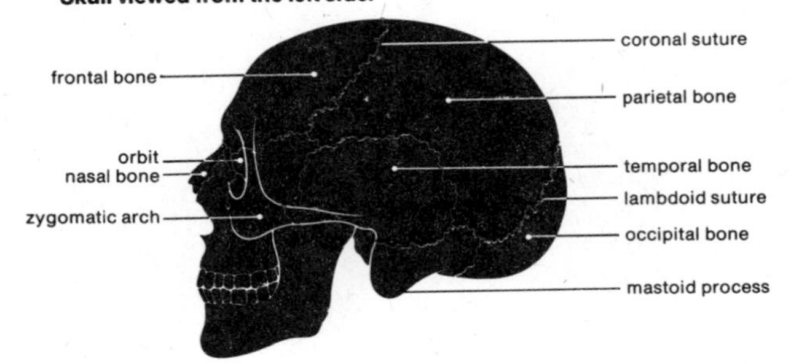

FIG. 9-58.
Fracture due to bursting action.

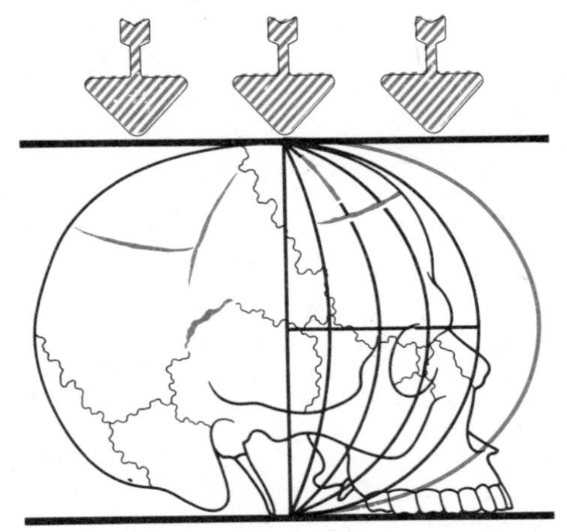

FIG. 9-59.
Pressure on the brain by hemorrhage.

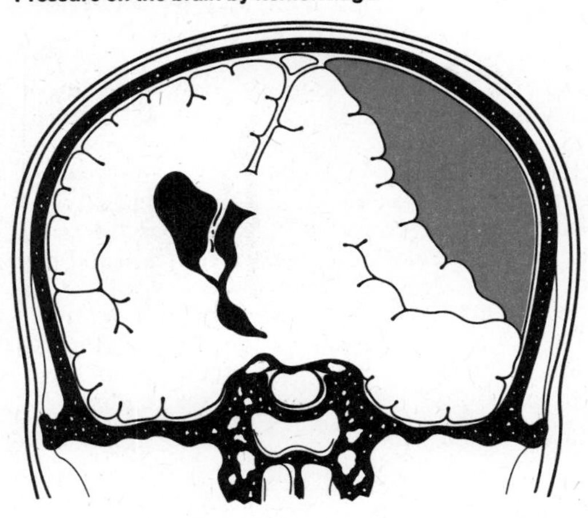

cranium are caused by hemorrhage from the ruptured temporal artery, so that an effusion of blood occurs under the dura mater (the outer membrane covering the brain). This blood cannot escape into cavities between adjacent tissues (as in an internal hemorrhage in some other part of the body), but instead builds up a pressure within the skull (Fig. 9-59). When this happens, the only remedy is to open the skull, drain out the blood and ligature the ruptured artery. Without such surgical intervention the patient will not regain consciousness and will die as a result of pressure on the brain.

If the violence suffered by the skull is of a more severe character, there will not only be concussion but also *contusion* of the brain. This involves damage to the nerve tissue as a result of numerous minor effusions of blood or direct bruising of the brain substance. The symptoms and possible long-term consequences of such injuries depend mainly on the location and extent of the brain damage. In some cases of skull fracture, e.g., strictly circumscribed flexural fractures, the splintered inner plate of the bone will heal with a scar and adhesion to the dura mater. As a result of this the patient may have convulsions very similar to epilepsy (see pages 346 ff), particularly the type known as Jacksonian epilepsy, which is usually confined to certain groups of muscles or to one side of the body and is otherwise due to disease involving the cerebral cortex. Operational separation of the coalescence of brain substance, membrane and bone can sometimes cure the disorder.

Fracture of the Base of the Skull

Like the vault of the skull, the base is composed of a number of bones joined by sutures. As described in the preceding section, these bones form a three-tiered system of fossae which support the various parts of the brain. In some respects the base is the weakest part of the skull, at any rate in those areas where the bone is very thin around holes for nerves and blood vessels; in other areas the bone forms solid ridges which stiffen and strengthen the base. Thus the petrous part of the temporal bone is just about the hardest bone in the whole body. Experience shows that the lines of fracture tend to be located between these strengthening ridges. Not infrequently, however, the petrous part itself is involved in the fracture.

The great majority of fractures of the base of the skull are caused by severe violence to the cranium. They start at the site of the injury and spread by the shortest path to the base. For example, a heavy blow on the frontal bone, as may occur in a motor accident, is likely to cause fracture involving the anterior fossa; fracture of the parietal or temporal bones spreads into the middle fossa; and fracture of the occipital bone is likely to involve the posterior fossa (Figs. 9-60 and 9-61). Other cases of fracture of the base of the skull are caused indirectly when parts of the facial skeleton or the upper end of the spine are driven into the base, e.g., as a result of a fall in which the head, the buttocks or the legs forcibly strike the ground (Fig. 9-62). A heavy blow on the nose may drive the nasal bones back into the anterior fossa. In the region of the articulation of the jaw the bone of the skull is very thin, so that a blow on the chin may force the head of the mandible (lower jaw) into the cranial cavity. An accident of this kind may occur in boxing; but an uppercut to the chin is more likely to fracture the jaw itself, which thus absorbs the force of the blow.

Fracture of the base of the skull is always a serious injury, as it is usually associated with brain damage or damage to the cranial nerves. The patient is usually unconscious immediately after the accident. Despite the otherwise strict rule that a patient with a head injury should not be moved, it is essential to turn him on his side or place him face downward in the reclining position, or else he will be in danger of suffocation due to inhalation of vomit. If major blood vessels are ruptured, the resulting effusion of blood within the skull will exert

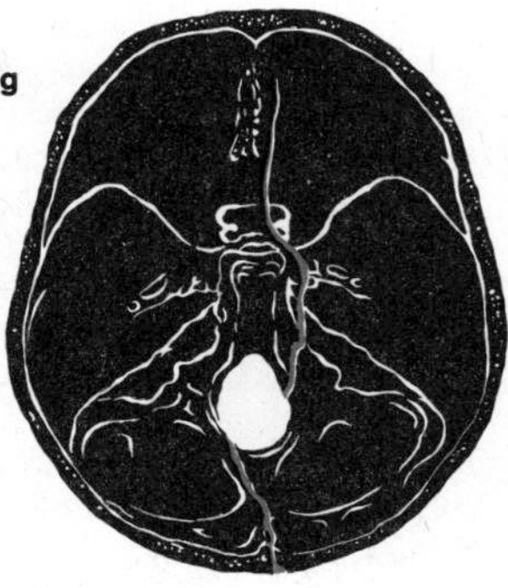

FIG. 9-60.
Fracture of the base of the skull in the longitudinal direction (e.g., due to falling on back of head).

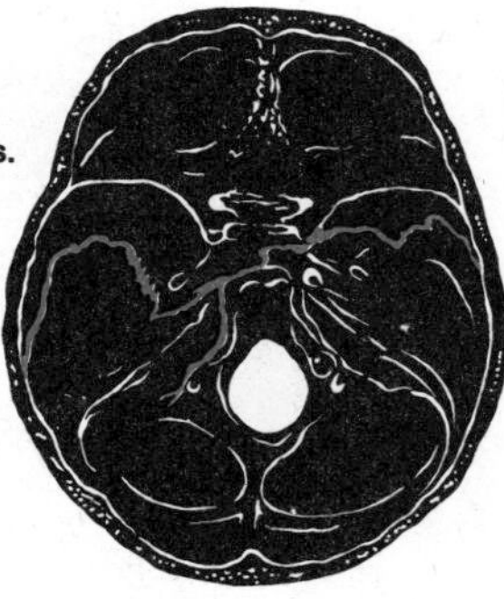

FIG. 9-61.
Fracture of the base of the skull involving the middle fossa due to a blow on the parietal and temporal bones.

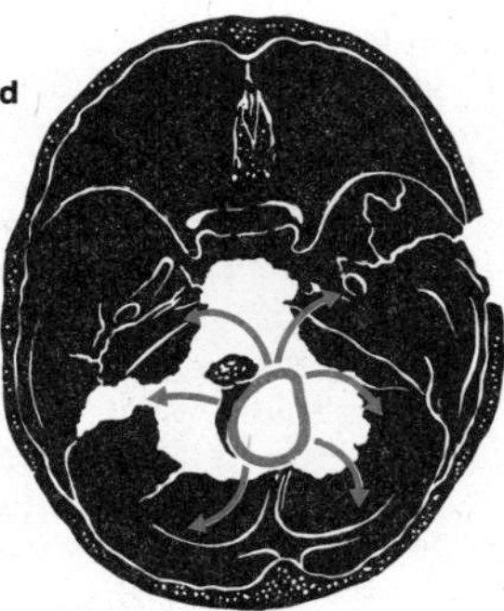

FIG. 9-62.
Fracture of the base of the skull due to a fall on the head (upper end of spine driven into skull).

pressure on the brain. The patient may regain consciousness, in which case there will be vomiting, severe headache, incoherent speech, spasms and paralysis; the heartbeat slows down alarmingly, and the patient relapses into unconsciousness. These phenomena are caused by the pressure exerted inside the skull, and the only way to save the patient is to open the cranial cavity and relieve the pressure. A sign of fracture of the base of the skull is the emergence of blood and cerebrospinal fluid from the mouth, nose and ears. If the orbits are injured by fracture affecting the anterior fossa, blood will make its way under the skin of the soft connective tissue surrounding the eyes.

If the coma is of long duration, the patient may have to be sustained by artificial respiration. If he survives the acute initial phase of maximum danger in a severe case, he may be left with permanent disabilities or paralysis. Such disorders may include tinnitus (buzzing sounds in the ears), possible deafness, and paralysis of the facial nerve (difficulty in closing the eyelids, sagging corner of the mouth). Also, the patient may suffer from tiredness, forgetfulness and lack of concentration for a long time after getting over the immediate effects of fracturing the base of the skull.

SPINAL COLUMN, SPINAL INJURIES

The *spinal column* or *vertebral column,* familiarly referred to as the *backbone* or the *spine* (Fig. 9-63), supports the trunk and protects the spinal cord. It consists of 33 small bones, the vertebrae, which are interconnected by joints and ligaments. There are 7 cervical, 12 thoracic, 5 lumbar, 5 sacral, and 4 coccygeal vertebrae. The sacral vertebrae are fused to form one bone (the sacrum), and the coccygeal vertebrae are rudimentary and partly fused to form the coccyx, a vestigial tail. With the exception of the first two cervical vertebrae, which are considerably modified to support the skull, all the vertebrae have the same

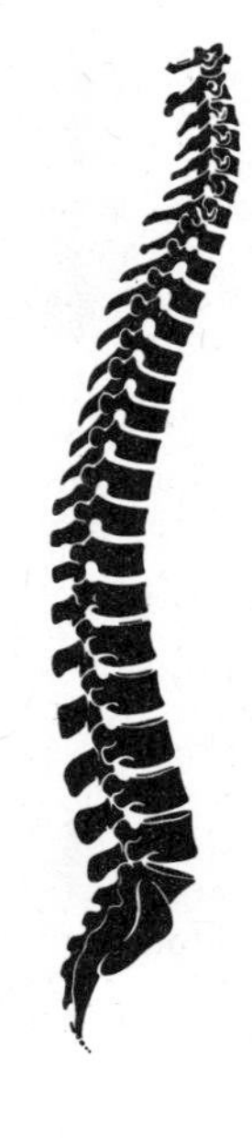

FIG. 9-63.
Spinal column: side view.

general shape. Each vertebra consists of the body (in front) and the neural arch (behind) (Fig. 9-64). The body is connected to the bodies of the adjacent vertebrae by cartilaginous joints and is the weight-bearing part. The neural arch, which encloses the spinal canal, has three projections to which the muscles of the back are attached. These projections are the spinous process (in the middle) and the transverse processes (at the sides). In the thoracic region the transverse processes articulate with the ribs and help to support them. In addition, there are upward projections and downward projections forming synovial joints with the vertebrae above and below. When the vertebrae are articulated, the neural arches of the successive vertebrae enclose openings, called intervertebral foramina, through which the spinal nerves emerge. The first cervical vertebra is called the *atlas,* the second is called the *axis* (Fig. 9-65).

The atlas is ring-shaped, having no body, but with bone masses at the sides; each mass articulates on its upper surface with the occipital condyle of the skull, and on its lower surface with the axis. The upper surface of the axis has a peglike projection (the dens) which protrudes into the opening of the atlas and articulates with the back of the anterior arch thereof. The dens is held in position by a transverse ligament which passes behind it; the spinal cord is immediately behind this ligament and may be injured if the latter is torn in an accident and the dens is displaced backward. Such injury is often fatal (broken neck).

The lumbar vertebrae have strong massive bodies for carrying the weight of the trunk.

The spinal column as a whole comprises four curves in the median plane of the body (Fig. 9-63). The thoracic curve and sacral curve are concave forward, while the cervical and lumbar curves are convex forward. These curves serve to increase the elasticity of the spinal column.

The cartilaginous joints between the bodies of adjacent vertebrae consist each of a flat round *intervertebral disc* (Fig. 9-66). This disc has an outer covering of fibrocartilage (annulus fibrosus) and a jellylike center (nucleus pulposus). The disc is attached by hyaline cartilage to the vertebral bodies. Fibrous ligaments in front and behind serve to strengthen the joint. Each joint has only a limited range of movement, but the large number of joints give the column as a whole a considerable amount of flexibility. Flexion, extension and lateral bending movements can be performed in the cervical and lumbar regions. The lumbar region al-

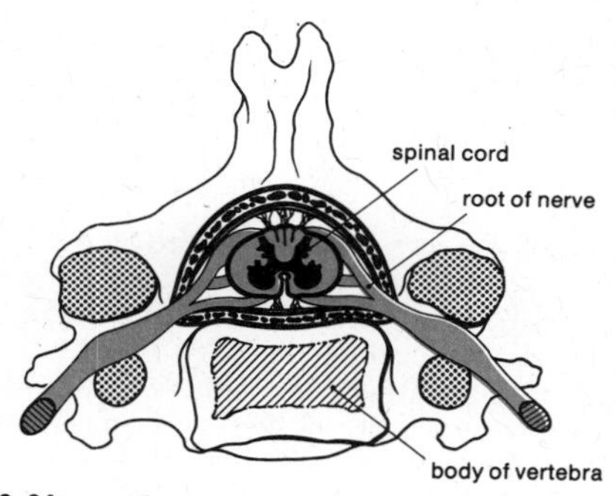

FIG. 9-64.
Section through a vertebra and spinal cord.

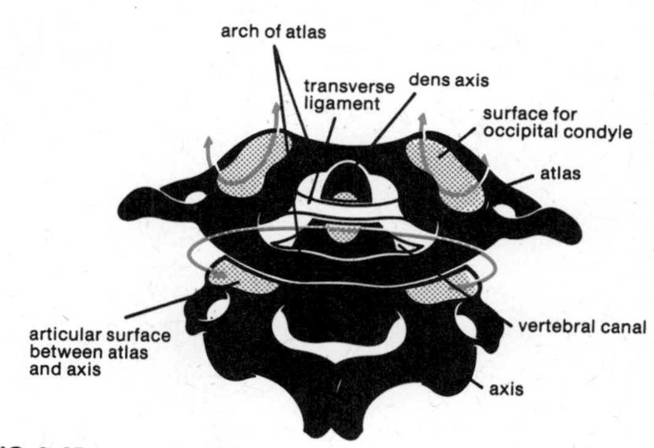

FIG. 9-65.
First and second cervical vertebrae (atlas and axis).

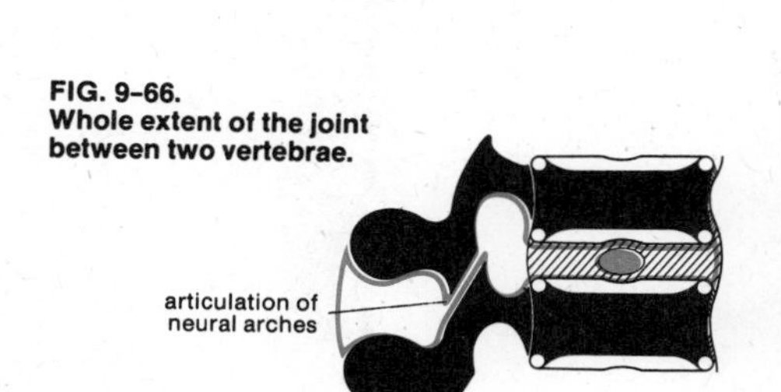

FIG. 9-66.
Whole extent of the joint between two vertebrae.

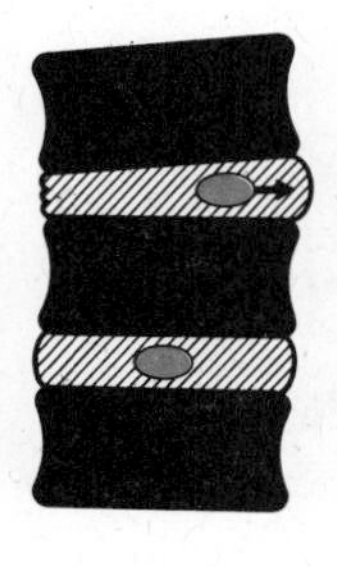

FIG. 9-67.
Deformation of the intervertebral disc due to bending movements of the spine.

lows some rotation, and more extensive rotation is possible in the cervical region. In the thoracic region the range of rotational movement is restricted by the ribs. Flexion of the spinal column by the force of gravity is counteracted by the pull of the extensor muscles of the back.

The intervertebral discs allow movement of the spinal column and carry the full load transmitted through it. The nucleus pulposus with its jellylike consistency is highly deformable, so that it can serve also to relieve and distribute local pressure in the event of sudden compressive loading, but also flexion and extension, of the spinal column. With flexion, i.e., when a bending movement is performed, the nucleus slides backward (Fig. 9-67), and the disc itself is compressed in front and increases in thickness behind. Conversely, when the spinal column is extended, i.e., a straightening movement is performed, the nucleus slides forward. With lateral bending, the nucleus similarly moves toward the convex side. If the nucleus loses its elasticity, the intervertebral disc loses its load-distributing capacity, resulting in damage to the vertebral bodies, while the capacity for movement by the spinal column is reduced. If the annulus fibrosus splits, the nucleus pulposus not only will move away from its normal at-rest central position when the spine is flexed, but will also move more or less unpredictably under any vertical loading and may protrude backward, so that it then presses on one of the nerve roots, causing acute pain. This condition is called "slipped disc" (see next section).

The spinal column derives its stability from a system of ligaments and muscles which counterbalance the high internal pressure of the nucleus pulposus. It can be compared to the mast of a boat, fixed in the pelvis (Fig. 9-68). With each movement of the body the intrinsically unstable equilibrium of the spinal column has to be restored and maintained. The erect posture is ensured by a system resembling the guying or staying of the mast of a sailing ship by the ropes and rigging (Fig. 9-69). The yards of the mast correspond to the processes (projections) of the vertebrae; the ropes represent the muscles, which are of various lengths—short ones from yard to yard, and longer ones extending down to the deck. The overall effect of these muscles is to prevent the spinal column from collapsing forward, sideways or backward (this latter collapse being prevented especially by the abdominal muscles).

Postural defects are associated with *abnormal spinal curvature*. They are due to weakness or failure of the muscles and ligaments attached to the vertebrae. Such effects can at first be compensated by voluntary muscular effort. Corrective exercises may be an effective countermeasure in certain cases. Some examples of faulty posture are illustrated in Figure 9-71. Abnormal curvature of the spine may take the form of lordosis, kyphosis or scoliosis. *Lordosis* is an exaggeration of the lumbar curve, i.e., abnormal forward curvature in the small of the back. It may be a congenital defect or be acquired through poor posture or metabolic disease. *Kyphosis* is an exaggeration

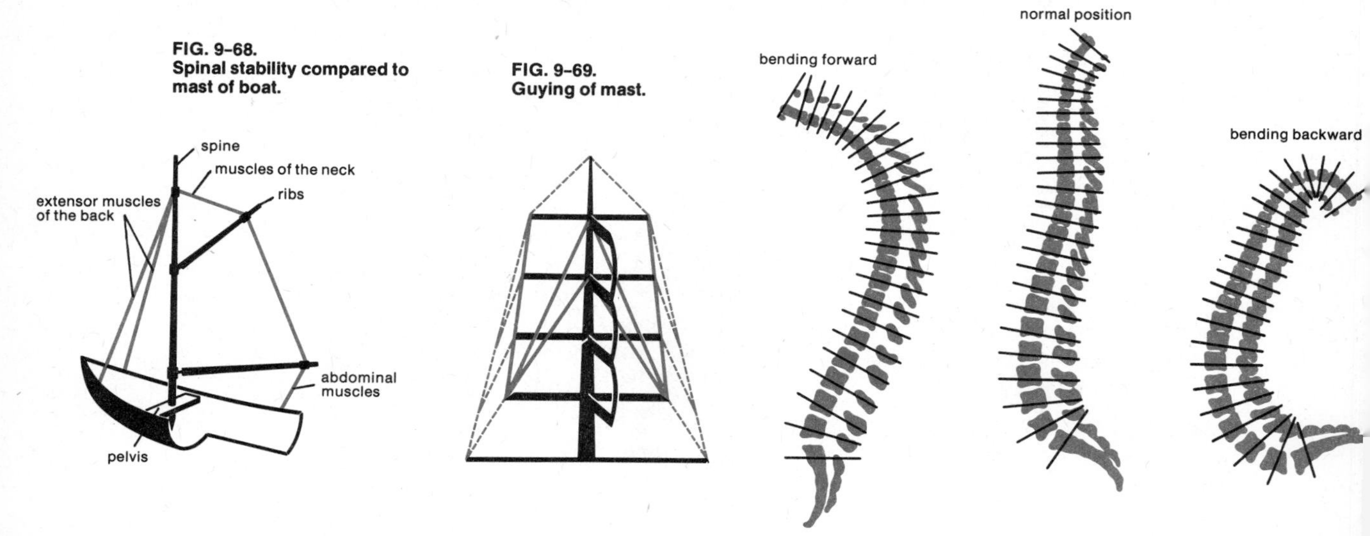

FIG. 9-68.
Spinal stability compared to mast of boat.

FIG. 9-69.
Guying of mast.

FIG. 9-70.
Movements of the spine.

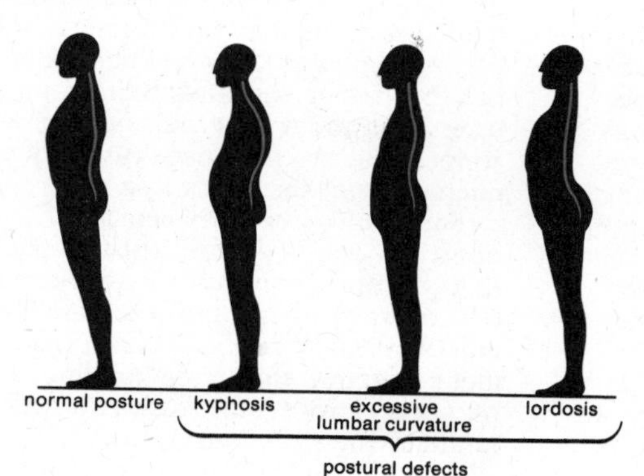

FIG. 9–71.
Examples of faulty posture.

FIG. 9–72.
Severe abnormal curvature of the spine due to rickets (vitamin D deficiency).

FIG. 9–73.
(a) Intervertebral disc penetrates into adjacent vertebral bodies.

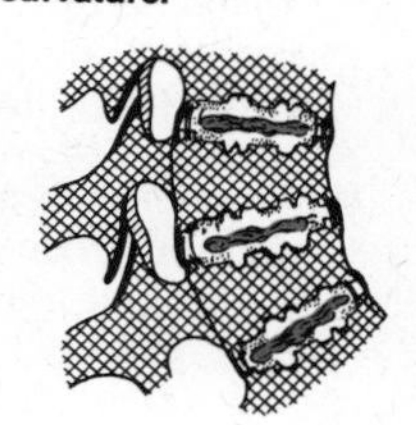

(b) As a result, the vertebral bodies become deformed (wedge-shaped) and the spine develops an abnormal curvature.

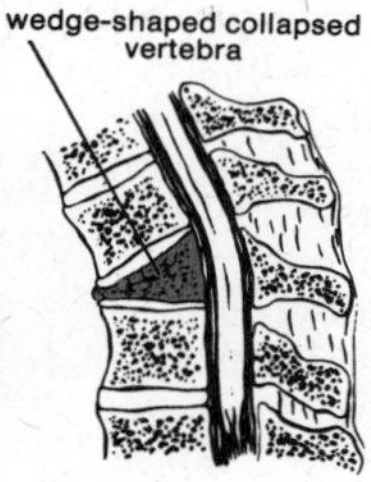

wedge-shaped collapsed vertebra

FIG. 9–74.
Abnormal spinal curvature due to calcium deficiency after the menopause.

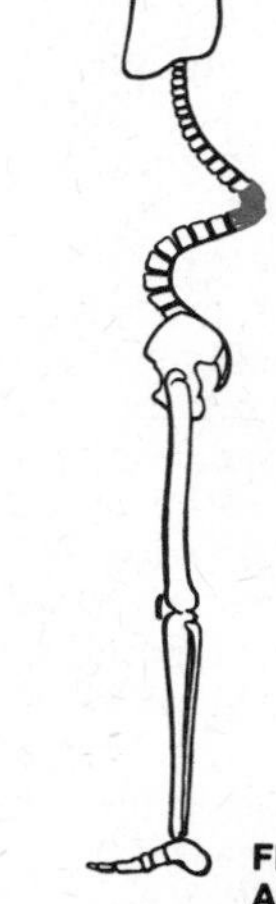

FIG. 9–75.
Abnormal spinal curvature due to tuberculosis of the vertebrae (Pott's disease).

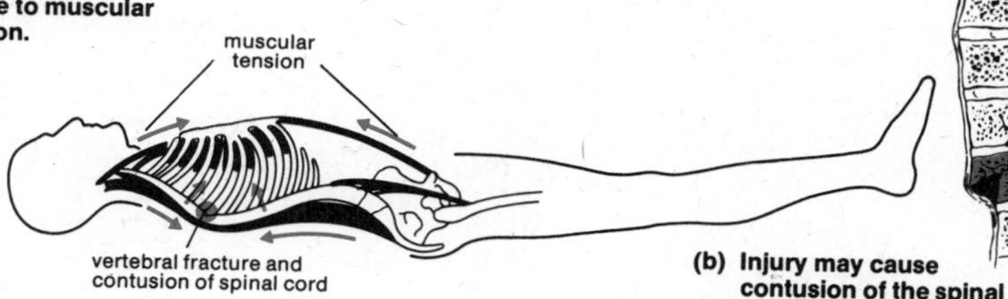

FIG. 9–76.
(a) Crushing fracture of the spine due to muscular convulsion.

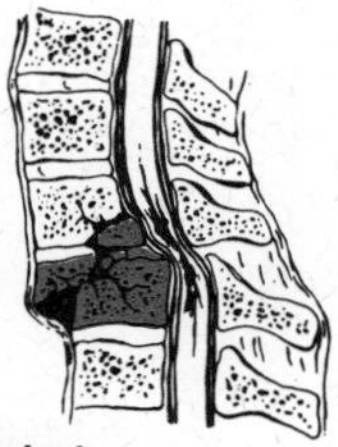

(b) Injury may cause contusion of the spinal cord and paralysis of the lower limbs.

of the thoracic curve, i.e., excessive curvature with convexity backward, which may be due to faulty posture, osteoarthritis or rheumatoid arthritis. In a pronounced form it is known as hunchback (also referred to by such terms as humpback, gibbus or Pott's curvature), which is frequently due to bone disease such as rickets, tuberculosis or syphilis.

Abnormal lateral curvature is called *scoliosis;* it may be due to congenital deformity, muscle paralysis (as in poliomyelitis), bone or muscle changes due to metabolic disease, or habitually poor posture. To maintain balance, any curvature of the spine is accompanied by a corresponding curve in the opposite direction. Thus scoliosis usually consists of two curves: the original one and a compensatory curve. It may be due to weakness of the muscle of one side of the back, occurring as a result of poliomyelitis, rickets, sciatica, etc. Severe scoliosis causes deformation of the thorax and may thus result in displacement of internal organs such as the heart and lungs. In some cases, as in rickets, for example, treatment will primarily be directed at the underlying disease (Fig. 9-72).

What is particularly important is to prevent subsequent further deterioration of a spinal deformity acquired in childhood. This may be achieved by corrective exercises, suitable reclining positions, the wearing of a special corset, etc. In extreme cases an operation to stiffen the spine may bring improvement. Children and adolescents may develop a stoop because portions of the intervertebral discs penetrate into the adjacent vertebral bodies (Fig. 9-73a). This results in disturbed growth and degenerative changes of the vertebral bodies, which become tapered (Fig. 9-73b).

Postmenopausal changes in the hormonal pattern in women may cause calcium deficiency and thus cause spinal curvature (Fig. 9-74). Collapse of vertebral bodies due to tuberculosis affecting the spine (Fig. 9-75) may cause curvature that is quite sharp or more deeply rounded, depending on whether only a few or a larger number of vertebrae are involved. Crushing fracture of the spinal column may occur, for example, as a result of a fall or jumping from a great height; violent convulsion of the extensor muscles of the back, as in epilepsy, may also cause it (Fig. 9-76a). Such serious injuries to the spine may constrict the spinal canal and cause contusion of the spinal cord, which could in turn result in paralysis of the lower limbs (Fig. 9-76b).

SLIPPED DISC

The intervertebral discs in the cartilaginous joints between the bodies of adjacent vertebrae function as shock-absorbers and allow a certain amount of movement between the vertebrae. Each disc has an outer covering of cartilage, called the annulus fibrosus, with a deformable gelatinous mass, the nucleus pulposus, in the center. From early middle age onward the intervertebral discs are prone to undergo wear caused by compression and aging (Fig. 9-77), this being associated with flattening of the discs, reduction of mobility and sometimes pain due to pressure on nerves. Also, the tendency of the annulus to rupture or split is increased. As a result, the overloaded nucleus may be displaced. This condition is popularly known as *slipped disc,* which is a misnomer, because in fact the disc ruptures, but does not slip. Other names for slipped disc are prolapsed disc or hernia of the nucleus pulposus.

The nucleus can be displaced in any direction (Fig. 9-67). If it is displaced backward, so that it protrudes into the neural canal, it is liable to exert pressure on the spinal cord or an adjacent nerve root. If the hernia of the nucleus is to one side of the midline, it will press on a nerve root and cause pain. If it is in the midline, the spinal cord itself will be affected (Fig. 9-78). Symptoms will depend on the location of the hernia. Slipped disc is especially common in the lumbar region, where the vertebrae are most severely loaded (Fig. 9-79). A frequent direct cause of slipped disc is strain while the spinal column is flexed, e.g., when lifting a heavy object while the back is bent and the legs are straight.

Instead of undergoing displacement in a plane parallel to the adjacent vertebral bodies, the nucleus pulposus may in certain cases penetrate into the vertebral body above or below it. The vertebral body penetrated in this way undergoes local breakdown of the bone; then the foreign material, i.e., the herniated nucleus, gradually becomes cartilaginous and eventually calcifies. It is believed that such cartilaginous nodes or knots occur in the spines of more than 30 percent of all people, especially in elderly men. These nodes may cause premature flattening and hardening of the intervertebral discs and thus result in restriction of movement and in spinal curvature.

Pressure exerted by a slipped disc on the spinal cord causes pain which is felt in those parts of the body where the afferent nerves affected by the pressure arise. Sometimes there is paralysis of the legs, the rectum and the bladder. Hernia of the nucleus to one side of the midline, causing pressure on a nerve root, is commoner; in such cases the pain radiates along the afferent nerves to the periphery of the body. As a result of cross-connections with the nerves of internal organs, internal pain may arise. For instance, a slipped disc in the cervical area may cause cardiac pain.

If cartilaginous nodes have not yet formed, i.e., if the hernia of the nucleus pulposus is still soft and yielding, it will sometimes slip back into its normal position without external inter-

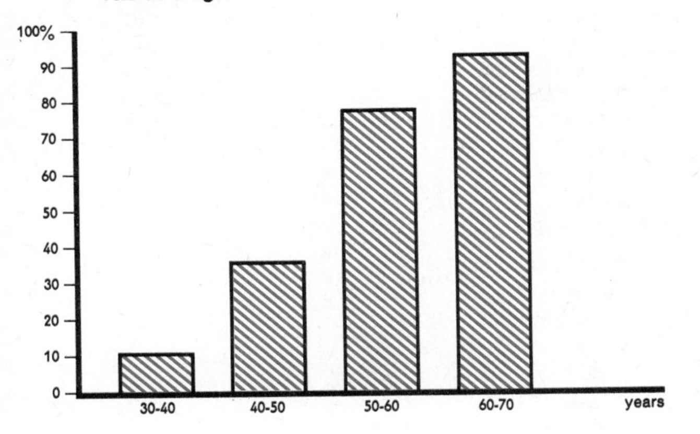

FIG. 9-77.
Frequency of defects due to wear of the spinal column as related to age.

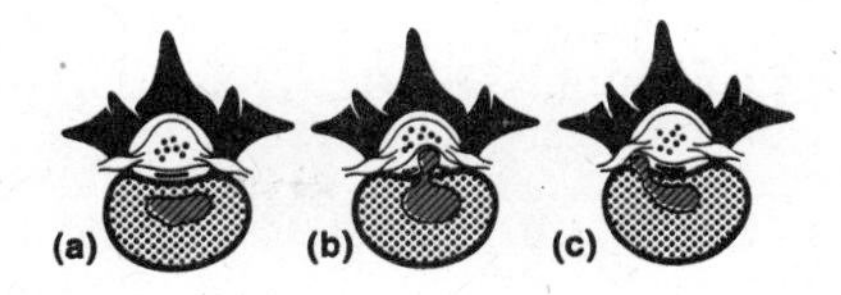

FIG. 9-78.
Slipped disc.

(a) Normal positon of nucleus pulposus.

(b) Nucleus displaced backward and exerting pressure on the spinal cord.

(c) Nucleus displaced sideways and exerting pressure on a nerve root.

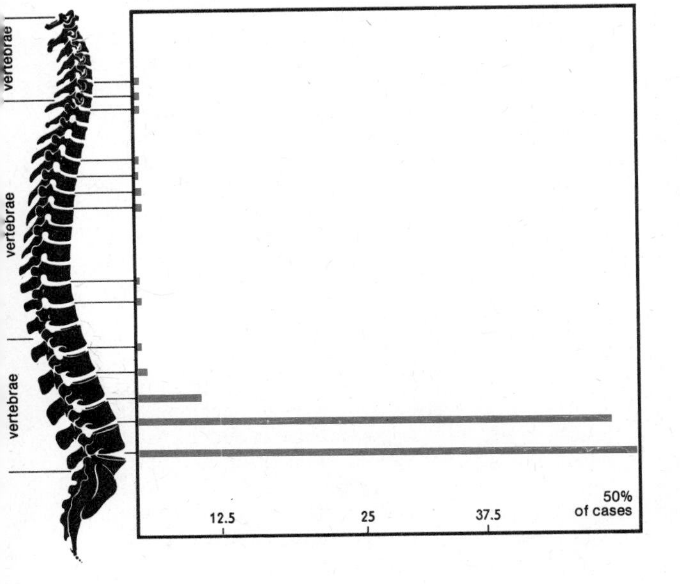

FIG. 9-79.
Frequency of slipped disc in the cervical, thoracic, and lumbar regions of the spinal column.

vention. However, the condition is likely to remain unstable in that the hernia may recur at irregular intervals, the direct cause being a sudden relatively violent bending or twisting movement of the trunk, e.g., when lifting a heavy load, or sometimes even merely stooping. A slipped disc that persists for a long time may cause sensations of tingling and formication ("crawling ants") in the lower part of the body. Stronger pressure on a nerve root will cause backache and pains resembling sciatica, which are intensified by coughing, sneezing, etc. In severe cases paralysis may develop. In this advanced stage of the condition the nucleus has turned into a permanently displaced cartilaginous or ossified node which may have to be removed by surgery.

In general, early intervention to correct a slipped disc has the best chance of succeeding. Manipulation and corrective exercises may suffice. In severe cases the patient may have to lie flat for some weeks. Occasionally the affected part of the spine will have to be splinted (immobilized) with a surgical collar or a special corset.

Part Ten

THE TEETH AND PALATE

All teeth are of similar structure in principle (Fig. 10-1). A tooth consists of the *crown,* which is the part that projects from the gums, and the *root,* which fits into a socketlike cavity (*alveolus*) in the jaw. The short portion between the crown and the root is called the neck of the tooth. The gum, a layer of dense connective tissue covered with mucous membrane and attached to the periosteum (the covering membrane) of the jaw, surrounds the neck. The main part of the tooth consists of a hard bonelike substance called *dentine* (ivory). Enclosed within this substance is the *pulp cavity,* or pulp chamber, extending from the crown down through the root, where a nerve and blood vessels enter the cavity. Nerves branch throughout the pulp cavity and into the dentine, which is thus sensitive to external stimuli from heat, cold, acid and sweetness. Dentine is formed by cells in the pulp which send out dentine fibers that subsequently become surrounded with a solid casing. It has a substantially higher calcium content than ordinary bone. When tooth growth has been completed, so-called secondary dentine can still be formed, which serves to protect the pulp cavity against tooth decay. The dentine of the crown is covered by a layer of *enamel,* which is the hardest and most resistant substance in the human body. Mastication causes wear of the enamel; tooth decay (dental caries) loosens and destroys the structure of enamel and dentine, spreading inward from the edges. Destroyed enamel is not replaced.

The root of the tooth is encased in a calcified substance, called *cement,* which bonds the tooth to the periosteum lining the alveolus; this lining is known as the *periodontal membrane.* The tooth is not completely fixed in the alveolus, but can perform slight rocking movements which stimulate the flow of blood to the tooth. The cement derives its nourishment from the periodontal membrane. So long as this nourishment is assured, the tooth will remain held in the alveolus even if the pulp is destroyed; but if the cement dies, the tooth is rejected. If the pulp cavity is penetrated from above, i.e., from the direction of the crown, bacteria can enter and eventually reach the apex of the root. This may give rise to an alveolar abscess or gumboil (see page 397). The gum and periodontal membrane form a protective collar round the neck of the tooth. It is here that periodontosis begins.

Two adjacent teeth form a functional unit with a tooth on the other jaw, with which they cooperate in mastication. Their biting surfaces fit together; this is called *occlusion* (Fig. 10-2); if the biting surfaces of the teeth of the upper and lower jaws are not in proper contact, the condition is called malocclusion (see page 399). Loss of a tooth disturbs the balance; in

FIG. 10-1.
Section through a tooth (molar).

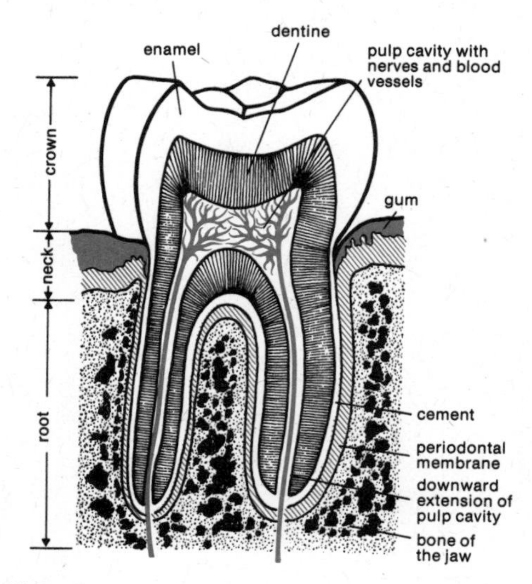

FIG. 10-2.
Each group of three teeth forms a functional unit.

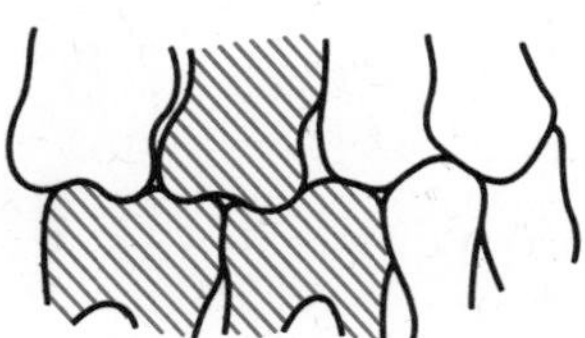

FIG. 10-3.
Temporary and permanent teeth.

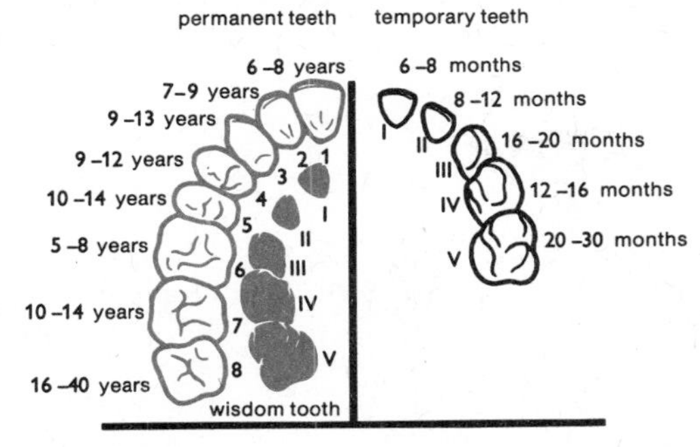

FIG. 10-4.
Development of permanent teeth, which dissolve the roots of the temporary teeth and displace the latter.

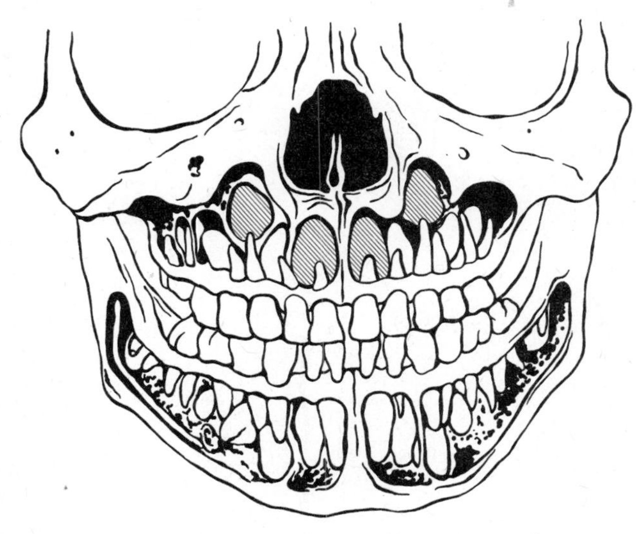

the course of time the positions of the adjacent teeth undergo a change in an effort to restore the balance and achieve something like correct occlusion.

According to their function, teeth are of three different kinds: *incisors* have a single root and a crown with a chisel-like edge for cutting; *canines* are long and pointed; *molars* (grinders) have two or three roots and a large flattish crown with rounded or pointed cusps. Man develops two sets of teeth: the *temporary teeth* (also known as milk teeth or deciduous teeth), which are shed in childhood, and the *permanent teeth* (Fig. 10-3). There are 20 temporary teeth, 5 in each half of each jaw; normally this set is complete by the age of 2–2½ years. The permanent set (Fig. 10-4) comprises 32 teeth, 8 in each half of each jaw: 2 incisors, 1 canine, 2 premolars and 3 molars. The molars have from 3 to 5 cusps and 2 roots (in the lower jaw) or 3 roots (in the upper jaw); each premolar has 2 cusps and a single root.

DENTAL CARIES

Dental caries (tooth decay) is progressive decalcification of the enamel and dentine. If untreated, it dissolves these substances until the pulp is exposed. Inflammation affecting the pulp may spread to the root and alveolus and may ultimately even result in blood poisoning. The teeth most likely to be infected by decay are the molars and the upper incisors (Fig. 10-5).

The underlying causes of tooth decay are not fully understood. There are factors that promote it and factors that initiate it. In the first place, the composition of the dental enamel itself is important; enamel that is poorly calcified in consequence of disturbed calcium and phosphorus metabolism increases the susceptibility to decay (e.g., as a secondary consequence of rickets, page 148). Chief among the local causes of decay is the presence of a coating (plaque) on the teeth. The parts where this more particularly tends to form are also those where decay is most likely to begin. Mainly these are crevices and depressions on the biting surfaces of the molars, which are difficult to clean; the interstices between the teeth and the neck areas of the teeth are likewise in the front-line of attack. Decay of the neck of a tooth usually begins at the border between the enamel and the cement and is most likely to occur if small lateral areas of the root lie exposed at the boundary with the gum (Fig. 10-6). The saliva is probably an important protective factor in the prevention of tooth decay; deficient saliva flow is often followed by rapid decay. Microbes present in the mouth, and fermentation processes associated with them, play an important part as initiators. The plaque that covers the teeth consists of mucus, saliva, epithelial cells and microbes; it forms a base in which carbohydrates (especially sugar) and caries-causing bacteria can accumulate. The time factor plays an important part in the process, for decay will start only if the caries-promoting agents contained in the plaque can act undisturbed on the tooth substance for a fairly long time.

Fermentation of sugar and other carbohydrates by certain bacteria (*Lactobacillus*) pro-

FIG. 10-5.
Frequency of decay affecting individual teeth.

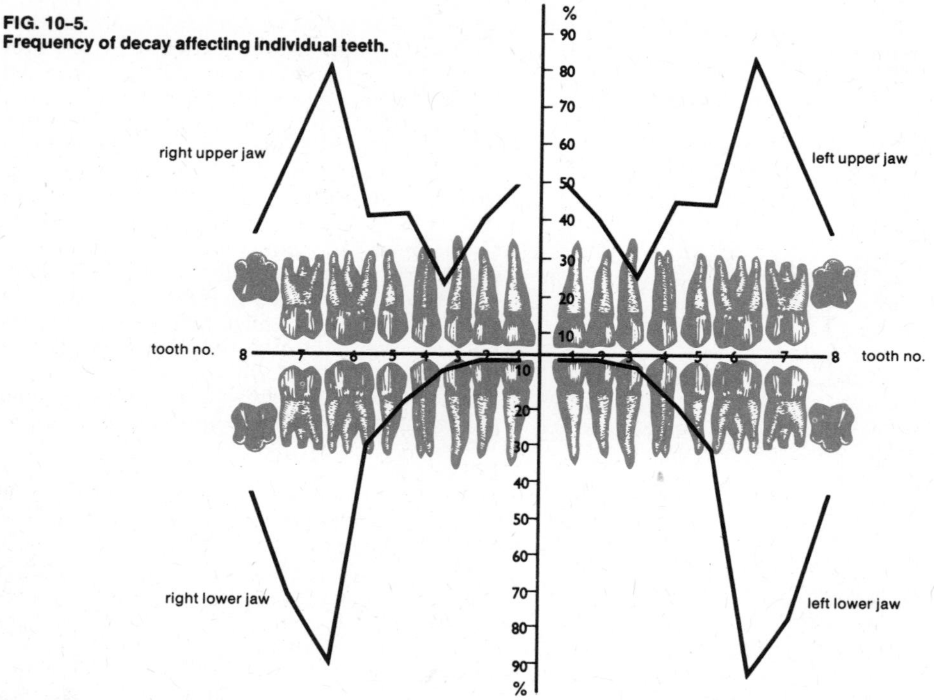

FIG. 10-6.
Areas where decay is likely to commence.

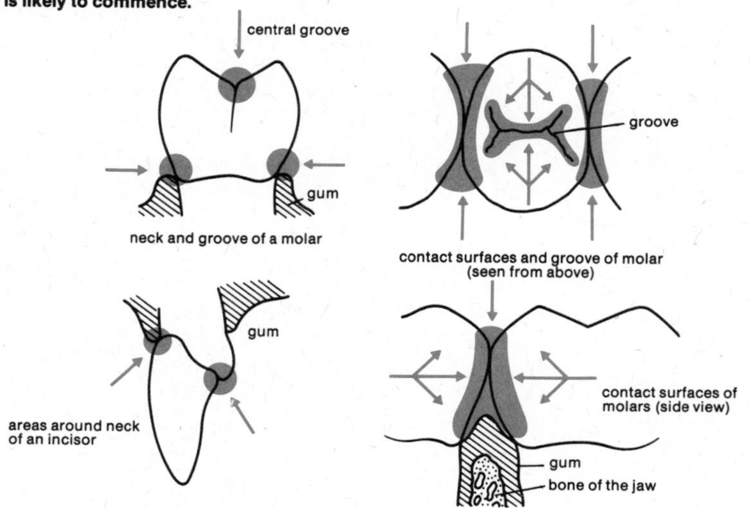

duces acids, notably lactic acid. It is this acid that is believed to be the real culprit: it dissolves the enamel and/or makes its way through tiny cracks in it and attacks the underlying dentine. The tooth may become sensitive to heat, cold or sweets when it contains a small cavity due to decay; as this cavity enlarges and approaches the pulp chamber of the tooth, the irritated nerves respond with toothache. Bacteria from the outside enter the cavity and attack the pulp, setting up inflammation. Sugar is especially harmful as a caries-promoting agent, particularly when taken between meals, because at such times there is little saliva flow.

Besides diet, there are many other factors involved in tooth decay and its causes: hereditary factors (a minority of fortunate people have teeth highly resistant to decay), mouth hygiene, endocrine gland activity, accumulation of tartar (dental calculus) on the teeth, character of the saliva, presence of crevices in the teeth, etc. Clean teeth are less likely to be attacked by decay than dirty ones; but toothbrushing alone (whatever brand of toothpaste is used) cannot guarantee protection from decay. Fluorine is an essential ingredient of bones and teeth. If the supply of this element is deficient, especially in the enamel, the teeth are more likely to decay. Small amounts of fluorine added to the drinking water supply can make up for the deficiency.

GUMBOIL

A gumboil is an acute inflammation of the periodontal membrane at the apex of the root of a tooth, developing into an abscess of the gum. The inflammation usually begins in the pulp cavity and spreads to the surroundings; eventually it also attacks the alveolus and bone of the jaw. A chronic alveolar abscess often starts as a small abscess or cystlike growth at the apex of the root, preceded by the formation of so-called granulation tissue (dental granuloma, Fig. 10-7); it may subsequently develop into a gumboil (Fig. 10-8). In general, the latter is an acute condition associated with a decayed tooth, or it may be the result of irritation or injury by a sharp denture.

The actual inflammation is often initiated by a catarrhal infection such as the common cold. First, the periodontal membrane around the apex of the root becomes inflamed. Very soon the adjacent bone and its marrow become involved. There is acute pain, and the surrounding gum is sore and sensitive to pressure. The pain is not only localized, but also radiates to the eye and ear on the affected side of the face. Soon there are throbbing pain, fever and general malaise. When the inflammation penetrates under the periosteum of the jaw, the tooth is loosened in its socket and the pain becomes very acute. The gum is swollen, tender and very painful. The patient has difficulty in opening his mouth, especially if one of the molars is affected. There is facial swelling, and in severe cases there is fever up to about 40°C (104°F). If the pus that forms in the abscess at the root of the tooth breaks out through the gum to the exterior, the pain subsides, though the swelling may continue. All this generally takes place within 24–48 hours.

The main risk associated with a gumboil is that it may develop into a more widespread infection. Thus a gumboil in the upper jaw may spread toward the nasal cavity, orbit or base of the skull (meningitis). Treatment will consist in draining the pus, which may require incision. The patient should be warned not to swallow the pus when the gumboil bursts. The tooth may have to be extracted. Antibiotics are given to prevent the infection from spreading.

The chronic alveolar abscess, on the other hand, at first usually causes little discomfort

FIG. 10-7.
(a) Normal tooth. (b) Dental granuloma.

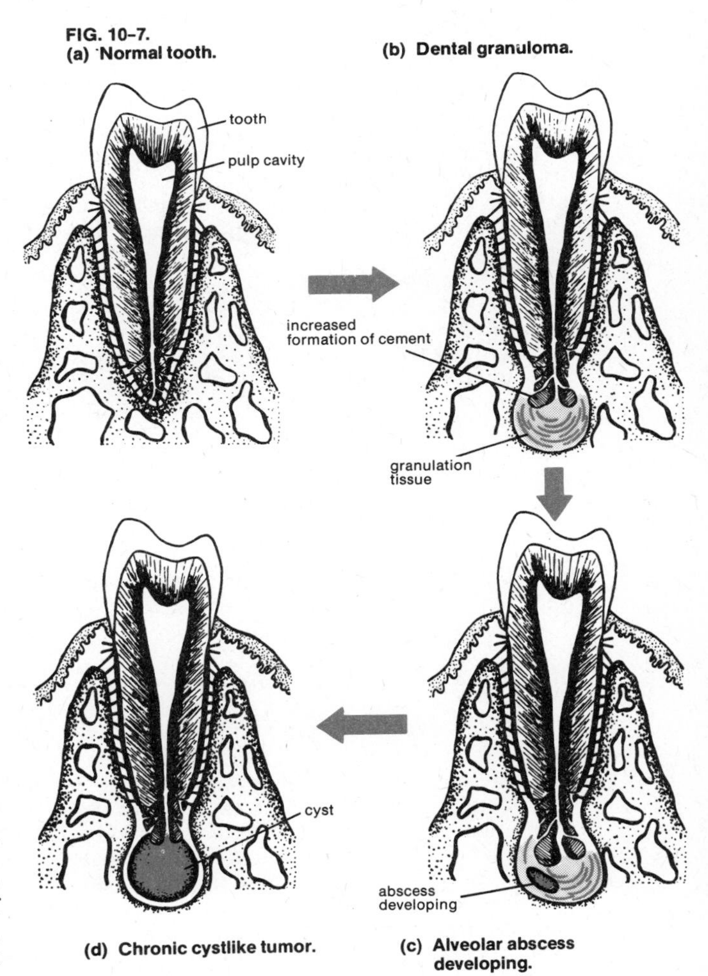

(d) Chronic cystlike tumor. (c) Alveolar abscess developing.

FIG. 10-8.
DEVELOPMENT OF A GUMBOIL.

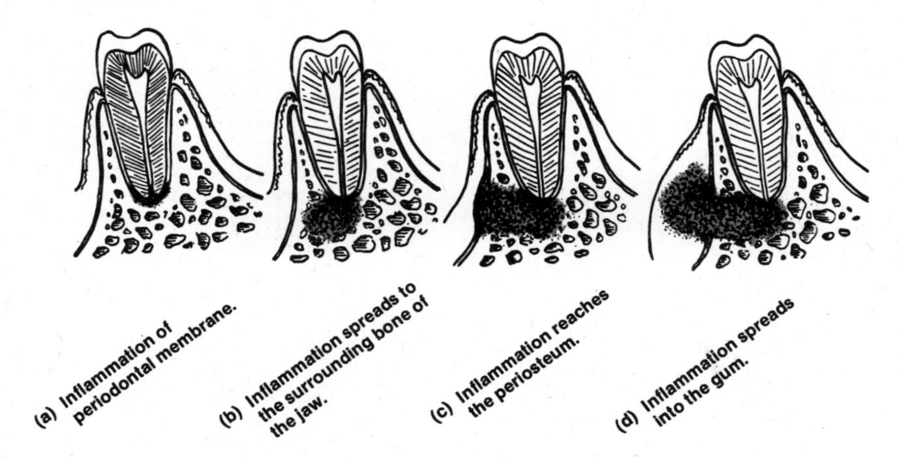

and may even remain undetected by the patient. Sometimes there is slight throbbing pain or tenderness at the root of the affected tooth. If the alveolar abscess or cyst at the root is an extensive one, the tooth may be loosened, and sometimes there is nerve pain radiating to the region of the eyes, ears, temples and back of the head. Quite often the patient does not suspect that these pains are in any way connected with a dental condition. Physical exertion or a catarrhal disorder aggravates the symptoms. The presence of the abscess can be detected by x-ray. In the early stages the condition can often be successfully treated through the root canal, i.e., the portion of the pulp cavity extending to the apex of the root. If this is unsuccessful, surgical removal of the affected tissue at the root may be necessary (resection of the apex: apicotomy). In some cases extraction of the tooth may be necessary to permit effective treatment.

DISEASES OF ALVEOLUS AND GUMS

Periodontopathic diseases affect the alveolus and gum (gingiva), i.e., the tissues that secure the tooth in the jaw (Fig. 10-9). They are associated with inflammation and/or with deterioration of these tissues. Noninflammatory degeneration is called *periodontosis* (or *parodontosis*), while inflammatory conditions are known by the general term of *periodontitis* (or *parodontitis*). These disorders are due to external and/or internal factors.

In periodontosis the gum and the bone of the jaw undergo gradual local degeneration, so that the teeth eventually look as though they had grown out of their sockets, this appearance being due to recession of the gums. There is no inflammation, however, and the affected teeth may remain firmly held in the jaw for a long time. Eventually, however, they will loosen, become displaced and perhaps drop out.

Inflammatory degeneration of the periodontal membrane, alveolar bone, cement and adjacent gum is called *pyorrhea alveolaris* (periodontitis). It is a common cause for loss of teeth after the age of about 35. The onset of pyorrhea (Fig. 10-10) is accompanied by inflammation of the gums (*gingivitis*), with recession of the bone of the jaw, so that the teeth become loosened. Factors that may cause pyorrhea include faulty bite (malocclusion), constant grinding of the teeth or clamping of the jaws (as a nervous habit), accumulation of tartar on the teeth, systemic diseases (e.g., diabetes and blood diseases), missing teeth, and faulty diet (lack of vitamin C). *Tartar* (dental calculus) is calcareous matter deposited on the teeth from the saliva; it shelters bacteria and will harm the gums if it is not regularly removed. *Concrement* is a tartarlike concretion which forms particularly in cavities between the gum and the neck of the tooth; together with tartar it is considered to be a major factor in causing these inflammatory conditions, besides constituting a serious hazard to the teeth. Pyorrhea causes the mouth to taste and smell unpleasant. In acute cases the gums are swollen and sore, bleed easily and are often puru-

FIG. 10-9.
Section through a tooth in position in the jaw.

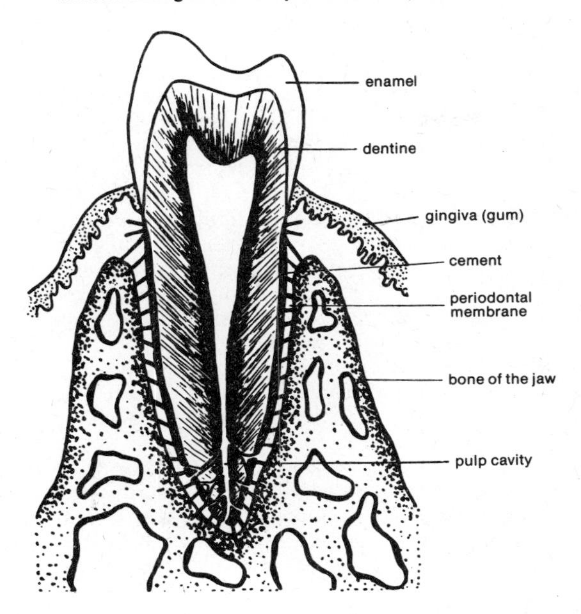

lent. The prevention and treatment of pyorrhea require regular removal of tartar, correction of malocclusion, replacement of missing teeth, medical treatment of underlying systemic disease, etc.

MALOCCLUSION

Malocclusion is imperfect occlusion of the teeth, i.e., the biting surfaces of the upper and lower teeth do not make contact properly. The branch of dentistry concerned with the correction and prevention of abnormalities in the position or alignment of teeth is called *orthodontics*. Normally the upper incisors fit in close contact over the lower incisors (this is known as *overbite*). Furthermore the cusps of the lower molars engage with the grooves of the upper molars, so that each pair of molars of one jaw forms a functional unit with a molar of the other jaw. Any deviation from this normal bite pattern is undesirable.

There are various forms that malocclusion can take: for example, one of the jaws may be narrower than the other (Fig. 10-11) or may be displaced sideways in relation to the other (crossbite, Fig. 10-12); the lower jaw may be incorrectly shaped so that the upper and lower front teeth cannot be brought into proper contact (open anterior bite, Fig. 10-13); the upper incisors may project excessively beyond the lower ones (close bite) or, alternatively, protrusion of the lower jaw (prognathism) causes the lower incisors to project in front of the upper ones.

Hereditary factors play an important part in causing malocclusion. External causes include dental caries (see page 395), systemic diseases, e.g., rickets (see page 148), and such infantile habits as thumb sucking, tongue sucking, lip biting, and continually pressing the tongue against the teeth. Also, many irregularities of the teeth are caused by premature loss of milk

FIG. 10-10.

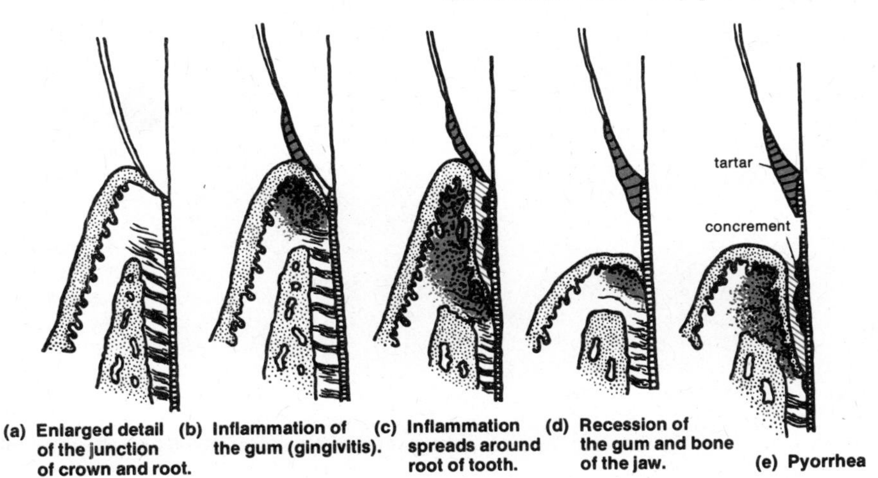

(a) Enlarged detail of the junction of crown and root. (b) Inflammation of the gum (gingivitis). (c) Inflammation spreads around root of tooth. (d) Recession of the gum and bone of the jaw. (e) Pyorrhea

teeth; then adjacent teeth will "move over" to fill the gap. Narrowness of one jaw in relation to the other is the most frequent cause of malocclusion and is a congenital deformity (Fig. 10-11). In the condition called crossbite (Fig. 10-12), the outer cusps of the upper molars engage with the longitudinal grooves of the lower molars, whereas with normal occlusion the outer cusps of the lower molars engage with the longitudinal grooves of the upper molars. Normally there is overbite, i.e., the upper incisors extend over the lower incisors, but this may be exaggerated (close bite) or reversed due to prognathism of the lower jaw, as already mentioned. However, this latter defect is often due also, and more particularly, to faulty guidance of the temporary incisors (milk teeth), resulting in subsequent faulty positioning of the permanent incisors.

In the case of open anterior bite (Fig. 10-13) it is not possible to bring the upper and lower incisors into contact. This defect may be caused by thumb sucking in babies or may occur as a consequence of rickets. If the habit is abandoned by the time the child is about 3 years old, no lasting defect is likely to develop. But if it is continued, the malocclusion will develop in the permanent teeth, particularly because the tongue, in speaking and swallowing, helps to perpetuate the defect. This may be a serious deformity when due to rickets. The incisors and canines then perform no function, so that mastication is impaired. There are often also defects of speech (lisping). The condition can be corrected by orthodontic treatment involving the wearing of a bridge or dental plate. In severe cases an operation for the removal of a wedge-shaped segment of bone from the jaw may be necessary (Fig. 10-13).

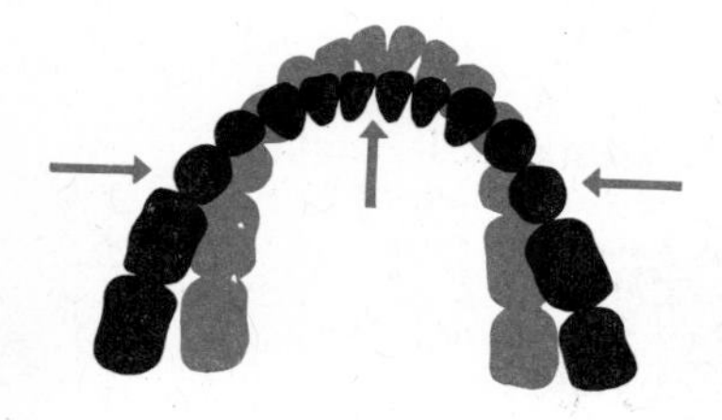

FIG. 10-11.
Malocclusion due to narrow jaw.

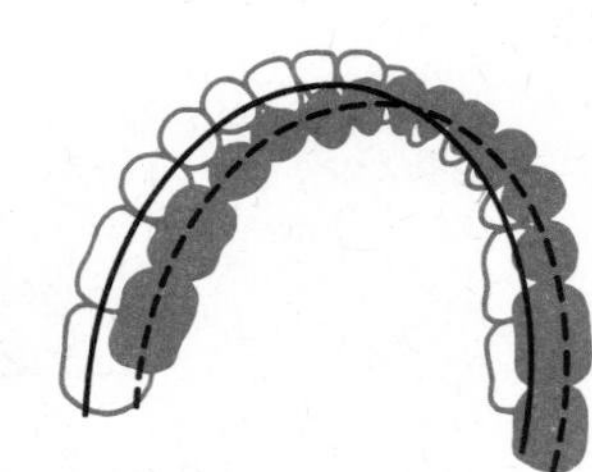

FIG. 10-12.
Crossbite.

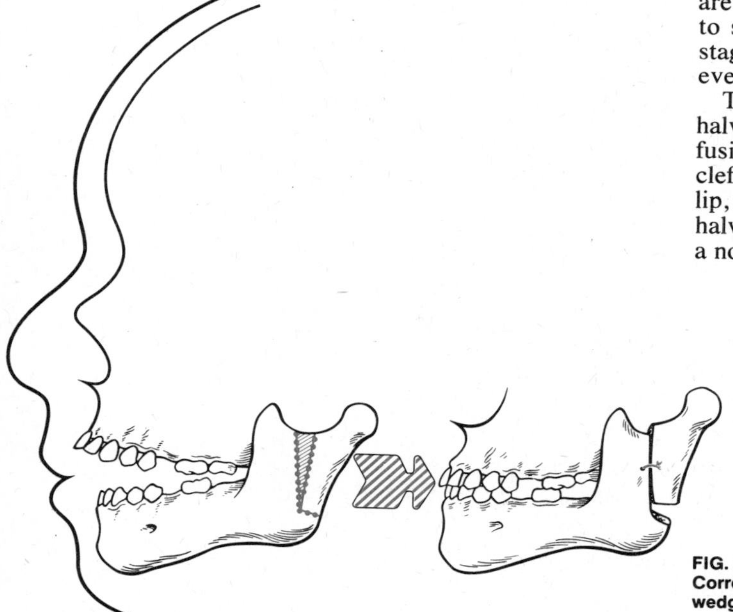

FIG. 10-13.
Correction of open anterior bite by surgical removal of a wedge of bone.

HARELIP, CLEFT PALATE

Harelip and cleft palate are congenital malformations occurring in about 1 baby in every 750–1000. It is believed that half of these cases are hereditary defects, while the others are due to some form of damage in an early prenatal stage. The precise cause is unknown, however.

The palate in the embryo develops in two halves which normally fuse in the midline. If fusion is incomplete, the condition is known as cleft palate. It is usually accompanied by harelip, which is due to a similar failure of the two halves of the upper lip to join, so that there is a notch giving it the appearance of a hare's (or

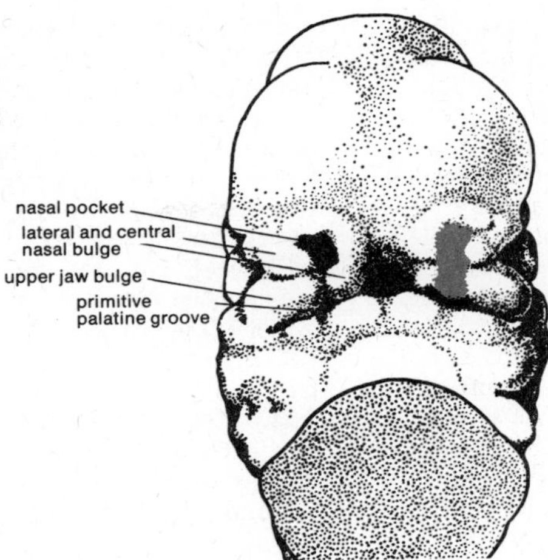

FIG. 10-14.
Diagram of human embryo in early stage of development (13 mm length). If a continuous cleft develops instead of the primitive palatine groove, it will remain as a cleft palate.

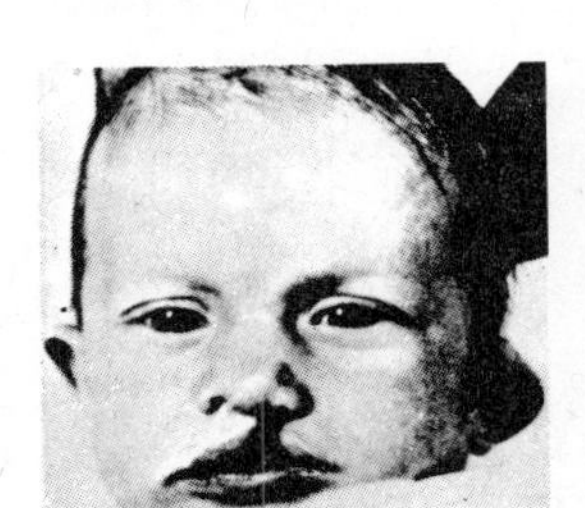

FIG. 10-15.
Harelip and cleft palate.

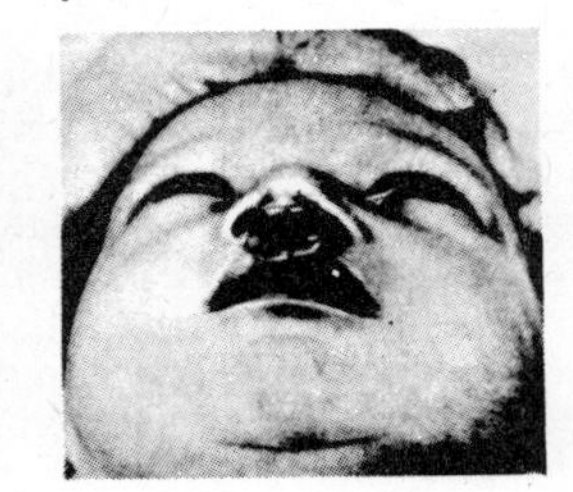

FIG. 10-16.
Bilateral harelip and cleft palate.

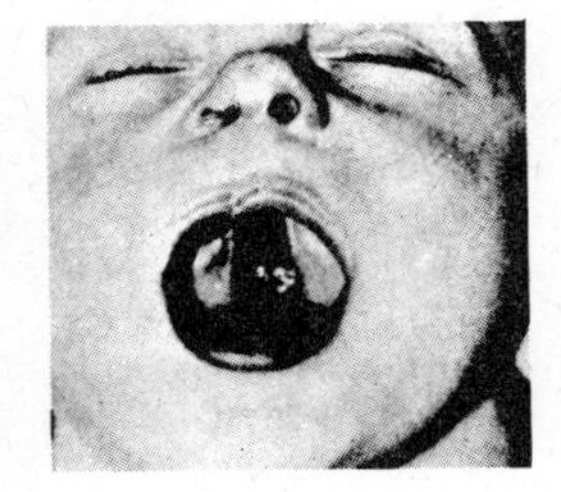

FIG. 10-17.
Cleft palate before corrective surgery.

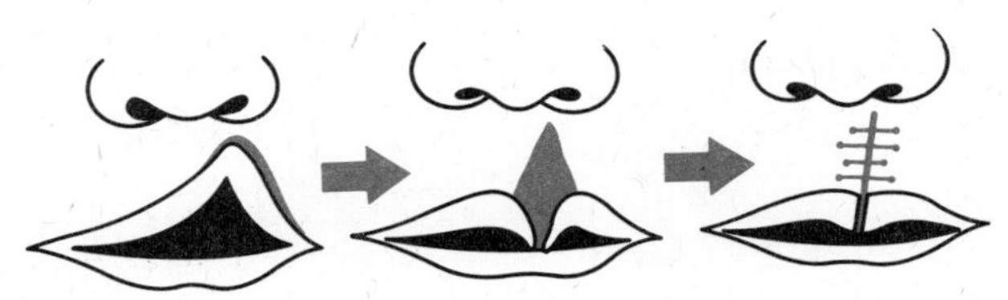

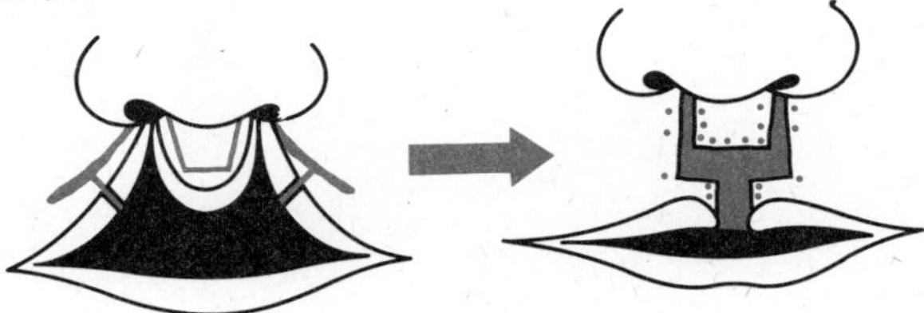

FIG. 10-18.
Operation to correct unilateral (upper diagrams) and bilateral harelip (lower diagrams).

rabbit's) lip. The defect is commoner in boys than in girls. These malformations are in part undoubtedly due to hereditary factors, as is evident from their reoccurrence in some families. The decisive stage in the development of the face occurs very early in prenatal life—in about the fourth to sixth week of pregnancy, when the embryo is only about 1 cm long. If a cleft develops in the embryo, a permanent gap will remain which can subsequently be corrected only by surgery (Fig. 10-14).

Harelip may be median, i.e., in the middle

of the lip, but is more often lateral; it may be unilateral or bilateral; it may be complete, when it extends into the nostril, or incomplete. The defect may range from a slight contraction of the upper lip to complete cleavage accompanied by deformity of the nose and by cleft palate (Fig. 10-15), so that the nasal and the oral cavity are in open communication with each other. The two conditions, i.e., harelip and cleft palate, may occur independently of each other, however.

Cleft palate may vary greatly in its severity. Thus, only the soft palate may be divided; or the soft palate and a portion of the hard palate may be divided. The cleft may extend through the hard palate to one side of the premaxillary portion of the palate, or it may extend to both sides thereof. Figure 10-16 illustrates a case of bilateral harelip and cleft palate, while Figure 10-17 shows the inside of a baby's mouth with cleft palate. These deformities make it difficult or even impossible for the baby to suck and thus obtain nourishment; special arrangements to feed it with milk may then be necessary, especially spoon-feeding, which involves a risk that the baby will "swallow the wrong way," so that milk enters the lung, with the attendant risk of pneumonia.

If the deformities are not corrected in infancy and childhood, the person affected by them will have to live with a permanent disfigurement, with some degree of speech defect which may be serious to the point of unintelligibility in a severe case of harelip and/or cleft palate. Development of the teeth is also likely to be defective. Plastic surgery in infancy and continuing through childhood can do much to improve the condition. Correct timing of such operations is very important. For instance, if surgery is deferred until the child has learned to speak, it may then be too late to correct the speech defect that has meanwhile developed.

Generally, treatment will comprise three main stages: (1) The clefts are closed by plastic surgery (Fig. 10-18). A corrective operation for the lips and nose is usually performed when the child is 4 or 5 months old; surgery on the cleft palate is performed much later, usually at 4 or 5 years. If the gaps are too wide for their edges to be united by stitching them together, a dental prosthesis (made of plastic) will have to be inserted; it will have to be altered and adjusted from time to time to keep pace with the growth of the skull. (2) Speech training is necessary to teach the child to speak intelligibly. This training should commence soon after surgical closure of the cleft palate and be continued consistently. If possible, speech should be intelligible when the child has reached school age; otherwise the child will have to attend a special school for the handicapped. (3) To prevent deformity of the jaw and malocclusion of the teeth, orthopedic treatment of the jaw will have to begin around the time that the temporary teeth are replaced by the permanent teeth and will continue until the child is 12–14 years old.

Part Eleven

THE SKIN

By means of the skin the body is brought into contact with the environment; at the same time the skin protects the body from harmful environmental influences (mechanical damage, pathogenic organisms, etc.). Its other main functions are sensation (through a large number of sensory nerve endings for touch, pressure, pain and temperature) and regulation of body temperature. The two principal layers of the skin are the epidermis, which is the superficial epithelial layer, and the dermis, which is the deeper layer of connective tissue (Fig. 11-1).

The *epidermis* is the cuticle or outer layer of the skin and consists of four layers or strata: germinal layer, granular layer, clear layer and horny layer. The deepest (innermost) of these is the germinal (or malpighian) layer, consisting of columnar cells which are constantly dividing and which contain granules of *melanin,* a dark brown pigment that protects these cells and the underlying tissues from ultraviolet light. The amount of melanin determines the color of the skin; in white-skinned people additional melanin is formed in response to exposure to sunlight (tanning). Melanin is also the pigment of the hair and the eyes. Abnormal absence of this pigment is called *albinism;* persons so afflicted are called *albinos*. They are unable to synthesize melanin. The cells of the granular layer of the skin contain granules which subsequently become *keratin,* a tough protein substance which is also present in hair and nails. These cells are flattened and dying. In the horny layer (stratum corneum) the cells are in fact dead; it consists of layers of keratin and forms a waterproof, tough protecting enclosure. The superficial layers are worn away and are constantly replaced by cells thrusting upward from the germinal layer; the regenerative power of the cells of this layer enables the epidermis to heal quickly after damage.

The *dermis,* or corium, is the layer of skin lying immediately under the epidermis and is composed of connective tissue. It comprises the papillary layer, whose surface is raised into tiny protuberances (papillae) which interlock with the epidermis (Fig. 11-2), and the reticular

FIG. 11-1.
Structure of the skin.

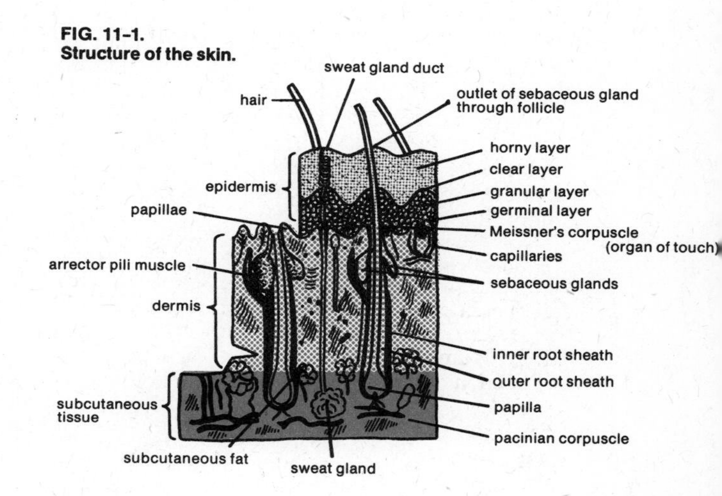

FIG. 11-2.
Blood vessels of the skin.

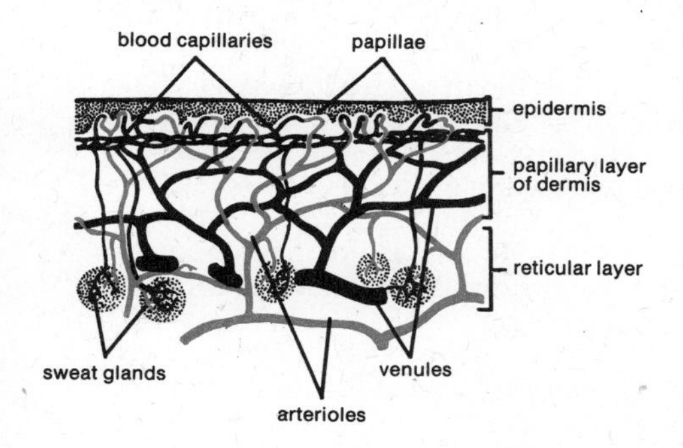

FIG. 11-3.
Glomus organ.

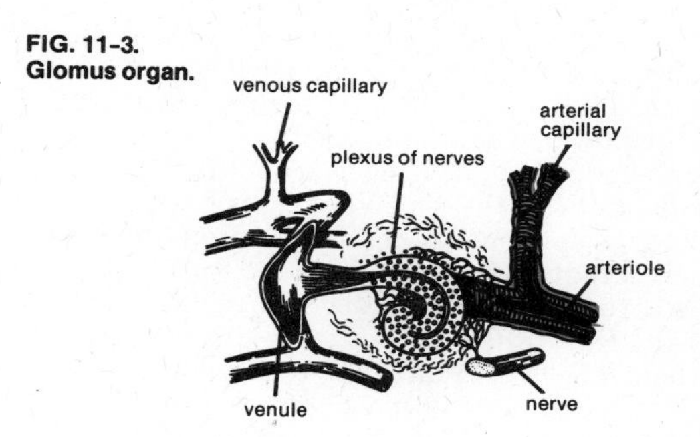

layer. A network of collagen fibers strengthens the dermis, which also contains numerous elastic fibers. The papillae contain an elaborate network of blood capillaries and nerve endings. Arterioles and networks of lymphatic capillaries are also present in the dermis. The papillary layer of the dermis determines the surface configuration of the skin, including the characteristic pattern of fine ridges on which the fingerprint technique of identification is based. Rupture of capillaries in the dermis causes an effusion of blood, appearing as a blue discolored area. The dermis also contains fibroblasts (cells that form connective tissue and play a part in the generation of the skin), histiocytes (cells that exhibit ameboid movement and phagocytic activity in biological defense processes), and mast cells (which are likewise important in defense, including blood coagulation). The reticular layer of the dermis contains the sweat glands as well as arterioles and nerves. In this part of the skin are located the so-called glomus organs (Fig. 11-3), which are well supplied with nerves and which control the peripheral blood circulation (particularly in the fingers and toes). Through the effect they exert upon the blood flow of the skin, they participate in the control of temperature and blood pressure. As they can respond very rapidly, thanks to their ample nerve supply, they can vary the blood flow as rapidly. They also play a part in blushing.

The dermis is separated from the underlying structures by a layer of loose connective tissue and adipose (fatty) tissue consisting of cells in which fat is stored in the form of globules (subcutaneous fat). This fat also acts as an insulating layer that reduces loss of heat from the body, besides having a mechanical protective effect (cushioning).

Hair follicles, sebaceous glands, sweat glands and nails are specialized structures derived from the epidermis and are known collectively as skin appendages. A *hair follicle* grows down into the dermis, where it forms a pit whose walls are the root sheath of the hair (Fig. 11-4). The bottom of the pit consists of a papilla which contains blood vessels and nerve endings. The hair itself is a thin flexible shaft of cornified cells which develops from the hair follicle and consists of a free portion (the shaft) and a root embedded within the follicle. There is a constant gradual loss and replacement of hair. Hair of the scalp lasts 2–5 years, that of the eyebrow 3–5 months. To each hair follicle is attached a small muscle (arrector pili) controlled by the sympathetic nervous system. When this muscle contracts, the hair stands up straight and the whole follicle rises, producing the appearance of "gooseflesh." Hair grows on average about 1.5 mm a week—somewhat faster in summer than in winter. The tendency to baldness appears to be inherited; sex hormones also play a part: women do not go bald unless their hormonal balance is upset.

FIG. 11-4.
Sebaceous gland and hair follicle.

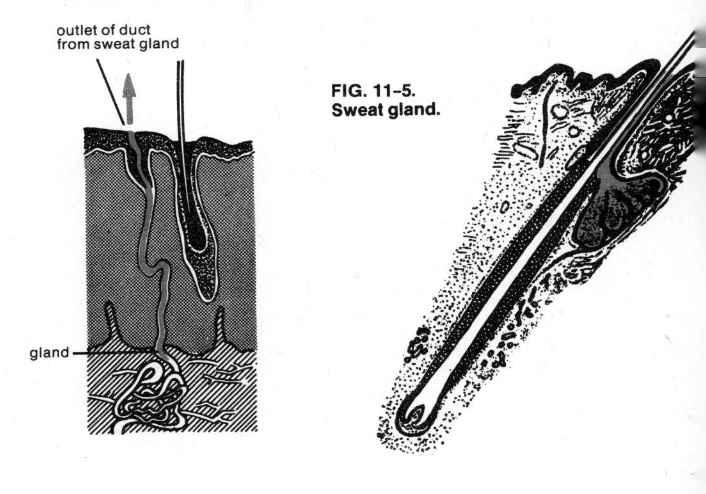

FIG. 11-5.
Sweat gland.

FIG. 11-6.
Temperature regulation.

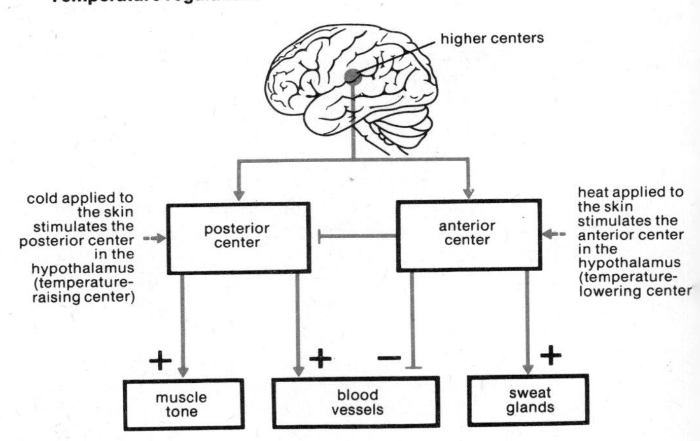

A *sebaceous gland* develops near the neck of every hair follicle and secretes an oily substance called *sebum* into the follicle. Sebum lubricates the hair and skin, protecting it from drying and cracking.

Sweat glands (Fig. 11-5), likewise derived from the epidermis, are coiled tubes laying deep in the dermis and communicating with the surface of the skin. Sympathetic nerves cause these glands to secrete a watery fluid which contains sodium chloride and some waste products of metabolism. The main function of the sweat glands is to facilitate heat loss from the body by the application of moisture to the skin. Evaporation of this moisture has a cooling effect.

Nails consist of keratin and are produced by multiplication of germinal cells under the *lunula*, the semilunar white arch near the root of the nail. The average growth rate of fingernails is 0.5–1 mm per week; it is somewhat slower in toenails.

RASHES

Rash is a general term applied to any eruption of the skin, especially associated with communicable disease and usually of a temporary character and of a red color. Related terms are eruption, lesion and roseola. *Eruption* refers more particularly to the appearance of a lesion, such as redness or spotting on the skin or mucous membrane. A *lesion* is a single infected patch in a skin disease (the term may also denote an injury or wound, however). *Roseola* is a skin condition marked by red spots of varying sizes (more specifically it refers to measles).

Macules (Fig. 11-7a) are small flat (nonraised) spots or colored areas on the skin which are caused by deposits of pigment embedded in the skin or by variations in its blood content. Pigments produced by certain groups of cells within the body and deposited in circumscribed areas may take the form of brown or black birthmarks or moles. *Petechiae* are small hemorrhagic spots which at first are red or purplish and then turn brown and continue for some time. They may be due to an abnormality of the blood clotting mechanism and are associated with certain severe fevers, such as typhus. *Purpura* denotes larger areas of discoloration, characterized by hemorrhages into the skin, which are dark red, then purple and finally brownish-yellow before disappearing. An example is idiopathic thrombocytopenic purpura, which is characterized by a reduction in the number of platelets in the blood, with bleeding from the mouth and skin upon slight injury. *Blood effusions* under the skin in consequence of bruising or similar injuries become discolored as a result of a local breakdown to blood pigment, changing in the course of a few days from red through blue or bluish-black to green and yellowish-green. *Papules* (Fig. 11-7c) are spots on the skin; their surface may be flat, convex or pointed; they generally leave no trace on healing. Larger circumscribed raised areas which extend into deeper layers of the skin may, however, leave scars, e.g., in smallpox. A *cyst* (Fig. 11-7d) is a closed sac or pouch with a wall that contains fluid, semifluid or solid matter. A *wheal* (pomphus) (Fig. 11-7b) is a pinkish-white elevation of the skin, usually more or less round but sometimes irregular in outline, often accompanied by itching. It is caused by the escape of blood serum from distended capillaries and can therefore be regarded as a circumscribed edema of the skin. Wheals may develop quite suddenly and may disappear very rapidly and without trace or may, in other cases, persist unchanged for a number of days. Urticaria (nettle rash, hives), insect bites, etc., are characterized by wheals. A *vesicle* (Fig. 11-7e) is a small breadlike blister containing serous fluid. Vesicles may be round, transparent, opaque or dark elevations of the skin. In herpes, for instance, they surmount an inflamed base. They occur also in chickenpox, smallpox and many other diseases. A *bleb* is a large vesicle; a *bulla* is a large irregular-shaped blister. A *pustule* (Fig. 11-7f) is a small elevation of the skin filled with pus, a pus blister. When the contents of a vesicle become purulent, it is called a pustule. Pustules occur in acne vulgaris, impetigo simplex and many other diseases. *Pimple* is a term somewhat loosely applied to any small papule or pustule. A *pock* is a pustule of an eruptive fever, particularly smallpox; it is characterized by its multilocular structure, i.e., consisting of many compartments. The eruption in smallpox illustrates the following evolution: macules become papules, which turn into vesicles and then pustules; finally scabs form. A *blister* (Fig. 11-7g) is a collection of fluid below the epidermis, usually the result of a burn and generally similar to a bleb or vesicle. A *callosity* (callus) (Fig. 11-7h) is a circumscribed thickening of the horny layer of the skin caused by friction, pressure or other irritation.

The foregoing types mostly occur as primary lesions, i.e., as the immediate consequence of an infection or an irritation. Some of them may also develop as secondary lesions in the course

FIG. 11-7.
Rashes and lesions.

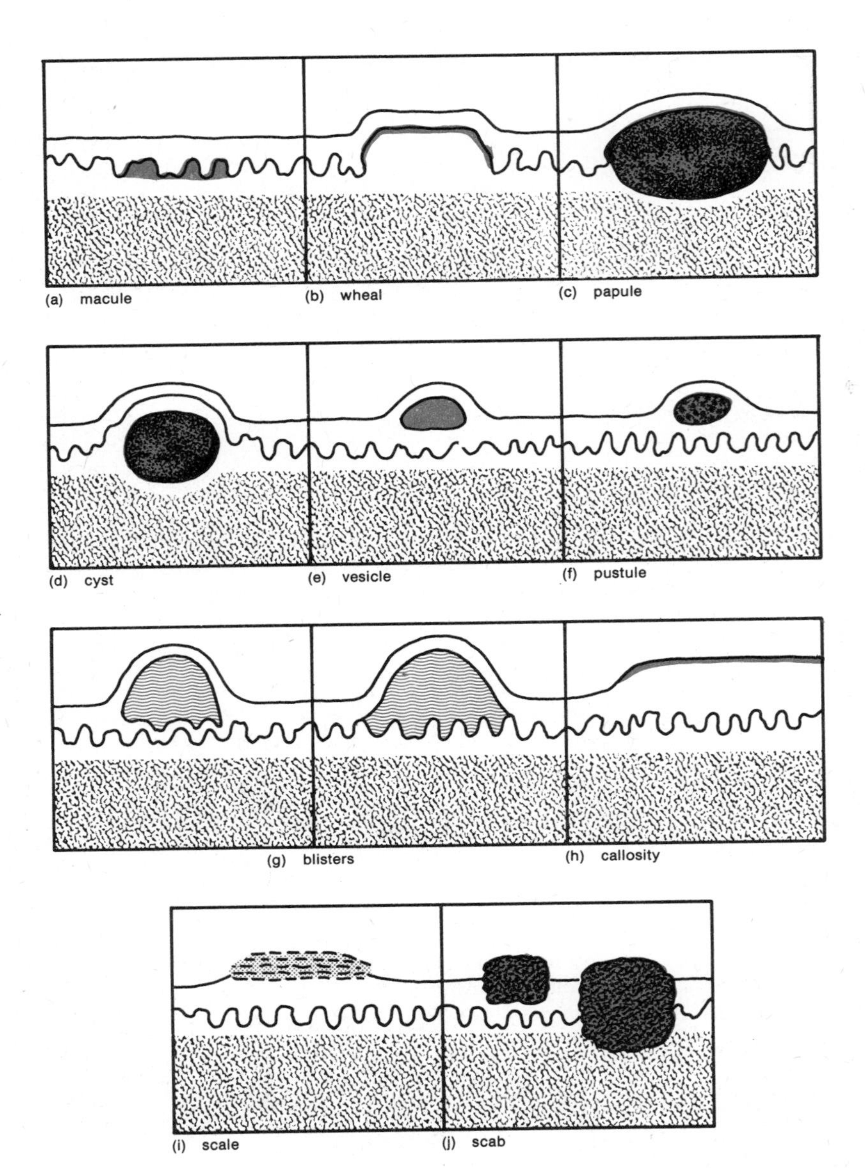

of healing or further progress of the disease, as exemplified by smallpox. The following lesions, etc., generally develop as secondary features. A *scale* (Fig. 11-7i) is a thin dry exfoliation shed from the upper layers of the skin and ranges in color from yellowish to grayish-black, depending upon its thickness. A common skin disease characterized by scaling is psoriasis (see page 408). A *scab* (crust) (Fig. 11-7j) is a secondary lesion that develops on a sore, wound, ulcer or pustule and is formed by drying up of the discharge. *Erosion* (Fig. 11-7k) is destruction of a surface layer by a physical or inflammatory process. Deeper erosion of the epidermis, occasionally extending into the dermis, is called *excoriation* (Fig. 11-7l). An *ulcer* (ulcus) (Fig. 11-7m) is an open superficial sore involving the epidermis and dermis accompanied by loss of substance and sometimes formation of pus. It leaves a scar when it heals. Ulcers may be classified in various ways, and there are many types. *Necrosis* (Fig. 11-7n) is death of an area of tissue; the dead matter (slough) becomes separated (sloughed) from the living tissue. It is caused by burns, frostbite, etc., but also by local de-

FIG. 11-7. (cont'd.) Rashes and lesions.

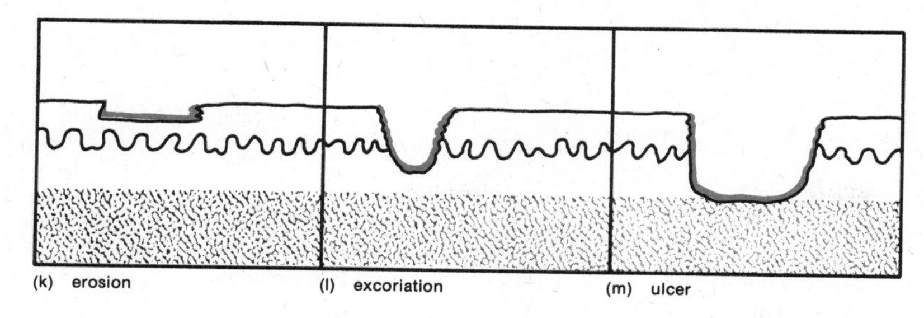

(k) erosion (l) excoriation (m) ulcer

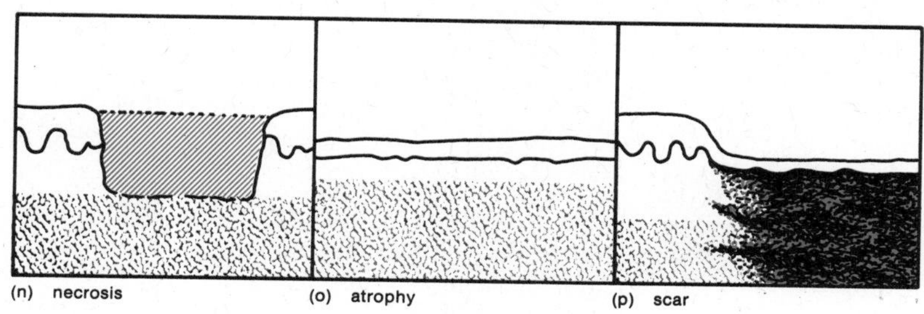

(n) necrosis (o) atrophy (p) scar

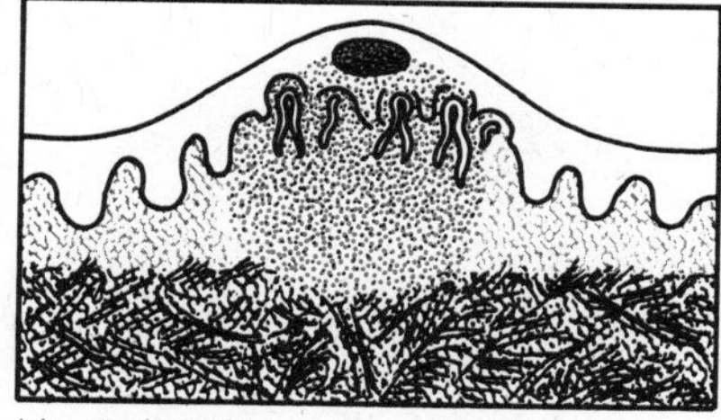

(q) papular vesicle

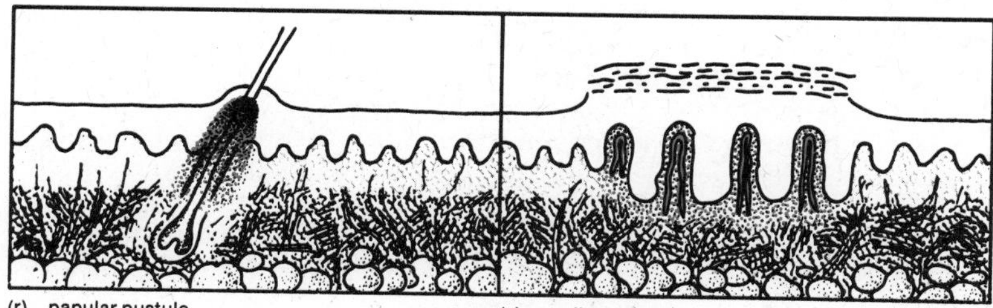

(r) papular pustule (s) scaling papule

ficiency in blood supply to the tissue. The term *necrosis* is usually applied to relatively small areas of tissue, while destruction of larger areas is called *gangrene*. A *scar* (cicatrix) (Fig. 11-7p) consists of fibrous connective tissue which replaces injured tissue when a wound or lesion heals. Scar tissue lacks the normal surface features of skin; it has neither hairs nor pores. At first it is red or purple, subsequently becoming whitish. A cicatricial scar is characterized by considerable contraction; a keloid scar is a red raised scar containing blood vessels. *Atrophy* (Fig. 11-7o) is a wasting of tissue due to lack of nutrition. On the skin it manifests itself in an overall reduction in thickness. *Lichenification* is thickening and coarsening of the skin caused by continued irritation. *Sclerosis* is thickening and hardening of the skin with loss of flexibility, due to excessive growth of fibrous tissue.

In addition to the various lesions described here, there are composite types, e.g., papular vesicles (Fig. 11-7q) as in eczema, papular pustules (Fig. 11-7r) as in inflammation of a hair follicle, or scaling papules (Fig. 11-7s) as in psoriasis.

DERMATITIS, ECZEMA

In present-day use the terms *dermatitis* and *eczema* are practically synonymous. They cover a variety of acute chronic disorders of the skin, many of which are of an allergic character. In a general way, eczema denotes red skin rashes that come in patches and which may or may not itch or burn. The skin may be swollen, blistered, scaly or oozing ("weeping eczema"). In many cases the cause is a food allergy, i.e., reaction to particular foods, especially in infants. There are two main groups of causes: (1) *external:* allergic contact, reaction to certain microorganisms, mechanical irritation (e.g., by rough clothing), irritation by chemicals; (2) *internal:* genetic factors, psychosomatic factors, disorders of internal secretions, etc.

Contact dermatitis is inflammation and irritation of the skin caused by contact with an irritating substance, e.g., a chemical. Apparently innocuous compounds such as perfume, deodorants and soaps may produce dermatitis in particularly sensitive individuals. Contact dermatitis is a common occupational disease in those who handle chemical irritants. In such cases it is usually the hands, forearm and face that are affected. Broadly speaking, the cause is either a direct toxic action of an irritant (toxic contact dermatitis) or an allergic reaction (allergic contact dermatitis). The causative agent (allergen) in allergy is generally a protein, but in contact dermatitis a simpler chemical agent acts probably by altering the human body's own proteins to become "foreign" proteins. A vast number of substances can cause contact dermatitis, but only in individuals who are susceptible to the causative agent in question. They include textiles, dyes, plants (e.g., poison ivy), drugs, cosmetics, lubricants, rubber (in gloves or clothing), metals (nickel, chromium), turpentine (in polish, varnish, etc.), and many others. More particularly in industry such substances as cement, leather, oils, greases, solvents, exotic woods, and synthetic resins are common causes of contact dermatitis. Cosmetics that may cause it include face lotions, after-shave lotions, hair dyes, mouthwashes, and bath oils. Antiseptics, ointments and many drugs can also give rise to contact dermatitis in certain persons.

Atopic dermatitis is a mild chronic inflammation of the skin, occurring as a patchy rash with itching. It may be accompanied by slight asthmalike difficulty in breathing and is attributable to an allergy, sometimes associated with a more definite allergic condition such as hay fever. Much of the skin trouble itself is due to scratching in response to itch rather than the actual disorder. This type of dermatitis, though unpredictable, generally clears up after a time.

There are many other types of dermatitis or eczema. For example, *seborrhea* (seborrheic dermatitis) is a kind of eczema characterized by overactivity of the sebaceous glands. It is liable to develop in greasy skin, especially on the scalp, forming rounded or irregular lesions covered with yellowish or grayish scales. On the scalp it may be dry with abundant scales or it may be oozing and crusted (*eczema capitis*). As a rule, seborrhea of the scalp is mild, with small scaly spots and excessive formation of dandruff. It can usually be treated quite effectively by frequent shampooing and lotions for removing grease and scales. Generalized seborrheic dermatitis requires careful attention, including skin hygiene and keeping the skin dry with the aid of dusting powders.

Babies sometimes have *infantile eczema,* which usually disappears in the course of time, but may develop into a dangerous complication when the infant is vaccinated. *Actinic dermatitis* is the reaction of the skin to sunlight, ultraviolet light or x-rays.

PSORIASIS

Psoriasis is a genetically determined and usually chronic skin disorder characterized by lesions (in the form of thickened red spots) covered with grayish or silvery scales of dead skin which drop off. The lesions tend to occur mainly on the elbows, knees and scalp, but may involve the entire back and thighs as well as other parts of the body. Also, they may merge with one another. The face is fairly seldom affected, however. Psoriasis occurs in both sexes and is not confined to any age group. It may begin at any age. Even without treatment the disease tends to come and go; it often improves in the summer, only to return in the autumn or winter, or vice versa.

Psoriasis starts with small red spots on the skin, which soon become covered with adhering whitish scales. If it is attempted to scratch these off, small bleeding spots (papillae) appear (Fig. 11-8). Older lesions shed their scales. The disease has no harmful effects, and there is usually no itching, but the lesions may

FIG. 11-8.
Bleeding spots caused by scratching the lesions.

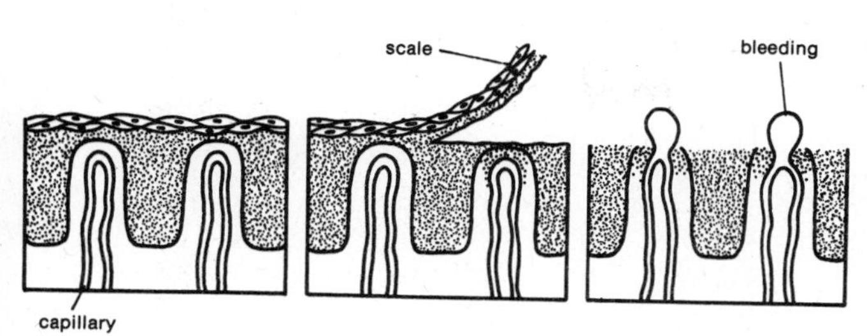

be distressing to the patient because of the disfigurement they cause. It may, however, be accompanied by a form of arthritis resembling rheumatoid arthritis. Also, the nails are often affected, becoming yellowish-gray, horny and thickened, so that they tend to lift off the nail bed; in other cases they may become pitted. As a rule, psoriasis affects the extensor side of the limbs, particularly the elbow and knee, but also the trunk and scalp. It may, however, occur "in reverse," i.e., affecting the flexor side of those joints, as well as the soles of the feet, the palms of the hands and folds of the skin generally. The disease is chronic in that it is liable to continue for years or indeed throughout the patient's lifetime, though varying in intensity. With treatment it can generally be cleared up for a time, but not permanently cured. The disease is not contagious and occurs only in individuals with a hereditary predisposition for it. Its precise cause is not known. An actual attack may be brought on by injury or acute illness (e.g., tonsillitis or pneumonia). It appears that in psoriasis the whole process of skin cell growth is abnormally speeded up: new cells are formed at a faster rate than dead ones are shed, so that local thickening occurs.

Sunlight and ultraviolet radiation can be effective in controlling psoriasis in some cases. Although there is no specific cure, treatment with coal tar, salicyl or sulfur ointments produces its effects by hastening the shedding of dead skin to keep pace with the overactivity of cell growth in the deeper layers. Other and more powerful agents producing the same effect have lately been introduced for the treatment of this disease. Drugs to discourage overproduction of cells have also been tried.

SHINGLES, FEVER BLISTERS, CHICKENPOX

Shingles (herpes zoster) and fever blisters (herpes simplex) are infectious diseases characterized by vesicular eruptions occurring on circumscribed areas of the body. Both are viral infections, but a close link between them has not yet been definitely established. On the other hand, it is known that shingles is caused by the same virus as chickenpox.

Herpes zoster is essentially a disease affecting the nerves. The virus attacks certain nerve ganglia, particularly the spinal nerves that serve the chest or one of the nerves that serve the face. The skin lesions appear in the area served by the affected nerves and usually take the form of small blisters (vesicles) on reddened bases. The condition is often very painful. The patient is likely to feel unwell and feverish for some days, especially if the vesicles become purulent. Although it may occur on any part of the body, herpes zoster usually appears on one side of the chest or face. The nerves affected on that side of the face and forehead may cause the eye to be involved, which can be a serious complication requiring special attention, preferably supervised by an ophthalmologist. Occasionally, too, shingles can cause deafness in one ear or loss of hair on one side of the scalp. These and similar complications can usually be prevented from becoming serious by prompt treatment. For the disease itself there is no specific treatment, however. The aim should be to make the patient as comfortable as possible and relieve the pain (drugs, pain-killing injections, etc.). The incubation period of herpes zoster is from one to three weeks; the total duration of the disease varies from about two to four weeks. It lasts longer and is usually more serious in adults than in children. If all the vesicles appear within about 24 hours, the total duration is likely to be short. In patients whose physical resistance is low, e.g., in those suffering from cancer, the vesicles may spread to all parts of the body. As a rule, however, the areas affected are circumscribed, and the disease involves little danger to life. In a few cases the patient continues to suffer neuralgic pains long after the infection has subsided.

Chickenpox (varicella) is an acute, highly contagious viral disease characterized by an eruption that appears in successive crops, developing from macules into papules, vesicles and crusts. As a rule, the disease involves no danger to life, except in a very severe type in which the lesions become gangrenous (which rarely occurs, however). The patient recovers in a few days, but remains infectious until all the spots have disappeared.

Herpes simplex (fever blisters) is an infectious, usually harmless, viral disease characterized by inflamed blisters (vesicles) which appear in most cases around the mouth. They may, however, sometimes occur on the genitals, where they are likely to be painful, or on other parts. It seems that many people carry the virus with them, which becomes active only in conjunction with some other infection, or on exposure to severe heat or cold, or as a result of even a quite minor injury. The blisters themselves do not cause fever, but are so named because they are often associated with feverish disorders such as influenza or the common cold. Some women regularly get fever blisters at each menstrual period. Strong ultraviolet radiation, as encountered by skiers and mountaineers in snowy surroundings at high altitudes, may cause fever blistering on the lips and nose.

None of the viral diseases described here confers immunity, and they cannot be prevented by protective vaccination.

BIRTHMARKS

Birthmark (*nevus*) is a general term denoting several types of congenital but nonhereditary blemishes associated more particularly with the pigment cells and blood vessels of the skin. The *mole* (pigmented nevus) is a discolored spot raised above the surface of the skin. Intradermal moles, i.e., wholly within the skin, are generally harmless unless irritated. Some types of mole may become malignant. Another form of birthmark is the *vascular nevus,* of which there are several types, characterized by enlargement of superficial blood vessels. These birthmarks are of variable size and shape, slightly elevated, reddish or purplish. The commonest vascular nevus is the *strawberry mark,* which may grow larger before shrinking and disappearing spontaneously. The *spider nevus* is characterized by red lines radiating from a central spot. It is due to dilation of capillaries and occurs mostly in later life, often associated with disease of the liver. *Nevus flammeus* is a circumscribed red or purplish discoloration of the face or neck, usually not elevated. It occurs quite commonly in babies, but often disappears in early childhood. Other, more persistent forms remain unchanged through life and can be a major disfigurement because of their size and color. Dark ones in particular are known as *portwine stains.* These are varieties of so-called flat angioma. The term *angioma,* essentially synonymous with nevus, more specifically denotes a congenital tumor composed of blood and lymph vessels; it is generally harmless and considered to be a remnant of misplaced fetal tissue. A *capillary nevus* consists of dilated capillary blood vessels elevated above the skin. *Nevus pilosus* is covered with hair; it consists of a conglomeration of pigment cells and closely spaced hair follicles. Small ones are generally no more objectionable than moles, but in a few cases they may cover large areas of skin and constitute a serious disfigurement. The *cavernous hemangioma* is a congenital tumor of larger dilated blood vessels, ranging in size from about half a centimeter to 10 cm or more. It commonly involves the subcutaneous tissue and may pulsate or throb. As a rule it is nonmalignant, but in rare cases it may increase so rapidly in size as to pose a threat to adjacent organs or bleed excessively on injury. Surgical treatment then becomes necessary. *Freckles* are small yellow and brown pigmented spots which are due to changes in the pigmented layer of the skin. Exposure to sun and wind stimulates this layer to produce spots of pigment.

These various types of nevus and related conditions are not diseases. For the most part they must be classed as cosmetic blemishes. Nevertheless, any mole or other dark-pigmented spot that suddenly begins to grow in size or change color should be viewed with suspicion for possible malignancy.

WARTS

A wart (*verruca*) is a small rough tumor on the skin resulting from hypertrophy (excessive growth) of the papillae and epidermis. Warts are caused by viral infection and differ from most other tumors in that they tend to disappear spontaneously, though the virus may lie dormant and subsequently produce another wart. Children and young women are sometimes affected by numerous so-called *fugitive warts,* which appear as reddish-yellow flat round spots; these may be particularly persistent and numerous, e.g., on the face, are spread by scratching and are communicable from person to person, but only to those who are susceptible to the virus.

The *common wart* is seen mostly on the skin of the hands and can be removed by surgery, freezing (with CO_2 snow) or the application of caustics. Such warts range from pinhead to pea size or larger, with a grayish-yellow, horny, irregular surface and a stem which penetrates into deeper layers of tissue. Following an injury, a number of secondary warts may develop around the original wart. As a rule, the common wart causes little or no discomfort, unless it is on a part of the body where it is subjected to pressure, particularly on the sole of the foot (plantar wart), where warts are painful and require treatment.

Seborrheic warts (senile warts) are flat greasy warts that occur about the face and neck of elderly persons. They are thought to be a form of seborrhea (overactivity of the sebaceous glands) and probably are not caused by a virus, so that in this sense they are not true warts. They occur frequently on the back, tending to spread downward. These warts are to be regarded as a normal accompaniment of aging; they are unobjectionable except as a blemish on the face.

The *fig wart* (verruca acuminata or condyloma acuminatum) is a reddish cauliflower-shaped tumor which occurs most frequently on mucous surfaces, especially the genitals, in which case it is known also as a venereal wart. It is caused by a virus which is believed to be the same as that which, in different circumstances, also causes the common wart. Fig warts can be transmitted from person to person by sexual intercourse, though they are not normally classed as venereal diseases in the more specific sense. They also occur in the anal region and, more rarely, in the mucous membrane of the mouth. These warts are liable to itch and to give off a foul-smelling secretion.

Molluscum contagiosum is an infectious disease of the skin caused by a virus and characterized by pinhead- to pea-sized hemispherical white tumor formations on the skin. These contain semifluid caseous (cheesy) matter and eventually heal without scarring, though they may suppurate. They occur singly or in rows and may be numerous, especially on the face and neck.

FUNGAL DISEASES OF THE SKIN

Dermatomycosis is a general term comprising a number of skin diseases which are caused by certain microscopic fungi or molds of the genera *Trichophyton, Epidermophyton* and *Microsporum.*

One such disease, caused by *Trichophyton schoenleini,* once common but now rarely encountered, is *favus,* which affects the scalp of children. It is characterized by cup-shaped yellowish crusts which form over the hair follicles and are accompanied by itching and a musty smell. Each crust (scutulum) is pierced by the hair around which it had developed. When the crusts fall off, the hairs beneath fall out, leaving bald patches.

Ringworm (*tinea*) is a term denoting infectious skin diseases caused by various species of fungus belonging to the genera *Microsporum* and *Trichophyton.* Ringworm may affect the scalp and is then known as tinea capitis, or the body (tinea corporis), or the beard (tinea barbae), or the nails (tinea unguium). In ringworm of the scalp (true ringworm) the fungus is in the hair follicles. There are reddish patches, more or less circular in shape, which are practically devoid of hair but covered with scales. The disease may also occur in a more deep-seated form, characterized by reddish flat tumors, sometimes discharging pus through dilated follicular openings. Ringworm of the body begins with red, slightly elevated scaly patches consisting of minute vesicles or papules. New patches develop from the edges, while the central portion clears up (hence the "ring"; there is, of course, no "worm" involved). In addition to the parts just mentioned, ringworm may occur in the groin (dhobie itch) or in the anal or genital areas, or it may occur between the fingers or the toes (athlete's foot). Personal hygiene can help to control the disease, and antiseptic footbaths may be beneficial in dealing with athlete's foot, but such measures cannot prevent the spread of the infection from person to person. Ringworm of the body may be transmitted to man by infected domestic pets or cattle. There are many ointments and lotions available for the treatment of ringworm, e.g., preparations of salicylic acid, antiseptic dyes, aluminum acetate, propionic acid. Different species of fungi are involved, and they respond to different remedies, so that a certain amount of trial and error may be necessary before the most effective treatment is established. Griseofulvin, an antifungal antibiotic for oral administration, can cure cases of ringworm, especially those of the *Trichophyton* type, which do not respond well to local treatment, in particular if the hair or nails are infected. Formerly, effective treatment of these conditions often necessitated removal of large areas of hair or infected fingernails.

Actinomycosis is caused by infection with *Actinomyces* (ray fungus), a microbe intermediate between fungi and bacteria. It usually invades the mucous membrane of the mouth, where it causes chronic suppurating lesions; other parts of the body may also be affected (lungs, gastrointestinal tract, brain). Infection may result from the chewing of stalks of grass or straw and is likely to center upon a decayed tooth as the focus of the disorder. The pus dischargd by the lesions is highly infectious. Prolonged treatment with penicillin is usually effective.

Sporotrichosis is a chronic infection of the skin and superficial lymph nodes, characterized by formation of abscesses, nodules and ulcers; it is caused by a fungus (*Sporotrichum*). It usually starts on the lower jaw, on the neck or below the ear, forming tumors which may grow in size to become as large as a hen's egg and which subsequently break down, with considerable wasting and loss of tissue. The disease is regional, occurring in France and the Middle West of the United States, for example, and is endemic in Rumania.

INSECT PARASITES OF MAN

Lice are small wingless insects which live as ectoparasites (i.e., on the outer surface of the body) on birds and mammals. There are numerous species, several of which live on man and are responsible for the transmission of disease, including typhus and relapsing fever.

The *head louse* (*Pediculus humanus capitis*) is about 2 mm long and gray with black abdominal markings (Fig. 11-9a–c). The female lays her eggs close to the scalp; they are small whitish objects called nits and are glued firmly to the hair, so that ordinary combing does not dislodge them. The louse feeds on blood sucked from the host. Louse bites itch and invite scratching, which may result in suppurating wounds. In a case of severe infestation (*pediculosis capitis*) there may be inflammation and secondary infection by bacteria, with formation of pustules, crusts and suppuration, while the hair may become matted and give off a disagreeable odor. There are many treatments for head louse infestation, e.g., benzyl benzoate emulsion rubbed into the scalp. In severe infestations the hair may have to be cut short to permit effective treatment.

The *body louse* (*Pediculus humanus corporis*) (Fig. 11-9d) is about 4 mm in length, i.e., a good deal larger than the head louse. It lives on or in the clothing worn next to the body, lays its eggs in the seams of the clothing, and

FIG. 11-9.
Lice (magnified 20 times).

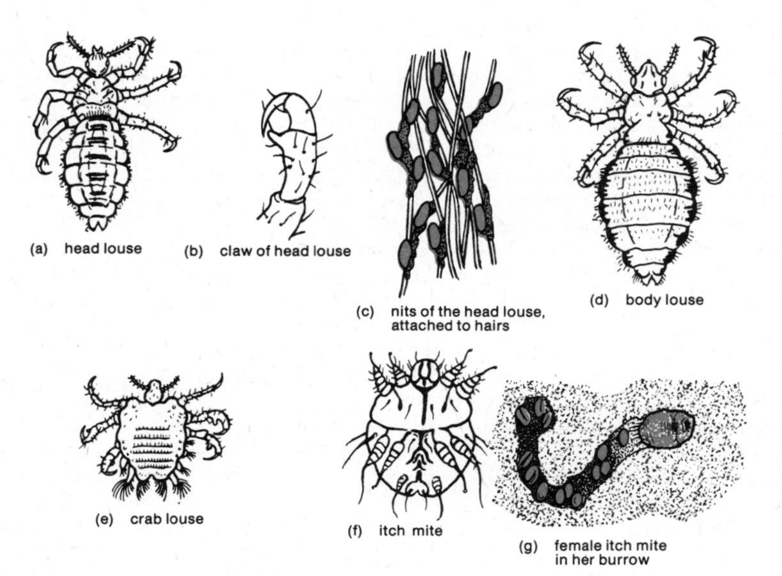

sucks the blood of its host, leaving small red spots (petechial marks) on the skin. The bites cause itching. Infestation with body lice is transmitted by direct contact or by the wearing of infested garments. In severe cases there may be a generalized skin eruption caused by the bites. Clothing and bedding should be sterilized by dry heat, hot water or dry cleaning. The patient should be given a hot bath, with thorough cleansing of the body with soap. A pediculicidal lotion should then be applied to hairy parts of the body.

The *crab louse* (*Phthirus pubis*) (Fig. 11-9e), known also as the *pubic louse,* attaches itself so firmly by means of little claws that forceps have to be used to pick it off. It lives principally in hair in the genital region, but is also found in the beard, eyebrows, eyelashes and armpits. The bite of this louse produces characteristic bluish spots resulting from the action of the insect's salivary secretion on hemoglobin. Treatment consists in washing thoroughly with soap and water, followed by application of a suitable pediculicide.

Fleas do not live in such intimate contact with their hosts as lice do. They are wingless bloodsucking insects with legs adapted for jumping. Different species live on different warm-blooded animals, including man. Fleas of the genus *Xenopsylla* transmit the bacillus of plague from rats to human beings. Certain other diseases are also transmitted by fleas. These insects lay their eggs in old furniture and clothing, where they live and hide. They can be destroyed by fumigation, insecticide sprays, etc. Flea attack can be prevented by treating the skin with a suitable insect-repellent. Dog or cat fleas occasionally have a meal of human blood. The *human flea* (*Pulex irritans*), once a very common parasite, has largely succumbed to modern hygiene, where this is consistently practiced.

The *bedbug* (*Cimex lectularius*) is about 5 mm long, flat, and gives off a peculiar musty odor. It is a bloodsucking insect whose saliva contains an irritating substance which, in some persons, sets up a degree of inflammation. The bugs live in furniture, particularly wooden furniture such as beds, in crevices of wooden walls or ceilings, etc. They have nocturnal habits. Fumigation does not destroy them, and it is necessary to kill them and their eggs with suitable insecticides in their hiding places.

The *itch mite* (*Sarcoptes scabiei*) (Fig. 11-9f,g) is not an insect, but an arachnid (a class of creatures which includes spiders), which causes the highly communicable skin disease called *scabies* (itch). The female mite burrows into the epidermis and lays her eggs there. The burrows appear as discolored lines up to several centimeters in length. Small pustules or vesicles are formed at the ends of the burrows. Parts most commonly affected are the webs between the fingers and toes, the front of the wrists, beneath the breasts, the genitals, the armpits, and the inside of the thighs. Treatment consists in the application of benzyl benzoate solution after the patient has taken a bath.

Ticks are small bloodsucking arachnids, certain species of which attach themselves to the human skin. They transmit various diseases collectively called tick fever and resembling typhus, also relapsing fever.

ACNE

There are several varieties of acne, the commonest being *acne vulgaris.* It is a chronic inflammation of the sebaceous glands, which are particularly numerous in the skin of the face, neck and shoulders. The glands may, especially at puberty, become overactive and produce so much fatty secretion (sebum) that their outlets become blocked. The dried secretion forms a plug which is visible as a so-called *blackhead* (*comedo*). In more severe cases, sebum from the obstructed glands escapes into the skin, where it forms lumps. As a rule, acne is a chronic condition characterized by successive crops of papules and pustules. These are surrounded by an area of inflammation; in more acute cases suppuration may occur. Various species of bacteria, including *Staphylococcus albus* (commonly found on the skin), may incubate in the trapped sebum and cause *pimples,* i.e., small superficial abscesses.

Acne occurs in both sexes, mainly between the ages of 14 and 20. The precise cause is not known. Hereditary factors and disturbances in the androgen-estrogen balance are involved. It appears that the male sex hormones and the "masculinizing" adrenal cortical hormones stimulate the secretion of sebum, whereas the female sex hormones tend to counteract this. For this reason acne is of less frequent occurrence in girls and of shorter duration. Specific factors causing acne in these circumstances may include excess fats and carbohydrates, endocrine factors, psychogenic disorders, contact with certain chemicals, and vitamin deficiencies. As a rule, acne disappears when puberty is over; but the endocrine balance may be disturbed in later life, e.g., by hormone treatment or other disorders, and acne may then reappear. In women it may also appear in the period just before the menopause.

Common acne can be treated with various nongreasy lotions and ointments, some containing sulfur. Sunlight and ultraviolet light generally have a beneficial effect. Noninflamed blackheads can be carefully squeezed out after the skin has been softened by washing or bathing. A special instrument can be used for the purpose. Papules can be emptied by squeezing under suitably hygienic conditions; pustules should be incised and drained, followed by application of an antiseptic. Unskilled or rough squeezing or pinching is likely to do more harm than good.

BOILS AND CARBUNCLES

A *boil* (*furuncle*) is a painful, acute, circumscribed inflammation due to infection of a hair follicle by staphylococci. There is no foundation for the popular belief that boils arise from "bad blood." In the condition called *blind boil* the inflammation subsides without suppuration. A boil usually begins as a round elevated reddening around a hair. Within a few days the reddened area turns dark red and is increasingly painful; a bead of pus forms on the summit, while the surrounding area becomes indurated (hardened). The pus is merely the tip of a capusulelike mass of pus that has formed in the hair follicle and has broken through the surface of the skin. In most cases the boil suppurates in this stage, discharging pus; a core of dead tissue is eventually extruded, leaving a cavity which heals. Boils may occur on any part of the body, but show a preference for the neck and upper back as well as the buttocks. Boils on the face, especially on the nose or the upper lip, are potentially dangerous because of the risk that infection may be carried into the skull and cause meningitis. In such cases penicillin or some other antibiotic will be given. Boils should not be injudiciously pinched or squeezed, especially those on the face, as such treatment may induce the infection to spread. Small boils should be protected with a gauze bandage and left to develop and heal naturally. Large ones may have to be lanced to let out pus. Intermittent application of moist heat (poultices) will generally bring a boil to a head more quickly. The area around a draining boil should be protected by an antiseptic ointment to prevent new boils from forming there. Persons with diabetes or kidney disease are especially subject to repeated attacks of boils (*furunculosis*). Although such attacks may also occur as a result of inadequate care of the skin and general lack of hygiene, they may alternatively be an indication of incipient or untreated existing diabetes. For this reason it is usual to perform blood tests to determine the blood sugar level in cases of furunculosis.

Like a furuncle (simple boil), a *carbuncle* is caused by staphylococci. It is actually a mass of interconnected boils close together and with several cores. They eventually coalesce and may then form a large craterlike ulcer. A carbuncle starts as a circumscribed, irregular and painful red area of swelling. The skin later becomes thin and perforates, discharging pus through several openings. Healing is often a slow process, leaving a deep scar. Carbuncles tend to develop in the same areas of the body as boils: neck, upper back and buttocks. Treatment is generally the same as for boils. Surgery or x-ray therapy is occasionally employed to assist healing; the necrosed tissue may be removed by electrocautery.

HAIR AND NAILS

The structure of a *hair* is quite complex (Fig.11-10). It comprises the shaft, the root (with a bulblike enlargment at the base) and the papilla (containing blood vessels that nourish the root). The root is implanted in the hair follicle, which is a pocket of epidermis that dips deeply into the dermis. The cells of the germinal layer of the papilla form the germinal matrix of the hair. These cells are pushed farther and farther away, die, and form the hair. The latter consists of keratin and comprises three layers of cells: the cuticle or outermost layer, the cortex forming the main horny portion of the hair, and the medulla or central axis. The color of the hair is due to a pigment (melanin) which is present in the cortex, but depends also on the surface condition of the hair, its grease content, and the presence of embedded minute air bubbles. Associated with each hair follicle is a sebaceous gland which discharges an oil secretion (sebum) into the follicle. It lubricates the hair and skin and also helps to kill bacteria.

On average, hair grows (Fig. 11-11) at a rate of 0.2–0.3 mm per day. Destruction of the germinal matrix results in permanent loss of hair. Hair growth proceeds cyclically and is closely associated with the structure and physiology

FIG. 11-10. Structure of a hair.

FIG. 11-11. Growth of hair.

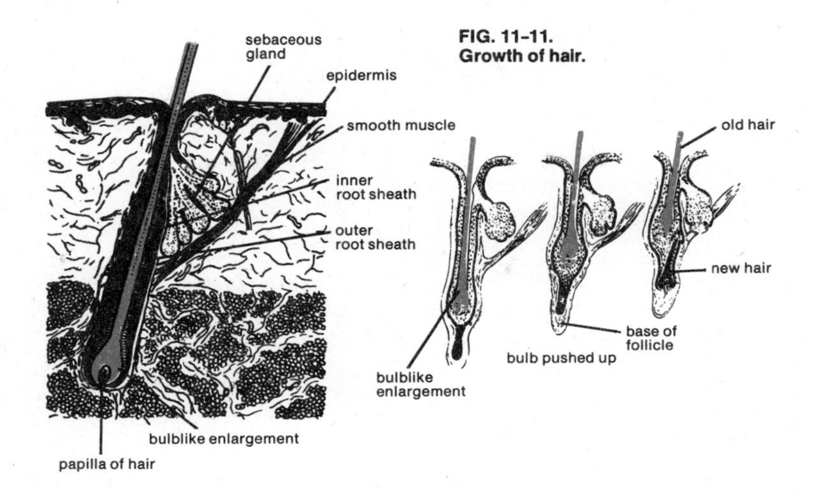

FIG. 11-12. Anatomy of a nail.

FIG. 11-13. Fingertip with nail and blood vessels.

FIG. 11-14. Nail and nail bed (one half of nail removed).

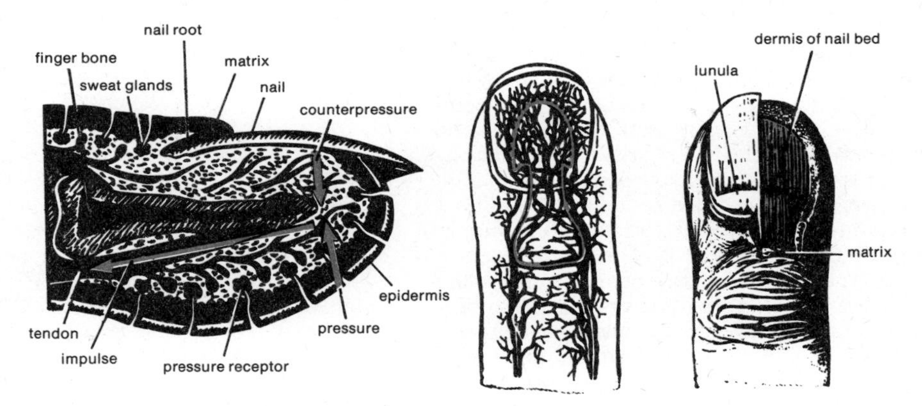

of the skin. An active phase, when the hairs increase in length, alternates with a resting phase, at the end of which the hairs are shed; new hairs then grow from the same follicles. In some animals there is seasonal molting, i.e., general shedding of hairs, but in man the process of hair growth and loss are not synchronized.

With advancing age, hair usually turns gray because the matrix cells stop producing pigment. If tiny air bubbles are present between the individual cells, the hair appears white. The earlier the hair darkens in childhood, the earlier does it usually turn gray in later life. Early graying is hereditary in some families; it is not a sign of premature senility.

Attached to each hair follicle is a small muscle (musculus arrector pili), consisting of smooth muscle fibers under the control of the sympathetic nervous system. When this muscle contracts, it causes the hair to stand up straight and pulls the whole follicle toward the surface, while it depresses the skin around it ("gooseflesh").

The *nails* (Figs. 11-12, 11-13, 11-14) consist of flat plates of keratin, the tough protein substance found also in hair and in the outermost layer of the skin, but the keratin in nails is harder. A nail consists of a body (the exposed portion) and a root (hidden by the nail fold), both of which rest on the nail bed. The crescent-shaped white area near the root is the *lu-*

nula. Under the nail fold is the matrix containing the germinal cells that produce nail growth at a rate of about 1 mm per week. The rate is slower in summer than in winter and also varies with age; toenails grow more slowly than fingernails. Disease and certain hormone deficiencies also affect nail growth. If the germinal matrix is damaged, the nail is shed and a new one has to grow. Ridges or other changes in nails may occur as a result of defective nutrition or following a serious illness, but splitting has no known cause. Discolorations of the nails may also be indicative of systemic disorders. Normally the nails themselves are fairly colorless and translucent, so that the apparent color is due to the nail bed.

LOSS OF HAIR, BALDNESS

Normal healthy hair is constantly being shed and new hair is formed in the follicles, pushing out the old hair. It has been estimated that children and adolescents lose about 30 or 40 hairs a day. This hair loss diminishes with advancing age. If new hairs fail to grow where old ones were shed, baldness (*alopecia*) develops. Most baldness is a natural process occurring in a certain proportion of men and is not a symptom of ill health or senility. The age at which men lose their hair varies from one individual to another, and some men do not develop baldness. Hereditary factors play a part here. In certain cases, however, baldness may occur as a result of a particular illness (for example, typhoid fever), endocrine disorders (e.g., hyperthyroidism), some forms of dermatitis, the action of certain drugs or poisons (such as arsenic or thallium), etc.

Temporary loss of hair may have various causes. For instance, it may be due to mechanical damage of the roots caused by excessive friction or certain techniques of setting the hair (with curlers, combs, inexpert permanent waving, etc.). In adolescent girls, and in adult women with unstable hormone balance, especially in the menopause, loss of hair may occur as a result of hormonal factors. At the same time, a tendency of the hair to be greasy induces such women to wash their hair very often, but this may be counterproductive in actually stimulating grease secretion and dandruff. Hormone therapy can be effective in some cases.

Circumscribed loss of hair, i.e., patches of baldness, may occur as a result of diseases of the scalp, e.g., fungal diseases (see page 41),

FIG. 11-15.
Three types of baldness in males.

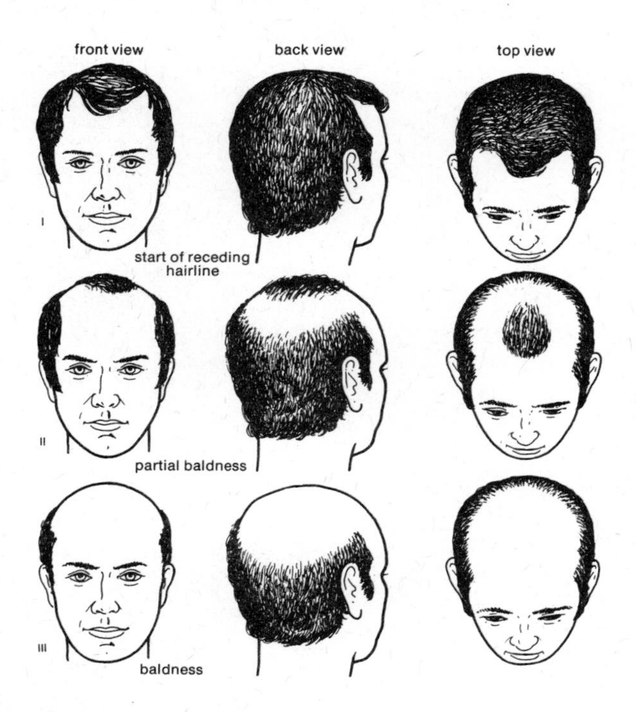

FIG. 11-16.
Baldness caused by disease.

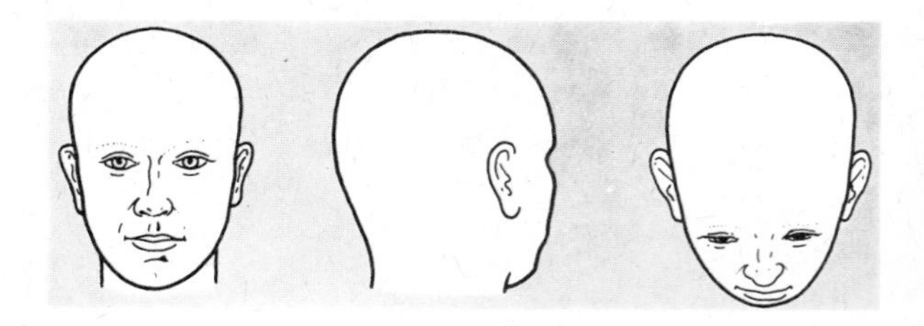

syphilis (page 320), dermatitis (page 408), shingles (page 409). *Alopecia areata* is baldness occurring in sharply defined circular patches which leave the scalp smooth and white. It occurs suddenly and has no known physical cause, though it is sometimes preceded by headache or neuralgic pains. It is believed to be brought on by severe mental strain in some cases. The hairs on the affected areas become detached from their roots and can eventually be pulled out easily and painlessly in bunches. Quite often, hair begins to grow again on the bald patches after a time. Occasionally the loss of hair is progressive, resulting in complete and permanent baldness (*alopecia maligna*), perhaps including the eyebrows and eyelashes and indeed all the hair on the body.

Like temporary loss, permanent loss of hair may be due to a number of causes. Apart from external physical causes such as burns, repeated exposure to x-rays, and action of chemical agents, there may be hormonal factors involved in causing baldness. The male sex hormone affects hair growth already in puberty; it induces increased secretion of grease and dandruff formation. In abnormal cases this may result in a loss of both scalp and body hair, together with abundant bran-like desquamation, i.e., shedding of scales from the surface of the epidermis (*alopecia pityroides*).

"Normal" baldness in men can occur in various typical forms (Fig. 11-15) which are hereditary and tend to recur in successive generations of a family. For this reason no hair tonic treatment will prevent these types of baldness or restore the hair once it has been lost. Whereas hereditary baldness is essentially confined to the male sex, progressive loss of hair sometimes resulting in baldness due to an unknown cause, characterized by a kind of cicatricial (scarring) contraction of the skin on the scalp, may occur in both sexes, usually in middle age, and more frequently in women than in men.

Part Twelve

THE SENSORY ORGANS

SENSE OF SMELL

The number of "pure" olfactory sensations, i.e., sensations of smell, is considerably greater than that of "pure" sensations of taste. Most odors produce mixed sensations in which the sense of taste participates. That the smell of food plays a very important part in determining its "taste" or "flavor" is a fact that can easily be verified by holding one's nose while eating: distinctions in flavor will then no longer be perceptible. And, of course, the distinctive character of a wine and its appreciation depends very largely on its "bouquet," i.e., its perfume. The acuteness of the sense of smell varies greatly from one animal species to another. Man has a relatively poorly developed sense of smell compared with that of, for example, a dog. Some dogs can smell millions of times more acutely than human beings. These animals live in a smell-oriented world of which man has only the dimmest conception. Dogs and pigs have traditionally been used to sniff out truffles buried underground, and in modern times dogs are used also for the detection of explosives and narcotics.

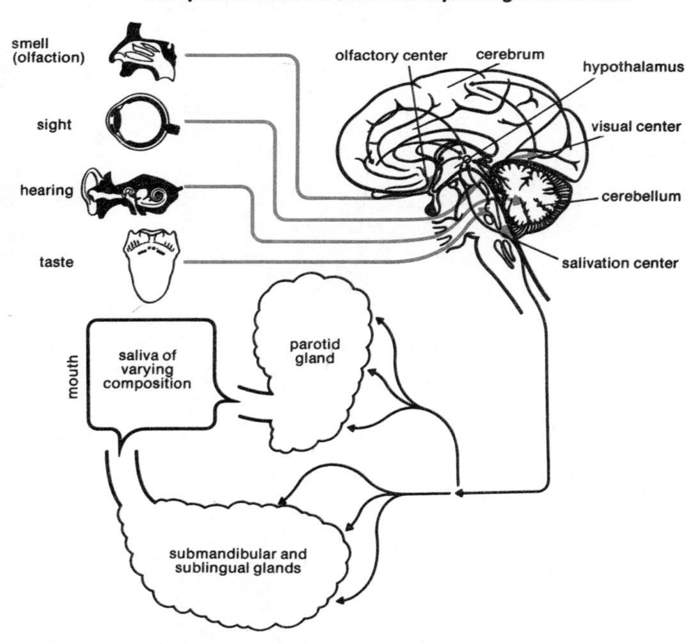

FIG. 12-1.
The special senses with their corresponding brain centers.

Only a small area in the roof of the human nasal cavity, the so-called regio olfactoria (Fig. 12-2), is provided with olfactory mucous membrane which contains sensory cells specialized to detect odors. In normal respiration only a small proportion of the odor-bearing air drawn into the lungs comes into contact with this membrane. Sniffing has the effect of increasing this proportion. Each sensory cell has dendrites (olfactory hairs) which project onto the surface and are stimulated by odors. The sensory cell bodies are situated between the epithelial cells of the mucous membrane; their axons pass through small holes in the ethmoid bone and synapse in the olfactory bulb, from where nerve fibers pass back in the olfactory tract to the centers for smell in the cerebral cortex. The finer molecular mechanism of the sense of smell is still largely unknown. A reaction occurs between the odor and the surface of the sensory cells (Fig. 12-3), which transform this stimulus into nervous impulses that are transmitted to the brain. How these sensory cells can detect and distinguish among thousands of odors and subtle differences between similar odors remains unexplained. Generally speaking, a sensation of smell is initiated by substances that volatilize, i.e., send tiny particles or individual molecules into the air. Nevertheless many volatile substances and gases are odorless. It appears that some degree of solubility of the substances in water and lipoids (fatlike substances such as cholesterol) is a further requirement in bringing about a sensation of smell.

The many thousands of odors and gradations of odor that the human nose can distinguish can be broadly subdivided into six some-

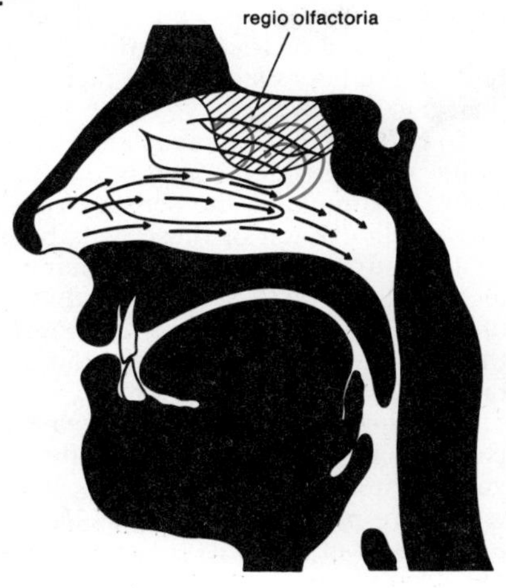

FIG. 12-2.
Area of smell reception in nasal cavity.

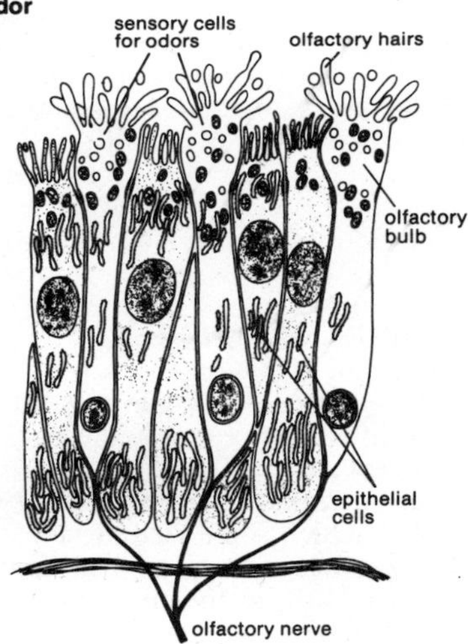

FIG. 12-3.
Apparatus for odor detection.

what arbitrary categories: (1) spicy (pepper, ginger); (2) fragrant or scented (jasmine oil); (3) fruity (malic ether); (4) resinous or balsamic (aromatic resins); (5) fetid (hydrogen sulfide); (6) tarry (tar).

The sense of smell may be affected by many conditions. Loss of the sense of smell, temporary or permanent, as a result of disease, injury or other causes is called *anosmia*. An increased sensitivity to odors is called *hyperosmia*. *Kakosmia* is the perception of bad odors where none exist. In some conditions the sense of smell is inverted: agreeable odors are found offensive, and offensive ones agreeable; this is known as *parosmia*.

SENSE OF TASTE

The overall phenomenon of "taste" is determined by the combined action and interaction of a variety of sensory perceptions such as temperature, pain and pressure, but more especially by sensations of smell. The "pungency" of any particular substance, such as pepper or curry powder, is determined by the involvement of pain receptors, for example. The effect of acids, causing the mouth to "draw together" by a kind of astringent action, manifests itself in a certain "dullness" of the mucous membrane, which is probably due to a slight degree of superficial damage to the pressure receptors. Superimposed upon these

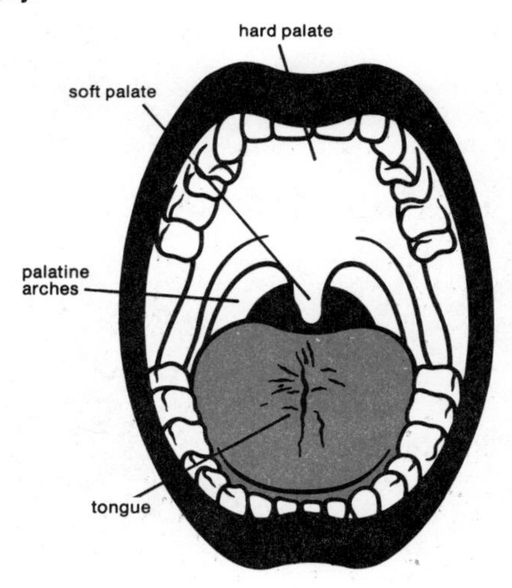

FIG. 12-4.
Oral cavity.

subsidiary "unspecific" sensations are the four basic types of taste sensation: sweet, sour (acid), salt and bitter. As Figure 12-5 shows, these perceptions are localized in certain areas of the tongue. The tip of the tongue is sensitive to sweet and salt, the edges to sour and salt, and the back to bitter (this is why a bitter "aftertaste" tends to persist). In a child the whole tongue is receptive to taste, but in an adult the central part ceases to function in this way.

The special sensory cells are found in the middle of the so-called *taste buds* (Fig. 12-6),

FIG. 12-5.
Areas of taste perception on tongue.

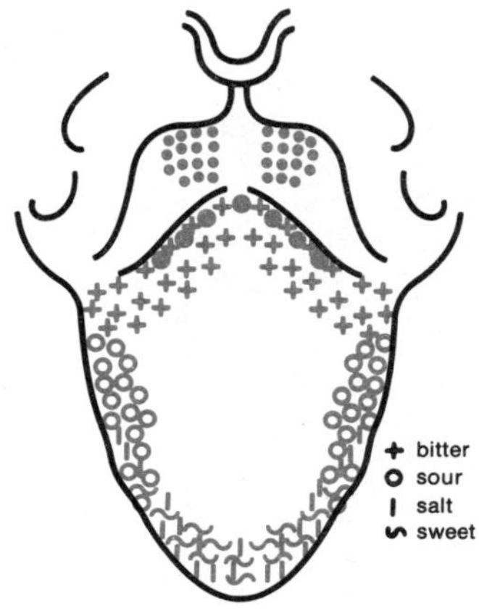

FIG. 12-6.
Taste bud and enlarged details of cells.

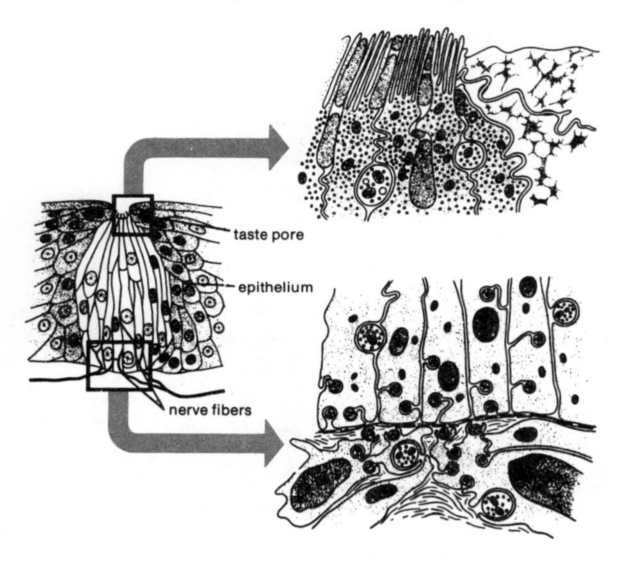

FIG. 12-7.
Papillae of tongue.

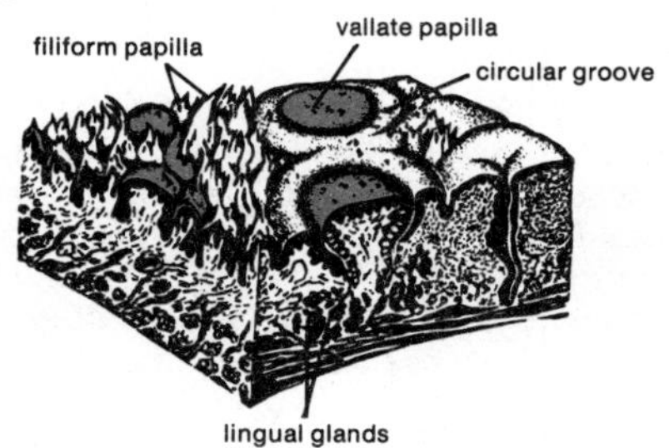

which are small groups of spindle-shaped cells. Most of the taste buds are situated in the epithelium of the tongue, but some are found in the palate and on the epiglottis. From the taste buds nerve fibers lead to the facial and glossopharyngeal nerves.

To be tasted, substances must be in solution in the saliva, which seeps into the taste buds where it stimulates the sensory receptor cells (taste cells). Each of these cells possesses on its free surface a short taste hair which projects into the taste pore, the external opening of the taste bud. It is here that the saliva with the dissolved substances is in contact with the taste hairs.

The surface of the tongue bears numerous papillae of three types: filiform, fungiform, and vallate (or circumvallate). The vallate papillae, which are present along the V-shaped groove on the back of the tongue, are each surrounded by a circular groove; taste buds are located mainly on the sides of these papillae (Fig. 12-7). The filiform papillae are slender ones at the tip of the tongue which respond only to heat and mechanical stimuli, while the fungiform papillae, which are broad and flat, are found chiefly at the edges of the tongue and are provided with taste buds.

HEARING, BALANCE

The ear has three main parts (Fig. 12-8): external (or outer) ear, middle ear, and internal (or inner) ear.

The *external ear* comprises the *auricle* (pinna), which is the visible part of the ear, and the *external auditory meatus* (or external ear canal) leading to the middle ear. The auricle consists mainly of elastic hyaline cartilage covered with skin; there is no cartilage in the lobe. The canal is about 2.5–3 cm long; its wall is of cartilage in the outer part and of bone farther inward; it ends at the eardrum (tympanic membrane). The skin that lines the external ear canal carries hairs and sebaceous glands; it also contains specialized sweat glands (ceruminous glands) which secrete a waxlike substance called *cerumen* that helps to trap foreign matter such as dust. An excess of this wax can cause temporary deafness. The *eardrum* is a thin membrane attached to the walls of the inner end of the external ear canal, separating it from the middle ear. It vibrates in

FIG. 12-8.
Structure of the ear.

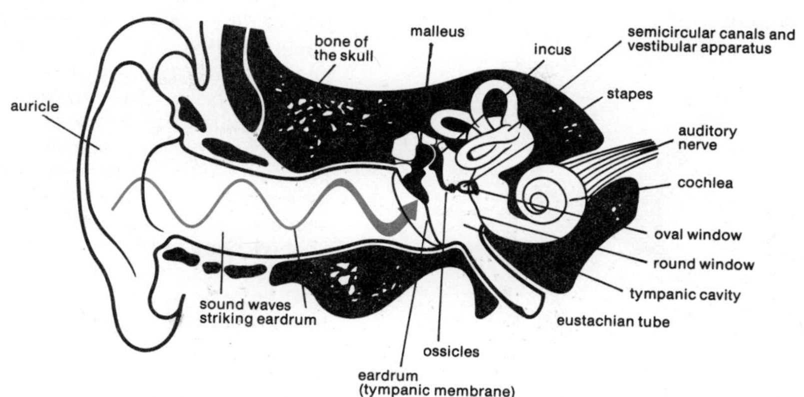

response to sound waves. The handle of the malleus is attached to it (Fig. 12-9).

The *middle ear* is a cavity in the temporal bone of the skull and is bridged by three tiny bones (*ossicles*): the *malleus* (hammer), the *incus* (anvil) and the *stapes* (stirrup). The handle of the malleus is attached to the eardrum, and its head articulates with the incus, which in turn articulates with the stapes; the footplate of the latter fits into the *oval window* which leads into the internal ear (Fig. 12-10). Below this is another small opening, the *round window*. At the back of the middle ear is an opening which leads to the tympanic antrum (or mastoid antrum), a hollow in the bone immediately behind the ear, which communicates with a network of small cavities, the mastoid air cells. The *eustachian tube* (pharyngotympanic tube) leads from the middle ear to the back of the nose. This tube and the middle ear are lined with mucous membrane which is continuous with that of the nose and throat.

The sound entering the ear is intensified by resonance in the external ear canal (external auditory meatus) and causes vibrations of the eardrum, which transmits them through the ossicles. These act as a lever system which further intensifies the pressure of the sound waves striking the eardrum and transmits this pressure to the oval window. The latter is much smaller than the eardrum, so that the intensity of the pressure is correspondingly increased (same force acting on reduced area).

The function of the ossicles is controlled by muscles (Fig. 12-12), which also serve as safety devices to protect the ear from harmful sound pressure. The muscles can rotate the ossicles so that less acoustic energy is transmitted from the eardrum to the internal ear; also, they make the eardrum more taut, so that

FIG. 12-9.
Eardrum.

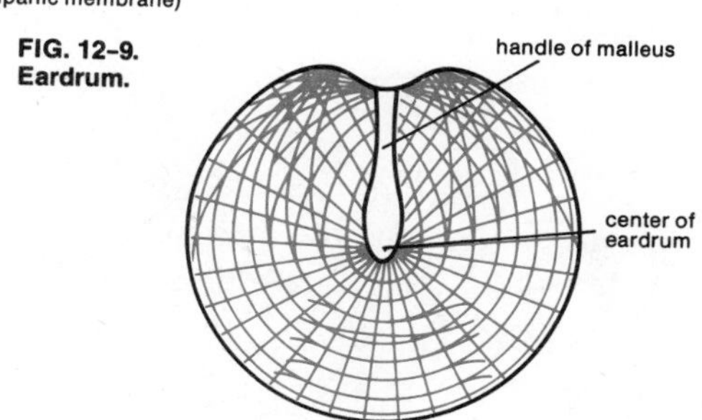

FIG. 12-10.
Middle ear structures.

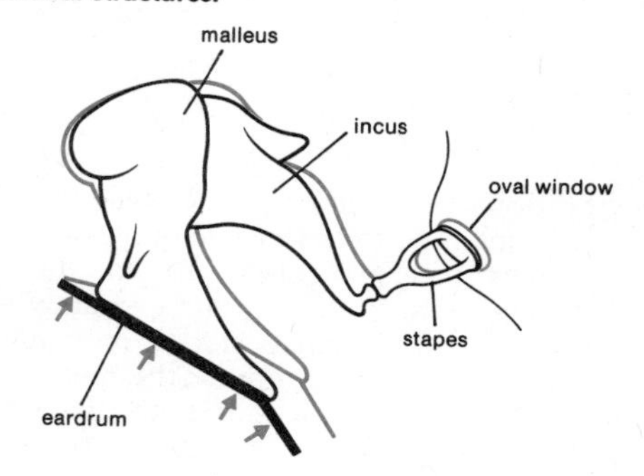

FIG. 12-11.
Transmission of pressure to oval window.

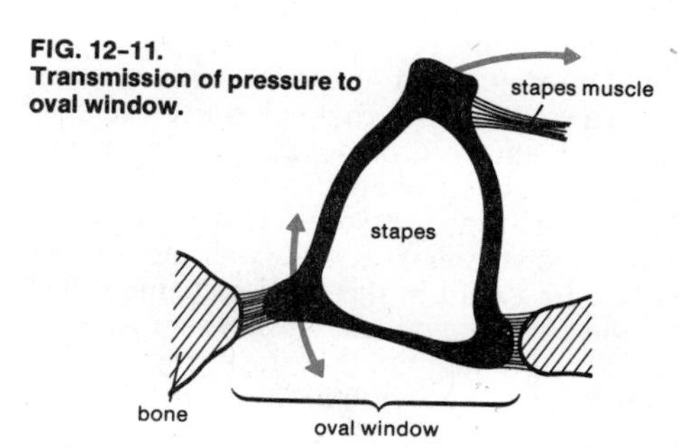

FIG. 12-12.
Stapes and its muscle.

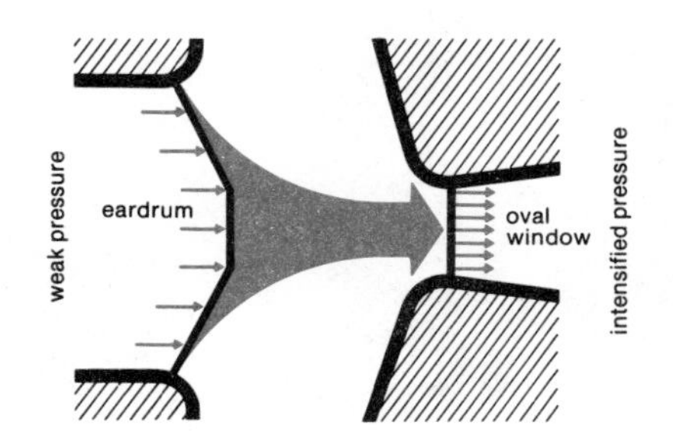

FIG. 12-13
Longitudinal section through the cochlea.

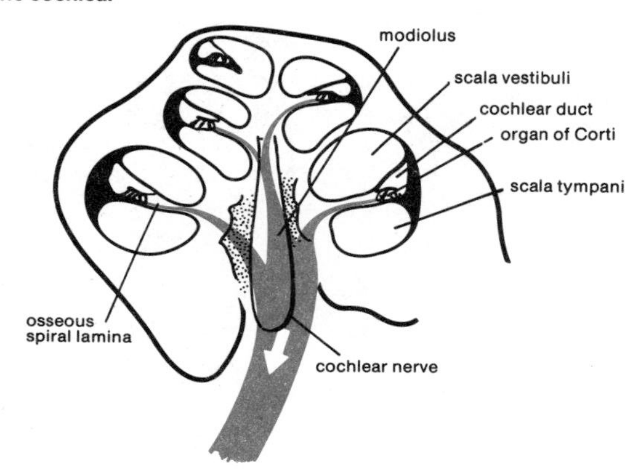

its vibrations are reduced. Since these muscles require some time to develop their action, they are effective in protecting the middle ear against noise that builds up fairly gradually in loudness, but not against sudden loud noise.

Through the eustachian tube the middle ear is normally in communication with the atmospheric air, so that the air pressure is the same on both sides of the eardrum; this ensures that the eardrum can respond easily to sound waves. Differences of pressure between the middle ear and the external ear cause the eardrum to bulge inward or outward. Swallowing movements help to open the eustachian tube and equalize the pressure. If the tube is blocked by swelling of its mucous membrane lining due to a cold in the head or some other cause, the air in the middle ear is absorbed and replaced by exudate, so that the eardrum cannot vibrate. Deafness results until the tube is opened again.

The *internal ear* consists of a delicate system of tubes called the membranous labyrinth, which is filled with *endolymph,* a watery fluid, and is situated inside the petrous part of the temporal bone. The *labyrinth* comprises two connected parts: the *cochlear duct* containing the receptor organ for hearing, and the vestibular part consisting of three *semicircular canals* and two cavities called the utricle and the saccule. The *cochlea* (Fig. 12-13) is the snail-shaped bone spiral tunnel in which the cochlear duct lies. The tunnel is divided lengthwise into three separate tunnels (Fig. 12-14) by the basilar membrane and the reticular membrane, which respectively form the lower and the upper boundary of the duct (Fig. 12-15) filled with endolymph. Above the duct is the *scala vestibuli,* below it the *scala tympani;* these two tunnels are joined to each other at the top and are filled with perilymph, which is similar to the cerebrospinal fluid. The lower end of the scala vestibuli terminates at the oval window, while the lower end of the scala tympani terminates at the round window, which is closed by a membrane. (In Figure 12-14 the cochlear duct, containing endolymph and communicating with the vestibular part of the inner ear, is colored red).

The cochlear duct contains the *organ of Corti,* consisting of highly specialized receptor cells in which sound vibrations produce nervous impulses for transmission to the brain. The receptor cells are arranged on the basilar membrane. The vibrations are transmitted by the ossicles from the eardrum to the oval window, where the footplate of the stapes communicates the vibrations to the perilymph in the scala vestibuli and scala tympani. The basilar membrane (Fig. 12-16) is thus made to vibrate in step with the sound vibrations entering the ear, so that the receptor cells of the organ of Corti are correspondingly stimulated. Different parts of the basilar membrane respond to different frequencies and stimulate a different part of that organ; high-pitched sounds are detected by its lower end, and low-pitched sounds by its upper end, where the fibers of the basilar membrane are longer than at the lower end. It used to be supposed that these fibers acted as acoustic resonators, but this theory is no longer accepted. Stimulation of the organ of Corti initiates electric currents whose voltage is higher as the movement is greater. High notes die out in the first part of the spiral, while lower notes travel farther up the length of the organ. The sum of the electric

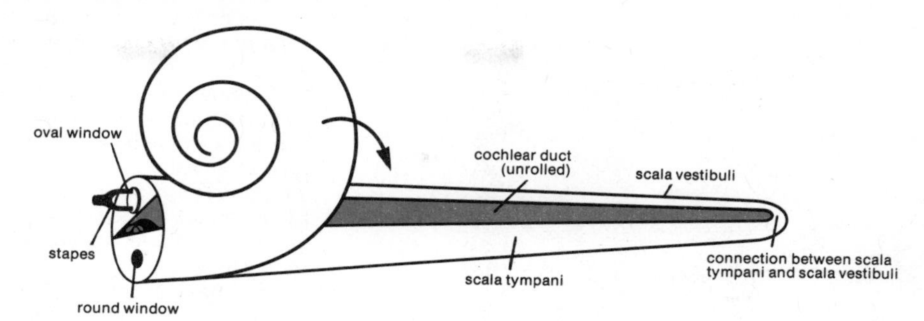

FIG. 12-14.
The cochlea (cochlear duct shown unrolled on right).

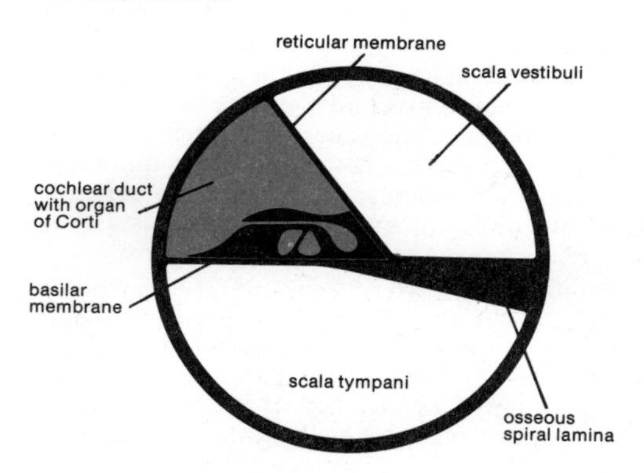

FIG. 12-15.
Cochlear division.

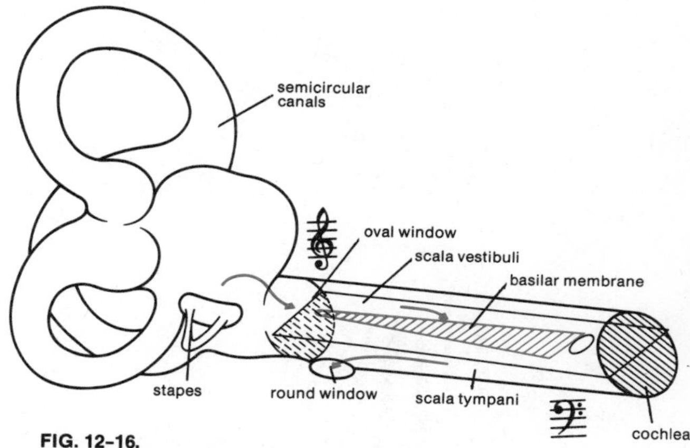

FIG. 12-16.
Diagram of basilar membrane with receptor cells for sound vibrations.

currents produced by a sound is an electrical replica of its vibrations, i.e., the organ of Corti behaves like a microphone. The brain does not, however, have this electrical record directly transmitted to it, but apparently identifies the distance that a particular sound travels along the organ of Corti. Nerve fibers from the receptor cells all along this organ communicate with the brain, so that different fibers from the ear transmit different frequencies. Lower notes, by traveling farther along the spiral of the cochlea, stimulate more cells of the organ of Corti than higher notes do. From here the nervous impulses are transmitted by the cochlear nerve to the junction of the medulla with the pons and then onward, after several synapses, to the auditory area of the cerebral cortex.

The *vestibular apparatus* (Fig. 12-17) is the

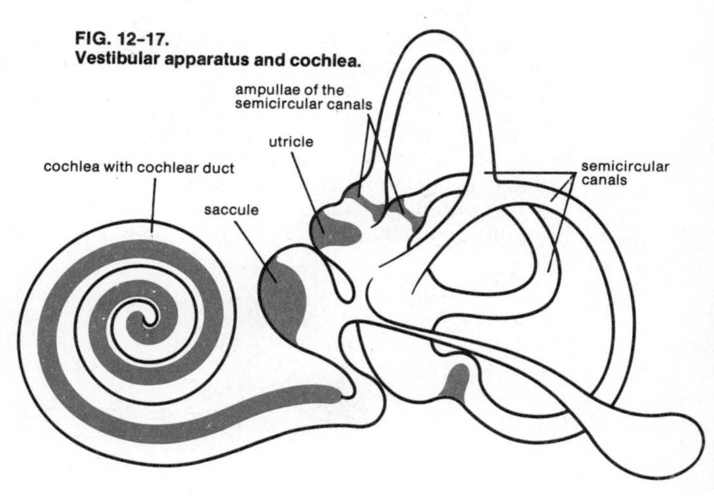

FIG. 12-17.
Vestibular apparatus and cochlea.

FIG. 12-18.
Semicircular canal lining.

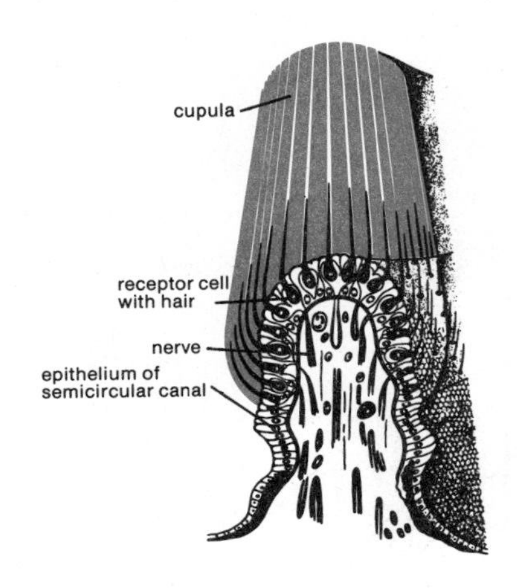

FIG. 12-19.
Receptor cells of macula.

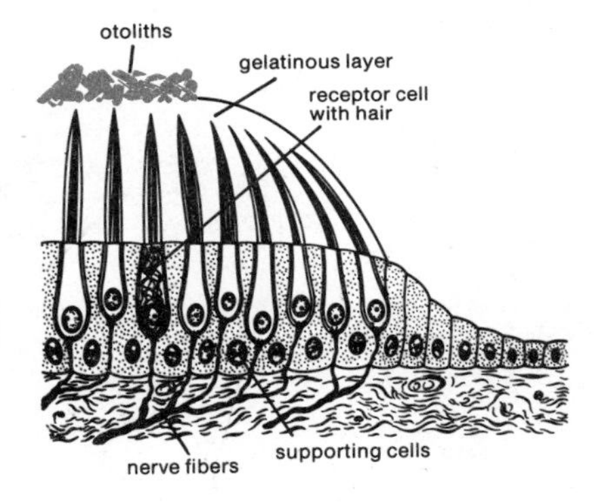

part of the internal ear concerned with balance. The larger central part is the *utricle,* which is connected by a narrow passage to the *saccule,* which is smaller than the utricle and is, in turn, connected to the cochlear duct. The three *semicircular canals* communicate at both ends with the utricle. They are located in mutually perpendicular planes. Each canal has an enlargement (ampulla) at one end, the walls of which contain receptor cells on localized thickenings of the membrane lining, called *cristae ampullares* (Fig. 12-18). The utricle and saccule also each have a smaller area of receptor cells, called a *macula,* on the surface of which is the otolithic membrane; these receptor cells have fine hairs with granules of calcium carbonate (otoliths) suspended from them (Fig. 12-19).

The maculae respond to gravitational pull, to linear acceleration and to position. When the head is upright, the macula of the utricle is horizontal and the macula of the saccule is vertical. Tilting movements of the head stimulate the maculae, so that the position of the head in relation to gravity is detected and is reported to the brain. The semicircular canals sense movements, especially rotary movements. When the head moves, the inertia of the endolymph stimulates the receptor cells in the ampullae. Since the canals are in three mutually perpendicular planes, different movements stimulate different ampullae. Each crista ampullaris is surmounted by a gelatinous mass (cupula) into which the sensory hairs of the receptor cells penetrate. The cupula projects into the endolymph and is unaffected by gravity, but responds to the relative movements of the endolymph. Prolonged or intense stimulation of these receptor cells causes sensations of dizziness and nausea (seasickness, etc.). (See Figs. 12-20 and 12-21.) Only a small proportion of the information about position and movements of the head reaches consciousness. Most of the impulses are transmitted to reflex centers in the brainstem and cerebellum which are concerned with monitoring and maintaining the body's balance.

FIG. 12-20.

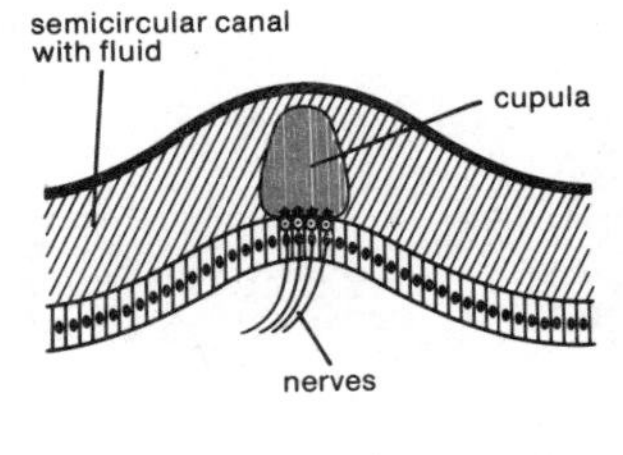

FIG. 12-21.

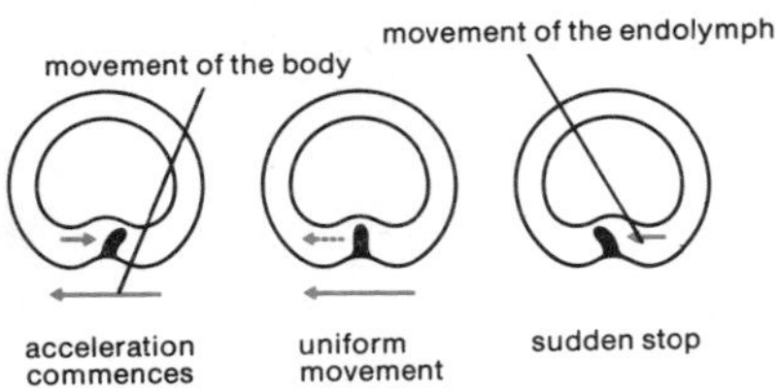

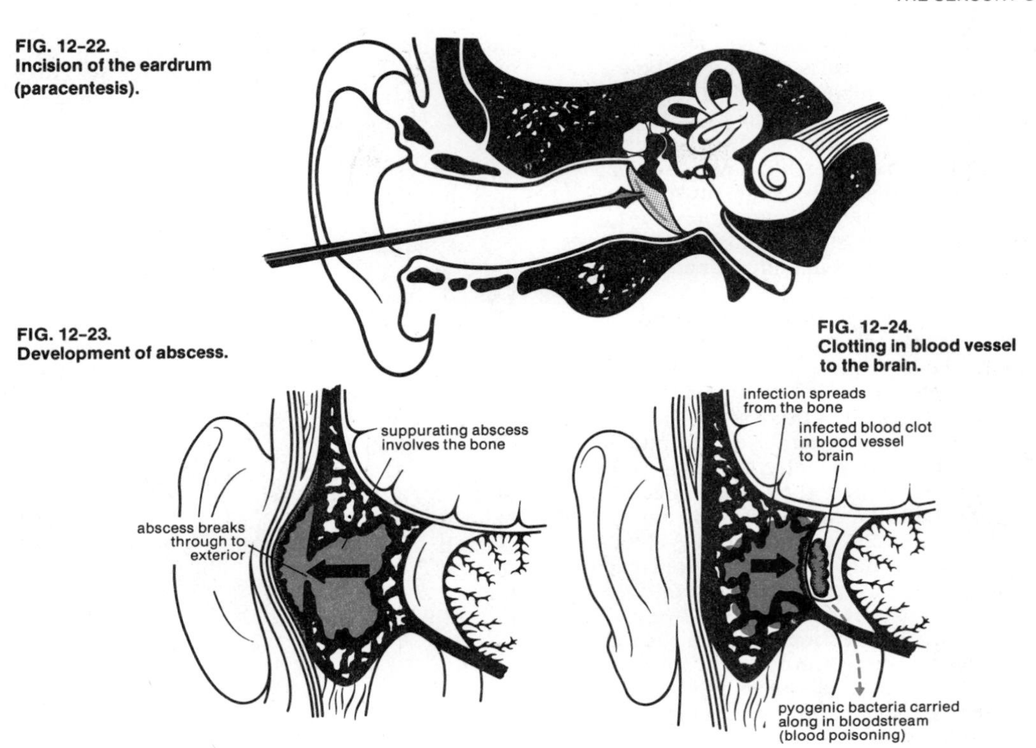

FIG. 12-22.
Incision of the eardrum (paracentesis).

FIG. 12-23.
Development of abscess.

FIG. 12-24.
Clotting in blood vessel to the brain.

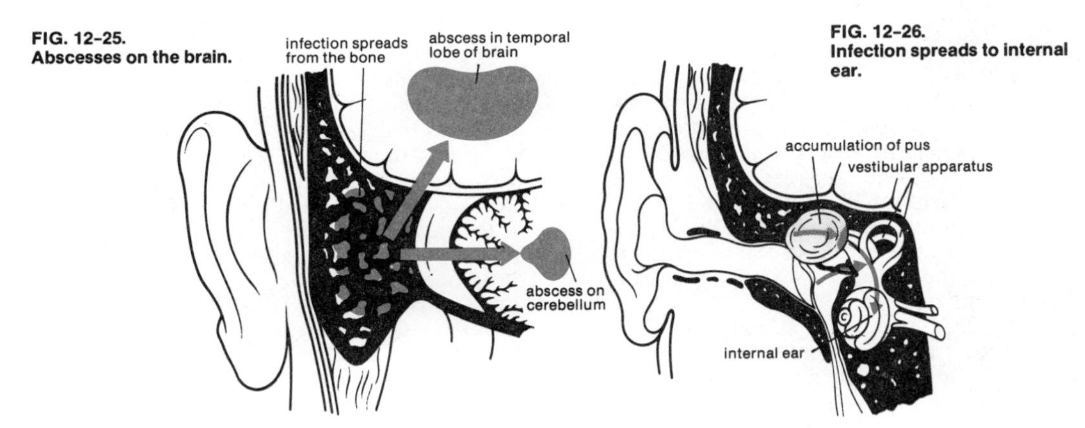

FIG. 12-25.
Abscesses on the brain.

FIG. 12-26.
Infection spreads to internal ear.

TUBAL CATARRH, EARACHE, OTITIS MEDIA

Acute *tubal catarrh,* which usually starts with a cold or sinusitis, is characterized by inflammation of the mucous membrane of the eustachian tube, so that it becomes temporarily obstructed and the pressure in the middle ear after a time becomes lower than the atmospheric pressure. As a result, the mucous membrane here is irritated, and catarrh spreads to the middle ear, though without the participation of pathogenic bacteria. Symptoms associated with this condition are hardness of hearing, tinnitus (buzzing or ringing sounds in the ear), and a sensation of pressure. When the acute swelling that blocks the tube eventually goes down spontaneously or as a result of treatment (nasal drops to reduce inflammation, application of heat, etc.), the eustachian tube can be finally cleared by blowing air through it. In general, a tube that is not properly clear is likely to cause trouble because it prevents equalization of pressure, especially when rapid changes of altitude, or similar causes, necessitate this. If the tubal catarrh persists, so that the eardrum continues to bulge inward because of reduced (lower-than-atmospheric) pressure in the middle ear, there is a risk that the hearing will be permanently impaired because the ossicles may stiffen.

Earache (otalgia) is a general term denoting pain in the ear. It may be due to a variety of causes, e.g., boils in the external ear, inflammation of the pharynx, wax or foreign bodies in the external ear, acute otitis media.

Otitis media is inflammation of the middle ear. This condition frequently develops from catarrh of the eustachian tube, with infection spreading up the tube, or, much more rarely, as a result of infection carried by the bloodstream or entering through a punctured eardrum. The entry of pyogenic bacteria into the middle ear is often preceded by a cold, tonsillitis, sinusitis or an infectious disease such as influenza, measles or scarlet fever. Eventually the mucous membrane of the middle ear becomes severely inflamed and discharges pus. The accumulation of pus presses against the eardrum, causing pain (earache), a sense of fullness in the ear, temporary loss of hearing, and fever. In infants the symptoms include drowsiness, stiffness of the neck, and vomiting. If the inflammation is relatively mild or if antibiotics or sulfonamides are promptly given, the pus may be absorbed. Otherwise its pressure may become so great that it bursts through the eardrum. If the fever and severe pain continue for two or three days, it will generally be preferable to incise the eardrum in order to drain the pus and relieve the pain (paracentesis tympani, Fig. 12-22).

In *mastoiditis,* which is a serious complication of otitis media, the infection advances from the middle ear to the air cells of the mastoid bone behind the ear and may even spread to the brain. A possible external symptom of this spread of the infection is the formation of a suppurating abscess behind the ear (Fig. 12-23). The main blood vessels to the brain may become blocked with infected blood clots, which may cause blood poisoning (Fig. 12-24). The pus may also infect the meninges and possibly cause abscesses on the brain itself (Fig. 12-25). Alternatively, the infection may spread from the middle ear to the internal ear (Fig. 12-26), disrupting the sense of hearing and balance. In general, these complications are attended by fever, headache, nausea, dizziness, vomiting, deafness, disturbed balance and possible other symptoms. Brain abscesses may also cause uncharacteristic general symptoms, together with specific ones such as speech difficulties and defective sense of balance.

Mastoiditis and associated complications require surgical treatment if they reach an advanced stage. The operation for removing part of the mastoid bone to drain the pus from the middle ear and brain is called *mastoidectomy.* Such operations were once common, but are now seldom necessary, because ear infections can be successfully treated with drugs, especially antibiotics.

A running ear (*otorrhea*) is a sign of continuing infection; the discharge indicates that inflammation in the middle ear has not been cleared up. Medical treatment should be sought. Plugging the ear with cotton wool and hoping that the trouble will eventually disappear of itself is ill-advised.

DEAFNESS, OTOSCLEROSIS, DEAF-MUTISM

Deafness in varying degrees can have a variety of causes. As a temporary condition it may be due to an accumulation of wax (cerumen) in the ear or to some foreign body wedged in the external ear canal. These disorders can usually be successfully treated by cleansing the canal with warm water. Catarrh of the eustachian tube or inflammation of the middle ear also causes temporary deafness. If these occur as chronic rather than acute conditions, the deafness is likely to be of longer duration. If infection spreads to the inner ear, deafness may be accompanied by disruption of the sense of balance. Deafness may also be due to damage to the inner ear resulting from injury or excessively loud noise, e.g., an explosion. Explosions cause permanent impairment of hearing, even though there may initially be some improvement. Occupational deafness may affect persons employed in noisy industries or occupations. Preventive measures such as the wearing of ear plugs and the use of appropriate sound insulation and sound attenuation, as well as general precautions against noise nuisance and harm to human hearing, are obviously essential in such cir-

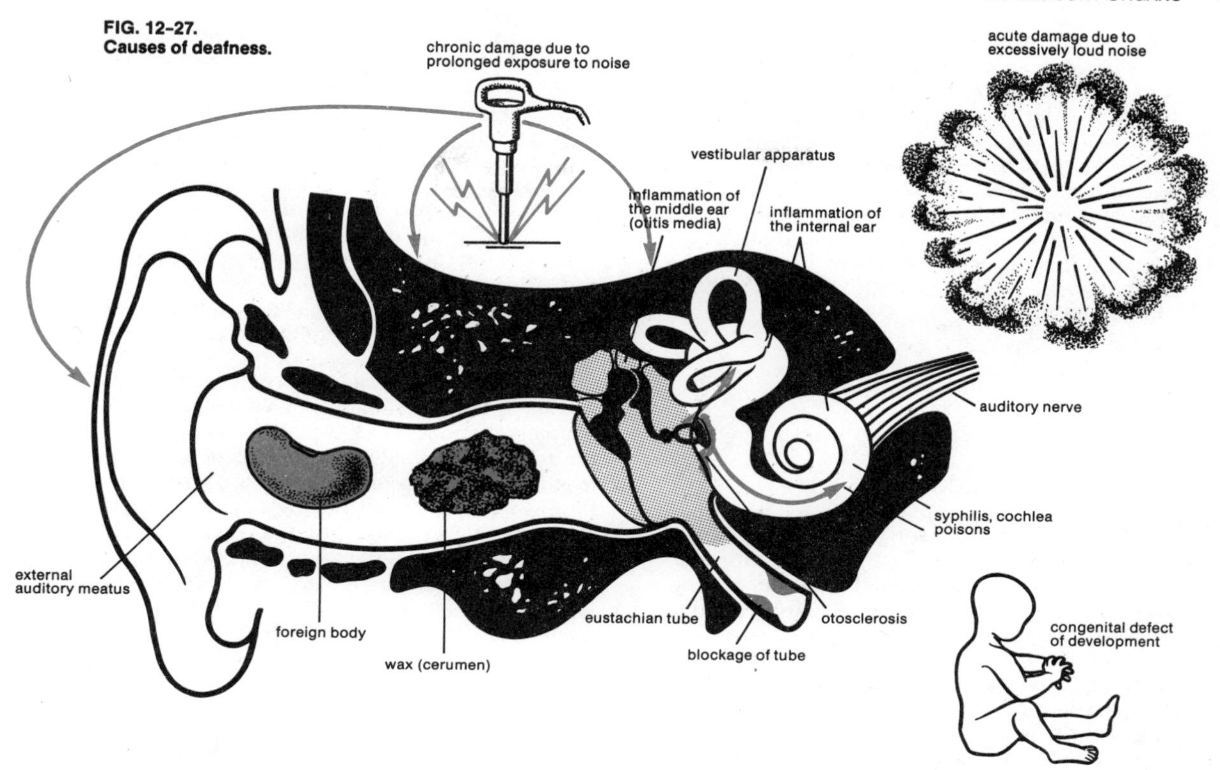

FIG. 12-27.
Causes of deafness.

cumstances. The inner ear can suffer direct injury from a fracture of the base of the skull (see page 385). Deafness may furthermore occur as the result of violence such as a blow or knock, usually associated with concussion.

Deafness is often due to *otosclerosis,* which is caused by the formation of spongy bone in the middle ear, especially around the oval window, so that the footplate of the stirrup bone (stapes) becomes fixed and thus no longer transmits sound. The disorder is characterized by progressive deafness, particularly for low tones. It is about twice as common in women as in men and may be worsened by pregnancy. Tinnitus occurs, and deafness develops gradually. Eventually there is deterioration of the sensory cells in the internal ear and of the nerve fibers, possibly resulting in total deafness. The cause of otosclerosis is unknown, but it does tend to run in families, so that a hereditary factor appears to be involved. There is no specific treatment. In some cases, however, considerable improvement can be achieved by surgery. The operation known as *fenestration* consists in making an artificial opening into the internal ear, so that sounds can be conducted into it. This operation has now, however, largely been superseded by an operation to remobilize the footplate of the stapes by intervention through an aperture in the eardrum. In other cases electronic hearing aids can be helpful.

Deaf-mutism is the state of being both deaf and unable to speak. It is generally due to deafness, congenital or acquired, in infancy. A child who cannot hear does not learn to speak. If the child has learned to speak, he usually loses the power of speech if deafness comes before the age of about 7 years. Deafness from birth can be caused by viral diseases, notably German measles, affecting the mother in early pregnancy. In general, the auditory nerves and internal ear then fail to develop properly. Diseases such as scarlet fever, typhoid and meningitis can cause deafness in infants, as can also hemorrhage in the internal ear and other injuries. Cretinism due to iodine deficiency is sometimes attended by deafness. In some cases, however, deafness and muteness (dumbness) are of psychogenic origin, not directly attributable to any physical defect. Any mental disturbance that impedes the normal process of learning is likely to affect a child's speech. Many deaf-mutes can be taught to speak intelligibly by specially skilled teachers. Basically, the training of such children consists in teaching them to imitate the speech positions of the mouth and make the sounds without, of course, being able themselves to hear and thus directly control the sounds. In cases where some vestigial sense of hearing remains, it is necessary and usually very helpful to take full advantage of this in teaching the child to speak.

VOICE AND SPEECH

The principal organ for producing the voice sounds, together with the tongue, mouth and throat, is the *larynx* (Figs. 12-28 and 12-29), the "voice box" situated in the front of the neck. It opens above into the pharynx and is continuous below with the trachea. The larynx appears to have evolved from a mere sphincter for preventing food from entering the air passages of the lung.

The skeleton of the larynx consists of several cartilages bound together by an elastic membrane and moved by muscles. The *cricoid cartilage* is ring-shaped and is situated at the entrance to the trachea; it is surmounted by the *thyroid cartilage,* comprising two halves which meet in front to form the Adam's apple. From behind the Adam's apple two ridges, the *vocal cords* (or vocal ligaments) extend rearward and are attached to the small triangular *arytenoid cartilages,* which slide and rotate on the cricoid cartilage, between the two halves of the thyroid cartilage. The vocal cords are the upper edges of the cricothyroid membrane which is attached below to the cricoid cartilage; they are not really "cords" at all, but the thickened upper edges of this membrane (Fig. 12-30). The gap between the vocal cords is the *glottis* (rima glottidis). A complex system of muscles moves the cartilages of the larynx.

The arytenoid cartilages alter their position during speech, so that the glottis varies from a V-shaped opening to a mere slit when the cords come together and the glottis is closed. This closure of the glottis occurs for swallowing, while at the same time a flap or cartilage (*epiglottis*) descends like a lid to cover the larynx. For speech the arytenoid cartilages may also be drawn together to close the glottis, but the epiglottis then remains in the raised position. When the vocal cords are close together, they are said to be adducted, and the glottis is then closed; when the arytenoid cartilages move apart, so that the glottis opens out and becomes V-shaped, the cords are said to be abducted. During quiet respiration the glottis is moderately open; in deeper respiration it is wider open.

Adduction of the vocal cords is necessary for phonation, i.e., for producing the sounds in speech. For speech, the glottis is narrowed, and air exhaled from the lungs makes the cords vibrate as it is expelled through the glottis. The intensity (loudness) of the sound depends on the force with which the air is expelled. The length and tension of the cords are altered by tilting movements of the thyroid cartilage in relation to the cricoid cartilage (Figs. 12-31 and 12-32), and this action controls the pitch of the voice: the greater the tension of the cords, the higher the pitch. The timbre, i.e., the resonant quality of the sound by which it is distinguished, depends on the shape and size of the upper air passages, including the nose. Changing the shape of the mouth alters the timbre. The differences between the vowel sounds, for example, are determined mainly by the timbre of the voice.

Speech is a complicated activity, involving the use of three groups of muscles: respiratory, laryngeal and oral. Their contractions are controlled and coordinated by the central nervous system. During speech the diaphragm is relaxed; contraction of the abdominal muscles pushes it upward, forcing air out through the larynx. The laryngeal muscles contract, reducing the glottis to a narrow slit. The air rushing through it causes vibration of the vocal cords, producing the primary sound of the voice (phonation). The cords vibrate only for short periods in speaking; in singing, the vibrations are of longer duration. The oral muscles include those of the tongue, lips, cheeks, palate, jaws and floor of the mouth. These control the shape of the upper air passages. If the palate is raised, so that air escapes only through the mouth, oral sounds are produced; a nasal sound is produced when the palate is lowered and the mouth shut, so that air escapes only through the nose.

Vowel sounds are phonated (produced in the larynx) and oral in character; the differences between vowel sounds are obtained by altering

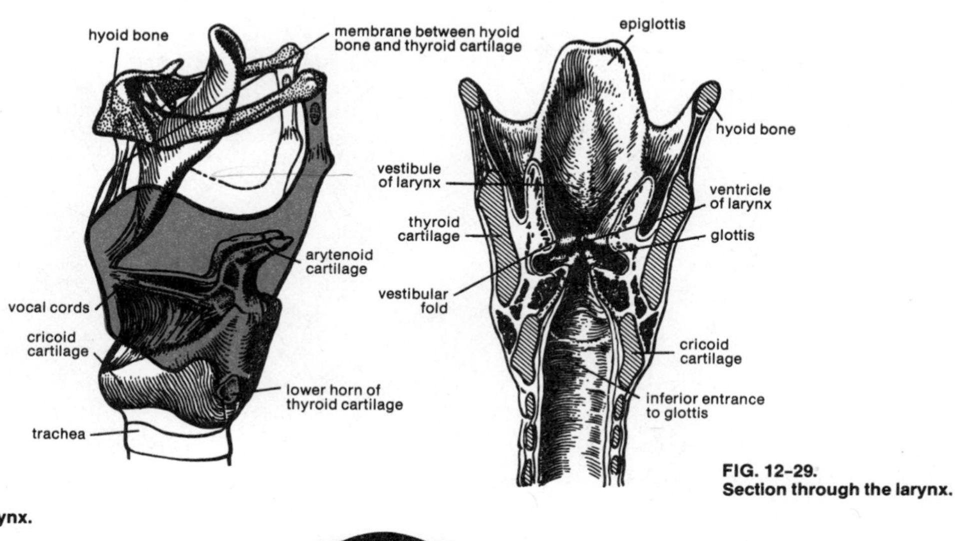

FIG. -12.28.
Side view of larynx.

FIG. 12-29.
Section through the larynx.

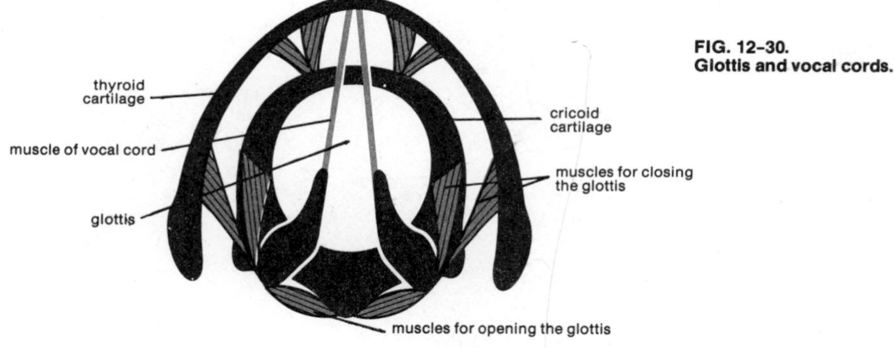

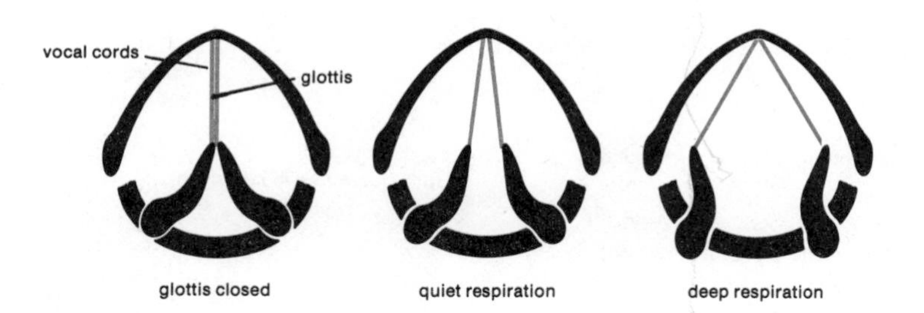

FIG. 12-30.
Glottis and vocal cords.

the shape of the mouth and varying the positions of the tongue and lips (Fig. 12-33). Consonants are produced by obstruction of the air flow at different places. Momentary closure of the lips forms the labial consonants (M, P, B); pressing different parts of the tongue against the palate forms lingual consonants (L, D, T, G, K, etc.). At the same time, there may or may not be phonation.

The mechanism of speech is controlled

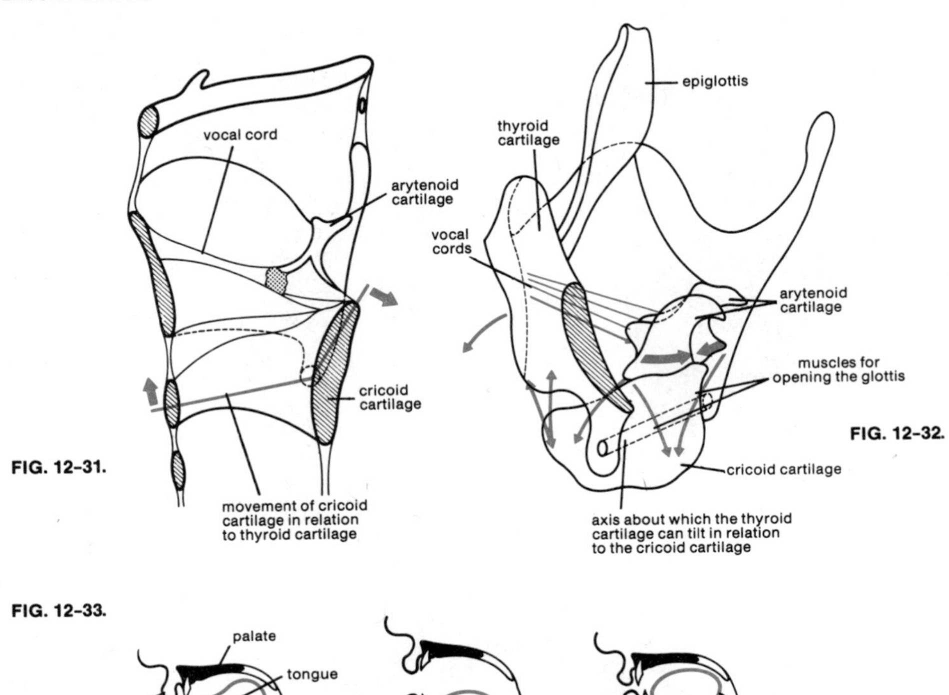

FIG. 12-31.

FIG. 12-32.

FIG. 12-33.

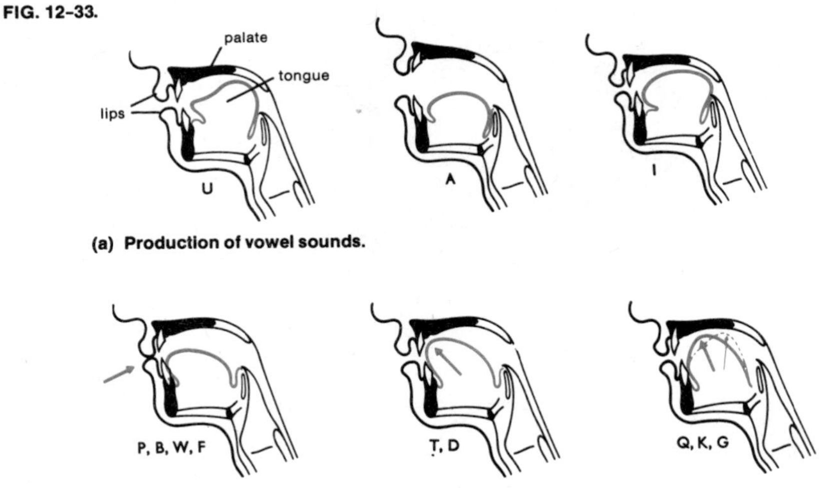

(a) Production of vowel sounds.

(b) Production of consonant sounds.

mainly by the motor speech center, which is located in the left cerebral hemisphere in a right-handed person. Words are formulated in the speech center, which is closely coordinated with centers for hearing (comprehension), vision (reading), and hand moving (writing). In a right-handed person the left cerebral hemisphere is dominant in relation to the right cerebral hemisphere; in a left-handed person the reverse is usually the case. This cerebral dominance, which to some extent applies also

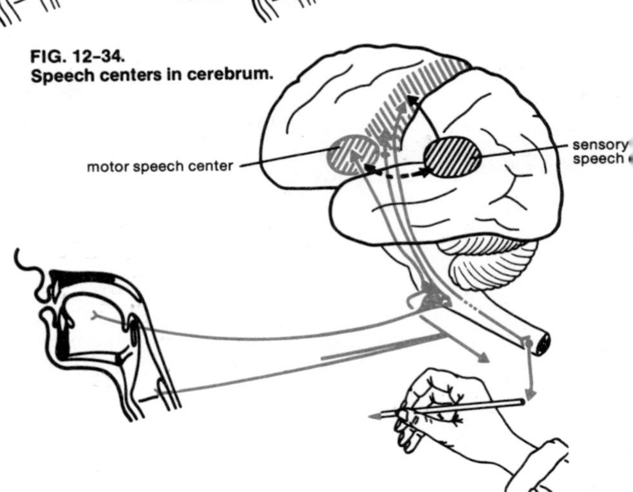

FIG. 12-34.
Speech centers in cerebrum.

FIG. 12-35.
Nerve tracts.

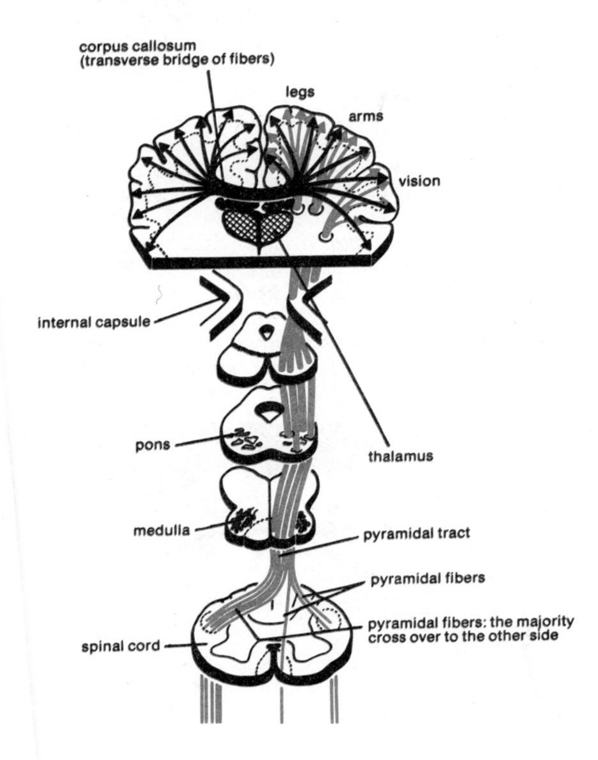

to sight and hearing, appears to be a specific characteristic of human beings, closely associated with the development of speech. A notable feature, too, is that the great majority of all nerve tracts that interconnect the brain and spinal cord cross over, from right to left or from left to right (Fig. 12-35), so that the left half of the brain is in close communication with the right half of the body, and vice versa.

SPEECH ABNORMALITIES

Abnormalities and defects of speech can have a wide variety of causes. They may be due to disturbed development of the ability to speak, or existing ability may subsequently be lost. Besides physical changes affecting the organs for producing the sounds of speech (lips, palate, larynx, nasopharyngeal cavity), psychogenic factors may cause abnormalities. Also, in some cases the cause may be located in one of the cerebral centers concerned with speech.

Deformity or disease of the organs of speech affect the sounds produced and thus impair intelligibility. Such defects are particularly those associated with harelip or cleft palate (see page 400), which are abnormalities caused by failure of the facial skeleton of the embryo to develop properly. Insofar as speech is concerned, they manifest themselves in disturbed pronunciation of consonants and in speech characterized by "open" nasal sounds due to the abnormal connection between the oral cavity and the nasopharyngeal cavity. Fairly common defects of the speech organs are due to local hypertrophy (excess growth) of mucous membranes or lymphatic tissue (enlarged tonsils in children, nasal polyps). In such cases the normal outflow of air through the resonance cavity of the nose is obstructed, so that speech is blurred and distorted, characterized by "closed" nasal sounds. Other organic defects are due to paralysis of the speech muscles resulting from absent or disturbed nerve supply of one or more muscles. The underlying cause may be degenerative changes along the course of individual nerves, in the medulla oblongata or in control centers located in the cerebrum; speech is lalling (babbling), indistinctly articulated or slurred. Paralysis of the laryngeal muscles manifests itself in a change of timbre; this may vary in degree; in severe cases the voice may be reduced to a hoarse whisper. Diseases of the cerebellum may, among various other abnormalities, cause *ataxic speech*—defective speech resulting from lack of muscular coordination, characterized by syllables and words uttered jerkily and with unequal loudness. Against this, disease affecting the root ganglia (e.g., in Parkinson's syndrome) is associated with monotonous, slow and barely articulate speech, sometimes (though rarely) accompanied by convulsive repetitions of the final syllables of individual words. These last-mentioned defects may occur as a result of general circulatory disorders.

Abnormalities due to disease of the main speech centers in the cerebellar cortex are collectively referred to as *aphasia*. In the condition called motor aphasia the patient knows what he wants to say, but cannot say it; he is unable to coordinate the muscles controlling speech, which is therefore incoherent because of his inability to articulate, although there is no paralysis of the muscles of articulation. In a mild case he may merely have difficulty in finding his words. Sensory aphasia, on the other hand, is characterized by inability to understand spoken words and/or written words (auditory and/or visual aphasia). The patient does not know what he is saying and produces a jumble of words, characterized by omissions, inversions and general mutilation of sentences. In extreme cases this becomes so-called gibberish aphasia ("word salad").

Defects due to disturbed speech development in children include stammering, stuttering, and lisping. These are usually of psychogenic origin and tend to arise particularly when the child starts school or (especially in boys) with the onset of puberty. In many cases these defects disappear of their own accord as the personality develops and matures. In severe cases it will be necessary to seek treatment (speech therapy).

Besides these mainly temporary abnormalities there are those associated with mental illness. For example, slowing of speech is generally encountered in depressed states; complete mutism may occur in schizophrenia. On the other hand, excessive talk flow occurs in mania and excited states.

EYE AND VISION

The human eye consists of the eyeball, which is approximately spherical in shape, and accessory parts for its protection and movement (eyelids, lacrimal apparatus, muscles). The *eyeball* (Fig. 12-36) is suspended by connective tissue in the orbit and is protected in front by the upper and the lower eyelids. The eyeball is enclosed in three layers (coats), the innermost of which is the retina, containing light-sensitive cells; around this is the choroid coat, containing numerous fine blood vessels for supplying the eyeball with blood; the outer coat consists of the sclera and, at the front, the cornea. The contents of the eyeball are the aqueous humor, the lens and the vitreous humor.

The *sclera* consists of fibrous connective tissue; its front part is the "white" of the eye, in the middle of which, in front of the lens, its place is taken by the cornea, which is transparent. The *cornea* also consists of fibrous connective tissue and is covered by a layer of epithelium which is continuous with the *conjunctiva* (the mucous membrane that lines the eyelids). There are no blood vessels in the cornea, and it can be grafted successfully because there are no blood cells to produce a defensive response resulting in rejection of a "foreign" cornea.

The *choroid coat* is joined to the iris by the *ciliary body,* a circle of tissue whose posterior surface has radial ridges (ciliary processes) to which is attached the ciliary zonule, i.e., the ligament by which the lens is suspended. The ciliary body consists of ciliary muscle: smooth muscle under the control of parasympathetic nerves. In front of the ciliary body the choroid coat continues as the *iris,* a ring of tissue which is the colored part of the eye and which rests against the front of the lens. In the middle of the iris is the *pupil,* the circular opening through which light enters the eye. The iris contains smooth muscle fibers, some of which are arranged circumferentially around the pupil and reduce its size by contracting; other fibers are arranged radially, so that their con-

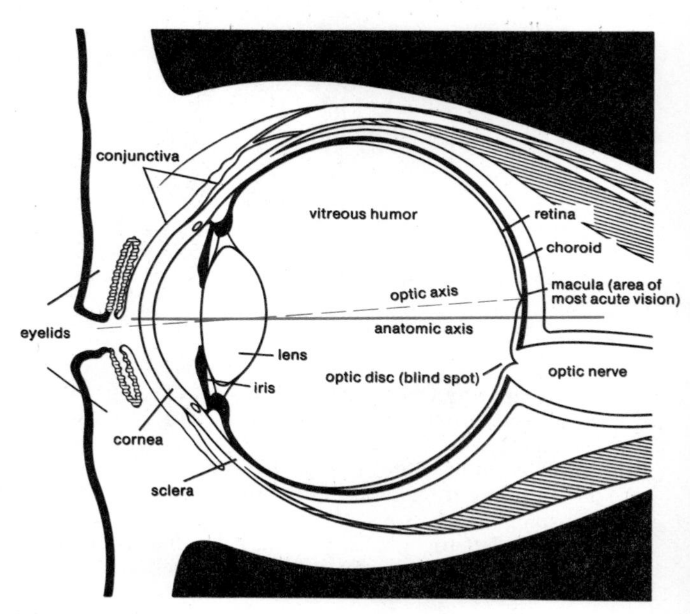

FIG. 12-36.
Section through the eyeball.

FIG. 12-37.
Section through front part of the eyeball.

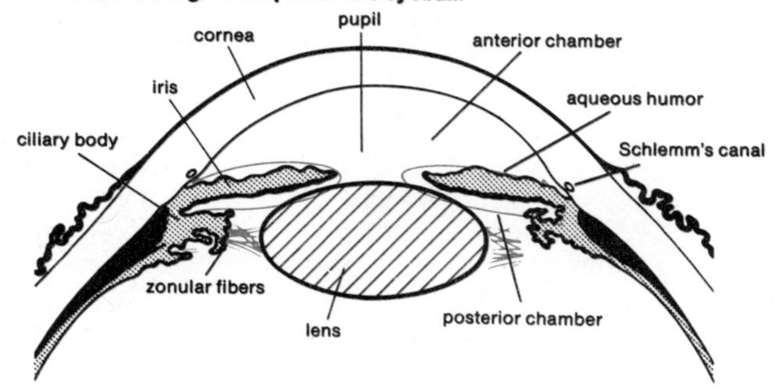

traction enlarges the pupil to let more light in. The pigment which is always present at the back of the iris looks blue; if there is additional pigment nearer the front, it modifies the color to brown; the color of the iris is darker in proportion to the amount of pigment it contains.

The posterior part of the *retina* (Fig. 12-38), the innermost coat of the eyeball, is light-sensitive. This part consists of three layers of cells, the outermost of which are the actual light receptors (rod and cone cells). The *cones* are stimulated only by bright light; their function is the perception of color and fine detail. The *rods* respond to dim light, but perceive only black-and-white images, i.e., they are color blind. Nerve fibers emerging from the rods and cones synapse with the bipolar cells, which in turn synapse with the ganglion cells (the third and innermost layer of the retina). The axons of the latter converge at the optic disc (blind spot) to form the optic nerve. About 2 mm sideways from the blind spot and directly opposite the pupil is the *macula lutea* (yellow spot), where the retina is reduced to a layer of closely packed cones functioning as the area of most acute vision.

The *lens* (Fig. 12-39) is biconvex and consists of lens fibers, which are transparent and elastic, and is surrounded by the lens capsule, the periphery of which is attached to the ciliary zonule. The latter exerts a pull on the lens capsule and thus tends to flatten the lens. When the eye accommodates, i.e., focuses on near objects, contraction of the ciliary muscle under the stimulation of parasympathetic nerve fibers eases the tension of the ciliary zonule, so that the pull it exerts on the lens capsule is reduced, thus allowing the lens to become thicker and more rounded.

The space between the cornea and the iris is the anterior chamber; the smaller space between the iris and the lens is the posterior chamber. The two chambers, which communicate through the pupil, are filled with a watery fluid called the *aqueous humor*. The eyeball behind the lens is filled with a transparent jellylike substance called the *vitreous humor*.

On traveling through the eye before reaching the retina the light passes through these various substances (refractive media), i.e., the cornea, aqueous humor, lens and vitreous humor. The lens is only a relatively minor component of the overall optical system of the eye. The main light-refracting structure is the cornea with the aqueous humor, while the lens provides the means of varying the focus.

In the normal eye, distant as well as near objects can be focused on the retina so as to produce a sharp image on it; such an eye is said to be *emmetropic*. A near object is

FIG. 12-38.
Section through the retina.

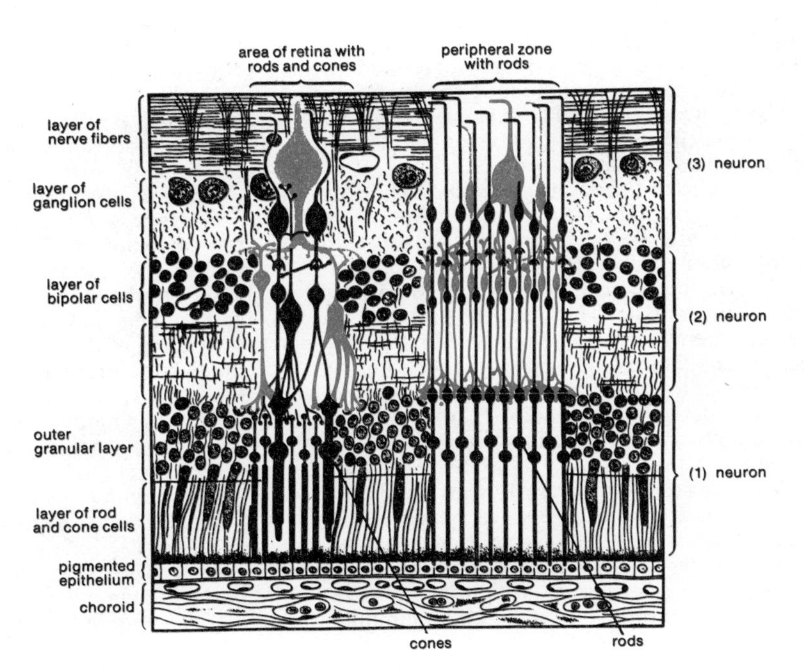

FIG. 12-39.
Section through the lens.

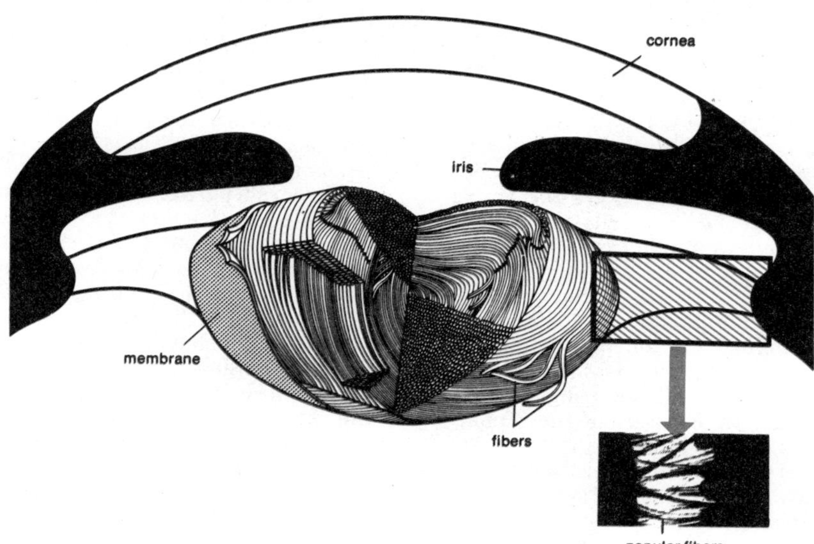

brought into focus on the retina by the process called *accommodation,* which occurs when the ciliary muscles contract and thus slacken the lens, so that it becomes more convex and its focal length shortens. At the same time there is reflex constriction of the pupil, providing greater depth of focus; also, there is convergence of the eyes, i.e., they are rotated inward to direct the gaze at the object looked at. Errors of refraction include hypermetropia (farsightedness), myopia (nearsightedness), presbyopia (farsightedness developing in middle age) and astigmatism.

The *cones* of the retina are stimulated only by bright light. Their function is to perceive color and fine detail. They are sensitive to three colors: red, yellow-green and blue-violet, more commonly referred to simply as red, green and blue. These are the primary colors; all other colors can be produced by mixing the primary colors in varying proportions, as in color photography and color television. An inherited defect causes color blindness, usually an inability to distinguish between red and green.

The *rods* respond to very low levels of light intensity and enable objects to be perceived even in dim twilight conditions (night vision). They contain a light-sensitive pigment called *visual purple* (*rhodopsin*), which becomes inactive on exposure to bright light; adaptation to night vision is brought about by slow regeneration of rhodopsin, a process that requires vitamin A. On passing from brightly illuminated surroundings into the dark, vision at first is poor, but improves after a time, though full adaptation of the rods may take as much as an hour, during which time the amount of rhodopsin in them steadily increases. This pigment consists of a protein (opsonin) and a chemical resembling carotene, which is stored in the liver and is there converted to vitamin A. When rhodopsin is regenerated, this vitamin is obtained from the blood; a deficiency of the vitamin results in night blindness.

Sharpness of vision depends not only on the proper focusing of the image on the retina, but also on the resolving power of the eye, i.e., its ability to distinguish between objects close together. This, in turn, depends on the closeness of spacing of the receptor cells in the retina. In the human retina there are about 125 million of these cells, of which there are something like 20 times as many rods as cones. In the area of maximum resolving power there are about 170,000 receptor cells per mm^2, so that points whose angular distance apart is about half a minute of arc can still be distinguished as separate points. In some birds of prey the acuteness of vision is about six times as good as that of the human eye.

The *optic nerve* is not a true peripheral

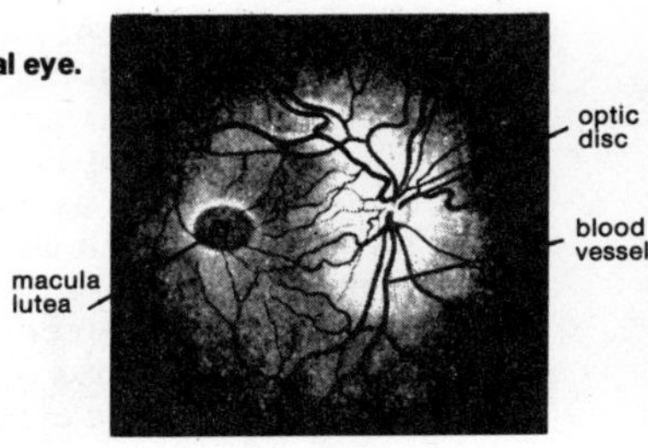

FIG. 12-40.
Fundus of the normal eye.

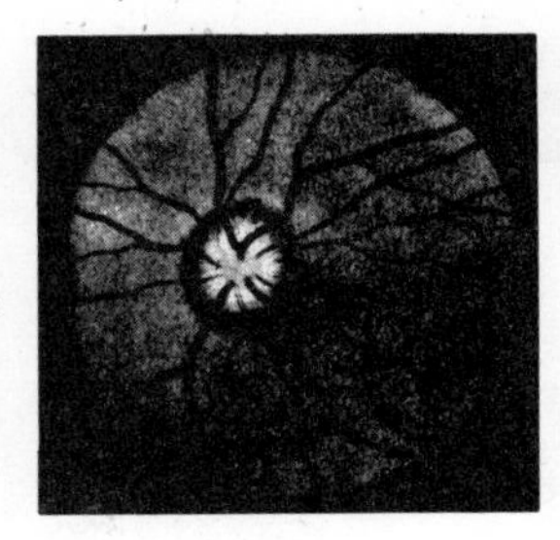

FIG. 12-41.
Fundus with enlarged optic disc and altered venous pattern.

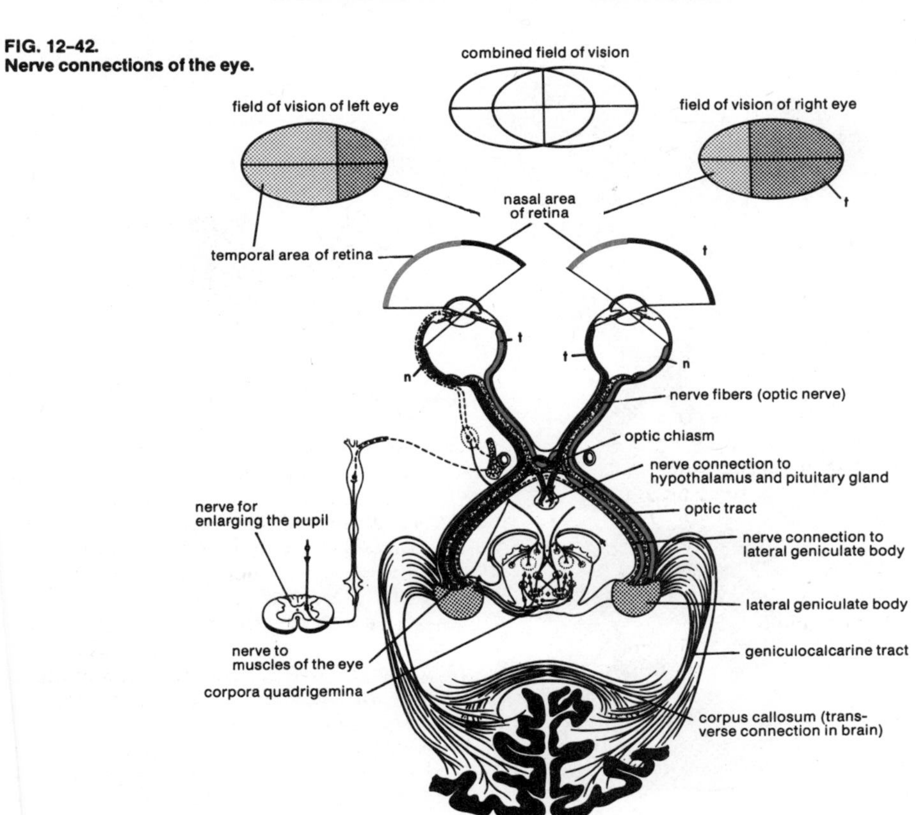

FIG. 12-42.
Nerve connections of the eye.

nerve, but is really a tract of the brain. Thus the fundus of the eye (the back of the eyeball), as viewed with the aid of an ophthalmoscope, is the only part of the human body whose examination can give direct indications about the condition of the brain. Normally the optic disc (blind spot), where the optic nerve emerges from the eyeball, is seen as a light-colored circular area (Fig. 12-40). Brain tumors often cause changes, more particularly enlargement, of the optic disc and of the pattern of the surrounding veins (Fig. 12-41); these can provide a valuable diagnostic indication. The optic nerve leads from the back of the eyeball and meets the optic nerve from the other eye at the *optic chiasm,* which is an X-shaped crossing of the optic fibers that originate in the inner half of the retina of each eye, whereas those optic fibers that originate in the outer half of each retina end up on the same side of the brain (Fig. 12-42). Past the chiasm the fibers form what are known as the two *optic tracts,* each of which terminates at the so-called lateral geniculate body, which acts as a relay station. From there the geniculocalcarine tract passes to the visual center in the occipital lobe of the cerebral cortex. Impulses are also transmitted from the lateral geniculate body to the superior colliculus in the midbrain, which plays an important part in visual reflexes.

Normally the two eyes are rotated reflexly so that both pupils point at the object being looked at; this is known as *convergence*. The images formed on the retinas of the two eyes are not quite identical; they are, however, "fused" in the visual center so that they appear as one (binocular vision), this being associated with a stereoscopic effect that produces

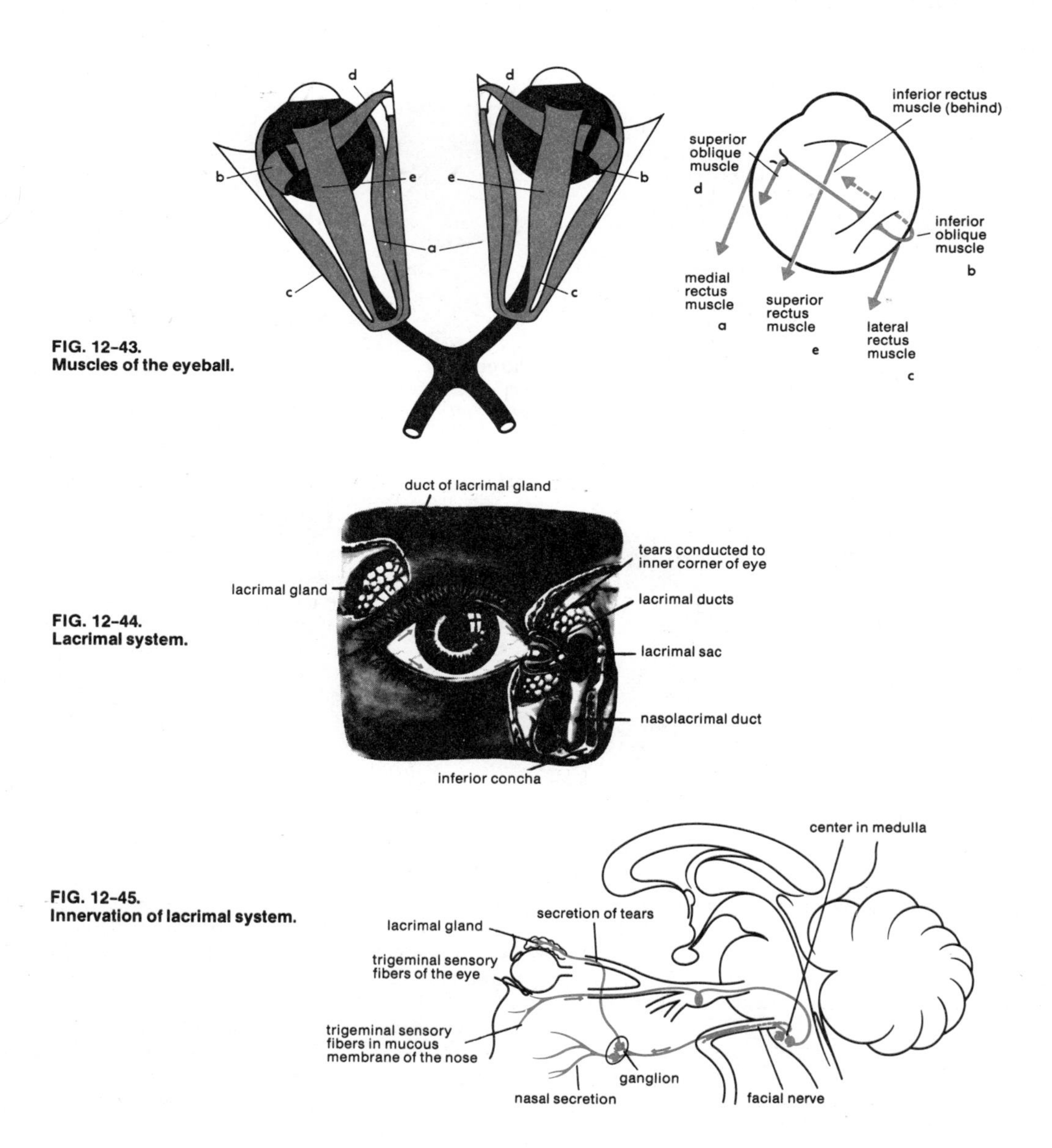

FIG. 12-43. Muscles of the eyeball.

FIG. 12-44. Lacrimal system.

FIG. 12-45. Innervation of lacrimal system.

a visual sensation of solidity and depth. If only one eye is used, only one image reaches the visual center; the sense of depth is then lacking. If convergence is defective, the images are not formed on corresponding points of the two retinas, and the visual center cannot "fuse" them properly; this defect is known as *diplopia*. *Squint* (*strabismus*), which is generally due to lack of balance of the muscles controlling the movement of the eyes, causes diplopia in which one eye may become virtually blind through lack of use because the images from it are ignored by the visual center. For this reason a squint should be treated as early as possible in childhood. A squint that develops in adult life causes double vision. Movements of the eyeball are brought about by six striped muscles: the superior, inferior, medial and lateral rectus muscles, and the superior and inferior oblique muscles (Fig. 12-43). The last two rotate the eyeball, while the four rectus muscles produce the up-and-down and side-to-side movements.

The so-called *adnexa* (accessory parts) of the eye comprise the *eyelids* and the *lacrimal apparatus*. Each eyelid is supported by a tarsal plate of fibroelastic tissue; the outer surface is covered with thin skin and the inner surface is lined with a thin epithelium, the *conjunctiva*. The eyelashes are small hairs that grow from the margin of the eyelid. Between the tarsal plate and the conjunctiva lies a row of sebaceous glands (tarsal glands); their secretion lubricates the eyelids and prevents tears from oozing out from under them. When the lids are closed, the space between them and the eyeball is called the conjunctival sac. When they are open, the cornea and conjunctiva are exposed to the air and have to be kept moist to survive, which is achieved by continuous secretion of tears by the lacrimal gland, which is located in a shallow cavity in the outer upper part of the orbit. Secretion of tears is stimulated by parasympathetic nerves. The tears are moved over the front of the eyeball by blinking of the eyelids and conveyed to the lacrimal sac (Fig. 12-44), from where the nasolacrimal duct leads to the nose, where the tears are utilized for moistening the inhaled air.

CATARACT

The lens of the eye is located behind the iris and pupil, in a concave recess of the vitreous humor. In the adult human being the lens consists of a nucleus enclosed by layers of fibers which, in turn, are surrounded by an elastic membrane (Fig. 12-39). In elderly people some degree of opacity (clouding) is nearly always found to have developed at the edges of the lens. If such opacity spreads farther inward, so that it affects the part directly behind the pupil and impairs vision, the condition is called *cataract*.

Initially the opacity usually appears in a spoke-shaped pattern developing at the front and rear of the lens capsule (Fig. 12-46). As the nucleus of the lens consists of dead material without metabolism of its own, it does not turn opaque. The opacity of the lens fibers is due to a kind of coagulation of the proteins of which they consist, comparable to what happens to the white of an egg when it is boiled. The spread of the opacity toward the center of the lens can be estimated by observing the shadow cast by the iris: the opaque regions light up when subjected to lateral illumination; the farther foward they lie, the narrower is the shadow, and vice versa (Fig. 12-47a,b). In this way it can usually be established that gradually larger and larger areas of the cortex (the peripheral parts of the lens) are affected by opacity until the iris casts no shadow at all (Fig. 12-47c). The cataract is then said to be *mature*. At that stage vision is completely lost in the affected eye, which can only distinguish between light and dark and detect the direction from which light is coming.

The mature cataract is treated by a comparatively simple operation, under local anesthesia, in which the lens is removed. First, the cornea is partly detached from the sclera around some two fifths of its circumference (Fig. 12-48). Next, the front membrane of the lens capsule is removed, and the lens is eased out with the aid of special instruments. Often the nucleus comes away first, leaving the cortex behind, which then has to be removed separately. The rear membrane of the capsule is left intact to serve as a partition between the aqueous humor (in the anterior chamber) and the vitreous humor. If this membrane is torn during the operation, the vitreous humor may prolapse, and this in turn may result in detachment of the retina. If portions of lens fibers remain adhering to the capsule in the cataract operation, there may be subsequent recurrence of opacity, sometimes necessitating a second operation (Fig. 12-49). As an alternative to the operation just described, the whole lens and capsule may be removed. Though it offers some advantages, this technique has the disadvantage that the rear membrane serving as a partition to retain the aqueous humor is also removed. A modern refinement consists in freezing the lens by means of a thin metal rod which is cooled to a temperature of −25°C, enabling the lens to be taken out as an intact solid whole.

Most cases of cataract are due to aging, but some are caused by injury or by faulty development of the lens in early childhood. Cataract in children can often be rectified merely by opening the anterior chamber to allow extrusion of the still-soft opaque material (Fig. 12-50).

In the refractive system of the eyeball, the lens is a relatively minor component. When it becomes an obstruction to the entry of light, as in cataract, its removal restores vision, though there is, of course, a loss in that the eye can now no longer accommodate, i.e., focus on near objects. This defect is compensated by the wearing of special corrective spectacles.

FIG. 12-46.
Initial stage of cataract.

FIG. 12-47.
(a) Cataract in a more advanced stage.

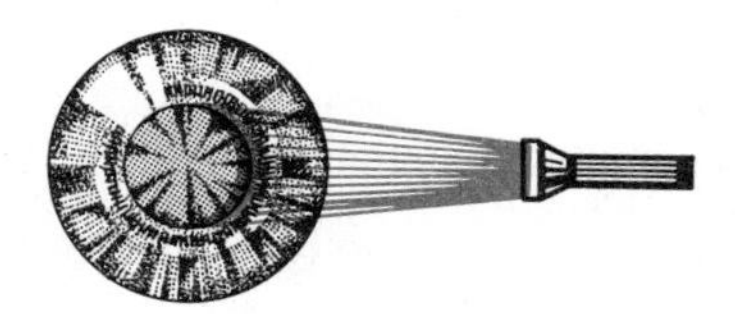

(b) Cataract in a very advanced stage.

(c) Mature cataract.

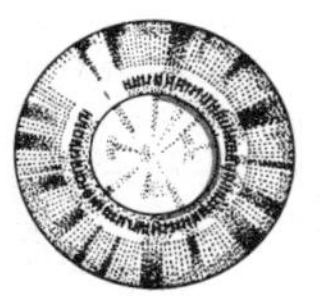

FIG. 12-48.
First step in cataract operation (partial detachment of the cornea).

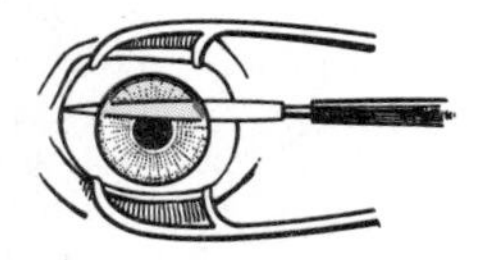

FIG. 12-49.
Second operation to treat recurrence of cataract.

FIG. 12-50.
Operation for cataract in children.

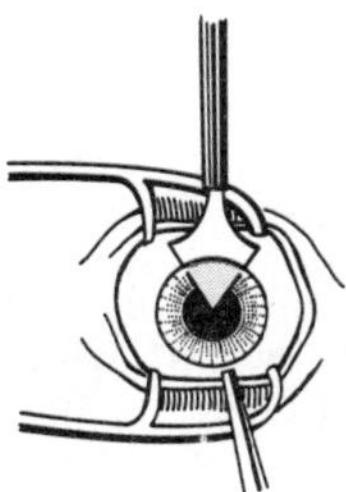

GLAUCOMA

The aqueous humor, the watery fluid between the cornea and the lens of the eye, is secreted in the posterior chamber behind the iris and circulates through the pupil into the anterior chamber, where it is absorbed in the angle between the rim of the iris and the back of the cornea (Fig. 12-51). From here it drains into the canal of Schlemm and seeps into the veins of the orbit. Its functions are to nourish the cornea and lens and to regulate the intraocular pressure, i.e., the pressure inside the eyeball, by the balance of its secretion and absorption. If this balance is upset, particularly by obstruction of the discharge or absorption of the fluid, an increase in pressure may have serious effects through compressing the blood vessels inside the eyeball, thus depriving the retina of an adequate supply of blood. Abnormally high intraocular pressure causes the condition called *glaucoma,* with disturbances of vision and, ultimately, blindness if it is left untreated.

The cause of reduction or obstruction of the discharge of fluid is often unknown. In some persons, however, the angle between the iris and the cornea in the anterior chamber is abnormally narrow so that even the normal thickening of the iris, when the pupil is fully dilated, may obstruct the outflow (Fig. 12-52b). In other cases the actual pores through which the fluid seeps away are deficient, or inflammation of the iris causes obstruction. Glaucoma may occur as an acute (sudden) and painful condition, but more often it is chronic, developing insidiously and perhaps remaining undetected until the optic nerve suffers permanent damage.

Besides the aqueous humor, the blood supply to the eyeball may play an important part in determining the intraocular pressure. The vortex veins from the choroid coat of the eyeball pass obliquely through the sclera (Fig. 12-53). When these veins are compressed as a result of increased intraocular pressure caused by obstructed outflow of the aqueous humor, the flow of blood to the eyeball through the arteries continues, thus adding to the intraocular pressure. This *acute glaucoma* is accompanied by pain, which may radiate from the eye to all parts of the head. Vision is greatly impaired because the deficient blood supply to the eye results in damage to the retina. The conjunctiva is inflamed, and the cornea becomes clouded or opaque.

Of more frequent occurrence is *chronic glaucoma* (glaucoma simplex chronicum), in which the angle of the anterior chamber is not abnormally narrow, but the outflow of fluid is nevertheless obstructed as a result of organic changes due to aging which block the canal of Schlemm. At first there are hardly any symptoms. The patient may have a sensation of a film before his eyes, and focusing is impaired, so that a book or newspaper has to be held farther away. Lights appear to have colored halos around them. Eventually, parts of the field of vision become sightless. The increase in intraocular pressure is not very great and at first produces only a slight sensation of tension; sometimes a dull pressure behind the forehead is felt.

Since increased pressure within the eyeball will in the long run cause permanent damage, possibly resulting in total loss of sight (glaucoma absolutum), early diagnosis of glaucoma is important. Measurement of the intraocular pressure with an instrument called a *tonometer* (Fig. 12-54) is a valuable aid. Only when the condition has reached a fairly advanced stage does it become possible to detect visible changes in the fundus of the eye by ophthalmoscopy.

In adults the sclera of the eyeball is tough and inelastic, so that the increased internal pressure most severely affects the least rigid part, i.e., the part with the most "give"—namely, the optic disc where the optic nerve enters the retina. As a result of this pressure on the optic disc a depression is formed in it; this condition is known as excavation of the optic disc or optic nerve, characterized by kinks in the blood vessels at its edge (Fig. 12-55).

In a young child the sclera is still resilient, so that increased intraocular pressure may instead cause considerable enlargement of the eyeball (hydrophthalmos: infantile glaucoma). This is generally a congenital condition, affecting both eyes and apparently due to extreme narrowness of the angle between the iris and the cornea, so that the outlet of fluid is blocked.

In addition to these various types of primary glaucoma, mostly occurring without precisely known cause, there is secondary glaucoma, in which there is an increase in intraocular pressure due to other eye disease, e.g., displacement of the lens (lenticular luxation). It may thus occur that the lens is displaced backward, whereby the cells secreting the aqueous humor are stimulated to excessive production.

FIG. 12-51.
Horizontal section through left eyeball (seen from above).

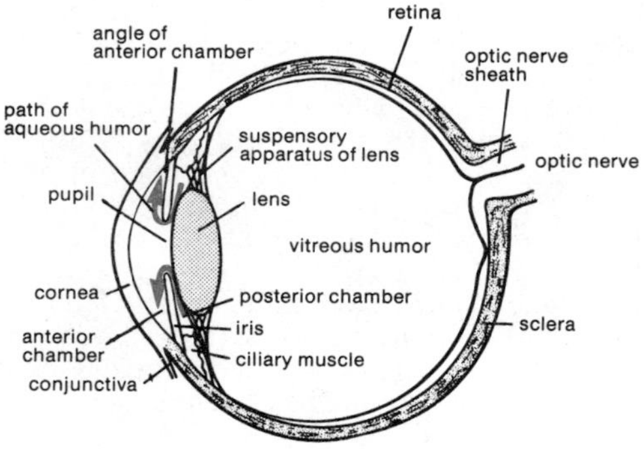

FIG. 12-52.

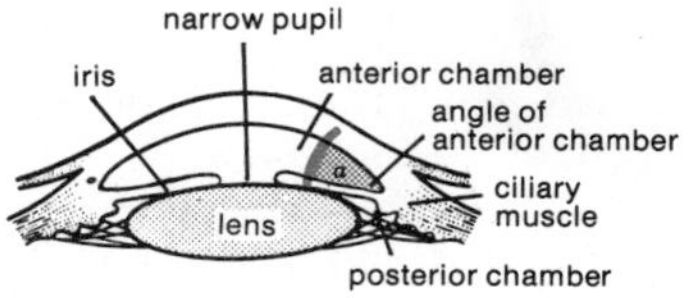

(a) In the normal eye the aqueous humor readily drains away.

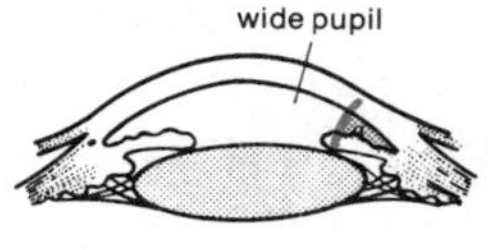

(b) If the angle of the anterior chamber is abnormally narrow, drainage is obstructed.

FIG. 12-53.
Blood vessels of the eye.

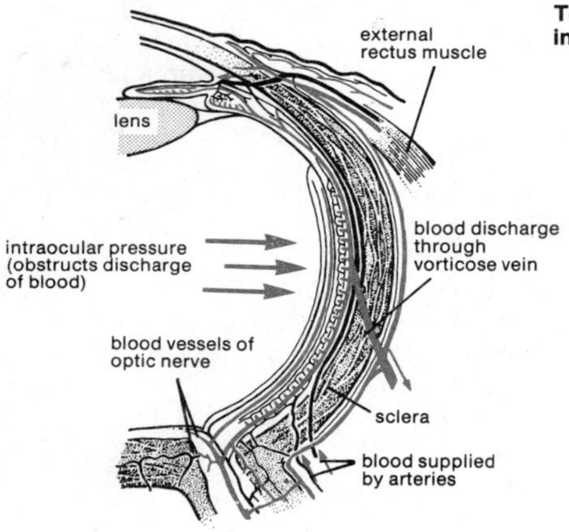

FIG. 12-54.
Tonometer for measuring intraocular pressure.

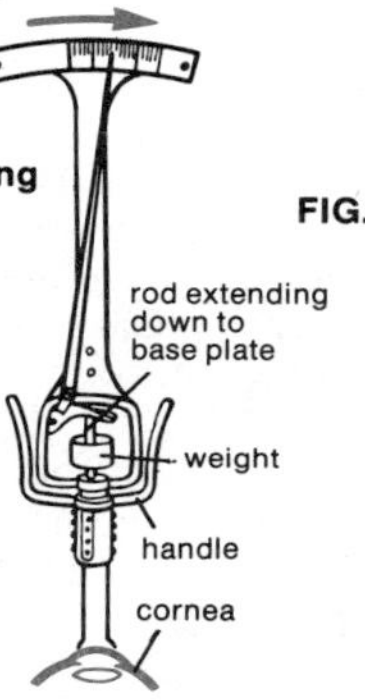

FIG. 12-55.

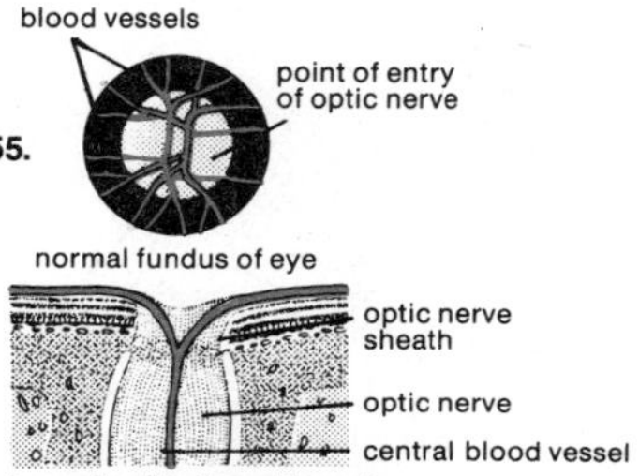

(a) Longitudinal section at point of entry of optic nerve into eyeball.

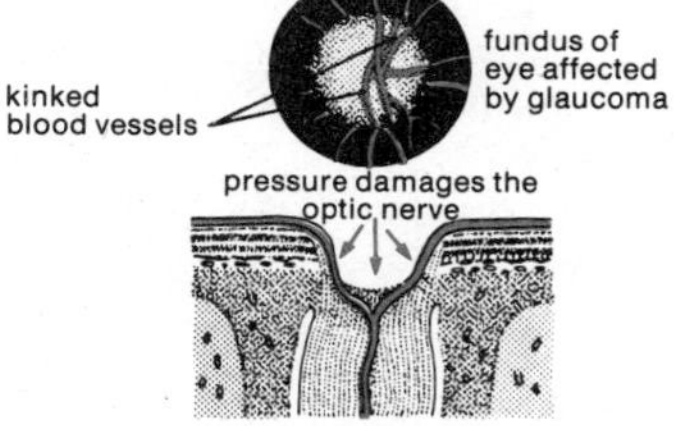

(b) Excavation of the optic disc due to intraocular pressure.

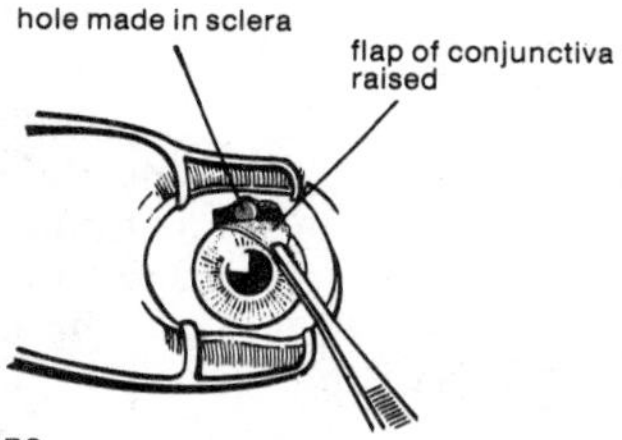

FIG. 12-56.
Operation for the treatment of glaucoma by perforation of the sclera (cyclodialysis).

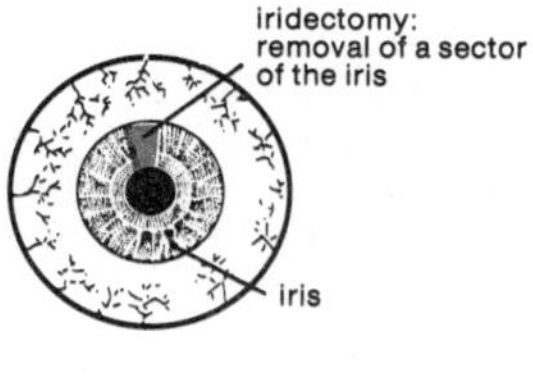

FIG. 12-57.
Iridectomy.

Or the lens may be displaced forward, causing obstruction of the fluid outlets in the anterior chamber. Inflammation of the iris or the use of drugs to enlarge the pupil (e.g., atropine) may disturb the balance between fluid production and absorption. Eye injuries and tumors in the eye may also cause glaucoma.

Initial treatment of glaucoma is generally

nonoperative, with special drugs (miotics) which cause the pupil to contract and widen the angle between the iris and the cornea in the anterior chamber (Fig. 12-52a). Such drugs are, for example, pilocarpine, eserine and physostigmine. At the same time, enlargement of the pupil by drugs (atropine, scopolamine, etc.) or darkness should be avoided as much as possible. In an acute attack of glaucoma, diuretics (particularly acetazolamide) may additionally be given. Severe cases of glaucoma will require operative treatment to relieve intraocular pressure. In the operation known as *cyclodialysis* (Fig. 12-56), a flap of the conjunctiva is raised and a hole is made in the sclera close to the edge of the cornea, enabling the aqueous humor to escape under the conjunctiva. Another operation is *iridectomy,* i.e., removal of part of the iris, which is equivalent to an artificial enlargement of the angle of the anterior chamber (Fig. 12-57). A third operation is *iridencleisis,* in which a portion of the iris is gripped (incarcerated) in a wound in the sclera, thereby forming a fistula through which the aqueous humor can drain to the exterior.

ERRORS OF REFRACTION, SPECTACLES

The basic requirement for sharpness of vision (visual acuity) is that the object observed forms a sharp image on the retina of the eye (Fig. 12-58). This is achieved by *accommodation,* i.e., by varying the curvature of the lens to focus the image. When the normal eye looks at anything more than about 5 meters away from it (the so-called far point: *a* in Figure 12-59) the eye forms a sharp image on the retina at *a'* without having to accommodate, i.e., the eye is fully relaxed. An object that is closer to the eye than 5 meters—at *b,* for example—will form its image at *b'* behind the retina if the eye is still fully relaxed (Fig. 12-59a). In order to see this nearer object clearly, the eye must accommodate, which it does by increasing the curvature of the lens. This is achieved by a reflex action causing the ciliary muscles to contract and slacken the ciliary zonule, so that the lens bulges. Increased curvature shortens the focal length of the lens, bringing the image sharply into focus on the retina (Figs. 12-59 and 12-60).

In a 10-year-old child the range of accommodation is large, so that an object only 8 cm from the eye (the near point) can still be seen sharply when the eye is at maximum accommodation. As a person grows older, the lens progressively loses its elasticity, so that it cannot bulge and thus increase its curvature so much. By the age of about 60 the lens has become almost completely inelastic. This is accompanied by *presbyopia* (''oldsightedness''), i.e., the farsightedness (or longsightedness) that normally comes with advancing age and which requires correction—by a convex lens (Fig. 12-62) for close work or reading—after the age of about 45. Eventually, the power of accommodation is so diminished (Fig. 12-63) that the unaided eye cannot focus sharply on anything closer to it than about 1 meter. The convex lenses of spectacles correct this, so that the focusing distance is reduced to about 33 cm (the normal distance for reading). These lenses may be of the bifocal type, with only the bottom half of the glass ground as a convex lens for close vision.

The optical capacity of the eye—i.e., the refractive power of its optical system comprising the cornea and aqueous humor, the lens, and the vitreous humor—is usually expressed in *diopters.* A lens with a focal distance of 1.00 m has a refractive power of 1 diopter. The shorter the focal distance, the greater is the power: thus a lens with a focal distance of 0.50 m has a refractive power of 2 diopters, and one with a focal distance of 0.20 m has a refractive power of 5 diopters (Fig. 12-64). The normal (*emmetropic*) eye of a young person has a refractive power of 58 diopters when fully relaxed (i.e., focused on the far point or beyond), with a range of 14 additional diopters for focusing by accommodation. If the lens has lost its elasticity, as in persons over the age of 60, spectacles with convex lenses of about 3 diopters will be needed to enable the person to read at about 33 cm. The lens of the eye itself has a refractive power of about 16 diopters. So if the lens has to be removed, as in an operation for cataract, spectacles of about 13 diopters power will be needed for distant vision, while reading will require spectacles of about 15–16 diopters.

In *hypermetropia* (farsightedness, longsightedness), which should not be confused with presbyopia (which is a normal effect of aging), the eyeball is too short, so that rays from the far point and beyond (''infinity'') are focused behind the retina when the eye is at rest (Fig.

FIG. 12-58.
Formation of the image in a camera (left) and in the eye (right).

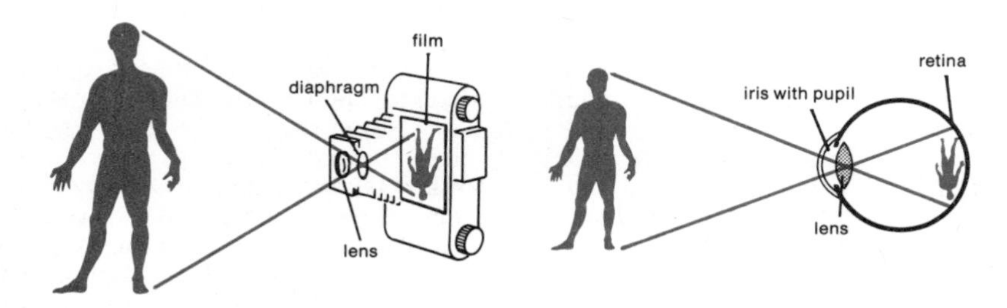

FIG. 12-59.

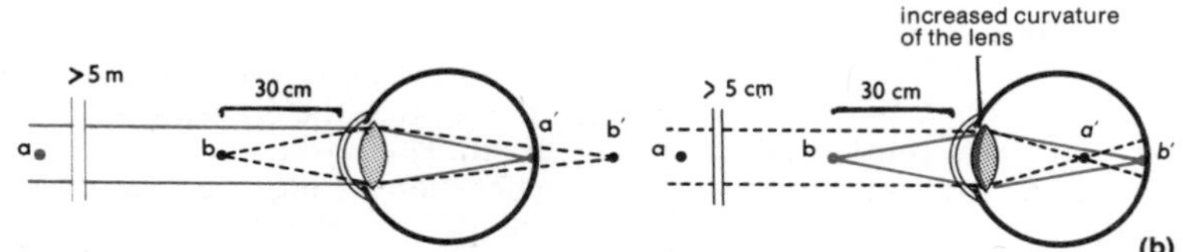

(a) When the normal eye looks at distant objects, any object farther than about 5 m will form a sharp image on the retina (such as the image *a′* of the object *a*), whereas a near object (such as *b*) will form its image behind the retina (at *b′*) and thus appear out of focus.

(b) When the eye focuses on a near object (this is known as accommodation), this will form a sharp image on the retina, while now the distant object will be out of focus.

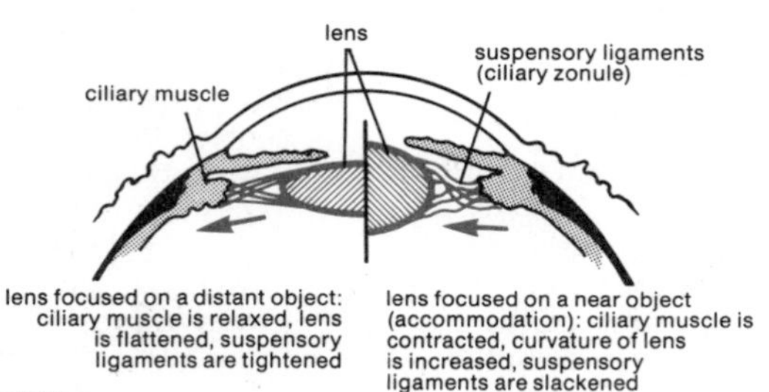

FIG. 12-60.
Lens curvature change in accommodation.

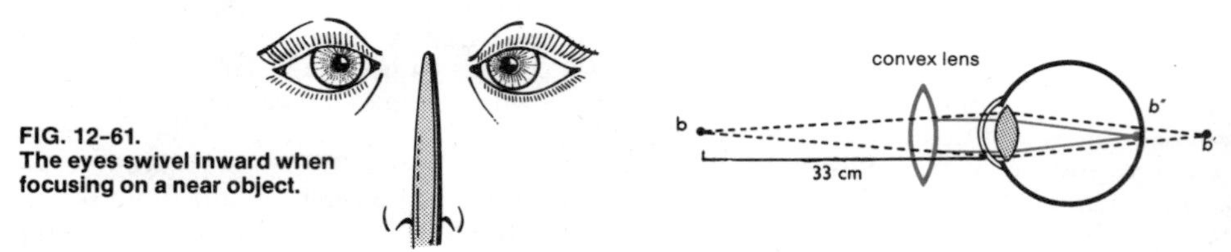

FIG. 12-61.
The eyes swivel inward when focusing on a near object.

FIG. 12-62.
If the refractive power of the eye is inadequate to form a sharp image of the object *b* on the retina, a suitable convex lens will compensate for the deficiency and will bring the image forward from *b′* to *b″* so that it is now formed on, instead of behind, the retina.

12-65a). To see distant objects clearly, it is necessary to accommodate; to avoid this necessity and to compensate for the deficient length of the eyeball, a corrective convex lens is required (Fig. 12-65b). The presbyopic eye only has to be aided by spectacles for reading or other close work, whereas the farsighted person requires spectacles in all circum-

FIG. 12-63.
Diminishing power of accommodation with advancing age (presbyopia).

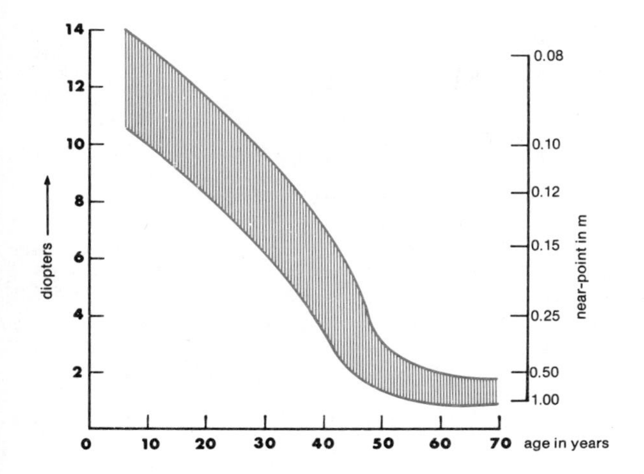

FIG. 12-64.
Refractive power of lenses expressed in diopters.

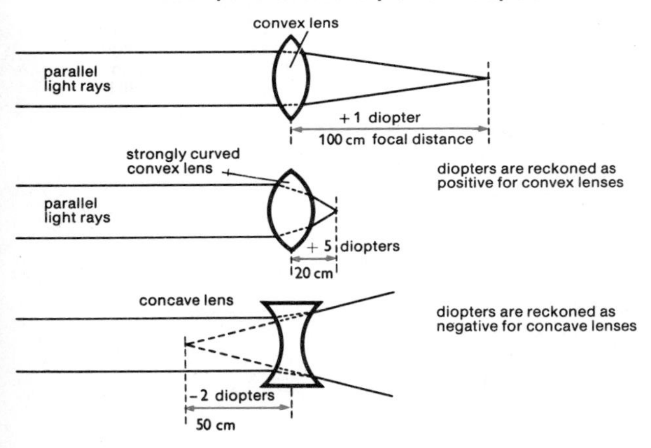

stances, including distant viewing. In a farsighted young person the automatic (reflex) accommodation of the eye compensates for the defect; but this compensation involves constant effort (contraction of the ciliary muscles), which may be harmful in the long run, resulting in undesirable changes in the fundus of the eye. With advancing age, the farsighted person, who may until then not have suspected that his vision was defective, will sooner have difficulty with reading than a person with normal eyesight which is developing into presbyopia. The farsighted eye uses its reserves of accommodation even for distant vision, so that the diminishing capacity of the lens to increase its curvature for close vision manifests itself sooner than in the normal eye.

In *myopia* (nearsightedness, shortsightedness) the eyeball is too long, so that rays from distant objects are focused in front of the retina; such objects therefore cannot be seen clearly (Fig. 12-66a) unless the deficiency is corrected with a concave lens (Fig. 12-66b). The only objects that the myopic eye can see clearly are those that are near enough to produce a sharp image on the retina when the lens is fully relaxed. Accommodation obviously cannot compensate for the myopic condition: it merely increases the curvature of the lens and further shortens the focal distance. On the other hand, the nearsighted person has no problem with reading and other close work; advancing age improves his condition in that he can then see objects farther away with less effort, and he is unlikely to need reading glasses.

Hypermetropia and myopia, due to the eyeball being relatively too short or too long, are to some extent the manifestation of mere random variation of eyeball shapes from the average condition of "normal" vision. In Figure 12-67 the mean or average is at +0.5 diopter, not 0 diopter, as might theoretically be expected. This would suggest that statistical "normal" actually corresponds to a slight degree of hypermetropia. The limit of statistical myopia, represented by the dotted curve, is bordered by a zone of very severe myopia, called malignant (or pernicious) myopia, represented by the red shaded area.

Whereas ordinary myopia, which develops in children of school age (though it generally has nothing to do with schoolwork as such), ceases to grow worse after adolescence, the malignant type is progressive, i.e., it steadily worsens during adult life. It is probably due to an abnormal elongation tendency of the eyeball, which continues with advancing age. The far point, i.e., the greatest distance at which an object can be seen sharply by the unaided eye, which is at approximately 5 meters for the normal eye, may diminish to as little as 7 cm, or even less, in an advanced case of malignant myopia. Correction of such extreme nearsightedness is possible, if at all, only with very powerful spectacles of 15 diopters or more. The progressive elongation of the eyeball causes the choroid coat and retina to suffer damage, which impairs vision, especially if

FIG. 12-65.
(a) If the eyeball is too short, the focal point of the lens is behind the retina (at a), so that distant objects are not seen clearly (hypermetropia).

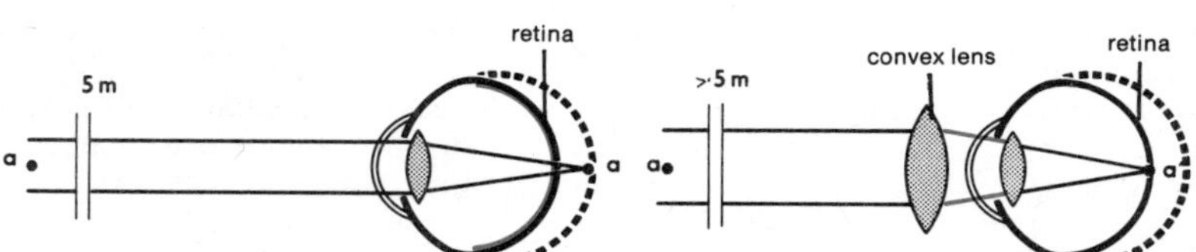

(b) A suitable convex lens increases the overall refractive power, so that the focal length is shortened and images of distant objects are formed on the retina.

FIG. 12-66.
(a) If the eyeball is too long, the focal point of the lens is in front of the retina (at a), so that distant objects are not seen clearly (myopia).

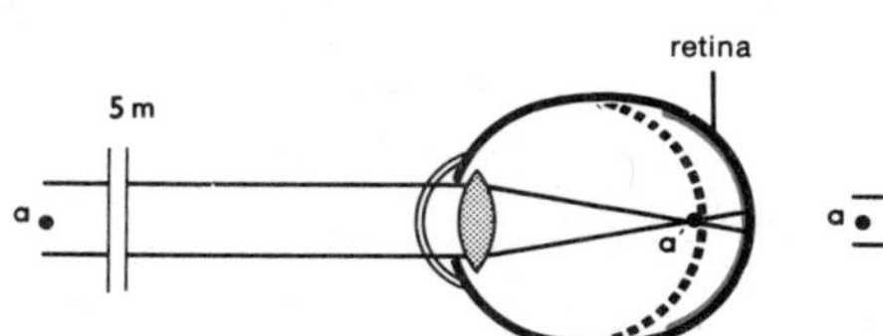

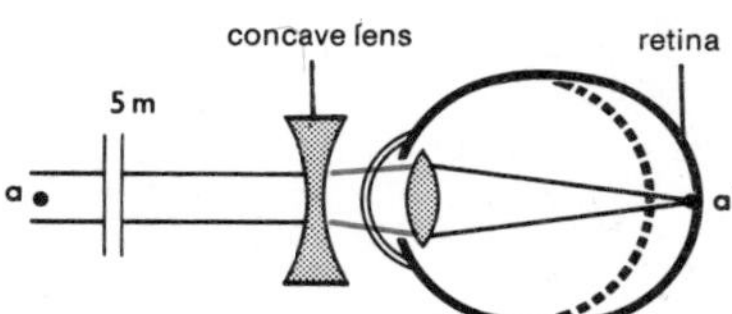

(b) A suitable concave lens decreases the overall refractive power, so that focal length is increased and images of distant objects are formed on the retina.

FIG. 12-67.
Curves showing frequency of hypermetropia and myopia.

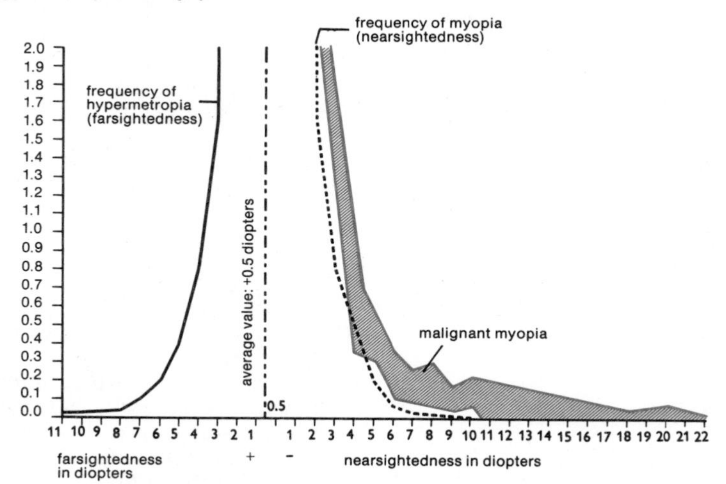

it—as it frequently does—occurs at the macula lutea in the center of the retina. Elongation of the eyeball also involves a risk of detachment of the retina.

Astigmatism is another fairly common defect of vision. In an optical lens, astigmatism is caused by the curvature being different in two mutually perpendicular planes. Thus, rays in one plane may be in focus, while those in the other plane are out of focus, producing distortion. In the eye, astigmatism is due to imperfectly spherical curvature, mostly of the cornea. Correction of the defect is achieved by means of so-called cylindrical lenses, which are ground to such a curvature that they compensate for the optical distortion of the eye. If astigmatism occurs in combination with hypermetropia or myopia, as it often does, so-called compound lenses are used to compensate for both types of defect.

COLOR BLINDNESS

Color blindness is the inability to distinguish between certain colors. Color is not a physical reality, but a subjective perception; it does, however, have an objective (physical) counterpart in electromagnetic waves of certain wavelengths, which are referred to as light waves, or visible light. These can stimulate the cones of the retina to transmit impulses to the brain where they come into consciousness as sensations of color. Radiation with a wavelength shorter than that of the violet end of the spectrum or longer than that of the red end, i.e., ultraviolet or infrared rays, is not experienced as "light" and is invisible. Visible light constitutes only a certain narrow range of the overall spectrum of electromagnetic waves (Fig. 12-68).

FIG. 12-68. Electromagnetic spectrum.

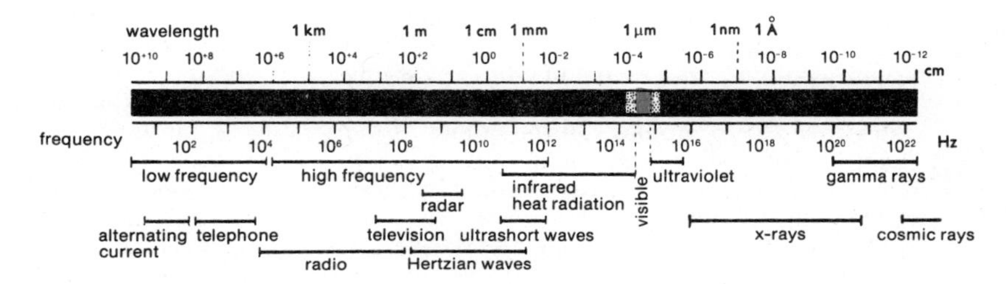

FIG. 12-69. Mixtures of three primary colors (red, green, blue) can produce all other colors and tints.

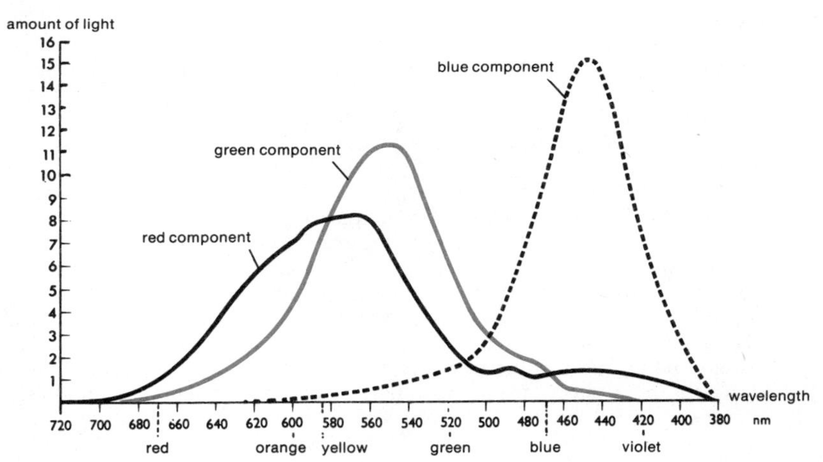

FIG. 12-70. Frequency of defective color vision.

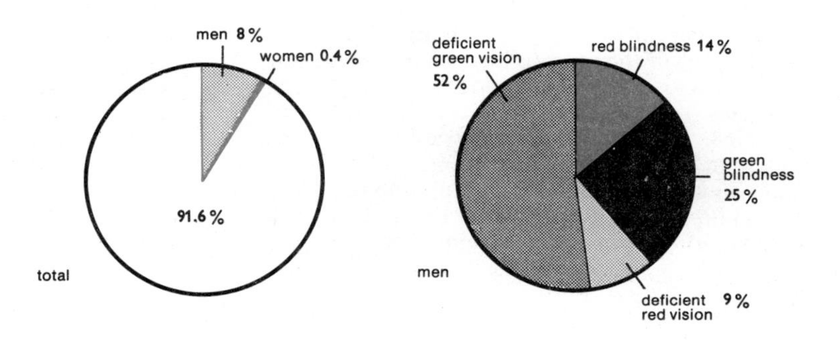

Normal color vision appears to be a composite function comprising three different physiological components involving three visual pigments in the color-sensitive receptor cells. These pigments are respectively sensitive to light of different wavelengths and utilize the energy of the absorbed light to change their own chemical structure. This change causes (electrical) stimulation of the receptor cells, which transmit their impulses to the brain. Thanks to the three-component pattern of color vision, the normal eye can produce any number of colors and tints by mixing the three primary colors (red, green and blue) in varying proportions. The proportions of the primary colors needed to produce all the colors of the spectrum of visible light in normal (trichromatic) color vision can be represented by three curves (Fig. 12-69).

Defects of color vision, collectively referred to as color blindness, arise from the absence or nonfunctioning of one of the three components, generally the red or the green. The condition is known as *red blindness* (*protanopia*), green blindness (*deuteranopia*) or *blue blindness* (*tritanopia*). Persons with one such component missing are said to have dichromatic vision, i.e., being able to see only two colors. Persons with red or green blindness are likely to confuse red, orange, yellow and yellow-green; a certain shade of red or a certain shade of green appears gray. The designation *red-green color blindness* is sometimes applied to this defect, denoting that red and green are the colors most likely to be confused with each other; actually the defect relates to the nonperception of only one of these two primary colors. Besides complete blindness for certain colors, which is a comparatively rare condition, there is more frequently partial color blindness—namely, deficient red vision (*protanomalia*) or deficient green vision (*deuteranomalia*). There is also a category of persons who have *monochromatic vision,* i.e., are totally color blind; they see only one color, though in different intensities or shades of light and dark. In congenital total color blindness there is usually a complete functional failure of the cones of the retina. Such a person relies on the rods also for seeing under bright light conditions.

Color blindness is considerably more common in men than in women (Fig. 12-70). It is a sex-linked character—especially the red-green variety—which women may transmit to their sons without themselves being affected. Red-green color blindness can occur in a woman, but only if her father is affected and her mother is a carrier of the defect. Other (rarer) forms of color blindness affect both sexes with approximately equal frequency.

SQUINT

Squint (strabismus) is a disorder of the eyes in which the optic axes cannot be directed to the same object. A squint may be inward (*convergent squint*), outward (*divergent squint*) or, though rarely, *vertical.* A squint that develops in an adult is usually due to paralysis of one of the muscles of the eye, resulting in lack of muscular coordination. Much more common is congenital squint, in which the muscles are healthy, but the eyes incorrectly coordinated. This is called *concomitant squint,* because both eyes move freely but retain their faulty relationship to each other, whereas in paralytic squint the degree of misalignment varies with the direction of the gaze. Concomitant squint in a child is often due to faulty accommodation of the eyes. If the child is farsighted, his eyes have to accommodate (i.e., focus by increasing the curvature of the lens) at all times, even for seeing distant objects. Accommodation is reflex-controlled by the eye muscles, which cause the eyes automatically to turn inward when they are focused for near vision. In the farsighted child, focusing on a distant object may wrongly produce the reflex that turns the eyes inward as though for close vision, so that he squints and sees double. He may then habitually ignore the images transmitted to the brain by one eye, so that this eye becomes useless and virtually blind.

In a person with normally aligned eyes, their optic axes are parallel when he looks at a distant object—in practice, a distant object is one that is at least 5 meters from the eyes (Fig. 12-71a). If the object approaches to within a short distance of the eyes, they turn inward while remaining directed to the object. This movement of the eyes is called convergence and takes place automatically as part of the function of accommodation (Fig. 12-71b). Any departure from this normal pattern of behavior results in squint (Fig. 12-71c,d).

The proper positioning and alignment of the eyes is of major importance for normal binocular vision, i.e., the fusion of the images formed by the two eyes to produce a sensation of a single image in the visual center of the

FIG. 12-71.

(a) The optic axes of normally aligned eyes are parellel when looking at a distant object.

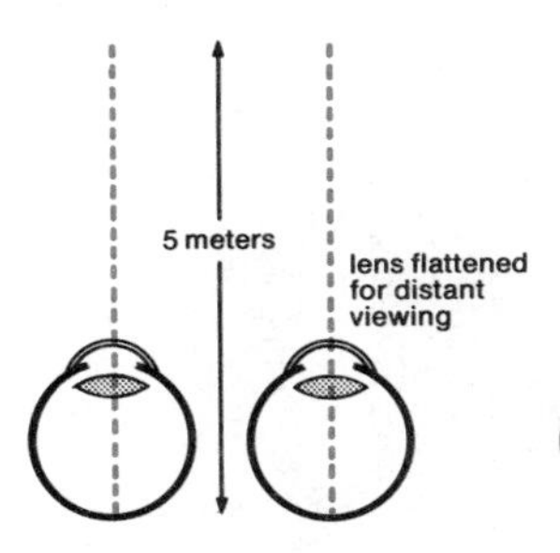

(b) When looking at a near object the optic axes converge on the object.

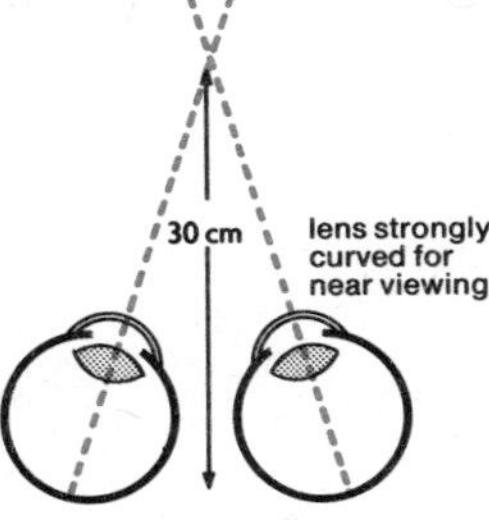

(c) Convergent squint of right eye.

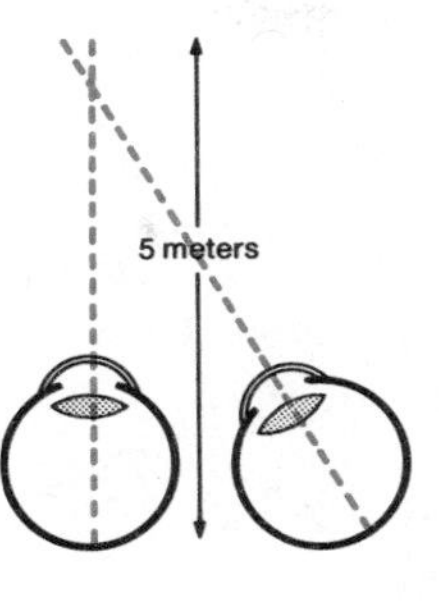

(d) Divergent squint of right eye.

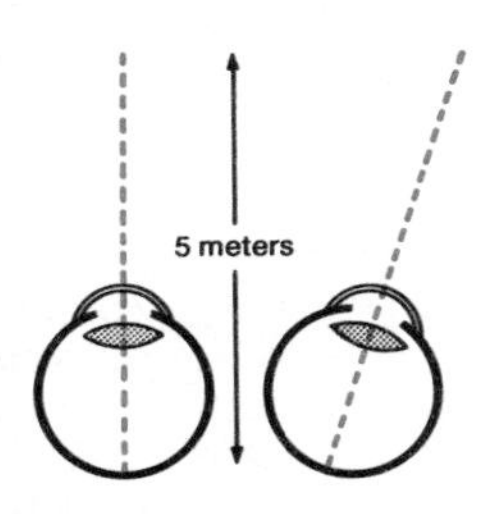

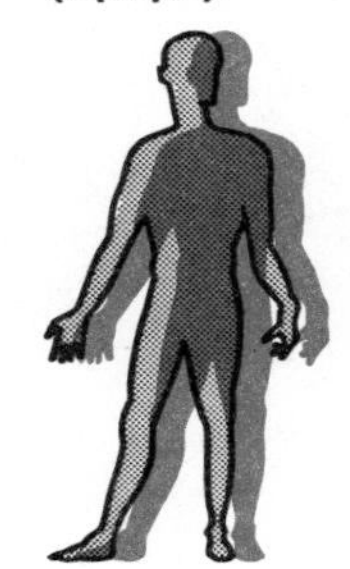

FIG. 12-72. Double vision (diplopia).

FIG. 12-73. Focus on corresponding parts of retina.

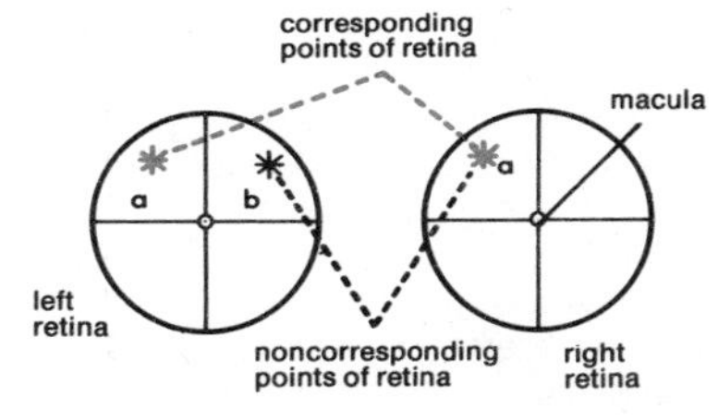

FIG. 12-74. The images from corresponding points of the two retinas are "fused" in the visual center of the brain so that they appear as one (binocular vision).

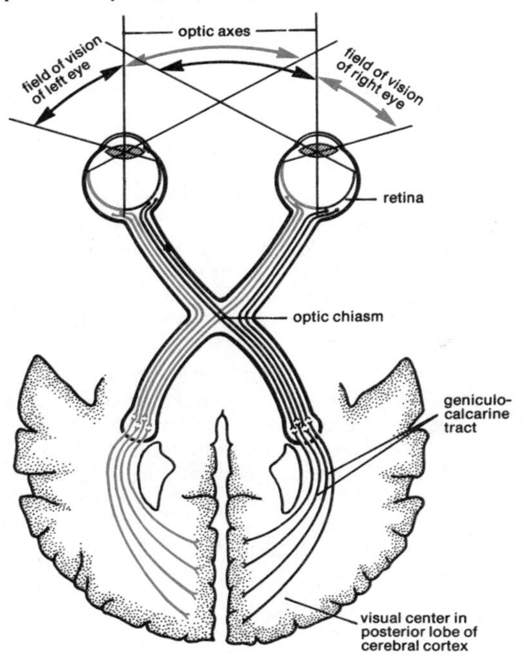

brain. Otherwise there is double vision (*diplopia*, Fig. 12-72). Normally the visual center ensures that the two images, formed in the left and the right eye respectively, are focused on corresponding areas of the retina (Fig. 12-73). This correspondence of retinal areas has its physiological and anatomic basis in the crossing of optic nerve fibers at the optic chiasm (Fig. 12-74, see also page 435). When both eyes focus on an object, the images that each eye forms of it are located on an area around the macula, i.e., the area of the retina directly opposite the pupil and composed of cones only; it is the area of most acute and most highly differentiated vision. This object will then produce a single visual image, and so will all objects located on a circle passing through the object on which the eyes are focused and through the two pupils (*horopter circle*). The images of points on this circle are formed on the aforementioned corresponding areas of the retina. Objects that are closer to, or farther from, the eye than the horopter circle are imaged on noncorresponding areas of the retina

FIG. 12-75.
Horopter circles.

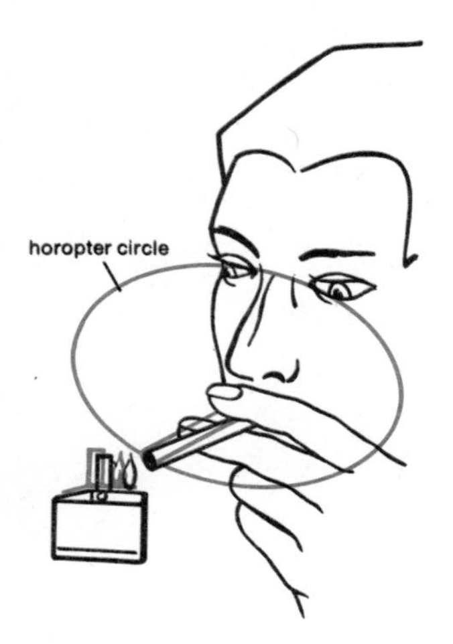

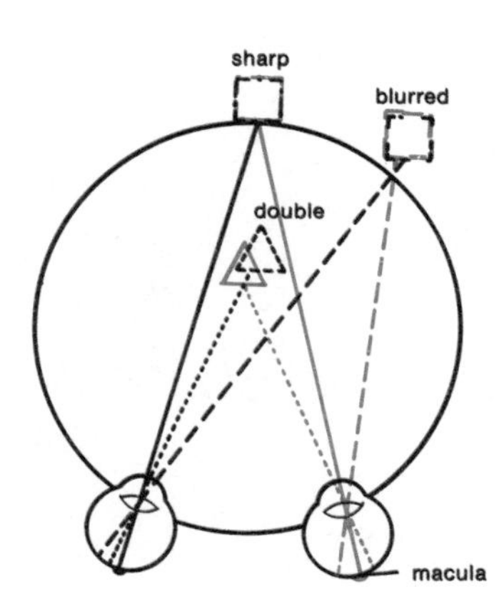

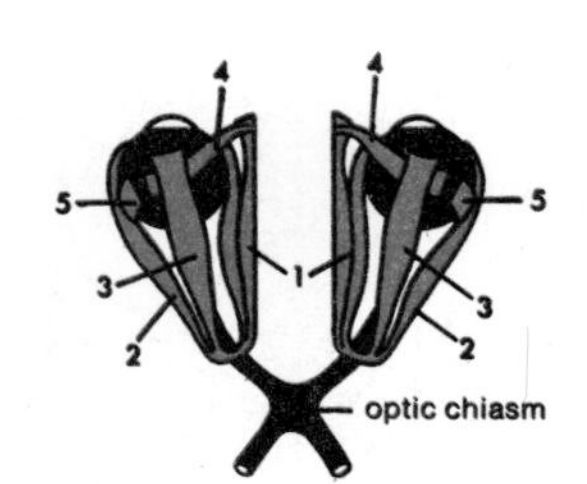

FIG. 12-76.
Eyeball muscles.

1= medial rectus muscle
2= lateral rectus muscle
3= superior rectus muscle
4= superior oblique muscle
5= attachment of inferior oblique muscle

and are seen double or unsharp (Fig. 12-75). In reality, only a small number of objects are located on the horopter circle, so that even in a normal field of vision a great many objects are in fact seen double. Yet they are not noticeable nor objectionable; this is because the brain concentrates its attention on the images of objects on the circle and suppresses all others. This capacity to suppress and eliminate images that would be confusing if they were allowed to obtrude themselves not only is important in connection with normal vision but is also a significant factor associated with squint. In a case of squint there is no correspondence of images on the two retinas, so that all objects are seen double.

The various forms of squint can be divided into two main categories: those cases where the person actually does "see things double," and those where the image from one of the two retinas is suppressed in the visual center, so that only one eye is really in use.

Paralytic squint, i.e., squint due to paralysis of a muscle, usually appears suddenly. The brain has no time to "learn" to suppress one of the retinal images. The patient then experiences very objectionable double vision, and his sense of spatial orientation is disturbed; this may be attended by severe nausea. Movements of the eyeball are brought about by six muscles: the superior, inferior, medial and lateral rectus ocular muscles, and the superior and inferior oblique ocular muscles (Fig. 12-76). The first four muscles control the up-

FIG. 12-77.
Function of rectus muscles.

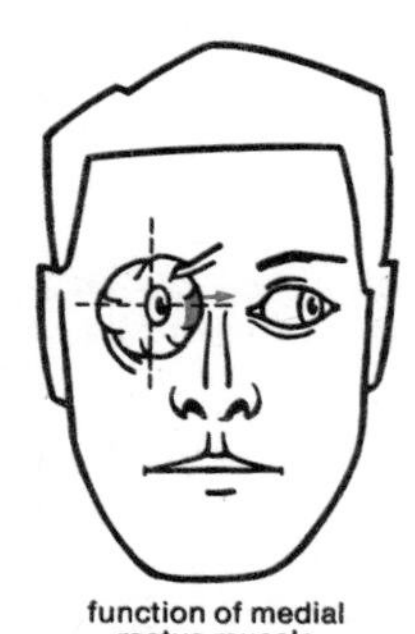

function of medial rectus muscle

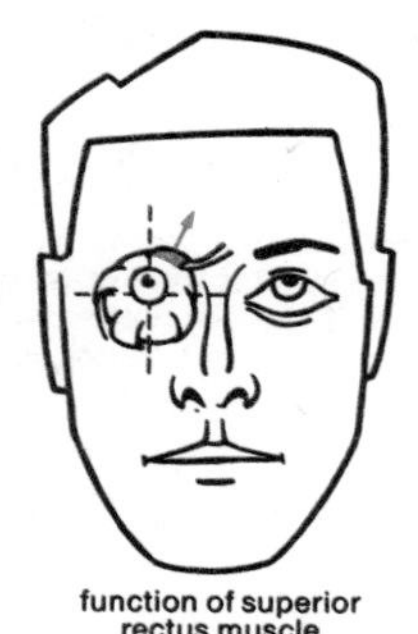

function of superior rectus muscle

and-down and side-to-side movements of the eyeball (Fig. 12-77), while the last two rotate it. Paralytic squint is due to paralysis of one or more of these muscles; it may be an indication of serious cerebral disease or some constitutional disease. In adults this paralysis is often due to syphilis involving the nerve centers or nerve trunks, to rheumatism, to brain tumors, to apoplexy, to multiple sclerosis or to injury. Muscular weakness, as distinct from paralysis, may also cause a condition similar to paralytic squint. Treatment of paralytic squint should be directed at the underlying disease. The patient can to some extent compensate for the defect by holding his head in certain positions so as to minimize the need for eyeball movements

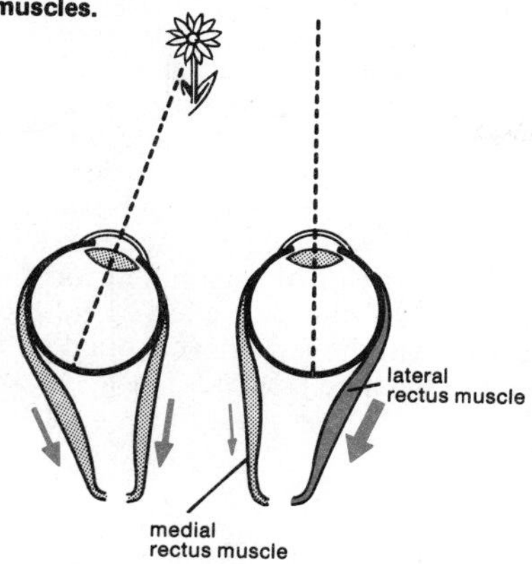

FIG. 12-78.
Squint due to unequal action of ocular muscles.

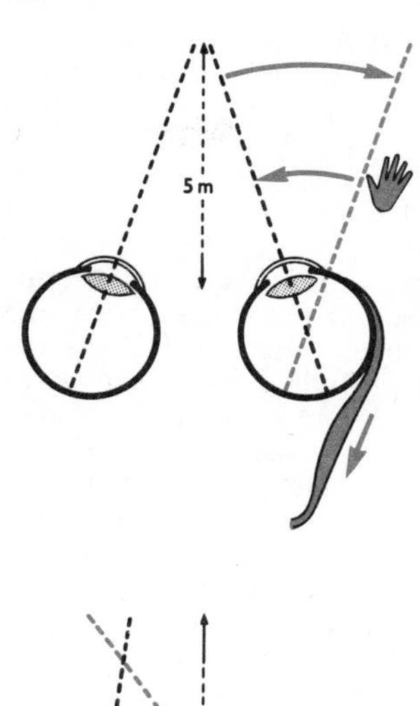

FIG. 12-79.
"Latent" squint.

FIG. 12-80.
(a) Divergent squint of the right eye in a nearsighted person.

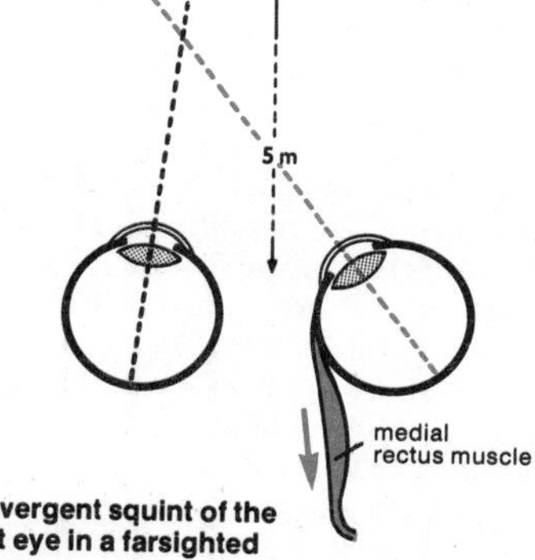

(b) Convergent squint of the right eye in a farsighted person.

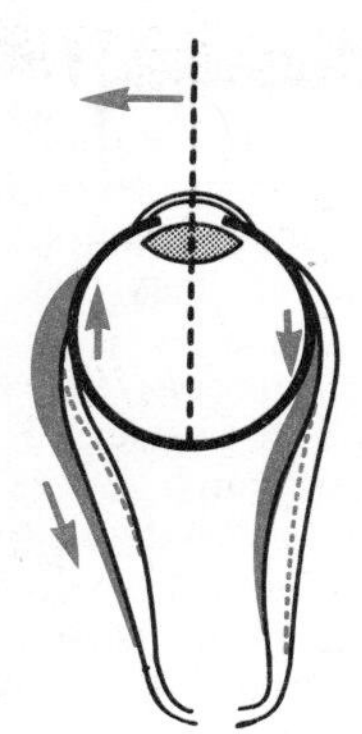

FIG. 12-81.
Operation to correct squint by shifting the points of attachment of the muscles to the eyeball.

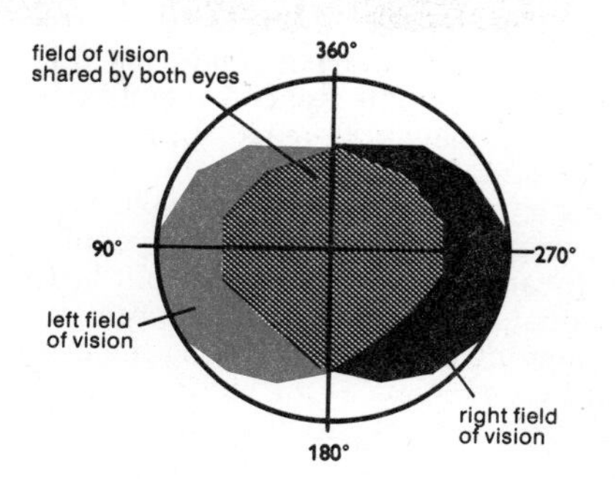

FIG. 12-82.
Field of vision.

that would normally be performed by the affected muscle. Also, he may learn to suppress the image received from the retina of the defective eye. If he cannot achieve this, he will have to wear an eyeshade or spectacles with one opaque glass to stop visual stimuli from reaching that eye.

Squint may also be caused by one of the ocular muscles of one eyeball developing a larger force than the opposite muscle. This imbalance causes faulty alignment, and the two eyes then cannot simultaneously be directed to the same object (Fig. 12-78). The defect will arise particularly in a person in whom coordinated muscular control of both eyes is disturbed. If this control is still functioning properly, however, the affected eye is automatically swung to the correctly aligned position by an effort that overcomes the imbalance of the muscular pull, in which case the squint will be suppressed and the objectionable phenomenon of double vision will not occur. The latent presence of the defect can be detected by letting the person look at a distant object, so that the optic axes of the two eyes are parallel or very nearly so. Now if the affected eye is covered, it will swivel to its abnormal (deviated) position because it is temporarily relieved of the compulsion to keep directed to the distant object. If the eye is then uncovered, it will be seen to swing back to the straight-ahead direction (Fig. 12-79).

Besides disturbed coordination, another factor liable to cause squint is defective vision such as myopia or hypermetropia. These defects, which are common causes of squint in children, can be corrected by suitable spectacles. If these are provided in good time, the squint will be prevented or, if it already exists, may be put right without surgery. It is generally not safe to assume that a child will "outgrow" a squint; left untreated, the condition may result in permanent damage. In the normal eye, accommodation (focusing the lens for near vision) is automatically associated with inward swiveling motions of the two eyeballs. A myopic (nearsighted) person does not have to accommodate for near vision; accordingly, the accompanying action of swiveling the eyeballs inward (convergence) for viewing a near object may be deficient, so that the inward-pulling ocular muscles are underdeveloped. The outward-pulling muscles then dominate. As a result of defective coordination of muscular control, the abnormality may become more pronounced in one eye than in the other, resulting in so-called *divergent squint* (Fig. 12-80a), in which the deviating eye turns outward. The opposite condition, *convergent squint,* may arise as a result of hypermetropia (farsightedness, Fig. 12-80b).

The wearing of spectacles and other nonoperative interventions may suffice to correct squint. If such methods fail, the defect can usually be corrected by an operation that lengthens, shortens or reattaches the eye muscles to restore proper binocular vision. By shifting the point of attachment of the muscle to the eyeball the pull exerted by the muscle can be strengthened or weakened so as to restore the balance (Fig. 12-81).

CONJUNCTIVITIS

Conjunctivitis is inflammation of the conjunctiva, which is the mucous membrane covering the inside of the eyelids and the front of the eyeball up to the edge of the cornea. It forms a sac around the edge of the eye. Normally the conjunctiva is clear, smooth and glistening with moisture; it contains small blood vessels which are distinctly visible where the conjunctiva overlies the sclera on the eyeball. Any irritation of the conjunctiva is characterized by increased flow of blood through these vessels, causing the eye to appear swollen and bloodshot, e.g., after a fairly long period of crying. Inflammation of the conjunctiva can be due to various causes. Sometimes purely physical irritants such as wind, draft, cold weather, smoke, dust or too much exposure to strong sunlight may give rise to conjunctivitis. Small objects trapped under the eyelid and swept into the recesses of the conjunctiva occasionally also cause acute inflammatory irritation. So also can chemical agents such as acids, alkalis, irritant fumes, and chemical dusts. Another possible cause of conjunctivitis is chlorinated water in swimming pools. Allergies, such as hay fever, may also cause it in some persons.

Infectious conjunctivitis is caused by specific bacteria and viruses. Thus gonorrhea in the mother can cause severe conjunctivitis in the newborn baby, with risk of blindness, unless preventive action is taken.

As the nasal cavity is in direct communication with the conjunctival sac through the nasolacrimal duct, a severe cold often involves the conjunctiva, causing irritation there, though generally of a harmless character. Occasionally, though rarely, the conjunctiva may be infected by bacteria or viruses conveyed by the bloodstream or lymph vessels. An early sign of conjunctivitis is a sense of dryness and friction in the eyes, as though some foreign body, such as a grain of sand, had got in. With sudden acute onset of conjunctivitis—due to streptococci, staphylococci or pneumococci—the lacrimal sac is also involved. The eyelids and the area around the eye are swollen; exuded pus causes the edges of the lids and the eyelashes to stick together; the eyes are bloodshot, hurt and are intolerant of light. *Pinkeye* (acute contagious conjunctivitis) is caused by the Koch-Weeks bacillus. It may, for example, be spread by the communal use of a towel.

Treatment of conjunctivitis includes bathing the eyes with a mild antiseptic solution, application of antibiotic lotions or ointments, shielding the eyes from excessive light, and rest.

Trachoma (granular conjunctivitis) is a chronic contagious form of conjunctivitis caused by a viruslike organism which is readily transmitted, especially in the early stages of the disease, either by direct contact or through shared towels, handkerchiefs, etc. In certain parts of the world, notably in Asia and Africa, this disease is very common and frequently causes blindness. It is characterized by "granulated eyelids," with granular elevations which ulcerate and form scars, so that the lids may not close properly. In some cases an opaque membrane of inflammatory tissue (pannus) may spread over the cornea. Trachoma responds well to treatment with sulfonamides or antibiotics; in this respect it is unlike true viral diseases. Advanced cases may need surgical treatment to correct deformity of the eyelids or to apply corneal grafting in order to prevent blindness.

EYELID INFECTIONS

A *stye* (hordeolum) is a small pimple or boil due to an acute bacterial infection of a hair follicle of an eyelash (external style) or suppuration of a tarsal gland at the root of an eyelash (internal style). The swelling is small, hard and painful, and there is edema of the eyelid and sometimes of adjacent tissues. It occurs in persons in poor health and sometimes in those who need spectacles. The stye usually softens in a few days, discharges some infectious matter, and recovers. It can be treated by hot compresses to speed it in coming to a head. When suppuration has taken place, the stye may be incised to remove the contents. Antibiotics can hasten healing. Certain vaccines can be helpful in combating recurrent styes.

Chalazion (meibomian cyst) is a small hard swelling due to inflammation and distension of a tarsal (or meibomian) gland, i.e., a sebaceous gland opening on the margin of the eyelid. It is caused by obstruction of the secretion and may be up to about 1 cm in diameter. In contrast with a stye, a chalazion is not a painful condition; on the other hand, it does not disappear of itself, but has to be removed by surgery. Similar sebaceous cysts, sometimes growing to a much larger size, may develop on other parts of the body; such a cyst is known as a *steatoma*.

Part Thirteen

INFECTIOUS DISEASES

PATHOGENIC BACTERIA

Bacteria are unicellular plantlike microorganisms. An average rod-shaped bacterium is about 4 microns in length and 1 micron in diameter (1 micron = 1/1000 millimeter), but many species are smaller (Fig. 13-1) and some others are larger, especially in length. There are three principal forms of bacteria: spherical (*coccus*), rod-shaped (*bacillus*) or spiral-shaped (Fig. 13-2). Single *cocci* are called monococci; diplococci occur in pairs; staphylococci form irregular clusters; streptococci form chains (Fig. 13-3). Bacteria may also lie chainwise end-to-end and are then referred to as *streptobacilli*. Rigid spiral bacteria are called *spirilla; spirochetes* are flexible spiral bacteria; *vibrios* are in the shape of a curved rod. Cocci are incapable of movement, but most bacilli and spiral bacteria can move independently. Locomotion is by means of one or more fine motile hairs (*flagella*) (Fig. 13-4). Some bacteria are surrounded by a capsule, a layer of slimy substance. A number of species can form *spores*, i.e., they can develop into an encysted or resting stage; this occurs especially when the bacteria encounter unfavorable environmental conditions. The spores are highly resistant to heat, drying and disinfectants. Among the pathogenic bacteria the number of spore-forming species is relatively small; they include the bacteria of anthrax and tetanus. Reproduction of bacteria is generally by fission (asexual reproduction in which the cell divides into two parts, each of which develops into a new bacterium), sometimes by budding or branching. Bacteria usually grow in groups called *colonies*, in which all the bacteria are descendants of a single cell. The type or shape of the colony is often characteristic of the species.

Unlike most higher plants, bacteria contain no green coloring matter (chlorophyll) and therefore cannot carry on photosynthesis, i.e., they are unable to utilize light energy to synthesize carbohydrates from carbon dioxide and water. As a rule, bacteria derive their

FIG. 13-1.
Relative sizes of various microorganisms. The large enclosing circle represents the size of a red blood cell drawn to the same scale (diameter: 7 microns = 0.007 mm).

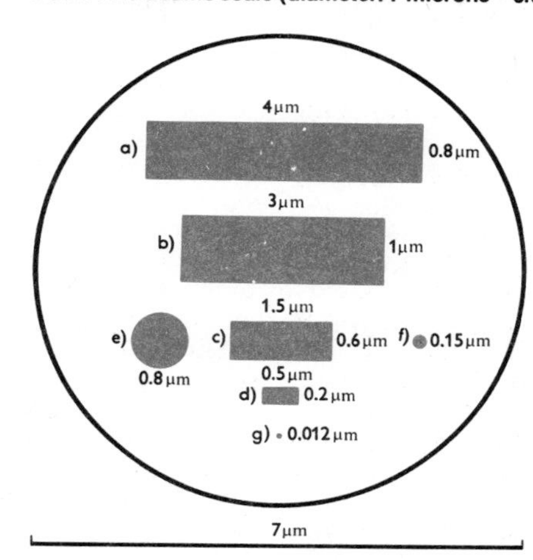

a gas gangrene bacillu
b anthrax bacillus
c typhoid bacillus
d large virus
e coccus
f smallpox virus
g virus of foot-and-mo disease

FIG. 13-2.
Basic forms of bacteria.

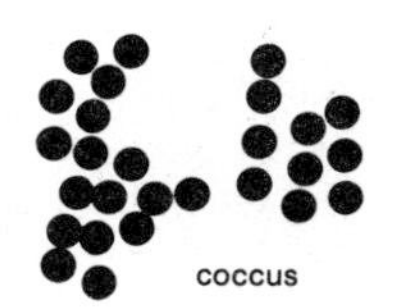

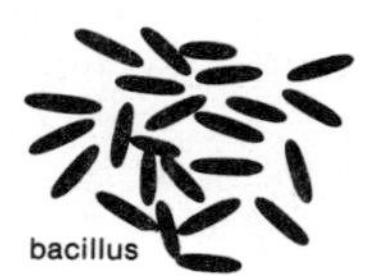

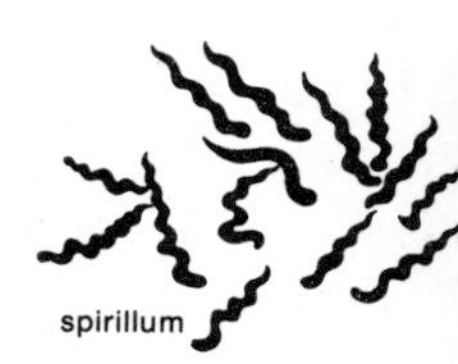

FIG. 13-3.
(a) Staphylococci. (b) Streptococci. (c) Diplococci. (d) Spirilla.

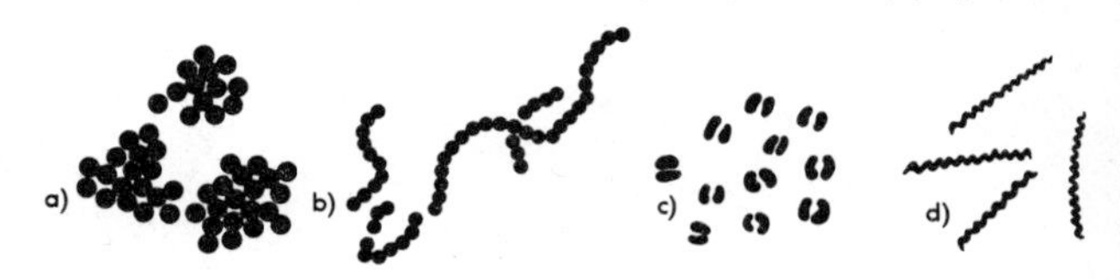

FIG. 13-4.
Various forms of flagella on bacteria.

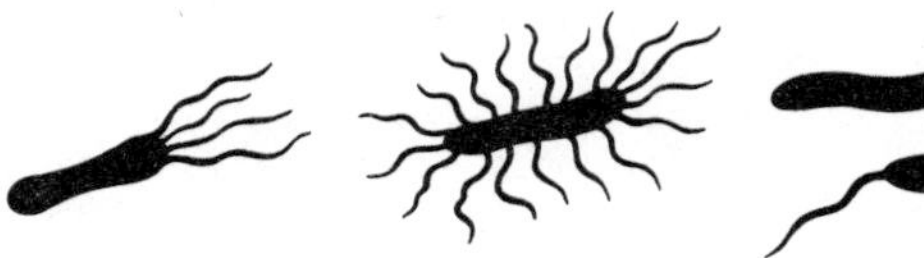

nourishment from organic material. Those that live on living organisms are called *parasites; saprophytes* are bacteria that obtain their food from dead organic material. *Pathogens* (pathogenic bacteria) cause disease in their host; only a relatively small number of the very many species of bacteria are pathogens. The majority of bacteria are *aerobic,* i.e., they require atmospheric (free) oxygen; some species live without atmospheric oxygen and are known as *anaerobic* bacteria. There are also species which, though basically aerobic, can exist also in the absence of atmospheric oxygen. Moderate temperatures provide the best environment for most bacteria—25°C (77°F) for saprophytes, 37°C (98.6°F) for pathogens—but some species prefer lower and others prefer higher temperatures.

Bacteria produce *enzymes,* which break down complex organic molecules into simpler ones which they can absorb as food. For example, carbohydrates (sugar, in particular) are broken down to carbon dioxide and alcohol, this process being known as *fermentation.* Some of the effects of diseases are due to enzymes produced by pathogenic bacteria and promoting changes unfavorable to the host. *Putrefaction* is due to the decomposition of organic substances, especially proteins, in the absence of air (by anaerobic bacteria). The gradual decomposition of organic material by bacteria under aerobic conditions is called *decay.* Some bacteria produce *toxins,* poisonous substances that bring about the symptoms of infection. They are of two general kinds: *endotoxins,* which are released from the bacterium only when it dies, and *exotoxins,* which are released by the living bacterium and diffuse into the tissues or are carried to other parts of the body in the bloodstream.

Bacteria can be examined for identification under the microscope in a so-called hanging drop culture with dark-field illumination. In most cases, however, bacteria are stained with suitable pigments or dyes before microscopic examination. This makes them easier to see (general staining), and also offers advantages in identification because different species of bacteria have affinities for particular dyes (differential staining, i.e., stains for specific bacteria or for specific parts of bacteria). For example, Gram's differential staining divides bacteria into two main groups: Gram-positive (which retain the stain) and Gram-negative (which are decolorized by alcohol). Another method of studying bacteria is by growing them on various culture media, such as broths, gelatin and agar. Other methods of bacteriological study include immunological methods, sterilization methods, and animal inoculation.

Most bacteria are killed by ultraviolet radiation, and all bacteria are destroyed by heat, particularly by boiling (certain spores are resistant to boiling, however). Refrigeration inhibits bacterial growth, but does not kill bacteria already present. Chemicals that kill bacteria are broadly divided into *antiseptics,* which do not greatly damage human tissues, and *disinfectants,* which are destructive to all living matter. Antiseptics may be further subdivided into *bacteriostats* (inhibiting bacterial growth) and *bactericides* or *germicides* (which kill bacteria). Chemotherapeutic agents are substances used in the treatment of bacterial diseases; they include antibiotics and sulfonamide compounds (sulfa drugs).

TYPHOID FEVER, FOOD POISONING, TYPHUS, CHOLERA

Typhoid fever (or enteric fever) is an acute infectious disease caused by a bacillus (*Salmonella typhi*), which, like other *Salmonella* species, is very resistant to drying and to low temperatures. The source of infection is contamination of food or drink by a person suffering from typhoid or by an apparently healthy carrier of the disease, i.e., someone harboring the bacillus (generally in the gallbladder) without symptoms (Fig. 13-5). The bacteria enter the body through the mouth and penetrate into the wall of the small intestine, where they multiply in areas of lymphoid tissue called *Peyer's patches,* but as yet produce no symptoms. After a week or two the now very numerous bacteria enter the bloodstream and are thus carried to the spleen, liver, bone marrow and other organs. At this stage the symptoms appear, starting with headache, weakness, indefinite pains, and fever (Fig. 13-6). There may also be nosebleeding and constipation. Within a few days to a week the temperature may rise to 40°C (105.8°F). In many cases the patient develops a rash of pink spots, particularly on the abdomen. During the following days the fever may show daily fluctuations, with evening temperatures 0.5°–1.5°C (0.9°–2.7°F) higher than morning temperatures. Relapses

FIG. 13-5.
Typhoid infection and transmission.

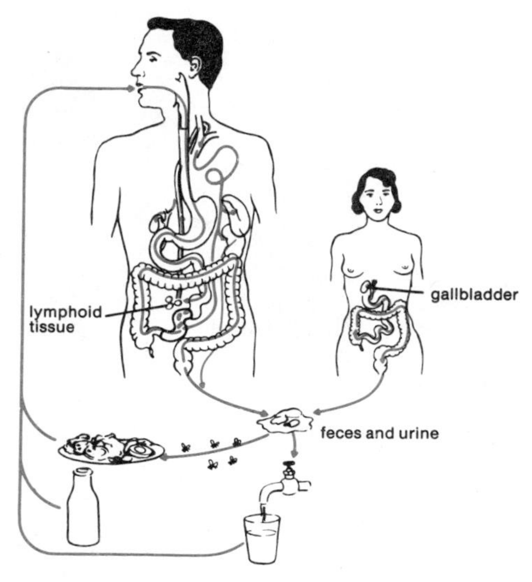

FIG. 13-6.
Pulse rate and fever during the course of typhoid.

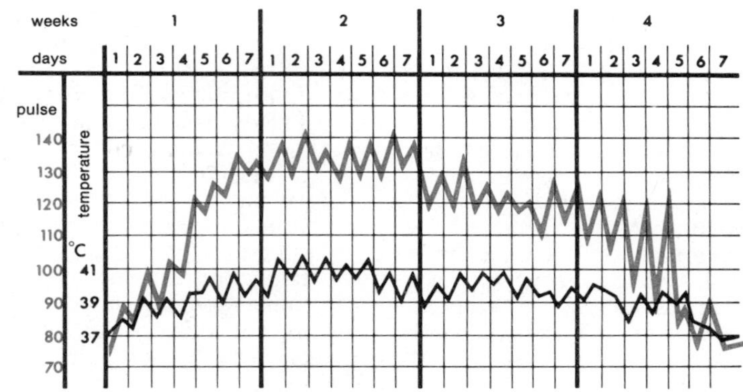

FIG. 13-7.
Intestinal ulcers in typhoid.

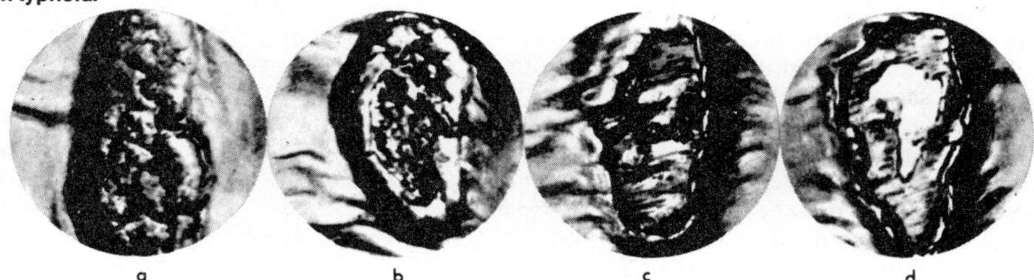

are not infrequent. Complications occur in about 25 percent of cases and are the main cause of death in this disease, which has a mortality rate of at least 10 percent in the absence of proper treatment. The most serious danger is from intestinal perforation and intestinal hemorrhage. A symptom indicating this is a sudden fall of several degrees in temperature, especially in the third or fourth week. The bacteria cause bleeding ulcers in the intestine, which are especially large and deep at the junction of the small and the large intestine, and it is here that the greatest risk of perforation exists (Fig. 13-7). Prevention of typhoid is achieved in modern communities by controlled water supply, proper sewage disposal, and hygiene, especially in the handling of food. Individual protection is provided by vaccination with killed typhoid bacilli, though the effectiveness of this has not been definitely established. Treatment of the active case will consist in careful nursing and the administration of an antibiotic such as chloramphenicol or ampicillin.

Paratyphoid fever comprises infectious diseases also caused by bacteria of the genus *Salmonella*. There are three types: A, B and C; only paratyphoid B occurs in western Europe, while A and C are confined mainly to the tropics and subtropics. In temperate climates the disease is spread by contaminated food or drink. In some cases the symptoms generally resemble those of typhoid, but are less severe; in others, the disorder is intestinal, with diarrhea as the main symptom. Complications involving intestinal ulcers are rarer than in typhoid fever. The same antibiotics as those used for the treatment of typhoid are also used in paratyphoid fever.

Bacterial food poisoning is mostly caused by *Salmonella* species. Food may be contaminated by a carrier of these bacteria who handles it. Animals may also transmit the disease to man. Duck eggs are a frequent source of the bacteria. Symptoms, which may occur between three hours and three days after eating contaminated food, include vomiting and/or diarrhea. The attack is not often dangerous, but may have to be combated by antibiotics or sulfonamides. In some types of food poisoning the symptoms are caused, not by bacterial infection of the intestine, but by toxins formed by bacteria (mostly staphylococci) in the food before it is eaten. In general, cooking kills bacteria in food and makes it safe, so far as bacterial infection is concerned, but may not destroy toxins. *Botulism* is a very dangerous type of food poisoning caused by a bacterium (*Clostridium botulinum*), which is commonly found in soil and in the intestines of domestic animals. It may contaminate improperly canned food, especially home-canned vegetables, where the bacteria can develop under anaerobic conditions; it may also occur in ham, sausage, etc., but is fortunately rare. The high temperatures normally employed in factory-canned foods destroy the bacteria. The toxin has a selective action on the central nervous system. Even a very small dose is fatal, causing cardiac and respiratory paralysis. Cooking destroys the toxin.

Typhus fever, or simply typhus, comprises a group of acute infectious diseases caused by microorganisms of the genus *Rickettsia*. These microorganisms occupy a position intermediate between viruses and true bacteria. They are usually transmitted by arthropods (lice, fleas, mites, etc.) and are responsible for what are collectively known as *rickettsial diseases,* the commonest of which are the spotted fevers (including Rocky Mountain spotted fever, tick fever, etc.), epidemic typhus, endemic typhus, trench fever, and others. *Louse-borne typhus* is transmitted by body lice and is therefore most likely to occur under unsanitary conditions, e.g., in wartime or during natural disasters when refugees are crowded together. These diseases cause a prostrating fever, with severe headache, rash, and neurological involvement. The onset is sudden, with fever rising rapidly (to 40.5°C or 104.9°F) in two or three days and remaining high for about ten days. The tongue may be covered with whitish fur; in severe cases it may turn black and roll up like a ball in the back of the mouth. The intensity of the rash is an indication of the severity of the attack. There are constipation and scanty urine. Complications may include bronchopneumonia, congestion of the lungs and nephritis. Treatment will require isolation of the patient and good nursing. Broad-spectrum antibiotics such as chloramphenicol and tetracylines are effective.

Cholera is an acute infectious disease caused by the "comma bacillus" (*Vibrio cholerae*) and contracted from food or water contaminated by feces. The disease is transmitted by cholera patients; there are no unaffected carriers as with, for example, typhoid fever. The disease is characterized by diarrhea, with severe loss of fluids and salts, the symptoms being due mainly to a potent endotoxin and to an enzyme produced by the bacteria. The incubation period can range from a few hours to about five days. The first symptoms are malaise, headache, diarrhea and loss of appetite. Sometimes these early symptoms are largely absent, and the onset is sudden, with violent diarrhea and vomiting, and watery stools containing mucus and blood. Muscular cramps occur, starting in the extremities. There is unquenchable thirst; hiccups sometimes develop. This stage lasts only a few hours and is followed by a state of collapse, with almost complete arrest of circulation, weak pulse, rapid respiration, suppression of urine. This stage lasts from a few hours to two days and usually results in death. There may,

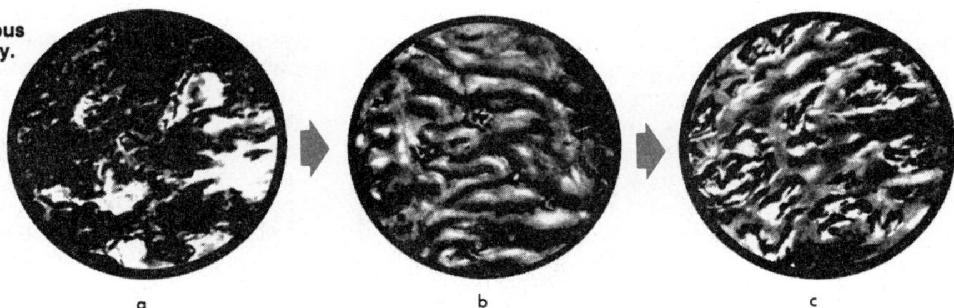

FIG. 13-8.
Changes in intestinal mucous membrane due to dysentery.

however, be a reaction leading to recovery. The danger associated with cholera arises from dehydration, loss of potassium, and acidosis (alteration of the body's acid-alkali balance toward greater acidity). Loss of potassium salts disturbs the conduction of nervous impulses, especially those controlling the heart muscle. In untreated cholera the death rate is at least 50 percent. This is dramatically reduced by treatment in which the losses of water, potassium and alkali are corrected, including replacement of fluid and electrolytes by intravenous injections of saline solution. Antibiotics such as tetracyline can be helpful. Vaccination can give protection from the disease for a limited period. The disease has been overcome in modern times by proper sanitation, since human excrement is the only source of infection. Contamination of drinking water supplies with sewage used to be the cause of epidemics in the past.

DYSENTERY

Dysentery comprises several intestinal disorders, especially of the colon (large intestine). More particularly, the term denotes either bacillary dysentery or amebic dysentery.

Bacillary dysentery is caused by bacteria of the genus *Shigella*. The infection by these bacteria is confined to the intestine, as distinct from *Salmonella,* which may cause widespread infection (see preceding section). The bacteria continually produce toxins which irritate and damage the intestinal mucous membrane (Fig. 13-8). They are transmitted by contaminated water or food, inadequately washed cups or plates, lavatory seats, or flies. The severity of the attack depends on the species of *Shigella* concerned. The most virulent form of the disease is known as *Shiga dysentery* and is caused by *Shigella dysenteriae*. It occurs

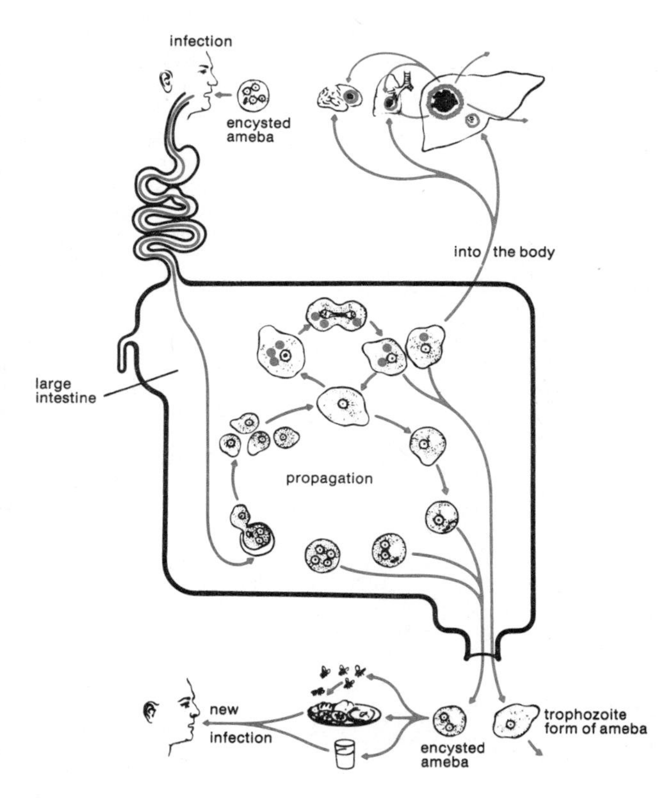

FIG. 13-9.
Dysentery infection and transmission (amebic dysentery).

mainly in the Far East and is characterized by severe diarrhea, bleeding and dehydration. There are acute abdominal pain, persistent desire to empty the bowel with spasmodic contraction, and the passage of slimy stools containing mucus and blood. The temperature rises, and there may be persistent nausea and vomiting. In general, the most serious effect of the disease is the loss of water and salts. The patient should drink freely; intravenous injections of saline solution may have to be given. *Flexner dysentery,* caused by *Shigella flexneri,* occurs in warm countries and is less severe than the Shiga variety. The type occurring in temperate climates is usually *Sonne dysentery,* caused by *Shigella sonnei.* It is a much milder disorder, with symptoms confined to diarrhea lasting no more than a few days. Most cases of dysentery, including the more violent types, are not fatal even without special treatment. In young children, however, the disease can be very dangerous because it depletes their reserves of water and salts, especially in tropical climates. A chronic form of bacillary dysentery is characterized by alternate diarrhea and constipation with coliclike pains. Such patients continually discharge bacteria in the feces, so that under unsanitary conditions they are a constant source of infection (Fig. 13-9). Bacillary dysentery can be effectively treated with sulfonamide drugs and especially with chloramphenicol (an antibiotic).

Amebic dysentery is caused by *Entamoeba histolytica,* a protozoan, which penetrates the lining of the intestine. The disease, which occurs mainly in the tropics and subtropics, is transmitted by contaminated food or water. In its encysted (encapsulated) form the ameba can be carried by flies or by wind. Many people who carry the infection show no symptoms, but can spread the disease. The disease may present a more or less chronic form, continuing for long periods, even years. In acute cases the symptoms are abdominal pains and diarrhea, with stools containing mucus and blood. The patient's temperature is subnormal. The disease may be complicated by ulceration and perforation of the intestinal wall and formation of amebic abscesses in the liver and other organs. It is treated with the drug emetine; antibiotics (e.g., tetracylines) may additionally be given, particularly against secondary bacterial infection of intestinal ulcers caused by the amebae.

TUBERCULOSIS

About a century ago something like one in six or seven of all people in Europe over 16 years of age died of tuberculosis. At the present time the figure is under 1 percent. A number of factors have helped to bring about this dramatic reduction in the frequency and severity of the disease (Fig. 13-10): the improved standard of life, better hygienic conditions, organized action to prevent and combat tuberculosis, vaccination, public health measures

FIG. 13-10.
Deaths from tuberculosis per 100,000 inhabitants in West Germany since 1878 (deaths from cancer are shown for comparison).

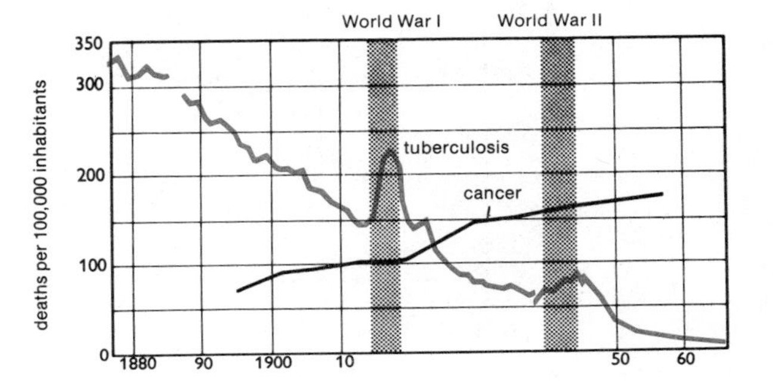

FIG. 13-11.
Number of male patients with active tuberculosis per 100,000 inhabitants in West Germany (in the older age groups there is only a small reduction in the number of cases between 1953 and 1960).

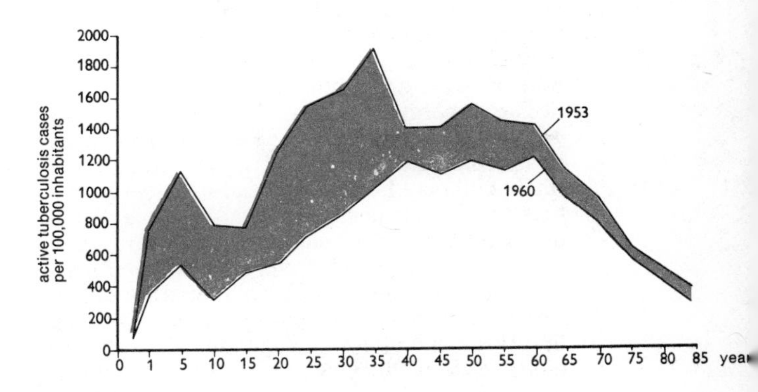

(including particularly the testing and control of milk), modern drugs for treating the disease, and improved operating techniques for cases requiring surgical treatment. Tuberculosis has by no means been eradicated, however. In recent years, the number of cases of the disease has not decreased as much as the number of deaths caused by it. Especially in the elderly, tuberculosis is still a not uncommon cause of death (Figs. 13-11 and 13-12), and it is estimated that throughout the world there are still about 15 million people affected with tubercu-

FIG. 13-12.
Deaths from tuberculosis per 100,000 inhabitants in West Germany in 1961, according to age and sex.

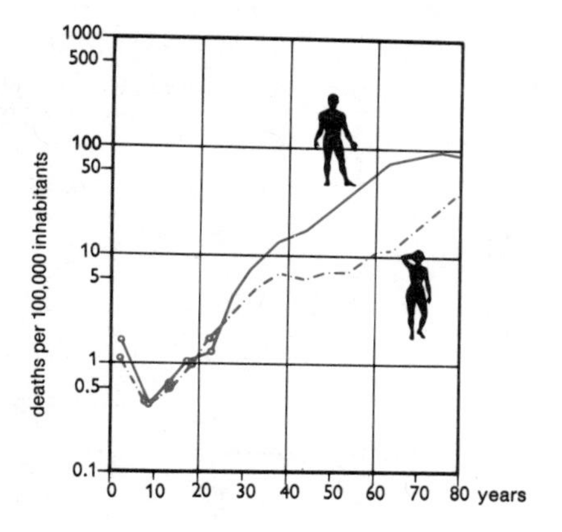

losis and that some 3 million a year die of it. In densely populated areas, even at the present time, 95 percent of all people have been or will be infected with tuberculosis at one time or other. The infection, though successfully resisted by the body's defenses in most cases, frequently leaves "inactive" tubercle bacilli behind which may be reactivated under what are, to them, favorable conditions, e.g., when the person's physical resistance is low due to other disease or to malnutrition.

The tubercle bacillus (*Mycobacterium tuberculosis*) is particularly resistant to dryness, heat, cold and vacuum. To assist diagnosis, the bacilli may be grown in cultures, or guinea pigs (which are particularly susceptible to the disease) may be inoculated to see how they respond. There are five *variants of the bacillus:* human, bovine (affecting cattle), avian (birds), murine (mice) and piscine (fish). The first three can affect man, though avian tuberculosis seldom does. Murine tubercle bacilli do not produce disease in man, but can confer immunity to the harmful variants. Bovine tuberculosis is transmitted to man chiefly through the drinking of milk; the infection usually spreads from the intestine of the infected person and may be carried to any part of the body, especially the bones and the lymph nodes. Under modern hygienic conditions of dairy farming, pasteurization and bottling of milk, etc., this source of tuberculous infection has been virtually eliminated in the developed countries. The human variant of the tubercle bacillus is transmitted mainly through the air from person to person, either in droplets of moisture (expelled by coughing or sneezing) or as "dust," i.e., the dried bacilli are carried by air currents and are inhaled into the lungs. Thus about 90 percent of cases of first infection with tuberculosis primarily affect the lungs.

Tuberculosis is not a highly contagious disease. Occasional exposure to a limited number of bacilli is unlikely to produce the disease in a person of normal health. Overcrowded living conditions, lack of hygiene, low resistance brought about by other disease (e.g., malaria) and especially by malnutrition appear to be major factors in promoting the spread of tuberculosis in a population. While it is true that the disease "runs in families," the supposed hereditary predisposition appears to have been exaggerated in the past. Resistance to tuberculosis is low at birth and in early infancy; instances of direct prenatal transmission of the disease to babies are very rare, however. Resistance develops when the baby is a few months old, but diminishes in adolescence. It subsequently increases again, but finally declines in old age.

As a chronic infectious disease, tuberculosis can run a complex and varied course. It can occur at any age, in any degree of severity, and can attack any organ of the body. Tuberculosis that has been "dormant" for many years may suddenly turn "active." The great variety of behavior presented by the disease is due to the fact that not only do the number and aggressiveness of the invading bacilli play a part, but also the human body's resistance—depending on age, general health, nutrition, predisposition—is an important factor in determining the pattern and course of the disease. Malnutrition, physical exhaustion, other diseases (whooping cough, measles, influenza, diabetes, silicosis, etc.) are all factors that lower a person's resistance to tuberculosis. In a healthy person, infection with the bacillus, when successfully overcome by the body's defenses, will increase his resistance to tuberculosis. Although this does not constitute absolute immunity such as that conferred by certain other diseases, it is nevertheless effective enough to ensure that tubercle bacilli that survive in the body after a first infection become inactivated and may well remain in that harmless condition all through that person's life. However, there is a risk that with subsequent lowering of the body's resistance, such bacilli are reactivated. The course run by the disease is also affected by hypersensitization reactions due to allergy with regard to the tox-

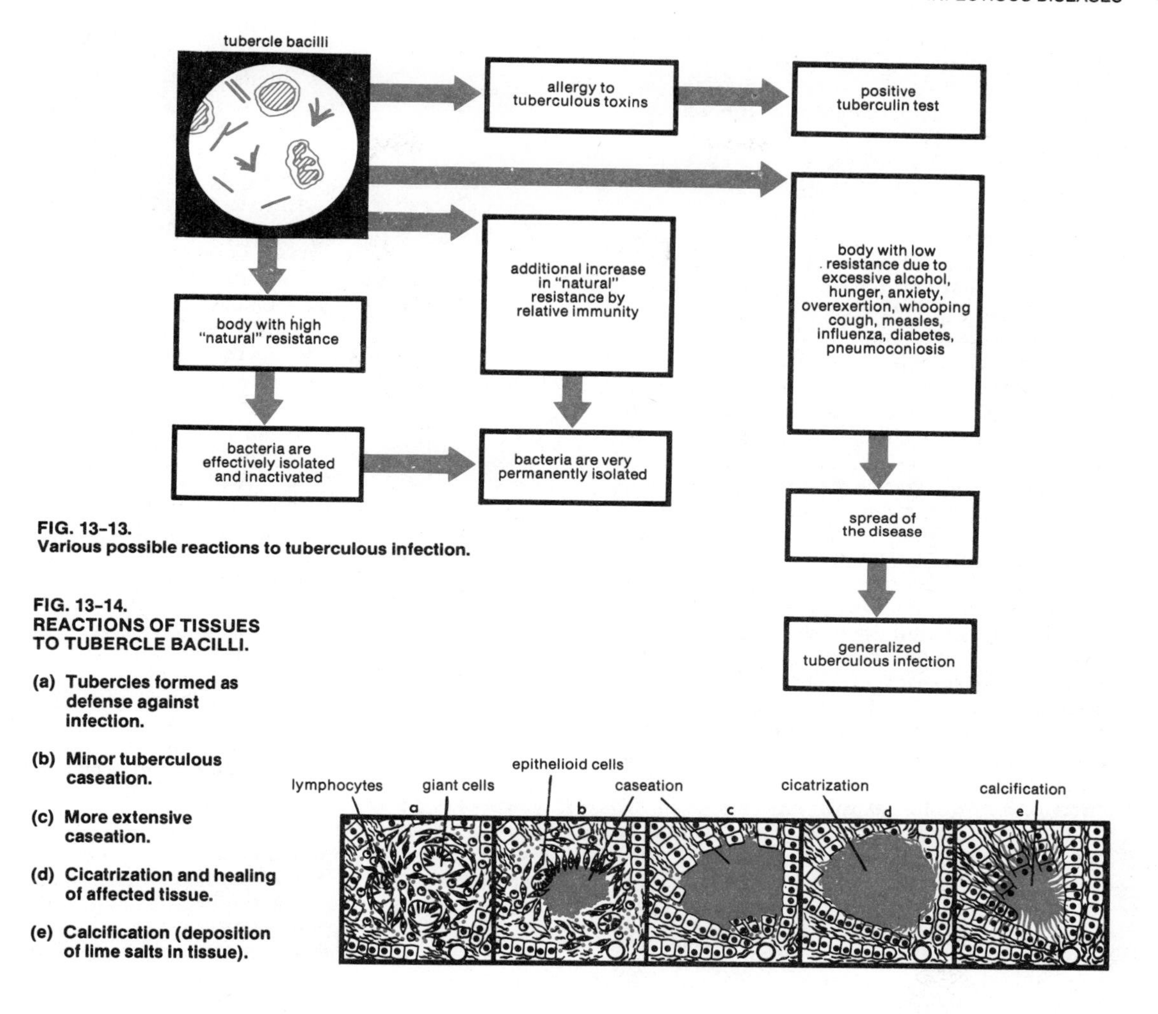

FIG. 13-13.
Various possible reactions to tuberculous infection.

FIG. 13-14.
REACTIONS OF TISSUES TO TUBERCLE BACILLI.

(a) Tubercles formed as defense against infection.

(b) Minor tuberculous caseation.

(c) More extensive caseation.

(d) Cicatrization and healing of affected tissue.

(e) Calcification (deposition of lime salts in tissue).

ins produced by the tubercle bacilli. This allergy manifests itself, for example, in a more or less distinct inflammatory reaction of the skin when the person is injected with toxin from the tubercle bacillus, more particularly tuberculin, which is a soluble substance prepared from the bacillus. The so-called *tuberculin test* is a test for determining the presence of a tuberculous infection based on positive or negative reaction to tuberculin applied to or injected into the skin. A positive result of the test merely shows that the person has been infected with tuberculosis, but does not indicate what degree of resistance there is to the disease (Fig. 13-13).

Reaction of tissues to tubercle bacilli depends on, among other factors, the local resistance of the affected tissues. Healthy tissue strives to fight off the bacilli and render them harmless by encapsulation. This process gives rise to the formation of tubercles, the characteristic lesions resulting from infection by the bacillus. A tubercle is, in fact, a small center of inflammation and tissue destruction; it consists of so-called granulation tissue and gives the disease its name (Fig. 13-14a). If the defensive mechanism of the affected tissue is deficient, there is limited local destruction of tissue, accompanied by what is known as tuberculous caseation, i.e., conversion of necrotic tissue into a granular amorphous mass somewhat resembling cheese (Fig. 13-14b,c). If these areas of caseation are extensive, particularly in the lungs, ejection of the degenerated material causes the formation of cavities, which are in effect chronic abscesses (exudative tuberculosis). Another type of response to infection is not really characteristic of tuberculosis. It consists in transformation of the affected tissue into fibrous tissue, with contraction and cicatrization (cirrhotic form of tuberculosis, Fig. 13-14d). Old centers of infection may calcify (Fig. 13-14e) or even ossify and nevertheless still contain tubercle bacilli.

Depending on the structure of the affected organ, tuberculosis spreads from the primary focus of infection either by advancing in the same tissue (Fig. 13-15a), by penetrating into passageways or tubes, e.g., the bronchial tree (Fig. 13-15b), by invading the lymph vessels (Fig. 13-15c), or by invading the blood vessels and being carried along with the bloodstream (Fig. 13-15d). From pulmonary cavities the tuberculous material may enter not only the bronchial tubes but also the pleural cavity (Fig. 13-15e). Whereas the bloodstream can spread the disease to any part of the body, its progress in the lymphatic system is generally arrested in the lymph nodes, and these tend to be involved in most forms of tuberculosis. Bovine tuberculosis transmitted by milk may spread from the throat to lymph nodes in the neck or, more usually, from the intestine to abdominal lymph nodes. In such cases the disease may remain confined to these nodes.

The commonest form of the disease is *pulmonary tuberculosis* (also known as consumption or phthisis), and about 90 percent of all cases of primary infection are of this type. The bacilli are inhaled into the lungs as dust or droplets expelled by coughing or sneezing by infected persons. The bacilli may sometimes, though rarely, enter the body through the alimentary canal, the skin, the conjunctiva, the tonsils, or the mucous membrane of the mouth or throat. About five or six weeks after first contact with the infection, the patient gives a positive result to the tuberculin test. A small area of inflammation develops in the lung, usually near the top, with caseation. From here the tubercle bacilli, carried by the lymph flow in the lung, spread to the lymph nodes at the hilus, i.e., the root of the lung through which the main bronchus enters. These nodes likewise degenerate by caseation. In this way a localized area of inflammation—the primary focus—develops in the lung (Fig. 13-16a); a

FIG. 13-15. Various ways in which pulmonary tuberculosis may spread.

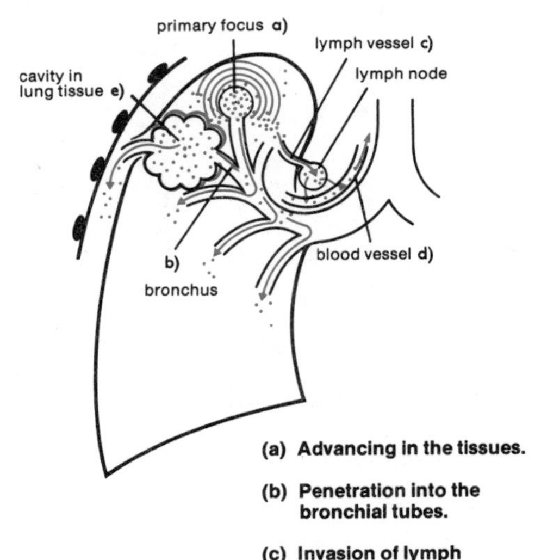

(a) Advancing in the tissues.

(b) Penetration into the bronchial tubes.

(c) Invasion of lymph vessels.

(d) Invasion of bloodstream.

(e) Tuberculous cavity in diseased tissue spreads to involve bronchus and pleural cavity.

small abscess forms here. The symptoms associated with this limited extent of the disease are slight and may pass unnoticed or be mistaken for a mild attack of influenza. The patient may feel slightly unwell, with assorted indefinite symptoms such as tiredness, loss of appetite, headache, chest pains, slight fever, some coughing. If no complications occur, the symptoms disappear before long. The focus diminishes in size, leaving a small scar which calcifies in about two years' time. This is what happens in the majority of cases. Tubercle bacilli remaining encapsulated in the healed focus may continue to be potentially capable of reactivation for many years. Complete healing of the disease sometimes occurs, however. The scar can be detected by x-ray examination, and the person will continue to respond positively to the tuberculin test.

In the minority of cases the infection spreads from the primary focus. This constitutes the second stage of tuberculosis. A hacking cough, breathlessness and a rise in temperature may be indications of extensive infection

FIG. 13-16. The stages of tuberculosis.

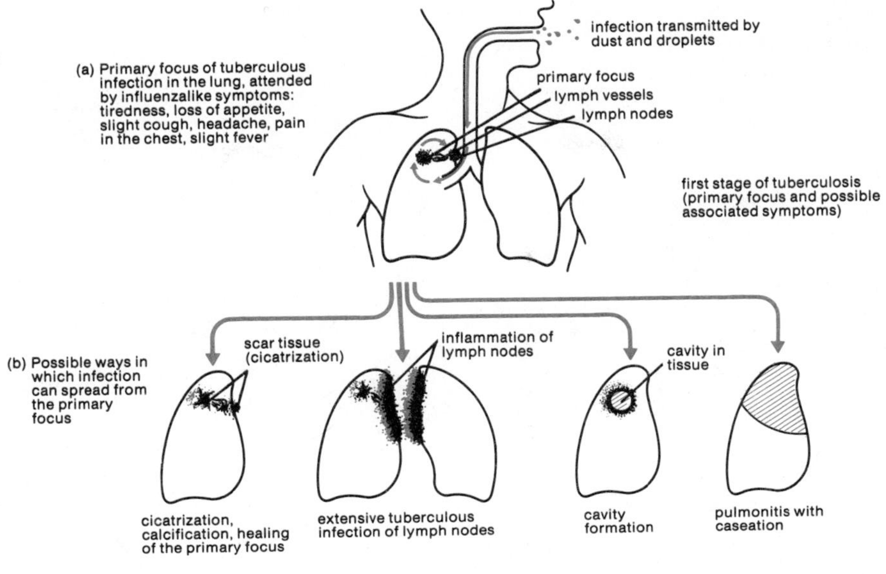

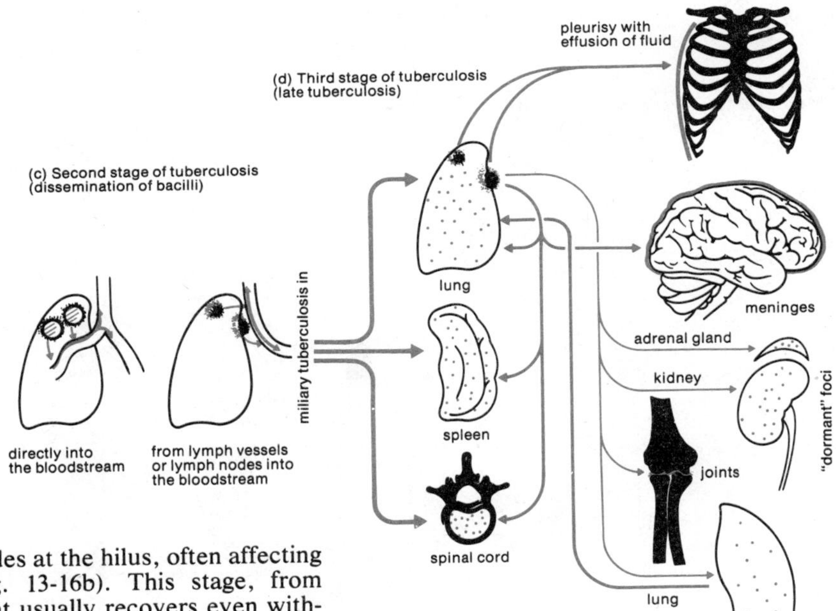

of the lymph nodes at the hilus, often affecting both lungs (Fig. 13-16b). This stage, from which the patient usually recovers even without any special treatment, is nevertheless potentially dangerous, because in some cases the disease may spread quickly through the lungs, resulting in early death ("galloping consumption"), or the bacilli may be widely disseminated through the body by the bloodstream. In the latter case, known as *miliary tuberculosis*, the tubercle bacilli form tubercles in various organs, where they can subsequently be detected by x-ray examination. The symptoms associated with this condition are high fever, coughing, breathlessness, headache, vomiting, cyanosis and respiratory trouble. The spleen and bone marrow are often involved. Some secondary foci of infection at first remain inactive, but may later develop into organic tuberculosis in the lungs, kidneys, adrenal glands, bones and joints (Fig. 13-16c). Miliary

FIG. 13-17.
The stages of pulmonary tuberculosis.

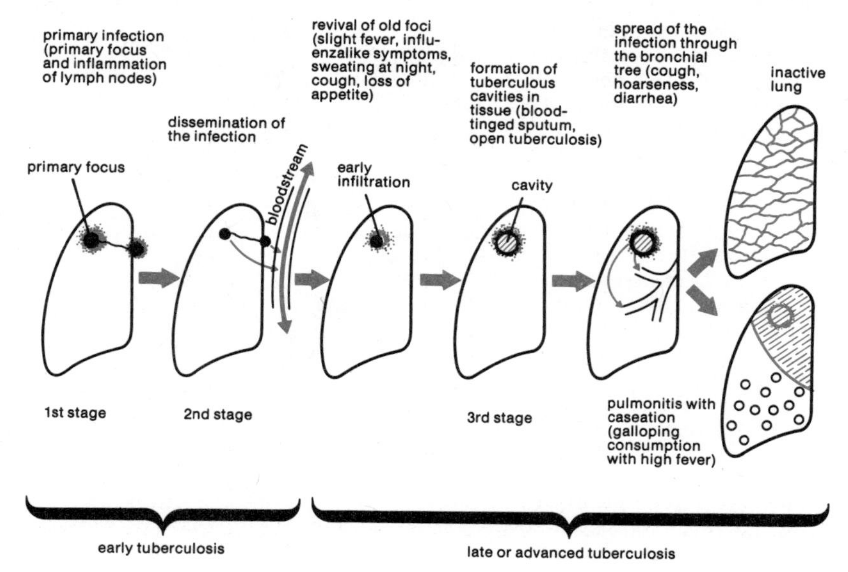

FIG. 13-18.
Tuberculosis localized in various organs and parts of the body.

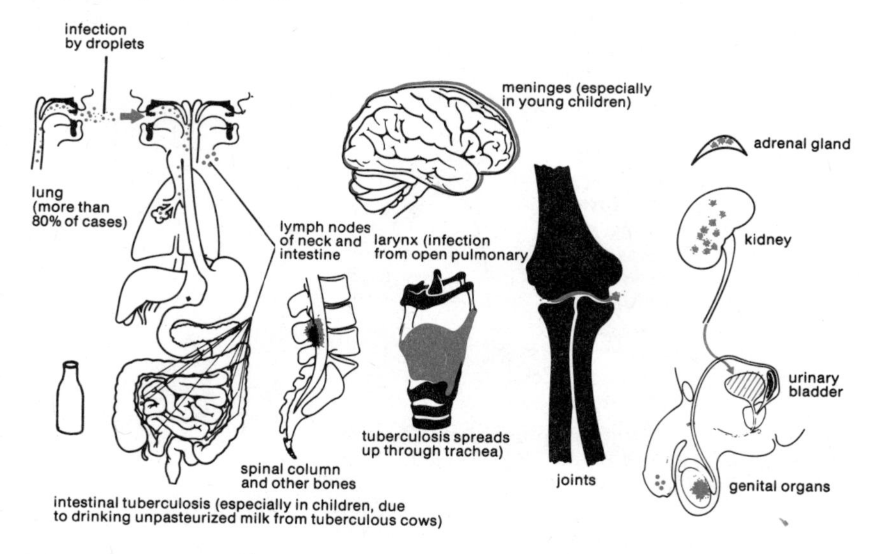

tuberculosis derives its name from the numerous tubercles that spread to the various parts of the body and somewhat resemble millet seeds (milia).

Tuberculous pleurisy may occur as a sequel to a late primary infection. In contrast with most other forms of early tuberculosis, which generally do not produce very pronounced symptoms, it soon manifests itself by stabbing respiratory pains in the chest. The initially dry pleurisy usually turns into pleurisy with effusion of fluid, which accumulates in the pleural cavity. When this happens, the pain diminishes, but the fluid may interfere with breathing. The effusion may become purulent, i.e., containing pus, the condition then being known as *empyema*. This clears gradually, often leaving adhesions between the pleural membranes. Something like 80 percent of all cases of pleurisy occurring in young adults are believed to be of tuberculous origin.

The first two stages of the disease are categorized as early tuberculosis. The third stage is that of *late* or *advanced tuberculosis* (Figs. 13-16d and 13-17). It may not begin—if at all—until months or indeed years after the primary infection; it may be a revival of the primary focus or of other foci. These may have formed in childhood and remained inactive. Their subsequent reactivation marks the onset of the disease in the adult patient. At first, just as at the time of the primary infection, the symptoms are easily mistaken for those of protracted influenza, with slight fever, loss of appetite, some loss of weight, sweating at night, coughing, and blood-spitting. Not all of these symptoms need necessarily be present. Early diagnosis is important, because the disease can be quite effectively treated if it is detected before extensive damage to the pulmonary tissue has occurred with formation of cavities (chronic abscesses). When that happens, the healing process is much slower and more difficult. The patient develops a severe cough, bringing up tenacious yellowish-gray sputum, sometimes blood-tinged, containing tubercle bacilli and fibers (open tuberculosis). The cavity in the lung tissue may develop a thickened wall and become chronic. The disease is then likely to spread through the air passages throughout the lung and invade the larynx, causing hoarseness; swallowed bacilli may carry the disease to the intestine, causing diarrhea. The infection of the lungs can still be arrested at any stage. However, it often spreads step by step downward from the top of the lung, giving rise to fresh foci, the symptoms being high fever, shortness of breath, and blood-spitting. At the new foci there is further destruction of lung tissue until eventually the whole lung is put out of action. When this stage is reached, severe bleeding from the lung is likely to occur and blood will be expectorated (hemoptysis). If the body's resistance is low, the condition develops into *pulmonitis,* with extensive caseation of tissue, high fever, and general deterioration of the patient's condition. This "galloping consumption" used to be almost invariably fatal, but is now rare and can moreover be arrested and reversed with modern chemotherapy administered in time.

Besides the lungs, other organs of the body may be affected by tuberculosis, either as a primary infection or as a result of spread of the disease from the lungs (Fig. 13-18). *Tuberculous meningitis* is acute inflammation of the cerebral meninges (the membranes surrounding the brain) caused by the tubercle bacillus. It arises as a complication of other forms of tuberculosis especially in children. Before the introduction of streptomycin it was nearly always fatal. Bones and joints affected by tuberculosis slowly degenerate and are replaced by unstable fibrous tissue. For instance, in *Pott's disease* there is tubercular inflammation of the bodies of the vertebrae: the destruction and collapse of the affected vertebrae result in angulation of the spinal column (kyphosis), with compression of the spinal cord and nerves. Tuberculosis of the intestine and abdominal lymph nodes, possibly affecting the whole peritoneal lining, can be caused by infected milk or by the patient's own sputum which he has swallowed. In many such cases the symptoms are vague and uncharacteristic, so that this form of the disease may remain undetected for a long time. The infection may involve any of the *abdominal organs,* particularly the kidneys, bladder and reproductive organs. Caseation and resulting degeneration of tissue can eventually render a kidney useless, necessitating its removal. Among the endocrine glands, the adrenal glands are most often affected. The skin may also be affected by tuberculosis, the commonest form being known as *lupus vulgaris,* characterized by patches which break down and ulcerate, leaving scars on healing. The disease may destructively attack the skin of the nose and cheeks.

As already stated, the early stages of tuberculosis are often mistaken for influenza or chronic bronchitis. Poor general health, slight fever, bronchial catarrh persisting for more than three weeks or so, especially when accompanied by loss of appetite and loss of weight, should be regarded with suspicion. Diagnosis of pulmonary tuberculosis is facilitated by modern radiographic (x-ray) techniques, such as tomography. Also important is the bacteriological examination of sputum, gastric juice, or swab specimens from the throat. The tuberculin test is suitable for mass examination, but merely indicates whether a primary infection has occurred; it does not show how active the condition is.

The traditional treatment of tuberculosis aimed at improving the patient's natural resistance by careful nursing, bed rest, well-balanced diet and favorable climatic conditions. Recovery was sometimes incomplete, however, and relapses were not infrequent. If the disease had firmly established itself or if it developed into galloping consumption or tuberculous meningitis, there was virtually no chance of recovery. Tuberculosis affecting a kidney, a joint or lymph nodes in the neck could be treated by removal of the kidney or nodes or by immobilization of the joint. Such cases used to be referred to as surgical tuberculosis. Although rest and proper care are still important cornerstones in the treatment of the disease, the tubercle bacilli can now be directly resisted with various tuberculostatic drugs. The first important breakthrough came with the introduction of streptomycin in 1944, followed by para-aminosalicylic acid (PAS) in 1946 and isoniazid in 1952. They may be given in combination with one another to overcome resistance that the bacillus may have developed to one of them. There are various other drugs available as well, e.g., pyrazinamide and ethionamide. Drug treatment usually enables even severe infections to be arrested, so that in the developed countries, where treatment is readily available, deaths from tuberculosis are comparatively rare. Surgical treatment, still applied in certain severe cases, has also become more effective. The lung can be temporarily immobilized (pneumothorax, see page 134); gas or air may be injected into the peritoneal cavity to treat tuberculous peritonitis or where pneumothorax is not possible (pneumoperitoneum). Thoracoplasty is the removal of portions of the ribs in stages in order to allow the wall of the lung to collapse and close the cavity of a large abscess. Modern drug therapy has now largely superseded this operation and is now seldom necessary. Removal of lobes of the lung or indeed a whole lung likewise used to be a routine treatment in certain advanced cases.

Vaccination with a modified bacillus that confers immunity or with the vole bacillus (murine tubercle bacillus) is an important preventive aid. Vaccinations are performed primarily on persons who are found to be negative in the tuberculin test, i.e., those who have acquired no immunity to the disease.

WHOOPING COUGH

Whooping cough (pertussis) is an acute infectious disease caused by a bacillus (*Hemophilus pertussis*) and chiefly affecting children. It is transmitted by droplets discharged by coughing or sneezing. In babies it is a serious disease, with risk of complications such as pneumonia and damage to the lungs. There is also a danger of suffocation. The disease is seldom dangerous in older children. It is highly infectious; even a short period spent in the same room as an infected person is liable to transmit it. Whooping cough confers a degree of immunity, though not of life-long duration. However, because of immunity acquired in childhood, the disease is rare in adults.

The incubation period is usually from one to two weeks. The first symptom is an ordinary cough which tends to be worse at night. The child is already infectious in this stage. After another week or two of this so-called catarrhal stage, the attacks of coughing become more frequent and more violent, developing into the typical paroxysmal cough from which the disease derives its name (paroxysmal stage). The attacks of coughing come on so suddenly that the patient is taken unawares, has no time to breathe in, and has a feeling of suffocation, which he counteracts by drawing in air with a whooping sound caused by spasmodic contraction of the glottis. In an attack of coughing the patient coughs repeatedly, becoming blue in the face, sometimes with saliva and mucus pouring from the mouth and nose and with tongue protruded. The attacks may come in fairly rapid succession until the patient is exhausted. The number of paroxysms in a 24-hour period may vary from 3 or 4 to as many as 40 or 50. They often end in vomiting. Sleep is disturbed and the patient looks pale and puffy. There is generally no fever, however. The number of white blood cells is characteristically increased. After about four weeks the coughing grows less frequent and less violent, vomiting ceases, the patient's appetite improves, and he sleeps better. About six weeks after the onset of the disease he is no longer infectious.

If there are no complications and no fever develops, the patient—usually a child—should not be kept in bed, but should, under suitable weather conditions, be allowed out into the fresh air, provided he can be kept warm and is, of course, not brought into contact with other children. A light, easily digestible, high-calorie diet is desirable. The infection generally responds to antibiotics. Passive and active immunization are available and, though not always completely effective, are nevertheless helpful in reducing the risk or the severity of the disease.

SCARLET FEVER

Scarlet fever (scarlatina) is an infectious disease, mainly of childhood, caused by a bacterium (*Streptococcus pyogenes*), of which numerous strains exist. Streptococci are also responsible for a number of related diseases (Fig. 13-19). The body may respond to the bacterial toxin by producing a specific antitoxin, thus acquiring antitoxic immunity. Independently of this response, the body may also, after contact with the bacteria, develop immunity against the bacteria themselves (antibacterial immunity). Whether a streptococcal invasion will, in fact, produce nothing worse than a septic sore throat or will produce scarlet fever (which is essentially sore throat with the additional feature of a rash) will depend on the type of streptococcus and on the patient's state of immunity: if he has antibacterial immunity, the bacteria are unable to establish themselves, and there is no illness; if there is merely antitoxic immunity, the infection may produce, for example, septic tonsillitis or otitis media. In the absence of both types of immunity, the purulent inflammation is likely to take the form of scarlet fever (Fig. 13-19). For these reasons the disease can be transmitted not only by a person actually afflicted with scarlet fever, but also by one with septic sore throat.

Scarlet fever occurs mostly in children between about 2 and 10 years of age. In older children and adults the disease is rarer because of the immunity that most people acquire against streptococcal toxic action. The onset of the disease is sudden, with sore throat, fever and vomiting, followed within 12–36 hours by a rash of tiny raised spots spreading over the body (Fig. 13-20). In this stage the palate and throat are bright red; the tonsils are swollen. As a rule, there is hardly any cough or catarrh. The tongue at first has a white coat with enlarged papillae standing out distinctly; later the white coat disappears, leaving a bright red surface (strawberry tongue). The face is flushed, but the skin around the mouth remains pale. The skin rash is most pronounced on the chest and back, in the groin and on the insides of the thigh. In an average case the duration of the rash is two days; the temperature then returns to normal, and recovery is uneventful. In mild cases the rash

FIG. 13-19. Development of scarlet fever.

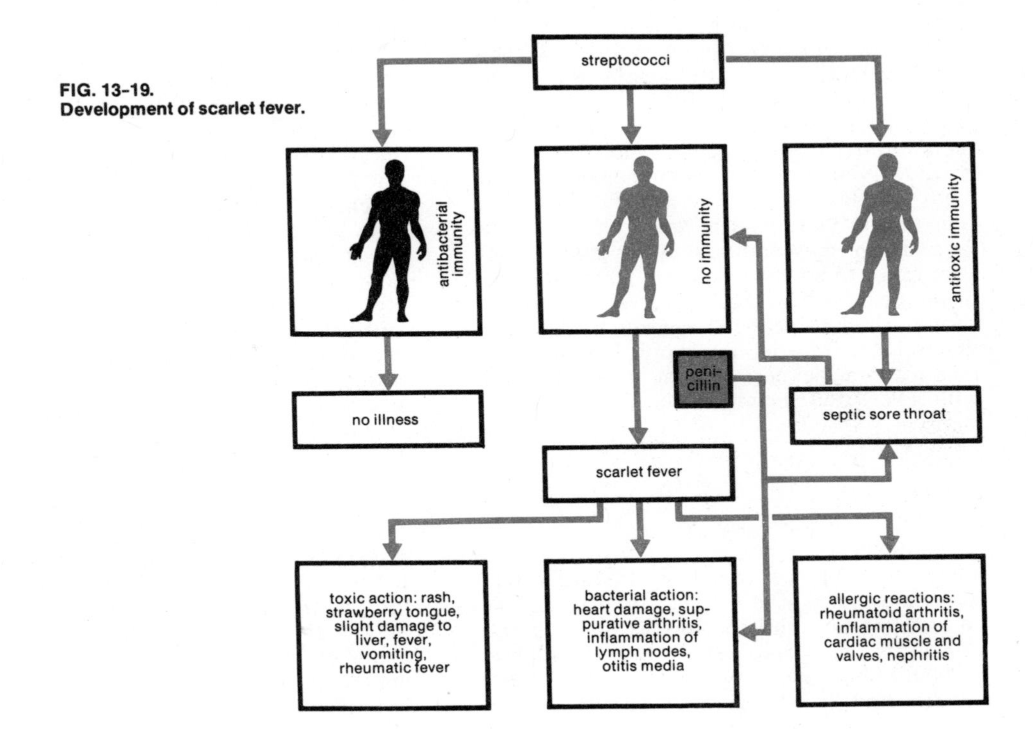

FIG. 13-20.
Course of the disease (scarlet fever).

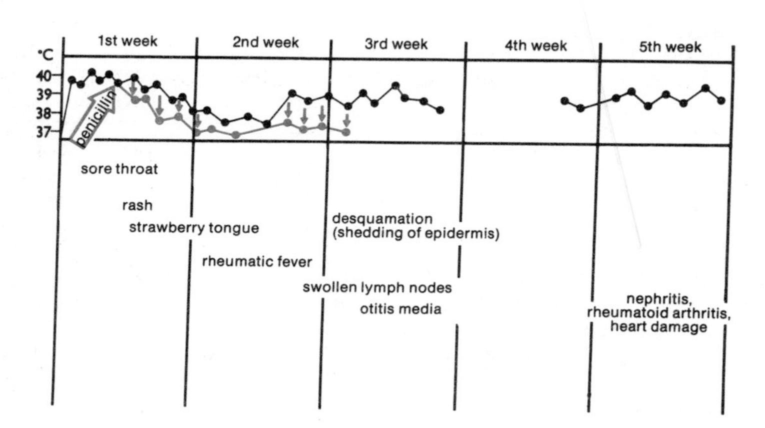

may be slight and of short duration or may not occur at all.

In the absence of treatment there is liable to be a recurrence of fever after a period varying from a few days to several weeks. The lymph nodes of the jaw become swollen and purulent. Otitis media with suppuration may develop, spreading rapidly and causing deafness, with the attendant risk of involving the meninges and blood vessels. Other complications that may result from untreated scarlet fever are allergic disorders such as rheumatic fever (page 381) or glomerulonephritis (page 204).

Isolation, diet and rest are important factors in the general treatment of scarlet fever. The patient should be kept in bed during the acute phase. Penicillin is given as the specific treatment, sometimes augmented with other antibiotics. Even in mild cases the penicillin treatment should continue for a sufficiently long time to prevent complications. Immune serum or antitoxin may additionally be given in severe cases to combat the toxic manifestations of the disease. In modern practice the use of serum, which is ineffective against the bacteria themselves, is generally avoided because of risk of serum sickness, except in certain cases with severe toxic manifestations.

DIPHTHERIA

Diphtheria is an acute infectious disease of mucous membranes, usually of the throat, but sometimes those of other parts such as the larynx (laryngeal diphtheria) or the nose. Occasionally the skin may be infected. Diphtheria of the throat begins about 2–5 days after infection, the first signs being reddened and swollen tonsils, with fever and sore throat. Spots soon appear which merge to form a continuous white membrane that adheres firmly to the underlying mucous membrane of the tonsils, as distinct from the purulent coating in ordinary tonsillitis. The formation of a false membrane (pseudomembrane) over mucous membrane is characteristic of the disease. Any attempt to remove the false membrane causes the mucous membrane to bleed. In a severe case the false membrane, now taking on a brownish color, spreads over all parts of the throat and may involve the larynx and nose. The lymph nodes at the sides of the neck are also affected, but the infection does not spread farther in the lymph system or into the bloodstream. The diphtheria patient looks pale, this pallor being due to the effect of the bacterial toxins on the circulatory system, for although the bacteria themselves do not spread far from their point of entry, the toxins they produce are absorbed into the blood and are highly poisonous to heart muscle, kidneys, nerve cells and other tissues. The heart muscle loses its contractility, slackens, becomes enlarged. The rhythm of the heartbeat may also be disturbed. The blood pressure is lowered and the patient has difficulty in breathing. Vomiting and enlarged pupils are serious danger signs, accompanied by blood pressure falling very low. The toxins may also produce localized paralysis. There is a more immediate danger in that the false membrane around the entrance to the trachea may obstruct the air passage and cause asphyxia (suffocation). In such cases it may be necessary to apply tracheotomy, i.e., to cut into the trachea (windpipe) below the obstruction for the insertion of a breathing tube, or to apply intubation, i.e., the insertion of a breathing tube through the glottis.

The disease is caused by the diphtheria bacillus (*Corynebacterium diphtheriae*), known

FIG. 13-21.
Course of diphtheria.

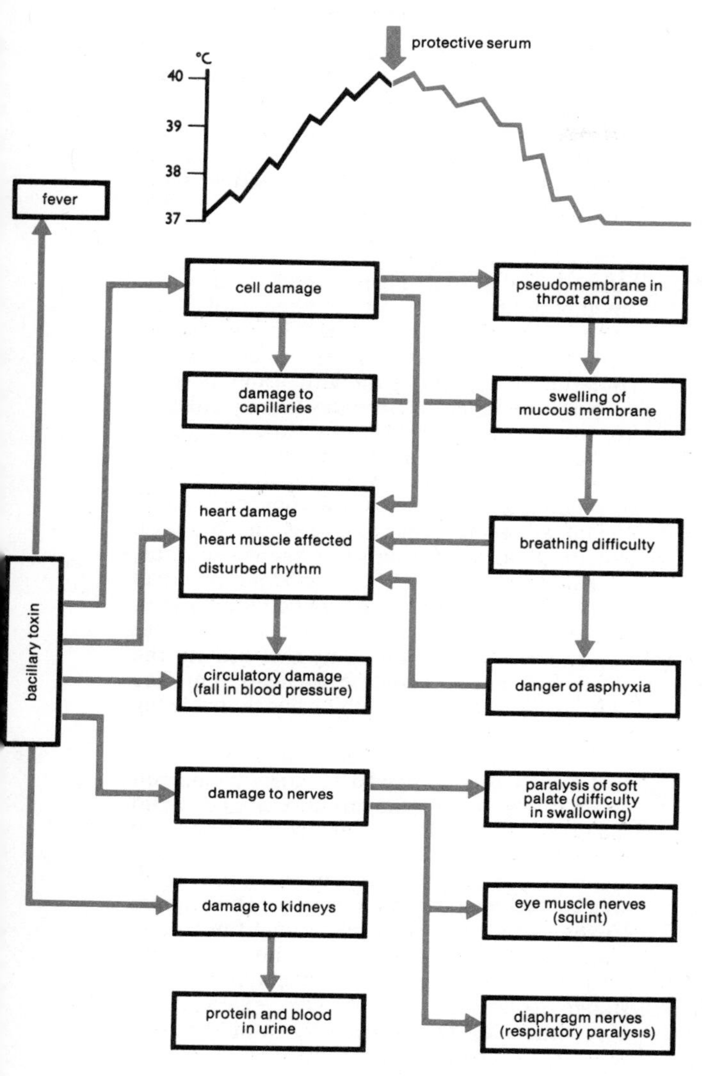

also as the Klebs-Löffler bacillus, and is contracted by airborne infection (coughs, sneezes) from a patient or from a carrier who harbors the bacilli but has no ill effects. The great majority of cases occur in young children, below the age of 10, but older children and adults are not exempt. The disease is rare in infants under 1 year of age.

Diphtheria may be complicated in many ways. In about 10–20 percent of cases of inadequately treated diphtheria of the throat, postdiphtherial paralysis is likely to develop. This frequently affects the soft palate, causing liquids (taken as drinks) to flow back through the nose, besides causing slurred speech. Paralysis of the muscles of the eyes may also occur, resulting in squint. Or there may be pharyngeal paralysis, with inability to swallow. Most dangerous is paralysis of the respiratory muscles, especially those of the diaphragm, involving a risk of asphyxia unless artificial respiration is applied. Damage to the kidneys is another possible complication. Once the patient has survived the disease, the damage to internal organs is usually reversed and disappears without permanent ill effects.

Laryngeal diphtheria (diphtheritic croup) is a complication that is liable to occur in very young children. It is characterized by the formation of a false membrane on the larynx, with hoarseness, loss of voice, and a hard dry cough. The patient has difficulty in breathing due to obstruction of the larynx. He lies quiet and pulseless, with a pale clammy face and blue lips, and is in danger of dying from asphyxia unless relief can be obtained by tracheotomy or intubation.

Routine immunization of infants has reduced diphtheria to a rare disease in many countries. Some individuals already possess immunity from the disease, so that they do not require immunization. The presence of immunity can be determined by the Schick test. Active immunization is achieved by injecting the child with diphtheria toxoid (detoxified diphtheria toxin), which is harmless but evokes antitoxin formation in the body.

Specific treatment of a patient affected by the disease is by injection of diphtheria antitoxin, i.e., the antibody that counteracts the toxins produced by the bacteria. Strict bed rest during the acute and convalescent stages of the disease is essential. In cases of myocardial involvement, i.e., where the heart muscle is affected, prolonged rest in bed is especially important.

LEPROSY

Leprosy is a communicable, but not a highly contagious, chronic disease caused by *Mycobacterium leprae,* which closely resembles the tubercle bacillus. The skin and nerves are attacked by the infection. The two main forms of the disease are lepromatous leprosy and tuberculoid leprosy.

The *lepromatous* form is characterized by skin lesions (raised areas and lumps) and symmetrical involvement of peripheral nerves, with muscular weakness and paralysis. The lesions, which may ulcerate, sometimes give the face a repulsive lionlike expression (leonine facies). They may spread into the nose and throat, causing respiratory difficulties. The sexual organs, kidneys and intestine are also attacked. The patient may live for 5–10 years with this form of the disease.

Tuberculoid leprosy also affects the nerves; the skin areas served by these nerves become white and hairless and lose all sensation or only the sense of pain. The absence of sensation in the areas of skin affected may cause the patient to burn or injure himself without being aware of it. Joints may be damaged by paralysis of muscles. In advanced cases of this form of the disease the hands and feet become paralyzed, fingers and toes shrivel and fall off; whole limbs may even be so affected. Yet the disease is generally a very slow killer, and patients may live for as long as 20–30 years.

Lepromatous leprosy is much more contagious than the tuberculoid type, but is, even so, relatively difficult to contract. Indeed, the disease is spread only between persons living in prolonged and intimate contact, particularly from diseased older members of a family to children living in overcrowded and primitive conditions. The disease may manifest itself after an incubation period that may range from 1 to as much as 30 years. In many respects the infection resembles tuberculosis. The onset is very gradual; the first signs are usually skin changes, but the true nature of the disease may remain undetected for years.

Leprosy used to be treated with chaulmoogra oil, but this has now largely been superseded by sulfone drugs, notably dapsone. Treatment is effective, but has to continue for at least two years. The leprosy bacteria may develop resistance to the drug, which for this reason is sometimes alternated with another type of drug, e.g., thiambutasone. Drug therapy makes the patient noninfectious, so that segregation in a special institution (leprosarium) or leper colony is now no longer considered necessary or desirable in countries where modern treatment is readily available. Because of their greater susceptibility to the disease, however, children should be removed from contact with leprosy patients. Leprosy is still quite common in many parts of the world, especially central Africa, southern Asia and South America.

VIRUSES AND DISEASE

Viruses are minute organisms not visible under an ordinary microscope. They are of very simple structure, on the borderline between living and dead matter, ranging in size from about 30 to 300 millionths of a millimeter and therefore visible only with the aid of an electron microscope. Viruses differ from bacteria not only in being much smaller but also in being totally parasitic: they can multiply only in the cells of other organisms (which may be human beings, animals, plants or indeed bacteria), but not in blood or in tissue fluid. An isolated viral particle is apparently dead, without the ability to feed, grow or multiply; but when it invades a living cell, it diverts the cell's functions to its own needs, as it were. Under these conditions the virus can multiply. It is not equipped to provide its own energy, but has to rely on what it can obtain from the host cell to which it attaches itself. Viruses do, however, contain nucleic acids; some consist of little else. These acids are found in the nuclei of all living cells and form the basis of genetics, the most notable of them being deoxyribonucleic acid (DNA). Under favorable conditions, a nucleic acid can produce a replica of itself. When a cell is infected by a virus, its resources are exploited by the virus to reproduce itself and invade other cells. This activity is harmful to the cells; besides, some viruses produce toxins.

The question of whether a virus is "alive" or "dead" is a somewhat academic one because it is difficult to decide whether the virus really reproduces itself in the manner of a living creature or is, in fact, merely an inanimate chemical component that is utilizing the favorable environment offered by the cell to make copies of itself or perhaps compel the cell to make the copies. Some viruses can actually be crystallized like many other chemical substances.

A virus (Fig. 13-22) comprises a thin protein membrane which encloses the mainly nucleic acid contents. When the virus invades a host cell, the nucleic acid of the virus mingles with that of the cell, where it insinuates itself into the metabolism of the latter, so that more viral nucleic acid is produced. The viral particles collect in crystal-like accumulations within the cell (Fig. 13-23). Eventually, when the nutrition provided by the host cell has been used up, the viral particles become enveloped with a protein membrane again, emerge from the

FIG. 13-22.
Various types of virus.

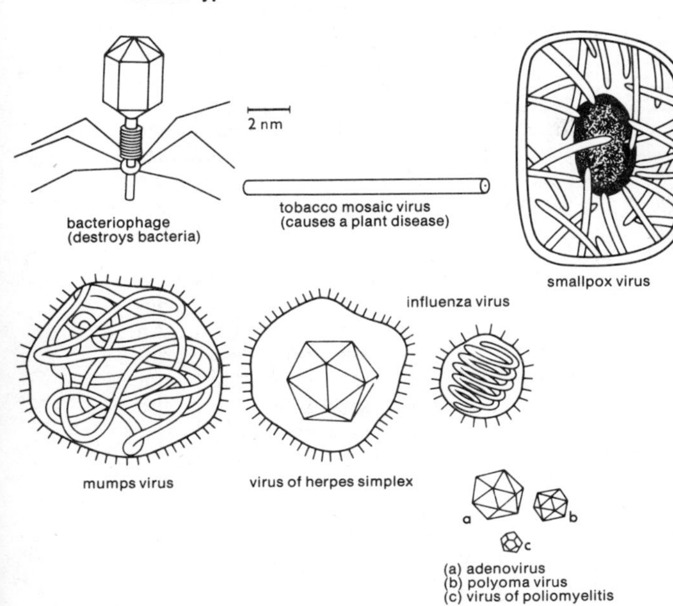

FIG. 13-23.
Viral particles in cell.

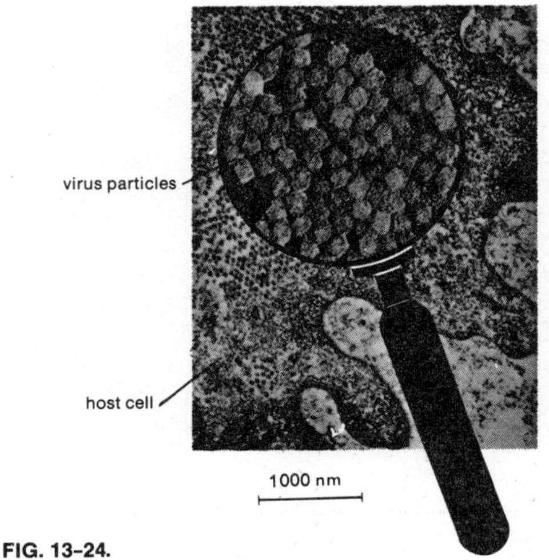

FIG. 13-24.
Virus particles invade cell.

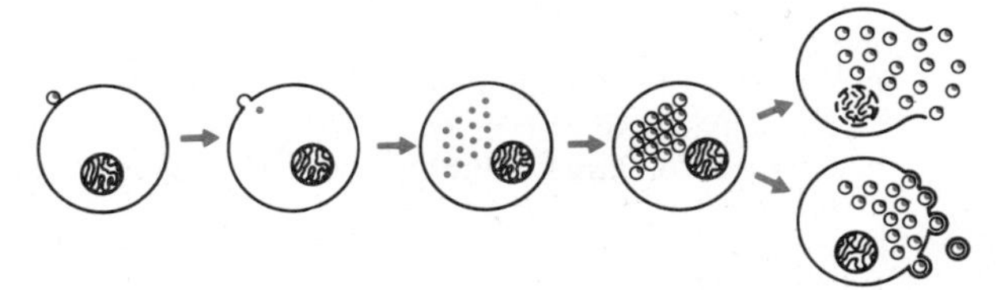

cell and invade neighboring cells. In some cases this emergence of the virus destroys the host cell; in others it does not (Fig. 13-24). The nucleic acid contained in this virus, though actually produced by the host cell, is an integral part of the virus. The latter could therefore be compared to a gene of a higher organism, but a parasitic gene that compels the host cell to manufacture more and more copies of the invader. Viruses that cause tumors may become permanently incorporated into the nucleic acid of the host cells and thereby incite them to continuous proliferation. Bacteriophages are viruses that attach themselves to bacteria as their host cells, which they eventually destroy. The bacteriophage consists of a head composed of DNA and a tail by which it attaches itself to the host cell of its choice.

Because of the simple structure of viruses and their resemblance to human genetic substances it is understandable that they can reproduce themselves only in living cells and that they are highly resistant to chemotherapeutic drugs, such as antibiotics, which are not harmful to the patient but have a specific and toxic effect on disease-causing microorganisms. Only a few of the "large" viruses respond to chemotherapy, e.g., the virus of psittacosis (parrot fever, a type of pneumonia transmitted to man by birds of the parrot family, including budgerigars).

Well-known viral diseases are smallpox, chickenpox, measles, German measles, poliomyelitis, rabies and certain types of pneumonia and hepatitis. They are transmitted mainly by interpersonal contact, but sometimes also indirectly or through animals. Many viral diseases confer lifelong immunity; active immunization by means of attenuated or inactivated virus can therefore be beneficial.

MEASLES, GERMAN MEASLES

Measles (rubeola, morbilli) is a highly contagious viral disease characterized by catarrhal symptoms and by a rash on the skin and mucous membranes (Koplik's spots). The virus is spread by airborne droplets or by direct contact. The human body is virtually 100 percent sensitive to the virus, i.e., no one who has not had the disease can hope to escape infection upon coming into contact with it. On the other hand, an attack of measles almost invariably confers permanent immunity. Infants up to the age of about 6 months are generally immune

FIG. 13-25.
Measles: course of the disease.

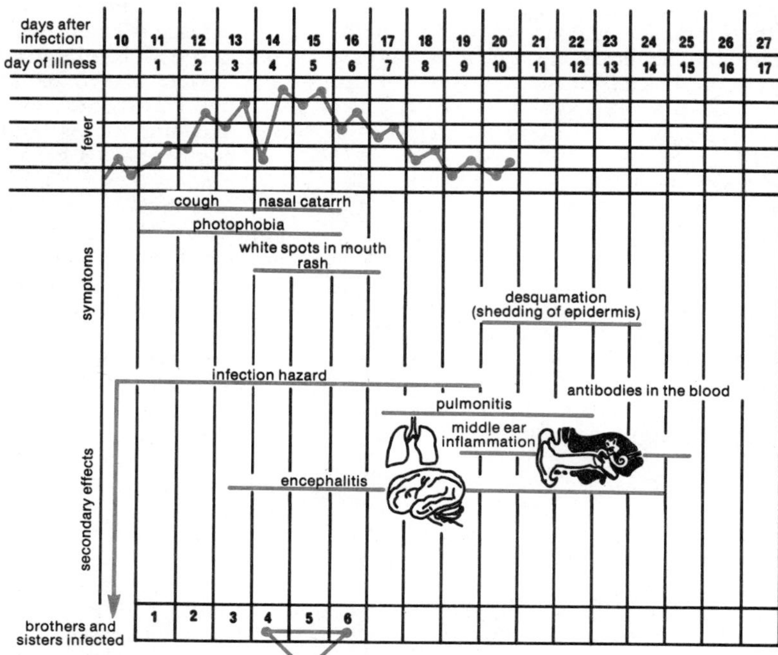

FIG. 13-26.
Frequency in relation to age for measles (left) and German measles (right).

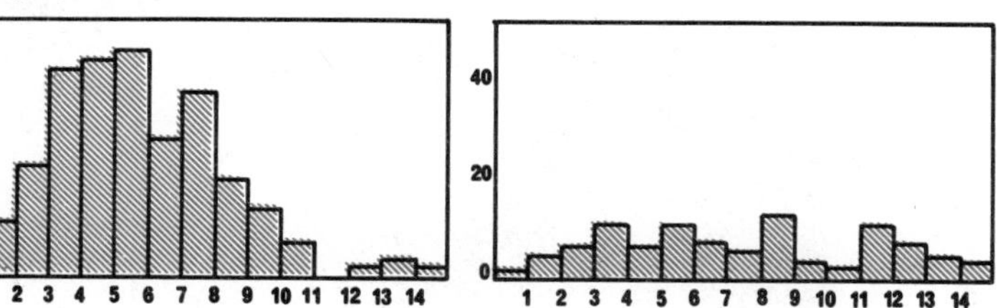

FIG. 13-27.
Symptoms and effects of measles and German measles.

measles	German measles
cough, catarrh, sore throat white spots in mouth rash, 6–8 days fever lymph nodes not swollen secondary diseases, risk of complications	cough, catarrh, sore throat no white spots rash, 3–4 days fever lymph nodes on neck are swollen usually no secondary diseases

because they still have the mother's protective substances in their blood. Active immunization can be provided by a measles vaccine containing the inactivated virus or, more effectively, the attenuated live virus. Gamma globulin can provide passive immunization.

The incubation period averages about 10 days. The onset of the disease is gradual, characterized by running nose and eyes, fever, and sore throat. A day or two later the spots (red with white centers), appear on the mucous membrane of the mouth, followed in a few days by the skin rash, which first appears on the face as small spots, but these soon coalesce to form large blotchy areas that may cause a mottled, swollen appearance. The rash spreads to the trunk and extremities. There is often photophobia (intolerance of light) and a cough which is due to the inflamed condition of the bronchial mucous membrane. If there are no complications, the rash lasts four or five days, after which it subsides. The whole attack lasts about a week. Measles is much more contagious before than after the rash has appeared.

Measles itself is generally not a dangerous disease, but while it lasts the patient is vulnerable to various kinds of secondary bacterial infection. The most serious complication of measles is encephalitis, i.e., inflammation of the brain, involving considerable risk of permanent damage to the central nervous system. Bronchopneumonia is another serious risk, as is also otitis media with the possible further complication of mastoiditis and brain abscess. Eye complications sometimes also occur.

Antibiotics are ineffective against the virus, but reduce the danger of secondary infection. The patient should be kept isolated in a well-ventilated room, which may have to be darkened if he complains of photophobia. He should be given plenty of fluids (water, fruit juice, milk).

Although measles is seldom dangerous to healthy well-nourished children, it is more likely to pose a threat to very young ones (under about 3 years old). Infants should be protected from the disease by active immunization at about 12 months of age.

German measles (rubella) is an acute contagious disease of short duration, resembling true measles but much milder (Fig.13-27). It is characterized by headache, fever and a pink rash, but the symptoms generally last no more than three or four days. The disease is nevertheless important because of the serious effects it has on the developing embryo in the first three months of pregnancy, with a high risk of the child being born with severe congenital defects, e.g., defective eyesight, hearing or heart function.

SMALLPOX, CHICKENPOX

Smallpox (variola) is an acute and highly contagious disease caused by a virus, acquired mainly by direct contact with a patient or with articles contaminated by a patient. Once common in all parts of the world, it has now become virtually a tropical disease (southern Asia, central Africa, South America), because it is in those parts that routine mass vaccination of the population has not yet been established. Natural immunity from the disease is believed to be almost nonexistent; immunity can be conferred only by the disease itself or by proper vaccination. The incubation period is about 10–12 days on average. The onset is abrupt, with high fever, chill, headache, nausea, vomiting, and intense backache or lumbar pains. The fever remains high until the third or fourth day and then falls sharply. When that happens, the rash appears, first on the face and soon afterward on the extremities and trunk. It may at first be difficult to distinguish from chickenpox. One difference, however, is that the spots in smallpox all begin more or less simultaneously and are all at the same stage of development, besides being of more uniform size and more deep-seated in the skin. In both diseases the lesions (spots) evolve from macules (flat spots) to papules (raised spots), then to vesicles (fluid-filled blisters) and then pustules (containing pus; these usually develop about one week after the appearance of the rash). The lesions are multilocular, i.e., composed of many compartments, and do not rupture easily nor collapse when pricked by a needle. There is usually some permanent scarring as a result of smallpox. Though the disease is potentially dangerous, requiring careful nursing and other precautions, most patients recover in a week or two. A particularly dangerous complication arises when the rash is extremely dense and extends to the air passages. Tracheotomy may save such patients. Another major hazard in some cases is presented by secondary bacterial infection of the lesions or by pneumonia; these complications can be prevented by treatment with penicillin. Vaccination provides reliable protection from smallpox infection; it should, however, be repeated during an epidemic or when a person has been exposed to known infection. The severity of an attack in susceptible (unvaccinated) persons can be reduced by passive immunization with a vaccine or by certain drugs if given within 24 hours after exposure to infection. Treatment of smallpox will consist in keeping the patient in strict isolation in an adequately ventilated room. This should continue until all scabs have disappeared. Plenty of liquids should be given (water, fruit juices, milk). Precautions to keep the eyes clean and free from infection are important.

Alastrim (variola minor) is a mild variant of smallpox, seldom dangerous.

Chickenpox (varicella) is also an acute and highly contagious viral disease characterized by a rash. The lesions (spots) make their appearance in successive crops, passing through the stages of macules, papules, vesicles and crusts. The virus of chickenpox also causes herpes zoster (shingles, see page 409). Although no age is exempt from the disease, chickenpox is relatively uncommon in adults. It is estimated that about 75 percent of children have had the disease by age 15 and have acquired lifelong immunity. After an incubation period of about two weeks, an itching rash appears on the back and chest, with crops of spots occurring at intervals of two or three days. They are superficial and rupture easily, seldom leaving any scars. The patient remains infectious until the spots have completely disappeared. In general, the disease is not dangerous. Complications may arise from secondary infections due to scratching. The comparatively rare variety called *varicella gangrenosa* is dangerous, however: necrosis develops around the lesions, causing gangrenous ulceration. There is no specific treatment for chickenpox.

INFLUENZA

Influenza is an acute contagious respiratory infection caused by a virus and characterized by sudden onset, chill, headache and myalgia (pain in the muscles). Sore throat, cough and catarrhal inflammation of the nasal mucous membrane ("cold in the head") also commonly occur. The disease is self-limited, usually lasting no more than a week, sometimes only a few days. It is spread by airborne droplets (coughing, sneezing) and may occur sporadically, epidemically or pandemically. A pandemic is a major epidemic affecting the majority of a population and spreading rapidly to, or occurring simultaneously in, many parts of the world.

There are many types and subtypes (strains) of influenza virus. An attack of influenza confers immunity to the strain of virus that caused it, but each new epidemic is due to a new strain to which there is no widespread immunity. These new strains apparently develop as modifications of existing forms of the virus; it may also be that certain types of the virus, which were previously confined to animals, evolve into a form that attacks man. So-called Asian influenza, which assumed pandemic proportions in 1957/1958, was caused by a variant of the "type A" influenza virus.

The virus itself can be dangerous, but most deaths from influenza are due to complications, especially pneumonia, e.g., in the notorious pandemic of 1918–1920, when 22 million deaths occurred in an estimated 500 million cases of the disease. Other complications due to secondary bacterial infections may affect the nasal sinuses or the middle ear. Death from influenza itself is rare except in babies, in elderly people and in those with certain chronic diseases.

Protection against known strains of the virus can be provided by vaccines, which are reckoned to be about 80 percent effective against such strains. In recent years, epidemics have been caused by new strains against which existing vaccines may be ineffective and which require new vaccines for their control. There is no specific treatment for influenza once it has gained a hold. Treatment should consist in bed rest in an adequately ventilated room, proper care, and a suitable diet. Symptomatic treatment aimed at making the patient as comfortable as possible may include aspirin, cough-control medicines and sleeping pills. Although recovery from influenza is fairly rapid, some patients experience lassitude (weariness) for weeks after.

Besides infants and old people, certain other categories of the population are particularly vulnerable to the disease and its attendant effects: persons with high blood pressure, arteriosclerosis, diabetes or asthma. Influenza can be somewhat arbitrarily subdivided into the common or respiratory type, the febrile type, the gastrointestinal type (characterized by vomiting, headache, abdominal pain, diarrhea), and the nervous type (headache, mental depression). Quite often the characteristics of these various types occur simultaneously in varying degrees.

MUMPS, GLANDULAR FEVER

Mumps (infectious parotitis) is an acute, febrile viral disease characterized by inflammation of the parotid glands and other salivary glands (Fig. 13-28). The disease is prevalent in the winter months; children are particularly susceptible to mumps (Fig. 13-29). The virus, resembling that of influenza, is airborne: conveyed by droplets of infected saliva expelled by a patient's coughing or sneezing. After an incubation period of 2–4 weeks, the onset of the disease is gradual, with malaise, headache,

FIG. 13-28.
Anatomy of the cervical region.

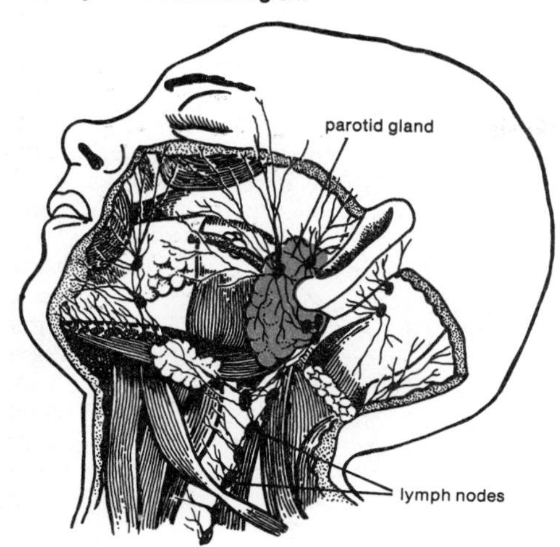

FIG. 13-29.
Frequency of mumps in children of different ages.

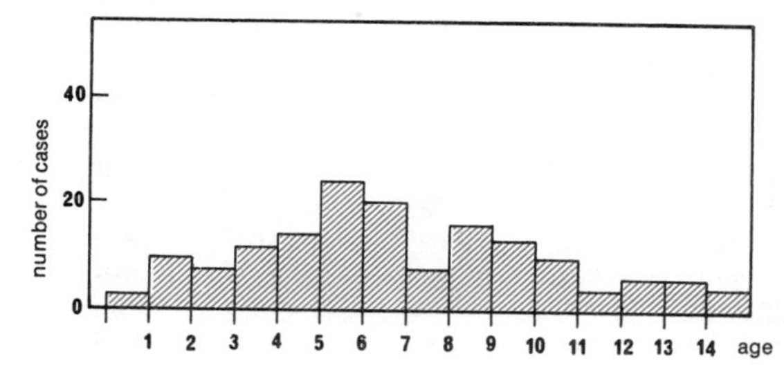

pain below the ears, and moderate fever, followed by swelling of one or both parotid glands, below and in front of the ear (Fig. 13-30). As a rule, mumps is a harmless disease, treatment being confined to bed rest with liquid diet. The most common complication in the adult male is orchitis (inflammation of a testis), which may result in sterility; in females there may be oophoritis (inflammation of an ovary). In some cases the disease involves the pancreas and is then characterized by vomiting and abdominal pain. Meningitis is another possible complication.

FIG. 13-30.
Swelling of parotid glands in mumps.

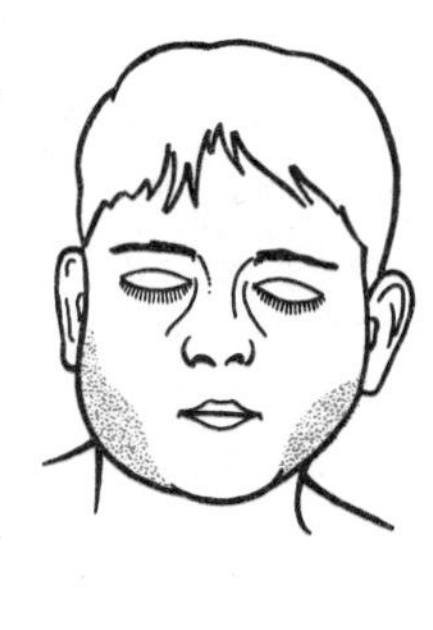

FIG. 13-31.
Enlarged lymph nodes in infectious mononucleosis.

Infectious mononucleosis (*glandular fever*) is an acute infectious disease believed to be caused by a virus which affects lymphoid tissues. It occurs mainly in young people and is characterized by enlarged lymph nodes in the neck (Fig. 13-31) and in other parts of the body and an enlarged spleen. There is a great increase in the number of leukocytes (white blood cells), especially the abnormal mononuclear type. The symptoms are headache, sore throat, rise in temperature, and pain in the neck. The tonsils are red and swollen; swallowing is difficult and painful. In some cases a rash appears. The fever subsides after a week or two, but the swollen condition of the lymph nodes and the abnormal characteristics of the blood may continue for a much longer time, and convalescence from infectious mononucleosis tends to be protracted. As a rule, there are no aftereffects.

PROTOZOA AS PATHOGENIC AGENTS

Protozoa belong to the animal kingdom and are mostly microscopic unicellular creatures, ranging in size from 0.005 to 0.5 mm. They depend on other living creatures for their food. Reproduction is usually asexual by fission; conjugation and sexual reproduction sometimes also occur, however. A typical protozoan consists of a blob of protoplasm surrounding a nucleus which contains chromosomes. In addition, the protoplasm contains so-called organelles, which are parts specialized to perform specific functions (propulsion, metabolism).

Protozoa of the genus *Trichomonas* are somewhat pear-shaped flagellate creatures, i.e., provided with flagella (hairlike motile projections) (Fig. 13-32a). A typical trichomonad is about 0.015 mm in size and has forward-facing flagella which perform propulsion movements. At the front end of the protozoan is a slitlike "mouth" aperture. Multiplication is by fission; there is no sexual reproduction. Trichomonads occur in the human intestine and, in a certain proportion of people, also in the oral cavity, where they exist as harmless

FIG. 13-32.
Pathogenic protozoa.

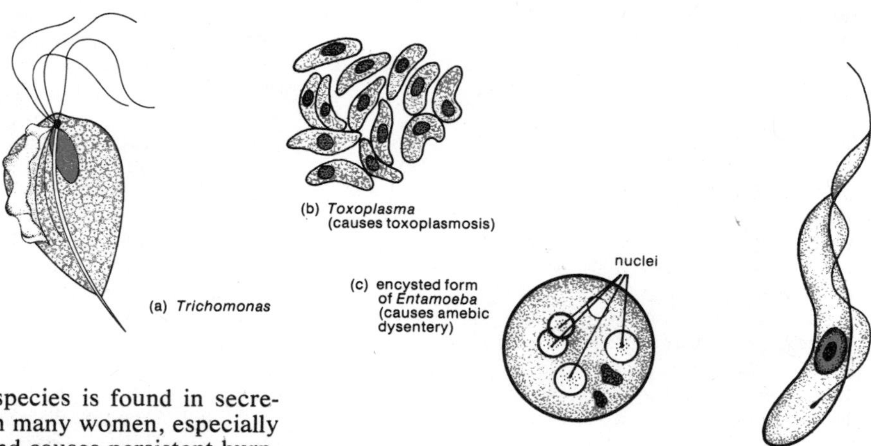

parasites. Another species is found in secretions of the vagina in many women, especially during pregnancy, and causes persistent burning and itching of the vulva, with white frothy discharge. The infection can be communicated by sexual intercourse and thus sometimes also occurs in the male urethra, where it occasionally causes inflammation.

Protozoa of the genus *Toxoplasma* comprise a number of species, ranging from crescent-shaped to approximately spherical. One end is rounded, while the other is somewhat pointed (Fig. 13-32b). They are 0.004–0.006 mm in length and reproduce by fission. The disease called *toxoplasmosis* is caused by *Toxoplasma gondii* and affects the central nervous system. In most cases it is fairly harmless, but it can be transmitted from the pregnant woman to her unborn child and seriously affect it.

An *ameba* is a protozoan of semifluid consistency and of irregular and constantly changing shape, moving about and feeding by sending out fingerlike protrusions of protoplasm (pseudopodia). It feeds by engulfing its prey into the protoplasm. Reproduction is by fission. There are many species of ameba in nature, especially in pools and ponds. *Entamoeba histolytica* (Fig. 13-32c) lives in the human colon and is about 0.025 mm in diameter. It causes *amebic dysentery* (see page 457); it feeds by engulfing particles of tissue, especially red blood cells. Amebae discharged in

FIG. 13-33.
Life cycle of *Plasmodium*, the protozoan that causes malaria.

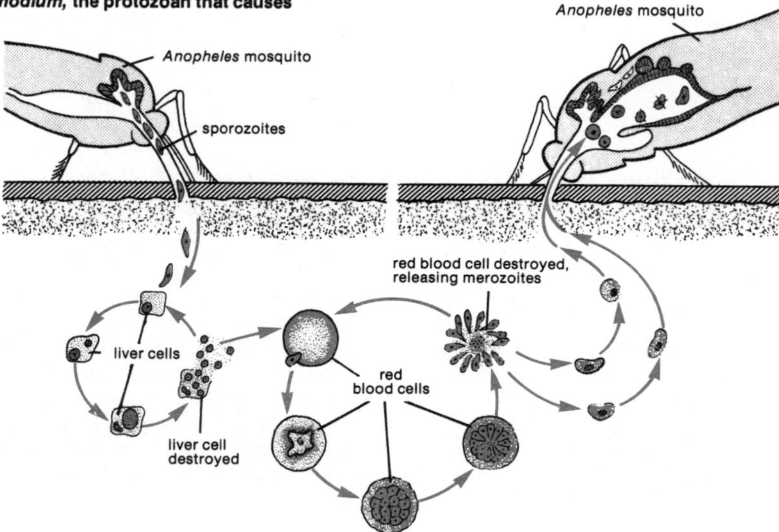

the feces of a dysentery patient become cysts, an inert form of the creature, which are transmitted to other persons, e.g., in contaminated food or water.

The protozoa of the genus *Plasmodium* include the causative agents of *malaria* (see next section), while those of the genus *Trypanosoma* (Fig. 13-32d) are flagellate protozoa which include the causative agent of West African *sleeping sickness* (*Trypanosoma gambiense*), transmitted by the tsetse fly. The latter disease is usually of a chronic type. The trypanosomes eventually invade the nervous system, causing the drowsiness that gives the disease its name. It is curable in its early stages. There is also an East African variety of sleeping sickness, caused by an allied species of *Trypanosoma*, which runs a more rapid course. *Chagas' disease*, which occurs in South America and is caused by *Trypanosoma cruzi*, is transmitted by certain species of bloodsucking bugs. It is often fairly harmless except when it attacks the heart muscle.

MALARIA

Malaria is an infectious disease caused by protozoan parasites of the genus *Plasmodium* within the red blood cells. It is transmitted by mosquitoes. The parasites undergo an asexual cycle of development in man and a sexual cycle in the mosquito, the bite of which injects the parasite in the form of so-called *sporozoites* from its salivary glands into the human bloodstream. The sporozoites enter tissue cells, such as liver cells, where they go through two reproductive divisions and then reenter the bloodstream and infect red blood cells. In these the parasites undergo further divisions, forming what are known as *merozoites*. These break free and in turn invade other red blood cells. The destruction of the blood cells and the resulting liberation of pigment and waste products cause the characteristic symptoms of violent shivering, headache, fever and sweating. These occur at intervals of 48 hours in tertian and of 72 hours in quartan malaria. After several generations, some merozoites develop into *gametocytes* (sexual forms of the parasite), which, when sucked up by a mosquito when it bites the patient, undergo further development within the insect. More particularly, there are two types of gametocytes—microgametocytes and macrogametocytes—which unite sexually, each pair forming a zygote which penetrates into the stomach wall of the mosquito and forms an oocyst in which sporozoites develop. When the oocyst matures, it bursts and liberates the sporozoites, which then make their way to the mosquito's salivary glands.

There are four types of malaria parasite: *Plasmodium malariae* causes quartan malaria and is common in subtropical regions. *Plasmodium vivax, Plasmodium ovale* and *Plasmodium falciparum* cause tertian malaria. The last-mentioned species causes the most dangerous type of the disease (*malignant tertian malaria*), which is encountered mainly in tropical Africa, Asia and South America. Here the fever occurs in more frequent and irregular attacks and is more persistent. Congestion of blood vessels in affected organs by damaged blood cells produces various symptoms and enlargement of the spleen. With each attack of fever there is increasing anemia from loss of red blood cells. Malignant tertian malaria can, moreover, have various complications. Thus it may give rise to a dysenterylike condition or it may attack the brain, causing cerebral malaria, which is particularly dangerous to children. *Blackwater fever*, fortunately rare, is another dangerous complication of this type of malaria and is characterized by fever, enlarged spleen and liver, dark urine (hemoglobinuria), abdominal pain, vomiting and jaundice.

In general, an attack of malaria does not confer immunity, but in the survivors of repeated attacks of the disease the symptoms become progressively less severe with each attack. The death rate from malaria is due mostly to anemia and general lowering of resistance. Even without treatment, most patients recover after undergoing two, three or more (sometimes as many as a dozen) episodes of fever.

Quinine, the traditional drug for treating malaria, has toxic side effects and is now seldom used. The modern synthetic drugs are safer: chloroquine, pyrimethamine, etc. They usually stop an attack and cure malignant tertian malaria (falciparum malaria). In other types of the disease, notably quartan malaria, the parasites may "hibernate" in the liver and cause subsequent attacks, possibly even after several years. The aforementioned drugs are not very effective against the "hibernating" parasites, but the drug primaquine usually destroys them. In some cases the malarial parasites are particularly resistant to drugs and then have to be treated with combinations of quinine and several of the modern drugs (Fig. 13-34).

FIG. 13–34.
Treatment of malaria.

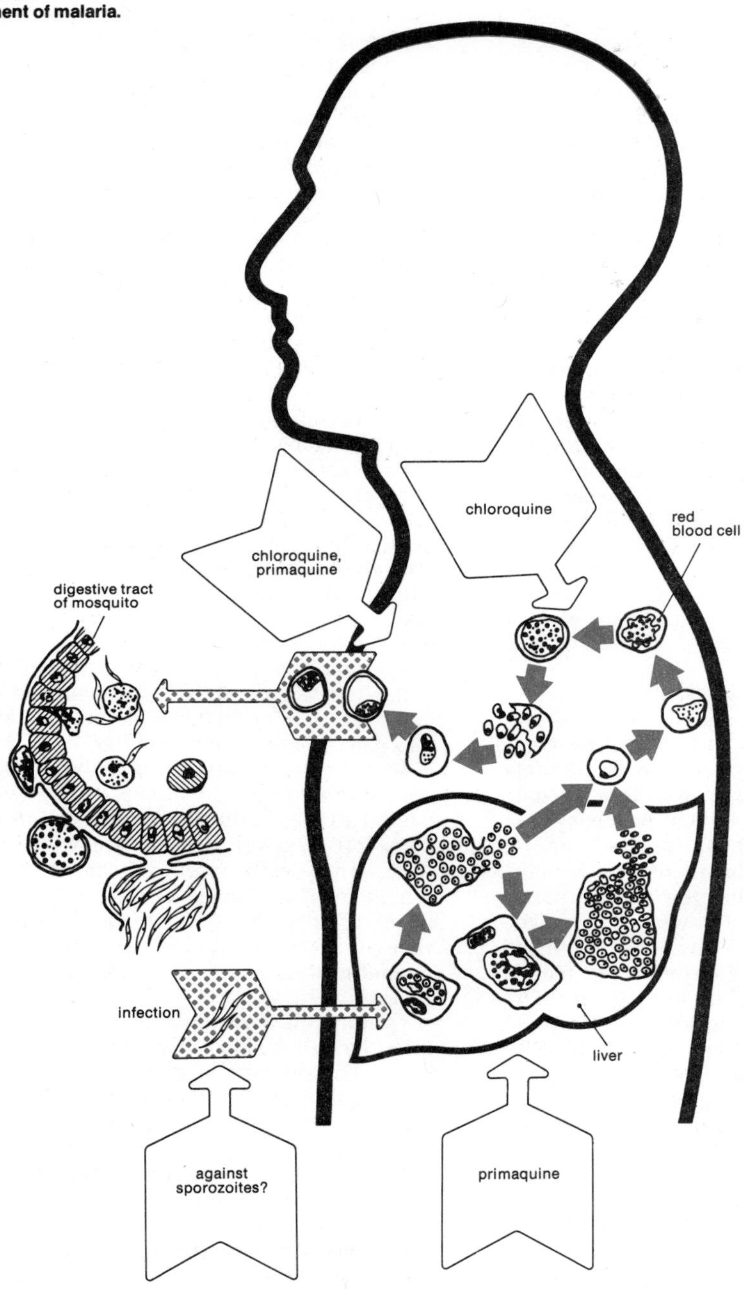

Part Fourteen

TUMORS

BENIGN TUMORS

Any swelling or enlargement may be designated as a *tumor*, but the term is more particularly applied to a spontaneous new growth of tissue (*neoplasm*) forming an abnormal mass which performs no useful physiological function and grows at the expense of healthy neighboring tissue. Such tumors are usually of unknown cause and noninflammatory. They do not obey the normal laws of tissue growth. A broad distinction is drawn between benign tumors and malignant tumors (cancer). A benign tumor is essentially harmless and becomes objectionable only if it obstructs the functions of adjacent tissues or organs or if it looks unsightly.

Cell division and cell differentiation characterize the development and growth of living organisms. The various biological processes that occur during the development of the form of the body and its organs is called *morphogenesis*. Once the organism has attained a certain level of development and maturity, the cells lose their capacity for differentiation into various types of tissue, but retain their capacity for divison—for the replacement of dead cells, regeneration of damaged tissue, healing of wounds, etc.—throughout the individual's life. These cells obey the laws of morphogenesis and are under the control of some kind of signal that starts the process of cell division and subsequently stops it when the repair or regeneration has been completed. A tumor does not obey these laws. Newly formed tumor cells tend to be less highly differentiated than the original tissue cells from which they arise, so that they cannot, or can only imperfectly, perform the function of that tissue. Tumor growth can start in any organ. If it remains localized, keeping within its own bounds, as it were, it is regarded as benign; but if it tends to infiltrate into surrounding tissues and to spread to other parts of the body through the lymph vessels and blood vessels, it is called malignant.

Benign tumors occur mainly in relatively young persons, under the age of 40 or so. A tumor of this general type is often enclosed within a capsule of connective tissue and can therefore be completely removed by surgery. Any symptoms of illness caused by a benign tumor are due to displacement of adjacent tissues. In an enclosed space, as in the skull, the pressure exerted by any tumor, whether benign or not, can pose a threat to life because it compresses a vital organ, in this case the brain.

A distinction can be drawn between tumors that originate in connective tissue and those that originate in epithelium. The commonest benign tumor of connective tissue is the *lipoma* (Fig. 14-1), which is composed of fat cells and is enclosed in a capsule, parts of which frequently extend into the tumor and subdivide it into lobes. Such tumors develop mostly under the skin, in the abdominal cavity, sometimes also in the bursae of joints. They vary in size; particularly large ones may grow to the size of a football. Sometimes the lipoma protrudes on a stalk (Fig. 14-1a). Lipomas do not turn malignant; their surgical removal is usually undertaken more for cosmetic than for health reasons. A *fibroma* is a fibrous encapsulated connective tissue tumor; it is slow-growing, of irregular shape and firm consistency. A tumor containing smooth muscle fiber is called a *myoma*. A commoner type, a combination of these two, is the *fibromyoma*, often found in the uterus and familiarly referred to as a "fibroid" (see page 480). A cartilaginous tumor of slow growth, called a *chondroma*, may arise from hyaline cartilage in any part of the body; it is often localized on the surface of bones, where it forms what is more particularly known as an *ecchondroma* (Fig. 14-2). Though benign, by thrusting into surrounding muscles such tumors may interfere with movement. Another type of chondroma is the *enchondroma*, which usually de-

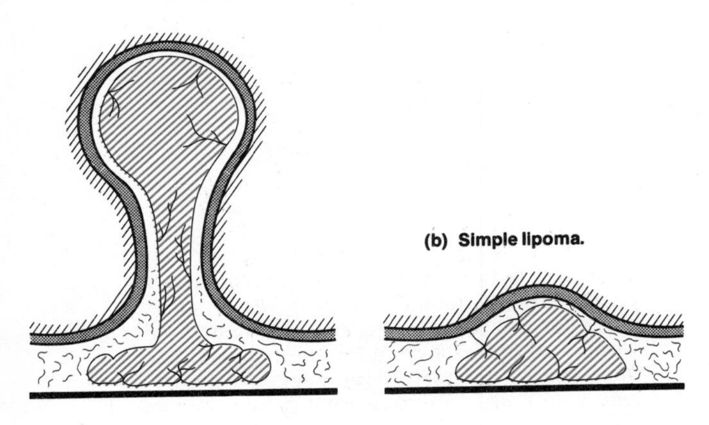

FIG. 14-1.
(a) Lipoma on a pedicle (stalk).
(b) Simple lipoma.

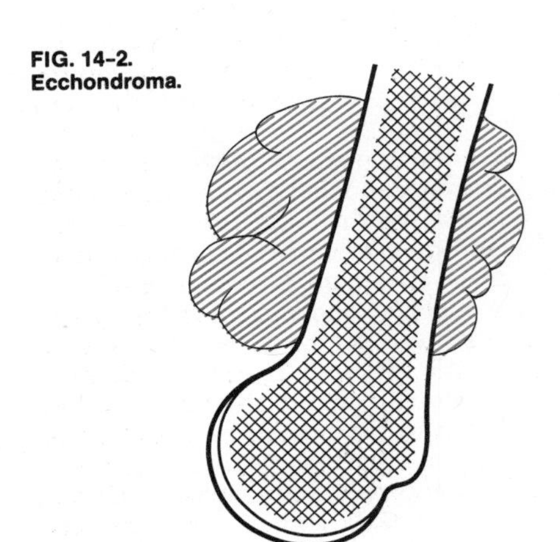

FIG. 14-2.
Ecchondroma.

velops inside one of the smaller long bones, e.g., in the fingers or toes (Fig. 14-3). A number of these tumors may occur simultaneously and cause unsightly deformities of the hands and feet. With tumors of the chondroma type there is a risk that they will become malignant, and it is therefore necessary to remove them surgically. An *osteoma* is a benign bony tumor, generally developing on the inside or outside of a bone and often extremely hard. If such a tumor grows inward from the inner surface of the skull, it displaces and compresses the brain substance and thus presents a serious threat to life (Fig. 14-4). Another type of benign connective tissue tumor is the *angioma*, consisting mainly of dilated blood vessels (hemangioma, including birthmarks: see page 410) or lymph vessels (*lymphangioma*). A tumor along the course of a nerve, more particularly arising from the nerve sheath, is called a *neuroma*.

The term *papilloma* may denote any benign epithelial tumor, but more specifically refers to a tumor of the skin or mucous membrane consisting of hypertrophied papillae covered by epithelium; warts (see page 410), polyps and tumors of the condyloma type come within this group. A benign neoplasm of glandular epithelium is called an *adenoma*. These epithelial tumors are not encapsulated; some of them tend to develop into malignancy. A papilloma may occur on the tongue or the vocal cords, or on the transitional epithelium of the urinary system (bladder, ureter, pelvis, kidney), where such tumors may obstruct the urinary passages and possibly degenerate into malignancy. It is therefore essential to remove them as soon as possible. Adenomas may arise from the salivary glands, the mucous glands of the gastrointestinal tract, the prostate and the mammary gland. Such tumors affecting the endocrine glands (e.g., thyroid, pancreas, pituitary, adrenal gland) cause proliferation of the glandular tissue and may thus result in hyperfunction

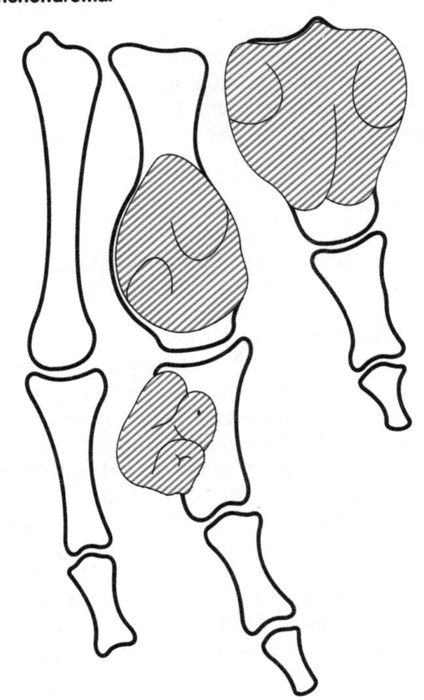

FIG. 14-3.
Enchondroma.

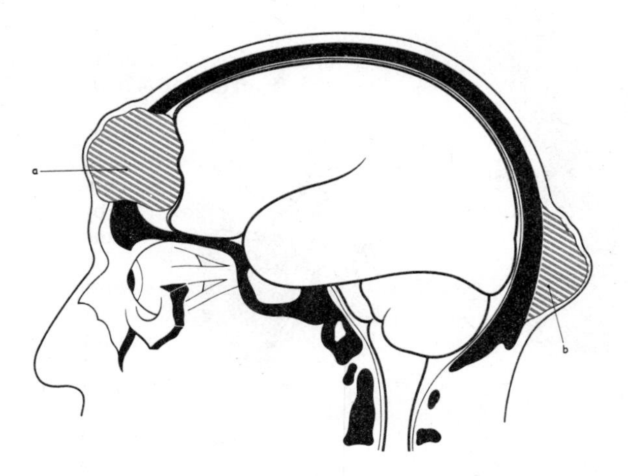

FIG. 14-4.
(a) Osteoma growing inward from the skull.

(b) Osteoma growing on the outside of the skull.

of the gland concerned, i.e., overproduction of the hormone it secretes. Treatment will generally consist in surgical removal of the tumor. A *fibroadenoma* is an adenoma with fibrous tissue; these tumors occur, among other places, in the prostate and in the female breast. Such a tumor, distinctly localized, can be felt as a painless lump. Although it is harmless, there is some slight risk that it will turn malignant. An *odontoma* is a tumor of a tooth or dental tissue.

BRAIN TUMORS

The term *brain tumor* is used rather loosely to denote any abnormal growth developing inside the skull, whether benign or malignant, occurring as neoplastic (new and abnormally formed) tissue, an abscess, a cyst, etc. Brain tumors are rare in comparison with tumors in other parts of the body: statistically only about 1 person in 20,000 is affected by a brain tumor. About half these tumors are of a neuroepithelial character, arising from the nervous tissue; about a quarter of all brain tumors affect the meninges (Fig. 14-5) and blood vessels of the skull, and about an eighth are tumors of the pituitary gland. The remainder include secondary growths (metastases) of lung cancer in men and of breast cancer in women (Fig. 14-6). The general symptoms of a brain tumor are due to an increase in intracranial pressure—headache, vomiting, characteristic changes in the retina of the eye (swollen optic disc) as observed by ophthalmoscopic examination, together with mental symptoms such as irritability, drowsiness, lethargy, dullness, weakness of memory, sometimes giddiness or convulsions. X-ray examination is often helpful in diagnosing a brain tumor. The technique of injecting air into the ventricles of the brain prior to radiography is known as *pneumoventriculography*. The location and outline of a tumor may also be determined by means of *scintiphotography*, i.e., photographing the scintillations emitted by radioactive substances injected into the tissues. Information gained from *electroencephalography* (the record of the electrical activity of the brain) can also be of great diagnostic value. In general, the earlier the tumor is diagnosed and located, the better are the chances of successful treatment.

The commonest brain tumor is the *glioma* (or glioblastoma), which arises from the neuroglia, the connective tissue of the brain. It is a fast-growing malignant tumor which may spread from one half of the brain to the other, but not to other tissues or organs. This type of tumor may cause paralysis and speech defects and is dangerous. A tumor may also develop at the junction of the cerebellum and pons (Fig. 14-7) and is characterized by tinnitus, progressive deafness, disturbed balance, followed by facial paralysis. A comparatively common brain tumor affecting younger per-

FIG. 14-5.
Meningioma: brain tumor which originates in the arachnoidal tissue of the meninges.

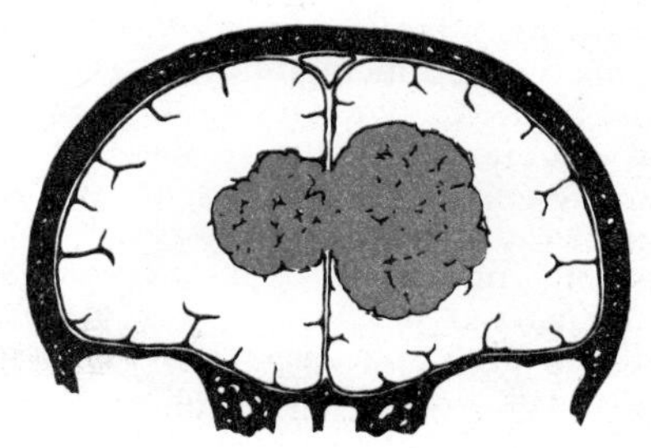

FIG. 14-6.
Secondary growths in the brain.

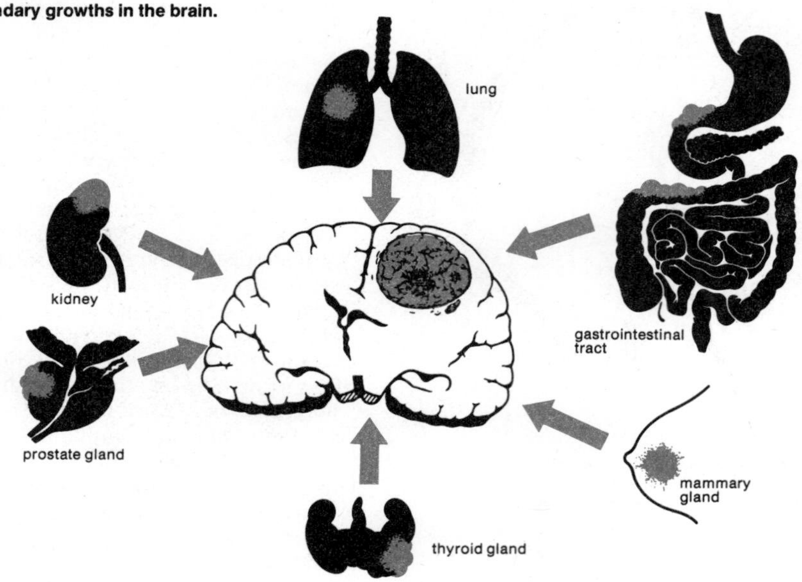

FIG. 14-7.
Cerebellopontine tumor.

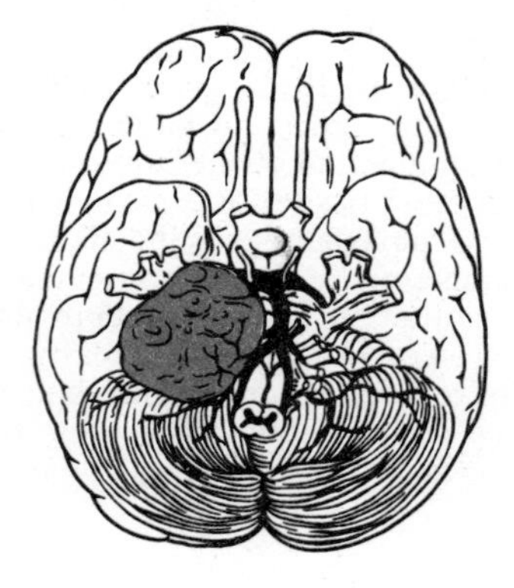

sons is the *astrocytoma*, which arises from the neuroglia of the cerebellum. This tumor, though causing disturbances of movement and equilibrium, is relatively benign and slow-growing. *Meningeal tumors* (*meningiomas*), affecting the membranes enclosing the brain, are also usually benign and slow-growing, with a preference for the region of the olfactory nerve and the pituitary gland, so that they are liable to cause disturbances in the sense of smell, one-sided impairment of the optic nerve, and paralysis of the eye muscles. A tumor of the pituitary gland (see pages 358 ff) disturbs the hormone balance; it may also press on the optic chiasm, causing partial loss of vision.

In general, the prognosis of a brain tumor will depend on its accessibility for surgery and on its malignancy. Success of surgical intervention often depends on preoperative diagnosis and determination of the exact location of the tumor.

UTERINE FIBROIDS

The wall of the uterus consists mainly of smooth muscle, which may be the site of tumors, mostly of a benign type called a *fibromyoma* (or *fibroma*), more familiarly known as a *fibroid*. These tumors are very common, being found in something like one in every four women over the age of 30–35. In a substantial proportion of these women they produce no symptoms; but in about one in every two they cause menstrual disorders. Less frequently the tumor is felt as a "foreign" presence in the abdominal cavity or is so located that it causes discomfort during evacuation of the bowel or during urination. The location of the tumor is important with regard to the symptoms it may produce. Thus, fibroids encroaching on the bladder region may cause painful or difficult urination (dysuria) or frequent urination.

Those pressing on the rectum may cause spasmodic contraction of the anal sphincter with persistent desire to empty the bowel (rectal tenesmus). As a rule, fibroids are painless, except when they grow very rapidly or when they degenerate as a result of deficient blood supply. Some fibroids may cause excessive menstrual bleeding (menorrhagia) and thus bring about an anemic condition, though this is of relatively rare occurrence.

Fibroids develop only under the influence of follicular hormones, i.e., only in sexually mature women. They are seldom found before the age of 25. After the menopause no more fibroids are formed, and existing ones decline because the supply of hormone on which they depend ceases. As a rule, uterine tumors that form, or continue to grow, after the menopause are not fibroids and probably not benign. For unknown reasons the menopause comes 5–7 years later in women with uterine fibroids than in women without them. These tumors may form in any part of the uterus; in 5–10 percent of cases they develop in the cervix. Quite often several fibroids are present. The various positions in which they may occur are indicated in Figure 14-8. They may develop within the muscular wall of the uterus, or just under the mucous membrane lining (endometrium), or on the outside of the uterus. This last-mentioned type of fibroid may be entirely symptomless even if it is a large one. Sometimes the tumor develops on a thin stalk (Fig. 14-9). If the stalk becomes twisted or if the tumor grows to a large size, it may degenerate because its blood supply is then inadequate to sustain it. Cavities may form in such a tumor (Fig. 14-10). In comparatively rare cases a fibroid may harden by calcification or may undergo fatty degeneration.

Fibroids vary in size from a few millimeters to large growths that may fill the abdominal cavity (Fig. 14-10). They are encapsulated in fibrous connective tissue containing blood vessels that supply the tumor. Various forms of benign degeneration occur. Occasionally a fibroid will undergo sarcomatous (malignant) degeneration, but this happens only in a small fraction of cases.

Fibroids that produce no symptoms are usually left in place, unless they show unusually rapid growth. Those that do produce symptoms require treatment. What form such treatment will take must depend on the age of the patient, the location and size of the tumor, and the symptoms it causes. In general, the presence of fibroids in the uterus is likely to pre-

FIG. 14-8.
Various positions of uterine fibroids.

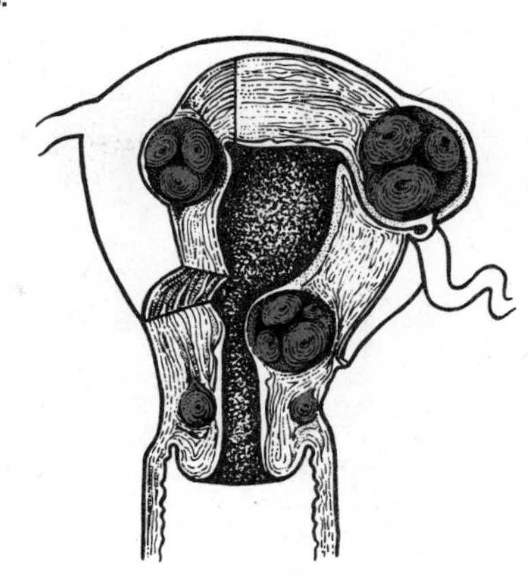

FIG. 14-9.
Development of a fibroid attached by a stalk to the fundus of the uterus.

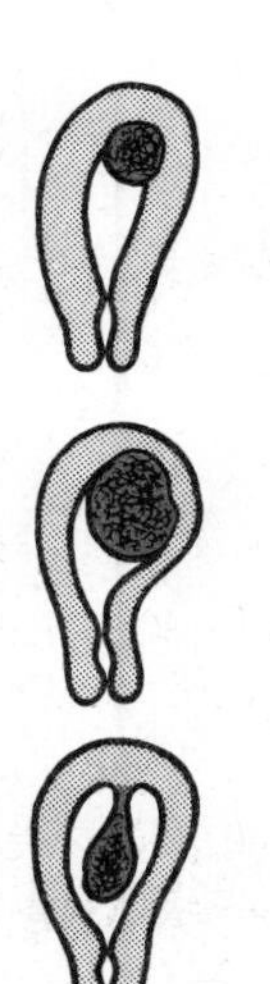

FIG. 14-10.
Cavity formed by degeneration of a large fibroid.

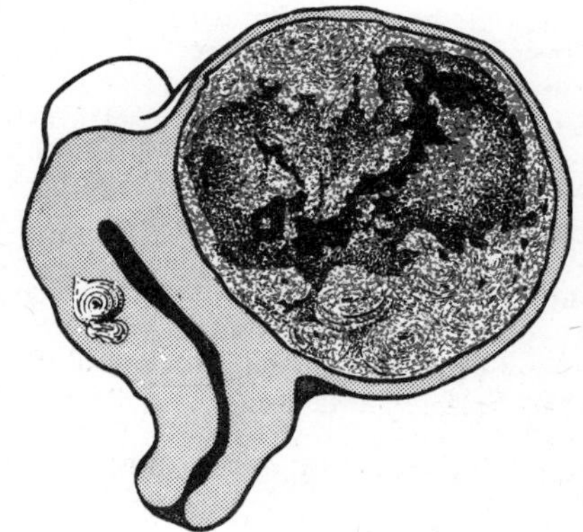

FIG. 14-11.
A large fibroid low down in the uterus may prevent normal childbirth.

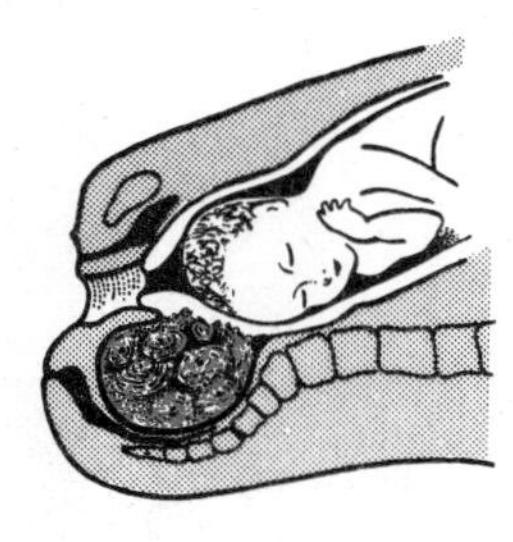

FIG. 14-12.
Surgical removal of fibroids.

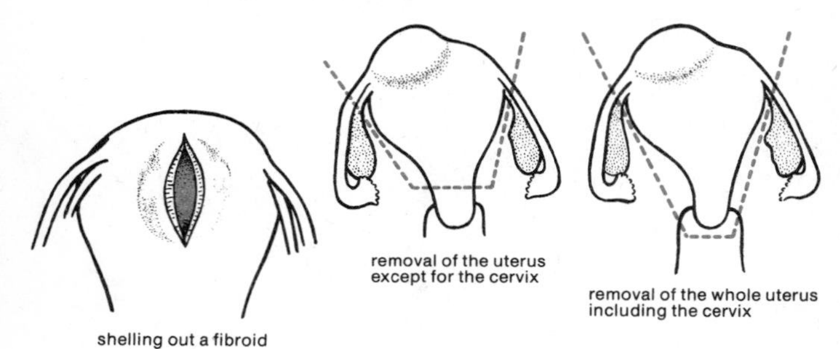

vent normal pregnancy (Fig. 14-11). Fibroids can be removed by surgery. Some fibroids can be shelled out, leaving the uterus capable of childbearing; in other cases the whole uterus may have to be removed (Fig. 14-12).

PROSTATE TROUBLE

The prostate is a gland surrounding the neck of the bladder and the urethra in the male. It is partly glandular, with ducts opening into the urethra, and partly muscular. Any condition causing obstruction of the prostate gland and resulting in retention of urine in the bladder is known by the general term of *prostatism*. The commonest cause is benign hypertrophy, i.e., excessive growth of the gland, occurring in middle or old age and familiarly known simply as enlarged prostate. Actually, hypertrophy does not involve the whole gland, but only the glandular elements which are controlled not only by the male sex hormone but also by the female sex hormone of the male. With advancing age, disturbed hormonal balance is thought to induce overgrowth of this hormone-dependent part of the prostate; the precise cause is not known, however. Enlarged prostate is very common, affecting a large proportion of elderly men, though in at least half these cases there are no objectionable symptoms.

The effects of prostatic hypertrophy are primarily mechanical. Pressure exerted by the enlarged gland produces an urge to urinate, and the encroachment on the urethra makes the flow of urine difficult. Obstruction of the flow of urine from the bladder, if left untreated, may result in damage to the bladder and kidneys. In the early stage of the disorder there is frequent urination urge, but the flow is feeble. Frequent small amounts of urine are discharged, also during the night. In the second stage the bladder can no longer be completely emptied; a residual quantity or more than 100 ml is retained in it, so that urination fails to relieve the urination urge. Bacteria are liable to infect the stagnant urine remaining in the bladder, causing purulent inflammation. In the third stage the urethra becomes so constricted that the bladder muscles are unable to overcome the resistance to the flow of urine. Under the action of the pressure that now builds up in the bladder the urine forces its way out dropwise in a slow but incessant discharge. Retention of urine is harmful to the kidneys and may cause their failure (Fig. 14-13).

Prostate trouble can be diagnosed by rectal examination, especially if the hypertrophy is located below the bladder (Fig. 14-14). If the growth or tumor, known as a prostatic *adenoma*, grows into the fundus of the bladder (Fig. 14-15), cystoscopic examination may be necessary for diagnosis. When the growth has been diagnosed as benign, hormone treatment may be tried, consisting in the administration of a combination of male and female sex hormones. However, completely successful treatment can be achieved only by surgery (Fig. 14-16). Removal of the enlarged part of the prostate, if undertaken before complications due to urine retention occur, is a fairly simple operation. It is called *prostatectomy*. In inoperable cases the patient can be taught to make regular use of a catheter himself (autocatheterism), or a permanent catheter for the discharge of urine may be fitted.

In a minority of cases, enlarged prostate is due to cancer (*prostatic carcinoma*). The early symptoms are similar to those of nonmalignant hypertrophy, but secondary tumors are liable to develop in other parts of the body (Fig.

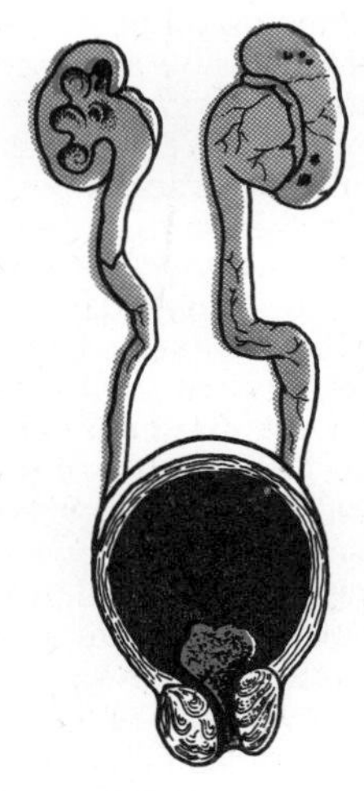

FIG. 14-13.
Retention of urine due to enlarged prostate, resulting in kidney damage.

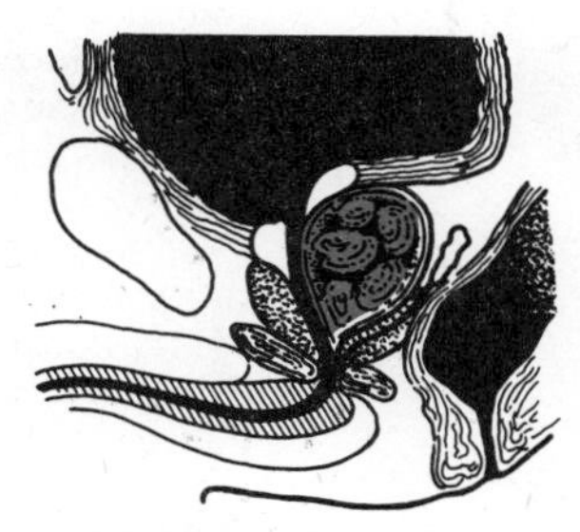

FIG. 14-14.
Prostatic tumor below the bladder.

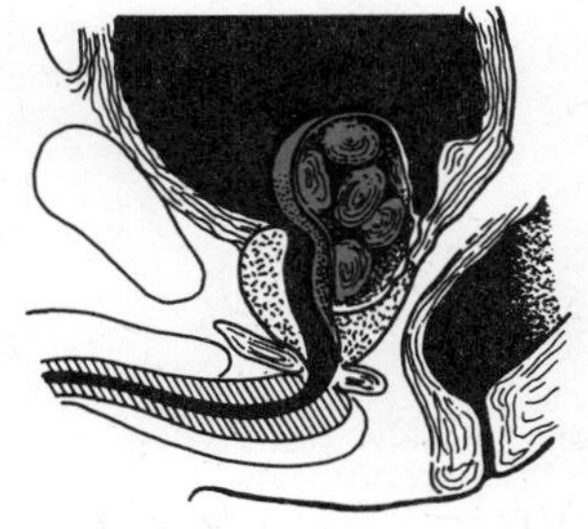

FIG. 14-15.
Prostatic tumor growing into the bladder.

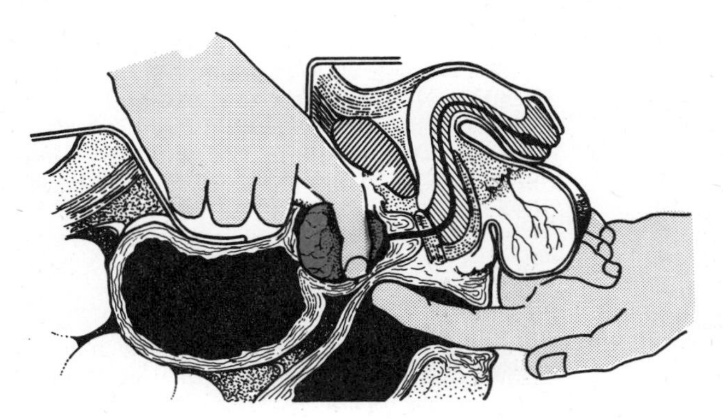

FIG. 14-16.
Surgical removal (shelling out) of the tumor.

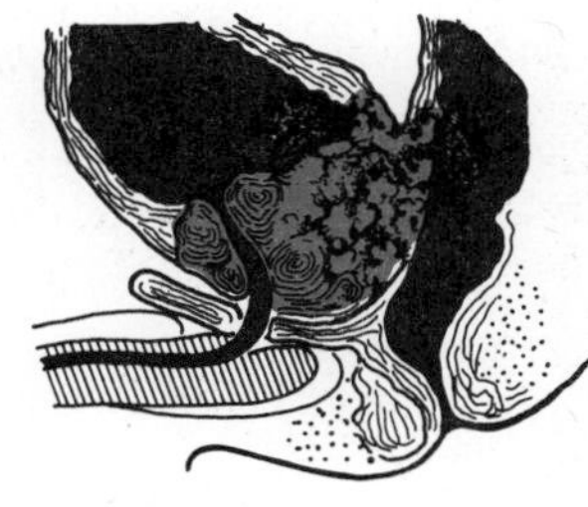

FIG. 14-17.
Cancer of the prostate spreading to bladder and rectum.

14-17). Early diagnosis is possible by rectal examination; the tumor is felt by the examining physician as an irregular hard lump. Microscopic examination of tissue specimens and laboratory tests may be necessary to decide with certainty whether the condition is benign or malignant. Prostate cancer is usually treated by prostatectomy. If this is not advisable, the cancer can be arrested by hormone therapy, which aims at altering the balance of the hormones that regulate the prostate, or by means of drugs such as stilbestrol, a synthetic preparation possessing estrogenic properties.

CANCER

Nearly all tissues of the human body can develop disorders of cell growth that give rise to tumors which can, broadly speaking, be divided into benign (nonmalignant) tumors (see page 477) and malignant tumors, the latter commonly known as *cancer*. Malignant tumors, like benign ones, are what are known as *neoplasms*—spontaneous new growths of tissue forming an abnormal mass with no physiological function—but differ from the benign ones in that they show uninhibited growth,

FIG. 14-18.
Cancer of the uterus may spread into the adjacent tissues and organs and form secondary tumors in other parts of the body.

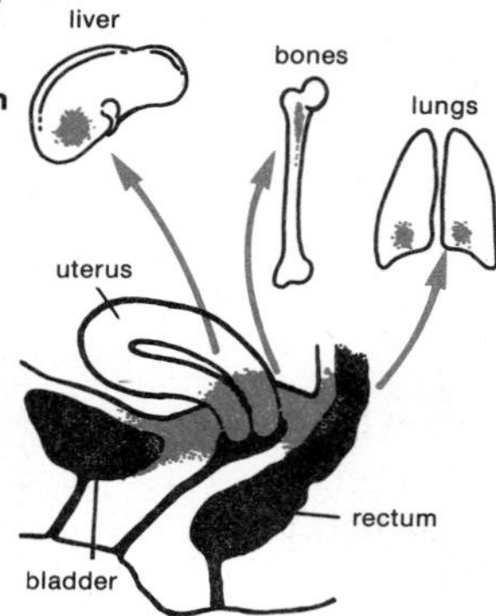

FIG. 14-19.
Location of primary cancer and of secondary tumors arising from it (malignant cells carried by blood and/or lymph).

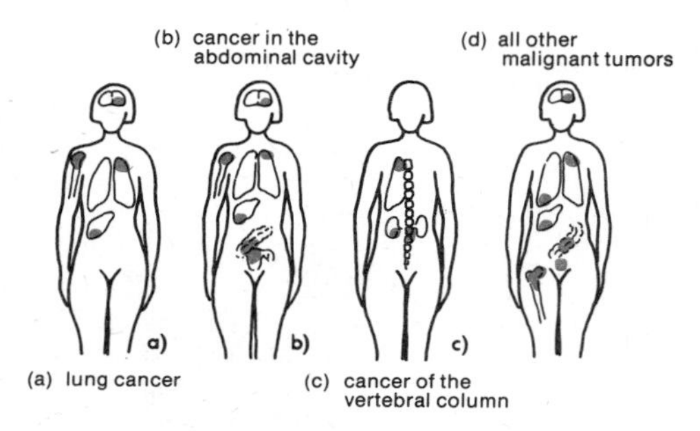

rapidly infiltrating adjacent tissues and organs, disrupting them and their blood vessels so that bleeding occurs, obstructing hollow organs, and being spread by groups of cells which are carried in blood vessels or lymph vessels to other parts of the body where they form secondary growths (*metastases*). Cancer of the uterus, for example, may spread downward into the vagina, rearward into the rectum, or forward into the bladder; at the same time, secondary tumors may arise in the lungs, bones or kidneys (Fig. 14-18). Cancer cells that have entered the bloodstream are found in about 50 percent of all cancer patients. Fortunately, only a small proportion of these migrant cancer cells actually establish themselves as secondary tumors. Surgical treatment of cancer will generally aim at removing it in order to prevent metastasization, taking care not to damage the tumor during the operation.

The location of secondary tumors will depend on the path of migration of the cancer cells from the organ in which the primary growth is located. Lung cancer cells, for example, establish themselves particularly in organs well supplied with blood by the systemic circulation (Fig. 14-19a). Cells originating from cancer in the abdominal cavity and in the region of the spinal column metastasize to particular preferred organs (Fig. 14-19a,b). Cancer that arises in certain organs—lungs, kidneys, breast, thyroid gland, stomach—tends to spread more especially to the bones (Fig. 14-20). All other cancers transmit their cells through the blood circulation to the lungs (Fig. 14-19d). Secondary tumors are seldom found in the spleen and pancreas, which are evidently able to destroy most cancer cells arriving there. Cancer cells can also migrate along the lymph vessels and tend to become trapped in the lymph nodes, where they thus chiefly establish themselves. In assessing how far a cancerous tumor has progressed, it is therefore usual to check whether secondary growths have formed in the adjacent lymph nodes.

Metastasization and uninhibited growth of cancer cells are among the reasons why these malignant tumors are so difficult to eradicate. If the original (primary) tumor is not completely removed by operation or destroyed by radiation therapy, or if cancer cells have already migrated to other parts, a recurrence of the disease is liable to occur. For this reason a cancer is not regarded as cured until five years

FIG. 14-20.
Malignant tumors which often produce secondary growths in the bones.

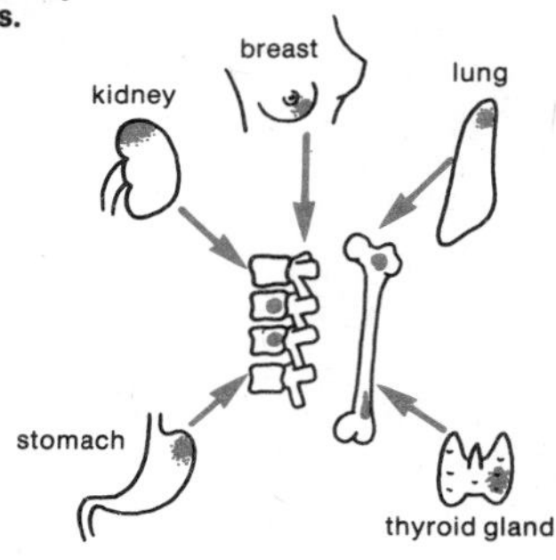

free from recurrence have passed. Figure 14-21 illustrates the aggressively uninhibited growth of a cancerous neoplasm, which starts as a small group of cells or perhaps a single cancer cell. If the neoplasm develops so rap-

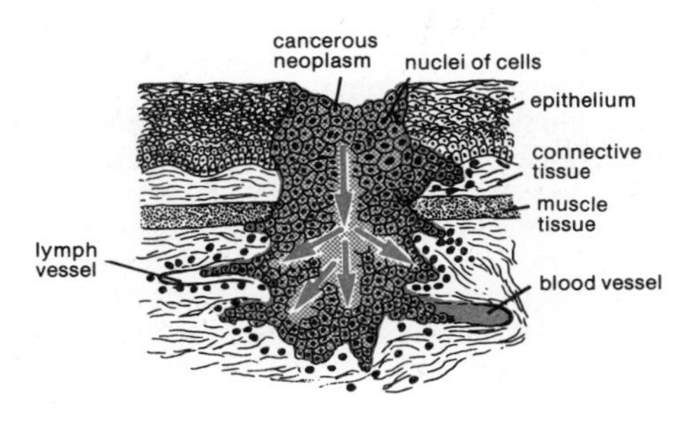

FIG. 14-21.
Malignant proliferation of a tumor, destroying adjacent connective and muscle tissue and penetrating into the blood vessels.

idly that the blood vessels "enslaved" by it are unable to supply its needs, it degenerates and turns into a cancerous ulcer. When that happens, progressive loss of health and general symptoms such as loss of weight and anemia begin to manifest themselves in addition to the local destruction of tissue. Precisely why the cancer should have these serious consequences is not known.

There are two main types of cancer: the *carcinoma,* arising in epithelial tissue of skin, glands and membranes, and the much less common *sarcoma,* arising in connective tissue—bone, muscle, etc. Within each of these two main groups there are a number of varieties.

Although cancer has been known from ancient times, having even been diagnosed in Inca mummies 5000 years old, and is also found among primitive tribes, as well as in animals, it appears to be on the increase in civilized modern societies. There are, however, two factors to be considered in connection with cancer statistics. Cancer is a disease whose prevalence increases with advancing age: between the ages of 40 and 80 the risk of getting cancer roughly doubles with every additional 10 years of age. In the past hundred years the average life expectation of children born in Western countries has almost doubled. With the increasing proportion of older people in the population, the statistical incidence of cancer is bound to increase as well. The second factor is that the diagnosis of cancer has become more reliable; older statistics may be somewhat misleading in that deaths really caused by cancer have been attributed to other causes simply because the cancer remained undetected or was wrongly diagnosed. Nevertheless, allowing for these factors, the frequency of cancer is indeed increasing, especially among men in the 45–65 age group.

The first question that comes to mind is whether this real and largely age-independent rise in the incidence of the disease is attributable to an increase in external causes linked to certain features of life in modern technological societies, or whether there is an increase in inherent (congenital) predisposition to cancer. Systematic efforts are being made to analyze the complex pattern and find the answer to these and allied questions. To this end, it is necessary not only to discriminate between the sexes, but also to draw distinctions as to the location of the cancer and its accessibility and amenability to treatment. As a result of the improved diagnosis and treatment of cancer of the female genital organs, for example, the overall death rate due to this disease in women has been reduced in the past few decades. Cancer of the stomach, liver and bile ducts is also on the decline. Against this, cancer of the respiratory passages and lungs has considerably increased. For example, in central Europe, the incidence of this form of the disease doubled in the decade between the mid-1950s and the mid-1960s. In men it is about six times as frequent as in women and accounts for something like a third of all male deaths from cancer (see also next section). Thus the notable increase in fatal cancer in men is to a substantial extent due to the increase in difficult-to-cure male lung cancer (Fig. 14-25). Smoking is regarded as the main factor in this increase. On the other hand, the consumption of potent alcoholic liquors is believed to promote cancer of the esophagus. Cancer of the penis is said to be less frequent in circumcised men. Cancer of the cervix is allegedly less frequent in women who do not have sexual intercourse and indeed, rather less credibly, in women who cohabit only with circumcised men. It is obviously very difficult to establish real proof for such suppositions and theories. The main difficulty is that there are always a number of variables involved, e.g., congential disposition, incidence of cancer in the community at large, climatic conditions, eating and drinking habits.

As for the causes of cancer, the precise factors are not fully known nor completely understood. For example, it has long been considered that various forms of cancer can be caused by chronic irritation, a term that can be interpreted in a fairly wide sense. This was the prevalent view at one time, but its general validity has been called into question. Ionizing radiation and certain chemical poisons such as coal tar products have also long been known as causal agents of cancer and are, for that reason, referred to as *carcinogenic agents* or *carcinogens*. Tobacco is now considered by many to be the most dangerous chemical carcinogen, statistically speaking, of all.

According to the chronic irritation theory of cancer, the disease is liable to develop as a result of persistent and frequent irritation of certain tissues. Stimuli such as inflammations, ulcers, chemicals, pressure, chafing and heat often give rise to a local increase in cell division even when such stimuli or irritants act only for a comparatively short time. Chronic irritation is, it is thought, liable eventually to trigger off cancerous proliferation of the cells affected. In the disease called schistosomiasis (or bilharziasis), caused by blood flukes (parasitic worms) and occurring in tropical regions, the worm eggs make their way into the bladder, where they set up chronic inflammation which in about 5 percent of cases gives rise to cancer of the bladder. The secretion of the preputial glands of the penis is supposed to cause chronic irritation in uncircumcised men, possibly resulting in cancer of the penis. Inflammatory irritation due to stones in the kidneys or gallbladder is held responsible for cancer in those organs, while a large gastric ulcer that fails to heal properly may be the forerunner of cancer of the stomach. In India so-called dhoti cancer develops on the left-hand side of the body, where people tie the knot securing the garment called a dhoti. Chubta cancer among the natives of Venezuela is caused by cigar burns in the mouth, while changri cancer affecting the Tibetans is caused by repeated burning, year after year, by a small stove which people carry about attached to the body. The pressure caused by parts of the harness on the bodies of draft animals sometimes also produces cancerous degeneration of the tissues so affected. The chronic irritation theory holds that cancer is liable to develop wherever tissues are continually damaged or irritated and are thus compelled constantly to regenerate and renew themselves. According to this view, cancer develops when this renewal process eventually gets out of hand and degenerates into uncontrolled and uninhibited proliferation of cells, e.g., in a chronic gastric ulcer, in skin damaged by tuberculosis (lupus vulgaris), in the tissues of the liver damaged by cirrhosis, or in the cervix of the uterus (Fig. 14-47).

Cancer has many features in common with young tissue, as though the biological clock had been put back: the tissue undergoes rapid growth at a relatively primitive level of cellular development. This fits in with the fact that cancer often occurs in "islands" of immature young tissue that became scattered in the early stages of embryonic growth—got left behind, as it were—and are prone to respond to subsequent growth stimuli by cancerous proliferation. For example, tumors of the testis or ovary are of this general type, also tumors of vestigial bronchial cells (rudimentary embryonic gills), tumors of the connective tissue of the nervous system (neuroglia), and tumors that develop more particularly in the brain, retina, bone marrow and lymph nodes of children (Fig. 14-22).

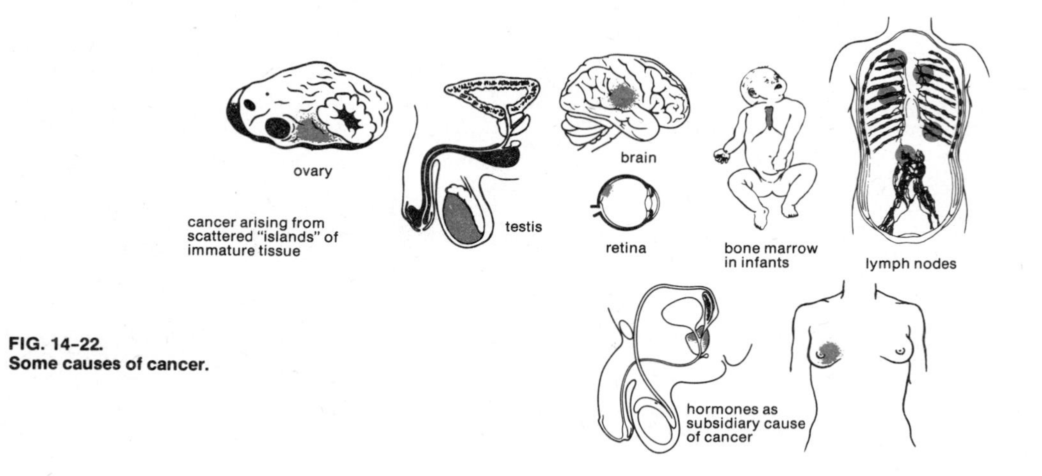

FIG. 14-22.
Some causes of cancer.

Radiation can be applied to the cure of cancer, but it can also cause cancer. For example, overexposure to x-rays or other ionizing radiation such as that from radioactive matter may cause leukemia (blood cancer); in the early days of radiography, before suitable precautions were introduced, many x-ray workers were affected by cancer of the skin. Excessive exposure to strong sunlight may also increase skin cancer susceptibility (see page 498). Whether radiation will cause cancer or will arrest it and thus have a beneficial effect depends on how large the dose is. In general, the cells and tissues that grow and multiply most rapidly are the most susceptible to radiation, which disturbs the growth-regulating system of cells or, if the dose is large enough, suppresses growth altogether. Such cells and tissues are particularly the sex glands, mucous membranes and skin. By the same token, the constantly multiplying cancer cells are more vulnerable to radiation than healthy tissue, which is the reason they can often be successfully controlled by radiation treatment.

According to the hormone theory of cancer, some tumors are hormone-dependent. This is particularly true of cancer of the male prostate gland (see page 482) and probably also, though less specifically, of female breast cancer (see page 503).

Cancer itself is certainly not hereditary, but a congenital predisposition to the disease may be present in certain individuals, families, sexes, age groups, or races of mankind. Such observations apparently suggest that susceptibility to cancer is to some extent hereditary. So far, however, a hereditary pattern has been established only for retinal cancer.

The infection theory of cancer bases itself on the fact that certain virus-infected cells in experimental animals may undergo cancerous degeneration. Since a virus contains nucleic acids and in many respects resembles and behaves like the genetic material in the nuclei of living cells, it may insinuate itself into the cells in such a way as to alter their genetic makeup, resulting in disordered or uninhibited growth.

The chemical theory of cancer is likewise based on animal experiments which have shown that hundreds of chemicals can cause malignant tumors in animals. In human beings cancer of the bladder, for example, is relatively frequent among workers in synthetic dye factories; the tar products of tobacco are liable to cause cancer of the lip in smokers of pipes which are not properly cleaned, etc. Occasional contamination with such carcinogens will not cause cancer; this results from constant or frequently repeated exposure over a long period of time.

In recent years an endeavor has been made to fit these various causes of cancer into an overall theory that can explain cancer in terms of changes in the genetic information passed on by each cell to its successors in the process of repeated cell division. It is supposed that the effects of the various causes of cancer are brought about by falsification of the genetic makeup of the cell, more particularly its genetic information content, thus giving rise to what is known as a mutation, a sudden change in genetic character and behavior patterns. Alternatively, cancer may be due to wrong interpretation of the genetic message.

Every malignant tumor starts as a small hard lump which grows slowly and may subsequently undergo ulcerative degeneration. Pain does not occur until the tumor begins to exert pressure on adjacent tissue. From the age of about 40 onward there is a steadily increasing risk of epithelial cancer (carcinoma)—true cancer—whereas the malignant tumors occasionally occurring in children below the age of about 10 are of the type arising from connective tissue such as muscle, bone, etc. (sarcoma).

The *signs of cancer* (Fig. 14-23) vary according to the location of the tumor and are, of course, more likely to be noticed where the cancer visibly manifests itself on the skin or mucous membranes (see page 498). Medical advice should promptly be sought if suspicious indications occur and especially if the symptoms are persistent or recurrent and will not respond to simple treatments.

Any birthmark or other darkly pigmented area that grows in size or even itches must be regarded as suspect; so must any lump that grows slowly, undergoes ulcerative degeneration and refuses to heal properly; growths of this latter type most often develop on the face or on the back of the hand (cancer caused by overexposure to strong sunlight, sometimes encountered in farmers, seamen, etc.; see pages 498 ff). A cancerous growth on the lip often gives rise to whitish discoloration and thickening of the mucous membrane. The habitual smoking of a dirty pipe or the regular and frequent tasting of hot foods or drinks over long periods of time are also believed to promote lip cancer. Early signs of cancer of the tongue include whitish discolorations, fol-

FIG. 14-23.
Early signs of cancer.

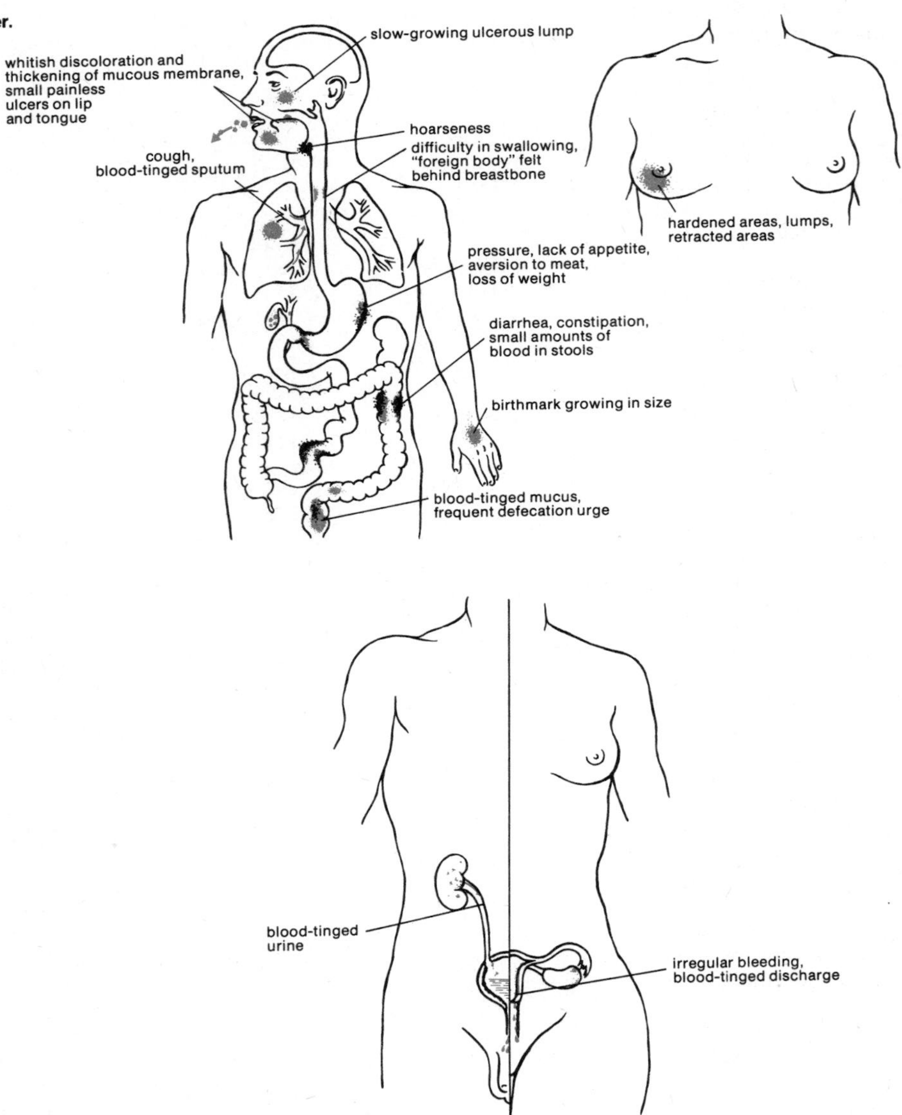

lowed by small painless ulcers which develop particularly in areas subject to chronic rubbing action and which are reluctant to heal. Difficulty in swallowing solid foods may indicate cancer of the esophagus, together with a sensation as of some "foreign body" behind the breastbone and, subsequently, also pain. This type of cancer supposedly occurs most frequently in persons who regularly consume strongly alcoholic liquors.

Cancer of the stomach can arise in consequence of a chronic gastric ulcer. Symptoms include a sensation of pressure and repletion quite soon after the start of a meal, lack of appetite, eructation (belching), aversion to meat and bread, followed eventually by anemia and loss of weight. Early x-ray examination and the earliest possible diagnosis are especially important in dealing with stomach cancer.

Intestinal cancer may cause discharge of mucous matter and, more particularly, also diarrhea or constipation, sometimes in alternation; discharge of blood can often be detected only by laboratory tests.

Cancer of the rectum is often mistaken for bleeding piles. Discharge of reddish mucus-containing fluid with the stools or with flatus, frequent defecation urge without emptying the bowel, and discharge of pencil-thin feces must be viewed with suspicion (see page 496). Cancer of the gallbladder and bile ducts is at first difficult to distinguish from the pain caused by ordinary gallstones, a condition that, incidentally, may be a forerunner of cancer (see page 196).

Malignant tumors in the kidney often cause the presence of blood in the urine; this is a symptom for which medical advice should be sought without delay. Cancer of the bladder is liable to be mistaken for a very persistent bladder catarrh with frequent urination (see page 504). Prostate cancer is a typical disease of elderly men, particularly after the age of 50. It causes increased urination urge, with frequent discharge of small amounts of urine, and is liable to be mistaken for benign enlargement of the prostate gland (see page 482). Ulcers of the penis may occasionally degenerate into cancerous tumors.

Abdominal cancer in women usually occurs as cancer of the uterus, causing irregular menstrual discharge, brownish discharge containing blood, and bleeding after sexual intercourse. Uterine bleeding and discharges occurring after the menopause are particularly suspect (see page 499). Cancer of the breast manifests itself as a hard lump in the mammary gland, causing distortion of the nipple and sometimes also producing a running sore (see page 503).

Cancer of the thyroid gland must be suspected when parts of a goiter become hard (see page 353). The larynx may also be affected by cancer, the symptoms being persistent hoarseness and coughing, often mistaken for "smoker's cough" or attributed to overstraining the voice. Later, there is a sensation of pressure.

Lung cancer (actually: bronchial cancer) causes coughing in about 80 percent and blood-tinged expectoration in about 60 percent of cases; respiratory trouble and pain are less frequently encountered.

If cancer is diagnosed in an early stage and treated without delay, the prospects of a cure are very much better than in a later stage. An important factor in the prevention of cancer is precautionary examination. A second factor would, ideally, be true prophylaxis by elimination of all causes of cancer and prevention of cancerous degeneration. Unfortunately, this is not possible, because the cancer hazards are not fully known and because the actual development process of the disease is only imperfectly understood. Hence cancer prevention must, for the present, aim at avoiding known or suspected cancer hazards. These include chronic irritation of tissues, overexposure to harmful radiation, many chemicals, overindulgence in tobacco, and probably also the increasingly heavy pollution of the atmosphere. Public health legislation in developed countries goes a long way toward eliminating some of these hazards: laws for the prevention of pollution (clean air, control of harmful wastes and effluents, etc.), laws for protection against radiation hazards, industrial hygiene regulations, regulations controlling the use of preservatives in food, etc. For instance, the observed decrease in cancer of the stomach, esophagus, mouth and throat is attributed to, among other factors, the enforcement of legal control over food preservatives. Preventive operative removal of chronic irritants—ulcers (especially gastric ulcers) or organs affected by stones (e.g., the gallbladder)—also reduces the risk of cancer.

Precautionary examination has long shown itself to be beneficial in circumstances where reliable and convenient cancer detection methods can suitably be applied to certain sections of the population. Foremost among such precautionary procedures are checks for cancer of the uterus (see page 499), which claims many thousands of victims each year. It is, of course, important to look out for early warning signs; these include irregular and/or increased menstrual bleeding, blood-tinged and possibly purulent discharge between menstrual periods, bleeding after sexual intercourse, or bleeding that occurs after the menopause. These symptoms, however, often do not manifest themselves until the cancer has firmly established itself. Early detection of cervical cancer is of particular importance in this context (see page 499). In almost 1 in every 100 women who, without having any symptoms, are examined for cancer there is found to be cancer in a preliminary or sufficiently early stage to enable the disease to be cured with a high degree of probability by surgical intervention. Hence it is particularly advisable that women of high-risk age should have themselves examined for cancer. Such an examination can reveal vaginal cancer, cervical cancer and—somewhat less reliably—cancer of the uterine cavity; early warning signs of breast cancer may also be detectable. With suitable instruction, women can also examine their own breasts for cancer (presence of a lump, contraction of the skin, etc.; see also page 503). Many early cancers can also be found in the course of general medical checkups: blood and urine tests, x-ray examination of the chest, bones and digestive system, tests on the gastric juice and feces, rectal examinations, etc. Even so, such tests and examinations are less reliable and the ratio of results to effort is lower than in tests for cervical cancer in women.

With modern treatment, about 20–30 percent of all cases of cancer, considered statistically as a whole, are curable. The most important precondition for successful treatment is early diagnosis of the disease so as to render it harmless before it has time to spread to tissues and organs other than those already affected. As relapses and recurrences are common in cancer, the guiding principle of treatment has to be "as radically as necessary, but as sparingly as possible." The most effective weapons for combating the disease still are surgery and radiation treatment (radiotherapy). In recent years chemotherapy has also come to the fore. In many cases a combination of these methods will offer the best prospects of arresting cancer.

The object of *surgical treatment* is to eliminate the cancer by its complete removal from the healthy tissues. When the cancer is detected in an early stage, the aim will be, if possible, to remove the entire affected organ together with adjacent lymph nodes. With modern surgical techniques and anesthetics it is possible to perform major cancer operations even on old people. Typical operations are the removal of the uterus (hysterectomy) for dealing with uterine cancer (see page 499); partial removal of the stomach for dealing with stomach cancer, which shows a 40–50 percent cure rate for cases treated in an early stage, but an average rate of only 5–8 percent (page 494); removal of a breast for breast cancer, with an 80 percent success rate for early diagnosis and 50 percent if the disease is detected late (page 504); removal of a lobe of a lung in lung cancer, an operation that is successful in about 60 percent of early cases, but offers an average cure rate of only 7–10 percent (see next section); removal of the rectum for the treatment of rectal cancer, with formation of a new outlet, offering up to a 70 percent chance of success (page 496). Sometimes a second operation will be necessary in order to remove secondary tumors or to deal with recurrence of the cancer; more often, however, such cases are treated by radiotherapy and chemotherapy.

Radiotherapy (radiation treatment) utilizes the fact that rapidly growing young tissues, more particularly cancerous tissues, are more sensitive to radiation than mature tissues. The difference is not very great, however, and for this reason the treatment of cancer often requires high radiation doses which are to some extent also harmful to normal healthy tissue. With special arrangements it is possible to direct and focus the rays to various depths within the body. With x-rays this is achieved by using high-voltage x-ray tubes, betatrons or linear accelerators which produce deep-penetrating "hard" rays. Radioactive cobalt can similarly provide a source of high-energy radiation. Radioactive isotopes can alternatively be used internally. For example, slender containers for radium (so-called radium needles) or other radioactive substances are inserted into tissue or into cavities of the human body in order to kill malignant cells (cf. cancer of the uterus, page 499). Even better selective and directed action is achieved when a radioactive isotope, on being injected into the bloodstream, for example, is absorbed into and concentrated in the organ or gland affected by cancer. Thus radioactive iodine is concentrated in the thyroid gland and combats cancer of this gland; it is also absorbed into secondary tumors arising from such cancer.

The third general type of treatment for cancer is by *chemotherapy*, i.e., the application of chemical agents which are not harmful to the patient but which have a specific suppressive or destructive effect on cancer cells. The problem is similar to that of radiotherapy: cancer cells are not bacteria with peculiarities of metabolism that make them susceptible to drugs, but are degenerated cells of the human body itself which differ little in their intrinsic character from normal cells. Cancer cells appear to be most vulnerable in their tendency to undergo rapid cell division. Not only radiation but also toxic substances that suppress division of cells can strike at this aspect of cancer. The drawback of cancer treatment with such cytostatic drugs is that they also attack normal healthy tissue in process of rapid growth, such as the blood-forming bone marrow, the intestinal mucous membrane, and the gonads. It is nevertheless hoped that, with improved understanding of the nature and causes of cancer, it will eventually become possible to produce drugs that can distinguish between healthy tissue and malignant tumors. Already some notable successes are achieved with drugs, e.g., in the treatment of leukemia in children and of hormone-dependent cancer of the prostate.

Lung Cancer

Lung cancer is on the increase in all countries. Its incidence is estimated to have gone up about 40-fold in comparison with the year 1900. In the main, the victims of lung cancer are men above the age of 50 (with 60 as the age of maximum frequency). Whereas cancer of the bladder, rectum, liver, stomach and prostate used to be the prevalent types, lung cancer has outstripped them all as the commonest male cancer. This form of the disease is less frequently encountered in women. Notable, too, are the great differences in lung cancer incidence in different parts of the world. Leading in this particular statistic is West Berlin, where in 1965 the disease claimed 111 deaths per 100,000 inhabitants (Fig. 14-24). This high incidence is presumably due to the fact that this city has a population in which the older age groups are disproportionately large. At the other end of the scale are countries like Spain, Portugal and Poland, the last with a mere 4 lung cancer deaths per 100,000. Such geographic differences are attributed mainly to differences in tobacco consumption and to differences in the methods of processing the tobacco for smoking.

Although the overall picture is not without some conflicting features, most medical experts now accept the existence of a close relationship between lung cancer and cigarette smoking. It is indeed remarkable that the increase in the disease follows a curve that closely reflects the increase in smoking (Fig. 14-25). The relatively low frequency of the disease in women is attributed to the fact that up to about 15 or 20 years ago they smoked less than men, so that they have, statistically speaking, not been exposed to the hazard to the same extent as men. In the United States, it is reported that lung cancer is increasing in women, too. Also, it is asserted that women do not inhale so deeply as men, smoke fewer cigarettes per day, and start smoking later in life. It has moreover been suggested, though without reliable evidence, that women are simply less predisposed to cancer. Inhaling is undoubtedly an important factor, as appears from the much less frequent occurrence of lung cancer in cigar and pipe smokers, who seldom inhale. Carcinogens have been isolated from the combustion products of tobacco and of cigarette paper. Filter tips give some protection, but not enough to make a major difference in the cancer risk. In general, it is advisable not to smoke cigarettes down to the smallest possible butt end before throwing them away. In persons who smoke up to 20 cigarettes a day the statistical incidence of lung cancer is about 4 times as high as in nonsmokers, while in those who smoke upward of 40 a day the cancer hazard shows a 40-fold increase. Smokers as a group run a 30 times higher risk of lung cancer than nonsmokers. Atmospheric pollution is also blamed for causing lung cancer, but according to some experts it accounts for no more than about 10 percent of such cases. The lower incidence of lung cancer in rural populations is attributed to lower tobacco consumption.

According to other investigators, there are, besides cigarette smoking, various contributory factors to lung cancer in industrialized countries, especially among city dewllers, e.g., carbon particles in the atmosphere (smoke), gases (especially the exhaust gases from motor vehicles), carcinogenic constituents in certain foods and food preservatives. Even the high rate of meat consumption in Western societies has come under suspicion as a cause of cancer, though such relationships based on statistics must always be viewed with skepticism until underlying causes have been established. On the other hand, there is little doubt that ionizing radiation, including x-rays and the emissions from radioactive substances, can cause cancer; for instance, quarry workers who have to deal with rock containing uranium run a high risk of lung cancer from inhaled particles. Chronic bronchial catarrh is also believed to be a potential cause of lung cancer as a result of continual irritation of the bronchial mucous membrane (Fig. 14-26).

Whereas lung cancer used to be almost invariably incurable and fatal, at present about 7–10 percent of those affected can be saved. If

FIG. 14-24.
Deaths from lung cancer per 100,000 inhabitants (1965).

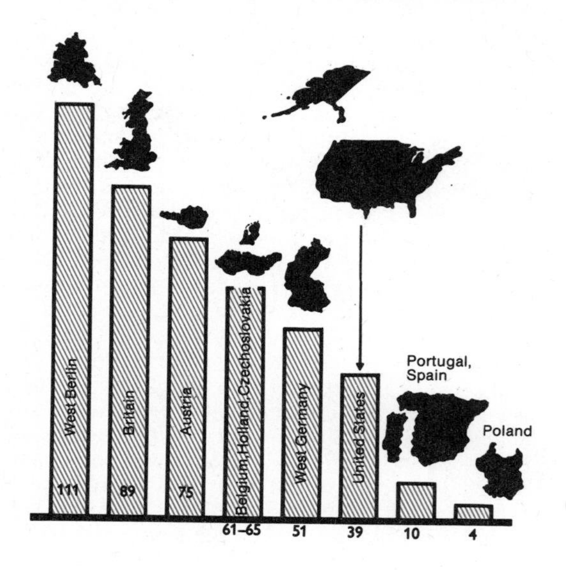

FIG. 14-25.
Lung cancer and cigarette production (according to Swiss statistics).

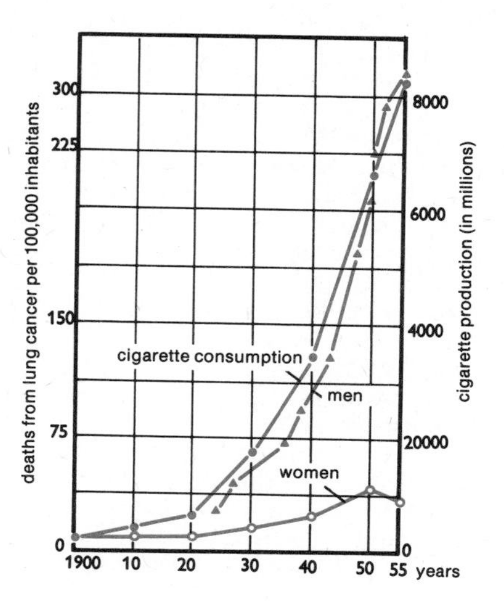

FIG. 14-26.
Lung cancer risks.

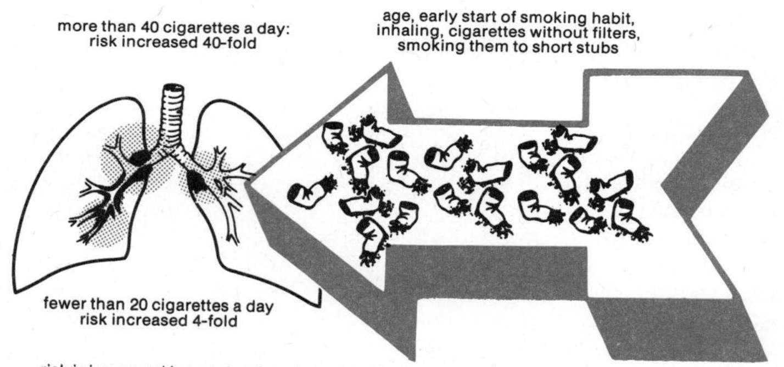

the disease is diagnosed in an early stage, e.g., in the course of a routine x-ray examination, the prospects of successful treatment are reasonably good, with a cure rate of up to 60 percent. Early diagnosis is therefore very important. Unfortunately, this is made difficult by the location and the manner of development of the disease. In 95 percent of lung cancer cases the condition is actually bronchial cancer arising at the terminal bronchioles, where it proliferates malignantly and uninhibitedly, destroying alveoli, invading alveolar ducts and penetrating into lymph vessels and blood vessels (Fig. 14-27). Secondary tumors, which continue to grow independently, develop in adjacent lymph nodes (in nearly 80 percent of cases) and in organs located downstream in the blood circulation—in the liver, the brain, the kidneys, the adrenal glands, the bones and especially the spinal column (in more than 60 percent of all cases) (Fig. 14-28). Displacement and destruction of alveolar ducts deprive adjacent tissues of communication with the air entering the lungs, so that such tissues become

FIG. 14-27.
(a) Various forms and consequences of lung cancer.

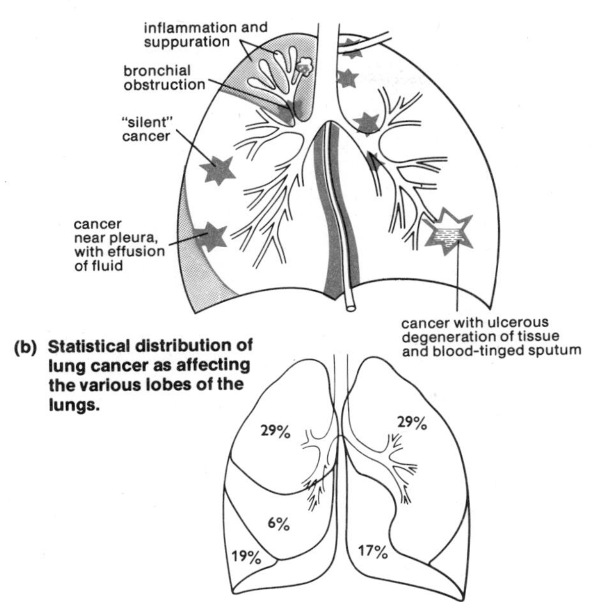

(b) Statistical distribution of lung cancer as affecting the various lobes of the lungs.

FIG. 14-28.
Secondary tumors associated with lung cancer.

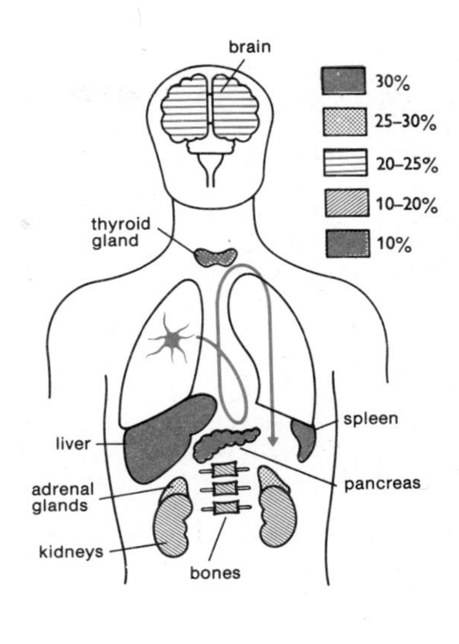

chronically inflamed and purulent. Death occurs as a result of internal hemorrhage, asphyxia, general wasting, or the effects of secondary tumors.

The symptoms of lung cancer correspond to the course that the disease generally runs. Coughing, usually the first sign of lung cancer, is caused by the irritation set up by the cancer in the bronchial tree. A dry cough without apparent cause, a persistent cough that continues after influenza or a cold, or a severe "smoker's cough" must be viewed with suspicion. Another sign of possible lung cancer is purulent or blood-tinged expectoration, which usually causes the patient to seek medical advice. Pain does not occur until the cancer reaches the pleura or exerts pressure on nerve tracts. Comparatively early warning signs also include a sensation of constriction, discomfort, an indefinite feeling of pressure or the presence of a "foreign body" in the chest. Secondary infections of the lungs may be mistakenly attributed to tuberculosis by producing a slight fever in combination with coughing. Shortness of breath does not become very noticeable until a fairly substantial portion of the lungs has been put out of action. In an advanced stage of the disease there are fever, a harassing

FIG. 14-29.
Symptoms of lung cancer: relative frequency of occurrence and time taken for them to develop.

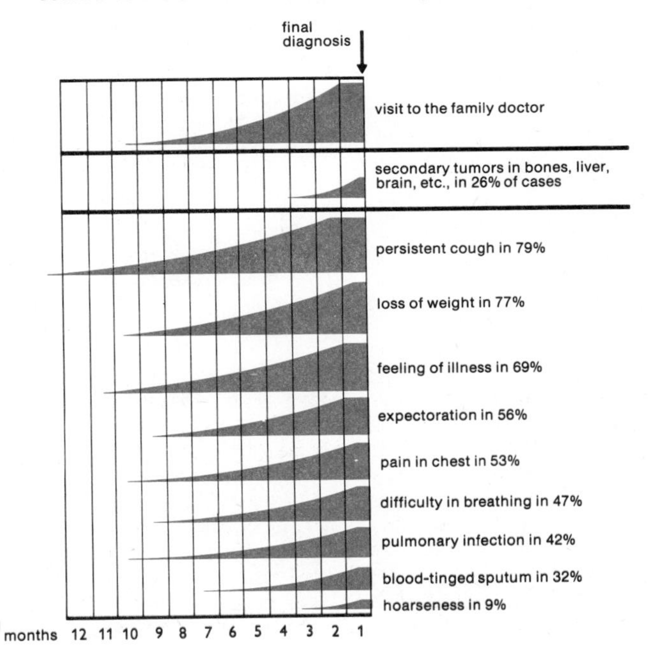

cough, expectoration, weariness, loss of weight, and general physical deterioration.

Various routine checks are available for the early detection of lung cancer. X-ray examination, in particular, can often reveal small centers of cancerous activity, which can be surgically removed, offering a fair chance of recovery. More precise information is provided by tomographic x-ray techniques, showing detailed images of structures in a selected plane of tissue by blurring the images of structures in other planes. If cancer is suspected, the bronchial tree can be made to show up clearly in x-ray photographs by suitable preparation with contrast media (radiopaque substances) or be directly inspected with the aid of a bronchoscope, an instrument for visual examination of the interior of a bronchus, which also has attachments for the painless removal of tissue specimens for microscopic examination.

When lung cancer has definitely been diagnosed, treatment will generally take the form of an operation or radiotherapy. Chemotherapy is used more particularly for dealing with secondary tumors. Surgery, which is still the preferred method of treating lung cancer, consists in the removal of the affected lobes.

Stomach Cancer

Cancer of the stomach (gastric cancer) occurs mainly between the ages of 40 and 60; it is twice as common in men as in women. It used to be the commonest form of male cancer, but has since been outstripped by lung cancer. In women, abdominal and breast cancer are far more frequent.

As with malignant tumors in general, the causes of stomach cancer are by no means definitely known. Some carcinogenic factors have been identified, however. Thus, cancer frequently attends inflammation of the gastric mucous membrane associated with contraction of tissue; persons afflicted with pernicious anemia (see page 107) run a high cancer risk (up to 15 percent); chronic gastric ulcers may be forerunners of cancer. It is estimated that 10 percent of gastric ulcers subsequently undergo cancerous degeneration and, conversely, that about 17 percent of all stomach cancers arise from such ulcers (see page 161). A congenital predisposition to cancer is believed to exist in some persons.

Unfortunately, the early signs of stomach cancer are so inconspicuous that in many cases the disease remains undetected until it has reached an advanced stage, when it is generally too late to operate effectively. For this reason, persons of high-risk cancer age should pay attention to even comparatively minor signs that could be an indication of stomach cancer. If a person who has had a chronic gastric ulcer for several years unaccountably loses much weight, has attacks of vomiting, suffers from tiredness, loss of appetite and aversion to certain foods (meat, for instance), he should seek medical advice without delay. Cancer of the stomach may, however, also

FIG. 14-30.

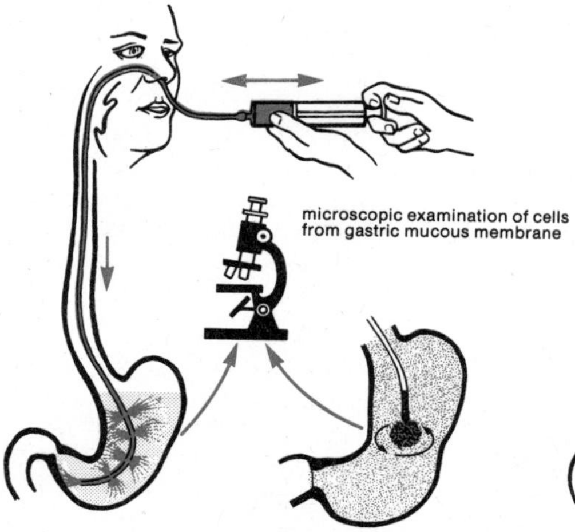

microscopic examination of cells from gastric mucous membrane

(a) Gastric lavage. **(b) Swabbing the stomach with foam-rubber swab.**

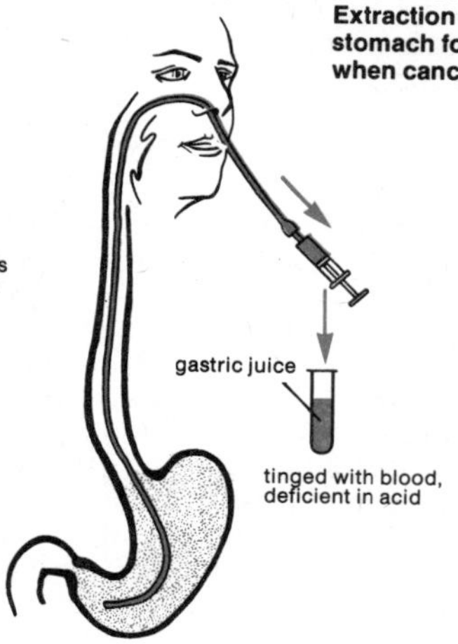

FIG. 14-31. Extraction of contents of stomach for examination when cancer is suspected.

gastric juice

tinged with blood, deficient in acid

FIG. 14-32.
Sites of stomach cancer.

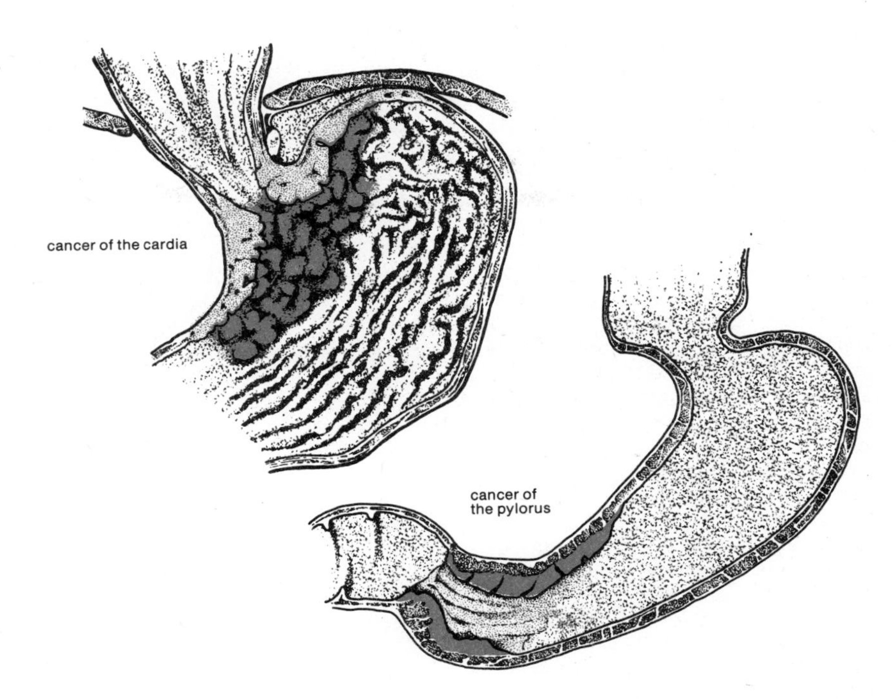

arise without being preceded by an ulcer and may at first produce no specific symptoms other than dyspepsia, poor appetite, aversion to meat, and a sensation of pressure and repletion. In a more advanced stage, vomiting and loss of weight occur as additional and more acute symptoms.

Stomach cancer is diagnosed by x-ray examination (with the aid of contrast media) or by gastroscopy, i.e., visual examination of the interior of the stomach with an instrument called a gastroscope. It may, however, not be easy to distinguish between a chronic ulcer and a cancer that has undergone ulcerative degeneration. In such cases it may be necessary to obtain a tissue specimen, either by swabbing (Fig. 14-30b) or by gastric lavage (Fig. 14-30a). On extraction, the contents of the stomach are, in a case of cancer, found to be deficient in acid and to be tinged with blood (Fig. 14-31). A test for the presence of blood in the stools will be positive if cancer occurs.

Treatment of stomach cancer will consist in surgical removal of the cancer, if possible. For operable cases the cure rate is over 20 percent; if the disease is diagnosed in an early stage, when the cancer has not had time to spread to the adjacent lymph nodes, it is over 45 percent. Depending on the location of the cancer, different portions of the stomach are removed (Fig. 14-32). Since many cancers are located in the vicinity of the pylorus, treatment will often consist in the surgical removal of about two thirds of the stomach (see Fig. 5-25).

Cancer of the Large Intestine

Nearly half of all malignant tumors arise in the alimentary tract, particularly the esophagus, the stomach and the large intestine. About 13 percent of all cancers are intestinal, and the great majority of these are located in the large intestine (cecum, colon and rectum), very seldom in the small intestine (Fig. 14-33). Men are more frequently affected than women, especially between the ages of 50 and 70.

If the cancer is located high up in the large intestine, the first symptoms are generally due to the narrowing of the intestine caused by the tumor. Stools become irregular; constipation alternates with diarrhea. At first, there is no pain. When severe blockage of the intestine develops, gases and feces accumulate in it, giving rise to colic-type pains in the region of the cecum, irrespective of the location of the actual tumor. Since ordinary constipation often produces similar symptoms, they can be regarded as possibly indicative of cancer only when they arise in persons of late middle age who have hitherto had regular bowel habits. If blood, pus and mucus are discharged, medical advice will in any case have to be sought. In a more advanced stage of the disease the symptoms are loss of appetite, loss of weight and general malaise. In some cases sudden occlusion of the intestine occurs as a result of an annular carcinoma which completely closes the passage, with acute symptoms. Cancer high up in the large intestine is treated by surgical removal of the diseased portion and reuniting the two ends.

The commonest intestinal cancer is cancer of the rectum (rectal carcinoma) (Fig. 14-33). The first symptom is small discharges of mucus or frequent urge to evacuate the bowel. In a more advanced stage the mucus discharges contain blood, the evacuation urge becomes very strong, with alternate diarrhea and constipation. The feces may emerge pencil-thin or ribbonlike. In most cases of cancer of the rectum there is no pain as an early sign; only if the tumor is located low down in the rectum is there sometimes pain in an early stage of the disease. Discomfort in sitting and nerve pain develop later, when the cancer penetrates into surrounding tissues. Occlusion of the rectum occurs in some cases. In contrast with hemorrhoids (piles), with which the early symptoms are sometimes confused, the symptoms of rectal carcinoma grow progessively more severe and acute as time goes by. Periods free from objectionable symptoms, such as usually occur in cases of hemorrhoids, are rare in persons affected by cancer.

The presence of blood in the stools should never be ignored; medical advice should be obtained. About 70 percent of tumors can be detected by rectal examination by insertion of the physician's finger, and nearly all tumors of the rectum can be detected by inspection with the aid of a rectoscope. Microscopic examination of a tissue specimen enables a definite diagnosis to be established.

Treatment is by surgery. If the tumor is more than about 15 cm inward from the anus, it will not be necessary to remove the anal sphincter (Figs. 14-34 and 14-35). Operative treatment of a tumor lower down in the rectum (Fig. 14-36) will, however, generally involve removal of the sphincter, in which case an artificial anus has to be formed by making a permanent fistula between the bowel and the abdominal wall; this operation is called *colostomy* (Fig. 14-37). A plastic bag is worn against the artificial anus, held in position by a

FIG. 14-35.
Frequency of cancer in various parts of the intestine.

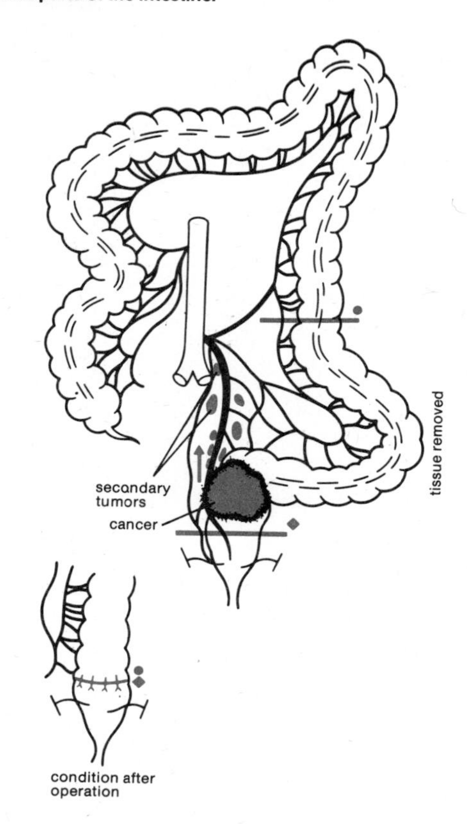

FIG. 14-33.
Cancer of the rectum about 20 cm from the anus.

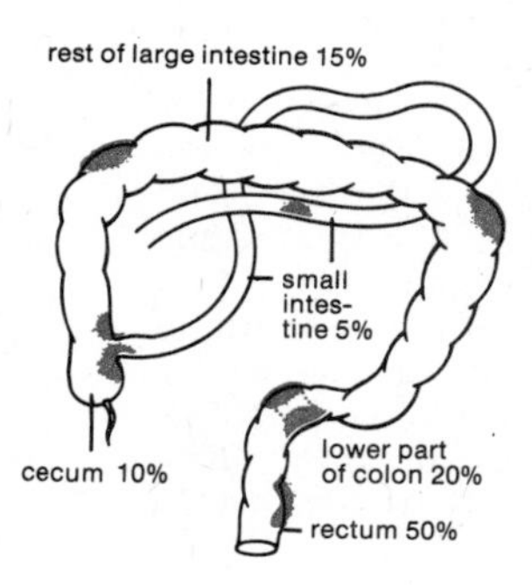

FIG. 14-34.
Operation for cancer high up in the rectum.

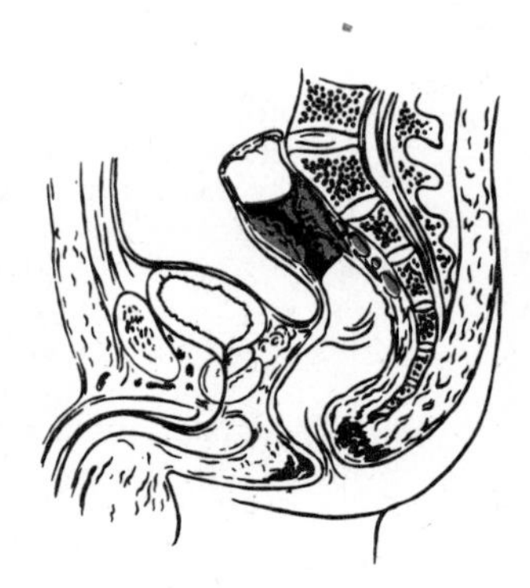

FIG. 14-36.
Operation for cancer low down in the rectum.

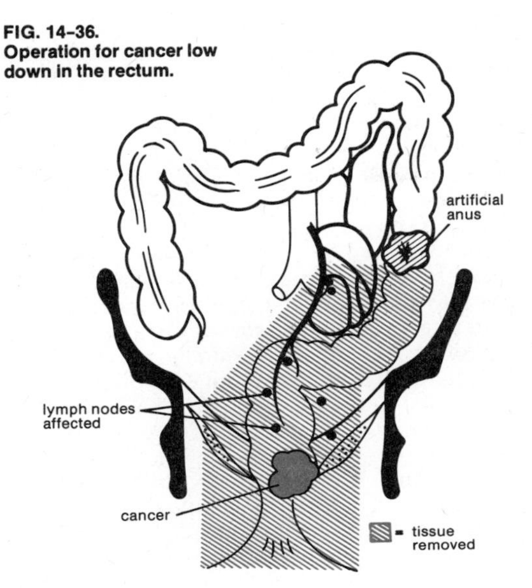

FIG. 14-37.
Colostomy, with artificial anus and plastic bag to receive feces.

special appliance, and receives the discharge of feces. With proper care this arrangement need give no offense or discomfort. Many patients learn to control the bowel action, so that discharge takes place only at fairly regular intervals.

Cancer of the Larynx

Something like half the malignant tumors arising in the ear, nose and throat region are laryngeal carcinomas. This form of cancer is encountered chiefly in older men, and has become increasingly common in the past fifty years or so, which is due not merely to more reliable diagnostic methods, but certainly also to external toxic factors, most important among which is the carcinogenic effect of cigarette smoking. Cancer of the larynx may be preceded by benign tumors and chronic inflammation of the larynx.

Visual inspection of the larynx is possible by laryngoscopy. Direct laryngoscopy is done under local anesthetic with an optical instrument called a laryngoscope, which is equipped with an illumination and magnification system; the patient is in the reclining position during the examination (Fig. 14-38). A simpler technique is that of indirect laryngoscopy with the aid of a mirror (Fig. 14-39) while the patient is seated. When a provisional diagnosis has been established by laryngoscopy, examination of a tissue specimen can provide the information for a definite diagnosis.

The symptoms and the possibilities of treatment of cancer of the larynx vary greatly according to the location of the tumor. Cancer of the vocal cords has a relatively favorable prognosis, as its presence is usually manifested in an early stage by persistent hoarseness. Besides, this part of the larynx contains few lymph vessels, so that spreading of the cancer by metastasis (migration of cancer cells) does not occur until relatively late. Treatment con-

FIG. 14-38.
Direct laryngoscopy.

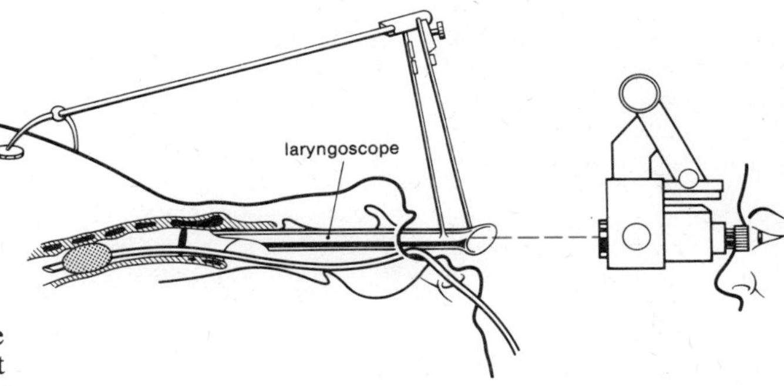

FIG. 14-39.
Indirect laryngoscopy.

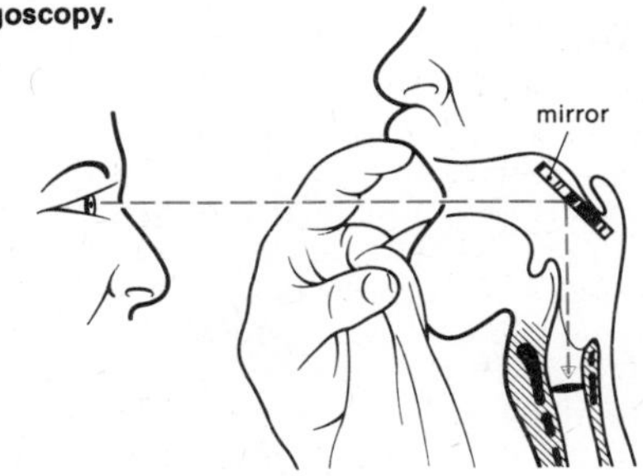

FIG. 14-40.
Possible sites of cancer of the larynx.

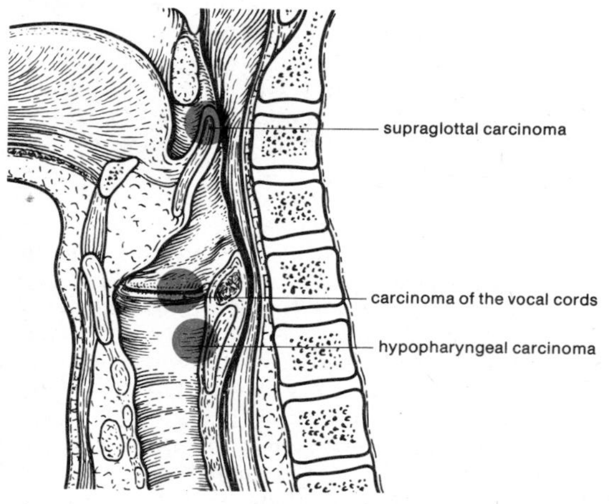

sists in the surgical removal of the vocal cord affected, partial removal of the larynx, or radiotherapy with radioactive cobalt. The symptoms of supraglottal carcinoma (cancer above the vocal cords) (Fig. 14-40) manifest themselves only in a relatively late stage (uncharacteristic sensation of pressure, hoarseness, husky voice), whereas formation of secondary growths in the lymph nodes of the throat occurs quite early. Treatment will generally consist in removal of one side of the larynx and, in advanced cases, additional removal of adjacent tissues of the throat. Hypopharyngeal carcinoma (cancer in the lower part of the larynx) (Fig. 14-40) has an unfavorable prognosis. In about 70 percent of such cases secondary growths will already have formed when the cancer first manifests itself by difficulty in swallowing and a sensation as though a "foreign body" were present in the esophagus. Treatment will involve total removal of the larynx and adjacent tissues or radiotherapy with radioactive cobalt.

Skin Cancer

Cancer of the skin is characterized by proliferation of skin cells causing displacement of normal tissue. Rapidly growing cancerous tissue degenerates, giving rise to ulcers, sometimes complicated by infections, while the breakdown products of the destroyed cells may have toxic effects. There are various types of skin cancer, differing in their origin and dangerousness. They are usually preceded by certain precancerous degenerative changes which, over a period of months or years, will possibly, probably or certainly develop into malignant tumors. The degree of probability with which malignancy will arise differs from one type of precancerous condition to another.

Changes of the skin caused by x-rays, by overexposure to strong sunlight over a period of years, or by incurable wounds and ulcers may be forerunners of cancer. Thus, the brown wartlike discolorations on the hands of seamen or agricultural workers may degenerate into malignancy; ulcerous varicose veins may likewise become cancerous after a number of years. Other potentially precancerous conditions are whitish cornified (horny) stains, especially on the lip, the tip or the edge of the tongue; they may occur in pipe smokers as a result of pressure exerted by the stem of the pipe. Scars that are frequently injured or scratched open, even the familiar brown birthmarks, may become malignant. Truly precancerous conditions of the skin are the discolorations that occur with advancing age. In some persons they may appear already at the age of 40 or so, but mostly they come in old age, and especially occur on those parts of the skin that are exposed to the light, i.e., the face, back of the hand, edge of the ear, neck, bald head, etc. These aging discolorations are surrounded by an inflamed reddish area, itch frequently, are often scratched open, and form scabs under which the skin becomes thickened. Eventually a small ulcer, at first harmless, develops in the middle of the thickened area and degenerates into skin cancer.

Another, though less frequent, precancerous condition of the skin is *erythroplasia,* which occurs as a small scaling node or lump on the face, trunk or genital organs. Lesions of this type may, for instance, occur on the penis even in fairly young men and, after developing slowly for a number of years, suddenly become more active in middle age. The risk associated with erythroplasia is that the condition may not be taken seriously and effectively treated until malignancy is well established, with secondary growths and involvement of adjacent lymph nodes. Treatment will then require extensive surgery, possibly with amputation of the penis. A somewhat similar condition, called *Paget's disease of the nipple,* may develop on the areola of the female nipple and is characterized by eczemalike reddening, scaling and slight weeping (exudation). Such lesions will not yield to the usual treatments with ointments, etc.; instead, they spread from the areola to the surrounding skin. The nipple itself becomes retracted, while the affected breast may be enlarged. This condition eventually becomes cancerous.

A *melanoma* is a malignant pigment spot, usually of a brown color, occurring in elderly people, often on the cheek. It develops from a precancerous condition characterized by brown pigment areas of the skin. These can be successfully treated by radiotherapy. If left untreated, the melanoma that may then develop is dangerous because cancer cells from this tumor spread quickly to other parts of the skin and may also metastasize to vital organs. Treatment of the melanoma will require radical surgery supplemented by radiotherapy.

Other types of skin cancer are the *basiloma,* which arises from the basal cells of the germinal layer and is relatively benign, and the *spinaloma,* arising in the prickle cell layer of the epidermis. The latter type of tumor may grow rapidly and is relatively unresponsive to radiotherapy; it may be caused by frequent contact with carcinogens such as bituminous materials (tar, pitch), crude paraffin, or soot, and occurs as an occupational disease in persons working with such materials.

Prevention of skin cancer will consist in the treatment by surgery or radiotherapy of any suspect birthmark, mole or discolored area that undergoes changes and/or increases in size.

Cancer of the Uterus

About 25 percent of all malignant tumors in women are located in the genital organs, and the uterus is the commonest site of such cancer. Something like 70 percent of genital cancers take the form of cervical carcinoma, i.e., cancer of the cervix of the uterus; a further 20 percent occur in the body of the uterus; the remaining 10 percent are in the ovary, oviduct, vagina or vulva (Fig. 14-41). Cervical cancer occurs mainly between the ages of 35 and 55, with its peak frequency around 45. Cancer affecting the body of the uterus is found mostly in women past the menopause.

The causes of cancer of the body of the uterus are not fully known. Its increasing incidence is attributed to the rising percentage of older women in the population. This type of cancer at first confines itself to the endometrium (mucous membrane lining); it then penetrates gradually into the wall of the uterus, eventually growing in size to fill the entire uterine cavity. The cancer is then likely to spread to other genital organs (Fig. 14-43). According to the extent of the cancerous growth, four stages are to be distinguished: cancer arising in the body of the uterus (stage I) invades the cervix (stage II), then affects other genital organs (stage III) and spreads also to the bladder and rectum; finally, cancer cells are carried by the blood and lymph, giving rise to secondary tumors in more distant organs (stage IV).

Early detection of malignant tumors is of vital importance to effective treatment. The first, most frequent and often only sign of cancer of the body of the uterus is the discharge of blood other than normal menstrual bleeding. In women before the menopause (about 20 percent of cases of this type), there are irregular intermediate discharges which are due to degenerative changes in the maligant tumor. Also after the menopause, when menstruation has ceased, cancer in the uterus will cause bleeding. Any such abnormal discharges, which may moreover contain pus, should be regarded as suspect. Other symptoms, such as pain, anemia, and loss of weight, do not occur until a more advanced stage of the disease. A reliable diagnosis of early cancer can be established by obtaining a tissue specimen (endometrial biopsy) for microscopic examination.

Treatment of cancer of the body of the uterus now achieves a cure rate of about 60 percent. It consists in the surgical removal of the uterus, oviducts, ovaries and part of the vagina. With radiotherapy to back up this treatment, a 70–80 percent success rate is attained if the cancer is diagnosed and dealt with in stage I. Radium or radioactive cobalt is used, contained in small capsules which can be inserted into the uterine cavity. Chemotherapy includes drugs (cytostatics or carcinostatics) and hormone treatment.

Cancer of the cervix of the uterus appears to be caused by irritation at the boundary of the glandular mucous membrane of the cervical canal and the epithelial lining of the vagina. This form of cancer is comparatively rare in women who have never borne a child. Hence it would appear that the effects of childbirth are a contributory factor. Discharge from the cervical canal which to some extent neutralizes or weakens the lactic acid secreted in the vagina may also promote the cancerous interaction of the two types of tissue. The woman's age affects the location of the cancer, because the tissue boundary moves upward from the external os into the canal with advancing age (Fig. 14-44).

FIG. 14-41.
Frequency distribution of cancer in various parts of the female genital organs.

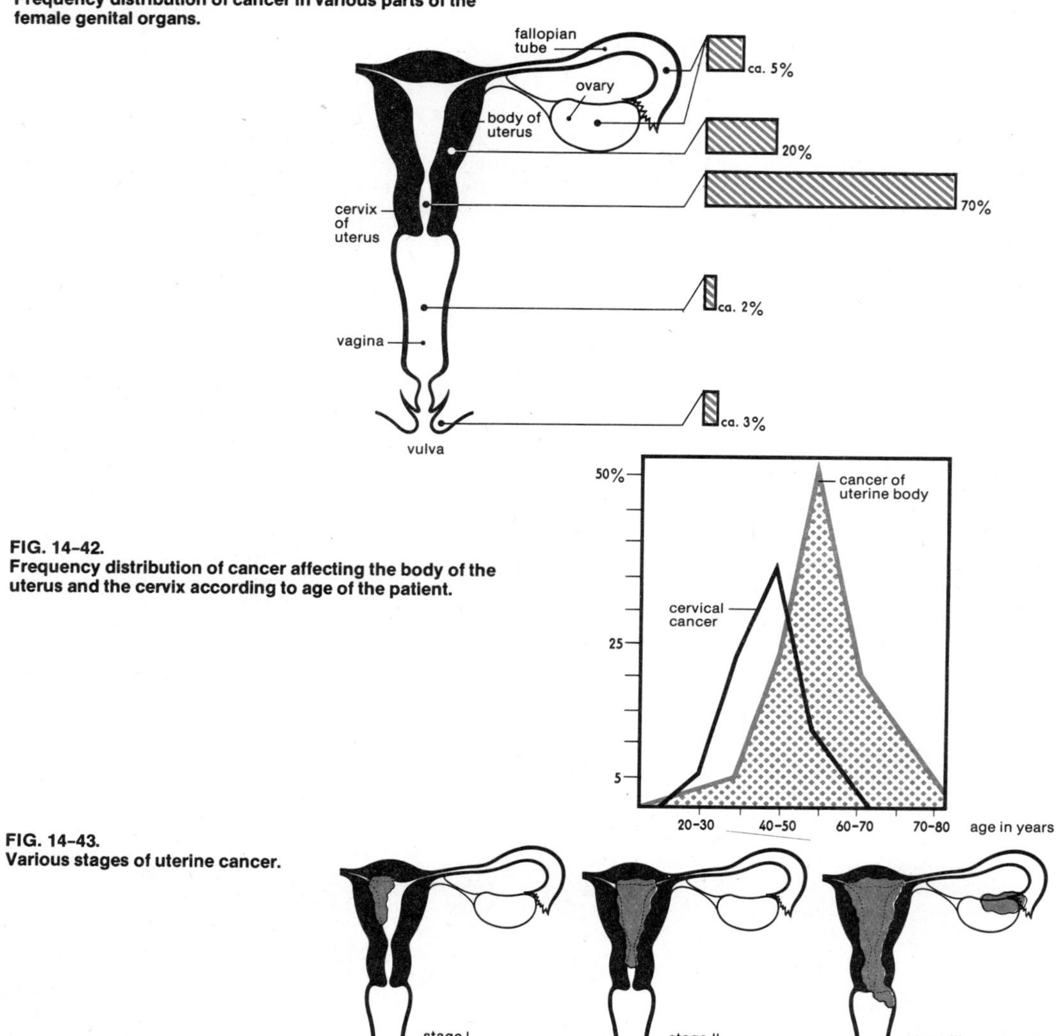

FIG. 14-42.
Frequency distribution of cancer affecting the body of the uterus and the cervix according to age of the patient.

FIG. 14-43.
Various stages of uterine cancer.

Cervical cancer is always preceded by a precancerous stage in which the disease is superficial, being confined to the lining, where it may remain semidormant for anything up to 15 years without causing any symptoms. If this precancerous condition is detected and treated, further development is arrested and a cure almost certainly effected (see Fig. 14-47). However, as there are no early warning signs, the superficial early carcinoma can be discovered only by a precautionary routine check (cervical screening), which should advisably by carried out at least once a year on every woman over the age of 30. For this check a small specimen of cells is obtained from the region of the external os of the cervix, and also from within the cervical canal, for microscopic examination (Fig. 14-45). With this technique the existence of a precancerous condition of the cervix can be detected with a high degree

FIG. 14-44.
Age related to site of cervical cancer.

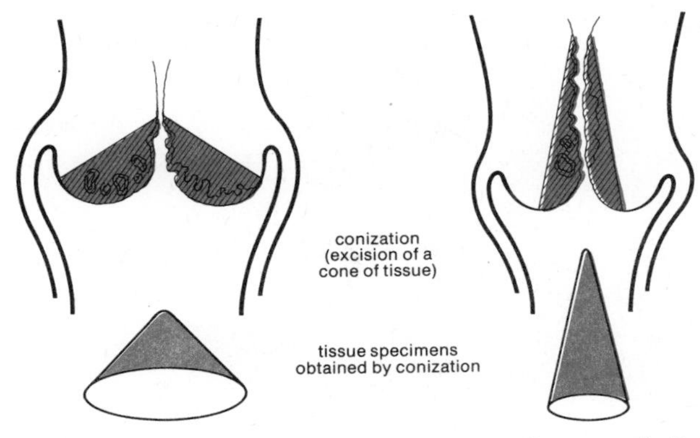

FIG. 14-45.
Precautionary check for cervical cancer.

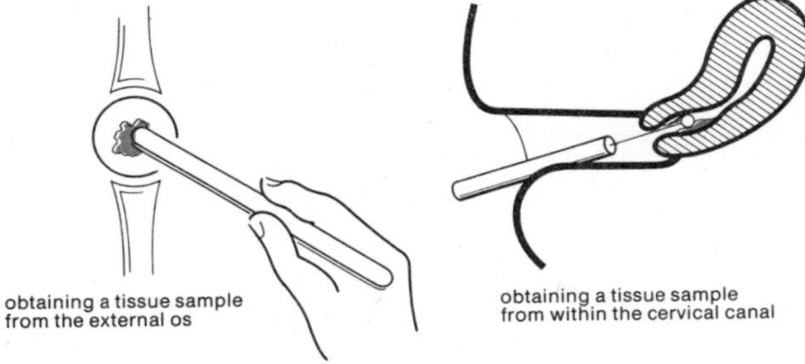

FIG. 14-46.
CHANGES AFFECTING THE EPITHELIUM AT THE EXTERNAL OS OF THE CERVIX.

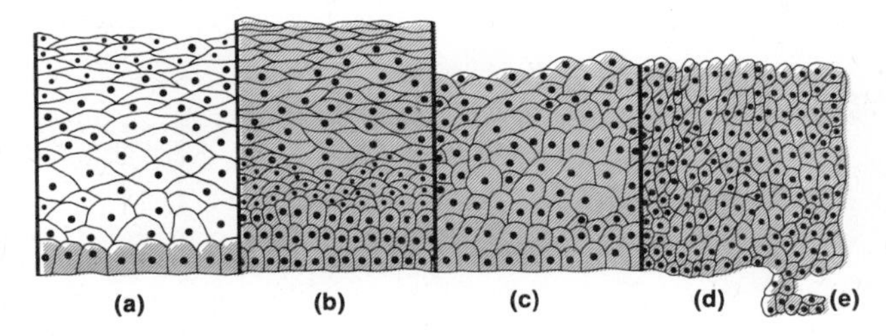

(a) Normal.

(b) Rapidly growing and "restless" epithelium.

(c) Characterized by nuclei varying in size, shape and color.

(d) Degeneratively changed epithelium.

(e) Proliferation into underlying connective tissue.

of probability, which can be further increased by direct inspection of the external os with the aid of an instrument called a colposcope. If additional evidence is required, a cervical biopsy is performed, consisting in the excision of a cone of mucous membrane from around the external os for microscopic examination. This minor operation, called *conization,* for obtaining a specimen of tissue also constitutes adequate treatment of the condition itself in women under the age of 40.

After the precancerous early stage, which may continue for months or years, the cervical cancer eventually invades the underlying tissue (Fig. 14-46e); this marks the start of the cancer proper. It is in this stage, when the prospects of a cure are already diminished, that the first distinct signs of the disease manifest themselves. Proliferation of the cancer may now proceed in various directions. It may spread up the cervical canal and undergo ulcerative degeneration. Discharges of blood

FIG. 14-47.
Characteristics and stages of development of uterine cancer.

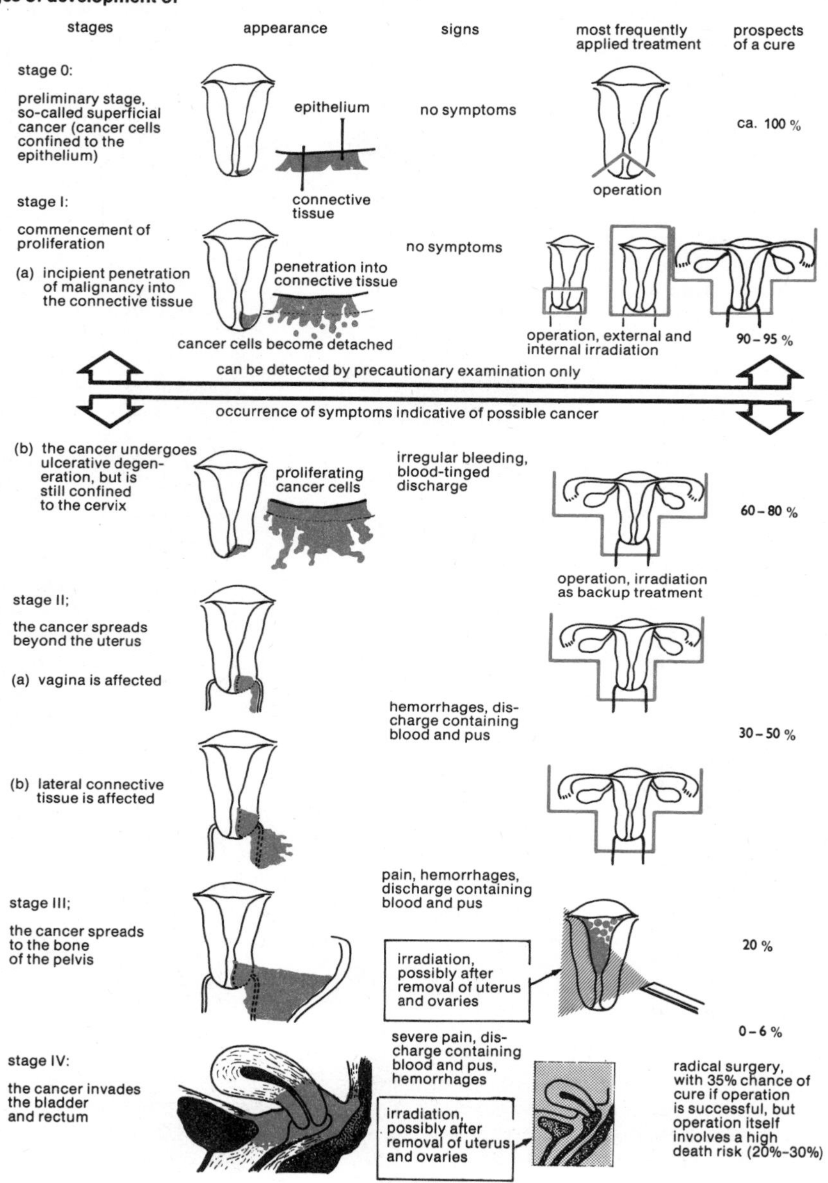

occur in this stage (Fig. 14-47). Alternatively, the cancer may spread cauliflowerlike into the vagina, or it may distend the cervix. Forward and rearward penetration of the cancer causes it to reach the bladder and/or the rectum, respectively. It may eventually exert pressure on nerve tracts and cause severe pain, though this can generally be made bearable by powerful analgesics such as morphine. The cancer cells make their way through the lymph vessels and establish themselves in the lymph nodes; others travel with the bloodstream to form secondary growths in distant parts (e.g., in the bones, liver, lungs, kidneys and brain).

The earliest symptoms of uterine cancer are discharge and bleeding. These are often overlooked because there is usually no pain in this stage of the disease, or are not taken seriously by the patient herself. A brownish, reddish or watery pink discharge is particularly suspect. In addition, there are irregular intermediate hemorrhages which occur spontaneously or sometimes after exertion, sexual intercourse or evacuation of the bowels. Hemorrhages occurring some time after the menopause must be viewed with particular suspicion.

Pain, discharge of fragments of tissue, fever, and retention of urine and feces are late symptoms of cancer. In this stage the physician can usually feel and see the cancer at the external os. If the growth is located farther up the cervical canal, a tissue specimen will have to be obtained for microscopic examination. The treatment and its prospects of lasting success will depend on the spread of the cancer.

For estimating the chances of a cure, an internationally adopted classification of the stages of cervical cancer has been established (Fig. 14-47). In stage 0, before the cancer has spread, there is very nearly 100 percent probability of a cure. As soon as the growth has penetrated into the underlying tissue, the chances of success are reduced, but are still quite fair (60–80 percent), so long as the cancer is confined to the uterus itself. The prospects are poor in cases where the cancer has been detected late, so that it has invaded adjacent organs and formed secondary tumors in other parts of the body. The minor operation known as conization provides adequate treatment in stage 0. If the cancer has spread only a little way into the cervical canal, surgical intervention may—depending on the patient's age—be confined to removing the cervix or the whole uterus. A better safeguard against relapse is to remove also the ovaries, the upper part of the vagina and surrounding tissues, together with the lymph nodes. In certain circumstances, removal of the bladder or the rectum may be considered, but in such advanced cases it is usually preferred to apply radiotherapy in an attempt to arrest the disease. Radiotherapy is often also applied as a follow-up to surgery.

Breast Cancer

Cancer of the breast is the commonest form of cancer in women between the ages of about 45 and 60. It occurs in something like 1 percent of all women, and in about 1 in 10 of these the cancer develops in both breasts. Breast cancer is rarer in women who have borne and suckled babies than in those who have not. With precautionary examination by a doctor or, with suitable instruction, by the woman herself, this form of cancer can usually be detected in an early stage. Any new node or lump that manifests itself on the breast must be viewed with suspicion. It may turn out to be a benign tumor, such as a fibroma or papilloma, but it would be unwise to take any chances or to assume that an initially benign tumor will not subsequently degenerate into malignancy.

As a rule, the cancer starts imperceptibly inside the breast. Less frequently it may start on the nipple, which looks reddened, cracked and wartlike (Fig. 14-48a). Another comparatively rare type is rapid cancerous growth in the whole breast, which becomes inflamed, with hardened skin somewhat resembling orange peel (Fig. 14-48b). The typical cancerous lump is hard, painless, usually irregular in shape and with an indefinite boundary; because of its intergrowth with its surroundings, it is not easily movable in relation to the glandular tissue of the breast. The overlying skin may subsequently coalesce with the cancer and will then look sunken, as does sometimes also the nipple (Fig. 14-48c). Eventually the tumor breaks through the skin and degenerates into an offensive cancerous ulcer. In some cases cancer gives rise to proliferation of connective tissue, associated with contraction. The affected breast then appears smaller, hard, lumpy and retracted. In others the cancer may spread like a hard platelike growth over a large area in the skin of the thorax. Adjacent lymph nodes at the outer edge of the major pectoral muscle and in the armpit become involved. Cancer cells then spread to more distant lymph nodes in the region of the collarbone, under the shoulder blade and, though rarely, in the interior of the thorax. From these affected lymph nodes there is often a further spread of cancer cells traveling in the bloodstream and giving rise to secondary tumors, particularly in the bones, e.g., in the vertebral bodies, the pelvis, the thighbone and the bone of the upper arm.

Any suspicious-looking growth or lump on or in the breast will, as a routine precaution, be surgically removed and subjected to microscopic examination. If malignancy is detected, the operation will be extended, generally involving removal of the whole affected breast, together with the major and minor pectoral

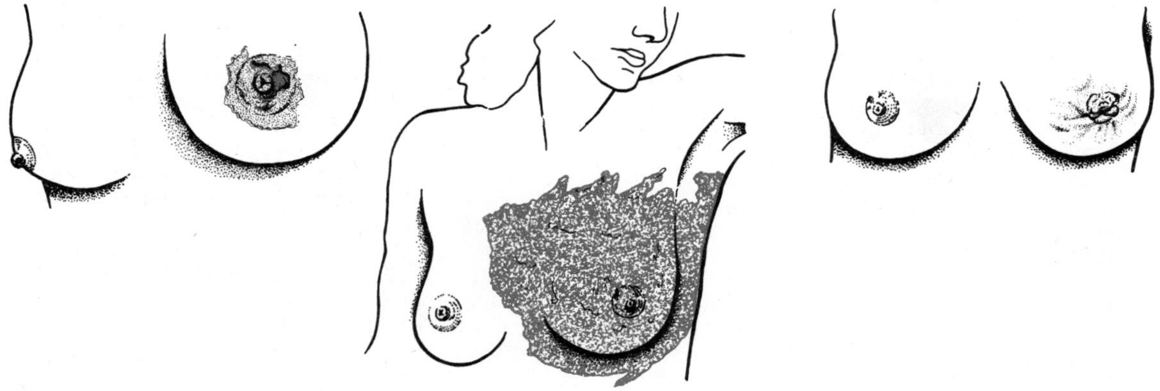

FIG. 14-48.
(a) Cancer spreading from the nipple.
(b) Rapidly spreading breast cancer with typical changes of the skin.
(c) Retraction of nipple due to coalescence with cancerous growth.

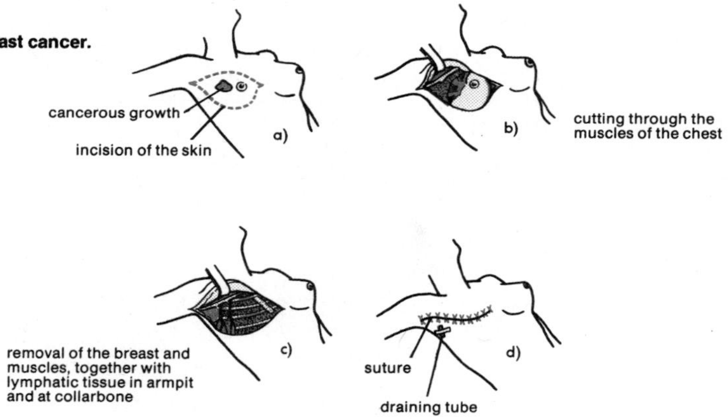

FIG. 14-49.
Operation for breast cancer.

muscles and lymphatic tissue in the armpit and at the collarbone (Fig. 14-49). This is backed up by radiotherapy. If the operation is performed before the cancer has spread to the lymph nodes, the prospects of a cure are between 75 and 90 percent. In a more advanced stage the cure rate diminishes to under 30 percent. About 40 percent of breast cancers are detected too late for an operation to offer much hope of lasting success. In such cases radiotherapy is usually applied, sometimes in combination with hormone treatment. In women below the age of 60, removal of the ovaries, sometimes also removal of the adrenal glands and suppressive intervention in the pituitary gland, can have a growth-arresting effect on breast cancer.

Cancer of the Bladder

Cancer affecting the urinary bladder is rather uncommon in comparison with certain other forms of the disease, such as lung cancer or stomach cancer. It is about three times more frequent in men than in women. It usually takes the form of malignant degeneration of the mucous membrane lining, i.e., it belongs to the *carcinoma* class of cancer. Malignancy arising from the muscular or connective tissue of the bladder (*sarcoma*) is rare. External factors play an important part in causing some cases of bladder cancer. For example, it has long been known that infection with *Schistosoma* (*Bilharzia*), a small parasitic worm found mainly in the tropics, is likely to be followed by cancer of the bladder. These parasites produce toxic substances which irritate the mucous membrane and eventually cause its cancerous degeneration. Another causal factor of bladder cancer is the action of certain chemicals, especially aromatic hydrocarbons such as aniline and other chemicals used in the dyestuff and rubber industries. These substances may enter the body with contaminated food and are eventually excreted through the kidneys and bladder.

Cancer of the bladder is often preceded by a precancerous condition which may occur as whitish thickened areas (leukoplakia) on the mucous membrane or in the form of an epithelial tumor (papilloma). Unfortunately, metastasis—i.e., spread of cancer cells through the lymph and blood to form secondary tumors—occurs in an early stage of bladder cancer. Such secondary tumors develop particularly in the lymph nodes of the pelvis and also in the liver. A common early symptom of cancer of the bladder is the presence of blood in the urine. Further symptoms are a strong urge to urinate, pains felt during urination and tending to radiate through the abdomen, and eventually retention of urine. Diagnosis of cancer is based primarily on examination of the urine for the presence of blood and especially also of cancer cells. In advanced cases the tumor can be located by rectal examination. The interior of the bladder can be directly examined with an optical instrument called a cystoscope, which is introduced through the urethra. A definite diagnosis can be established by microscopic examination of tissue specimens.

If the examination shows the tumor to be a papilloma, i.e., an initially benign tumor, it is usually possible to remove or destroy it by the technique of electrocision or electrocautery performed with an instrument passed up the urethra. If the condition has, however, already degenerated into cancer, this treatment will not suffice; a major surgical operation will then be required. If the cancer is not in a very advanced stage, part of the bladder and often also the adjacent lymph nodes will be removed. The ureters can then be connected to the bowel so that they discharge the urine into it. By the time such an operation is performed, the cancer has invariably reached an advanced stage, and for this reason the average survival period of such patients is not more than about five years.

Index